International
REVIEW OF
Neurobiology

Volume 51

Glucose Metabolism in the Brain

International

REVIEW OF

Neurobiology

Volume 51

Glucose Metabolism IN THE Brain

EDITED BY

DONARD DWYER
Departments of Psychiatry and Pharmacology
LSU Health Sciences Center
Shreveport, Louisiana

ACADEMIC PRESS
An imprint of Elsevier Science
Amsterdam Boston London New York Oxford Paris
San Diego San Francisco Singapore Sydney Tokyo

This book is printed on acid-free paper. ∞

Academic Press
An imprint of Elsevier Science.
525 B Street, Suite 1900, San Diego, California 92101-4495, USA
http://www.academicpress.com

Academic Press
84 Theobalds Road, London WC1X 8RR, UK
http://www.academicpress.com

International Standard Book Number: 0-12-366852-2

PRINTED IN THE UNITED STATES OF AMERICA
02 03 04 05 06 07 MM 9 8 7 6 5 4 3 2 1

CONTENTS

Energy Metabolism in the Brain

Leif Hertz and Gerald A. Dienel

The Cerebral Glucose-Fatty Acid Cycle: Evolutionary Roots, Regulation, and (Patho)physiological Importance

Kurt Heininger

Expression, Regulation, and Functional Role of Glucose Transporters (GLUTs) in Brain

Donard S. Dwyer, Susan J. Vannucci, and Ian A. Simpson

Insulin-Like Growth Factor-1 Promotes Neuronal Glucose Utilization During Brain Development and Repair Processes

CAROLYN A. BONDY AND CLARA M. CHENG

CNS Sensing and Regulation of Peripheral Glucose Levels

BARRY E. LEVIN, AMBROSE A. DUNN-MEYNELL, AND VANESSA H. ROUTH

Glucose Transporter Protein Syndromes

DARRYL C. DE VIVO, DONG WANG, JUAN M. PASCUAL, AND YUAN YUAN HO

Glucose, Stress, and Hippocampal Neuronal Vulnerability

Lawrence P. Reagan

Glucose/Mitochondria in Neurological Conditions

John P. Blass

Energy Utilization in the Ischemic/Reperfused Brain

John W. Phillis and Michael H. O'Regan

Diabetes Mellitus and the Central Nervous System

Anthony L. McCall

Diabetes, the Brain, and Behavior: Is There a Biological Mechanism Underlying the Association between Diabetes and Depression?

ALAN M. JACOBSON, JACQUELINE A. SAMSON, KATIE WEINGER, AND CHRISTOPHER M. RYAN

Schizophrenia and Diabetes

DAVID C. HENDERSON AND ELISSA R. ETTINGER

Psychoactive Drugs Affect Glucose Transport and the Regulation of Glucose Metabolism

Donard S. Dwyer, Timothy D. Ardizzone, and Ronald J. Bradley

CONTRIBUTORS

Numbers in parentheses indicate the pages on which the authors' contributions begin.

Timothy D. Ardizzone (503), Departments of Psychiatry and Pharmacology, LSU Health Sciences Center, Shreveport, Louisiana 71130[1]

John P. Blass (325), Dementia Research Service, Burke Medical Research Institute, Weill Medical College of Cornell University, White Plains, New York 10605

Carolyn A. Bondy (189), Developmental Endocrinology Branch, National Institute of Child Health and Human Development, National Institutes of Health, Bethesda, Maryland 20892

Ronald J. Bradley (503), Departments of Psychiatry, LSU Health Sciences Center, Shreveport, Louisiana 71130

Clara M. Cheng (189), Developmental Endocrinology Branch, National Institute of Child Health and Human Development, National Institutes of Health, Bethesda, Maryland 20892

Darryl C. De Vivo (259), Department of Pediatric Neurology, Colleen Giblin Research Laboratories for Pediatric Neurology, Columbia Presbyterian Medical Center, New York 10032

Gerald A. Dienel (1), Department of Neurology, University of Arkansas for Medical Sciences, Little Rock, Arkansas 72205

Ambrose A. Dunn-Meynell (219), Neurology Service, VA Medical Center, East Orange, New Jersey 07018 and Department of Neurosciences, New Jersey Medical School, Newark, New Jersey 07103

Donard S. Dwyer (159, 503), Departments of Psychiatry and Pharmacology, LSU Health Sciences Center, Shreveport, Louisiana 71130

Elissa R. Ettinger (481), Schizophrenia Program, Massachusetts General Hospital, Harvard Medical School, Boston, Massachusetts 02114

Kurt Heininger (103), Department of Neurology, Heinrich Heine Universität, D-40597 Düsseldorf, Germany

David C. Henderson (481), Schizophrenia Program, Massachusetts General Hospital, Harvard Medical School, Boston, Massachusetts 02114

Leif Hertz (1), Hong Kong DNA Chips, Ltd., Kowloon, Hong Kong, China

[1]Current affiliation: Department of Neurology, Vontz Center for Molecular Studies, University of Cincinnati Medical Center, Cincinnati, Ohio 45267.

Yuan Yuan Ho (259), Department of Neurology, Colleen Giblin Research Laboratories for Pediatric Neurology, Columbia Presbyterian Medical Center, New York 10032

Alan M. Jacobson (455), Joslin Diabetes Center, Harvard Medical School, Boston, Massachusetts 02215

Barry E. Levin (219), Neurology Service, VA Medical Center, East Orange, New Jersey 07018 and Department of Neurosciences, New Jersey Medical School, Newark, New Jersey 07018

Anthony L. McCall (415), James M. Moss Professor in Diabetes in Internal Medicine, University of Virginia School of Medicine, Division of Endocrinology & Metabolism, Department of Internal Medicine, UVA Health System, Charlottesville, Virginia 22908

Michael H. O'Regan (377), Biomedical Sciences, School of Dentistry, University of Detroit Mercy, Detroit, Michigan 48219

Juan M. Pascual (259), Department of Neurology, Colleen Giblin Research Laboratories for Pediatric Neurology, Columbia Presbyterian Medical Center, New York 10032

John W. Phillis (377), Department of Physiology, Wayne State University School of Medicine, Detroit, Michigan 48201

Lawrence P. Reagan[2] (289), Laboratory of Neuroendocrinology, The Rockefeller University, New York 10021[1]

Vanessa H. Routh (219), Department of Pharmacology and Physiology, New Jersey Medical School, Newark, New Jersey 07103

Christopher M. Ryan (455), Harvard Medical School, Joslin Diabetes Center, Boston, Massachusetts 02215[3]

Jacqueline A. Samson (455), McLean Hospital, Harvard Medical School, Boston, Massachusetts 02215

Ian A. Simpson (159), Department of Neuroscience and Anatomy, Penn State University College of Medicine, Hershey, Pennsylvania 17033

Susan J. Vannucci (159), Departments of Pediatrics and Neuroscience and Anatomy, Penn State University College of Medicine, Hershey, Pennsylvania 17033

Dong Wang (259), Department of Neurology, Colleen Giblin Research Laboratories for Pediatric Neurology, Columbia Presbyterian Medical Center, New York 10032

Katie Weinger (455), Joslin Diabetes Center, Harvard Medical School, Boston, Massachusetts 02215

[2]Current affiliation: Department of Pharmacology, Physiology, and Neuroscience, University of South Carolina School of Medicine, Columbia, South Carolina 29208.

[3]Current affiliation: Department of Psychiatry, University of Pittsburgh School of Medicine, Pittsburgh, Pennsylvania, 15213.

PREFACE

Glucose is an essential energy source for the adult human brain. When blood glucose levels fall by half from a normal fasting value (to about 2–3 mM), there is some cognitive impairment and, at glucose levels below 1 mM, mental confusion is evident and coma may result from sustained glucose deprivation. Recent studies have demonstrated that optimum neuronal function and even cell survival are determined by the efficiency of energy metabolism. Thus, cells that use glucose as their primary energy source are less susceptible to stress and harmful conditions than cells that use other energy sources or that exhibit higher rates of glycolysis. Although our knowledge about the critical role of glucose metabolism in the maintenance of high level brain function has grown considerably in recent years, the various factors that regulate glucose uptake and utilization in the CNS are not well understood. Moreover, the brain must regulate the relative use of glucose, glutamine, and ketone bodies for energy under normal circumstances and especially during development and aging.

The regulation of glucose metabolism in humans is a complex process that proceeds via a two-way street connecting the brain and peripheral tissues. The CNS monitors glucose levels and exerts indirect control over both the utilization and production of glucose via hormone secretion and innervation. Peripheral tissues, including muscle, fat, pancreas, and liver, are responsible for direct control over the transport, synthesis, storage, and metabolism of glucose. Specific signals are conveyed back to the brain and nonspecific information, such as the overall metabolic state, can produce a significant impact on brain function. For example, injury to brain tissue in stroke is much more extensive in patients with hyperglycemia or diabetes as compared to patients with normal levels of blood glucose. Similarly, tardive dyskinesia, a serious movement disorder induced by neuroleptic drugs, occurs more frequently in patients with elevated blood glucose levels. Furthermore, a number of neurologic and psychiatric conditions are associated with alternations in energy metabolism in both the brain and peripheral tissues. Research efforts aimed at providing insights into the relationship between glucose metabolism and brain dysfunction should be a high priority.

The purpose of this volume is to bring together the latest research related to both basic and clinical aspects of glucose metabolism in the brain. The first two chapters by Drs. Hertz and Dienel and Dr. Heininger cover general aspects of energy metabolism in the CNS, including the use of alternative sources of energy and the evolutionary basis for these metabolic processes. Drs. Dwyer, Vannucci, and Simpson provide an update on glucose transporter proteins that are expressed in the brain in relation to the regulation of glucose uptake. Drs. Bondy and Cheng discuss the important role of insulin-like growth factors in neuronal development and neuroprotection. The mechanisms by which the brain senses glucose and integrates other metabolic signals are summarized by Drs. Levin, Dunn-Meynell, and Routh. The clinical significance of glucose transport and glucose transporter proteins is elegantly illustrated in the chapters by Dr. De Vivo and colleagues (on glucose transporter deficiencies) and by Dr. Reagan (on the effects of stress in the CNS). The next three chapters describe neurological aspects of energy metabolism specifically, the role of mitochondria in a variety of disorders (Dr. Blass), the interrelationship between ischemic injury and glucose metabolism (Drs. Phillis and O'Regan), and the clinical implications of diabetes for brain function (Dr. McCall). The final three chapters concern the interface between behavioral disorders and glucose metabolism. Dr. Jacobson and colleagues describe various associations between major depression and diabetes, Drs. Henderson and Ettinger summarize the evidence for biological connections between schizophrenia and abnormal glucose metabolism, and last, Drs. Dwyer, Ardizzone, and Bradley discuss the drugs used to treat various psychiatric disorders in relation to their effects on glucose transport. In each chapter, there is special emphasis on possible future directions for research on the various topics covered here.

I wish to thank Dr. Noelle Gracy and others at Academic Press for their encouragement and support for this project. I also want to thank all of the contributors who ultimately made this volume possible. Tremendous creative energy was harnessed for this effort and the finished product represents a state-of-the-art summary of the field that will serve as an indispensable reference for many years to come. Finally, to the reader, I express our collective appreciation for your interest in this area of research and I hope that this information stimulates further exchange of ideas, challenges to the prevailing dogma, and the design of additional experiments that take the field to new vistas.

Donard S. Dwyer

ENERGY METABOLISM IN THE BRAIN

Leif Hertz

Hong Kong DNA Chips, Ltd., Kowloon, Hong Kong, China

Gerald A. Dienel

Department of Neurology, University of Arkansas for Medical Sciences
Little Rock, Arkansas 72205

0074-7742/02 $35.00

I. Introduction

Studies of glucose metabolism in the brain reflect a dichotomy due to the fact that the complex, integrating functions of the brain can only be studied in the intact, functioning brain in the conscious individual (human or animal), whereas properties of brain cells, cell-cell interactions, and mechanisms are most readily evaluated *in vitro* under controlled conditions using brain slices, subcellular fractions or purified, isolated cells of different types. *In vivo* studies have most commonly been done in studies with labeled 2-deoxy-D-glucose (DG) or 2-fluorodeoxyglucose (FDG). "Imaging" with DG revolutionized investigations of correlations between brain function and brain metabolism (Sokoloff *et al.*, 1977), because this glucose analog enables local functional analysis of hexokinase activity *in vivo*, from which local rates of glucose utilization can be calculated under steady-state conditions. On the other side, it is becoming overwhelmingly clear that such studies represent only one aspect of brain function, i.e., the "big picture," identifying the pathways and magnitude of functional metabolic activities; the underlying contributions of different cells and cell types in the brain are not identified and quantified, and the character of the energy-requiring processes are not determined. Brain cells can behave metabolically in very different manners in response to various stimuli and interact so that one cell type may generate a glucose metabolite (e.g., glutamate or lactate), which then undergoes "metabolic trafficking" to sustain function, to be further metabolized in a different cell type, or even to leave the activated area. These heterogeneous interactions have the consequence that imaging of overall brain metabolism cannot provide a picture of glucose metabolism at the cellular level.

A variety of *in vitro* methods have been used to assess metabolic activities in different brain cell types and in subcellular structures. Apart from the difficulty that these methods provide no direct information about metabolic activities in the functioning brain *in vivo*, they are almost all encumbered with potential methodological problems. Immunohistochemical studies of enzymes and substrate carriers in intact brain tissue have given much useful information, but with one notable exception (see SectionII.D.1), they *only* provide information about the amount of enzyme or transporter protein, *not* about the dynamic, condition-dependent activity of the enzyme, or the transporter. Information about enzyme and transporter activities in different cell types can be obtained in cellularly homogenous preparations, today most often cultured cells, derived from immature tissues, but differentiating during the culturing. Well-differentiated primary cultures of neurons and astrocytes have similar rates of oxidative metabolism and similar contents of adenine nucleotides as the brain *in vivo* (Hertz and Peng, 1992a; Silver and Erecinska, 1997). However, tissue culture methodology has the potential source of error that the cultured cells may differ in metabolic characteristics from their *in vivo* counterparts, in part because of the very feature that makes them attractive for metabolic studies, namely their homogeneity and ensuing lack of cellular interactions and exposure to a temporal sequence of trophic factors, known to play critical roles in the development of the central nervous system (CNS). This source of error does not apply to preparations of different cell types or subcellular fractions obtained by dissociation of intact brain tissue followed by gradient centrifugation, but the resulting cellular or subcellular (e.g., synaptosomes, mitochondria) fractions have been rendered ischemic (i.e., exposed to severe energy failure and accompanying autolytic processes from which they may never fully recover), removed from their natural surrounding, and physically damaged, especially in older studies, due to exposure to more or less harsh treatment during their isolation. Accordingly, these preparations as well as brain slices show lower metabolic activities and contents of ATP than intact brain (Hertz and Schousboe, 1986), although the ATP/ADP ratio in carefully prepared synaptosomal preparations approaches that in the brain (Erecinska *et al.,* 1996), suggesting the presence of a functional, metabolically intact component.

Nevertheless, by combining different methodologies and continuously maintaining the *in vivo* situation as the general standard to which results obtained with different cellular and subcellular techniques must be compared, a picture of cellular interactions in glucose metabolism has emerged, and information has been obtained about the identity of energy-requiring and energy-yielding processes. Perhaps even more importantly, these studies have triggered the development of *in vivo* methods, primarily utilizing nuclear magnetic resonance imaging and spectroscopy, which have confirmed and further expanded many observations made *in vitro.* In this review, we will first discuss pathways and regulation of glucose metabolism in the functioning brain in the conscious human or animal during rest and during stimulation; this will be followed by a description of mechanisms which increase glucose metabolism *in vitro.* Combination of these two approaches allows a tentative determination of not only the quantitative contributions to glucose metabolism by some of the major cell types, but also identification of mechanisms creating a demand for metabolically generated energy and their relationships to functional activation and neurotransmission.

II. Pathways and Regulation of Glucose Utilization

A. Oxidative and Nonoxidative Metabolism

Metabolism of glucose is tightly regulated to generate ATP and provide carbon for biosynthetic reactions in conjunction with local functional

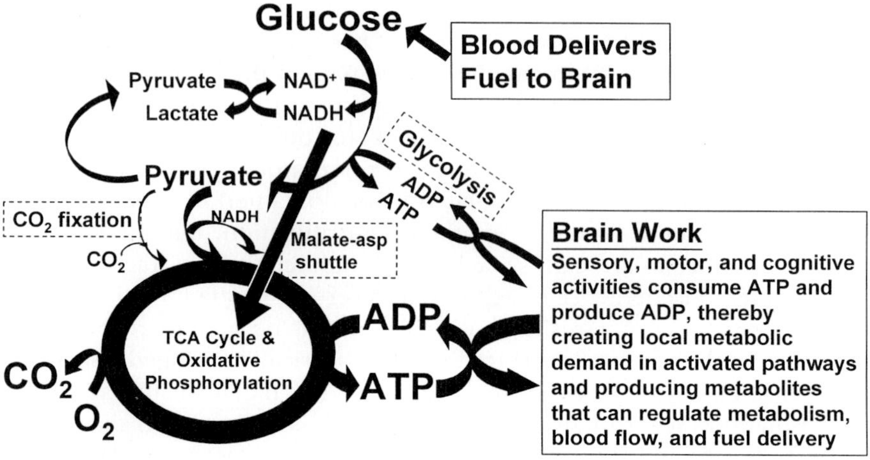

FIG. 1. ATP-ADP cycling links brain function and glucose metabolism. Functional tasks activate neuronal signaling and consumption of neuronal and glial ATP, thereby stimulating glucose utilization (CMR_{glc}) in specific brain structures. By-products of metabolism stimulate local blood flow to increase local delivery of glucose and oxygen. Cytoplasmic NADH is oxidized via lactate dehydrogenase and/or the malate-aspartate (asp) shuttle (see Fig. 4 and text), depending on conditions in the cell. Both the glycolytic pathway (glucose to pyruvate) and pyruvate oxidation in the tricarboxylic acid (TCA) cycle generate ATP for working brain. The glycolytic pathway can be rapidly activated, whereas the TCA cycle has the highest energy yield (see Fig. 2). (Adapted from G. A. Dienel. Energy generation in the central nervous system. *In* "Cerebral Blood Flow and Metabolism, 2nd ed." (L. Edvinsson and D. Krause, eds.), 2002, Lippincott Williams & Wilkins.©)

activities of the brain (Fig. 1). The catabolic process has nonoxidative (glycolytic) and oxidative components, and branch points can divert a portion of the glucose carbon from energy production toward other uses. Oxidative metabolism of pyruvate via the tricarboxylic acid (TCA) cycle produces ATP in high yield via the electron transport system and links bioenergetics to the large amino acid pools. In whole brain at steady state >90% of the glucose is oxidatively degraded as can be concluded from a ratio between rates of utilization of glucose (CMR_{glc}) and of oxygen (CMR_{O_2}) of at least 5.5, which is close to the theoretically expected ratio of 6. In the resting (i.e., not specifically stimulated) human brain, CMR_{glc} is 0.3 μmol/min/g wet wt., compared to 0.7 μmol/min/g wet wt. in the rat brain (Sokoloff, 1986).

B. Glycolysis

1. *Glycolytic Pathway*

Glucose enters the cytoplasmic compartment of brain cells from a capillary or the extracellular space via an equilibrative glucose transporter.

Glucose breakdown takes place in "stages," beginning with its phosphorylation at the C6 position by hexokinase, metabolically "primed" by hydrolysis of one molecule of ATP. Most glucose-6-phosphate (glucose-6-P) is converted to pyruvate (Table I), but glucose-6-P can be diverted from the glycolytic pathway by entry into the pentose-P shunt pathway to produce NADPH and five-carbon compounds, or it can be converted to glucose-1-P and utilized for synthesis of glycogen, galactose, glycoprotein, and glycolipids (Fig. 2; see color insert). Myo-inositol is also synthesized from glucose-6-P and serves as the precursor for the phosphatidylinositide signaling molecules. The second glycolytic step, formation of fructose-6-P also produces a branch point product for biosynthetic pathways; small quantities are converted to mannose-6-P (for synthesis of fucose, and complex carbohydrates via GDP-mannose) or glucosamine-6-P (a precursor for sialic acid). Thus, the initial phase of glucose metabolism requires ATP to "prime" each glucose molecule, and the first two metabolic steps yield "branch point" metabolites that are precursors for important but quantitatively minor metabolic pathways.

The controlling and most highly regulated reaction of the glycolytic pathway is the second ATP-dependent phosphorylation to fructose-1,6-bisphosphate (fructose-1,6-P_2) carried out by 6-phosphofructo-1-kinase (Passonneau and Lowry, 1964), the activity of which is governed by many downstream metabolites (see the next section). Formation of fructose-1,6-P_2 is followed by splitting of the 6-carbon compound into two triose phosphates (triose-P), dihydroxyacetone-phosphate (dihydroxyacetone-P), and glyceraldehyde-phosphate (glyceraldehyde-P). This sets the stage for a series of oxidation-reduction reactions that generate cytoplasmic NADH and ATP. Of the two triose phosphates, only glyceraldehyde-P is oxidized, but new glyceraldehyde-P is generated from dihydroxyacetone-P, catalyzed by triose-P isomerase. Glyceraldehyde 3-P dehydrogenase produces 1,3-bisphosphoglycerate plus NADH, which must be reoxidized to NAD^+, either via the malate-aspartate shuttle (MAS) and associated with generation of ATP (Section II.E.1), or by conversion of equimolar amounts of pyruvate to lactate, without any ATP synthesis (Section II.E.3). Two molecules of ATP per molecule of glucose are generated by the next step which is carried out by phosphoglycerate kinase. The 3-P-glycerate undergoes a mutase reaction to shift the phosphate group to the two position, followed by dehydration by enolase to form phosphoenolpyruvate (PEP); conversion of PEP to pyruvate by pyruvate kinase produces two more molecules of ATP per molecule glucose. Pyruvate is also a branch point metabolite (Fig. 2); it can either (1) enter mitochondria for conversion to acetyl-CoA and serve as substrate for oxidative metabolism or biosynthesis of fatty acids or acetylcholine; (2) be reduced to lactate in the cytosol for later oxidation and/or export from the cell; (3) be converted to alanine by transamination; or (4) be converted to oxaloacetate in the

TABLE I
ENZYMATIC STEPS OF THE GLYCOLYTIC PATHWAY

Sequential enzymatic step	Reaction[a]	Maximal velocity[b] (μmol min^{-1} g wet wt^{-1})	
		Mouse	Human
1. Hexokinase	Glucose + ATP → glucose-6-P + ADP	11	4
2. Phosphohexose isomerase	Glucose-6-P ↔ fructose-6-P	55	58
3. 6-Phosphofructo-1-kinase	Fructose-6-P + ATP → fructose-1,6-P_2 + ADP	9	1
4. Aldolase	Fructose-1,6-P_2 ↔ glyceraldehyde-3-P + dihydroxyacetone-P	5	4
5. Triose phosphate isomerase	Dihydroxyacetone-P ↔ glyceraldehyde-3-P		747
6. Phosphoglyceraldehyde dehydrogenase	2Glyceraldehyde-3-P + $2P_i$ + $2NAD^+$ ↔ 2glycerate-1,3-P_2 + 2NADH	52	2
7. 3-Phosphoglycerate kinase	2Glycerate-1,3-P_2 + 2ADP + $2P_i$ → 2glycerate-3-P + 2ATP	167	
8. Phosphoglyceromutase	2Glycerate-3-P ↔ 2glycerate-2-P	39	41
9. Enolase	2Glycerate-2-P ↔ 2phosphoenolpyruvate + $2H_2O$	30	
10. Pyruvate kinase	2Phosphoenolpyruvate + 2ADP + $2P_i$ → 2pyruvate + 2ATP	118	70
Net reaction	Glucose + 2ADP + $2NAD^+$ + $2P_i$ → 2pyruvate + 2ATP + 2NADH		
Cytoplasmic oxidation of NADH by lactate dehydrogenase	2NADH + 2pyruvate ↔ $2NAD^+$ + 2lactate	59	66
Mitochondrial oxidation of NADH[c]	2NADH + O_2 + 4-6ADP + 4-6P_i → $2NAD^+$ + 4-6ATP + $2H_2O$		

[a] Reactions do not include hydrogen ions.

[b] Rate data from mouse brain were compiled by McIlwain and Bachelard (1985); values from human brain were calculated from data summarized by Sheu and Blass (1999), assuming 100 mg protein (g brain tissue)$^{-1}$. Note that maximal velocities of all steps in the glycolytic pathway greatly exceed the average rate of glucose utilization, i.e., about 0.7 μmol g^{-1} min^{-1} in rat brain and about 0.3 μmol g^{-1} min^{-1} in human brain (Sokoloff, 1986, 1996), demonstrating very high capacity to increase fuel consumption with an abrupt rise in energy demand.

[c] Either 2 or 3 ATP can be formed from each cytoplasmic NADH, depending on the shuttle system that brings the reducing equivalents into the mitochondria. The glycerol-3-P shuttle activity is low in brain, and provides electrons at the level of $FADH_2$, with a total yield of 4 ATP. The malate shuttle is predominant, and transfers electrons to mitochondrial complex 1, yielding a total of 6 ATP (see text). (Modified from G. A. Dienel. Energy generation in the central nervous system. *In* "Cerebral Blood Flow and Metabolism, 2nd ed." (L. Edvinsson and D. Krause, eds.), 2002, Lippincott Williams & Wilkins.©)

mitochondria by pyruvate carboxylase. The moment-to-moment energy status of the cell, tissue oxygen level, relative fluxes of the glycolytic pathway and tricarboxylic acid cycle, and the cell type determine the fate of pyruvate.

To summarize, the glycolytic pathway of glucose metabolism uses two ATP to prime one molecule of glucose and produces two molecules of NADH and four ATP via substrate-level phosphorylation reactions, for a net gain of two ATP per molecule glucose. Oxidation of NADH to NAD^+ by MAS, is under oxygenated conditions, followed by oxidation of NADH in the mitochondria, creating another six molecules of ATP, whereas lactate formation is not associated with ATP formation or utilization.

2. *Metabolic Control by Energy Demand and Levels of Intermediates*

ATP production is closely coupled to brain work, due, in part, to the requirement for ADP as a substrate for the energy-producing reactions (Figs. 1 and 2). If glycolysis were not regulated, metabolism of all available glucose that entered the brain would simply consume glucose and ATP, trap phosphate as triose-P, and produce lactate. Major sites for control of glycolytic flux are hexokinase, phosphofructokinase (PFK), and pyruvate kinase; PFK is the key enzyme.

Type I hexokinase, the predominant isozyme in brain, is normally saturated with substrate and strongly inhibited by its product (glucose-6-P); its kinetic properties are altered by reversible binding to mitochondria. In rat brain, the apparent K_m of hexokinase for glucose is ~0.05 m*M*, which is well below the intracellular glucose concentration in brain, i.e., 2–3 m*M* (Siesjö, 1978; Pfeuffer *et al.*, 2000). The K_m for ATP-Mg^{2+} is about 0.4 m*M*, and brain ATP level is 2–3 m*M*. Brain hexokinase is inhibited by ADP as well as by glucose-6-P ($K_i \cong 10\ \mu M$) and fructose-1,6-P_2, an inhibition which is antagonized by phosphate (P_i). Comparison of glycolytic flux in normal rat brain to maximal hexokinase activity assayed *in vitro* (Table I) indicates that the enzyme is normally inhibited by more than 95%, so brain has high capacity to increase glycolytic flux as needed. Hexokinase I binds to the outer mitochondrial membrane via a pore-forming protein (porin) through which ATP and ADP cross the mitochondrial membrane, thereby giving hexokinase preferential access to mitochondrially generated ATP (Cesar and Wilson, 1998). Binding to mitochondria alters specific epitopic regions in the hexokinase molecule (Hashimoto and Wilson, 2000). The phosphorylation of glucose by brain hexokinase bound to mitochondria is not only *able to* use mitochondrially generated ATP but *selectively uses* ATP formed in mitochondria, which may help coordinate glycolysis and TCA cycle activity (BeltrandelRio and Wilson, 1992a,b; Cesar and Wilson, 1998). When bound to mitochondria, the K_i of hexokinase for glucose-6-P is increased and the K_m for ATP is reduced, suggesting that the bound form is more active.

In smooth muscle preparations, it has been shown that hexokinase association with mitochondria is reduced from 70 to 40% of total hexokinase activity by treatment with DG to increase glucose-6-P content fourfold, suggesting that hexokinase binding to mitochondria is regulated by the metabolic state of the cells (Lynch *et al.,* 1991, 1996). Besides moment-to-moment regulation of activity by metabolic effectors (Wilson, 1995), functional activity modulates hexokinase amount; thus, hexokinase activity increases over several days in structures involved in body fluid regulation in response to water deprivation, diabetes, and aortic baroreceptor denervation (Turton *et al.,* 1986; Krukoff *et al.,* 1986).

All three isoenzymes (muscle [M], liver [L], and brain [C]) of PFK occur in brain, and quantitative differences in allosteric properties are found between the different isozymes (Zeitschel *et al.,* 1996). Regulation of PFK activity is a major control point for glycolysis (Passonneau and Lowry, 1964). For example, activation of metabolism by ischemia increases the rate of glycolysis greater than four-fold, and causes the concentrations of metabolites upstream of fructose-6,1-P_2 (glucose, glucose-6-P, and fructose-6-P) to fall, whereas the levels of those downstream (between fructose-1,6-P_2 and lactate) rise, indicating rapid PFK activation. PFK is inhibited by compounds that accumulate when the energy charge is high, and it is activated by products of functional metabolic activity (Fig. 2). Energy charge is the relative ATP level of the adenine nucleotide pool, calculated as [ATP + 0.5ADP]/[ATP + ADP + AMP]. The cellular concentration of ATP (2–3 mM) greatly exceeds those of ADP (about 0.2–0.6 mM) and AMP (about 0.05 mM), and the normal energy charge is slightly less than one. AMP is produced by adenylate kinase (myokinase) when ADP is used to regenerate ATP; due to the low concentration of AMP, small changes in ATP level are amplified and reflected by larger fractional changes in the amount of AMP. PFK activators include NH_4^+, K^+, P_i, AMP, cAMP, ADP, and fructose-2,6-P_2; inhibitors include ATP, phosphocreatine, 3-P-glycerate, 2-P-glycerate, 2,3-P_2-glycerate, phosphoenolpyruvate, citrate, hydrogen ion (low pH), and Mg^{2+}. Fructose-6-P, fructose-1,6-P_2, ADP, AMP, P_i, and NH_4^+ are all increased in ischemia; these compounds overcome the inhibition of PFK by ATP. Inhibition of PFK by citrate helps to coordinate TCA cycle activity with glycolytic flux. Inhibition by ATP of both PFK and pyruvate kinase slows glycolysis when energy supplies are high. Many regulatory mechanisms act concertedly to fine-tune glucose metabolism to meet local energy demand. Typical metabolite levels obtained in rat brain (Table II) show that brain reserves of energy metabolites are low relative to flux through catabolic pathways (0.3 μmol glucose/min/g wet wt in man; 0.7 μmol glucose/min/g wet wt in the rat). A continuous supply of glucose is, therefore, required to sustain brain function.

TABLE II
REPRESENTATIVE LEVELS OF ENERGY METABOLITES IN FREEZE-BLOWN RAT BRAIN

Compound	Concentration (μmol/g wet wt)
Glycogen	2.8
Glucose	1.6
Glucose-6-P	0.2
Fructose 1,6-P_2	0.01
Dihydroxyacetone-P	0.02
α-Glycerol-P	0.11
Pyruvate	0.09
Lactate	1.4
Citrate	0.28
α-Ketoglutarate	0.22
Malate	0.32
Glutamate	12
Aspartate	3
Glutamine	6
ATP	2.5
ADP	0.6
AMP	0.07
Creatine-P	4

Data are from Veech, R. L., 1980.

3. *Immunohistochemistry of Glycolytic Enzymes and Transporters*

a. Glucose Transporter. Transit of glucose from blood across the capillary endothelium and ultimately into brain cells requires the action of several isoforms of the glucose transporter family. Endothelial cells constituting the blood-brain barrier express the glucose transporter GLUT1, whereas neurons express GLUT3. Neuronal perikarya and proximal dendrites have little immunochemically visualized glucose transporter, but the adjacent neuropil is intensely stained for the glucose transporter (Bagley *et al.*, 1989; Mantych *et al.*, 1993; McCall *et al.*, 1994; Gerhart *et al.*, 1995; Fattoretti *et al.*, 2001), which is densely expressed both pre- and postsynaptically (Leino *et al.*, 1997). GLUT3 is not expressed by astrocytes, oligodendrocytes, or endothelial cells (Nagamutsi *et al.*, 1993; Morro and Yamada, 1994). These cells instead express the glucose transporter GLUT1, which is concentrated in astrocytic end feet and astrocytic processes surrounding synapses (Morgello *et al.*, 1995; McCall *et al.*, 1996; Yu and Ding, 1998), although it is also present in astrocytic cell bodies (Leino *et al.*, 1997). GLUT1 is expressed in the choroid plexus and ependymal cells (Hacker *et al.*, 1991; Cornford *et al.*, 1998; de los A Garcia *et al.*, 2001), but not in microglia, which express GLUT5 (Payne *et al.*, 1997; Yu and Ding, 1998). There is relatively good regional correlation between staining for glucose transporter and local CMR_{glc} (Wree *et al.*, 1988; McCall *et al.*, 1994; Gronlund *et al.*, 1996). Both GLUT1 and GLUT3 immunostaining increase in abundance in a region-specific manner following chronic seizures (Gronlund *et al.*, 1996).

b. Hexokinase. Some, although not perfect, correlation is found between density of glucose transporter sites and expression of hexokinase, which can be observed in the cytoplasm of neuronal, astrocytic, and choroid plexus cells as well as in the neuropil and purified synaptosomes (Wilkin and Wilson, 1977; Fields *et al.*, 1999). The distribution of hexokinase has been especially well examined in the cerebellar cortex (Kao-Jen and Wilson, 1980). Extensive staining of cytoplasmic regions, with some increased density at mitochondrial profiles was found in most types of neurons and their processes and in astrocytes, whereas oligodendrocytes showed no staining. The expression of dense staining for hexokinase in both neurons and astrocytes is consistent with the finding of almost identical values for hexokinase activities in cultured neurons and astrocytes (Lai *et al.*, 1999); the deficient staining in oligodendrocytes is mirrored by very low activity of hexokinase in cultured oligodendrocytes (Rust *et al.*, 1991), and a much lower CMR_{glc} in white than in gray matter (Sokoloff *et al.*, 1977). An exception to intense neuronal staining was Purkinje cells and part of their dendrites, which showed only little hexokinase expression. Granule cell dendrites were well stained in their proximal parts but void of stain in their terminal digits, which form part of the cerebellar glomeruli; in contrast, the mossy fiber terminals of brain stem neurons, with which the granule cells synapse, exhibited intense staining, as did synaptic vesicles adjacent to the mitochondria. Endothelial cells in brain microvessels express hexokinase activity (Djuricic and Mrsulja, 1979; de Cerqueira Cesar and Wilson, 1995).

c. PFK. All three isotypes of PFK have been found by immunohistochemistry in both neurons and astrocytes. M-type PFK is preferentially found perinuclearly, L-type PFK shows a characteristic staining in the cytoplasm and the processes of cells, whereas the C-type antibodies almost homogeneously stain whole cell bodies as well as large dendrites; because the PFK isoenzymes differ with respect to their allosteric properties, their differential distribution in different cell constituents might be of importance for the regulation of brain glycolysis in the different cellular compartments of the brain (Zeitschel *et al.*, 1996).

d. Pyruvate Kinase. Pyruvate kinase is expressed in both neurons and astrocytes, but appears to be especially prominent in large neurons and in nerve terminals (Gali *et al.,* 1981); pyruvate kinase staining may be absent in oligodendrocytes and microglia (van Erp *et al.,* 1988).

e. Lactate Dehydrogenase and Lactate Transporters. There are different isoforms of lactate dehydrogenase. The H4 tetramer, which shows much greater inhibition by a pyruvate/NAD^+ complex at the active site, is predominant in aerobic heart tissue, raising the possibility that the efflux of pyruvate as lactate is minimized and lactate is mainly converted to pyruvate in these tissues. On the other hand, muscle, which has mainly the M4 isoform, can operate anaerobically and needs to produce lactate from pyruvate.

High levels of lactate dehydrogenase reactivity are found in the neuropil in certain, specific afferent terminal fields (Borowsky and Collins, 1989a,b). It has been reported that M4, i.e., the isoenzyme favoring conversion of pyruvate to lactate is enriched in astrocytes compared to neurons (Bittar *et al.,* 1996; Pellerin *et al.,* 1998); however, this observation is not consistent with early work using *in vivo* immunofluorescense, which demonstrated approximately equal distribution of the H and M forms of lactate dehydrogenase in astrocytes and neurons in the CNS (Brumberg and Pevzner, 1975; Pevzner, 1979).

Pyruvate and lactate are transported across cell membranes via an equilibrative monocarboxylic acid transporter (MCT), which exists as different isotypes. It has been suggested that the MCT isotypes expressed by brain cells favor lactate formation and release from astrocytes and lactate uptake into neurons (Broer *et al.,* 1997; Pellerin *et al.,* 1998; Pierre *et al.,* 2000), but this proposal is controversial, since different MCT isoform distributions have been demonstrated by Gerhart *et al.* (1997, 1998) and Hanu *et al.* (2000). Moreover, it should be kept in mind that MCT mediates facilitated diffusion, and that sustained flux of lactate across a cell membrane in a given direction is determined by the transmembrane lactate concentration gradient (together with the H^+ gradient). Maintained net uptake of lactate from extracellular fluid therefore will be governed by rate of metabolism of lactate, which is much slower than the equilibrative transport process (Dienel and Hertz, 2001).

f. Synopsis of Immunochemistry. With the exception of oligodendrocytes, hexokinase is readily detectable in brain parenchymal cells, in brain endothelial cells, and in choroid plexus; at least in some pathways the density is greater pre- than postsynaptically. There is intense staining in the neuropil. PFK and pyruvate kinase are also expressed in both neurons and glial cells, but the level of pyruvate kinase appears to be low in oligodendrocytes. LDH is expressed in the neuropil, especially in specific afferent fields, and differences exist within both neurons and astrocytes according to the pathways with which they are associated.

C. Formation of Acetyl Coenzyme A (Acetyl-CoA)

1. *Acetyl-CoA Formation from Pyruvate*

Pyruvate oxidation is initiated by pyruvate entry into the mitochondrion, mediated by an MCT. The participation of MCTs both in transmembrane transport of lactate and pyruvate and in the entry of pyruvate from the cytosol into the mitochondria renders it difficult to utilize an MCT inhibitor in order to draw any conclusions about the importance of lactate (or pyruvate) as a metabolic fuel.

Inside the mitochondria, the pyruvate dehydrogenase (PDH) complex (PDHC) catalyzes the first step of pyruvate utilization to produce acetyl-CoA plus CO_2 and NADH from pyruvate, coenzyme A (CoASH) and NAD^+; in this thiamine-dependent step, carbons three and four of glucose (carbon one of pyruvate) are converted to CO_2, whereas the remaining carbon atoms are introduced into the TCA cycle (Fig. 2). Pyruvate dehydrogenase has a K_m for pyruvate of $\sim$0.05 m*M* (Ksiezak-Reding *et al.*, 1982), which is approximately equal to the pyruvate concentration in brain (Siesjö, 1978). The PDH multienzyme complex is composed of pyruvate dehydrogenase tetramers (each with two decarboxylase and two dehydrogenase sites), transacetylase, and lipoamide dehydrogenase. Activity of the PDH complex is regulated by phosphorylation at a serine residue on the pyruvate decarboxylase polypeptide to make PDH *inactive.* A Mg^{2+}- and Ca^{2+}-dependent phosphatase dephosphorylates and activates the PDH complex. Acetyl-CoA and NADH inhibit the active dephosphorylated form of PDH and are also positive effectors of the kinase, which will inactivate the enzyme. CoASH, NAD^+, and pyruvate are all PDH substrates that inhibit the PDH kinase and thereby activate PDH, as does ADP. Thus, metabolic demand regulates pyruvate utilization: increased precursor supply reduces inactivation of the PDH complex by the kinase, and products of the reaction both inhibit the active PDH complex and activate the kinase. Overload of the TCA cycle will cause acetyl-CoA and NADH to rise, thereby turning off pyruvate utilization, whereas increased energy demand raises the ADP level and activates the flux of pyruvate into the TCA cycle. Another stimulus for activation of PDH is an increase in intramitochondrial Ca^{2+}, resulting from transmitter-induced increase in free cytosolic Ca^{2+} concentration ($[Ca^{2+}]_i$) (McCormack and Denton, 1990), as will be discussed in Section IV.D.8. Because PDH is a regulated enzyme and its Km for pyruvate is similar to that of the brain pyruvate concentration, this metabolic step can be a transient "bottle-neck," in which flux of pyruvate into the oxidative pathway is limited compared to the rate of glycolysis, causing an increase in lactate formation in order to regenerate NAD^+ and maintain glycolytic flux, especially when brain work is suddenly increased.

Pyruvate dehydrogenase immunoreactivity has been observed in neuronal cell bodies (with pronounced differences between different neurons), proximal cell processes, and at several locations in the neuropil (Milner *et al.*, 1987; Bagley *et al.*, 1989; Calingasan *et al.*, 1994). No information is available about the cellular distribution of pyruvate dehydrogenase within the neuropil, but cortical mouse astrocytes in primary cultures show higher rates of pyruvate dehydrogenase-mediated flux from [1-^{14}C]pyruvate than cultures of cortical neurons (Hertz *et al.*, 1987).

2. *Acetyl-CoA Formation from Other Sources*

Although brain acetyl-CoA is mainly derived from pyruvate, it can also be formed from fatty acids, ketone bodies, and monocarboxylic acids like acetate. Lipid is not a major energy source in brain, and astrocytes are the only cell type to oxidize fatty acids as primary fuel, whereas neurons, astrocytes, and oligodendrocytes can metabolize ketone bodies (Edmond *et al.*, 1987; Auestad *et al.*, 1991). Formation of acetyl-CoA from acetate is of little physiological importance (although ethanol and the neurotransmitter acetylcholine are metabolized to acetate), but it is of considerable experimental interest, because acetate is preferentially transported into glial cells (Fig. 3A). Autoradiographic studies have localized acetate uptake to neuropil and astrocytes, whereas it is not accumulated into perikarya (Muir *et al.*, 1986). This finding is corroborated by the demonstration that [^{14}C]acetate is taken up much more rapidly in primary cultures of astrocytes than in primary cultures of neurons, as shown in Fig. 3B (O'Dowd, 1995; Waniewski and Martin, 1998). Therefore, acetate can be utilized as a "reporter molecule" of astrocyte metabolism. The first step in acetate metabolism in mammalian brain is conversion to acetyl-CoA by acetate thiokinase (acetyl-CoA synthase) in the presence of CoASH and ATP; this enzyme is present in both cultured astrocytes and synaptosomes (see in the next section). Label from the astrocytically accumulated [^{14}C]acetate can, after formation of acetyl-CoA, become incorporated into TCA-cycle-derived amino acids (Fig. 3A), and may eventually be transported to neurons due to cycling of glutamate, glutamine, and GABA between neurons and astrocytes (Section II.H.2–4). Since acetyl-CoA formation from pyruvate does not proceed in the opposite direction, acetate is not a precursor for pyruvate or for oxaloacetate, which is formed by pyruvate carboxylation (Section II.F.2).

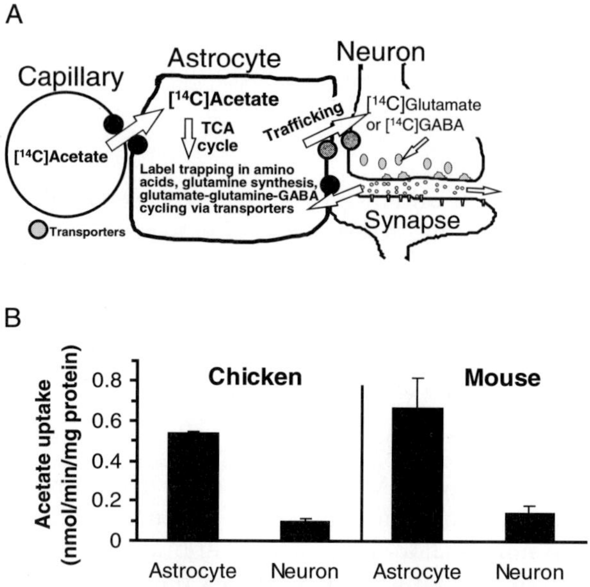

FIG. 3. Acetate is a "glial reporter molecule." (A) Preferential entry into the astrocyte and metabolic trapping in the amino acid pools provides a means for autoradiographic detection of a local increase in astrocytic activity and for NMR assays of astrocyte TCA cycle activity. This schematic drawing illustrates preferential uptake of blood-borne [^{14}C]acetate into astrocytes via a monocarboxylic acid transporter, incorporation into TCA-cycle-derived amino acids in astrocytes, and local trafficking of labeled compounds due to cycling of glutamate, glutamine, and GABA between neurons and astrocytes (see text). (Adapted from G. A. Dienel. Energy generation in the central nervous system. *In* "Cerebral Blood Flow and Metabolism, 2nd ed." (L. Edvinsson and D. Krause, eds.), 2002, Lippincott Williams & Wilkins.©) (B) Rates of [^{14}C]acetate uptake in primary cultures of chick and mouse astrocytes. Acetate uptake was measured during a 10-min period of incubation with 50 μM [2-^{14}C]acetate in tissue culture medium containing 6 mM glucose. The uptake was rectilinear and was calculated from accumulated radioactivity per mg protein and the specific activity of the incubation medium. SEM values are shown by vertical bars. In both chick and mouse cultures acetate uptake is significantly higher ($p < 0.05$ or better) in astrocytes than in neurons. (From O'Dowd, 1995, with the permission of O'Dowd.)

3. *Acetate and "Metabolic Compartmentation"*

a. Metabolic Compartmentation. The preferential uptake of acetate (and some other monocarboxylic acids) into astrocytes is consistent with pioneering tracer labeling experiments carried out between the late 1950s and the 1970s (reviewed in Berl *et al.*, 1975). The patterns of labeling of glutamate and glutamine in whole brain tissue by different radioactively labeled precursors were studied during an experimental period when the specific activities (μCi/mmol) of the compounds of interest were increasing *in vivo.*

When [^{14}C]glucose was used as the labeled precursor, the specific activity of an obligatory precursor, glutamate (formed from glucose via the TCA cycle intermediate α-ketoglutarate—see Sections II.D.1 and II.H.1), was always higher than that of its product, glutamine. This behavior is characteristic for a normal precursor-product relationship, because the specific activity of any compound will fall as the tracer is further metabolized and mixes with the pool of initially unlabeled product in the tissue. However, after administration of [^{14}C]acetate or several other compounds, the specific activity of glutamine was higher than that of its precursor, glutamate. This intriguing result led to the following conclusions: (1) at least two metabolically distinct compartments coexisted in brain (a "small" and a "large" metabolic compartment); (2) glutamate pools existed in both compartments; (3) glucose had equal access to both compartments; (4) acetate only entered one of them; and (5) the compartment into which acetate entered was the only compartment that synthesized glutamine from glutamate (Fig. 3A). Accordingly, administration of labeled acetate as the precursor leads to labeling of the entire glutamine pool in the tissue but to labeling of only one of the two (or more) glutamate pools in the tissue. Determination of specific activities by analysis of disrupted brain tissue does not distinguish between the different glutamate pools, and accordingly the specific activity of the labeled glutamate pool will be "diluted" with nonradioactive glutamate from the pool(s) into which acetate has no access, whereas little or no corresponding dilution will occur in glutamine, all of which had been formed in the pool labeled by acetate (Balazs and Cremer, 1972; Berl *et al.*, 1975; Berl and Clarke, 1969). Initially the small compartment was thought to represent presynaptic structures, but with the demonstration of glutamine synthetase as a glial-specific enzyme (Norenberg and Martinez-Hernandez, 1977) it became obvious that astrocytes at the least are included in the small compartment. Conceivably, kinetic and anatomical pools might not be the same, the large compartment may consist of several different metabolic pools, and more than one anatomical compartment (e.g., neuronal presynaptic structures and their adjacent astrocytic processes) might correspond to a single kinetic compartment.

b. Metabolic Studies with [^{13}C]- and [^{14}C]Acetate. Since acetate is not taken up into neurons and converted to acetyl-CoA it is possible to distinguish between neuronal and glial metabolism by following the metabolic fate of [^{13}C]acetate, using nuclear magnetic resonance spectroscopy (Cerdan *et al.*, 1990; Sonnewald *et al.*, 1993; Cruz and Cerdan, 1999). This method allows *in vivo* tracking of the metabolic fate of acetate, of the incorporation of its carbon into a neurotransmitter precursor (glutamine), and of the turnover of excitatory (glutamate) and inhibitory (GABA, γ-aminobutyric acid) transmitters, formed from glutamine not only under physiological conditions, but also under such pathophysiological conditions as brain ischemia (Pascual *et al.,* 1998; Haberg *et al.,* 1998a, 2001). Also, enhanced accumulation of label from [^{14}C]acetate in brain tissue during activation is an indication of increased astrocytic metabolism (see also Section III,F and Figs. 11,13).

D. The TCA Cycle and Electron Transport Chain

1. *The Cycle Pathway*

Acetyl-CoA enters the TCA cycle by condensation with oxaloacetate to form citrate, which is a branch point for carbon efflux from the TCA cycle (Fig. 2, Table III). Citrate can be exported from mitochondria and serve as a precursor of the neurotransmitter acetylcholine and cytoplasmic oxaloacetate. Acetylcholine synthesis accounts for only about 1% of the pyruvate decarboxylated (Gibson *et al.,* 1976). The inhibition of PFK by citrate (Section II.B.2) influences pyruvate availability for the synthase. Purified citrate synthase is inhibited by ATP, NADH, and succinyl-CoA. The next TCA cycle steps are catalyzed by aconitase, and involve successive dehydration and rehydration steps to form isocitrate. Similar to acetate, the highly toxic aconitase inhibitor fluoroacetate is preferentially taken up by astrocytes (Muir *et al.,* 1986); in the astrocytes it is converted to fluorocitrate which strongly inhibits aconitase. For this reason, fluoroacetate is an inhibitor of the astrocytic TCA cycle and, at appropriately low doses, has no effect on the neuronal TCA cycle (Clarke *et al.,* 1970; Fonnum *et al.,* 1997). Isocitrate dehydrogenase carries out the first oxidative decarboxylation step in the TCA cycle and produces CO_2, NADH, and α-ketoglutarate (α-KG). This enzyme is stimulated by ADP and inhibited by ATP and NADH, thereby reducing its activity when high-energy phosphate levels are adequate; it is also stimulated by an increase in $[Ca^{2+}]_i$ (McCormack and Denton, 1990). α-KG is a major branch point metabolite in the TCA cycle (Fig. 2), and its carbon rapidly exchanges with the glutamate amino acid pool, which has the highest concentration (about 12 μmol/g) of all amino acids in brain (Table II).

The second oxidative decarboxylation step of the TCA cycle is carried out by the α-KG dehydrogenase complex (KGDHC), which like PDH is a thiamine-dependent reaction; it converts α-KG to succinyl CoA and produces NADH. From Table III it can be seen that this step is rate limiting for TCA cycle activity, at least in humans (Sheu and Blass, 1999). Regulation of KGDHC activity is very complex, involving substrate, co-factors, products, energy metabolites, and Ca^{2+}. ATP, GTP, NADH, and succinyl-CoA inhibit the enzyme, whereas ADP, P_i, and Ca^{2+} enhance its activity, although it is inhibited by high levels of Ca^{2+}. Inhibition of KGDHC secondary to thiamine deficiency leads to selective cell death, which is associated with the Wernicke-Korsakoff syndrome in chronic alcoholism (Gibson *et al.,* 2000).

TABLE III
ENZYMATIC STEPS OF THE TRICARBOXYLIC ACID (TCA) CYCLE

Sequential enzymatic step	Reaction[a]	Maximal velocity[b] (μmol min^{-1} g wet wt^{-1}) Various species	Human
1. Pyruvate dehydrogenase	Pyruvate + NAD^+ + CoA-SH → acetyl CoA + NADH + CO_2	2–3.6	1.1
2. Citrate synthase	Acetyl-CoA + oxaloacetate → citrate + CoASH	2	4 (12,biopsy)
3. Aconitase	Citrate ↔ [*cis*-aconitate] ↔ isocitrate	1.5	1
4. Isocitrate dehydrogenase	Isocitrate + NAD^+ → α-ketoglutarate + NADH + CO_2	2	1
5. α-Ketoglutarate dehydrogenase	α-Ketoglutarate + CoASH + NAD^+ → succinyl-CoA + NADH + CO_2	1.7	0.2
6. Succinyl-CoA synthetase	Succinyl-CoA + GDP + Pi → succinate + GTP + CoASH		
7. Succinate dehydrogenase	Succinate + FAD → fumarate + $FADH_2$	9.7	10
8. Fumarase	Fumarate + H_2O ↔ L-malate	33	
9. Malate dehydrogenase	L-Malate + NAD^+ → oxaloacetate + NADH	88	18
Net reaction	1Pyruvate + GDP + 4NAD^+ + FAD + Pi → 3CO_2 + GTP + 4NADH + $FADH_2$		
ATP production by oxidative phosphorylation[c]	4NADH + 12ADP + 12Pi + 2O_2 → 4NAD^+ + 12ATP + 4H_2O		
	$FADH_2$ + 2ADP + 2Pi + 1/2O_2 → FAD + 2ATP + H_2O		

[a] Reactions do not include hydrogen ions.

[b] Rate data from mammalian brain were compiled from literature sources by McIlwain and Bachelard (1985).Values from human brain were calculated from data summarized in Sheu and Blass (1999), assuming 100 mg protein (g brain tissue)$^{-1}$; values for human electron transport Complex I, Complex II/III, and Complex IV are 2, 98, and 285 μmol min^{-1} (g wet wt)$^{-1}$, respectively (Sheu and Blass, 1999). At steady state, the rate of the TCA cycle is about twice that of glucose utilization due to formation of two moles of pyruvate per mole glucose.

[c] The total maximal yield of ATP per mole glucose (2 moles pyruvate) is 36–38, i.e., 28 ATP (24 via NADH + 4 via $FADH_2$) + 2GTP = 30ATP from TCA cycle oxidation of pyruvate, plus 4–6 ATP from TCA cycle oxidation of two moles of cytoplasmic NADH, plus 2ATP from glycolysis.

(Modified from G. A. Dienel. Energy generation in the central nervous system. *In* "Cerebral Blood Flow and Metabolism, 2nd ed." (L. Edvinsson and D. Krause, eds.), 2002, Lippincott Williams & Wilkins.©)

The α-KGDH complex shows immunochemical staining in neurons, glia, and neuropil throughout the brain, but some regions express denser perikaryal staining than other areas (Calingasan *et al.*, 1994). However, reduction in α-KGDH activity is not accompanied by a reduction of staining (Sheu *et al.*, 1998). Therefore a method has been developed, which provides quantitative staining of α-KGDH activity rather than of the expression of the enzyme protein; the distribution of the α-KGDH activity in the resting brain differs from that of the protein (Park *et al.*, 2000).

Succinyl-CoA is derived mainly from α-KG, but it can also be produced from degradation of odd-chain length fatty acids and branched-chain α-keto acids derived from amino acid catabolism. Succinyl-CoA synthase (thiokinase) forms succinate and phosphorylates GDP to GTP, which energetically equals formation of one ATP. Succinate is also a catabolite of the neurotransmitter GABA (γ-aminobutyrate); an aminotransferase converts GABA to succinic semialdehyde (SSA), and SSA dehydrogenase forms succinate plus NADH. Succinate is oxidized to fumarate in a reaction with FAD, instead of NAD^+, by succinate dehydrogenase. Succinate dehydrogenase is inhibited by oxaloacetate (and malonate, a competitive inhibitor), and is activated by P_i, and succinate. Fumarate is symmetric around a double bond, which is of importance for labeling of individual carbon atoms during TCA cycle activity, because specifically labeled fumarate (e.g., C1) becomes symmetrically labeled (e.g., C1 and C4) malate. In its hydration the OH group adds to only one side of the double bond and produces L-malate, which is oxidized by malate dehydrogenase (L-malate:NAD^+ oxidoreductase) to form oxaloacetate and produce the third molecule of NADH per turn of the cycle. Oxaloacetate can be transaminated to form aspartate, coupling the TCA cycle with a second large amino acid pool (about 3 μmol/g; Table II). Since each turn of the TCA cycle is initiated by condensation of oxaloacetate with acetyl-CoA and the cycle terminates by regenerating another molecule of oxaloacatetate, there is no net consumption or generation of TCA cycle intermediates during the process of oxidation of pyruvate (Fig. 2).

2. *Turnover Rate*

The formation of glutamate from α-KG is so rapid and the glutamate pool so large that incorporation of radioactivity from [^{14}C or ^{13}C]glucose initially may occur rectilinearly with time and at a rate which is equal to the rate of turnover of the TCA cycle (Mason *et al.*, 1992, 1995; Rothman *et al.*, 1999), provided all label is trapped in the tissue (Section II.H.2). By using [^{13}C]acetate as the precursor, it is possible to determine TCA cycle turnover rates in the neuronal and glial TCA cycle, and Cruz and Cerdan (1999) calculated turnover rates in the rat brain of 1.0 μmol/min/g and

0.4 μmol/min/g, respectively, under resting conditions. Since the TCA cycle turns over once for every pyruvate molecule metabolized and since two molecules of pyruvate are formed from one molecule of glucose, the sum of neuronal and glial metabolism (1.4 μmol/min/g) is consistent with the CMR_{glc} of 0.7 μmol/min/g.

3. *The Electron Transport Chain*

Pyruvate utilization is coupled to reduction of NAD^+ and FAD to produce NADH and $FADH_2$ (Fig. 2, Table III). These cofactors transfer electrons to oxygen via the electron transport chain. Complex I is the NADH dehydrogenase component that accepts electrons and transfers them to coenzyme Q. Complex II is the succinate dehydrogenase flavoprotein component that transfers electrons to coenzyme Q. Electrons are sequentially transferred from coenzyme Q to Complex III (cytochrome bc_1), cytochrome c, and Complex IV (cytochrome oxidase), where oxygen is the terminal acceptor. The electron transport process is coupled to pumping of protons across the inner mitochondrial membrane at three points (Complex I, III, IV) to produce an electrochemical proton gradient (the proton motive force); movement of protons back across the membrane is used to drive ATP synthesis. Complete oxidation of pyruvate in respiring mitochondria (Fig. 2, Table III) produces 3 CO_2 + 4 NADH + $FADH_2$ + GTP. If there is perfect coupling of electron transport with ATP synthase, each NADH produces 3 ATP, each $FADH_2$ gives 2 ATP, and GTP is equivalent to ATP, for a net yield of 15 ATP/molecule pyruvate or 30 ATP/molecule glucose. If reducing equivalents generated by oxidation of the NADH derived from the cytoplasmic glyceraldehyde-3-P dehydrogenase reaction are shuttled into the mitochondria and oxidized, an additional 4–6 ATP is obtained, depending upon which shuttle is utilized (see Section II.E). Thus, oxidative metabolic steps that are linked to the electron transport chain produce most of the ATP generated by complete oxidative glucose catabolism.

Detailed studies of the histochemical localization of cytochrome oxidase in hippocampus were initiated by Kageyama and Wong-Riley (1982) and Borowsky and Collins (1989a). Marked differences were observed not only between different types of neurons but also between dendrites, neuropil (which generally is densely labeled; Borowsky and Collins, 1989b), cell perikarya, and axonal endings (which often show very little labeling). Among the glial cells, astrocytes constituting the glia limitans are heavily stained, but most other glial cell bodies show only little staining. However, high activities of cytochrome oxidase in astrocytes from glutamatergic regions have been demonstrated by Aoki *et al.* (1987b). A correlation has also been observed between cytochrome oxidase and Na^+,K^+-ATPase expression in monkey hippocampus and striate cortex (Hevner *et al.*, 1992; Wong-Riley *et al.*, 1998), suggesting that regions with high energy requirements for ion pumping have high capacity for oxidative metabolism.

The striking differences in cytochrome oxidase staining between different neurons have been further examined in the primate striate cortex, where GABAergic neurons receive mainly glutamatergic, excitatory input, whereas glutamatergic neurons receive GABAergic inhibitory innervation. The GABAergic, glutamatergically innervated neurons, were found to have three times as many mitochondria darkly reactive for cytochrome oxidase activity as the glutamatergic, GABAergically innervated neurons (Nie and Wong-Riley, 1996). However, strongly cytochrome oxidase-positive areas in the striate monkey cortex also frequently coincide with areas expressing high GFAP immunoreactivity, i.e., reflecting astrocytic localization (Colombo *et al.*, 1999).

E. Regeneration of Cytosolic NAD$^+$

1. *Malate-Aspartate Shuttle*

NADH and NAD$^+$ are present in brain at low concentrations (~20 and 300 nmol/g, respectively) and act catalytically. The NADH produced by cytoplasmic glyceraldehyde-3-P dehydrogenase (Table I) must, therefore, be continuously reoxidized to supply NAD for the glycolytic rate to be maintained; this can occur by coupling to respiration or to lactate production. NADH cannot traverse the mitochondrial membranes, and instead reducing equivalents are transferred from the cytosol into mitochondria. In brain, this transfer occurs mainly via the malate-aspartate shuttle (MAS) (Fig. 4). In this shuttle *cytoplasmic* NADH is oxidized to NAD$^+$ by reduction of cytosolic oxaloacetate to malate, which traverses the mitochondrial membrane in exchange for α-KG. Mitochondrial malate dehydrogenase converts malate to oxaloacetate and in the process generates NADH, which is oxidized via the electron transport chain, generating three ATP. Oxaloacetate is transaminated intramitochondrially with glutamate, catalyzed by aspartate aminotransferase to form aspartate together with α-KG. The mitochondrial aspartate is then exchanged for cytoplasmic glutamate via another antiporter, and cytoplasmic aspartate aminotransferase regenerates oxaloacetate by transamination of aspartate with α-KG, which is converted to glutamate.

The MAS is critical not only for transfer of reducing equivalents from cytoplasm to mitochondria, but also for exchange of glucose-derived carbon between the mitochondria and cytoplasm. In metabolic experiments using labeled precursors, such as glucose or acetate, this shuttle process should facilitate mixing of labeled products derived from TCA cycle intermediates

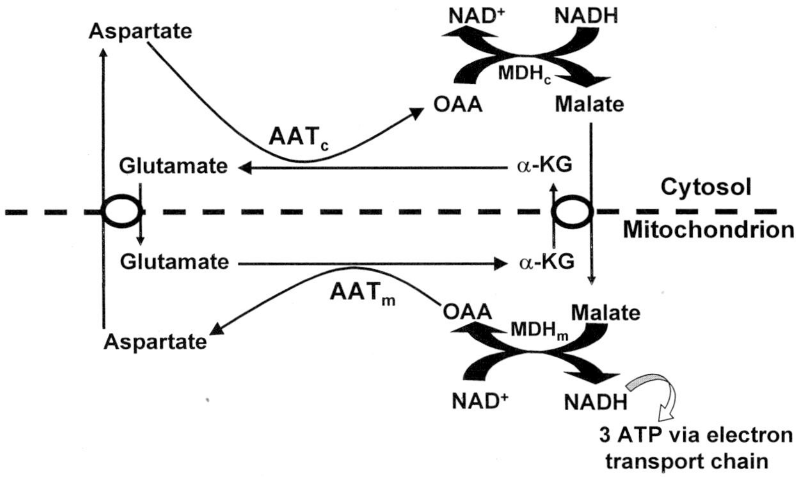

FIG. 4. The malate-aspartate shuttle (MAS) links oxidation-reduction reactions in the cytosol with electron transport and oxidative phosphorylation in the mitochondrion. Because NAD^+ and NADH cannot cross the mitochondrial membrane, reducing equivalents are transferred into the mitochondria by coupled transport of metabolites. The cytosolic malate dehydrogenase (MDH_c) oxidizes NADH produced by glycolysis (glyceraldehyde phosphate dehydrogenase step) and produces malate (top right region of schematic drawing), which enters the mitochondria via an antiporter in exchange for α-ketoglutarate (α-KG). The mitochondrial malate dehydrogenase (MDH_m) oxidizes malate to oxaloacetate (OAA) and produces NADH, which yields three ATP via the electron transport chain. The OAA is then transaminated to form aspartate by the mitochondrial aspartate aminotransferase (AAT_m). Aspartate is exported to the cytoplasm via an antiporter in exchange for glutamate, which supplies the other substrate for the mitochondrial transaminase reaction, producing the α-KG which is exported as malate is imported. Cytoplasmic aspartate aminotransferase (AAT_c) then regenerates OAA and glutamate to complete the cyclic process. Note that α-KG and OAA are also intermediates in the tricarboxylic acid (TCA) cycle, so if labeled precursors (derived from labeled glucose or acetate tracers in metabolic assays) enter the TCA cycle, the label can be exported to cytoplasm via the metabolite exchange processes. Thus, the highly labeled TCA cycle metabolites can mix with the (presumably) much larger unlabeled amino acid pools in the cytoplasm and thereby help to trap label by reducing the specific activity of labeled products produced by the TCA cycle. It is not known whether the efficiency of metabolic trapping of glucose-derived label in the large amino acid pools is dependent upon mitochondrial-cytoplasmic exchange of intermediates; conceivably an increase in lactate production and export from the cell might impair this exchange, as well as eliminate a high specific activity metabolite (i.e., lactate) from an activated cell.

with the larger cytoplasmic amino acid pools (see legend, Fig. 4), suggesting that coupling of NADH oxidation to lactate production instead of MAS activity might reduce label trapping in amino acids for two reasons: (1) glucose-derived label would be lost when labeled lactate is cleared from the activated tissue; and (2) there would be less mixing of the glucose-derived label that did enter the TCA cycle, due to reduced exchange of labeled amino and keto acids between the mitochondrial and cytoplasmic pools by means of this shuttle system.

2. *The Glycerol Phosphate Shuttle*

An alternative pathway, the glycerol phosphate shuttle, has low activity in brain (Siesjö, 1978). This shuttle, which interconverts dihydroxyacetone-P and glycerol-3-phosphate, yields less ATP (two molecules per cytoplasmic NADH), because it generates $FADH_2$, not NADH in the mitochondria.

3. *Lactate Formation*

When the rate of glycolysis exceeds the rate of triose entry into the TCA cycle or when oxygen is limiting, NADH oxidation is achieved in the cytoplasm by reduction of pyruvate to lactate by lactate dehydrogenase (LDH), without production or utilization of ATP. Rapid clearance of lactate from the cells in which it has been produced must be a necessary and integral component of this process, because local lactate accumulation would otherwise become an opposing driving force that would influence many reversible NAD^+/NADH-coupled redox reactions, including continued conversion of pyruvate to lactate under oxygenated conditions (Dienel and Hertz, 2001). This problem is overcome by release of lactate from the cells, resulting in lactate overflow and/or lactate release to blood and cerebrospinal fluid, as will be discussed in Section III.D.2.

F. TCA Cycle Expansion and Elimination of TCA Cycle Constituents

1. *Anaplerotic and Cataplerotic Reactions*

An anaplerotic reaction is a biosynthetic process leading to expansion of the total pool of constituents in a pathway, e.g., by *de novo* synthesis of TCA cycle intermediates from pyruvate. The previously described PDH-mediated conversion of pyruvate to acetyl-CoA and the subsequent condensation between the acetate moiety of acetyl-CoA with the four-carbon TCA cycle intermediate, oxaloacetate, to form one molecule of citrate, is *not* an anaplerotic reaction, because it does not give rise to an enlargement of the pool of TCA cycle intermediates. Since TCA cycle intermediates are consumed in biosynthetic reactions (e.g., by diversion of α-KG and oxaloacetate away from the TCA cycle to form glutamate and aspartate, respectively), these compounds must be replenished, and anaplerosis is essential for brain function. The most important anaplerotic reaction in brain is condensation of CO_2 with pyruvate to generate oxaloacetate (Patel, 1974), catalyzed by the pyruvate carboxylase (PC). Other routes by which glycolytic intermediates lead to an expansion of the pool of TCA cycle intermediates, or which allow TCA cycle intermediates to be taken out of the TCA cycle (a process known as cataplerosis) and converted to glycolytic intermediates, are interconversions between pyruvate plus CO_2 and malate, catalyzed by malic enzyme, and between phosphoenolpyruvate plus CO_2 and oxaloacetate, catalyzed by phosphoenolpyruvate carboxykinase.

2. *Pyruvate Carboxylase*

Pyruvate carboxylase (PC) is the major brain enzyme catalyzing CO_2 fixation, and thus net formation of TCA cycle intermediates from pyruvate (Patel, 1974). Besides depending upon the substrates pyruvate and CO_2 (bicarbonate), pyruvate carboxylation requires hydrolysis of one molecule of ATP. PC is a widespread biotin-dependent enzyme, which consists of four identical subunits, arranged in a tetrahedron-like structure. Each subunit contains three functional domains: biotin carboxylation, transcarboxylation, and biotin carboxyl carrier. Pyruvate carboxylase is a tightly regulated allosteric enzyme, and acetyl-CoA is a positive modulator that is required for synthesis of oxaloacetate. Following pyruvate carboxylation, the newly formed oxaloacetate (Fig. 2) reacts with acetyl CoA to form citric acid, from which any other TCA cycle constituent can be synthesized.

Since pyruvate is introduced differently into the TCA cycle by PDH and by PC, exposure of cells to [1-^{14}C]glucose or [3-^{14}C]lactate leads to labeling of different carbon atoms in the citrate, α-KG, glutamate, and glutamine molecule when they are formed by PC compared to labeling by PDH activity. The differential labeling patterns can be detected in the brain *in vivo* by NMR analysis, allowing determination of the relative rate of pyruvate carboxylation, which consistently has been found to correspond to 10–20% of total TCA cycle activity, with higher percentage values for pyruvate carboxylation in human than in rodent brain (Lapidot and Gopher, 1994; Aureli *et al.*, 1997; Gruetter *et al.*, 1998, 2001; Sibson *et al.*, 2001). Enhanced pyruvate carboxylation in humans probably reflects a higher density of glial cells in human brain (Bass *et al.*, 1971), since pyruvate carboxylase has been biochemically (Yu *et al.*, 1983) and histochemically (Shank *et al.*, 1985) demonstrated in astrocytes but is absent in neurons. The selective astrocytic localization of PC activity is of major importance for brain function because this enzyme is required for *de novo* synthesis of transmitter glutamate, thereby requiring coordinated metabolic interactions between neurons and astrocytes and substrate transport ("metabolic trafficking") between these two cell types.

3. *Malic Enzyme-Mediated Flux*

Malic enzyme (decarboxylating L-malate:oxidoreductase) differs from malate dehydrogenase by catalyzing combined decarboxylation/carboxylation and oxidation/reduction reactions between malate and pyruvate. The enzyme exists in two isoforms (mitochondrial and cytoplasmic [Bukato *et al.*,

1995]), of which the cytosolic is strictly dependent upon $NADP^+$, whereas the mitochondrial isoform can operate with either $NADP^+$ or NAD^+. In the presence of high concentrations of pyruvate, malic enzyme may catalyze an anaplerotic CO_2-fixation reaction coupled with reduction of $NADP^+$ to produce malate from pyruvate (Hassel and Brathe, 2000). However, it is generally assumed that this enzyme in the brain mainly catalyzes formation of pyruvate from malate, i.e., takes TCA cycle intermediates out from the TCA cycle by oxidizing and decarboxylating them to form pyruvate (pyruvate recycling), which can then be completely oxidized to CO_2 via PDH and TCA cycle activity, with or without intermittent formation of lactate. Cruz and Cerdan (1999) have concluded that pyruvate recycling in rat brain amounts to 0.3 μmol/min/g wet wt, or about 20% of a turnover rate in the cycle (1.4 μmol/min/mg protein), i.e., a value which is reasonably close to the estimate of entry into the TCA cycle by pyruvate carboxylation. The cytosolic malic enzyme is abundant in astrocytes (Martinez-Rodriguez *et al.,* 1989), where it catalyzes formation of pyruvate/lactate from glutamate and other TCA cycle intermediates and their derivatives (Sonnewald *et al.,* 1993, 1997; Bakken *et al.,* 1998b; Haberg *et al.,* 1998b); this isoform appears to be absent in neurons (Kurz *et al.,* 1993; McKenna *et al.,* 1995). Most pyruvate recycling seems to occur in astrocytes (Waagepetersen *et al.,* 2002).

4. *Phosphoenolpyruvate Carboxykinase-Mediated Flux*

The phosphoenolpyruvate carboxykinase reaction requires GTP as a phosphate donor and operates exclusively to convert oxaloacetate to phosphoenolpyruvate. This enzyme is important for gluconeogenesis from TCA cycle intermediates and from pyruvate because the interconversion between pyruvate and phosphoenolpyruvate, for thermodynamic reasons, is not able to generate phosphoenolpyruvate directly from pyruvate (Fig. 2). Accordingly, pyruvate is initially carboxylated to oxaloacetate in an ATP-dependent step, and oxaloacetate is then converted in a GTP-dependent step to phosphoenolpyruvate. This energy-dependent synthetic reaction is favored when the ATP/ADP ratio is high and pyruvate is present in excess amounts.

G. Glycogen Turnover

1. *Glycogen Synthesis*

Glycogen is the major reserve of glucose in the brain and is located mainly in astrocytes (Ibrahim, 1975). It is derived from glucose-6-P via glucose-1-P and the nucleotide sugar, UDP-glucose (Fig. 2). Whole brain glycogen levels are about 2–5 μmol glucose equivalent per gram, and there are regional differences in glycogen content. The equilibrium of the phosphoglucomutase reaction between glucose-6-P and glucose-1-P is such that the level of glucose-1-P in brain is about 7% that of glucose-6-P, and carbon can be drawn from the glycolytic pathway as needed. The role of glycogen in brain energy homeostasis is not well-established; it probably serves as a reservoir to rapidly supply fuel during the interval between activation of metabolism and increased delivery.

In addition to synthesis of glycogen from glucose, gluconeogenesis, i.e., formation of glycogen from pyruvate, alanine, or TCA cycle constituents has been demonstrated in brain tissue (Prokhorava *et al.*, 1978; Bhattacharya and Datta, 1993) and in astrocytes (Hevor *et al.*, 1986; Dringen *et al.*, 1993; Huang *et al.*, 1994; Schmoll *et al.*, 1995a), but the quantitative contribution of this pathway to glycogen synthesis is not known. The first steps of gluconeogenesis from pyruvate are formation of oxaloacetate by PC and decarboxylation to phosphoenolpyruvate. Phosphoenolpyruvate is then converted by a reversal of the glycolytic process to fructose-1,6-P_2 (Fig. 2), which is dephosphorylated in the one position to form fructose-6-P by fructose-1,6-bisphosphatase, a different enzyme than that catalyzing the catabolic glycolytic reaction. Fructose-1,6-bisphosphatase is present in brain; it is an allosteric enzyme, strongly inhibited by AMP (i.e., it requires high energy charge) and stimulated by 3-phosphoglycerate and citrate. It appears to be be astrocyte-specific (Hevor *et al.*, 1986; Schmoll *et al.*, 1995b), although fructose-1,6-bisphosphatase mRNA isoforms have been demonstrated in neurons (Loffler *et al.*, 2001). Fructose-6-P can be converted via glucose-6-P to glycogen.

2. *Glycogenolysis*

Glycogenolysis is catalyzed by phosphorylase, a cytosolic enzyme which is present as an inactive form, phosphorylase b, and an active form, phosphorylase a. Phosphorylase b is converted to phosphorylase a by phosphorylation, mediated by phosphorylase kinase which, in turn is converted from a low-activity form to a high-activity form by phosphorylation, catalyzed by protein kinase A, or by an increase in $[Ca^{2+}]_i$. Therefore, glycogenolysis can be stimulated either by transmitters acting by stimulating adenylyl cyclase or by transmitter or depolarizing procedures increasing $[Ca^{2+}]_i$ as will be discussed in Sections IV.D.3 and IV.D.8. Turnover of glycogen is tightly regulated, but in contrast to metabolism of glucose, glycolytic breakdown of glucose equivalents in glycogen is not dependent upon initial "priming" with ATP. Glycogen is quickly mobilized in response to abnormally high demand for glycolytically derived energy (e.g., hypoxia or ischemia [Siesjö, 1978]), during normal sensory stimulation (e.g., whisker movement in rodents [Swanson *et al.*, 1992]), and during specific stages of learning in day-old chicks (O'Dowd *et al.*, 1994; O'Dowd, 1995). However, chronic deafferentiation of barrel cortex by clipping of the whiskers also elevates the expression of glycogen phosphorylase (Dietrich *et al.*, 1982). Glycogenolysis in the brain is increased in a Ca^{2+}-dependent manner by elevated extracellular concentrations of K^+ (Hof *et al.*, 1988) and by many transmitters (Section IV). On account of the very low activity of glucose-6-phosphatase in brain (Nelson *et al.*, 1987), glycogenolysis in cultured cells gives rise to release of lactate, rather than of glucose (Wiesinger *et al.*, 1997). This is in contrast to the situation in liver where the level of glucose-6-phosphatase is high. Glycogen phosphorylase has been demonstrated in astrocytes, ependymal cells, and retinal Muller cells, whereas oligodendrocytes and most neurons are negative, and choroid plexus cells stain poorly or not at all (Pfeiffer *et al.*, 1990, 1992, 1994). In the neuropil, immunopositive astrocytic processes are frequently observed close to synaptic structures (Richter *et al.*, 1996). Thus, turnover of glycogen, the major carbohydrate energy reserve in brain, is probably linked to astrocytic demands for energy and/or glucose carbon as they interact with neurons during functional changes in neurotransmission.

H. Synthesis of Amino Acids

1. *Metabolic Pathways*

The nonessential amino acids which also serve as neurotransmitters (e.g., aspartate, glutamate, GABA, glycine) cannot readily cross the blood-brain barrier and must be synthesized in the brain from glucose plus an amino group donor. Three amino acids are produced from glycolytic intermediates, and four others from TCA cycle intermediates (Fig. 2). Serine is produced from 3-P-glycerate in three steps. Glycine, an inhibitory neurotransmitter, is synthesized from serine via serine hydroxymethyltransferase. Alanine is formed by transamination of pyruvate. Glutamate and aspartate are formed from α-KG and oxaloacetate, respectively, and thus require *de novo* synthesis of TCA cycle constituents; glutamine is formed from glutamate in astrocytes (Norenberg and Martinez-Hernandez, 1977) and other glial cells (D'Amelio *et al.*, 1990; Tansey *et al.*, 1991). GABA is formed in neurons from glutamate by the action of the pyridoxal phosphate-dependent enzyme, glutamate decarboxylase. GABA release from cultured cerebral cortical interneurons is considerably slower than release of glutamate from cultures of cerebellar granule cell neurons (Hertz and Schousboe, 1987); therefore synthesis of transmitter glutamate may create a greater demand for pyruvate carboxylation than synthesis of transmitter GABA. It is important to distinguish between [^{14}C]- or [^{13}C]glucose incorporation into glutamate, which mainly reflects isotope exchange between α-KG and glutamate, due to high activity of the aspartate aminotransferase (AAT; the reaction that allows determination of TCA cycle activity by incorporation of label from

[^{13}C]glucose into glutamate), and net synthesis of glutamate, accompanied by pyruvate carboxylation and by an increased pool size of glutamate.

2. *Rates of Glutamate and Glutamine Synthesis and Glutamate-Glutamine Cycling*

a. Determination of the Turnover Rate in the TCA Cycle by NMR. The AAT-mediated isotope exchange between α-KG and glutamate in brain is so rapid that determination of the flux of the ^{13}C label from [1-^{13}C]glucose into glutamate has been used to provide an *in vivo* measurement of the cerebral TCA cycle rate, from which the turnover rate in the TCA cycle rate, and thus CMR_{O_2} can be calculated (Mason *et al.*, 1992, 1995; Rothman *et al.*, 1999). In support of the validity of this approach, the average rate of resting oxidative metabolism in the nonstimulated human brain calculated in this manner (Sibson *et al.*, 1998; Gruetter *et al.*, 2001) corresponds well to independent measurements of CMR_{glc}. Such correspondence can, however, only be expected if there is quantitative trapping of ^{13}C by glutamate, a condition that may not always be fulfilled during increased functional activity (see Section III.D).

Since PDH activity during the first turn of the TCA cycle specifically incorporates ^{13}C from [1-^{13}C]- or [2-^{13}C]-labeled glucose into carbon 4 or C3, respectively, in the glutamate molecule, determination of label associated with these carbon atoms reflects specifically PDH-mediated entry into the TCA cycle, whereas PC-mediated entry of [1-^{13}C]glucose is reflected by labeling of glutamate carbon 2.

b. Formation of Glutamate by de novo Synthesis from Glucose. Because the concentrations of the TCA cycle intermediates in brain are quite low (totaling less than 1 μmol/g [Table II], versus ~25 μmol/g for all glucose-derived amino acids) and glutamate turnover is rapid, net synthesis of TCA-cycle-derived amino acids is the major reason for an intense pyruvate carboxylation in brain; without continuous *de novo* synthesis of oxaloacetate, glutamate synthesis would quickly drain the catalytic pool of TCA cycle intermediates and reduce the capacity of the cycle. The rate of *de novo* synthesis of glutamate (plus aspartate) can be approximated by the rate of pyruvate carboxylation, which as already mentioned, amounts to 10–20% of total entry of pyruvate into the TCA cycle. Since synthesis of one molecule of oxaloacetate requires not only formation of one molecule oxaloacetate but also condensation of oxaloacetate with one molecule acetyl-CoA, *de novo* synthesis of glutamate in astrocytes in the resting brain must account for 20–40% of total brain glucose metabolism. Since pyruvate carboxylation is absent in neurons, a precursor of glutamate's carbon skeleton (perhaps α-KG or glutamine, formed in astrocytes from α-KG via glutamate by glutamine synthetase [Peng *et al.*, 1993; Hassel *et al.*, 1997]) must be transported from astrocytes to neurons (Fig. 5) by metabolic trafficking (reviewed by Hertz *et al.*, 1999, 2000). Glutamine synthetase is ubiquitously expressed throughout brain in astrocytes and other glial cells (Norenberg and Martinez-Hernandez, 1977; D'Amelio *et al.*, 1990; Tansey *et al.*, 1991). The glutamate/glutamine α-amino nitrogen group can be supplied by various amino acids, e.g., alanine (Westergaard *et al.*, 1993; Erecinska *et al.*, 1994) and branched chain amino acids (Yudkoff, 1997; Hutson *et al.*, 2001), whereas the amide nitrogen of glutamine comes from ammonia (Cooper and Plum, 1987).

c. Glutamine Formation and the Glutamate-Glutamine Cycle. The rate of glutamine synthesis can be determined in the intact brain by incorporation of [^{15}N]ammonia into glutamine in *in vivo* MRS labeling studies; it is higher that that of *de novo* glutamate synthesis from glucose because an additional process contributes to glutamine synthesis: continuous glutamate-glutamine cycling linked to neurotransmission. The glutamate-glutamine cycle involves conversion of transmitter glutamate into glutamine after glutamate is released from neurons and accumulated in astrocytes. This glutamine is then released for uptake by neurons, where its hydrolysis replenishes the transmitter glutamate and GABA pools; this process does not require synthesis of the carbon skeleton, whereas *de novo* glutamate synthesis does (Fig. 5). Detoxification of ammonia entering brain from blood also contributes to glutamine synthesis, particularly under hyperammonemic conditions (Cooper and Plum, 1987; Lapidot and Gopher, 1997; Sibson *et al.*, 2001). The rate of glutamine synthesis in resting human brain corresponds to ~40% of the turnover rate of glucose in the TCA cycle, and the rate of glutamine synthesis via the glutamate-glutamine cycle is about twice that of the glutamate synthesis via the anaplerotic CO_2 fixation step (Gruetter *et al.*, 2001). In lightly anesthetized rat brain a higher rate of glutamine synthesis relative to TCA cycle turnover has been reported (Sibson *et al.*, 1998).

Modeling and determination of the *in vivo* rates of the TCA cycle in neurons and astrocytes, anaplerotic reactions in astrocytes, and glutamate-glutamine cycling between neurons and astrocytes are currently under intensive study because they are critically important to understanding metabolic interactions between the two major brain cell types during functional activity and the cellular basis for metabolic brain imaging. One of the unparalleled strengths of *in vivo* NMR studies is that the time courses of labeling of all detectable brain metabolites from a common precursor can be simultaneously measured under steady-state conditions in each subject, so uncertainties arising from animal-to-animal variations are minimized when metabolic fluxes in different pathways are calculated by means of biochemical-mathematical models. Unfortunately, experimental animals have to be anesthetized during these assays to minimize movement artifacts,

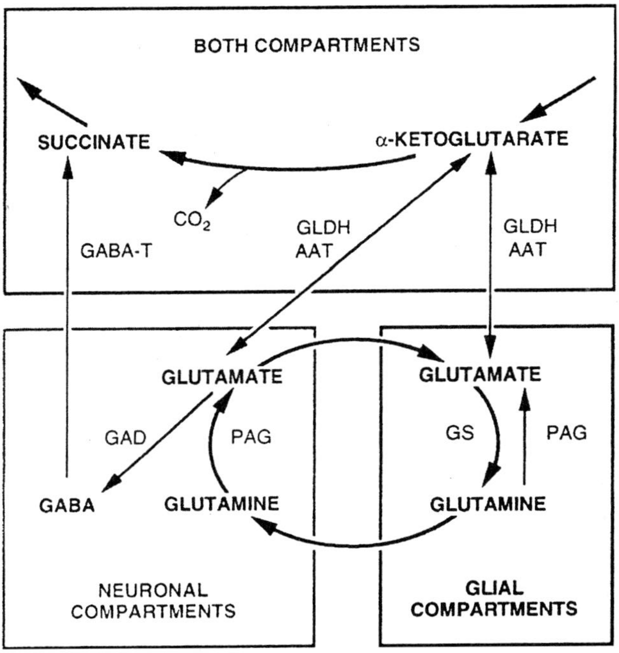

FIG. 5. Schematic illustration of glutamate and GABA carbon cycling via glutamine: *de novo* synthesis of glutamate and GABA from glucose, their degradation to TCA cycle intermediates, and neuronal "recovery" of transmitter glutamate, which after release to the extracellular space and uptake into astrocytes is returned to neurons as glutamine (glutamate-glutamine cycle). The top panel shows a section of the TCA cycle from α-ketoglutarate (α-KG) toward succinate, operating in both neuronal and astrocytic (glial) metabolic compartments, and the lower panels show astrocyte-specific (right panel) and neuron-specific (left panel) reactions involved in the formation, trafficking, and degradation of glutamate, GABA, and glutamine. Glutamate can be formed from α-ketoglutarate by reductive amination catalyzed by glutamate dehydrogenase (GLDH), or by transamination catalyzed by aspartate aminotransferase (AAT); glutamate can be reconverted to α-KG in both neurons or astrocytes by reversal of either of these reactions. Glutamine can be formed from glutamate in astrocytes in an irreversible reaction catalyzed by glutamine synthetase (GS) and reconverted to glutamate in another irreversible reaction catalyzed by phosphate-activated glutaminase (PAG), which is present in both astrocytes and neurons; neurons can accumulate glutamine after its release from astrocytes and reuptake into neurons via the glutamate-glutamine cycle. GABA can be formed via glutamate (glutamine is a good precursor of this glutamate) in neurons in an irreversible reaction catalyzed by glutamate decarboxylase (GAD). GABA is metabolized to succinate in irreversible reaction catalyzed sequentially by GABA transaminase (GABA-T) and succinic semialdehyde dehydrogenase (not shown). GS is a glial-specific enzyme, whereas GAD is neuronal-specific; the other enzymes involved in these reactions are not cell-type-specific. However, net formation of α-KG from glucose occurs in astrocytes and requires the glial-specific pyruvate carboxylation reaction; this α-KG might be directly transferred to neurons then converted to glutamate (by transamination or reductive amination), or it can be first converted to glutamate in astrocytes, followed by synthesis of glutamine, which is then transferred via the glutamate-glutamine cycle to neurons. (From Robinson *et al.*, 1997, with modifications. Reprinted with the permission of Cambridge University Press).

whereas studies in human brain are performed in normal, conscious individuals; anesthesia, due to its consciousness-suppressive actions, can markedly alter blood flow and metabolism in many unidentified ways, depending on the anesthetic, its dose, and, perhaps, duration. Thus, important issues that are integral to interpretation of *in vivo* MRS studies in experimental animals are the influence of anesthesia on (1) the magnitude, pathways, and regulation of glucose and oxygen metabolism during rest and stimulation; and (2) extrapolation of data obtained during various anesthesia regimens to the unanesthetized, conscious, activated state (Shulman *et al.*, 1999). Activation-dependent responses of blood flow and metabolism and the mechanisms that regulate these processes differ during anesthetic and conscious states, and marked regional differences in anesthetic effects might arise, in part, from the different transmitters and signaling systems employed as sensory signals are processed and integrated within the central nervous system (Nakao *et al.*, 2001).

3. *Glutamate and Glutamine Transporters*

Extracellular glutamate, especially the neuronally released transmitter glutamate in the synaptic cleft, is accumulated to a minor extent by neuronal glutamate transporters and to a major extent by the astrocyte-specific glutamate transporters, GLAST and GLT, which are located ubiquitously on astrocytic membranes, particularly those in close vicinity of synapses (Danbolt, 2001). Regional differences exist between the distribution of GLAST and GLT, but in most regions extracellular glutamate is predominantly accumulated in astrocytes, and glutamate transporters are of major importance for brain function because they can control the duration, magnitude, and range of glutamate signaling (Bergles *et al.*, 1999; Danbolt, 2001). Since glutamate uptake by these transporters is metabolically driven by Na^+ entry along its concentration gradient, it is dependent upon continuous extrusion from astrocytes of accumulated Na^+ by the Na^+,K^+-ATPase. The metabolic consequences of this stimulation have mainly been studied in cultured astrocytes and will be discussed in Section IV.C.4. GABA is also partly accumulated into astrocytes, although the neuronal uptake of GABA is of relatively greater quantitative importance compared to that of glutamate (Hertz and Schousboe, 1986).

Different isoforms of glutamine transporters are essential for blood-brain barrier glutamine transport and the glutamate-glutamine cycling process. Glutamine transporters can be coupled to amino acid countertransport exchange or to ionic gradient. They are sensitive to physiologic changes in pH and are capable of adaptive changes; different transporters appear to be present in astrocytes and neurons (Broer and Brookes, 2001; Boullard *et al.*, 2002).

4. *Glutamate, Glutamine, and GABA Degradation*

At steady state, net synthesis of glutamate from glucose must be quantitatively matched by complete oxidative degradation of glutamate, or the glutamate content in the brain would continuously rise (the blood-brain barrier is poorly permeable to glutamate). Glutamate degradation is initiated by conversion of glutamate to α-KG (Fig. 5), a process that may occur either as a transamination, involving concurrent transamination of a keto acid to its corresponding amino acid (e.g., oxaloacetate to aspartate or pyruvate to alanine) or as an oxidative deamination, catalyzed by glutamate dehydrogenase and leading to concomitant release of NH_3. Glutamate dehydrogenase shows a low regionally homogenous immunoreactivity in neurons and intense heterogeneous labeling of astrocytes, which does not always coincide with regional glutamatergic innervation in rat brain (Aoki *et al.*, 1987a). In spite of the paucity of neuronal labeling, neurons express considerable mRNA for glutamate dehydrogenase (Schmitt and Kugler, 1999). Detailed studies of glutamate dehydrogenase immunoreactivity have been performed in the cerebellar cortex, where labeling of mitochondria in Bergmann glia, other astrocytes, and oligodendrocytes is intense, especially in astrocytic processes. In contrast, staining of neuronal mitochondria amounts to only $\sim$15% of that in astrocytes, with no difference between glutamatergic and nonglutamatergic neurons (Madl *et al.*, 1988; Rothe *et al.*, 1994). These studies suggest that in some brain regions astrocytes might be the major site for both glutamate synthesis and degradation. However, synaptosomes also express glutamate dehydrogenase activity (McKenna *et al.*, 2000).

Glutamine is converted to glutamate by phosphate-activated glutaminase (PAG), and in spite of its role in the glutamate-glutamine cycle, glutaminase is not a neuronal-specific enzyme (Hogstad *et al.*, 1988), but expressed both in neurons and in astrocytes (Fig. 5). In astrocytes, glutaminase might be mainly involved in oxidative degradation of glutamine via glutamate. Astrocytically accumulated GABA is converted to glutamate (via formation of succinate and α-KG [Fig. 2]) and subsequently returned to the glutamate-glutamine cycle. Alternatively, malate (formed from succinate) may be converted to pyruvate and completely oxidized (II.F.3).

Glutamate is a good fuel that is oxidized at a high rate in astrocytes, but at lower rates in cerebral cortical GABAergic neurons and at the lowest rate in glutamatergic cerebellar granule cells (Hertz *et al.*, 1988). The rate of $^{14}CO_2$ glutamate formation from [1-^{14}C]glutamate (C1 is decarboxylated after entry into the TCA cycle by α-ketoglutarate dehydrogenase) is higher than that from [U-^{14}C]glutamate, where the label is presumably equally divided among the five carbon atoms, so fractional $^{14}CO_2$ evolution per molecule of glutamate metabolized in the first turn of the cycle is lower (Fig. 6A). In mouse astrocytes $^{14}CO_2$ formation from [1-^{14}C]glutamate has

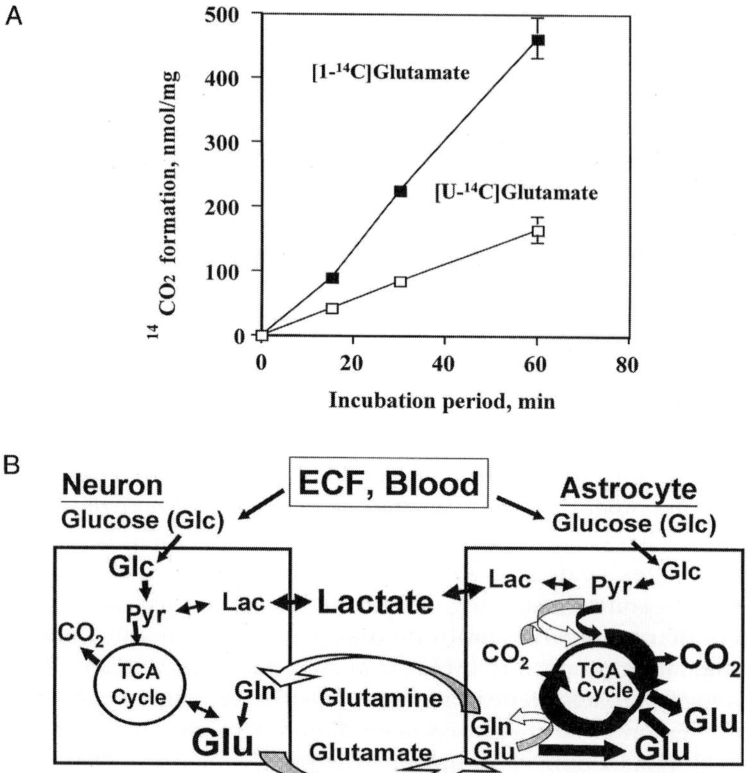

FIG. 6. Glutamate oxidation by astrocytes. (A) Formation of $^{14}CO_2$ from [1-^{14}C]- and [U-^{14}C]glutamate in primary cultures of astrocytes. The uptake and metabolism of glutamate (50 μM of either [1-^{14}C]- or [U-^{14}C]glutamate in tissue culture medium) was calculated from accumulated radioactivity per mg protein and the specific activity of glutamate in the tissue. The amount of $^{14}CO_2$ generated from [1-^{14}C]glutamate indicates the amount of glutamate (~7 nmol/min/mg protein), which after conversion to α-KG has been decarboxylated to succinyl-CoA, not the total amount of generated CO_2. If the products of glutamate beyond the succinyl-CoA step were not further decarboxylated, one would expect $^{14}CO_2$ formation from [U-^{14}C]glutamate to be five times lower than from [1-^{14}C]glutamate (glutamate is a five-carbon amino acid). The observation that the difference in $^{14}CO_2$ formation with the two substrates is smaller, two- to three-fold, suggests that a second decarboxylation step rapidly follows the formation of succinyl-CoA. The second decarboxylation of [U-^{14}C]glutamate might either have occurred during conversion of malate to pyruvate, taking glutamate-derived carbon out of the TCA cycle (Fig. 2), or glutamate may have remained in the TCA cycle to form oxaloacetate (i.e., transiently expanding the pool of TCA cycle intermediates) which then could react with acetyl-CoA, generating citrate and eventually α-KG, from which glutamate can be resynthesized, and the specific activity of labeled tracer diluted. In this process the carbon released during formation of α-KG from isocitrate during the first turn of the TCA cycle originates from [U-^{14}C]glutamate. (Yu and Hertz, Metabolic sources of energy in astrocytes, *In* "Glutamine, Glutamate and GABA in the Central Nervous System" (L. Hertz, E. Kvamme, E. G. McGeer,

consistently been found to occur at least as rapidly as glutamine synthesis (Yu *et al.,* 1982; Hertz and Schousboe, 1986, 1987). However, in rat astrocytes oxidative degradation is relatively slow at low glutamate concentrations, but increases when the extracellular glutamate concentration is increased (McKenna *et al.,* 1996). Utilization of glutamate or glutamine as a metabolic substrate is not restricted to cultured cells but has also been observed in brain slices and dissociated cell preparations and in intact brain (Yu and Hertz, 1983; Tildon and Roeder, 1984; Zielke *et al.,* 1998). As long as the glutamate utilized as a metabolic fuel originally is produced from glucose within the confines of the blood-brain barrier, its use as an alternative fuel is not a violation of the fact that the adult brain *in vivo* under normal conditions almost exclusively utilizes glucose as its substrate for energy metabolism.

The carbon skeleton of glutamate entering the TCA cycle can have several metabolic fates: (1) immediate complete oxidation to CO_2 and water (after conversion of malate to pyruvate, catalyzed by malic enzyme, and reentry of pyruvate into the cycle as acetyl-CoA), a process leading to utilization of oxygen without consumption of glucose (Fig. 2); (2) "storage" as glycogen (after conversion of oxaloacetate to PEP, catalyzed by phosphoenolpyruvate carboxykinase) for metabolic degradation at a later time; or (3) transient expansion of the quantity of TCA cycle intermediates.

During use of glutamate to transiently expand the catalytic capacity of the TCA cycle, the newly synthesized molecule of α-KG is converted to oxaloacetate, which combines with acetyl-CoA to form citrate; the citrate is cycled to α-KG, which can be retained in the cycle or be reconverted to glutamate (Fig. 6B). Although this process represents enhanced net utilization of one molecule of acetyl-CoA, the two carbon atoms released during one turn of the TCA cycle by decarboxylation are both from glutamate. Glutamate entry into the TCA cycle followed by cycling and resynthesis of glutamate from α-KG has been demonstrated in cultured astrocytes (Westergaard *et al.,* 1996), and rapid decarboxylation of two carbon atoms in the glutamate molecule is consistent with the observation that $^{14}CO_2$ formation rate from 50 μM [U-^{14}C]glutamate within the first hour of incubation is approximately

and A. Schousboe, eds.). Copyright © [1983], Wiley-Liss, Inc.) (B) Schematic illustration of transient expansion of the astrocytic pool of TCA cycle intermediates by glutamate. The sequential steps include uptake of glutamate into the astrocyte, conversion of glutamate to α-KG, release of the first CO_2 during the formation of succinyl-CoA, condensation of glutamate-derived oxaloacetate with acetyl-CoA (derived from pyruvate), release of second CO_2 during formation of α-KG. This α-KG can be retained in the cycle to transiently expand the capacity of the cycle, or be used for regeneration of glutamate after one or more turns of the cycle. This model for stimulation of glucose metabolism in astrocytes during functional activation by oxidation of glutamate and increasing the concentration of oxaloacetate has the following stoichiometry: 1 glu + 1 pyr + 15 ADP + 2.5O_2 = 1 glu + 15 ATP + 2H_2O + 3CO_2.

2.5 times lower than that from [1-^{14}C]glutamate (Fig. 6A). Also, the $^{14}CO_2$ production rate from labeled glucose is increased by 50–100% in the presence of 1 m*M* unlabeled aspartate in cultured astrocytes, but not in cultured neurons (Murthy and Hertz, 1988). Again, this is probably because the availability of oxaloacetate for condensation with acetyl-CoA to form citrate is increased, perhaps during several turns of the TCA cycle, thereby releasing the feedback inhibition of pyruvate dehydrogenation by acetyl-CoA.

To summarize, glutamate entering the TCA cycle can be used directly as an oxidative fuel and also to enhance the cell's capacity for oxidation of pyruvate by increasing the quantity of the catalytic components without the necessity for the ATP-dependent carboxylation of pyruvate to synthesize oxaloacetate; transamination of aspartate could also serve this purpose.

I. Formation of Fatty Acids and Cholesterol

Fatty acids are synthesized from acetyl-CoA (Fig. 2), with the rate-limiting enzyme being acetyl-CoA carboxylase, which plays an important role in supplying fatty acids for myelination. Rat brain acetyl-CoA carboxylase is indistinguishable immunologically from the isozyme in rat adipose tissue and liver. Its total activity and mRNA in brain decline from birth to 4 weeks of age, but unlike acetyl-CoA carboxylase in liver and adipose tissue, the brain enzyme is unaffected by nutritional state (Spencer *et al.,* 1993).

The central nervous system accounts for only 2% of the whole body mass but contains almost a quarter of the unesterified cholesterol present in the whole individual (Dietschy and Turley, 2001). Brain cholesterol is largely present in the plasma membranes of glial cells and neurons and in myelin. All cholesterol in myelin is synthesized in the brain from glucose, via acetyl-CoA (Morell and Jurevics, 1996). In addition, brain cholesterol is the precursor for neurosteroids (Majewska, 1992), agents that are mainly formed in glial cells and have neuromodulatory and behavioral effects (Baulieu, 1997).

J. The Pentose Phosphate Shunt Pathway

Glucose-6-P oxidation via the pentose phosphate shunt pathway is a two-step process yielding ribulose-5-P, CO_2, and 2 NADPH (Fig. 2). Glucose-6-P dehydrogenase is the regulatory enzyme for the pathway, governed mainly by the NADPH/$NADP^+$ ratio. A series of non-oxidative reactions catalyzed by isomerase, epimerase, transaldolase, and transketolase can subsequently transform the 5-carbon sugar-P to fructose-6-P according to the following net reaction: 6 glucose-6-P + 12 $NADP^+$ $\rightarrow$ $6CO_2$ + 5 fructose-6-P + 12 NADPH.

Glucose-6-P can be regenerated from fructose-6-P by reversal of this step in glycolysis. Pentose shunt pathway activity has often been assayed *in vitro* by comparing relative rates of decarboxylation of [1-^{14}C]glucose and [6-^{14}C] glucose; the greater the relative production of $^{14}CO_2$ from [1-^{14}C]glucose the higher the pentose shunt activity. The activity of the shunt pathway is about three-fold higher in developing compared to adult brain, presumably due to lipogenesis and myelin formation during development, which require NADPH for biosynthetic reactions. Estimates of the fraction of glucose oxidized by the shunt pathway in adult rat brain are in the range of 2–5%. However, when assayed in the presence of an artificial electron acceptor, phenazine methosulfate, the shunt pathway shows similar activity in brain slices at all ages, suggesting that adult brain has high capacity. The pentose shunt enzymes are enriched in synaptosomes, and in brain slices shunt activity is activated by electrical stimulation or addition of monoaminergic transmitters, H_2O_2, or glutathione to the assay medium. These results suggest linkage to neurotransmitter turnover and use of NADPH to metabolize aldehydes and peroxides produced by monoamine oxidase action on biogenic amines. Because glutathione reductase requires NADPH to regenerate glutathione (GSH) from glutathione disulfide (GSSG), a product of peroxide scavenging, the pathway might be important for protection against oxidative damage (Baquer *et al.*, 1988). NADPH is also used to reduce carbon one of an aldose (e.g., glucose and galactose) to an alcohol. Sorbitol is the alcohol of glucose, and is normally present in brain in very small amounts because aldose reductase has a high substrate K_m (e.g., 37 mM for glucose). Accordingly, the synthesis of sorbitol becomes biologically significant only under hyperglycaemic conditions like diabetes.

K. Summary

This section has discussed the complex interactions between glycolysis and TCA cycle activity and emphasized the regulation of multiple enzyme activities by upstream and downstream metabolites. Thus, glucose metabolism proceeds according to the needs for energy production and synthesis of glucose-derived metabolites, including fatty acids and nonessential amino acids. Pyruvate, the end product of glycolysis, enters the TCA cycle by two different routes: (1) via acetyl-CoA formation, catalyzed by the pyruvate dehydrogenase complex; and (2) by formation of oxaloacetate catalyzed by PC. The former pathway is followed during energy production, but only the latter gives rise to net synthesis of a TCA cycle intermediate. This section has also touched upon metabolic differences between different cell types or subcellular structures of the brain. Although actual metabolic fluxes generally are considerably lower than enzyme activities in brain (i.e., metabolic capacity exceeds normal demand), the activities provide estimates of maximum metabolic rates under activated conditions and they tentatively identify the rate-limiting reactions. The most consistent immunochemical observations are: (1) the high enzyme activities in the neuropil; (2) the uneven distribution of both glycolytic and oxidative enzymes between similar structures in different types of neurons and possibly also between astrocytes at different locations; and (3) the low activities of glycolytic enzymes, and therefore glycolytic capacity, in oligodendrocytes. In addition, acetate is preferentially accumulated in astrocytes and can be used as a "glial reporter molecule." There is no experimental demonstration that some cell *types* in the brain *in vivo* are fueled by glycolytically derived energy and others by energy generated in the TCA cycle. Examples of metabolic specialization in brain cells include four enzymes enriched in astrocytes and absent in neurons: PC, glycogen phosphorylase, fructose-1,6 P_2 phosphatase, and glutamine synthetase. This cell-type selectivity has major functional implications and necessitates transfer of metabolites between neurons and astrocytes (metabolic trafficking). Specifically, synthesis of the transmitters glutamate and GABA depends upon pyruvate carboxylation in astrocytes and metabolic trafficking of a glutamate precursor to glutamatergic and GABAergic neurons. Moreover, released glutamate is mainly, and released GABA partly, accumulated in astrocytes, necessitating further metabolic trafficking via the glutamate-glutamine cycle. Thus, there is considerable interchange of compounds between and among brain cells, depending on fuel availability, energy demand, and product requirement; various compounds synthesized from glucose are available for use as fuel, and their normal levels can be replenished when demand subsides. The ability to measure the position of label in metabolites after exposure to specifically labeled glucose (or other substrates, e.g., acetate) by NMR has allowed not only *in vivo* determination of the turnover rate of the TCA cycle but also distinction between pyruvate dehydrogenation/decarboxylation and pyruvate carboxylation.

III. Relation Between Glucose Utilization and Function

A. Functional Activity Governs Glucose Utilization

In the previous section "resting" brain metabolism was discussed. Although the resting brain is not really at rest, but shows continuous sensory and cognitive activity, resting brain activity is the activity without any specific activation. It is metabolically well defined, and it can be reduced as much as 50% by anesthesia, e.g., by barbiturates (see Section III.B.1). One of the major advances in the neurosciences during the last 30 years is that it has become firmly established that brain work increases utilization of glucose (CMR_{glc}) in specific brain areas, whereas reduced or impaired activity and neuronal loss diminish CMR_{glc} in the affected pathway. The brain is a very heterogeneous tissue not only at the microscopic level but also macroscopically, with specific functions localized to small nuclei and neuronal networks. Analytical methods that rely on gross tissue dissection for metabolite analysis can, therefore, lose vital information and obtain misleading results due to averaging of results obtained in adjacent structures that have unrelated function or different metabolic rates. Highly sensitive, precise microassays that were developed in Oliver Lowry's laboratory to determine levels of many metabolites and enzymes overcame this problem, but the labor-intensive analyses limited the number of samples and microregions from each brain that could be analyzed. Development of quantitative autoradiographic methods provided sufficient spatial resolution and even permitted simultaneous determination of local rates of blood flow and metabolism in all structures of the brain in a single subject. This approach established close linkage between functional activity (ATP demand), blood flow, and glucose utilization (e.g., Sokoloff, 1986, 1996).

B. Measurement of CMR_{glc} with 2-Deoxy-D-Glucose (DG)

1. *Macroscopic Level*

DG is a glucose analog lacking the hydroxyl group at carbon two. In 1954, Sols and Crane reported that DG isolated the hexokinase reaction, because it could be metabolized to glucose-6-P, but not further converted to fructose-6-P. In the late 1950s Tower (1980) used loading doses of unlabeled DG as a competitive inhibitor in assays of glucose and oxygen utilization, and showed accumulation of its phosphorylated derivative, DG-phosphate (DG-P) in tissue. Based upon the intracellular trapping of DG-P, which within a reasonable time frame is not converted to metabolites leaving the cells, the [^{14}C]DG methodology was elegantly developed by Sokoloff and co-workers (1977) as a tracer to determine local rates of glucose utilization autoradiographically in brains of experimental animals. The concentration of DG used for this purpose is so low that it is without significant inhibitory effect on CMR_{glc}. Synthesis of a positron-emitting analog, 2-[^{18}F]fluorodeoxyglucose (FDG) allowed application of the method to studies in primates and man by positron emission tomography (PET), (Phelps *et al.*, 1979). Corrections which must be made for kinetic differences between glucose and [^{14}C]DG in their transport across cell membranes and

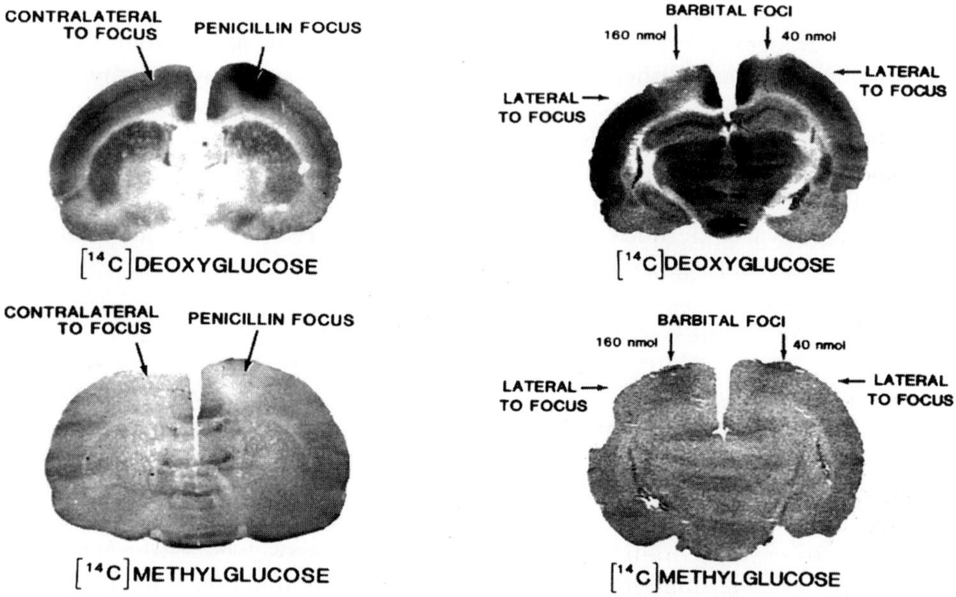

FIG. 7. Glucose supply and demand vary between regions and during altered activity and are closely matched over a wide range of rates of glucose utilization (CMR_{glc}). Local CMR_{glc} and glucose levels are illustrated during focal seizure and focal depression of metabolism in otherwise normal brain. [^{14}C]Deoxyglucose autoradiographs (top panels) illustrate the heterogeneity of CMR_{glc} throughout brain; the higher the optical density, the greater CMR_{glc}. Gray matter (especially cerebral cortical and hippocampal subregions with highly active synapses) has much higher CMR_{glc} than white matter. Topical application of penicillin produced a focal seizure and doubled CMR_{glc} (top left), whereas different topical doses of barbital depressed CMR_{glc} below the application sites by 40–50% (top right). [^{14}C]Methylglucose, which distributes in brain according to glucose concentration (Gjedde, 1982; Dienel *et al.,* 1999), is relatively uniform throughout the brain. Tissue glucose levels fell slightly at the seizure focus, and increased somewhat when CMR_{glc} was lowered (bottom panels). (From Nakanishi *et al.,* Influence of glucose supply and demand on determination of brain glucose content with labeled methylglucose, *J. Cereb. Blood Flow Metab.,* **16,** 439–449, Lippincott Williams & Wilkins, 1996.)

phosphorylation by hexokinase, as well as the detailed calculations and precautions needed in the use of these techniques are discussed by Sokoloff (Sokoloff *et al.,* 1977; Sokoloff, 1986, 1996).

The DG autoradiographs in Fig. 7 illustrate the heterogeneity of glucose metabolism throughout the brain and specific, local alterations in CMR_{glc} in response to drug-induced changes in functional activity. The heterogeneous optical densities of the DG autoradiographs reflect regional differences in metabolism: the higher the optical density, the greater the accumulation of product. Also illustrated are focal seizure activity and local depression of metabolism induced by topical application of penicillin and barbiturate, respectively. CMR_{glc} is increased by about 100% in the seizure focus and depressed by 40–50% in the barbital foci. In normal rat brain CMR_{glc} ranges from a low of about 0.3–0.4 μmol/min/g wet wt in white matter to about 0.5–1.8 μmol/min/g wet wt in various gray matter structures, with highest values in the auditory structures (Sokoloff *et al.*, 1977). In the rat the weighted average in brain is 0.7 μmol/min/g wet wt or, with a protein content of 10%, 7 nmol/min/mg protein. In primates, values for CMR_{glc} are about half those in the rat, but the rank order of metabolic rates in most brain structures is similar (Sokoloff, 1986, 1996), and the weighted average CMR_{glc} in whole brain is 0.3 μmol/min/g wet wt. This species difference is consistent with maximal capacities of enzymes in the glycolytic and TCA cycle pathways in human brain of about half of those in the rat (Tables I and III).

Many physiological and pharmacological studies have shown that CMR_{glc} increases when functional activity is stimulated and falls when function is depressed (Sokoloff, 1986, 1996). Neuroanatomic processing of physiological information, e.g., sensory input, is readily detected and quantified by the DG method, the effects of pharmacological intervention are easily visualized and determined simultaneously in all brain structures, and tumors that show intense glycolytic activity can be detected and localized. Electrical stimulation of afferent sensory nerves increases CMR_{glc} in the synaptic areas of the spinal cord in proportion to frequency of stimulation, whereas CMR_{glc} remains unchanged in dorsal root ganglia containing the nerve cell bodies (Kadekaro *et al.*, 1985), suggesting that the increased metabolic activity occurs in the neuropil, not in neuronal perikarya. Intracerebral administration of adrenergic antagonists leads to a decrease in CMR_{glc} in many regions (Savaki *et al.*, 1982). In posterior pituitary tissue *in vitro* the increase in CMR_{glc} evoked by electrical stimulation (or by exposure to veratridine) is abolished by ouabain (Mata *et al.*, 1980); this finding has been extrapolated to suggest that activation of brain metabolism *in vivo* is mainly or exclusively a metabolic manifestation of increased neuronal Na^+ pump activity (Sokoloff, 1996), an issue that will be discussed in more detail in Section IV.

Under steady-state conditions, assay of the rate of any single step in a multistep metabolic pathway yields the rate of the pathway, and the overall rate of glucose utilization can be calculated from determination of hexokinase activity by means of accumulation of [^{14}C]DG-P (Sokoloff *et al.*, 1977). The ultimate fate of glucose downstream of the glucose-6-P step cannot, however, be evaluated only by assay of the first reaction in a complex pathway. Interpretation of results obtained during non-steady state or during stimulation must take into account the likelihood of changes in the partitioning of glucose carbon into different pathways to meet new functional demands, and this shift in metabolism might include the fraction consumed by oxidative metabolism by different cell types. Information about glucose metabolism under these conditions is very difficult to obtain experimentally, because it requires a fully quantitative analysis and an accounting of all labeled products of glucose metabolism, and can only be gained by purification of labeled [^{14}C]glucose metabolites or by *in vivo* [^{13}C]glucose NMR studies; the analysis requires that all labeled products of glucose metabolism are quantitatively trapped in the tissue of interest or that those lost from tissue can be quantitatively recovered and assayed.

2. *Microscopic Level*

Little quantitative information is presently available regarding the relative rates of glucose consumption in neurons and glial cells during rest or brain activation *in vivo,* because ^{14}C autoradiography does not have the necessary spatial resolution (Smith, 1983). Quantitative ^{3}H autoradiography would provide greater resolution, but technical problems (e.g., loss or spreading of labeled metabolites during lipid extraction to minimize differential absorption in gray and white matter) have not been solved. However, early studies (reviewed by Sharp *et al.,* 1993) suggest labeling of both neurons and astrocytes by DG, not global, predominant labeling of only one major cell type. Recent elegant work has used individual trajectories of the electrons emitted by [^{14}C]DG, a method which allows the precise localization of the origin of the track to either a neuron or a glial cell (Wittendorp-Rechenmann *et al.,* 2001). Approximately one half of the electrons emitted by [^{14}C]DG originate in glial cells and the other half in neurons (Fig. 8), but quantification is limited by the fact that the recovery of labeled products is only about 30%; however, the distribution between neurons and glial cells did not seem to depend upon the degree of recovery. It is also of considerable interest that McCasland and Hibbard (1997) found a higher retention of [^{3}H]DG in glutamatergically innervated GABAergic neurons in the hamster striate cortex compared to nearby GABAergically innervated glutamatergic neurons; unfortunately, the recovery of label after immunocytochemistry was very low in these experiments (about 10%). A different approach has been to use microanalytical procedures and direct biochemical assays of nonradioactive DG and DG-6-P (i.e., to use the DG molecule as a tracer for glucose) to assay of CMR_{glc} in very small brain regions and single cells; results show (1) local differences during seizures in 0.1–10 μg dry weight samples (McDougal *et al.,* 1990), and (2) dissected anterior horn cell bodies (1.5–5 ng dry weight) had under resting conditions two-fold higher CMR_{glc} than adjacent neuropil and dorsal root ganglion cells (Akabayashi and Kato, 1993). This is not in disagreement with the conclusion that stimulation of glucose phosphorylation mainly occurs in the neuropil, but emphasizes that glucose metabolism is by no means negligible in neuronal

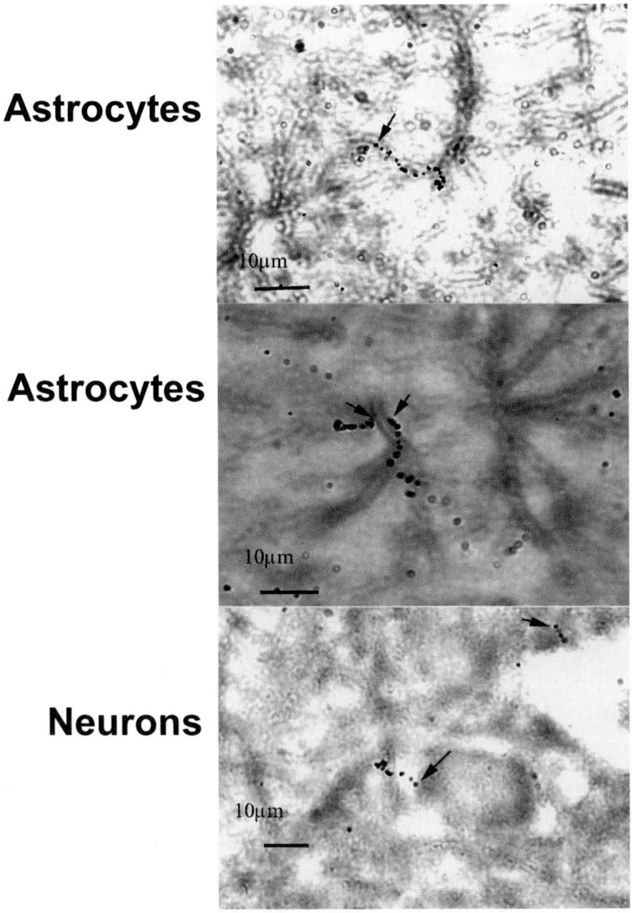

FIG. 8. Individual trajectories of electrons emitted from [^{14}C]deoxyglucose (DG) precisely localize the origin of the tracks to neurons and astrocytes in intact brain. A pulse of [^{14}C]DG was injected to adult rats through a venous femoral catheter, and the animals were euthanized 5 or 45 min after injection. The brains were fixed in paraformaldehyde or submitted to microwave fixation before classical paraffin histology and immunohistochemical demonstration of glial fibrillary acidic protein (GFAP) for identification of astrocytes or microtubule-associated protein 2 (MAP2) for identification of neurons in 5-μm-thick sections. Individual trajectories of the electrons emitted by [^{14}C]2DG were visualized by a track-autoradiographic method after a 5 to 10-day exposure at 4°C of the immunohistochemically treated sections placed in contact with a 15-μm-thick Ilford K5 or K2 nuclear emulsion followed by gold signal enhancement and amidol development and scanning of the immuno-[^{14}C]DG track-autoradiograms. (Modified from Wittendorp-Rechenmann *et al.*, First *in vivo* demonstration of the uptake of [^{14}C]deoxyglucose by astrocytes and neurons: a microautoradiographic study, *J. Cereb. Blood Flow Metab.*, **21,** Suppl 1: S321, Lippincott Williams and Wilkins, 2001.)

perikarya, probably reflecting "housekeeping" activities like maintenance of membrane potential, protein synthesis, etc.

To summarize, regional rates of glucose consumption vary widely at both the macroscopic and microscopic levels. CMR_{glc} is tightly linked to cellular activity, and rises significantly when functional activation requires ion pumping to reestablish transmembrane gradients. To date there is no evidence to indicate a generalized preference for glycolysis (as opposed to oxidative metabolism) by any given cell type, although various cells at a local level (glutamatergically or GABAergically innervated neurons and their surrounding astrocytes) certainly have different metabolic requirements and rates, and activation may alter the relative distribution between glycolytic and oxidative metabolism.

C. Dissociation between CMR_{glc} and CMR_{O_2} during Activation

An apparent uncoupling of glucose and oxygen metabolism, with an increased brain uptake or metabolism of glucose relative to that of oxygen, has been observed during brain activation in normoxic subjects and is often referred to as "aerobic glycolysis." The widely accepted concept of close "coupling" of oxidative metabolism and functional activity in brain was first challenged in studies of human brain. Good correlations between blood flow and CMR_{O_2} and CMR_{glc} were found during rest, but stimulation caused disproportionately greater increases in blood flow and/or CMR_{glc} (30–50%) with no or little change in CMR_{O_2}, suggesting increased lactate production (Fox and Raichle, 1986; Fox *et al.*, 1988; Madsen *et al.*, 1995; Fujita *et al.*, 1999). These findings have been corroborated in studies using NMR techniques for determination of blood flow and oxygenation level (Davis *et al.*, 1998) or of TCA cycle activity by determination of incorporation of label from [1-^{13}C]glucose into carbon 4 of glutamate (Kim and Ugurbil, 1997; Chen *et al.*, 2001). Many microdialysis studies have shown that extracellular lactate levels in brain rise about two- to three-fold during stress and handling (e.g., Kuhr and Korf, 1988; Kuhr *et al.*, 1988; De Bruin *et al.*, 1990; Takita *et al.*, 1992; Fellows *et al.*, 1993; Taylor *et al.*, 1994; Korf, 1996), and higher during seizures (Kuhr *et al.*, 1988; Hu and Wilson, 1997). However, recent studies of aerobic glycolysis in rat brain induced by generalized somatosensory stimulation have shown that (1) changes in the quantities of lactate and glycogen retained in the tissue do not fully explain the fall in the ratio of oxygen to glucose utilization and (2) lactate efflux to blood does not occur in this situation (see Section III.D.3; Dienel *et al.*, 1997b; Madsen *et al.*, 1999). In addition, labeling studies using [1- or 6-^{14}C]glucose (described in Section III.D.1) show that during activation of the visual or auditory pathways product trapping is incomplete and yield calculated metabolic rates that are too low compared to values obtained in parallel experiments with [^{14}C]DG. Taken together, all of these data suggest that the imbalance between CMR_{O_2} and CMR_{glc} during aerobic glycolysis probably mainly arises from rapid efflux of high-specific activity, nonoxidized glucose metabolites from the active tissue. If lactate or other metabolites were taken up and quantitatively metabolized locally, as suggested by Magistretti *et al.* (1999), label should be locally trapped in the large amino acid pools and oxidative metabolism would match glucose utilization. Thus, it is likely that spreading of incompletely oxidized metabolites of glucose within brain may contribute to aerobic glycolysis during brain activation.

Because the decreased CMR_{O_2}/CMR_{glc} ratio during brain activation is followed by an increased CMR_{O_2}/CMR_{glc} ratio after activation, some glucose metabolites (such as glycogen, lactate, and glutamate) that remain in the activated area might be oxidized during the "recovery" process. Hertz and Fillenz (1999) proposed that *de novo* synthesis of glutamate during the onset of glutamatergic activity may contribute to "anaerobic glycolysis," because synthesis of glutamate from glucose leads to generation of only 4 molecules of NADH (2 NADH during conversion of glyceraldehyde-3-P to 1,3-P_2 glycerate, one NADH during the PDHC step and the fourth NADH during the isocitrate DH step). Thus, there would be much less oxygen consumption than corresponding to oxidation of 2 mol of pyruvate (Fig. 2). Moreover, if excess glutamate is oxidized during the recovery there would be disproportionate utilization of oxygen compared to glucose. In conclusion, the biochemical and cellular basis of aerobic glycolysis is not fully understood, but it is a very complex phenomenon involving synthesis and, presumably, efflux of nonoxidized metabolites such as lactate, increased glycogen turnover, increased oxidative metabolism of glucose, altered amino acid levels, and perhaps increased biosynthesis of material from glucose.

D. Comparison between CMR_{glc} Determined with Labeled Glucose and DG

1. *Labeling with Glucose Is Much Less Than with DG*

Brain activation studies in normal conscious rats that assayed CMR_{glc} in parallel with [1- or 6-^{14}C]glucose and [^{14}C]DG found that local increases in CMR_{glc} were much too low when [^{14}C]glucose was the metabolic tracer. These findings suggest a major shift in the predominant pathway(s) of glucose metabolism during brain activation, because most of the labeled products of glucose metabolism that correspond to the additional glucose consumed by the activated tissue over and above that in the resting state are not retained in the stimulated structure. Establishing the identities of the labeled metabolites lost from and retained within the activated tissue and understanding processes that contribute to underestimation of CMR_{glc} with labeled glucose are key to elucidating metabolic demands and interactions of working neurons and astrocytes.

During graded unilateral visual stimulation of the normal conscious rat, CMR_{glc} in the dorsal superior colliculus increased in proportion to the on-off frequency of pattern stimulation to a maximum of about twice control when [^{14}C]DG was the tracer, but rose to a plateau of only about 30% above control over the same stimulus range when [6-^{14}C]glucose was used to track metabolism (Collins *et al.*, 1987). The inability of [6-^{14}C]glucose to accurately register visual activation, illustrated in Fig. 9, was reflected by an increasingly larger difference between CMR_{glc} determined with [^{14}C]DG and [6-^{14}C]glucose as the magnitude of stimulus rose, and was ascribed to [^{14}C]lactate loss (Collins *et al.*, 1987; Ackermann and Lear, 1989). Unfortunately, the identity of exiting labeled products is extremely difficult to determine *in vivo* due to inaccessibility of the venous drainage of most brain structures.

Underestimation of CMR_{glc} and local metabolite spreading of products of glucose beyond the activated area(s) also occur in a larger structure, the inferior colliculus, during unilateral stimulation of the auditory pathway. When normal, conscious rats were exposed to an 8-kHz tone, the activated inferior colliculus showed two (tonotopic) bands labeled by [^{14}C]DG, with peak values 2.2- and 1.6-fold higher than the contralateral tissue (Fig. 10, top panel; see color insert). In contrast, ^{14}C levels in the activated colliculus labeled by [1-^{14}C]glucose did not exhibit this striking bimodal pattern (Fig. 10, middle panel), and the highest ^{14}C levels were only 1.3-fold higher than the contralateral tissue. During spreading cortical depression, accumulation of products of [6-^{14}C]glucose in the activated tissue (16% greater than control cortex) is one-third that of [^{14}C]DG (51%), indicating rapid loss of labeled metabolites (Adachi *et al.*, 1995). Also, a laminar distribution in the [^{14}C]DG-P autoradiograms (Fig. 11; see color insert) was not obvious in the [^{14}C]glucose autoradiographs, indicating metabolite spreading within cerebral cortex (Cruz *et al.*, 1999). Spreading cortical depression is a peculiar electrophysiological phenomenon (Leao, 1944), during which a wave of suppression of electrical activity, preceded by brief electrical hyperactivity, slowly spreads from its point of origin across the brain cortex. This wave is accompanied by a very substantial release and subsequent active reaccumulation of K^+ and there are large increases in CMR_{O_2}, tissue lactate, and local cerebral blood flow (Bures *et al.*, 1974; Rosenthal and Somjen, 1973; Shinohara *et al.*, 1979; Mayevsky and Weiss, 1991; Martins-Ferreira, 1994; Kager *et al.*, 2000; Somjen, 2001). Thus, low labeling of activated tissue, failure to detect or resolve tonotopic bands, and more label spread with [^{14}C]glucose indicate that (1) glucose metabolites do not accumulate

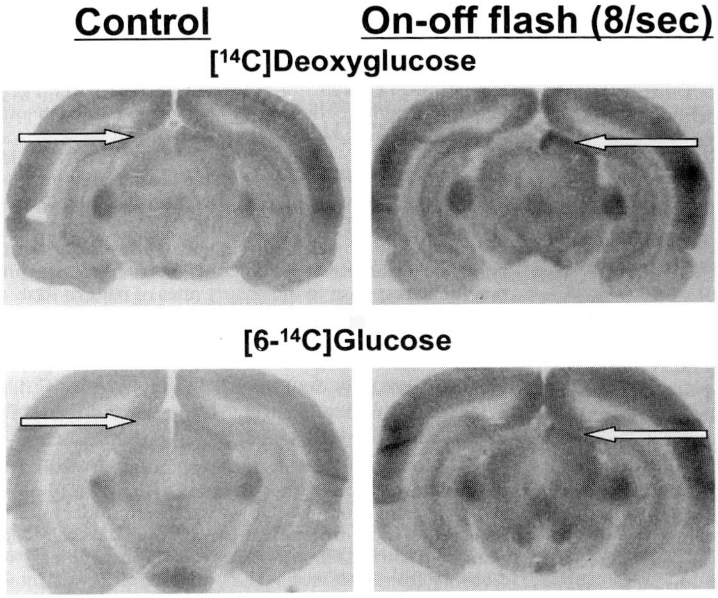

FIG. 9. Glucose utilization in dorsal superior colliculus of conscious rats at rest and during on-off photic stimulation measured with [^{14}C]deoxyglucose (DG) and [6-^{14}C]glucose. [^{14}C]Deoxyglucose (DG) and [6-^{14}C]glucose (Glc) autoradiographs are modified from data in Collins *et al.* (1987), with modifications, with the permission of Blackwell Science Ltd. Rats were unilaterally enucleated under anesthesia prior to the experiment; because about 90% of the retinal input to the dorsal superior colliculus is derived from the contralateral eye, the dorsal superior colliculus corresponding to the enucleated eye has reduced neuronal signaling activity and a much lower metabolic rate. Under control conditions (flash rate = 0), calculated CMR_{glc} for both DG and Glc fell about 30% in the left (arrows, left panels) compared to right dorsal superior colliculus due to removal of retinal input. Functional metabolism of glucose is increased in the right superior colliculus by 8-Hz on-off photic stimulation (arrows, right panels). The largest metabolic increase was obtained with [^{14}C]DG, which rose about 40% at 8 Hz and progressively increased with higher flash rate over the range 4–33 Hz to a peak that was twice control, whereas maximal calculated increases with [6-^{14}C]glucose reached a plateau of about 20–30% above control over the same stimulus range; tissue glucose levels in superior colliculus were the same during rest and activation, indicating matching of supply and demand (With modifications, from Collins *et al.,* Cerebral glucose utilization: comparison of [^{14}C]deoxyglucose and [6-^{14}C]glucose quantitative autoradiography. *J. Neurochem.* **49,** Blackwell Science Ltd., 1987.)

quantitatively in stimulated areas; and (2) [^{14}C]metabolite loss from these areas exceeds any *local* ^{14}C trapping that might arise from lactate metabolism and trafficking.

2. *Identification of Labeled Metabolites That Exit the Activated Tissue*

a. Potential Metabolites. Loss of diffusible products that are rapidly labeled, by glucose, have high-specific activities, or normally participate in metabolite

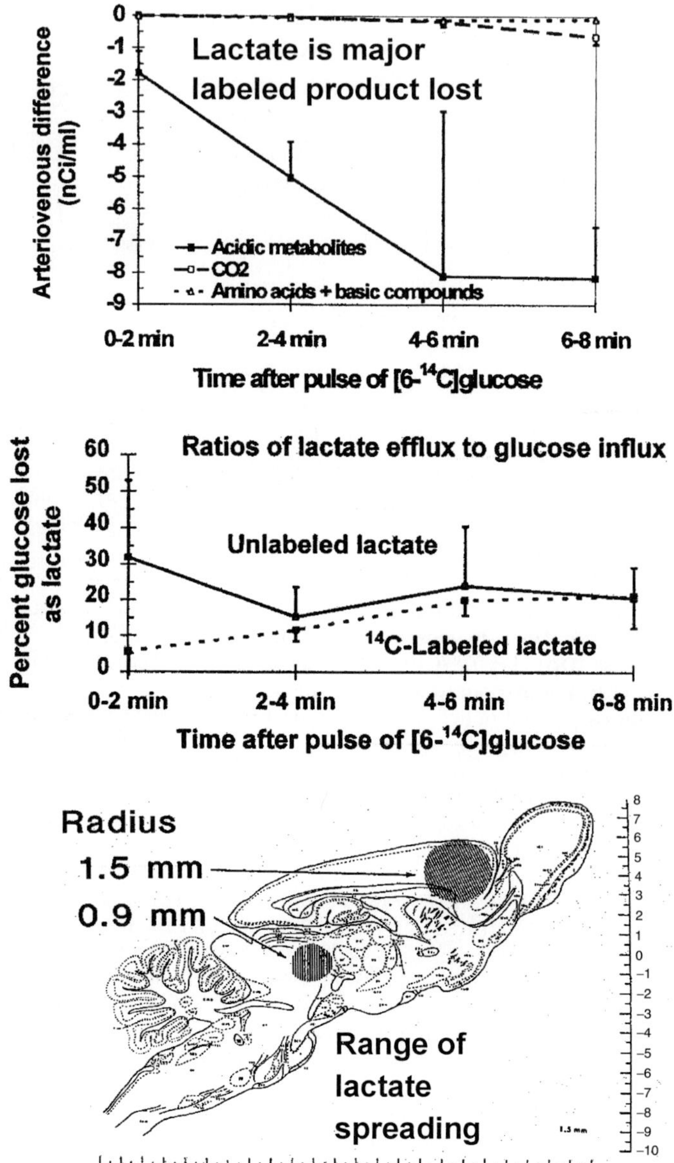

FIG. 12. Rapid clearance of lactate to blood during spreading cortical depression and spreading within brain after injection into tissue. Top panel: Arteriovenous (A-V) differences across the cerebral cortex of conscious rats were assayed during spreading depression, and labeled products in paired (A-V) samples were fractionated to identify major metabolites lost to blood from brain; a negative (A-V) difference indicates net loss from brain, i.e., a higher concentration in venous blood. Lactate efflux was detectable within 2 min after pulse labeling

trafficking, are most likely to contribute to underestimates of CMR_{glc}. These metabolites include CO_2, lactate, and amino acids.

b. CO_2. Eventually, most of the label from [^{14}C]glucose will end up as $^{14}CO_2$, depending on the turnover rates of the pools into which the labeled glucose enters; some compounds (e.g., protein and lipid) turn over very slowly compared to intermediary metabolites and are sparsely labeled but retain that label for a longer time. It is generally assumed that appearance of radioactive $^{14}CO_2$ is substantially delayed by trapping of labeled glucose metabolites in the large pools of glutamate, glutamine, and aspartate, which rapidly exchange with the small quantities of TCA cycle constituents (Fig. 2, Table II), thereby greatly reducing the specific activities of labeled metabolites. Loss of labeled CO_2 via the TCA cycle is further delayed by use of C1- or C6-labeled glucose; the carbon atoms in these positions are not oxidized during the formation of acetyl coenzyme A from pyruvate or during the first turn of the TCA cycle, although loss from C1-glucose is slightly higher than from C6-glucose due to the pentose phosphate shunt pathway. Nevertheless, loss of labeled CO_2 could be accelerated by entry of high-specific activity compounds into small, extremely active compartment(s) (perhaps located in synaptic structures or astrocytes) that do not have large unlabeled metabolite pools to trap ^{14}C or do not quickly mix with total brain glutamate. Glutamate pools with different turnover rates have been detected in cerebrocortical brain slices and *in vivo* by ^{13}C NMR studies (Badar-Goffer *et al.*, 1992; Shank *et al.*, 1993; Cruz and Cerdan, 1999). Retarded loss of $^{14}CO_2$ is evident in PET studies in normal human brain (Blomqvist *et al.*, 1990) and also during spreading depression (Fig. 12, top panel), induced

---

with [6-^{14}C]glucose, and [^{14}C]lactate accounted for about 95% of the total ^{14}C lost from brain within 8 min. $^{14}CO_2$ loss was delayed, becoming detectable between 6–8 min, and was about 5% of the total ^{14}C lost to blood. Efflux of labeled amino acids was negligible. Middle panel: Assay of (A-V) differences across the cerebral cortex of conscious rats during spreading depression shows continuous efflux of similar amounts of labeled and unlabeled lactate from about 2–8 min after the pulse intravenous injection of [6-^{14}C]glucose. The quantity of lactate exiting brain was approximately equal to 20% of the glucose influx to brain during this interval. The lag before the quantity of labeled and unlabeled lactate loss from brain became equal is due to the time required for entry into and mixing of the [^{14}C]glucose with the unlabeled brain metabolite pools. Bottom panel: Spreading of lactate and its labeled metabolites within brain can reach up to about 1.5 mm from a point source in the halothane-anesthetized rat (see Table IV); even a range of 60% of this distance (i.e., 0.9 mm) is large compared to the size of many rat brain structures, indicating that spreading of lactate in brain can contribute to loss of resolution of activated tissue if lactate is produced and exported from the cell (see Fig. 10, lack of tonotopic bands in the inferior colliculus in the autoradiographs derived from labeled glucose and acetate compared to the defined bands obtained with DG). (From Cruz *et al.*, Rapid efflux of lactate from cerebral cortex during K^+-induced spreading cortical depression, *J. Cereb. Blood Flow Metab.*, **19**, 380–392, 1999, Lippincott Williams & Wilkins.)

by topical application of KCl, where only 5% of the radioactivity released to blood during an 8-min labeling period was recovered as $^{14}CO_2$ when [6-^{14}C]glucose was the precursor (Cruz *et al.*, 1999).

c. Lactate. In normal, conscious rats, brain and plasma glucose specific activities are similar within a few minutes after an intravenous injection of labeled glucose (Cremer *et al.*, 1978; Adachi *et al.*, 1995), and in adult brain, lactate quickly attains a high-specific activity, the expected value of about half that of [6-^{14}C]glucose (Adachi *et al.*, 1995). This is a much higher specific activity than those of the TCA-cycle-derived compounds (primarily glutamate, glutamine, and aspartate), which are diluted due to mixing with large unlabeled endogenous pools. Rapid loss of labeled lactate from activated tissue (Fig. 12, middle panel) would, therefore, have a disproportionately high negative impact on labeled product accumulation and calculated CMR_{glc} compared to loss of equivalent molar quantities of other labeled metabolites. Astrocytes contribute to the release of lactate and may under some conditions be the major source of extracellular lactate (Elekes *et al.*, 1996; Korf, 1996). As discussed in more detail in Section IV, astrocytic lactate production may be associated with K^+ clearance from the extracellular space, at least initially since the initial phase of K^+ clearance is impaired if glycolysis is inhibited, but not when oxidative metabolism is inhibited, whereas the opposite is true for the later component of K^+ clearance (Raffin *et al.*, 1992; Roberts, 1993). It has also been suggested that glutamate uptake may be associated with astrocytic glycolysis (Magistretti *et al.*, 1999); however, glutamate recycling can be calculated to use only a small fraction of the total energy consumption by brain (Attwell and Laughlin, 2001), and oxidative metabolism of glutamate itself appears to be able to provide fuel for its uptake (Section IV.C.4).

When aerobic glycolysis (i.e., lactate production during adequate oxygenation) occurs in normal, normoxic brain, removal of the lactate from the cells in which it is continuously produced is essential in order to maintain intracellular redox (oxidation/reduction) conditions favorable for net conversion of pyruvate to lactate. The equilibrium constant of the LDH reaction strongly favors production of lactate from pyruvate, and in tissues with high LDH activity that allows the reaction to be close to equilibrium, the direction of the reaction will be governed by the relative concentrations of the reactants. Thus, if the rate of NADH oxidation to regenerate NAD^+ (via shuttling reducing equivalents into mitochondria for oxidation) is slow compared to the rate of pyruvate and NADH formation, then lactate synthesis will be favored. However, subsequent accumulation of intracellular lactate to very high levels would eventually cause this process to reverse, possibly also affecting other reversible reactions coupled to NAD^+/NADH. Therefore, lactate must be exported from the activated cell and also quickly removed from the surrounding extracellular fluid. Once outside the cell clearance from the surrounding extracellular space could occur via different mechanisms: (1) local, short-distance diffusion within extracellular fluid and uptake into other cells; (2) intermediate-distance spreading via an astrocytic syncytium coupled by gap junctions (Yamamoto *et al.*, 1990; Lee *et al.*, 1994; Blomstrand *et al.*, 1999; Rouach *et al.*, 2000); (3) longer range movement within tissue via flow along paravascular spaces, extending along arteries entering from the subarachnoidal space and eventually reaching venules and veins and brain lymphatics (Rennels *et al.*, 1985; Weller *et al.*, 1992; Ichimura *et al.*, 1991); (4) dispersal by flow of the cerebrospinal fluid (Ghersi-Egea *et al.*, 1996), which turns over with a half-life of 1–2 h (Davson, 1962); and (5) efflux to blood with clearance from brain.

Spread of labeled lactate to neighboring regions of the brain may contribute to the loss of labeled metabolites of [^{14}C]glucose from the activated areas. The likelihood of lactate spreading beyond the activated area is emphasized by studies in which movement of lactate from a point source was studied by intracerebral injection of 0.5 μl saline with [^{14}C]lactate or [^{14}C]inulin in halothane-anesthetized rats (Cruz *et al.*, 1999). Within 10 min label from these tracers had become distributed within an area reaching up to 1.5 mm from the injection site for lactate and 2.4 mm for inulin, a distance about half the thickness of the cerebral cortex (Fig. 12, bottom panel); the volume of labeled brain was 17 times that of the injectant for lactate, and 100-fold greater for inulin (Table IV). Transport distance and volume of labeled tissue were greater for inulin, a macromolecule restricted to the extracellular space, suggesting that lactate enters cells surrounding the injection

TABLE IV
SPREADING OF LABELED LACTATE (AND ITS METABOLITES) AND INULIN WITHIN BRAIN

Tracer	Labeled tissue volume (mm^3)	Maximum distance (mm)
[U-^{14}C]lactate	8.4 ± 2.2	1.5 ± 0.2
[^{14}C]inulin	51.0 ± 13.7	2.4 ± 0.9

Halothane-anesthetized rats were given an intracerebral injection of 0.5 μl of labeled compounds into brain over a 2-min interval; the rats were killed at 10 min, and labeled tissue was assayed by autoradiography. The volume of tissue labeled by lactate and its labeled metabolites were much less than that of inulin, an uncharged, nonmetabolizable polymer (MW $\sim$ 5000), which is restricted to the extracellular space. Values are mean $\pm$ SD (n = 6). Data from Cruz *et al.* (1999).

site, where it can be metabolized or cleared to the blood, thereby limiting its movement. It is an indication of involvement of gap junction conductivity that the tissue volume labeled in conscious rats (many anesthetics, including halothane, block gap junctions) by unidentified metabolites synthesized after injection of [1-^{14}C]glucose can be reduced by about half by prior infusion of gap junction inhibitors (Dienel *et al.*, 2001a).

A major reason for the extra- and intracellular spreading of lactate in brain tissue is that the rates of sustained lactate uptake and metabolism in both cultured neurons and astrocytes are about equal and not fast enough to keep pace with the rapid release in the stimulated area. Net lactate uptake occurs by two sequential processes, transporter-mediated facilitated diffusion (catalyzed by the monocarboxylate transporter [MCT]) and oxidative metabolism, and the rate of maintained net uptake depends upon the rate at which lactate is metabolized. The rate of MCT-mediated diffusion across the plasma membrane is high enough to cause rapid equilibration of intracellular with extracellular lactate. Oxidative metabolism decreases the concentration of unmetabolized lactate within the cell, thereby maintaining an extracellular/intracellular concentration gradient which allows the continuation of lactate uptake by facilitated diffusion of lactate into the cell. In cultured cells that were incubated in media containing lactate concentrations relevant to working brain *in vivo* (1–3 m*M*) the rate of lactate metabolism corresponds at most to one quarter of the rate of glucose utilization, eliminating the possibility than lactate can replace glucose as primary fuel (Dienel and Hertz, 2001).

During certain circumstances lactate efflux from brain to blood can be rapid and considerable, even in adult brain, which has a very low level of the blood-brain barrier monocarboxylic acid transporter compared to suckling animals (Cremer *et al.*, 1976). For example, during spreading depression as much as 20% of the radioactivity accumulated in brain was recovered in blood leaving the brain, with [^{14}C]lactate accounting for 95% (Fig. 12, top panel). Intense seizure activity also leads to accumulation of lactate in brain (Beresford *et al.*, 1969; Bolwig and Quistorff, 1973), to release of lactate to cerebrospinal fluid (Calabrese *et al.*, 1991), and to a large underestimation of metabolic labeling of seizing tissue with labeled glucose compared to DG (Ackermann and Lear, 1989). Both spreading depression and seizures are associated with large increases in $[K^+]_e$ in brain (reviewed by Walz and Hertz, 1983; Hertz, 1986a). Hyperammonemia is a third condition that is accompanied by an increase in brain lactate content and release of lactate to blood. Ammonia is detoxified in astrocytes, and it increases glycolysis specifically in astrocytes (Kala, 1991) by direct stimulation of PFK (Sugden and Newsholme, 1975). After an acute ammonia load that caused lactate levels in brains of conscious rats to rise modestly from 1.3 to

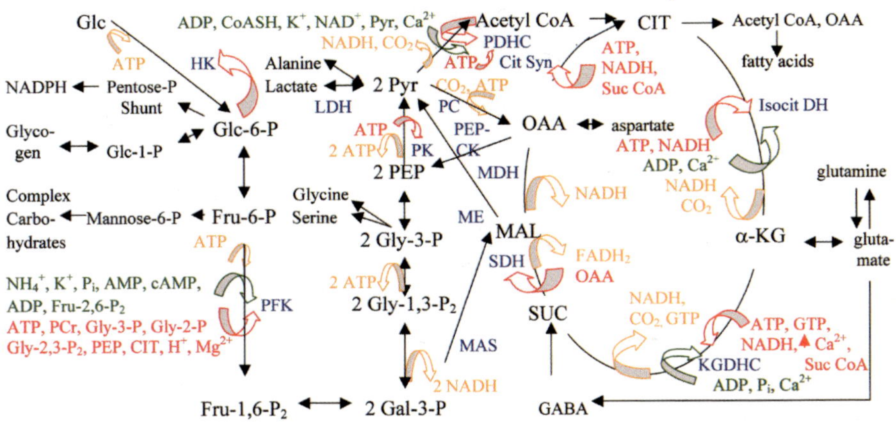

FIG. 1.2. Pathways, major branch points and representative regulators of glucose metabolism. Pathways are shown in black, enzymes and transporters in blue, input and output of ATP, NADH, $FADH_2$ and CO_2 in yellow, inhibitory factors in red, and stimulatory factors in green (with arrows). One molecule of glucose (Glc) generates two molecules of pyruvate (Pyr), which can be introduced into the TCA cycle (upright oval) either via acetyl CoA (for energy production by oxidation of NADH [3 ATP/NADH] and $FADH_2$ [2 ATP/$FADH_2$]) or by pyruvate carboxylation (for biosynthesis). Abbreviations are Glc-6-P: glucose-6-phosphate; Fru-6-P: fructose-6-P; Fru-1,6-P_2: fructose-1,6-bisphosphate; Gal-3-P: glyceraldehyde-3-phosphate; Gly-1,3-P_2: 1,3-bisphosphoglycerate; Gly-3-P: 3-phosphoglycerate; PEP: phosphoenolpyruvate; Acetyl CoA: acetyl coenzyme A; CIT: citrate; α-KG: α-ketoglutarate; SUC: succinate; MAL: malate; OAA: oxaloacetate; HK: hexokinase; PFK, phosphofructokinase; PK: pyruvate kinase; LDH: lactate dehydrogenase; PDHC, pyruvate dehydrogenase complex; PC, pyruvate carboxylase; Cit Syn: citrate synthase; Isocit DH: isocitrate dehydrogenase; KGDHC: α-keto-glutarate dehydrogenase complex; SDH, succinate dehydrogenase; MDH: malate dehydrogenase; ME: malic enzyme; PEP-CK: phosphoenolpyruvate carboxykinase; MAS, malate-aspartate shuttle with associated transporters (see Fig. 4); P_i: inorganic phosphate; Fru-2,6-P_2: fructose-2,6 bisphosphate; Gly-2-P: 2-phosphoglycerate; Gly-2,3-P_2: 2,3-bisphosphoglycerate; CoASH: coenzyme A; Suc CoA: succinyl coenzyme A.

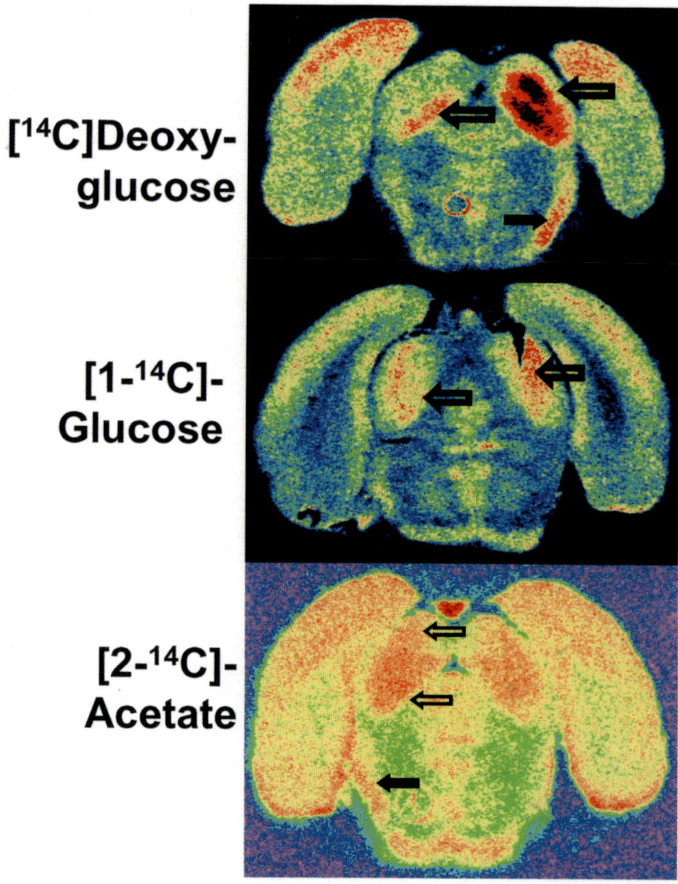

FIG. 1.10. Utilization of glucose (CMR_{glc}) and acetate utilization in inferior colliculus in rats with unilateral auditory (8-kHz tone) stimulation. Autoradiographs show activation of the lateral lemniscus (solid arrows at the lower right and lower left border in the [^{14}C]deoxyglucose (DG) and [^{14}C]acetate autoradiographs, respectively) and in the inferior colliculus (open arrows). Preferential unilateral stimulation was achieved by plugging one auditory canal with wax during the preparative surgical procedure; the right inferior colliculus was stimulated in the DG and glucose studies, whereas the left was activated in the acetate study. The mean rate of glucose utilization determined with [^{14}C]DG (top panel) increased 48% during stimulation, from 0.73 and 1.08 μmol/g/min in the control (left) and activated (right) inferior colliculus, with peak values about 40 and 70% higher than the contralateral tissue in the two tonotopic activation bands in the right colliculus (black indicates highest metabolic rate, followed by red, with progressively lower rates represented by yellow, green, and blue). When ^{14}C-labeled glucose (middle panel) and acetate (lower panel) were used as tracers, the increase in the activated inferior colliculus over the control inferior colliculus was 17% for glucose and 15% for acetate, and the tonotopic bands were not detectable with these two tracers. (Data are from Dienel *et al.,* 2000; Figure reproduced from Glucose and lactate metabolism during brain activation, Dienel and Hertz, *J. Neurosci. Res.*, **66**, Copyright © [2002], John Wiley & Sons, Inc.)

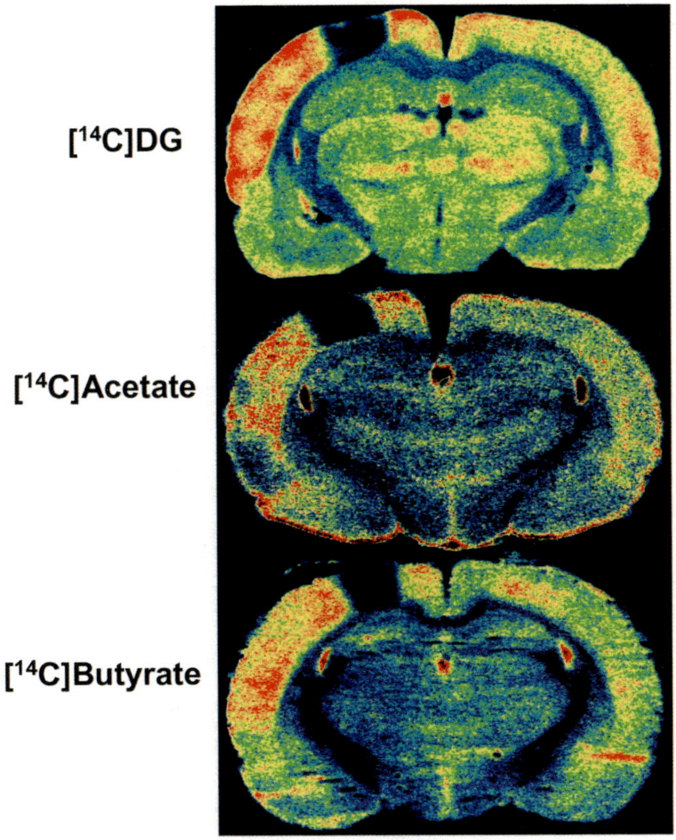

FIG. 1.11. Metabolic imaging of unilateral spreading cortical depression. An intravenous pulse of [^{14}C]tracer was injected at 20 min after induction of unilateral spreading depression by topical application of a cotton ball soaked with 5 M KCl to the intact dura of left cerebral cortex of the conscious rat. The labeling period was 5 min for all tracers, and autoradiographs were prepared from serial coronal sections. Spreading depression caused heterogeneous increases in labeling of the left compared to the untreated right cerebral cortex with [^{14}C]DG, [1-^{14}C]acetate, and [1-^{14}C]butyrate; red indicates high metabolic rate, with progressively lower rates represented by yellow, green, blue, and black. The dark area in the left cortex is the cortical tissue below the KCl site, which had very low uptake of all tracers, presumably due to the high KCl level. Labeling with DG, acetate, and butyrate was highest near the KCl application site, and tended to be higher than average in the most dorsal and most ventral layers of left cerebral cortex. Butyrate, like acetate, is an astrocyte reporter molecule (Berl *et al.,* 1975). In contrast, labeling by [6-^{14}C]glucose was relatively homogeneous throughout the layers of K^+-activated cerebral cortex, there was no intense labeling adjacent to the KCl application site, and left-right differences were small, (Data not shown, See Adachi *et al.,* 1995; Cruz *et al.,* 1999; and Dienel *et al.*, 2001c). Failure of [6-^{14}C]glucose to show the same labeling pattern as [^{14}C]DG indicates incomplete trapping of [^{14}C]metabolites (probably mainly lactate) in activated cells (acetate is not a precursor for pyruvate/lactate), with rapid loss of [^{14}C]metabolites from the activated cortex and also [^{14}C]metabolite spreading within the activated cortex. Note that labeling by acetate and butyrate was heterogeneous in gray matter in both hemispheres, and corpus callosum (white matter) had lower levels compared to gray matter structures. (Data from Local uptake of ^{14}C-labeled acetate and butyrate in rat brain *in vivo* during spreading cortical depression, Dienel *et al., J. Neurosci. Res.,* **66**, Copyright © [2002], John Wiley & Sons, Inc.)

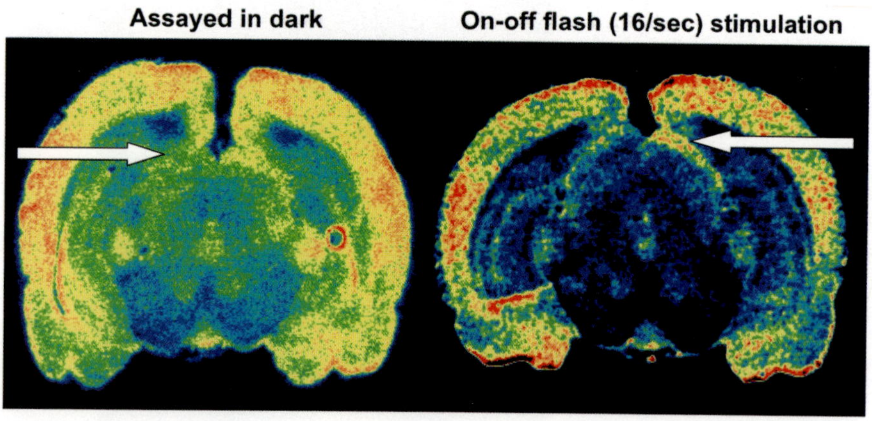

FIG. 1.13. Acetate uptake in dorsal superior colliculus in conscious rats at rest and during on-off photic stimulation. Autoradiographs are from Dienel *et al.* (1999). In rats with the right eye removed under anesthesia during preparative surgical procedures, the calculated [^{14}C]acetate uptake fell about 10% (the basal net uptake rate was 0.040 ml/g/min, calculated by dividing the tissue ^{14}C concentration [nCi/g] by the plasma time-activity integral [μCi/ml][min]) in the left (arrow, left panel) compared to the right superior colliculus, when assayed in the dark. In contrast, under the same conditions CMR_{glc} assayed with [^{14}C]DG fell about 40% from a basal value of about 0.70 normal μmol/g/min (not shown). Acetate uptake in the right dorsal superior colliculus increased about 20% by 16 Hz on-off photic stimulation of the left eye (arrow, right panels), whereas the same stimulus increased CMR_{glc} by 60% (not shown). (Reproduced from Glucose and lactate metabolism during brain activation, Dienel and Hertz, *J. Neurosci. Res.*, **66**, Copyright © [2002], John Wiley & Sons, Inc.)

2.4 μmol/g (i.e., similar to the 2- to 3-fold rise in extracellular lactate level observed during normal physiological stimulation), the rate of lactate efflux to blood quickly increased from about 4% of the glucose entering the brain in control animals to 15% in ammonia-injected rats (Hawkins *et al.*, 1973).

3. *Identification of Labeled Metabolites Remaining in Tissue During Activation*

a. Potential Metabolites. In addition to identification of products released from activated tissue, evaluation of energetics of *in vivo* brain stimulation must also account for shifts in the fraction of glucose metabolized by different pathways (e.g., to amino acids or glycogen), and this analysis can help to identify involvement of major energy or carbon demands, and, perhaps, the contributions of different cell types. In addition, labeled but unmetabolized glucose and lactate may remain in the tissue during the period of stimulation.

b. Glucose, Lactate, and Glycogen. Generalized sensory stimulation of conscious normoxic rats by gentle brushing of the head, whiskers, face, back, paws, and tail with a soft paint brush leads to a 25% increase in labeling of brain metabolite pools within the activated cortical areas 5 min after the pulse of the [6-^{14}C]glucose tracer (Dienel *et al.*, 1997b). About 20% of the label in brain represented nonmetabolized glucose both during rest and activation. Sensory stimulation caused labeling of lactate to increase 2- to 3-fold, but due to the small pool size (1.7 μmol/g), lactate only accounted for about 8% of the ^{14}C recovered in glucose metabolites. Close to one-quarter of the total increase in labeled metabolites in the tissue was in nonidentified glucose metabolites, presumably glycolytic and TCA cycle intermediates and their derivatives. Turnover of glycogen is enhanced during the 5-min activation period and for at least 15 min after stimulation, indicating prolonged changes in the metabolic activities of astrocytes.

c. Amino Acids. One-half of the increased labeling during generalized sensory stimulation occurs in glutamate, which showed a 50% increase in labeling 5 min after the pulse, but there is also a significant increase in GABA and alanine labeling; glutamine labeling was not statistically significantly altered (Dienel *et al.*, 1997b). The increase in glutamate labeling is accompanied by a small (6%) but statistically significant increase in glutamate pool size (from 12.5 to 13.3 μmol/g) and a doubling of the size of the alanine pool (from 0.2 to 0.4 μmol/g), whereas the aspartate pool is significantly decreased (from 4.4 to 3.3 μmol/g). The main reason for the increase in glutamate labeling must be a rapid exchange between α-KG and glutamate, as discussed in Section II.H.2. This increase may occur in both neurons and astrocytes, and will reflect their relative degree of stimulation of

TCA cycle activity. The increase in glutamate pool size indicates net synthesis of glutamate, but the simultaneous decrease in aspartate pool size prevents any definitive conclusion whether there was an increase in anaplerotic activity or whether aspartate had been used for glutamate synthesis, or both. However, the increase in specific activity was higher in aspartate than in any other amino acid, which would be consistent with *de novo* synthesis of aspartate. All amino acid changes, with the exception of the increased pool size of alanine, were reversed 15 min after the cessation of the 5-min stimulation period. Thus, the complexity of metabolic shifts in astrocytes and neurons during sensory stimulation is underscored by simultaneous increases in oxidative metabolism and aerobic glycolysis; the biochemical and cellular basis for these changes are not understood.

E. Correlation between Glucose Supply and Demand

1. *Glucose Supply-Demand Relationships*

Matching of glucose supply with local energy demand under normal and elevated neuronal firing activity is evident in parallel experiments that show local changes in CMR_{glc} and glucose level (Collins *et al.*, 1987). [^{14}C]Methylglucose competes with glucose for transport to and from cells, and it distributes within tissue according to the blood and tissue glucose concentration, so it can be used to determine local tissue glucose levels when glucose supply and demand are altered (Gjedde, 1982; Dienel *et al.*, 1997a). Distribution of [^{14}C]methylglucose in normal brain is relatively uniform (Fig. 7, bottom panels), indicating that glucose levels are similar throughout the brain and glucose supply and utilization are closely matched in normal tissue even though there are large regional differences in energy demand, indicated by metabolism of [^{14}C]DG (Fig. 7, top). There is a high correlation between capillary density, cerebral blood flow, distribution of glucose transporters, and CMR_{glc} (Sokoloff, 1982; Gross *et al.*, 1987; Zeller *et al.*, 1997). Furthermore, glucose delivery nearly meets metabolic demand during prolonged focal seizures, and there is only a small decrease in the [^{14}C]methylglucose level (Fig. 7, bottom) and, therefore, glucose level. However, transient and larger decreases in glucose level occur at seizure onset, and the magnitude of the shift depends upon seizure intensity (Siesjö, 1978). On the other hand, inhibition of CMR_{glc} with barbital caused a small, local increase in tissue glucose level, reflected by the higher optical density in the [^{14}C]methylglucose autoradiograph (Fig. 7, bottom). Short-term changes in glucose supply are achieved via alterations in blood flow, whereas prolonged shifts in functional activity over days or weeks alter glucose transporter gene expression and protein amount (see the next section).

2. *Glucose Transport Supports Altered Utilization*

Up- and downregulation of the blood-brain barrier glucose transporter (GLUT1) in adult rat brain occurs in response to long-term changes in metabolic demand or chronic pharmacological intervention, indicating capacity for adaptation of fuel transport to prolonged local shifts in energy requirements. For example, physiological activation of osmoregulatory structures by water deprivation enhances CMR_{glc} and increases the levels of GLUT1 and GLUT3, with a rise in the mRNA level of only GLUT3; these changes normalize after rehydration (Koehler-Stec *et al.,* 2000). In the visual pathway selective decreases in GLUT1 and GLUT3 glucose transporter density, 3-*O*-methylglucose transport, and glucose utilization occur after one week of visual deprivation (Duelli *et al.,* 1998). Chronic hypoglycemia increases glucose uptake, GLUT1 mRNA, and total GLUT1 protein, and it redistributes more GLUT1 to the luminal surface of the microvessels; chronic hyperglycemia did not, however, change glucose transport, GLUT1 protein level or distribution, even though GLUT1 mRNA levels were substantially increased (Simpson *et al.,* 1999). Thus, long-term changes in neuronal functional activity and energy demand, as well as brain development and aging, might regulate glucose transporter expression (Vannucci *et al.,* 1998). It is notable that changes in mRNA levels did not necessarily parallel those of the transporter protein level or glucose utilization. A neurological syndrome characterized by infantile seizures, developmental delay, and acquired microcephaly underscores the impact of failure to match glucose supply with demand. These infants have insufficient glucose transport across the blood-brain barrier, apparently due a mutation in GLUT1 that does not alter its affinity for glucose, but lowers maximal transport capacity to $<30\%$ that of the parents (Klepper *et al.,* 1999).

F. Acetate Utilization as a Tool to Assay Astrocyte TCA Cycle Activity

1. *Rationale*

Preferential transport of acetate by astrocytes compared to neurons (Section II.C.2 and 3) by the MCT constitutes the basis for the use of ^{14}C- or ^{13}C-labeled acetate to achieve "biochemical isolation" of the glial TCA cycle (Fig. 3A). These substrates can be used as "reporter molecules" to track and visualize glial activity and glial-neuronal metabolic interactions under various normal and pathological conditions by NMR (e.g., Cerdan *et al.,* 1990; Badar-Goffer *et al.,* 1992; Hassel *et al.,* 1997; Cruz and Cerdan, 1999) and autoradiography (Fig. 13; see color insert). Moreover, fluoroacetate can be used as a specific inhibitor of TCA cycle activity in astrocytes. Even if the metabolites formed in astrocytes are subsequently transferred to neurons, labeling with acetate indicates that they have been generated in astrocytes. The selectivity of [^{14}C]acetate autoradiography for astrocytes is shown by the observation that autoradiographs obtained with [^{3}H]acetate in rat brain indicate greater labeling of neuropil compared to neuronal perikarya; hippocampal pyramidal neurons and neuronal cell bodies in cerebellum and retina are essentially unlabeled, whereas the retinal Müller (glial) cells are highly labeled (Muir *et al.,* 1986). Cellular transport/metabolic substrate specificity can also be exploited to localize, detect, and characterize brain tumors, since [^{14}C]acetate preferentially labels glial and meningial brain tumors (Dienel *et al.,* 2001b).

2. *Effects of Stimulation on Local [^{14}C]Acetate Uptake*

Functional activity (e.g., evoked by visual or auditory stimulation) is detectable by autoradiography with [2-^{14}C]acetate, although it is considerably less than the rise in glucose utilization. Net acetate uptake in the dorsal superior colliculus increased only about 20% with 8- to 16-Hz on-off visual stimulation (Fig. 13), whereas CMR_{glc} assayed with [^{14}C]DG rose 60% in parallel studies (Dienel *et al.,* 1999). With unilateral auditory stimulation, the [^{14}C]acetate utilization was about 15% higher in the activated compared to contralateral lateral lemniscus and inferior colliculus (Fig. 10; Dienel *et al.,* 2000). During spreading depression, the maximal increases in acetate (and butyrate, another glial reporter molecule) uptake were much higher (Fig. 11), about 40% (two-thirds of the increase observed with [^{14}C]DG). Moreover, the pattern of increased acetate and butyrate labeling in the tissue surrounding the KCl application site and the dorsal and ventral layers of cerebral cortex resembled that of [^{14}C]DG (Fig. 11), not [1-^{14}C]glucose (Adachi *et al.,* 1995; Dienel and Cruz, 1997; Cruz *et al.,* 1999; Dienel *et al.,* 2001c), probably reflecting that the trapped labeled products of acetate (i.e., TCA-cycle-derived amino acids) are less diffusible (and less labeled) compared to lactate; nevertheless, the patterns of labeling by glucose and acetate are similar to each other during auditory stimulation; neither tracer shows the tonotopic bands detected with DG. Thus, enhanced acetate uptake *in vivo* tentatively suggests that astrocytic oxidative metabolism is increased during normal physiological stimulation and that the stimulation may be specially pronounced during such a pathophysiologic condition as spreading depression. However, interpretation of upward shifts in acetate metabolism is complicated by the possibility that enhanced lactate production in astrocytes might contribute to increased uptake of acetate by heteroexchange between lactate release and acetate uptake due to transacceleration of transport (Waniewski and Martin, 1998), so increased metabolism might reflect either or both more precursor and higher activity of the astrocytic TCA cycle;

paradoxically, a rise in glycolysis might be identified by use of an oxidizable substrate.

G. Activation of TCA Cycle Turnover Determined by NMR

Visual stimulation of humans increases the rate of TCA cycle turnover by ~30% (Chen *et al.*, 2001) determined by incorporation of label from [1-^{13}C]glucose into glutamate carbon as a result of PDH activity. Forepaw stimulation in the anesthetized rat causes an even larger stimulation of oxidative metabolism in the somatosensory cortex (Hyder *et al.*, 1996), but anesthesia may change the response to stimulation (Nakao *et al.*, 2001). Unfortunately as of yet there does not appear to be any determination in the brain *in vivo* whether pyruvate carboxylation is affected during brain activation.

H. NAD^+/NADH Ratio as an Indication of Relative Oxidative Metabolism

Mitochondrial NAD^+ is reduced to NADH by the TCA cycle dehydrogenases and regenerated to NAD^+ in the electron transport chain and ultimately by oxygen. Thus, changes in the cellular NAD^+/NADH ratio during "brain work" provide information about the relative alterations in overall oxidation/reduction processes in the cell, although they do not allow determination of the exact quantity of ATP or oxygen utilized (Lothman *et al.*, 1975). Stimulation of electron flow through the respiratory chain under well-oxygenated conditions is triggered by a decrease in energy charge (i.e., a rise in the concentration of ADP due to activation of energy-requiring processes) and in some cell types also by an increase in intramitochondrial Ca^{2+} (see Section IV.D.1). Electron flow causes oxidation of NADH and thereby increases the NAD^+/NADH ratio. Because extracellular K^+ ($[K^+]_e$) is altered during brain activity, changes in $[K^+]_e$ might be associated with shifts in the NAD^+/NADH ratio mediated by Na^+,K^+-ATPase activity or by depolarization-induced, calcium-dependent changes in enzyme activity.

Microelectrode studies have shown that $[K^+]_e$ rises slightly during functional activity but dramatically under pathological conditions. For example, $[K^+]_e$ rises a few millimoles/l (e.g., from the normal resting level of 3 to 4 m*M*) during normal nerve cell activity, up to 8–10 m*M* during intense afferent stimulation, to a "ceiling level" of 10–12 m*M* during seizures, and to exceedingly high levels (more than 50 m*M*) during energy failure and spreading depression (reviewed by Sykova, 1983; Walz and Hertz, 1983). In most other tissues such alterations in extracellular ion concentrations would rapidly be reduced or abolished by diffusion into the vascular system, but

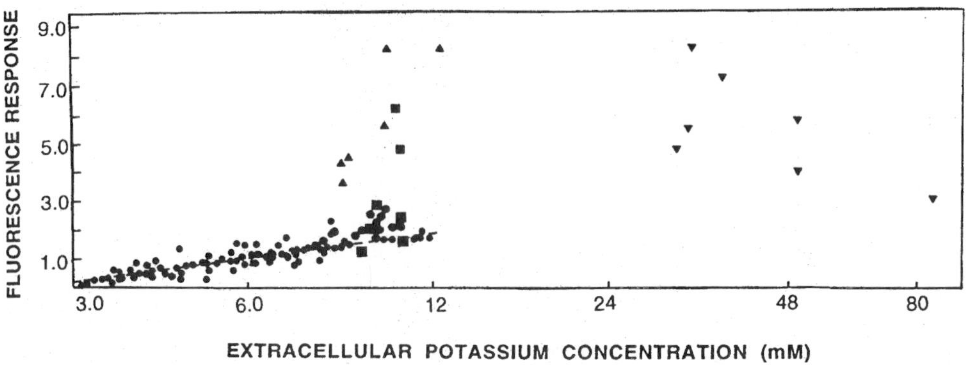

FIG. 14. Changes in oxidative metabolism reflected by shifts in $NAD^+/NADH$ ratios in the rat brain cortex *in vivo* as a function of extracellular potassium level ($[K^+]_e$). The $NAD^+/NADH$ ratios are indicated as fluorescence response, which cannot be converted to absolute values. The increases in $[K^+]_e$ were brought about by stimulus trains of varying duration (closed circles), paroxysmal after-discharges (filled squares), pentylenetetrazol-induced seizures (closed triangles), or spreading depression (inverted filled triangles); each point represents an individual simultaneous measurement of fluorescence ($NAD^+/NADH$ ratio) and $[K^+]_e$. (Modified from *Brain Res.*, **88**, Lothman *et al.*, Responses of electrical potential, potassium levels and oxidative metabolic activity of cerebral neocortex of cats., 15–36,© (1975), with permission from Elsevier Science.)

this does not occur in the CNS due to the presence of a blood-brain barrier. Figure 14 shows that intense stimulation of energy metabolism in the cat neocortex during spreading depression causes an increase in $[K^+]_e$ above 8–10 m*M*, and leads to a huge increase in $NAD^+/NADH$ ratio. There is no major difference between the effects of 10 and of 50 m*M* K^+ (Lothman *et al.*, 1975). In contrast, elevation of $[K^+]_e$ from about 3 up to 8–10 m*M* caused by stimulus trains of varying duration is correlated with much smaller increases in $NAD^+/NADH$ ratio. These experiments allowed the conclusion that "during activation of the cerebral cortex, the oxidative metabolic activity is increased as a function of $[K^+]_e$ activity" (Lothman *et al.*, 1975). This does not necessarily imply that the stimulus for the increased respiratory activity is the $[K^+]_e$ per se, since $[K^+]_e$ also is an indication of the extent of neuronal activity. The abrupt change in the correlation between $NAD^+/NADH$ ratio around 10 m*M* $[K^+]_e$ indicates a particularly large increase in energy metabolism, possibly reflecting concomitant depolarization-induced increases in intracellular Na^+ and/or Ca^{2+}. Thus, *in vivo* determinations of $NAD^+/NADH$ ratios clearly indicate a complex correlation between glucose metabolism and $[K^+]_e$, but they do not provide precise quantitative information about the metabolic changes, about the cell type(s) involved, or about the mechanism(s) by which energy metabolism is activated.

I. Summary

Brain activation is associated with an increase in glucose utilization that can sometimes be larger than the concomitant increase in oxygen consumption. The mechanisms and cell types involved in this metabolic uncoupling are not understood, but probably include accumulation and/or release of lactate and generation of other incompletely oxidized glucose metabolites. The rates of lactate release by unidentified cells, under some conditions perhaps mainly astrocytes, in stimulated areas may become high, but a limited rate of cellular uptake of lactate causes overflow into adjacent brain regions and even into blood and cerebrospinal fluid. The rate of a *possible* astrocytic-to-neuronal flow and metabolism of lactate is very unlikely to exceed one-quarter of the rate of glucose metabolism under physiological conditions. The regions with the highest rates of functional activity and CMR_{glc} have the greatest capacity for fuel delivery, and the components of the system (e.g., transporters and enzymes) that enable matching local fuel delivery with local rates of consumption can adapt to changes in demand. NMR and autoradiographic studies using labeled acetate are beginning to evaluate activation of glial metabolism in response to stimulation of neuronal activity in the brain *in vivo*. The combined use of different metabolic tracers, especially in NMR studies, which can track the fate of carbon atoms, will help to separate the metabolic interactions between neurons and astrocytes. The relationship between oxidative metabolism and extracellular K^+ concentration suggests that two quantitatively different changes in $NAD^+/NADH$ ratio occur: (1) an increase in $NAD^+/NADH$ ratio that is a function of the K^+ concentration per se within the range up to 10–12 m*M*; and (2) a second, much larger increase in $NAD^+/NADH$ ratio, probably brought about by changes in membrane potential and other K^+-induced events, which dramatically alter the energy demands in brain tissue.

IV. *In vitro* Studies of Stimulatory Mechanisms

A. *In vivo* versus *In vitro* Studies

The *in vivo* studies discussed in Sections II–III have given a substantial amount of information about glucose metabolism in the brain under resting and activated conditions, and they have begun to provide knowledge about metabolic capabilities and responses in different brain structures and cellular participation in the bioenergetics of brain activation. However, *in vivo* studies have given very little information about the mechanisms which lead to stimulation of brain metabolism when cerebral activity is enhanced. Studies described below using *in vitro* preparations have helped fill this void.

B. "Classical" and "Emerging" Concepts of Metabolic Regulatory Mechanisms

The classical concept is that energy demand and energy production are linked mainly by stimulation or inhibition of key enzymes by products generated as a result of work or metabolism, e.g., ATP hydrolysis to ADP (which is a substrate for glycolytic reactions and oxidative phosphorylation) and AMP, and that the individual rates of the many metabolic processes involved in complete degradation of glucose are adjusted and fine tuned by various feedback mechanisms (Figs. 1 and 2). A good example of classical linkage between ADP production and energy metabolism is activation of Na^+, K^+-ATPase by Na^+-dependent glutamate uptake in cultured astrocytes (see Section IV.C.4). Although this broad concept is undoubtedly true, regulation of glycolysis and oxidative phosphorylation by ADP, AMP, and other metabolites may not fully explain stimulation of energy metabolism, and signaling processes and second messengers might also be important modulators that act alone or in conjunction with metabolic regulators.

During the last decade, studies in other tissues (e.g., muscle and liver) have shown that an increase in free intramitochondrial Ca^{2+}, secondary to a rise in $[Ca^{2+}]_i$, can cause a direct stimulation of the mitochondrial dehydrogenases, pyruvate dehydrogenase, isocitrate dehydrogenase, and α-ketoglutarate dehydrogenase (McCormack and Denton, 1990; Rutter *et al.*, 1996), as well as of glutaminase activity (Halestrap, 1989). This effect has mainly been studied following administration of transmitters leading to a stimulation of the phosphatidyl inositide second messenger system and resulting release of Ca^{2+} from binding sites on the endoplasmic reticulum. In cultured hepatocytes, a rapid increase in intramitochondrial Ca^{2+} is followed within a few seconds by a transient increase in the concentration of reduced nucleotide cofactors (NADH and NADPH), reflecting stimulation of the mitochondrial dehydrogenases; however, NADH/NADPH was reoxidized during the period of continued activation of the PDH, reflecting an additional Ca^{2+}-dependent increase in both the mitochondrial membrane potential component and the proton gradient component of the proton motive force (Robb-Gaspers *et al.*, 1998). The metabolic effects of transmitter-induced increase in $[Ca^{2+}]_i$ have been studied most intensively, and the proximity between the release site and mitochondria is assumed to be critical for the increase in free mitochondrial Ca^{2+}. However, opening of voltage-sensitive Ca^{2+} channels can trigger an increase in free intramitochondrial Ca^{2+} in epithelial cells (Lawrie *et al.*, 1996), and might accordingly be able to induce similar metabolic effects as transmitter-induced release of Ca^{2+}. The roles of Na^+,K^+-ATPase- and calcium-mediated functions leading to changes in metabolic activity of neurons and astrocytes will be discussed in detail in the following sections. For non-stimulating cells oxidative metabolism in astrocytes is as intense as in neurons (Peng *et al.*, 1994).

C. Na^+,K^+-ATPase-Mediated Stimulation of Glucose Metabolism

1. *Altered Extracellular or Intracellular Levels of Na^+ or K^+*

Although utilization of ATP by stimulation of the Na^+,K^+-ATPase has powerful regulatory effects on glucose metabolism in many types of cell cultures, it should be emphasized that some cultures, e.g., rat astrocytes in homogenous cultures, often show a deficient maturation of the isozyme pattern; they may therefore fail to show specific stimulatory effects (e.g., Sokoloff *et al.*, 1996), whereas corresponding cultures of mouse astrocytes and rat astrocytes in neuronal-astrocytic co-cultures differentiate in culture in parallel with the development *in vivo* (Peng *et al.*, 1997, 1998; Hertz *et al.*, 1998b). Interpretation of stimulation of glucose phosphorylation and/or oxidation in brain cells as being due to activation of Na^+,K^+-ATPase activity is based on inhibition of these metabolic effects by ouabain, a relatively specific inhibitor of this ATPase. Activation of this ATPase in brain cells arises from various conditions (Table V), including (1) modest increases in extracellular K^+ concentrations (in the range 5–12 mM), which directly stimulate the extracellular, K^+-sensitive site of the astrocytic (but not the neuronal) Na^+,K^+-ATPase; (2) K^+-mediated depolarization and subsequent increase in intracellular Na^+ in neurons; (3) electrical stimulation of brain slices (McIlwain, 1951), slices of the posterior pituitary (Mata *et al.*, 1980) or synaptosomes (De Belleroche and Bradford, 1972), which increases the intracellular Na^+ concentration. Energy metabolism in astrocytes can also be stimulated by increased intracellular Na^+ concentration, e.g., by exposure to glutamate, which is avidly accumulated in astrocytes together with Na^+ (see Section IV.C.4).

2. *Stimulation of Extracellular K^+-Sensitive Na^+,K^+-ATPase Site in Astrocytes*

a. Enzyme Activity. A moderate elevation of the concentration of extracellular K^+ causes a stimulation of Na^+,K^+-ATPase activity at its extracellular K^+-sensitive site in cultured astrocytes and in astrocytes from mature brain obtained by gradient centrifugation (Henn *et al.*, 1972; Grisar *et al.*,

TABLE V
METABOLIC EFFECTS ARISING FROM CHANGES IN EXTRACELLULAR K^+ CONCENTRATION OR ELECTRICAL STIMULATION

		Na^+,K^+-ATPase-mediated		$[Ca^{2+}]_i$-mediated	
Preparation	K^+ level (m*M*)	Glycolysis	Glucose oxidation	Glycogenolysis	Glucose oxidation
Astrocytes	5–12	Yes[a]	?		
Astrocytes	>10			Yes[c]	Yes[d]
Neurons	>10	Yes[a]	Yes[a]		
Synaptosomes	Electrical stimulation	Yes	Yes		
Synaptosomes	>10				Yes[d,e]
Brain slices	Electrical stimulation	Yes	Yes[b]		
Brain slices	>10			Yes[e]	Yes[d,e,f]
Posterior pituitary	Electrical stimulation	Yes[a]			Yes[d,e]

[a] Ouabain-sensitive.
[b] Tetrodotoxin-sensitive.
[c] Dihydropyridine-sensitive.
[d] Ouabain-resistant.
[e] Ca^{2+}-dependent.
[f] Tetrodotoxin-insensitive.

1979; Moonen *et al.,* 1980; Mercado and Hernandez, 1992; Hajek *et al.,* 1996), but not in corresponding preparations of neurons (Grisar *et al.,* 1979; Hajek *et al.,* 1996). In cultured astrocytes maximum Na^+,K^+-ATPase activity is reached at a K^+ concentration of ~12 m*M*, and the enzyme activity follows Michaelis-Menten kinetics with a K_m of 1.9 m*M* for K^+. In cultured neurons (Hajek *et al.,* 1996) and synaptosomes (Kimelberg *et al.,* 1978) the enzyme has a three- to fivefold higher affinity for K^+ compared to astrocytes, and is therefore not stimulated by above-normal $[K^+]_e$.

b. Astrocytic Glucose Metabolism. Fig. 15 and Table VI show that a rise of the $[K^+]_e$ from 5 to 12 m*M* increases glucose phosphorylation in mouse astrocytes in primary cultures and in neuronal-astrocytic co-cultures from the rat by 25–50%, whereas 12 m*M* K^+ does not enhance CMR_{glc} in neurons in primary culture (Peng *et al.,* 1994, 1996; Huang *et al.,* 1994; Honegger and Pardo, 1999). This stimulation is inhibited by ouabain, and there is no further increase in the stimulation of CMR_{glc} in astrocytes when the K^+ concentration is increased to 50 m*M* (Table VI). A K^+-induced stimulation of $^{14}CO_2$ production has also been observed after long incubation time (18 h) with [U-^{14}C]glucose, but not after 1 h of incubation, reflecting lack of

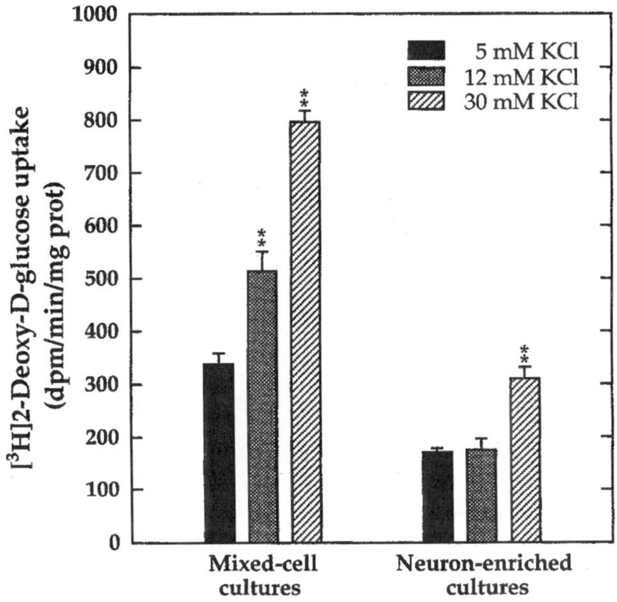

FIG. 15. Effect of elevated extracellular potassium level ($[K^+]_e$) on deoxyglucose (DG) phosphorylation in mixed neuronal-astrocytic aggregate cultures and in neuron-enriched aggregate cultures from rat brain. The rates of DG phosphorylation were measured during a 30-min incubation in tissue culture medium with 5.5 m*M* glucose and a fixed concentration of $[^3H]$DG. Note that both 12 and 30 m*M* $[K^+]_e$ stimulate the DG phosphorylation rate in the mixed neuronal-astrocytic cultures above the rate obtained with 5 m*M* $[K^+]_e$. On the other hand, only 30 m*M* $[K^+]_e$ has a stimulatory effect in the neuron-enriched cultures, presumably secondary to $[K^+]_e$-induced excitation. The absence of effect by 12 m*M* $[K^+]_e$ in the neuronal aggregates suggests an effect on astrocytes, which is consistent with the stimulatory effect of 12 m*M* $[K^+]_e$ on DG phosphorylation in astrocyte cultures shown in Table VI. Vertical bars denote SD. The stimulatory effects of 12 and 30 m*M* $[K^+]_e$ in mixed-cell cultures and of 30 m*M* $[K^+]_e$ in neuron-enriched aggregates are statistically significant, as is the difference between the effects of 12 and 30 m*M* $[K^+]_e$ in the mixed-cell cultures ($p < 0.05$ or better). (Modified from Honegger and Pardo, Separate neuronal and glial Na^+,K^+-ATPase isoforms regulate glucose utilization in response to membrane depolarization and elevated extracellular potassium, *J. Cereb. Blood Flow Metab.*, **19**, 1051–1059, Lippincott Williams & Wilkins, 1999.)

isotope equilibration in glucose metabolites (Yu and Hertz, 1983). Astrocytes accumulate K^+ by active transport (Walz and Hertz, 1982, 1983; Reichenbach *et al.*, 1992; Walz and Wuttke, 1999), and both glycolysis and oxidative metabolism appear to be able to sustain active K^+/Na^+ exchange (Rose *et al.*, 1998). K^+ uptake in the intact brain at highly elevated K^+ concentrations appears mainly to occur in astrocytes (Largo *et al.*, 1996; Xiong and Stringer, 1999; Ransom *et al.*, 2000); in brain slices, clearance of extracellular K^+ is dependent upon Na^+,K^+-ATPase activity (Xiong and Stringer, 2000).

TABLE VI
MAGNITUDE OF K^+-INDUCED STIMULATION OF GLUCOSE UTILIZATION AND OXIDATIVE METABOLISM AND INHIBITION OF THE K^+-INDUCED STIMULATION BY OUABAIN

Preparation	K^+ (mM)	K^+-induced stimulation (%) Glucose utilization	K^+-induced stimulation (%) Oxidative metabolism	Ouabain inhibition (%) Glucose utilization	Ouabain inhibition (%) Oxidative metabolism
Cerebellar granule cells[a]	50	87	500	100	75
Dendrite-impaired cerebellar granule cells[b]	50	269[f]	100	60	23
Synaptosomes[c]	40	121	169	0	0
Astrocytes[d]	12	26		82	
Astrocytes[d]	50	27	>28[g]		0
Brain slices[e]	50		41		29

[a] Peng *et al.*, 1994; Peng, 1995. Control [U-^{14}C]glucose to $^{14}CO_2$: 1.00 ± 0.16 nmol/min/mg protein; this value is an underestimate due to lack of isotope equilibration.

[b] Peng and Hertz, 1993; Peng, 1995. Control [U-^{14}C]glucose to $^{14}CO_2$: 0.32 ± 0.05 nmol/min/mg protein.

[c] Erecinska *et al.*, 1991; Erecinska and Dagani, 1990. Control CMR_{O_2}: 3.4 nmol O_2/min/mg protein.

[d] Peng *et al.*, 1994; Peng, 1995; Hertz *et al.*, 1998; and L. Peng and L. Hertz, unpublished experiments. Control [U-^{14}C]glucose to $^{14}CO_2$: 1.21 ± 0.27 nmol/min/mg protein.

[e] C. S. Kjeldsen and L. Hertz, unpublished experiments. Control CMR_{O_2}: 21.9 nmol O_2/min/mg protein.

[f] Control DG phosphorylation similar to conventional cerebellar granule cells.

[g] Rapidly declining effect, measured during a 10-min incubation.

3. *Stimulation of the Intracellular Na^+-Sensitive Na^+,K^+-ATPase Site in Neurons*

a. Excitation-Induced Na^+ Entry in Neurons but Not in Astrocytes. Increased tetrodotoxin-sensitive $^{22}Na^+$ uptake has been demonstrated in cultured hippocampal and striatal neurons during exposure to highly elevated K^+, whereas K^+-induced depolarization does not result in Na^+ uptake in corresponding cultures of astrocytes (Rose and Ransom, 1996, 1997; Takahashi *et al.*, 1997). Electrical stimulation of brain slices also leads to an increase in intracellular content of Na^+, which is inhibited by the Na^+ channel blocker tetrodotoxin (Varon and McIlwain, 1961; Bachelard *et al.*, 1962; Joanny and Hillman, 1963), suggesting that electrical stimulation opens Na^+ channels in neuronal populations, and thereby stimulates the Na^+,K^+-ATPase at its intracellular, Na^+-sensitive site (Keesey *et al.*, 1965).

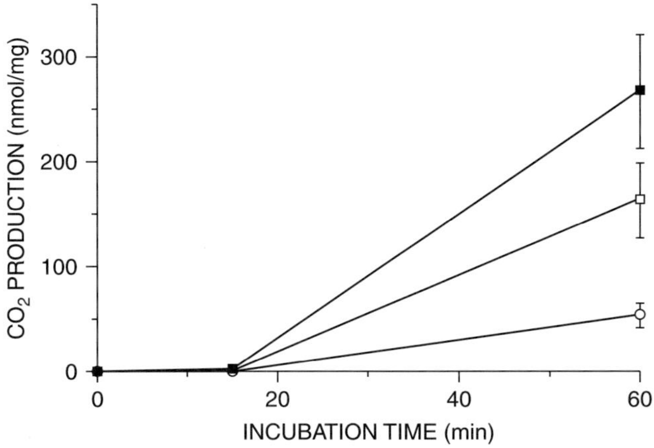

FIG. 16. Production of $^{14}CO_2$ from [U-^{14}C]glucose in cerebellar granule cell neurons as a function of the length of the incubation time. Cultures of cerebellar granule cell neurons were incubated for either 15 or 60 min at extracellular K^+ concentrations of 5 m*M* (open circles), 25 m*M* (open squares), or 50 m*M* (filled squares). All values are means ± SEM of 5–10 individual cultures. Note the negligible $^{14}CO_2$ production during the first 15 min, regardless of the K^+ concentration. (From Peng, L., 1995, with modifications, with the permission of Dr. Peng.)

b. Na^+ and K^+ Effects on Neuronal Glucose Metabolism. The K^+-induced intracellular increase in Na^+ concentration in cultured glutamatergic cerebellar granule cell neurons or hippocampal neurons is accompanied by a large increase in glucose oxidation (Fig. 16) and phosphorylation (Peng *et al.,* 1994), which is almost completely inhibited by tetrodotoxin (Takahashi *et al.,* 1995) and ouabain (Tables V and VI). The increase in CO_2 production is delayed by at least 15 min (Fig. 16), probably reflecting slow metabolic conversion through the neuronal glycolytic and oxidative pathways. However, there is very little K^+-induced stimulation of glucose oxidation in (1) cultures of the inhibitory GABAergic cortical interneurons (Peng *et al.,* 1994); and (2) glutamatergic cerebellar granule cells with severe dendritic degeneration, but histologically normal perikarya and presynaptic structures (Peng and Hertz, 1993); ouabain has also little effect on K^+-stimulated glucose oxidation in the dendrite-impaired cells, indicating that the stimulated energy requiring processes only to a minor extent include active exchange between Na^+ and K^+ (Table VI). This finding is in keeping with the conclusion that the capacity for glycolytic and oxidative energy metabolism is high in dendrites (Lowry *et al.,* 1954; Strominger and Lowry, 1955). However, the granule cells with dendritic degeneration show a high rate of glucose phosphorylation, which is further increased during exposure to high $[K^+]_e$ (Table VI) probably reflecting the large capacity of glycolysis to maintain

ATP in synaptosomes (Nicholls, 1993) and their high hexokinase activity (Section II.B.3).

Electrical stimulation of brain slices and synaptosomes is accompanied by an increase in the rates of aerobic glycolysis and oxygen consumption (McIlwain, 1951; McIlwain *et al.*, 1952; McIlwain and Bachelard, 1985; De Belleroche and Bradford, 1972). The stimulation of glucose metabolism in brain slices is inhibited by tetrodotoxin (Okamoto and Quastel, 1970) and ouabain (Wallgren, 1963), and therefore does, at least in part, reflect increased intracellular Na^+ level and a depolarization-induced stimulation of the intracellular Na^+-dependent site of the Na^+,K^+-ATPase. In contrast, K^+-induced stimulation of oxygen consumption in brain slices is only slightly inhibited by ouabain (Table VI), but is abolished by Ca^{2+} depletion (see Section IV.D.4). Accordingly, elevated K^+ does not appear to stimulate the neuronal populations within the slice by its depolarizing effect (perhaps due to inactivation of Na^+ channels or glutamate receptors), but exerts mechanistically different effects, which may not have the same cellular target.

4. *Stimulation of the Na^+-Sensitive Site of the Astrocytic Na^+,K^+-ATPase*

a. Processes Increasing Intracellular Na^+ in Astrocytes. The astrocytic Na^+,K^+-ATPase can be stimulated by an increase in intracellular Na^+, evoked by exposure (1) to a Na^+ ionophore (Silver and Erecinska, 1997) or veratridine, a drug that opens Na^+ channels (Sokoloff *et al.*, 1996; Peng, 1995), which are present in astrocytes, but not in sufficient density to make the cells excitable (Sontheimer, 1994); (2) to monensin, which facilitates exchange between intracellular H^+ and extracellular Na^+ (Mollenhauer *et al.*, 1990); or (3) to extracellular L-glutamate or D-aspartate, which are accumulated in astrocytes in co-transport with Na^+, and therefore require continuous extrusion of accumulated Na^+ by Na^+,K^+-ATPase. These stimuli cause an increase in glucose phosphorylation, oxygen consumption, and/or $^{14}CO_2$ production from labeled glucose in astrocytes (Yarowsky *et al.*, 1986; Peng *et al.*, 1994, 2001; Peng, 1995; Eriksson *et al.*, 1995; Hertz *et al.*, 1998a) although glutamate uptake may be fueled by glutamate oxidation (see below).

b. Glutamate-Induced Stimulation of Astrocytic Glycolysis and/or Oxidative Metabolism. The exact correlation between Na^+-mediated stimulation of the astrocytic Na^+, K^+-ATPase and activation of glycolysis and oxidative metabolism is disputed. Pellerin and Magistretti (1994) and Sokoloff *et al.* (1996) reported a very substantial ouabain-inhibited increase in glucose phosphorylation in cultured astrocytes exposed to extracellular glutamate, and this observation led to the concept that astrocytic glutamate uptake and glutamine formation require glycolytically derived energy (Magistretti *et al.*, 1999). However, the above studies did not examine oxidative metabolism, and Eriksson *et al.* (1995) observed that the uptake of glutamate in astrocytes in primary cultures is accompanied by a similar increase in oxygen

Astrocytic CMR_{O2} and CMR_{glc} are Increased by Exposure to 100 μM L-Glutamate *in vitro*

•1.5-fold ↑O_2 consumption

9.1→14.1 nmol O_2/min/mg
net: 5 nmol x 2-3 P/O x 2 = **20-30 ATP**
(Eriksson et al., *Glia* 15:152,1995)

•1.5-fold ↑Lactate production

10.7→16.3 nmol glc/min/mg protein
net: 5.6 nmol glc x 2 ATP/mole glc = **11 ATP**
(Pellerin & Magistretti, *Proc Nat Acad Sci* 91:10625, 1994)

•Total ↑ATP = 31-41 nmol ATP/min/mg
Glycolysis: ~30%
Oxidative phosphorylation: ~70%

FIG. 17. Glutamate transport into astrocytes increases oxidative metabolism which provides most of the ATP in the activated cells. Similar percent increases in oxygen consumption and glucose utilization were obtained in cultured astrocytes exposed to the same concentration of L-glutamate. ATP yields were calculated from the net change in oxygen consumption by astrocytes incubated with 100 μM L-glutamate by Eriksson *et al.* (1995) and using P/O ratios of 2 or 3; the P/O ratio is probably closer to 3 in brain. ATP yields were also calculated from net change in lactate production from data in Figs. 1 and 4 of Pellerin and Magistretti (1994); data from Fig. 4 in their paper were used to calculate the net change in glucose consumed to produce the measured amount of lactate released into the culture medium. These data, which were obtained in the presence of 200 μM L-glutamate, were then corrected by multiplying by 0.75, based on the lower [^{3}H]deoxyglucose (DG) uptake by astrocytes incubated with 100 μM compared to 200 μM glutamate, as shown in their Fig. 1. Thus, glycolysis supplies about one-third of the ATP produced during exposure of cultured astrocytes to 100 μM L-glutamate when lactate production and oxygen utilization are both increased 1.5-fold. In the cited studies, the increases in both [^{3}H]DG uptake and oxygen consumption were ouabain-sensitive, indicating dependence on Na^+-K^+-ATPase activity. (Adapted from Dienel and Hertz, Glucose and lactate metabolism during brain activation. *J. Neurosci. Res.*, **66,** Copyright © [2001], John Wiley & Sons, Inc.)

consumption, an observation, which we have been able to confirm (E. and L. Hertz, unpublished experiments). These observations suggest that glutamate uptake into astrocytes can be metabolically supported by either glycolytic or oxidative metabolism, or both. From the magnitudes of the glutamate-induced stimulations of glucose phosphorylation and of oxygen consumption, which have been reported by Pellerin and Magistretti (1994) and Eriksson *et al.* (1995), respectively, it can be calculated that ~70% of the energy is derived from oxidative metabolism and ~30% from glycolysis if both pathways are stimulated (Fig. 17).

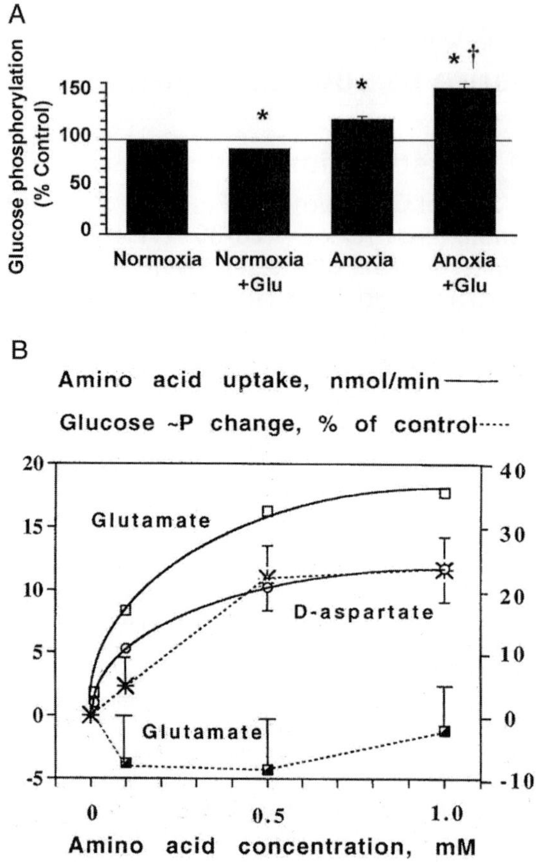

FIG. 18. Effect of L-glutamate uptake on astrocyte metabolism. (A) Effect of glutamate on rate of deoxyglucose (DG) phosphorylation. [^{14}C]DG utilization was determined in primary cultures of mouse astrocytes incubated during normoxic and anoxic conditions in tissue culture medium for 30 min in the absence and presence of 50 μM L-glutamate. All rates of [^{14}C]DG phosphorylation are expressed as percentages of the rate per mg protein during incubation under oxygenated conditions in the absence of glutamate and were calculated from the accumulated radioactivities in cells and the respective specific activities of the incubation media. SEM values are shown by vertical bars if extending beyond the symbols. The decrease in DG phosphorylation rate in the presence of glutamate is statistically significant ($p < 0.05$ or better), as are the increases during anoxia and the effect of glutamate under anoxic conditions. (B) D-Aspartate uptake into astrocytes increases glucose phosphorylation, whereas L-glutamate uptake does not. The solid, curved lines show amino acid uptake rates (nmol/min/mg protein, plotted on the left ordinate axis) of ^{14}C-labeled L-glutamate (open squares) and D-aspartate (open circles) in primary cultures of mouse astrocytes incubated in glucose-containing saline. Amino acid uptake was measured during a 5-min period and calculated from the respective accumulated radioactivity and specific activity of the incubation media; data are plotted as a function of the extracellular amino acid concentration (0–1.0 mM). The dotted, straight lines are plots of DG phosphorylation (glucose ~P, plotted as percent changes of control on the

The studies by Pellerin and Magistretti (1994) and by Sokoloff *et al.* (1996) were performed using cultures that had been grown at an occasionally used but very high glucose concentration (25 m*M*), which might have favored the expression of glycolytic mechanisms at the expense of oxidative metabolism in the cultured cells. For this reason, the correlation between glutamate uptake and [^{14}C]DG phosphorylation has been reinvestigated in cultures grown at a much lower glucose concentration (6–7.5 m*M* at the time of feeding) that is closer to the physiological level in rat brain tissue *in vivo* (i.e., about 2–3 m*M* [Siesjö, 1978; Pfeuffer *et al.*, 2000]). The re-investigation showed the major difference that glutamate uptake caused a small decline in DG phosphorylation under oxygenated conditions (between 3 and 30%, dependent on culturing conditions), and only increased DG phosphorylation under anoxic conditions (Fig. 18A), when the administration of glutamate increased DG phosphorylation over and above the increase caused by anoxia alone (Hertz *et al.*, 1998a). In contrast, an increase in glucose phosphorylation by D-aspartate (which is accumulated by the same carrier as glutamate, but differs from glutamate by not being metabolizable) increased DG phosphorylation (Fig. 18B) (Peng *et al.*, 2001), as previously reported by Pellerin and Magistretti (1994). From these observations, combined with the stimulation of oxygen consumption observed by Eriksson *et al.* (1995) and a well-established ability of most preparations of cultured astrocytes to degrade glutamate oxidatively (Yu *et al.*, 1982; McKenna *et al.*, 1996; Westergaard *et al.*, 1996; Sonnewald *et al.*, 1997), it was concluded that glutamate and/or glucose oxidation normally is able to fuel glutamate uptake, whereas the nonmetabolizeable D-aspartate cannot do so. In further support of this concept, it was found that glutamate decreases glucose utilization and oxidation in astrocytes (Peng *et al.*, 2001;

right scale of the ordinate) as function of L-glutamate (half-filled squares), and D-aspartate (stars) concentration, which ranged from 0.1–1.0 m*M*. Experiments were carried out in primary cultures of mouse astrocytes as described in panel A; results are expressed as percentage changes of DG phosphorylation in control cultures from the same batches that were measured in the same experiment (0,0 value). The scale showing DG phosphorylation was chosen so that maximum effects of unlabeled D-aspartate on its own uptake (solid line with open circles) and on DG phosphorylation (dotted lines with star symbols) are identical, allowing easy comparison of the shapes of the two curves. In conclusion, [^{14}C]DG phosphorylation was enhanced by D-aspartate, with similar concentration dependence as its uptake rate, whereas glutamate uptake did not increase glucose phosphorylation even though it was taken up at higher rates than aspartate. SEM values are shown by vertical bars if extending beyond the symbols. The apparent decrease in DG phosphorylation rate in the presence of glutamate is not statistically significant, but the increases in DG phosphorylation during incubation with 0.5 and 1.0 m*M* aspartate are significant ($p < 0.05$). (Modified from *Neurochem. Int.*, **38,** Peng, L., Swanson, R. A., and Hertz, L., Effects of L-glutamate, D-aspartate and monensin on glycolytic and oxidative glucose metabolism in mouse astrocyte cultures, 437–443, © (2001), with permission from Elsevier Science.)

Chen and Liao, 2001; Qu *et al.*, 2001). Glutamine synthesis from glutamate is an ATP-requiring step necessary for glutamate-glutamine cycling; in cultured astrocytes this reaction proceeds normally even during severe hypoglycemia (Bakken *et al.*, 1998a) and might also be metabolically fueled by glutamate oxidation (Peng *et al.*, 2001). Thus, the energetics of glutamate cycling (Na^+ extrusion and glutamine synthesis) can probably be supported by either or both glycolytic and oxidative metabolism of glucose and oxidative metabolism of some of the transported glutamate.

Noradrenaline stimulates glutamate uptake, both in cultured astrocytes (Hansson and Ronnback, 1992) and in intact brain tissue (Alexander *et al.*, 1997), thereby increasing Na^+,K^+-ATPase activity. In addition, noradrenaline (Hajek *et al.*, 1996) and serotonin (Mercado and Hernandez, 1992; Huang *et al.*, 1994) exert direct stimulatory effects on Na^+,K^+-ATPase activity in cultured astrocytes and in brain tissue.

To summarize, several effects of Na^+ and K^+ on astrocytic and neuronal metabolism are mediated by ADP production and are metabolic manifestations of the sensitivity of Na^+,K^+-ATPase at the respective intracellular and extracellular sites for these ions. Since ADP is formed from ATP during both glycolysis and oxidative metabolism, Na^+,K^+-ATPase-mediated metabolic effects can be exerted by a direct stimulation of either or both pathways. If energy demand exceeds the capability to increase oxidative metabolism (due to the failure of increasing PDH beyond a rather limited level), glycolysis may accordingly show a disproportionately large increase, as seen during spreading depression and seizure activity.

D. Ca^{2+}-Mediated Stimulation of Glucose Metabolism in Brain Cells

1. *Which Stimulatory Effects Are Ca^{2+}-Mediated?*

Ca^{2+}-dependent effects on glucose metabolism are not inhibited by ouabain or tetrodotoxin, but those Ca^{2+}-mediated effects evoked by Ca^{2+} entry into the cell are abolished in the absence of Ca^{2+} in the incubation medium. In some cases Ca^{2+} depletion is effective only combined with a simultaneous elevation of the concentration of Mg^{2+}, which competes with Ca^{2+} for its binding or uptake sites (including mitochondrial uptake) but does not exert similar physiological activities. As shown in Table V, stimulatory effects associated with an increase in $[Ca^{2+}]_i$ include (1) a K^+-mediated and very short-lasting increase in oxygen consumption in synaptosomes, which is inhibited by Ca^{2+} depletion, but not by ouabain (Erecinska *et al.*, 1991; Erecinska and Silver, 1994); (2) K^+-mediated glycogenolysis, which both in cultured astrocytes and in brain slices is dependent upon Ca^{2+} entry; (3) a transient ouabain-resistant, K^+-mediated stimulation of oxidative metabolism of glucose in cultured astrocytes and in brain slices; and (4) an increase in oxygen utilization by electrical stimulation or elevated $[K^+]_e$ in posterior pituitary *in vitro,* most of which is resistant to ouabain, but sensitive to Ca^{2+} depletion (Shibuki, 1989). A fifth metabolic effect of Ca^{2+} is that evoked by transmitters activating the phosphatidylinositide second messenger system, particularly noradrenaline, thereby increasing $[Ca^{2+}]_i$ by release of bound Ca^{2+} from the endoplasmic reticulum. Since the endoplasmic reticulum is not immediately depleted of bound Ca^{2+} in the absence of extracellular Ca^{2+}, transmitter-induced increase of $[Ca^{2+}]_i$ is resistant to short-lasting Ca^{2+} depletion, but the increase in free mitochondrial Ca^{2+} resulting from a rise in $[Ca^{2+}]_i$ is inhibited by omission of Ca^{2+} in the medium with concomitant elevation of Mg^{2+} (e.g., Chen and Hertz, 1999).

2. *K^+ Effect on Free Cytosolic Ca^{2+} Concentration ($[Ca^{2+}]_i$) in Cultured Cells*

K^+-mediated depolarization causes an increase in $[Ca^{2+}]_i$ in both neurons and astrocytes, albeit with different K^+ concentration dependence. The concentration dependence of the K^+-induced increase in $[Ca^{2+}]_i$ in glutamatergic cerebellar granule cell neurons is illustrated in Fig. 19, top panel. Between 3 and 15 m*M* K^+, there is a very small rise in $[Ca^{2+}]_i$ from its resting level of ~100–150 nM, whereas between 15 and 20 m*M* K^+ there is a steep increase in $[Ca^{2+}]_i$ to close to 1 μM due to depolarization and neuronal excitation, which activates Ca^{2+} entry through voltage-dependent Ca^{2+} channels; there is no further increase in $[Ca^{2+}]_i$ when extracellular K^+ is elevated above 20 m*M* (Zhao, 1992). In well-differentiated cultured astrocytes, elevated $[K^+]_e$ also enhances Ca^{2+} uptake (Hertz *et al.*, 1989) and increases $[Ca^{2+}]_i$ by opening L-type Ca^{2+} channels (Zhao *et al.*, 1996). A similar $[K^+]$-induced increase in $[Ca^{2+}]_i$ or Ca^{2+} influx has been observed in noncultured dissociated astrocytes (Duffy and McVicar, 1994; Thorlin *et al.*, 1998) and in astrocytes in intact brain tissue (Shao and McCarthy, 1997; Kulik *et al.*, 1999). In astrocytes, the slope of $[Ca^{2+}]_i$ as a function of $[K^+]_e$ is much shallower than in neurons (Fig. 19, bottom panel) because astrocytes are not excitable cells (Barres *et al.*, 1990; Sontheimer, 1994) and therefore do not show an abrupt depolarization when the membrane potential is lowered to a certain threshold. To sum up, Ca^{2+} is an important intracellular messenger and its free intracellular concentration can be increased by either Ca^{2+} entry into the cell or release of intracellularly bound Ca^{2+}; a rise in $[Ca^{2+}]_i$ exerts various effects in both neurons (e.g., Ca^{2+}-dependent transmitter release) and astrocytes (e.g., metabolic stimulation).

3. *Depolarization-Induced, $[Ca^{2+}]_i$-Mediated Glycogenolysis in Astrocytes*

Elevated $[K^+]_e$ enhances glycogenolysis in brain slices (Hof *et al.*, 1988), and this must be an astrocytic phenomenon, since most neurons normally

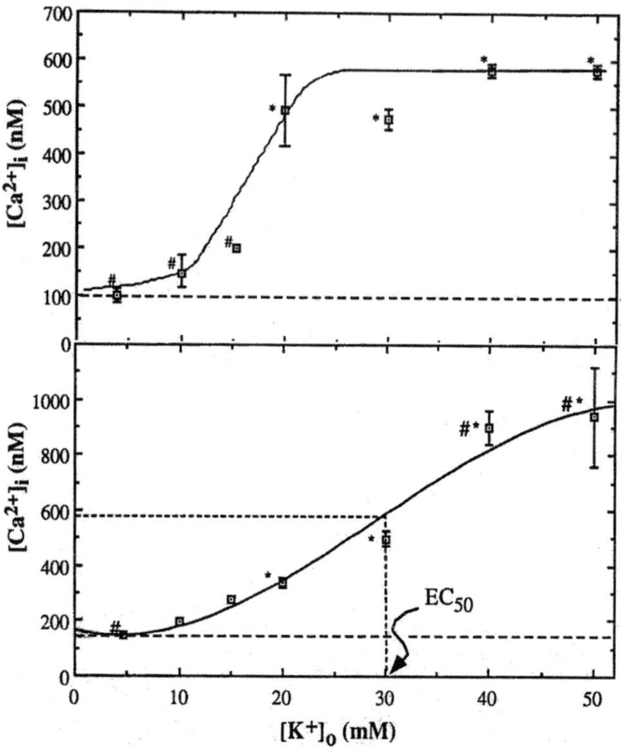

FIG. 19. Increase in intracellular calcium concentration ($[Ca^{2+}]_i$) in primary cultures of mouse cerebellar granule cell neurons (upper panel) and mouse cortical astrocytes (lower panel) as a function of elevation of $[K^+]_e$. $[Ca^{2+}]_i$ was measured by means of the fluorescent probe Indo-1 during incubation of cultured cells in phosphate-buffered saline containing 6 m*M* glucose. SEM values are shown by vertical bars if extending beyond the symbols. Statistically significant increases ($p < 0.05$ or better) are indicated by asterisks. (From Zhao, 1992, with the permission of thesis advisor [upper panel] and Zhao *et al.*, 1996 [lower panel], with modifications, with the permission of the National Research Council of Canada.)

neither contain glycogen (Ibrahim, 1975) nor express any activity of phosphorylase (Pfeiffer *et al.*, 1990, 1992, 1994), a glycogenolytic enzyme stimulated by Ca^{2+} (see Section II). K^+-stimulation of glycogenolysis occurs in dibutyryl cyclic AMP-treated astrocyte cultures but not in untreated cultures which do not express voltage-dependent, L-type Ca^{2+} channels (Subbarao *et al.*, 1995). K^+-induced glycogenolysis is (1) abolished by specific inhibition of L-type Ca^{2+} channels (by 100 n*M* nifedipine, a dihydropyridine Ca^{2+} channel blocker); and (2) enhanced by a benzodiazepine (midazolam), which augments the K^+-induced increase in $[Ca^{2+}]_i$ in cultured astrocytes (Subbarao *et al.*, 1995; Zhao *et al.*, 1996).

4. *K^+-Induced Stimulation of Astrocytic Glucose Oxidation Via Stimulation of the Na^+,K^+,Cl^- Co-Transporter*

a. Astrocytes in Primary Cultures. A substantial, but transient stimulation of oxygen consumption by high K^+ concentrations (>15 m*M*) occurs in microdissected glial cells (Hertz, 1966; Aleksidze and Blomstrand, 1969), glial cells obtained by gradient centrifugation (Haljamae and Hamberger, 1971), and astrocytes in primary cultures (Hertz *et al.*, 1973; E. Hertz and Hertz, 1979; E. Hertz *et al.*, 1986). Increased production of $^{14}CO_2$ from [U-^{14}C]glucose in primary astrocyte cultures (averaging 28% during a 10-min incubation [Table VI]) occurs *transiently* (i.e., lasting less than 30 min) after onset of exposure to elevated K^+ concentrations (i.e., a different effect that that seen after prolonged incubation times [see Section IV.C.2]). This effect is unaffected by ouabain, even at high concentrations (L. Hertz, unpublished experiments), but is abolished by furosemide, an inhibitor of co-transport of Na^+, K^+, and Cl^- (Hertz, 1986b), linking oxidative metabolism to activation of Na^+, K^+,Cl^- co-transporter activity, as discussed below.

b. K^+-Induced, Ca^{2+}-Dependent Stimulation of Brain Slices. High extracellular K^+ concentrations cause a large increase in oxygen consumption in brain slices with either glucose (Ashford and Dixon, 1935; Dickens and Greville, 1935) or pyruvate (Ghosh and Quastel, 1954) as the substrate. Maximum stimulation occurs at 35–50 m*M* K^+ with a threshold at 10–15 m*M* K^+ (Hertz and Schou, 1962; Hertz and Kjeldsen, 1985). The metabolic stimulation is associated with a considerable swelling of the tissue, i.e., fluid uptake, which shows a similar K^+ concentration dependence (Lund-Andersen and Hertz, 1970; Hertz and Kjeldsen, 1985). The rate of respiratory decline is enhanced after the stimulation (Hertz and Schou, 1962). Aerobic production of pyruvate and lactate is also enhanced by ~50 m*M* K^+ in both brain slices (Ashford and Dixon, 1935; Takagaki, 1972) and synaptosomes (De Belleroche and Bradford, 1972; Erecinska and Dagani, 1990).

Stimulation of oxygen consumption by excess K^+ in brain slices is not inhibited by tetrodotoxin, which, in contrast, does block the rise in CMR_{O_2} induced by electrical stimulation (Okamoto and Quastel, 1970). These data suggest that the K^+ effect is not secondary to excitation-induced entry of Na^+, i.e., may not primarily be a neuronal phenomenon. This conclusion is supported by the observation that production of $^{14}CO_2$ from [1-^{14}C]acetate, a "glial reporter substrate," is increased by excess K^+ (Gonda and Quastel, 1966), indicating that the K^+-induced stimulation of brain slices at least in part occurs in the astrocytic cell population, which is consistent with results of NMR studies by Badar-Goffer *et al.* (1992). The K^+-induced stimulation of oxygen consumption in brain slices is also relatively resistant to ouabain (Table VI; Hertz and Peng, 1992a), suggesting Ca^{2+}-dependence, which is consistent with inhibition of the K^+-induced stimulation of CMR_{O2} by Ca^{2+} depletion combined with an elevation of Mg^{2+} (Hertz and Schou, 1962). Together, these observations suggest that the K^+-induced stimulation of oxidative metabolism in brain slices is, at least partly, caused by a depolarization-induced entry of Ca^{2+} through voltage-sensitive channels. The stimulation is abolished by ethacrynic acid, an inhibitor of the Na^+, K^+, Cl^- co-transporter system, suggesting that a K^+-induced increase in swelling is a result of the joint ion uptake by the co-transporter. The relatively high threshold concentration of 10–15 m*M* $[K^+]_e$, required to elicit these effects make the question pertinent whether they are physiologically relevant during normal brain function. However, it should be kept in mind (1) that $[K^+]_e$ values in the brain *in vivo* are measured in relative large "pockets" of extracellular fluid, not in narrow spaces separating adjacent neuronal and astrocytic processes; (2) that K^+-induced depolarization in *in vivo* situations may be additive with transmitter-induced depolarization of astrocytes; noradrenaline, the α_1-adrenergic agonist phenylephrine, and glutamate cause membrane depolarization in cultured astrocytes (Hosli *et al.,* 1982; Bowman and Kimelberg, 1984, 1987); and (3) that most cell culture and brain slice experiments have been carried out using a $[K^+]_e$ of 5 m*M* as the control value, rather than the ~3 m*M* in brain extracellular fluid; therefore a doubling of $[K^+]_e$ may occur at lower levels in the brain, e.g., 6 m*M* $[K^+]_e$.

5. *Depolarization-Activated, $[Ca^{2+}]_i$-Mediated Na^+, K^+, and Cl^- Co-Transport*

Inhibition of K^+-induced CMR_{glc} in cultured astrocytes by furosemide (Hertz, 1986b) and oxygen consumption in brain slices by ethacrynic acid (Kjeldsen and Hertz, unpublished experiments) suggested a link between a Na^+, K^+, and Cl^- co-transporter and oxidative metabolism. The presence in astrocytes of a transporter, which jointly accumulates one K^+, one Na^+, and 2 Cl^- and is inhibited by furosemide, ethacrynic acid, and bumetanide is well established (Kimelberg and Fragakis, 1985; Tas *et al.,* 1987). In cultured rodent astrocytes this co-transporter is activated by high $[K^+]_e$ (Walz and Hertz, 1984), probably as a result of K^+-stimulated Ca^{2+} entry through L-channels, a conclusion based on inhibition of co-transporter activity by 0.5 μM of the dihydropyridine, nifedipine (Su *et al.,* 2000). Similar and/or related co-transporters are also present in neurons, but less information is available about their K^+ sensitivity and they might operate in the reverse direction, transporting the ions out of the cells. In contrast to cultured astrocytes, cultured cerebellar granule cell neurons and hippocampal neurons did not accumulate Na^+ by co-transporter activity (Chen *et al.,* 1992; Rose and Ransom, 1997).

6. *Transmitter-Induced, Ca^{2+}-Mediated Effects in Neurons*

It is likely that the K^+-induced, ouabain-insensitive stimulation of oxygen consumption in synaptosomes (Erecinska *et al.*, 1991), at least partly, is a manifestation of processes associated with transmitter release and normalization of $[Ca^{2+}]_i$. Both transmitter release and re-establishment of the low resting neuronal $[Ca^{2+}]_i$ (by Ca^{2+}-ATPase activity and Na^+/Ca^{2+} exchange, triggering subsequent stimulation of Na^+,K^+-ATPase activity by Na^+) are assumed to be associated with a relatively low, but undefined energy utilization, which is consistent with *very* slight increases in CMR_{glc} after K^+-induced depolarization of cultured GABAergic cortical interneurons, and the modest, largely ouabain-resistant stimulation in dendrite-impaired neurons.

7. *K^+-Stimulated, Ca^{2+}-Mediated Increase in CMR_{O_2} in Posterior Pituitary*

The increase in oxygen utilization during electrical or K^+-induced stimulation of the intact incubated pituitary is only slightly inhibited by ouabain, but abolished by Ca^{2+} depletion (Shibuki, 1989), a finding which appears to disagree with data of Mata *et al.* (1980), who found that glucose utilization in the same tissue is stimulated in an ouabain-sensitive manner by electrical stimulation. Unfortunately the two studies give no information about absolute rates of either CMR_{glc} or CMR_{O_2}. Nerve terminals releasing vasopressin and oxytocin might be the site of the Ca^{2+}-dependent stimulation of oxygen consumption. However, about 30% of the volume of the posterior pituitary consists of glial cells (Nordmann, 1977), and Ca^{2+}-dependent stimulation of oxygen consumption or use of endogenous fuel (e.g., glutamate) in astrocytes can not be ruled out.

8. *Transmitter-Induced, Ca^{2+}-Mediated Stimulation of Astrocytic CMR_{glc}*

a. Astrocytic Receptors. Receptor expression is not a neuronal prerogative, and astrocytes also express a wide spectrum of receptors. This attribute is shared by many cell types, e.g., muscle cells and epithelial cells, in which receptor activation regulates functions like energy metabolism and ion transport. In the CNS, transmitters released to the intimacy of a synapse convey private information between a pre- and a postsynaptic neuron, as well as to the astrocytes surrounding the synaptic cleft. In addition, diffusion of transmitters released by varicosities can reach all neighboring cells, so, depending on their location, astrocytes can receive signals from various sources. Most studies of receptor signaling in astrocytes have been carried out using cultured astrocytes, but it is now well established that astrocytes *in situ* also express functional receptors for many transmitters (Aoki 1997; Thorlin *et al.*, 1998; Kulik *et al.*, 1999; Kimelberg *et al.*, 2000). Adrenergic, especially β-adrenergic, receptors may be expressed on most cerebral astrocytes (i.e., astrocytes represent a major target for activation of locus coeruleus, the nucleus of origin for noradrenergic fibers to the brain [Stone and Ariano, 1989; Stone *et al.,* 1992]), whereas other receptors, e.g., serotonergic and certain adrenergic subtypes (such as the α_2-adrenergic receptor) are found only on subsets of astrocytes. Astrocytes also display metabotropic glutamate receptors and non-NMDA ionotropic glutamate receptors, whereas NMDA receptors are either absent on astrocytes or only expressed at a few locations.

In cultured astrocytes several transmitter agonists, including α-adrenergic and some serotonergic agonists, activate production of the two second messengers, inositol trisphosphate (IP_3) and diacylglycerol (DAG), from phosphatidylinositide diphosphate (PIP_2). DAG stimulates protein kinase C (PKC) activity, whereas IP_3 causes a release of Ca^{2+} from intracellular stores, leading to an increase in $[Ca^{2+}]_i$, which in the short term is independent of extracellular Ca^{2+}. An increase in $[Ca^{2+}]_i$ spreads in a waveform across an astrocytic syncytium, partly due to transport of IP_3 through gap junctions, eventually exciting adjacent neurons by release of astrocytic glutamate (Cornell-Bell *et al.,* 1990; Kim *et al.,* 1994; Parpura and Haydon, 2000).

b. Stimulation of Glycogenolysis. In primary cultures of astrocytes, noradrenaline and serotonin (5-HT) activation of postjunctional α_2-adrenergic and 5-HT_{2A} and 5-HT_{2B} receptors, respectively, leads to an increase in $[Ca^{2+}]_i$ and stimulation of glycogenolysis via Ca^{2+}-dependent activation of phosphorylase (Subbarao and Hertz, 1990a; Chen *et al.,* 1995; Chen and Hertz, 1999).

c. Stimulation of Mitochondrial Dehydrogenase and Glutaminase Activity. Mitochondrial dehydrogenases stimulated by noradrenaline in many tissues include PDH, α-KG dehydrogenase, and isocitrate dehydrogenase (McCormack and Denton, 1990). In astrocytes, metabolic fluxes through the reactions catalyzed by PDH and α-KG dehydrogenase are increased following the rise in mitochondrial Ca^{2+} concentration secondary to a neurotransmitter-induced increase in $[Ca^{2+}]_i$ (Subbarao and Hertz, 1991; Hertz and Peng, 1992b; Peuchen *et al.,* 1996; Chen and Hertz, 1999). Both noradrenaline and the α_2-specific agonists clonidine and dexmedetomidine increase $[Ca^{2+}]_i$ and $[1\text{-}^{14}C]$pyruvate decarboxylation (which mainly reflects PDH activity [Erecinska and Dagani, 1990; Kaufman and Driscoll, 1992]) in astrocytes (Chen and Hertz, 1999; Chen *et al.,* 2000), but dexmedetomidine has no effect on $[Ca^{2+}]_i$ in cerebellar granule cell neurons (Zhao *et al.,* 1992). Increased $^{14}CO_2$ formation from pyruvate is abolished in the absence of extracellular Ca^{2+}, combined with a high $[Mg^{2+}]$ (Hertz and Peng, 1992a; Chen and Hertz, 1999), and the biphasic dependence on dexmedetomidine concentration in astrocytes is similar for the $[Ca^{2+}]_i$ response and the increase in pyruvate decarboxylation (Fig. 20). No data are available regarding

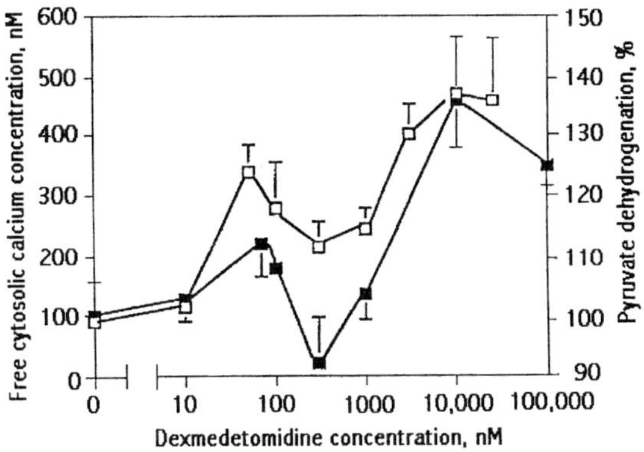

FIG. 20. Increases in intracellular calcium concentration ($[Ca^{2+}]_i$) (open squares) and pyruvate dehydrogenation (filled squares) in primary cultures of mouse astrocytes as functions of the concentration of dexmedetomidine, a highly specific α_2-adrenergic agonist. $[Ca^{2+}]_i$ was measured by means of the fluorescent probe Indo-1 during incubation in phosphate-buffered saline containing 6 m*M* glucose. The rate of pyruvate dehydrogenation was assayed as production of $^{14}CO_2$ from 1-[^{14}C]pyruvate during a 30 min of incubation in an air-tight chamber in glucose-free tissue culture medium containing 5 m*M* pyruvate; at the end of the experiment, the cells were acidified, and CO_2 was trapped and counted. SEM values are shown by vertical bars. The increase in $[Ca^{2+}]_i$ is statistically significant at all concentrations above 10 n*M* (except 500 n*M*), and $^{14}CO_2$ production was significantly greater than control at 70 and 100 n*M* and 10 and 100 μM. (Modified from Chen *et al.*, Correlation between dexmedetomidine-induced biphasic increases in free cytosolic calcium concentration and in energy metabolism in astrocytes, *Anesth. Analg.*, **91,** 353–357, Lippincott Williams & Wilkins, 2000.)

stimulation of isocitrate dehydrogenase by noradrenaline in astrocytes, but $^{14}CO_2$ production from [1-^{14}C]glutamate via the action of succinate dehydrogenase is stimulated by noradrenaline in cultured astrocytes but not in neurons (Subbarao and Hertz, 1990b, 1991). Stimulation of glutaminase activity might be unexpected, but glutamine is a good metabolic substrate (Section II.H.4), and flux from glutamine to glutamate is enhanced by noradrenaline in cultured astrocytes, whereas glutamine synthesis (catalyzed by glutamine synthetase) is unaffected (Huang and Hertz, 1995).

E. Metabolic Effects of Transmitters Activating Adenylyl Cyclase Activity

Activation of β-adrenergic receptors enhances adenylyl cyclase activity and increases the level of cAMP and the activity of protein kinase A. This leads to a stimulation of glycogenolysis in cultured astrocytes (Subbarao and

Hertz, 1990a), which is consistent with the activation by protein kinase A of phosphorylase kinase, the enzyme converting inactive phosphorylase to its active form. Vasoactive intestinal peptide (VIP), a neuropeptide stimulating adenylyl cyclase activity, has a similar effect (Magistretti *et al.,* 1983). In addition, noradrenaline and vasopressin stimulate glycogen synthesis (Sorg and Magistretti, 1992).

F. K^+-Stimulated Enzyme Reactions

Several enzymes are stimulated by elevated K^+ concentrations, including PC (Ruiz-Amil *et al.,* 1965; McClure *et al.,* 1971), and pyruvate carboxylation increases with a rise in the extracellular K^+ concentration from 2–25 m*M* in cultured astrocytes (Kaufman and Driscoll, 1992).

G. Summary

Graded increases in extracellular K^+ concentrations, similar to those which either occur during physiological brain activity or evoke neuronal excitation, stimulate glucose phosphorylation and oxidation in neurons and astrocytes at different $[K^+]_e$ levels by different mechanisms involving Na^+,K^+-ATPase activity, ATP hydrolysis, ADP production, and opening of Ca^{2+} channels. In the brain *in vivo,* there are different relationships between $[K^+]_e$ and NAD^+/NADH ratio (Fig. 14). *In vitro* studies help to explain these concentration-dependent effects of $[K^+]_e$ by identifying different sensitivities of the multiple mechanisms expressed in astrocytes and neurons. At the lower range, the effect appears to be mainly due to action of K^+ at the extracellular, K^+-stimulated site of the astrocytic Na^+,K^+-ATPase, which has sufficiently low affinity to be stimulated by 5–12 m*M* K^+; at this $[K^+]_e$ level, there is no corresponding stimulation at the K^+-sensitive site of the neuronal Na^+,K^+-ATPase. Because neurons, in contrast to astrocytes, are excitable cells, high levels of $[K^+]_e$ that exceed ~10 m*M* (as well as the actions of neurotransmitters) cause membrane depolarization, Na^+entry, and activation of the intracellular, Na^+-sensitive site of the Na^+,K^+-ATPase, thereby stimulating both CMR_{glc} and CMR_{O_2}. Although presynaptic events in glutamatergic and GABAergic neurons may elicit similar energy demands, a large K^+-induced stimulation of CMR_{O_2} was exclusively seen in the intact glutamatergic cerebellar granule cells (where all functional synapses are glutamatergic), but not in either GABAergic cerebral cortical neurons or cerebellar granule cells that had dendritic degeneration and intact presynaptic structures. These findings suggest that dendrites carrying excitatory, glutamatergic impulses may be the major site of stimulation of glucose metabolism during brain activation.

Astrocytes are nonexcitable cells and do not react to partial depolarization with an all-or-none depolarization evoked by Na^+ entry. $[K^+]_e$ above $\sim$10 m*M* stimulates both glycogenolysis and oxidative metabolism of glucose, the latter presumably due to increased Na^+,K^+,Cl^--co-transporter activity. Both K^+-induced glycogenolysis and co-transporter-linked metabolic effects appear to be dependent on opening of voltage-dependent Ca^{2+} channels in cultured astrocytes and in brain slices, with a threshold of $\sim$10 m*M* K^+. The noradrenaline-induced stimulation of a wide range of metabolic and enzymatic activities in cultured astrocytes is also associated with a marked increase in $[Ca^{2+}]_i$. Calcium-mediated activation of mitochondrial dehydrogenases is independent of altered energy charge or nucleoside phosphates. However, there is overlap and cooperation between Ca^{2+}-mediated and ADP- and AMP-mediated effects on glucose metabolism, perhaps especially in neurons, where Ca^{2+}-mediated transmitter release triggers energy-requiring processes. In general, elevated K^+ concentrations and transmitters stimulate different metabolic processes (e.g., a noradrenaline-induced stimulation of the PDH complex versus a K^+-induced stimulation of pyruvate carboxylation). Because (1) elevated K^+ concentrations and several transmitters stimulate glycogenolysis; and (2) pyruvate is the common precursor for both acetyl-CoA formation and pyruvate carboxylation, coordinated activation of glucose, glycogen, and TCA cycle metabolism in astrocytes might reflect simultaneous activation of both energy-producing steps and biosynthetic pathways for the excitatory amino acids during and after functional activation of metabolism in working brain.

V. Concluding Remarks

A. Contributions of Different Cell Types to Brain Glucose Metabolism

The metabolic capabilities and activities of each of the major brain cell types and their contributions to resting and functionally stimulated metabolism during normal physiological and disease states has been one of the major unresolved problems in the field of metabolic brain imaging for decades. This issue has received considerable attention in current models for neuronal-astrocytic interactions in working brain. Neurons and astrocytes are the two major cell types in brain cortex, with oligodendrocytes and microglia, as well as nonparenchymal cells like brain endothelial cells, accounting for considerably smaller fractions of the volume. Astrocytes may constitute 30% of brain cortical volume in man, but less in rodents (Bass *et al.*, 1971). Two lines of evidence, (1) *in vitro* assays showing that astrocytes and neurons have similar rates of glucose oxidation (Peng *et al.*, 1994); and (2) NMR studies of the relative contribution of astrocytic pyruvate carboxylation to total TCA cycle activity in brain, suggest that astrocytes account for roughly one-third of the total glucose phosphorylation and oxidation. The cells that produce lactate *in vivo* are not identified, but astrocytes (and perhaps also neurons) are able to release large amounts of lactate during brain activation. However, due to a relatively slow lactate uptake and metabolism in both neurons and astrocytes, the released lactate is highly unlikely to function as the *major* energy fuel in either type (Dienel and Hertz, 2001). Instead, an overflow of lactate is likely to occur beyond the activated regions, explaining the quantitative difference in stimulated metabolic activity when measured with [^{14}C]DG and with [^{14}C]glucose.

Almost two-thirds of glucose metabolism in cerebral cortex probably occurs in neurons because astrocytes account for only about one-third and oligodendrocytes, which are most prominent in white matter, have low expression of glycolytic enzymes and rates of glucose utilization in the adult brain *in vivo*. The distribution of metabolic activity between different neuronal constituents is uncertain, and the conclusion that stimulation of metabolic activity primarily seems to occur in the neuropil does not necessarily imply that resting metabolism also is uniformly low in neuronal perikarya; there is evidence to the contrary in isolated neurons. However, pronounced differences in metabolic activity and capacity in different neuronal pathways have been identified by direct assays of CMR_{glc} and immunochemical observations showing large regional and subcellular differences in enzyme localization and maximal activities; synaptic nerve endings have negligible energy reserves, and possess high hexokinase activity and high glycolytic capacity, as well as mitochondrial oxidative metabolism (Kauppinen and Nicholls, 1986a–c).

Comparison of intact glutamatergic (i.e., excitatory) cerebellar granule cell neurons in primary cultures with (1) similar cells with dendritic degeneration, but intact presynaptic structures; and (2) intact GABAergic (i.e., inhibitory neurons) suggest that the increase in oxidative metabolism mainly takes place in dendrites conveying excitatory input. This possibility is supported by the immunohistochemical demonstration of high cytochrome oxidase expression in cells receiving glutamatergic input and high rates of DG phosphorylation in such cells. Propagation of sodium-dependent action potentials along nonmyelinated dendrites (Martina *et al.*, 2000) will create high energy demand to restore resting levels of intra- and extracellular K^+ and Na^+.

B. Enhancement of Energy-Dependent Processes during Brain Activation

Na^+,K^+-ATPase has a predominant and widespread role in the direct linkage of functional activity to metabolic demand by consuming ATP and producing compounds, including ADP and AMP, that stimulate and are required substrates for the electron transport chain and glucose metabolism. However, the activity of Na^+,K^+-ATPase need not be the major or exclusive step that couples function and metabolism, and the same stimulus can cause activation of different metabolic pathways or utilization of different fuels. For example, electrical stimulation of CMR_{glc} in isolated posterior pituitary increases is completely inhibited by ouabain (Mata *et al.*, 1980), whereas only a small fraction of the stimulation of CMR_{O_2} is blocked by ouabain (Shibuki, 1989). Apparent discrepancies might arise because most experimental studies focus on specific processes with limited assays. The combination of ouabain-sensitive glucose phosphorylation and ouabain-resistant oxidative metabolism is completely feasible, for example, if a cell population within the tissue oxidized glutamate.

Stimulation of glucose phosphorylation and oxidation by depolarizing levels of K^+ is in some neurons (and perhaps all neurons with predominantly glutamatergic innervation) almost completely inhibited by ouabain, indicating that the energy metabolism is increased mainly by enhanced Na^+ and K^+ pumping following the excitation-induced disruption of ionic gradients. Astrocytes are nonexcitable cells, and their activation occurs either at the extracellular K^+-sensitive site of the ATPase, as a direct response to increases in $[K^+]_e$ above its normal resting level or at its intracellular site, due to increases in intracellular Na^+ concentration as a result of Na^+ entry, e.g., during Na^+-dependent uptake of transmitters such as glutamate. Both K^+ and glutamate are released to the extracellular space by excited neurons, so it is likely that stimulated astrocytic metabolism occurs mainly along dendrites receiving glutamatergic input. The relative contributions of the metabolic activation corresponding to Na^+,K^+-ATPase activity in neurons and astrocytes are, however, not known and may vary with the magnitude of the ambient $[K^+]_e$.

Ca^{2+}-mediated stimulation of metabolism occurs in both neurons and astrocytes, and the linkage between functional roles of $[Ca^{2+}]_i$ and metabolic activity is quite different in these two cell types. It is well-established that exocytosis due to Ca^{2+} entry into neurons must be correlated with some metabolic stimulation, as is extrusion and sequestering of Ca^{2+}. This energy may be modest compared to that used for reestablishment of Na^+ and K^+ gradients, since inhibitory neurons have a much smaller increase in metabolism compared to that in excitatory neurons when $[K^+]_e$ is raised. An emerging role for $[Ca^{2+}]_i$ in regulation of glucose metabolism in astrocytes is evident in several areas. Elevated K^+ concentrations stimulate glycogenolysis by a $[Ca^{2+}]_i$-mediated effect, regardless whether the increase in $[Ca^{2+}]_i$ is due to opening of L-channels or transmitter-dependent activation of the phosphatidyl inositide second messenger system. Noradrenaline-induced increase in $[Ca^{2+}]_i$ also leads to a substantial rise in intramitochondial Ca^{2+}, which, in turn, can stimulate intramitochondrial dehydrogenase activity, glutaminase activity, and even the electron transport chain, thereby causing a direct stimulation of oxidative metabolism, independent of any work-induced change in energy charge. Both "upstream" and "downstream" reactions in glucose metabolism and oxidative phosphorylation will be affected by the metabolic actions of Ca^{2+}, and the Ca^{2+} effects may operate in conjunction with increased production of ADP. Studies of the importance of the Ca^{2+}-mediated signaling system for brain function and stimulus-induced changes in metabolism (including the establishment of memory, which is almost always associated with increased noradrenaline at one or more specific stages) are still in their infancy. The recent demonstration that transmitter-induced increases in $[Ca^{2+}]_i$ are not a localized phenomenon, but may travel long distances through an astrocytic network, greatly enhances the potential importance of Ca^{2+}-mediated signaling in astrocytes and the contributions of astrocytes to metabolic brain imaging.

C. Future Directions

Glucose utilization studies firmly link energy generation to functional activity at a local level, but determination of oxygen consumption or assay of glucose flux through the hexokinase step does not identify the cells, pathways, or functional processes using the energy or the carbon derived from degradation of glucose. Intermediate steps directly linking glucose metabolism and functional activity, e.g., neurotransmitter cycling, have been implicated by determination of relationships between glucose flux into the glutamate pool and glutamate-glutamine cycling rates by NMR when functional and metabolic activity are challenged. However anesthetized animals were used in many of these experiments, and the influence of different types of anesthesia on the processes involved in activation, metabolic regulation, and the magnitude of substrate flux through different pathways during response to activation remains to be clarified. Many aspects of brain activation need to be validated in conscious subjects, and incorporation of other neurotransmitter systems exerting metabolic effects into the model is required to develop a more complete framework for functional metabolic activity in working brain. It was shown in both cultured cells and in intact tissue that working brain cells might activate different aspects or pathways

of glucose metabolism to satisfy functions that have relatively high energy needs that stimulate oxygen utilization. The actions of neurotransmitters on astrocytes stimulate energy demand, activate use of glycogen, the brain's major energy store, and might transiently enhance biosynthetic capacity to integrate the activities of neurons and astrocytes during brain activation. Elucidation of these processes in the brain *in vivo* should lead to a greater understanding of the functions and interactions of working astrocytes and neurons in the excitatory pathways. These interactions are likely to vary with physiological state and be "task-specific," and it will be necessary to identify potential differences between different modes of brain activation.

The fate of glucose in working brain, trafficking of metabolites within brain and from brain to blood, neuronal-glial interactions, including those regulating astrocytic $[Ca^{2+}]_i$, and cellular basis of brain images in working brain are key, unresolved, interrelated issues in the neurobiology of energy generation. Although bioenergetic processes are well understood at the molecular and cellular levels, a challenge for the future is to understand how the different brain cell types work together *in vivo* to carry out normal brain functions, and how gradual disruption of these processes leads to progressive neurological diseases and mental dysfunction.

Acknowledgments

This work was supported, in part, by grants IBN 9728171 from the National Science Foundation and NS 36720 and NS 38230 from the National Institutes of Health to G. A. Dienel.

References

Ackermann, R. F., and Lear, J. L. (1989). Glycolysis-induced discordance between glucose metabolic rates measured with radiolabeled fluorodeoxyglucose and glucose. *J. Cereb. Blood Flow Metab.* **9,** 774–785.

Adachi, K., Cruz, N. F., Sokoloff, L., and Dienel, G. A. (1995). Labeling of metabolic pools by [6-^{14}C]glucose during K^+-induced stimulation of glucose utilization in rat brain. *J. Cereb. Blood Flow Metab.* **15,** 97–110.

Akabayashi, A., and Kato, T. (1993). Glucose utilization rates in single neurons and neuropil determined by injecting nontracer amounts of 2-deoxyglucose. *J. Neurochem.* **60,** 931–935.

Aleksidze, N. G., and Blomstrand, K. (1969). Effect of potassium ion on the respiration of the neurons and neuroglia of the lateral vestibular nucleus in rabbits. *Dokl. Akad. Nauk. SSSR.* **186,** 1429–1430.

Alexander, G. M., Grothusen, J. R., Gordon, S. W., and Schwartzman, R. J. (1997). Intracerebral microdialysis study of glutamate reuptake in awake, behaving rats. *Brain Res.* **766,** 1–10.

Aoki, C. (1997). Differential timing for the appearance of neuronal and astrocytic beta-adrenergic receptors in the developing rat visual cortex as revealed by light and electron-microscopic immunocytochemistry. *Vis. Neurosci.* **14,** 1128–1142.

Aoki, C., Milner, T. A., Berger, S. B., Sheu, K. F., Blass, J. P., and Pickel, V. M. (1987a). Glial glutamate dehydrogenase: ultrastructural localization and regional distribution in relation to the mitochondrial enzyme, cytochrome oxidase. *J. Neurosci. Res.* **18,** 305–318.

Aoki, C., Milner, T. A., Sheu, K. F., Blass, J. P., and Pickel, V. M. (1987b). Regional distribution of astrocytes with intense immunoreactivity for glutamate dehydrogenase in rat brain: implications for neuron-glia interactions in glutamate transmission. *J. Neurosci.* **7,** 2214–2231.

Ashford, C. A., and Dixon, K. C. (1935). The effect of potassium on the glycolysis of brain tissue with reference to the Pasteur effect. *Biochem. J.* **29,** 157–168.

Attwell, D., and Laughlin, S. B. (2001). An energy budget for signaling in the grey matter of the brain. *J. Cereb. Blood Flow Metab.* **21,** 1133–1145.

Aureli, T., Di Cocco, M. E., Calvani, M., and Conti, F. (1997). The entry of [1-^{13}C]glucose into biochemical pathways reveals a complex compartmentation and metabolite trafficking between glia and neurons: a study by ^{13}C-NMR spectroscopy. *Brain Res.* **765,** 218–227.

Auestad, N., Korsak, R. A., Morrow, J. W., and Edmond, J. (1991). Fatty acid oxidation and ketogenesis by astrocytes in primary culture. *J. Neurochem.* **56,** 1376–1386.

Bachelard, H. S., Campbell, W. J., and McIlwain, H. (1962). The sodium and other ions of mammalian cerebral tissues, maintained and electrically stimulated *in vitro*. *Biochem. J.* **84,** 225–232.

Badar-Goffer, R. S., Ben-Yoseph, O., Bachelard, H. S., and Morris, P. G. (1992). Neuronal-glial metabolism under depolarizing conditions: ^{13}C-n.m.r.study. *Biochem. J.* **282,** 225–230.

Bagley, P. R., Tucker, S. P., Nolan, C., Lindsay, J. G., Davies, A., Baldwin, S. A., Cremer, J. E., and Cunningham, M. V. J. (1989). Anatomical mapping of glucose transporter protein and pyruvate dehydrogenase in rat brain: An immunogold study. *Brain Res.* **499,** 214–224.

Bakken, I. J., White, L. R., Unsgard, G., Aasly, J., and Sonnewald, U. (1998a). [U-^{13}C]glutamate metabolism in astrocytes during hypoglycemia and hypoxia. *J. Neurosci. Res.* **51,** 636–645.

Bakken, I. J., White, L. R., Unsgard, G., Aasly, J., and Sonnewald, U. (1998b). [U-^{13}C]aspartate metabolism in cultured cortical astrocytes and cerebellar granule neurons studied by NMR spectroscopy. *Glia* **23,** 271–277.

Balazs, R., and Cremer, J. E., eds. (1972). "Metabolic Compartmentation in the Brain." John Wiley & Sons, New York.

Baquer, N. Z., Hothersall, J. S., and McLean, P. (1988). Function and regulation of the pentose phosphate pathway in brain. *In* "Current Topics in Cellular Regulation" (B. A. Horecker and E. R. Stadtman, eds.), pp. 265–289. Academic Press, New York.

Barden, R. E., and Scrutton, M. C. (1974). Pyruvate carboxylase from chicken liver. Effects of univalent and divalent cations on catalytic activity. *J. Biol. Chem.* **249,** 4829–4838.

Barres, B. A., Chun, L. L. Y., and Corey, D. P. (1990). Ion channels in vertebrate glia. *Ann. Rev. Neurosci.* **13,** 441–474.

Bass, N. H., Hess, H. H., Pope, A., and Thalheimer, C. (1971). Quantitative cytoarchitectonic distribution of neurons, glia and DNA in rat cerebral cortex. *J. Comp. Neurol.* **143,** 481–490.

Baulieu, E. E. (1997). Neurosteroids: of the nervous system, by the nervous system, for the nervous system. *Recent Prog. Horm. Res.* **52,** 1–32.

Beresford, H. R., Posner, J. B., and Plum, F. (1969). Changes in brain lactate during induced cerebral seizures. *Arch. Neurol.* **20,** 243–248.

Bergles, D. E., Diamond, J. S., and Jahr, C. E. (1999). Clearance of glutamate inside the synapse and beyond. *Curr. Opin. Neurobiol.* **9,** 293–298.

BeltrandelRio, H., and Wilson, J. E. (1992a). Interaction of mitochondrially bound rat brain hexokinase with intramitochondrial compartments of ATP generated by oxidative phosphorylation and creatine kinase. *Arch. Biochem. Biophys.* **299,** 116–124.

BeltrandelRio, H., and Wilson, J. E. (1992b). Coordinated regulation of cerebral glycolytic and oxidative metabolism, mediated by mitochondrially bound hexokinase dependent on intramitochondrially generated ATP. *Arch. Biochem. Biophys.* **296,** 667–677.

Berl, S., and Clarke, D. D. (1969). Compartmentation of amino acid metabolism. *In* "Handbook of Neurochemistry" (A. Lajtha, ed.), Vol. 2, *Structural Neurochemistry,* pp. 447–472. Plenum Press, New York.

Berl, S., Clarke, D. D., and Schneider, D., eds. (1975). "Metabolic Compartmentation and Neurotransmission: Relation to Brain Structure & Function." Plenum Press, New York.

Bhattacharya, S. K., and Datta, A. G. (1993). Is brain a gluconeogenic organ? *Mol. Cell. Biochem.* **125,** 51–57.

Bittar, P. G., Charnay, Y., Pellerin, L., Bouras, C., and Magistretti, P. J. (1996). Selective distribution of lactate dehydrogenase isoenzymes in neurons and astrocytes of human brain. *J. Cereb. Blood Flow Metab.* **16,** 1079–1089.

Blomqvist, G., Stone-Elander, S., Halldin, C., Roland, P. E., Widen, L., Lindqvist, M., Swahn, C.-G., Langstrom, B., and and Wiesel, F. A. (1990). Positron emission tomographic measurements of cerebral glucose utilization using [1-^{11}C]glucose. *J. Cereb. Blood Flow Metab.* **10,** 467–483.

Blomstrand, F., Aberg, N. D., Eriksson, P. S., Hansson, E., and Ronnback, L. (1999). Extent of intercellular calcium wave propagation is related to gap junction permeability and level of connexin-43 expression in astrocytes in primary cultures from four brain regions. *Neuroscience* **92,** 255–265.

Bolwig, T. G., and Quistorff, B. (1973). *In vivo* concentration of lactate in the brain of conscious rats before and during seizures: a new ultra-rapid technique of the freeze-sampling of brain tissue. *J. Neurochem.* **21,** 1345–1348.

Borowsky, I. W., and Collins, R. C. (1989a). Histochemical changes in enzymes of energy metabolism in the dentate gyrus accompany deafferentation and synaptic reorganization. *Neuroscience* **33,** 253–262.

Borowsky, I. W., and Collins, R. C. (1989b). Metabolic anatomy of brain: a comparison of regional capillary density, glucose metabolism, and enzyme activities. *J. Comp. Neurol.* **288,** 401–413.

Boullard, H., Osen, K. K., Levy, L. M., Danbolt, N. C., Edwards, R. H., Storm-Mathisen, J., and Chaudhry, F. (2002). Cell-specific expression of the glutamine transporter SN1 suggests differences in dependence on the glutamine cycle. *Eur. J. Neurosci.* **15,** 1615–1631.

Bouzier, A.-K., Thiaudiere, E., Biran, M., Rouland, R., Canioni, P., and Merle, M. (2000). The metabolism of [3-^{13}C]lactate in the rat brain is specific of a pyruvate carboxylase-deprived compartment. *J. Neurochem.* **75,** 480–486.

Bowman, C. L., and Kimelberg, H. K. (1984). Excitatory amino acids directly depolarize rat brain astrocytes in primary culture. *Nature* **311,** 656–9.

Bowman, C. L., and Kimelberg, H. K. (1987). Pharmacological properties of the norepinephrine-induced depolarization of astrocytes in primary culture: evidence for the involvement of an alpha 1-adrenergic receptor. *Brain Res.* **423,** 403–407.

Broer, S., Rahman, B., Pellegri, G., Pellerin, L., Martin, J. L., Verleysdonk, S., Hamprecht, B., and Magistretti, P. J. (1997). Comparison of lactate transport in astroglial cells and monocarboxylate transporter 1 (MCT 1) expressing *Xenopus laevis* oocytes. Expression of two different monocarboxylate transporters in astroglial cells and neurons. *J. Biol. Chem.* **272,** 30096–30102.

Broer, S., and Brookes, N. (2001). Transfer of glutamine between astrocytes and neurons. *J. Neurochem.* **77,** 705–719.

Brumberg, V. A., and Pevzner, L. (1975). Cytospectrophotometric studies on the lactate dehydrogenase isoenzymes in functionally different neurone-neuroglia units. I. Optimal procedure conditions and topochemical comparisons in normal mice. *Acta Histochem.* **54,** 218–229.

Bukato, G., Kochan, Z., and Swierczynski, J. (1995). Different regulatory properties of the cytosolic and mitochondrial forms of malic enzyme isolated from human brain. *Int. J. Biochem. Cell Biol.* **27,** 1003–1008.

Bures, J., Buresova, O., and Krivanek, J. (1974). "The Mechanism and Applications of Leao's Spreading Depression of Electroencephalographic Activity." Academic Press, New York.

Calabrese, V. P., Gruemer, H. D., James, K., Hranowsky, N., and DeLorenzo, R. J. (1991). Cerebrospinal fluid lactate levels and prognosis in status epilepticus. *Epilepsia* **32,** 816–821.

Calingasan, N. Y., Baker, H., Sheu, K. F., and Gibson, G. E. (1994). Distribution of the alpha-ketoglutarate dehydrogenase complex in rat brain. *J. Comp. Neurol.* **346,** 461–479.

Cerdan, S., Künnecke, B., and Seelig, J. (1990). Cerebral metabolism of [1,2-^{13}C2] acetate as detected by in vivo and in vitro ^{13}C NMR. *J. Biol. Chem.* **265,** 12916–12926.

Cesar, M. de C., and Wilson, J. E. (1998). Further studies on the coupling of mitochondrially bound hexokinase to intramitochondrially compartmented ATP, generated by oxidative phosphorylation. *Arch. Biochem. Biophys.* **350,** 109–117.

Chen, C.-J., and Liao, S.-L. (2001). Effects of glutamate on glucose utilization and lactate output in cultured rat cortical astrocytes. *J. Cereb. Blood Flow Metab.* **21,** Suppl. 1, S254.

Chen, W., Zhu, X. H., Gruetter, R., Seaquist, E. R., Adriany, G., and Ugurbil, K. (2001). Study of tricarboxylic acid cycle flux changes in human visual cortex during hemifield visual stimulation using (1)H-(13)C MRS and fMRI. *Magn. Reson. Med.* **45,** 349–355.

Chen, Y., and Hertz, L. (1999). Noradrenaline effects on pyruvate decarboxylation-correlation with calcium signaling. *J. Neurosci. Res.* **58,** 599–606.

Chen, Y., McNeill, J. R., Hajek, I., and Hertz, L. (1992). Effect of vasopressin on brain swelling at the cellular level - do astrocytes exhibit a furosemide-vasopressin-sensitive mechanism for volume regulation? *Can. J. Physiol. Pharmacol.* **70,** S367–S373.

Chen, Y., Peng, L., Zhang, X., Stolzenburg, J.-U., and Hertz, L. (1995). Further evidence that fluoxetine interacts with a 5-HT_{1C} receptor. *Brain Res. Bull.* **38,** 153–159.

Chen, Y., Zhao, Z., Code, W. E., and Hertz, L. (2000). Correlation between dexmedetomidine-induced biphasic increases in free cytosolic calcium concentration and in energy metabolism in astrocytes. *Anesth. Analg.* **91,** 353–357.

Clarke, D. D., Nicklas, W. J., and Berl, S. (1970). Tricarboxylic acid-cycle metabolism in brain. Effect of fluoroacetate and fluorocitrate on the labelling of glutamate, aspartate, glutamine, and γ-aminobutyrate. *Biochem. J.* **120,** 345–351.

Collins, R. C., McCandless, D. W., and Wagman, I. L. (1987). Cerebral glucose utilization: comparison of [^{14}C]deoxyglucose and [6-^{14}C]glucose quantitative autoradiography. *J. Neurochem.* **49,** 1564–1570.

Colombo, J. A., Schleicher, A., and Zilles, K. (1999). Patterned distribution of immunoreactive astroglial processes in the striate (V1) cortex of New World monkeys. *Glia* **25,** 85–92.

Cooper, A. J., and Plum, F. (1987). Biochemistry and physiology of brain ammonia. *Physiol. Rev.* **67,** 440–519.

Cornell-Bell, A. H., Finkbeiner, S. M., Cooper, M. S., and Smith, S. J. (1990). Glutamate induces calcium waves in cultured astrocytes: long-range glial signaling. *Science* **247,** 470–473.

Cornford, E. M., Hyman, S., Cornford, M. E., and Damian, R. T. (1998). Glut1 glucose transporter in the primate choroid plexus endothelium. *J. Neuropathol. Exp. Neurol.* **57,** 404–414.

Cremer, J. E., Braun, L. D., and Oldendorf, W. H. (1976). Changes during development in transport processes of the blood-brain barrier. *Biochim. Biophys. Acta* **448,** 633–637.

Cremer, J. E., Sarna, C. S., Teal, H. M., and Cunningham, V. J. (1978). Amino acid precursors: Their transport into brain and initial metabolism. *In* "Amino Acids as Chemical Transmitters. NATO Advanced Study Institutes Series: Series A, Life Sciences" (F. Fonnum, ed.), Vol. 16, pp. 669–689. Plenum Press, New York.

Cruz, F., and Cerdan, S. (1999). Quantitative ^{13}C NMR studies of metabolic compartmentation in the adult mammalian brain. *NMR Biomed.* **12,** 451–462.

Cruz, N. F., Adachi, K., and Dienel, G. A. (1999). Metabolite trafficking during K^+-induced spreading cortical depression: Rapid efflux of lactate from cerebral cortex. *J. Cereb. Blood Flow Metab.* **19,** 380–392.

D'Amelio, F., Eng, L. F., and Gibbs, M. A. (1990). Glutamine synthetase immunoreactivity is present in oligodendroglia of various regions of the central nervous system. *Glia* **3,** 335–341.

Danbolt, N. (2001). Glutamate uptake. *Prog. Neurobiol.* **65,** 1–105.

Davis, T. L., Kwong, K. K., Weisskoff, R. M., and Rosen, B. R. (1998). Calibrated functional MRI: Mapping the dynamics of oxidative metabolism. *Proc. Natl. Acad. Sci. USA* **95,** 1834–1839.

Davson, H. (1962). "A Textbook of General Physiology," 4th edition, J. & A. Churchill, London.

De Belleroche, J. S., and Bradford, H. F. (1972). Metabolism of beds of mammalian cortical synaptosome. Response to depolarizing influences. *J. Neurochem.* **19,** 585–602.

De Bruin, L. A., Schasfoort, E. M. C., Steffens, A. B., and Korf, J. (1990). Effects of stress and exercise on rat hippocampus and striatum extracellular lactate. *Am. J. Physiol.* **259,** R773–R779.

de Cerqueira Cesar, M., and Wilson, J. E. (1995). Application of a double isotopic labeling method to a study of the interaction of mitochondrially bound rat brain hexokinase with intramitochondrial compartments of ATP generated by oxidative phosphorylation. *Arch. Biochem. Biophys.* **324,** 9–14.

de los A Garcia, M., Carrasco, M., Godoy, A., Reinicke, M., Montecinos, V. P., Aguayo, L. G., Tapia, J. C., Vera, J. C., and and Nualart, F. (2001). Elevated expression of glucose transporter-1 in hypothalamic ependymal cells not involved in the formation of the brain-cerebrospinal fluid barrier. *J. Cell. Biochem.* **80,** 491–503.

Dickens, F., and Greville, G. D. (1935). The metabolism of normal and tumour tissue (XIII). Neutral salt effects. *Biochem. J.* **29,** 1468–1483.

Dienel, G. A. (2002). Energy generation in the central nervous system. *In* "Cerebral Blood Flow and Metabolism, 2nd ed." (L. Edvinsson and D. Krause, eds.). Lippincott, Williams & Wilkins, Philadelphia, pp. 140–161.

Dienel, G. A., and Cruz, N. F. (1997). Altered incorporation of $H^{14}CO_3$ into brain metabolites during spreading cortical depression. *Abstr. Soc. Neurosci.* **23,** 1310.

Dienel, G. A., and Hertz, L. (2001). Glucose and lactate metabolism during brain activation. *J. Neurosci. Res.* **66,** 824–838.

Dienel, G. A., Cruz, N. F., Adachi, K., Sokoloff, L., and Holden, J. E. (1997a). Determination of local brain glucose level with [^{14}C]methylglucose: Effects of glucose supply and demand. *Am. J. Physiol.* **273,** E839–E849.

Dienel, G. A., Wang, R. Y., and Cruz, N. F. (1997b). Flux of [6-^{14}C]glucose into metabolic pools in neurons and glia *in vivo* during "aerobic glycolysis" induced by sensory stimulation. *J. Cereb. Blood Flow. Metab.* **17** (suppl.), S644.

Dienel, G. A., Liu, K., Popp, D., and Cruz, N. F. (1999). Enhanced acetate and glucose utilization during graded photic stimulation: Neuronal-glial interactions in vivo. *Ann. N.Y. Acad. Sci.* **893,** 279–281.

Dienel, G. A., Ball, K., Popp, D., and Cruz, N. F. (2000). Trafficking of metabolites of [^{14}C]glucose in brain during auditory stimulation of the conscious rat. *J. Neurochem.* **74**(Suppl.), S85A.

Dienel, G. A., Ball, K., Popp, D., and Cruz, N. F. (2001a). A role for gap junctions in metabolite spreading? *J. Neurochem.* **78**(Suppl.), 86.

Dienel, G. A., Popp, D., Drew, P. D., Ball, K., Krisht, A., and Cruz, N. F. (2001b). Preferential labeling of glial and meningial brain tumors with [2-^{14}C]acetate. *J. Nucl. Med.* **42,** 1243–1250.

Dienel, G. A., Liu, K., and Cruz, N. F. (2001c). Local uptake of ^{14}C-labeled acetate and butyrate in rat brain *in vivo* during spreading cortical depression. *J. Neurosci. Res.* **66,** 812–820.

Dietrich, W. D., Durham, D., Lowry, O. H., and Woolsey, T. A. (1982). "Increased" sensory stimulation leads to changes in energy-related enzymes in the brain. *J. Neurosci.* **2,** 1608–1613.

Dietschy, J. M., and Turley, S. D. (2001). Cholesterol metabolism in the brain. *Curr. Opin. Lipidol.* **12,** 105–112.

Djuricic, B. M., and Mrsulja, B. B. (1979). Brain microvessel hexokinase: kinetic properties. *Experientia* **35,** 179–181.

Dringen, R., Schmoll, D., Cesar, R., and Hamprecht, B. (1993). Incorporation of radioactivity from [^{14}C]lactate into the glycogen of cultured mouse astroglial cells. Evidence for gluconeogenesis in brain cells. *Biol. Chem. Hoppe Seyler* **374,** 343–347.

Duelli, R., Maurer, M. H., and Kuschinsky, W. (1998). Decreased glucose transporter densities, rate constants and glucose utilization in visual structures of rat brain during chronic visual deprivation. *Neurosci. Lett.* **26,** 49–52.

Duffy, S., and MacVicar, B. A. (1994). Potassium-dependent calcium influx in acutely isolated hippocampal astrocytes. *Neuroscience* **61,** 51–61.

Edmond, J., Robbins, R. A., Bergstrom, J. D., Cole, R. A., and de Vellis, J. (1987). Capacity for substrate utilization in oxidative metabolism by neurons, astrocytes, and oligodendrocytes from developing brain in primary culture. *J. Neurosci. Res.* **18,** 551–561.

Elekes, O., Venema, K., Postema, F., Dringen, R., Hamprecht, B., and Korf, J. (1996). Evidence that stress activates glial lactate formation in vivo assessed with rat hippocampus lactography. *Neurosci. Lett.* **208,** 69–72.

Erecinska, M., and Dagani, F. (1990). Effects of K^+, Na^+, and adenosine triphosphate in isolated synaptosomes. *J. Gen. Physiol.* **95,** 591–616.

Erecinska, M., and Silver, I. A. (1994). Ions and energy in mammalian brain. *Prog. Neurobiol.* **43,** 37–71.

Erecinska, M., Nelson, D., and Chance, B. (1991). Depolarization-induced changes in cellular energy production. *Proc. Natl. Acad. Sci. USA* **88,** 7600–7604.

Erecinska, M., Nelson, D., Nissim, I., Daikhin, Y., and Yudkoff, M. (1994). Cerebral alanine transport and alanine aminotransferase reaction: alanine as a source of neuronal glutamate. *J. Neurochem.* **62,** 1953–1964.

Erecinska, M., Nelson, D., and Silver, I. A. (1996). Metabolic and energetic properties of isolated nerve ending particles (synaptosomes). *Biochim. Biophys. Acta* **1277,** 13–34.

Eriksson, G., Peterson, A., Iverfeldt, K., and Walum, E. (1995). Sodium-dependent glutamate uptake as an activator of oxidative metabolism in primary astrocyte cultures from newborn rat. *Glia* **15,** 152–156.

Fattoretti, P., Bertoni-Freddari, C., Di Stefano, G., Casoli, T., Gracciotti, N., Solazzi, M., and Pompei, P. (2001). Quantitative immunochemistry of glucose transport protein (Glut3) in the rat hippocampus during aging. *J. Histochem. Cytochem.* **49,** 671–672.

Fellows, L. K., Boutelle, M. G., and Fillenz, M. (1993). Physiological stimulation increases nonoxidative glucose metabolism in the brain of the freely moving rat. *J. Neurochem.* **60,** 1258–1263.

Fields, H. M., Rinaman, L., and Devaskar, S. U. (1999). Distribution of glucose transporter isoform-3 and hexokinase I in the postnatal murine brain. *Brain Res.* **846,** 260–254.

Fonnum, F., Johnsen, A., and Hassel, B. (1997). Use of fluorocitrate and fluoroacetate in the study of brain metabolism. *Glia* **21,** 106–113.

Fox, P. T., and Raichle, M. E. (1986). Focal physiological uncoupling of cerebral blood flow and oxidative metabolism during somatosensory stimulation in human subjects. *Proc. Natl. Acad. Sci. USA* **83,** 1140–1144.

Fox, P. T., Raichle, M. E., Mintun, M. A., and Dence, C. (1988). Nonoxidative glucose consumption during focal physiologic neural activation. *Science* **241,** 462–464.

Fujita, H., Kuwabara, H., Reutens, D. C., and Gjedde, A. (1999). Oxygen consumption of cerebral cortex fails to increase during continued vibrotactile stimulation. *J. Cereb. Blood Flow Metab.* **19,** 266–271.

Gali, P., Hauw, J. J., Boutry, J. M., and Hartmann, L. (1981). Immunofluorescence and histochemical methods for neural M1 pyruvate kinase localization. *J. Neurochem.* **37,** 1377–1384.

Gerhart, D. Z., Leino, R. L., Borson, N. D., Taylor, W. E., Gronlund, K. M., McCall, A. L., and Drewes, L. R. (1995). Localization of glucose transporter GLUT 3 in brain: comparison of rodent and dog using species-specific carboxyl-terminal antisera. *Neuroscience* **66,** 237–246.

Gerhart, D. Z., Enerson, B. E., Zhdankina, O. Y., Leino, R. L., and Drewes, L. R. (1997). Expression of the monocarboxylate transporter MCT1 by brain endothelium and glia in adult and suckling rats. *Am. J. Physiol. Endocrinol. Metab.* **273,** E207–E213.

Gerhart, D. Z., Enerson, B. E., Zhdankina, O. Y., Leino, R. L., and Drewes, L. R. (1998). Expression of the monocarboxylate transporter MCT2 by rat brain glia. *Glia* **22,** 272–281.

Ghersi-Egea, J.-F., Finnegan, W., Chen, J.-L., and Fenstermacher, J. D. (1996). Rapid distribution of intraventricularly administered sucrose into cerebrospinal fluid cisterns via subarachnoid velae in rat. *Neuroscience* **75,** 1271–1288.

Ghosh, J. J., and Quastel, J. H. (1954). Narcotics and brain respiration. *Nature* **174,** 28–31.

Gibson, G. E., and Blass, J. P. (1976). Impaired synthesis of acetylcholine in brain accompanying mild hypoxia and hypoglycemia. *J. Neurochem.* **27,** 37–42.

Gibson, G. E., Park, L. C. H., Sheu, K.-F. R., Blass, J. P., and Calingasan, N. Y. (2000). The α-ketoglutarate dehydrogenase complex in neurodegeneration. *Neurochem. Int.* **36,** 97–112.

Gjedde, A. (1982). Calculation of cerebral glucose phosphorylation from brain uptake of glucose analogs *in vivo:* a re-examination. *Brain Res. Rev.* **4,** 237–310.

Gonda, O., and Quastel, J. H. (1966). Transport and metabolism of acetate in rat brain cortex in vitro. *Biochem. J.* **100,** 83–94.

Grisar, T., Frere, J. M., and Franck, G. (1979). Effects of K^+ ions on kinetic properties of the (Na^+-K^+)-ATPase (EC 3.6.1.3) in bulk isolated glia cells, perikarya and synaptosomes from rabbit brain cortex. *Brain Res.* **165,** 87–103.

Gronlund, K. M., Gerhart, D. Z., Leino, R. L., McCall, A. L., and Drewes, L. R. (1996). Chronic seizures increase glucose transporter abundance in rat brain. *J. Neuropathol. Exp. Neurol.* **55,** 832–840.

Gross, P. M., Sposito, N. M., Pettersen, S. E., Panton, D. G., and Fenstermacher, J. D. (1987). Topography of capillary density, glucose metabolism, and microvascular function within the rat inferior colliculus. *J. Cereb. Blood Flow Metab.* **7,** 154–160.

Gruetter, R., Seaquist, E. R., Kim, S., and Ugurbil, K. (1998). Localized in vivo ^{13}C NMR of glutamate metabolism in the human brain: initial results at 4 Tesla. *Dev. Neurosci.* **20,** 380–388.

Gruetter, R., Seaquist, E. R., and Ugurbil, K. (2001). A mathematical model of compartmentalized neurotransmitter metabolism in human brain. *Am. J. Physiol. Endicrinol. Metab.* **282,** E100–E112.

Haberg, A., Qu, H., Haraldseth, O., Unsgard, G., and Sonnewald, U. (1998a). *In vivo* injection of [1-^{13}C]glucose and [1,2-^{13}C]acetate combined with ex vivo ^{13}C nuclear magnetic resonance spectroscopy: A novel approach to the study of middle cerebral occlusion in the rat. *J. Cereb. Blood Flow Metab.* **18,** 1223–1232.

Haberg, A., Qu, H., Bakken, I. J., Sande, L. M., White, L. R., Haraldseth, O., Unsgard, G., Aasly, J., and and Sonnewald, U. (1998b). *In vitro* and ex vivo ^{13}C-NMR spectroscopy studies of pyruvate recycling in brain. *Dev. Neurosci.* **20,** 389–398.

Haberg, A., Qu, H., Saether, O., Unsgard, G., Haraldseth, O., and Sonnewald, U. (2001). Differences in neurotransmitter synthesis and intermediary metabolism between glutamatergic and GABAergic neurons during 4 hours of MCAO in the rat. The role of astrocytes in neuronal survival. *J. Cereb. Blood Flow Metab.* **21,** 1451–1463.

Hacker, H. J., Thorens, B., and Grobholz, R. (1991). Expression of facilitative glucose transporter in rat liver and choroid plexus. A histochemical study in native cryostat sections. *Histochemistry* **96,** 435–439.

Hajek, I., Subbarao, K. V., and Hertz, L. (1996). Stimulation of Na^+, K^+-ATPase activity in astrocytes and neurons by K^+ and/or noradrenaline. *Neurochem. Int.* **28,** 335–342.

Halestrap, A. P. (1989). The regulation of the matrix volume of mammalian mitochondria *in vivo* and *in vitro* and its role in the control of mitochondrial metabolism. *Biochim. Biophys. Acta* **973,** 355–382.

Haljamae, H., and Hamberger, A. (1971). Potassium accumulation by bulk prepared neuronal and glial cells. *J. Neurochem.* **18,** 1903–1912.

Hansson, E., and Ronnback, L. (1992). Adrenergic receptor regulation of amino acid neurotransmitter uptake in astrocytes. *Brain Res. Bull.* **29,** 297–301.

Hanu, R., McKenna, M., O'Neill, A., Resneck, W. G., and Bloch, R. J. (2000). Monocarboxylic acid transporters, MCT1 and MCT2, in cortical astrocytes in vitro and in vivo. *Am. J. Physiol. Cell Physiol.* **278,** C921–C930.

Hashimoto, M., and Wilson, J. E. (2000). Membrane potential-dependent conformational changes in mitochondrially bound hexokinase of brain. *Arch. Biochem. Biophys.* **384,** 163–173.

Hassel, B., and Brathe, A. (2000). Neuronal pyruvate carboxylation supports formation of transmitter glutamate. *J. Neurosci.* **20,** 1342–1347.

Hassel, B., Bachelard, H., Jones, P., Fonnum, F., and Sonnewald, U. (1997). Trafficking of amino acids between neurons and glia in vivo. Effects of inhibition of glial metabolism by fluoroacetate. *J. Cereb. Blood Flow Metab.* **17,** 1230–1238.

Hawkins, R. A., Miller, A. L., Nielsen, R. C., and Veech, R. L. (1973). The acute action of ammonia on rat brain metabolism in vivo. *Biochem. J.* **134,** 1001–1008.

Henn, F. A., Haljamae, H., and Hamberger, A. (1972). Glial cell function: active control of extracellular K^+ concentration. *Brain Res.* **43,** 437–443.

Hertz, E., and Hertz, L. (1979). Polarographic measurement of oxygen uptake by astrocytes in primary cultures using the tissue culture flask as the respirometer chamber. *In Vitro* **15,** 429–436.

Hertz, E., Shargool, M., and Hertz, L. (1986). Effects of barbiturates on energy metabolism by cultured astrocytes and neurons in the presence of normal and elevated potassium concentrations. *Neuropharmacology* **25,** 533–539.

Hertz, L. (1966). Neuroglial localization of potassium and sodium effects on respiration. *J. Neurochem.* **13,** 1373–87.

Hertz, L. (1986a). Potassium transport in astrocytes and neurons in primary cultures. *Ann. N.Y. Acad. Sci.* **481,** 318–333.

Hertz, L. (1986b). Potassium as a signal in metabolic interactions between neurons and astrocytes. *In* "Dynamic Properties of Glial Cells, Cellular and Molecular Aspects" (T. Grisar, G. Franck, L. Hertz, W. T. Norton, M. Sensenbrenner, and D. M. Woodbury, eds.), pp. 215–224. Pergamon Press, New York.

Hertz, L., and Fillenz, M. (1999). Does "the mystery of the extra glucose" during CNS activation reflect glutamate synthesis? *Neurochem. Int.* **34,** 71–75.

Hertz, L., and Kjeldsen, C. S. (1985). Functional role of the potassium-induced stimulation of oxygen uptake in brain slices studied with cesium as a probe. *J. Neurosci. Res.* **14,** 83–93.

Hertz, L., and Peng, L. (1992a). Energy metabolism at the cellular level of the CNS. *Can. J. Physiol. Pharmacol.* **70,** S145–S157.

Hertz, L., and Peng, L. (1992b). Effects of monoamine transmitters on neurons and astrocytes: correlation between energy metabolism and intracellular messengers. *Progr. Brain Res.* **94,** 283–301.

Hertz, L., and Schou, M. (1962). Univalent cations and the respiration of brain cortex slices. *Biochem. J.* **85,** 93–104.

Hertz, L., and Schousboe, A. (1986). Role of astrocytes in compartmentation of amino acid and energy metabolism. *In* "Astrocytes" (S. Fedoroff and A. Vernadakis, eds.), Vol. 2, pp. 179–208. Academic Press, New York.

Hertz, L., and Schousboe, A. (1987). Primary cultures of GABAergic and glutamatergic neurons as model systems to study neurotransmitter functions. I. Differentiated cells. *In* "Model Systems of Development and Aging of the Nervous System" (A. Vernadakis, A. Privat, J. M. Lauder, P. S. Timiras, and E. Giacobini, eds.), pp. 19–31. Martinus Nijhoff Publishers, Boston.

Hertz, L., Dittmann, L., and Mandel, P. (1973). K^+ induced stimulation of oxygen uptake in cultured cerebral glial cells. *Brain Res.* **60,** 517–520.

Hertz, L., Murthy, Ch. R. K., Lai, J. C. K., Fitzpatrick, S. M., and Cooper, A. J. L. (1987). Some metabolic effects of ammonia on astrocytes and neurons in primary cultures. *Neurochem. Pathol.* **6,** 97–129.

Hertz, L., Drejer, J., and Schousboe, A. (1988). Energy metabolism in glutamatergic neurons, GABAergic neurons and astrocytes in primary cultures. *Neurochem. Res.* **13,** 605–610.

Hertz, L., Bender, A. S., Woodbury, D., and White, S. A. (1989). Potassium induced calcium uptake in astrocytes and its potent inhibition by a calcium channel blocker. *J. Neurosci. Res.* **22,** 209–215.

Hertz, L., Swanson, R. A., Newman, G. C., Mariff, H., Juurlink, B. H. J., and Peng, L. (1998a). Can experimental conditions explain the discrepancy whether or not glutamate stimulates aerobic glycolysis? *Dev. Neurosci.* **20,** 339–347.

Hertz, L., Peng, L., and Lai, J. C. K. (1998b). Functional studies in cultured astrocytes. *Methods Comp. Methods Enzymol.* **16,** 293–310.

Hertz, L., Dringen, R., Schousboe, A., and Robinson, S. R. (1999). Astrocytes: Glutamate producers for neurons. *J. Neurosci. Res.* **57,** 417–428.

Hertz, L., Yu, A. C. H., Kala, G., and Schousboe, A. (2000). Neuronal-astrocytic and cytosolic-mitochondrial metabolite trafficking during brain activation, hyperammonemia and energy deprivation. *Neurochem. Int.* **37,** 83–102.

Hevner, R. F., Duff, R. S., and Wong-Riley, M. T. (1992). Coordination of ATP production and consumption in brain: parallel regulation of cytochrome oxidase and Na(+), K(+)-ATPase. *Neurosci. Lett.* **138,** 188–192.

Hevor, T. K., Delorme, P., and Beauvillain, J. C. (1986). Glycogen synthesis and immunocytochemical study of fructose-1,6-biphosphatase in methionine sulfoximine epileptogenic rodent brain. *J. Cereb. Blood Flow Metab.* **6,** 292–297.

Hof, P. R., Pascale, E., and Magistretti, P. J. (1988). K^+ at concentrations reached in the extracellular space during neuronal activity promotes a Ca^{2+}-dependent glycogen hydrolysis in mouse cerebral cortex. *J. Neurosci.* **8,** 1922–1928.

Hogstad, S., Svenneby, G., Torgner, I. A., Kvamme, E., Hertz, L., and Schousboe, A. (1988). Glutaminase in neurons and astrocytes cultured from mouse brain: Kinetic properties and effects of phosphate, glutamate and ammonia. *Neurochem. Res.* **13,** 383–388.

Honegger, P., and Pardo, B. (1999). Separate neuronal and glial Na^+,K^+-ATPase isoforms regulate glucose utilization in response to membrane depolarization and elevated extracellular potassium. *J. Cereb. Blood Flow Metab.* **19,** 1051–1059.

Hosli, L., Hosli, E., Zehnter, C., Lehmann, R., and Lutz, T. W. (1982). Evidence for the existence of alpha- and beta-adrenoceptors on cultured glial cells—an electrophysiological study. *Neuroscience* **7,** 2867–2872.

Hu, Y., and Wilson, G. S. (1997). A temporary local energy pool coupled to neuronal activity: fluctuations of extracellular lactate levels in rat brain monitored with rapid-response enzyme-based sensor. *J. Neurochem.* **69,** 1484–1490.

Huang, R., and Hertz, L. (1995). Noradrenaline induced stimulation of glutamine metabolism in primary cultures of astrocytes. *J. Neurosci. Res.* **41,** 677–683.

Huang, R., Peng, L., Chen, Y., Hajek, I., Zhao, Z., and Hertz, L. (1994). Signaling effect of monoamines and of elevated potassium concentrations on brain energy metabolism at the cellular level. *Dev. Neurosci.* **16,** 337–351.

Hutson, S. M., Lieth, E., and LaNoue, K. F. (2001). Function of leucine in excitatory neurotransmitter metabolism in the central nervous system. *J. Nutr.* **131,** 846S–850S.

Hyder, F., Chase, J. R., Behar, K. L., Mason, G. F., Siddeek, M., Rothman, D. L., and Shulman, R. G. (1996). Increased tricarboxylic acid cycle flux in rat brain during forepaw stimulation detected with $^1H[^{13}C]$NMR. *Proc. Natl. Acad. Sci. USA* **93,** 7612–7.

Ibrahim, M. Z. (1975). Glycogen and its related enzymes of metabolism in the central nervous system. *Adv. Anat. Embryol. Cell Biol.* **52,** 3–89.

Ichimura, T., Fraser, P. A., and Cserr, H. F. (1991). Distribution of extracellular tracers in perivascular spaces of the rat brain. *Brain Res.* **545,** 03–113.

Joanny, P., and Hillman, H. H. (1963). Substrates and the potassium and sodium levels of guinea pig cerebral cortex slices in vitro: Effects of application of electrical pulses, of inhibitors and of anoxia. *J. Neurochem.* **10,** 655–664.

Kadekaro, M., Crane, A. M., and Sokoloff, L. (1985). Differential effects of electrical stimulation of sciatic nerve on metabolic activity in spinal cord and dorsal root ganglion in the rat. *Proc. Natl. Acad. Sci. USA* **82,** 6010–6013.

Kageyama, G. H., and Wong-Riley, M. T. (1982). Histochemical localization of cytochrome oxidase in the hippocampus: correlation with specific neuronal types and afferent pathways. *Neuroscience* **7,** 2337–2361.

Kager, H., Wadman, W. J., and Somjen, G. G. (2000). Simulated seizures and spreading depression in a neuron model incorporating interstitial space and ion concentrations. *J. Neurophysiol.* **84,** 495–512.

Kala, G. Cerebrotoxic effects of ammonia at the cellular level. Ph.D. Thesis, University of Saskatchewan.

Kao-Jen, J., and Wilson, J. E. (1980). Localization of hexokinase in neural tissue: electron microscopic studies of rat cerebellar cortex. *J. Neurochem.* **35,** 667–678.

Kaufman, E. E., and Driscoll, B. F. (1992). Carbon dioxide fixation in neuronal and astroglial cells in culture. *J. Neurochem.* **58,** 258–262.

Kauppinen, R. A., and Nicholls, D. G. (1986a). Failure to maintain glycolysis in anoxic nerve terminals. *J. Neurochem.* **47,** 1864–1869.

Kauppinen, R. A., and Nicholls, D. G. (1986b). Synaptosomal bioenergetics. The role of glycolysis, pyruvate oxidation and responses to hypoglycaemia. *Eur. J. Biochem.* **158,** 159–165.

Kauppinen, R. A., and Nicholls, D. G. (1986c). Pyruvate utilization by synaptosomes is independent of calcium. *FEBS Lett.* **199,** 222–226.

Keesey, J. C., Wallgren, H., and McIlwain, H. (1965). The sodium, potassium and choride of cerebral tissues: maintenance, change on stimulation and subsequent recovery. *Biochem. J.* **95,** 289–300.

Kim, S. G., and Ugurbil, K. (1997). Comparison of blood oxygenation and cerebral blood flow effects in fMRI: estimation of relative oxygen consumption change. *Magn. Reson. Med.* **38,** 59–65.

Kim, W. T., Rioult, M. G., and Cornell-Bell, A. H. (1994). Glutamate-induced calcium signaling in astrocytes. *Glia* **11,** 173–184.

Kimelberg, H. K., and Frangakis, M. V. (1985). Furosemide- and bumetanide-sensitive ion transport and volume control in primary astrocyte cultures from rat brain. *Brain Res.* **361,** 125–134.

Kimelberg, H. K., Biddlecome, S., Narumi, S., and Bourke, R. S. (1978). ATPase and carbonic anhydrase activities of bulk-isolated neuron, glia and synaptosome fractions from rat brain. *Brain Res.* **141,** 305–323.

Kimelberg, H. K., Cai, Z., Schools, G., and Zhou, M. (2000). Acutely isolated astrocytes as models to probe astrocytic functions. *Neurochem. Int.* **36,** 359–367.

Klepper, J., Wang, D., Fischbarg, J., Vera, J. C., Jarjour, I. T., O'Driscoll, K. R., and De Vivo, D. C. (1999). Defective glucose transport across brain tissue barriers: a newly recognized neurological syndrome. *Neurochem. Res.* **24,** 587–594.

Koehler-Stec, E. M., Li, K., Maher, F., Vannucci, S. J., Smith, C. B., and Simpson, I. A. (2000). Cerebral glucose utilization and glucose transporter expression: response to water deprivation and restoration. *J. Cereb. Blood Flow Metab.* **20,** 192–200.

Korf, J. (1996). Intracerebral trafficking of lactate in vivo during stress, exercise, electroconvulsive shock and ischemia as studied with microdialysis. *Dev. Neurosci.* **18,** 405–414.

Krukoff, T. L., Turton, W. E., and Calaresu, F. R. (1986). Increased hexokinase activity in forebrain of water-deprived and diabetes insipidus rats. *Am. J. Physiol.* **251,** R268–R273.

Ksiezak-Reding, H., Blass, J. P., and Gibson, G. E. (1982). Studies on the pyruvate dehydrogenase complex in brain with the arylamine acetyltransferase-coupled assay. *J. Neurochem.* **38,** 1627–1636.

Kuhr, W. G., and Korf, J. (1988). Extracellular lactic acid as an indicator of brain metabolism: Continuous on-line measurement in conscious, freely moving rats with intrastriatal dialysis. *J. Cereb. Blood Flow Metab.* **8,** 130–137.

Kuhr, W. G., van den Berg, C. J., and Korf, J. (1988). In vivo identification and quantitative evaluation of carrier-mediated transport of lactate at the cellular levels in the striatum of conscious, freely moving rats. *J. Cereb. Blood Flow Metab.* **8,** 848–846.

Kulik, A., Haentzsch, A., Luckermann, M., Reichelt, W., and Ballanyi, K. (1999). Neuron-glia signaling via α_1 adrenoceptor-mediated Ca^{2+} release in Bergmann glial cells *in situ*. *J. Neurosci.* **19,** 8401–8408.

Kurz, G. M., Wiesinger, H., and Hamprecht, B. (1993). Purification of cytosolic malic enzyme from bovine brain, generation of monoclonal antibodies and immunocytochemical localization of the enzyme in glial cells of neural primary cultures. *J. Neurochem.* **60,** 1467–1474.

Lai, J. C. K., Behar, K. L., Liang, B. B., and Hertz, L. (1999). Hexokinase in astrocytes: Kinetic and regulatory properties. *Metab. Brain Dis.* **14,** 125–133.

Lapidot, A., and Gopher, A. (1994). Cerebral metabolic compartmentation. Estimation of glucose flux via pyruvate carboxylase/pyruvate dehydrogenase by ^{13}C NMR isotopomer analysis of D-[U-^{13}C]glucose metabolites. *J. Biol. Chem.* **269,** 27198–27208.

Lapidot, A., and Gopher, A. (1997). Quantitation of metabolic compartmentation in hyperammonemic brain by natural abundance ^{13}C-NMR detection of ^{13}C-^{15}N coupling patterns and isotopic shifts. *Eur. J. Biochem.* **243,** 597–604.

Largo, C., Cuevas, P., Somjen, G. G., Martin del Rio, R., and Herreras, O. (1996). The effect of depressing glial function in rat brain in situ on ion homeostasis, synaptic transmission, and neuron survival. *J. Neurosci.* **16,** 1219–1229.

Lawrie, A. M., Rizzuto, R., Pozzan, T., and Simpson, A. W. (1996). A role for calcium influx in the regulation of mitochondrial calcium in endothelial cells. *J. Biol. Chem.* **271,** 10753–10759.

Leao, A. A. P. (1944). Spreading depression of activity in the cerebral cortex. *J. Neurophysiol.* **7,** 359–390.

Lee, S. H., Kim, W. T., Cornell-Bell, A. H., and Sontheimer, H. (1994). Astrocytes exhibit regional specificity in gap-junction coupling. *Glia* **11,** 315–325.

Leino, R. L., Gerhart, D. Z., van Bueren, A. M., McCall, A. L., and Drewes, L. R. (1997). Ultrastructural localization of GLUT 1 and GLUT 3 glucose transporters in rat brain. *J. Neurosci. Res.* **49,** 617–626.

Loffler, T., Al-Robaiy, S., Bigl, M., Eschrich, K., and Schliebs, R. (2001). Expression of fructose-1,6-bisphosphatase mRNA isoforms in normal and basal forebrain cholinergic lesioned rat brain. *Int. J. Dev. Neurosci.* **19,** 279–285.

Lothman, E., Lamanna, J., Cordingley, G., Rosenthal, M., and Somjen, G. (1975). Responses of electrical potential, potassium levels and oxidative metabolic activity of cerebral neocortex of cats. *Brain Res.* **88,** 15–36.

Lowry, O. H., Roberts, N. R., Leiner, K. Y., Wu, M.-L., Farr, A. L., and Albers, R. W. (1954). The quantitative histochemistry of brain; III. Ammon's horn. *J. Biol. Chem.* **207,** 39–49.

Lund-Andersen, H., and Hertz, L. (1970). Effects of potassium and of glutamate on swelling and on sodium and potassium content in brain cortex slices from adult rats. *Exp. Brain Res.* **11,** 199–212.

Lynch, R. M., Fogarty, K. E., and Fay, F. S. (1991). Modulation of hexokinase association with mitochondria analyzed wioth quantitative three-dimensional confocal microscopy. *J. Cell Biol.* **112,** 385–395.

Lynch, R. M., Carrington, W., Fogarty, K. E., and Fay, F. S. (1996). Metabolic modulation of hexokinase association with mitochondria in living smooth muscle cells. *Am. J. Physiol.* **270,** C488–C499.

Madl, J. E., Clements, J. R., Beitz, A. J., Wenthold, R. J., and Larson, A. A. (1988). Immunocytochemical localization of glutamate dehydrogenase in mitochondria of the cerebellum: An ultrastructural study using a monoclonal antibody. *Brain Res.* **452,** 396–402.

Madsen, P. L., Hasselbalch, S. A. G., Hangman, L. P., Olsen, K. S., Bulow, J., Holm, S., Wildschcidtz, G., Paulson, O. B., and and Lassen, N. A. (1995). Persistent resetting of the cerebral oxygen/glucose uptake ratio by brain activation: Evidence obtained by the Kety-Schmidt technique. *J. Cereb. Blood Flow Metab.* **15,** 485–491.

Madsen, P. L., Cruz, N. F., Sokoloff, L., and Dienel, G. A. (1999). Cerebral oxygen/glucose ratio is low during sensory stimulation and rises above normal during recovery: Excess glucose consumption during stimulation is not accounted for by lactate efflux from or accumulation in brain tissue. *J. Cereb. Blood Flow Metab.* **19,** 393–400.

Magistretti, P. J., Manthorpe, M., Bloom, F. E., and Varon, S. (1983). Functional receptors for vasoactive intestinal polypeptide in cultured astroglia from neonatal rat brain. *Regul. Pept.* **6,** 71–80.

Magistretti, P. J., Pellerin, L., Rothman, D. L., and Shulman, R. G. (1999). Energy on demand. *Science* **283,** 496–497.

Majewska, M. D. (1992). Neurosteroids: Endogenous bimodal modulators of the $GABA_A$ receptor. Mechanism of action and physiological significance. *Prog. Neurobiol.* **38,** 379–395.

Mantych, G. J., James, D. E., Chung, H. D., and Devaskar, S. U. (1993). Cellular localization and characterization of Glut 3 glucose transporter isoform in human brain. *Endocrinology* **131,** 1270–1278.

Martina, M., Vida, I., and Jonas, P. (2000). Distal initiation and active propagation of action potentials in interneuron dendrites. *Science* **287,** 295–300.

Martinez-Rodriguez, R., Arenas, G., Hidalgo, M. M., and Carnicero, M. B. (1989). Light and electron microscope immunolocalization of cytosolic enzyme-like activity in the rat's cerebellar cortex. *J. Hirnforsch.* **30,** 291–296.

Martins-Ferreira, H. (1994). Spreading depression: a neurohumoral reaction. *Braz. J. Med. Biol. Res.* **27,** 851–863.

Mason, G. F., Rothman, D. L., Behar, K. L., and Shulman, R. G. (1992). NMR determination of the TCA cycle rate and alpha-ketoglutarate/glutamate exchange rate in rat brain. *J. Cereb. Blood Flow Metab.* **12,** 434–447.

Mason, G. F., Gruetter, R., Rothman, D. L., Behar, K. L., Shulman, R. G., and Novotny, E. J. (1995). Simultaneous determination of the rates of the TCA cycle, glucose utilization, alpha-ketoglutarate/glutamate exchange, and glutamine synthesis in human brain by NMR. *J. Cereb. Blood Flow Metab.* **15,** 12–25.

Mata, M., Fink, D. J., Gainer, H., Smith, C. B., Davidsen, L., Savaki, H., Schwartz, W. J., and Sokoloff, L. (1980). Activity-dependent energy metabolism in rat posterior pituitary primarily reflects sodium pump activity. *J. Neurochem.* **34,** 213–215.

Mayevsky, A., and Weiss, H. R. (1991). Cerebral blood flow and oxygen consumption in cortical spreading depression. *J. Cereb. Blood Flow Metab.* **11,** 829–836.

McCall, A. L., Van Bueren, A. M., Moholt-Siebert, M., Cherry, N. J., and Woodward, W. R. (1994). Immunohistochemical localization of the neuron-specific glucose transporter (GLUT3) to neuropil in adult rat brain. *Brain Res.* **659,** 292–297.

McCall, A. L., Van Bueren, A. M., Nipper, V., Moholt-Siebert, M., Downes, H., and Lessov, N. (1996). Forebrain ischemia increases GLUT 1 protein in brain microvessels and parenchyma. *J. Cereb. Blood Flow Metab.* **16,** 69–76.

McCasland, J. S., and Hibbard, L. S. (1997). GABAergic neurons in barrel cortex show strong, whisker-dependent metabolic activation during normal behavior. *J. Neurosci.* **17,** 5509–5527.

McClure, W. R., Lardy, H. A., and Kneifel, H. P. (1971). Rat liver pyruvate carboxylase, I. Preparation, properties and cation specificity. *J. Biol. Chem.* **246,** 3569–3578.

McCormack, J. G., and Denton, R. M. (1990). The role of mitochondrial Ca^{2+} transport and matrix Ca^{2+} in signal transduction in mammalian tissues. *Biochim. Biophys. Acta.* **1018,** 287–291.

McDougal, D. B. Jr., Ferrendelli, J. A., Yip, V., Pusateri, M. E., Carter, J. G., Chi, M. M., Morris, B., Manchester., J., and and Lowry, O. H. (1990). Use of nonradioactive 2-deoxyglucose to study compartmentation of brain glucose metabolism and rapid regional changes in rate. *Proc. Natl. Acad. Sci. USA* **87,** 1357–1361.

McIlwain, H. (1951). Metabolic response in vitro to electrical stimulation of sections of mammalian brain. *Biochem. J.* **49,** 382–393.

McIlwain, H., and Bachelard, H. S. (1985). "Biochemistry and the Central Nervous System," 5th ed. Churchill Livingstone, Edinburgh.

McIlwain, H., Anguiano, G., and Cheshire, J. D. (1952). Electrical stimulation in vitro of the metabolism of glucose by mammalian cerebral cortex. *Biochem. J.* **50,** 12–18.

McKenna, M. C., Tildon, J. T., Stevenson, J. H., Huang, X., and Kingwell, K. G. (1995). Regulation of mitochondrial and cytosolic malic enzymes from cultured rat brain astrocytes. *Neurochem. Res.* **20,** 1491–1501.

McKenna, M. C., Sonnewald, U., Huang, X., Stevenson, J., and Zielke, H. R. (1996). Exogenous glutamate concentration regulates the metabolic fate of glutamate in astrocytes. *J. Neurochem.* **66,** 386–393.

McKenna, M. C., Stevenson, J. H., Huang, X., and Hopkins, I. B. (2000). Differential distribution of the enzymes glutamate dehydrogenase and aspartate aminotransferase in cortical synaptic mitochondria contributes to metabolic compartmentation in cortical synaptic terminals. *Neurochem. Int.* **37,** 229–241.

Mercado, R., and Hernandez, J. (1992). Regulatory role of a neurotransmitter (5-HT) on glial Na^+,K^+-ATPase in the rat brain. *Neurochem. Int.* **21,** 119–127.

Milner, T. A., Aoki, C., Sheu, K. F., Blass, J. P., and Pickel, V. M. (1987). Light microscopic immunocytochemical localization of pyruvate dehydrogenase complex in rat brain: Topographical distribution and relation to to cholinergic and catecholaminergic nuclei. *J. Neurosci.* **7,** 3171–3190.

Mollenhauer, H. H., Morre, D. J., and Rowe, L. D. (1990). Alteration of intracellular traffic by monensin; mechanism, specificity and relationship to toxicity. *Biochim. Biophys. Acta* **1031,** 225–246.

Moonen, G., Franck, G., and Schoffeniels, E. (1980). Glial control of neuronal excitability in mammals. I. Electrophysiological and isotopic evidence in culture. *Neurochem. Int.* **2,** 299–310.

Morell, P., and Jurevics, H. (1996). Origin of cholesterol in myelin. *Neurochem. Res.* **21,** 463–470.

Morgello, S., Uson, R. R., Schwartz, E. J., and Haber, R. S. (1995). The human blood-brain barrier glucose transporter (GLUT 1) is a glucose transporter of gray matter astrocytes. *Glia* **14,** 43–54.

Morro, I., and Yamada, T. (1994). Immunohistochemistry using antibody to the glucose transporter 3 in human brainstem and cerebellar tissues. *No To Shinkei* **46,** 1039–1043.

Muir, D., Berl, S., and Clarke, D. D. (1986). Acetate and fluoroacetate as possible markers for glial metabolism in vivo. *Brain Res.* **380,** 336–340.

Murthy, Ch.R. K., and Hertz, L. (1988). Pyruvate decarboxylation in astrocytes and in neurons in primary cultures in the presence and the absence of ammonia. *Neurochem. Res.* **13,** 57–61.

Nagamutsi, S., Sawa, H., Kamada, K., Nakamichi, Y., Yoshimoto, K., and Hoshino, T. (1993). Neuron-specific glucose transporter (NSGT): CNS distribution of GLUT3 rat glucose transporter (RTG3) in rat central neurons. *FEBS Lett.* **334,** 289–295.

Nakanishi, H., Cruz, N. F., Adachi, K., Sokoloff, L., and Dienel, G. A. (1996). Influence of glucose supply and demand on determination of brain glucose content with labeled methylglucose. *J. Cereb. Blood Flow Metab.* **16,** 439–449.

Nakao, Y., Itoh, Y., Kuang, T. Y., Cook, M., Jehle, J., and Sokoloff, L. (2001). Effects of anesthesia on functional activation of cerebral blood flow and metabolism. *Proc. Natl. Acad. Sci. USA* **98,** 7593–7598.

Nelson, T., Dienel, G. A., Mori, K., Cruz, N. F., and Sokoloff, S. (1987). Deoxyglucose-6-phosphate stability in vivo and the deoxyglucose method: response to comments of Hawkins and Miller. *J. Neurochem.* **49,** 1949–1960.

Nicholls, D. G. (1993). The glutamatergic nerve terminal. *Eur. J. Biochem.* **212,** 613–631.

Nie, F., and Wong-Riley, M. T. (1996). Differential glutamatergic innervation in cytochrome oxidase-rich and -poor regions of the macaque striate cortex: quantitative EM analysis of neurons and neuropil. *J. Comp. Neurol.* **369,** 571–590.

Nordmann, J. J. (1977). Ultrastructural morphology of the rat neurohypophysis. *J. Anat.* **123,** 213–218.

Norenberg, M. D., and Martinez-Hernandez, A. (1977). Fine structural localization of glutamine synthetase in astrocytes of rat brain. *Brain Res.* **161,** 303–310.

O'Dowd, B. S. (1995). Metabolic activity in astrocytes is essential for the consolidation of one-trial passive avoidance memory in day-old chick. Ph.D. thesis. La Trobe University, Bundoora, Victoria, Australia.

O'Dowd, B. S., Gibbs, M. E., Ng, K. T., Hertz, E., and Hertz, L. (1994). Astrocytic glycogenolysis energizes memory processes in neonate chicks. *Dev. Brain Res.* **78,** 137–141.

Okamoto, K., and Quastel, J. H. (1970). Tetrodotoxin-sensitive uptake of ions and water by slices of rat brain in vitro. *Biochem. J.* **120,** 37–47.

Park, L. C., Calingasan, N. Y., Sheu, K. F., and Gibson, G. E. (2000). Quantitative alpha-ketoglutarate dehydrogenase activity staining in brain sections and in cultured cells. *Anal. Biochem.* **277,** 86–93.

Parpura, V., and Haydon, P. G. (2000). Physiological astrocytic calcium levels stimulate glutamate release to modulate adjacent neurons. *Proc. Natl. Acad. Sci. USA* **97,** 8629–8634.

Pascual, J. M., Carceller, F., Roda, J. M., and Cerdan, S. (1998). Glutamate, glutamine and GABA as substrates for the neuronal and glial compartments after focal cerebral ischemia in rats. *Stroke* **29,** 1048–1056.

Passonneau, J. V., and Lowry, O. H. (1964). The role of phosphofructokinase in metabolic regulation. *Adv. Enzyme Regul.* **2,** 265–274.

Patel, M. S. (1974). The relative significance of CO_2 fixing enzymes in the metabolism of rat brain. *J. Neurochem.* **22,** 717–724.

Payne, J., Maher, F., Simpson, I., Mattice, L., and Davies, P. (1997). Glucose transporter Glut 5 expression in microglial cells. *Glia* **21,** 327–331.

Pellerin, L., and Magistretti, P. J. (1994). Glutamate uptake into astrocytes stimulates aerobic glycolysis: A mechanism coupling neuronal activity to glucose utilization. *Proc. Natl. Acad. Sci. USA* **91,** 10625–10629.

Pellerin, L., Pellegri, G., Bittar, P. G., Charnay, Y., Bouras, C., Martin, J. L., Stella, N., and Magistretti, P. J. (1998). Evidence supporting the existence of an activity-dependent astrocyte-neuron lactate shuttle. *Dev. Neurosci.* **20,** 291–299.

Peng, L. (1995). "Metabolic Trafficking between Neurons and Astrocytes." Thesis. University of Saskatchewan, Saskatoon, Saskatchewan, Canada.

Peng, L., and Hertz, L. (1993). Potassium induced stimulation of oxidative metabolism of glucose in cultures of intact cerebellar granule cells but not in corresponding cells with dendritic degeneration. *Brain Res.* **629,** 331–334.

Peng, L., Hertz, L., Huang, R., Sonnewald, U., Petersen, S. B., Westergaard, N., Larsson, O., and Schousboe, A. (1993). Utilization of glutamine and of TCA cycle constituents as precursors for transmitter glutamate and GABA. *Dev. Neurosci.* **15,** 367–377.

Peng, L., Zhang, X., and Hertz, L. (1994). High extracellular potassium concentrations stimulate oxidative metabolism in a glutamatergic neuronal culture and glycolysis in cultured astrocytes, but have no stimulatory effect in a GABAergic neuronal culture. *Brain Res.* **663,** 168–172.

Peng, L., Juurlink, B. H. J., and Hertz, L. (1996). Pharmacological and developmental evidence that the potassium-induced stimulation of deoxyglucose uptake in astrocytes is a metabolic manifestation of increased Na^+-K^+-ATPase activity. *Dev. Neurosci.* **18,** 353–359.

Peng, L., Martin-Vasallo, P., and Sweadner, K. J. (1997). Isoforms of Na,K-ATPase alpha and beta subunits in the rat cerebellum and in granule cell cultures. *J. Neurosci.* **17,** 3488–3502.

Peng, L., Arystarkhova, E., and Sweadner, K. J. (1998). Plasticity of Na,K-ATPase isoform expression in cultures of flat astrocytes: species differences in gene expression. *Glia* **24,** 257–271.

Peng, L., Swanson, R. A., and Hertz, L. (2001). Effects of L-glutamate, D-aspartate and monensin on glycolytic and oxidative glucose metabolism in mouse astrocyte cultures. *Neurochem. Int.* **38,** 437–443.

Peuchen, S., Duchen, M. R., and Clark, J. B. (1996). Energy metabolism of adult astrocytes in vitro. *Neuroscience* **71,** 855–870.

Pevzner, L. (1979). "Functional Biochemistry of the Neuroglia," p. 87. Consultants Bureau, New York.

Pfeiffer, B., Elmer, K., Roggendorf, W., Reinhart, P. H., and Hamprecht, B. (1990). Immunohistochemical demonstration of glycogen phosphorylase in rat brain slices. *Histochemistry* **94,** 73–80.

Pfeiffer, B., Meyermann, R., and Hamprecht, B. (1992). Immunohistochemical co-localization of glycogen phosphorylase with the astroglial markers glial fibrillary acidic protein and S-100 protein in rat brain sections. *Histochemistry* **97,** 405–412.

Pfeiffer, B., Grosche, J., Reichenbach, A., and Hamprecht, B. (1994). Immunocytochemical demonstration of glycogen phosphorylase in Muller (glial) cells of the mammalian retina. *Glia* **12,** 62–67.

Pfeuffer, J., Tkac, I., and Gruetter, R. (2000). Extracellular-intracellular distribution of glucose and lactate in the rat brain assessed noninvasively by diffusion-weighted ^{1}H nuclear magnetic resonance spectroscopy in vivo. *J. Cereb. Blood Flow Metab.* **20,** 736–746.

Phelps, M. E., Huang, S. C., Hoffman, E. J., Selin, C., Sokoloff, L., and Kuhl, D. E. (1979). Tomographic measurement of local cerebral glucose metabolic rate in humans with (F-18)2-fluoro-2-deoxy-D-glucose: validation of the method. *Ann. Neurol.* **6,** 371–388.

Pierre, K., Pellerin, L., Debernardi, R., Riederer, B. M., and Magistretti, P. J. (2000). Cell-specific localization of monocarboxylate transporters, MCT1 and MCT2, in the adult mouse brain revealed by double immunohistochemical labeling and confocal microscopy. *Neuroscience* **100,** 617–627.

Prokhorova, M. I., Osadchaia, L. M., Vilkova, V. A., Eshchenko, N. D., and Zakharova, L. I. (1978). Participation of glutamate and alanine in amino acid biosynthesis, in lipogenesis, and in gluconeogenesis in the brain. *Vopr. Biokhim. Mozga.* **13,** 248–259.

Qu, H., Eloqayle, H., Unsgard, G., and Sonnewald, U. (2001). Effect of glutamate on [U-^{13}C]glucose metabolism in cerebellar astrocytes and granule neurons by ^{13}C MRS. *Chinese J. Neurosci.* **17** (Suppl.), 49.

Raffin, C. N., Rosenthal, M., Busto, R., and Sick, T. J. (1992). Glycolysis, oxidative metabolism, and brain potassium clearance. *J. Cereb. Blood Flow Metab.* **12,** 34–42.

Ransom, C. B., Ransom, B. R., and Sontheimer, H. (2000). Activity-dependent extracellular K^+ accumulation in rat optic nerve: the role of glial and axonal Na^+ pumps. *J. Physiol.* **522,** 427–442.

Reichenbach, A., Henke, A., Eberhardt, W., and Reichelt, W. (1992). K^+ ion regulation in retina. *Can. J. Physiol. Pharmacol.* **70,** S239–S247.

Rennels, M. L., Gregory, T. F., Blaumanis, O. R., Fujimoto, K., and Grady, P. A. (1985). Evidence for a "paravascular" fluid circulation in the mammalian central nervous system, provided by the rapid distribution of tracer protein throughout the brain from the subarachnoid space. *Brain Res.* **32,** 47–63.

Richter, K., Hamprecht, B., and Scheich, H. (1996). Ultrastructural localization of glycogen phosphorylase predominantly in astrocytes of the gerbil brain. *Glia* **17,** 263–273.

Robb-Gaspers, L. D., Burnett, P., Rutter, G. A., Denton, R. M., Rissuto, R., and Thomas, A. P. (1998). Integrating cytosolic calcium signals into mitochondrial metabolic responses. *EMBO J.* **17,** 4987–5000.

Roberts, E. L., Jr. (1993). Glycolysis and recovery of potassium ion homeostasis and synaptic transmission in hippocampal slices after anoxia or stimulated potassium release. *Brain Res.* **620,** 251–258.

Robinson, S. R., Schousboe, A., Dringen, R., Magistretti, P., Coles, J., and Hertz, L. (1998). Metabolic trafficking between neurones and glia. *In* "Glial Contribution to Behaviour"

(P. Laming, E. Sykova, A. Reichenbach, G. I. Hatton, and H. Bauer, eds.), pp. 83–106. Cambridge University Press, New York.

Rose, C. R., and Ransom, B. R. (1996). Intracellular sodium homeostasis in rat hippocampal astrocytes. *J. Physiol.* **491,** 291–305.

Rose, C. R., and Ransom, B. R. (1997). Regulation of intracellular sodium in cultured rat hippocampal neurones. *J. Physiol.* **499,** 573–587.

Rose, C. R., Waxman, S. G., and Ransom, B. R. (1998). Effects of glucose deprivation, chemical hypoxia, and simulated ischemia on Na^+ homeostasis in rat spinal cord astrocytes. *J. Neurosci.* **18,** 3554–3562.

Rosenthal, M., and Somjen, G. (1973). Spreading depression, sustained potential shifts, and metabolic activity of cerebral cortex of cats. *J. Neurophysiol.* **36,** 739–749.

Rothe, F., Brosz, M., and Storm-Mathisen, J. (1994). Quantitative ultrastructural localization of glutamate dehydrogenase in the rat cerebellar cortex. *Neuroscience* **62,** 1133–1146.

Rothman, D. L., Sibson, N. R., Hyder, F., Shen, J., Behar, K. L., and Shulman, R. G. (1999). *In vivo* nuclear magnetic resonance spectroscopy studies of the relationship between the glutamate-glutamine neurotransmitter cycle and functional neuroenergetics. *Phil. Trans. R. Soc. Lond. B* **354,** 1165–1177.

Rouach, N., Glowinski, J., and Giaume, C. (2000). Activity dependent neuronal control of gap-junctional communication in astrocytes. *J. Cell Biol.* **149,** 1513–1526.

Ruiz-Amil, M., De Torrontegui, G., Palacian, E., Catalina, L., and Losada, M. (1965). Properties and function of yeast pyruvate carboxylase. *J. Biol. Chem.* **240,** 3485–3492.

Rust, R. S., Jr., Carter, J. G., Martin, D., Nerbonne, J. M., Lampe, P. A., Pusateri, M. E., and Lowry, O. H. (1991). Enzyme levels in cultured astrocytes, oligodendrocytes and Schwann cells, and neurons from the cerebral cortex and superior cervical ganglia of the rat. *Neurochem. Res.* **16,** 991–999.

Rutter, G. A., Burnett, P., Rizzuto, R., Brini, M., Murgia, M., Pozza, T., Tavare, J. M., and Denton, R. M. (1996). Subcellular imaging of intramitochondrial Ca^{2+} with recombinant targeted aequorin: significance for the regulation of pyruvate dehydrogenase activity. *Proc. Natl. Acad. Sci. USA* **93,** 5489–5494.

Savaki, H. E., Kadekaro, M., McCulloch, J., and Sokoloff, L. (1982). The central noradrenergic system in the rat: metabolic mapping with alpha-adrenergic blocking agents. *Brain Res.* **234,** 65–79.

Schmitt, A., and Kugler, P. (1999). Cellular and regional expression of glutamate dehydrogenase in the rat nervous system: non-radioactive in situ hybridization and comparative immunocytochemistry. *Neuroscience* **92,** 293–308.

Schmoll, D., Fuhrmann, E., Gebhardt, R., and Hamprecht, B. (1995a). Significant amounts of glycogen are synthesized from 3-carbon compounds in astroglial primary cultures from mice with participation of the mitochondrial phosphoenol pyruvate carboxykinase isoenzyme. *Eur. J. Biochem.* **227,** 308–315.

Schmoll, D., Cesar, M., Fuhrmann, E., and Hamprecht, B. (1995b). Colocalization of fructose-1,6-bisphosphatase and glial fibrillary acidic protein in rat brain. *Brain Res.* **677,** 341–344.

Shank, R. P., Bennett, G. S., Freytag, S. O., and Campbell, G. L. (1985). Pyruvate carboxylase: An astrocyte-specific enzyme implicated in the replenishment of amino acid neurotransmitter pools. *Brain Res.* **329,** 364–367.

Shank, R. P., Leo, G. C., and Zielke, H. R. (1993). Cerebral metabolic compartmentation as revealed by nuclear magnetic resonance analysis of D-[^{13}C]glucose metabolism. *J. Neurochem.* **61,** 315–323.

Shao, Y., and McCarthy, K. D. (1997). Responses of Bergmann glia and granule neurons in situ to *N*-methyl-D-aspartate, norepinephrine and high potassium. *J. Neurochem.* **68,** 2405–2411.

Sharp, F. R., Sagar, S. M., and Swanson, R. A. (1993). Metabolic mapping with cellular resolution: c-fos vs. 2-deoxyglucose. *Crit. Rev. Neurobiol.* **7,** 205–228.

Sheu, K.-F. R., and Blass, J. P. (1999). The α-ketoglutarate dehydrogenase complex. *Ann. N.Y. Acad. Sci.* **893,** 61–78.

Sheu, K. F., Calingasan, N. Y., Lindsay, J. G., and Gibson, G. E. (1998). Immunochemical characterization of the deficiency of the alpha-ketoglutarate dehydrogenase complex in thiamine-deficient rat brain. *J. Neurochem.* **70,** 1143–1150.

Shibuki, K. (1989). Calcium-dependent and ouabain-resistant oxygen consumption in the rat neurohypophysis. *Brain Res.* **487,** 96–104.

Shinohara, M., Dollinger, B., Brown, G., Rapoport, S., and Sokoloff, L. (1979). Cerebral glucose utilization: Local changes during and after recovery from spreading cortical depression. *Science* **203,** 188–190.

Shulman, R. G., Rothman, D. L., and Hyder, F. (1999). Stimulated changes in localized cerebral energy consumption under anesthesia. *Proc. Natl. Acad. Sci. USA* **96,** 3245–3250.

Sibson, N. R., Dhankhar, A., Mason, G. F., Rothman, D. L., Behar, K. L., and Shulman, R. G. (1998). Stoichiometric coupling of brain glucose metabolism and glutamatergic neuronal activity. *Proc. Natl. Acad. Sci. USA* **95,** 316–321.

Sibson, N. R., Mason, G. F., Shen, J., Cline, G. W., Herskovits, A. Z., Wall, J. E., Behar, K. L., Rothman, D. L., and and Shulman, R. G. (2001). *In vivo* 13C NMR measurement of neurotransmitter glutamate cycling, anaplerosis and TCA cycle flux in rat brain during [2-^{13}C]glucose infusion. *J. Neurochem.* **76,** 975–989.

Siesjö, B. K. (1978). "Brain Energy Metabolism." John Wiley and Sons, New York.

Silver, I. A., and Erecinska, M. (1997). Energy demands of the Na^+/K^+-ATPase in mammalian astrocytes. *Glia* **21,** 35–45.

Simpson, I. A., Appel, N. M., and Hokari, M. (1999). Blood-brain barrier glucose transporter: effects of hypo- and hyperglycemia revisited. *J. Neurochem.* **72,** 238–247.

Smith, C. B. (1983). Localization of activity-associated changes in metabolism of the central nervous system with the deoxyglucose method: Prospects for cellular resolution. *In* "Current Methods in Cellular Neurobiology" (J. L. Barker, and F. F. McKelvy, eds.), Vol. 1, Anatomical Techniques, pp. 269–317. John Wiley and Sons, New York.

Sokoloff, L. (1982). The radioactive deoxyglucose method. Theory, procedure, and applications for the measurement of local glucose utilization in the central nervous system. *In* "Advances in Neurochemistry" (B. W. Agranoff, and M. H. Aprison, eds.), pp. 1–82. Plenum Press, New York.

Sokoloff, L. (1986). Cerebral circulation, energy metabolism, and protein synthesis: General characteristics and principles of measurement. *In* "Positron Emission Tomography and Autoradiography: Principles and Applications for the Brain and Heart" (M. Phelps, J. Mazziotta, and H. Schelbert, eds.), pp. 1–71. Raven Press, New York.

Sokoloff, L. (1996). Cerebral metabolism and visualization of cerebral activity. *In* "Comprehensive Human Physiology." (R. Greger and U. Windhorst, eds.), Vol. 1, pp. 579–602. Springer-Verlag, Heidelberg.

Sokoloff, L., Reivich, M., Kennedy, C., Des Rosiers, M. H., Patlak, C. S., Pettigrew, K. D., Sakurada, O., and Shinohara, M. (1977). The [^{14}C]deoxyglucose method for the measurement of local glucose utilization: theory, procedure, and normal values in the conscious and anesthetized albino rat. *J. Neurochem.* **28,** 897–916.

Sokoloff, L., Takahashi, S., Gotoh, J., Driscoll, B. F., and Law, M. J. (1996). Contribution of astroglia to functionally activated energy metabolism. *Dev. Neurosci.* **18,** 343–352.

Sols, A., and Crane, R. K. (1954). Substrate specificity of brain hexokinase. *J. Biol. Chem.* **210,** 581–595.

Somjen, G. G. (2001). Mechanisms of spreading depression and hypoxic spreading depression-like depolarization. *Physiol. Rev.* **81,** 1065–1096.

Sonnewald, U., Westergaard, N., Hassel, B., Muller, T. B., Unsgard, G., Fonnum, F., Hertz, L., Schousboe, A., and Petersen, S. B. (1993). NMR spectroscopic studies of ^{13}C acetate and ^{13}C glucose metabolism in neocortical astrocytes: Evidence for mitochondrial heterogeneity. *Dev. Neurosci.* **15,** 351–358.

Sonnewald, U., Westergaard, N., and Schousboe, A. (1997). Glutamate transport and metabolism in astrocytes. *Glia* **21,** 56–63.

Sontheimer, H. (1994). Voltage-dependent ion channels in glial cells. *Glia* **11,** 156–72.

Sorg, O., and Magistretti, P. J. (1992). Vasoactive intestinal peptide and noradrenaline exert long-term control on glycogen levels in astrocytes: Blockade by protein synthesis inhibition. *J. Neurosci.* **12,** 4923–4931.

Spencer, E. B., Bianchi, A., Widmer, J., and Witters, L. A. (1993). Brain acetyl-CoA carboxylase: Isozymic identification and studies of its regulation during development and altered nutrition. *Biochem. Biophys. Res. Comm.* **192,** 820–825.

Stone, E. A., and Ariano, M. A. (1989). Are glial cells target of the central noradrenergic system? A review of the evidence. *Brain Res. Rev.* **14,** 297–309.

Stone, E. A., John, S. M., and Zhang, Y. (1992). Studies of the cellular localization of biochemical responses to catecholamines in the brain. *Brain Res. Bull.* **29,** 285–288.

Strominger, J. L., and Lowry, O. H. (1955). The quantitative histochemistry of the brain. IV. Lactic, malic, and glutamic dehydrogenases. *J. Biol. Chem.* **213,** 635–646.

Su, G., Haworth, R. A., Dempsey, R. J., and Sun, D. (2000). Regulation of Na(+)-K(+)-Cl(−) cotransporter in primary astrocytes by dibutyryl cAMP and high $[K(+)]_O$. *Am. J. Physiol. Cell. Physiol.* **279,** 1710–1721.

Subbarao, K. V., and Hertz, L. (1990a). Effects of adrenergic agonists on glycogenolysis in primary cultures of astrocytes. *Brain Res.* **527,** 346–349.

Subbarao, K. V., and Hertz, L. (1990b). Noradrenaline induced stimulation of oxidative metabolism in astrocytes but not in neurons in primary cultures. *Brain Res.* **527,** 346–349.

Subbarao, K. V., and Hertz, L. (1991). Stimulation of energy metabolism in astrocytes by adrenergic agonists. *J. Neurosci. Res.* **28,** 399–405.

Subbarao, K. V., Stolzenburg, J.-U., and Hertz, L. (1995). Pharmacological characteristics of potassium-induced glycogenolysis in astrocytes. *Neurosci. Lett.* **196,** 45–48.

Sugden, P. H., and Newsholme, E. A. (1975). The effects of ammonium, inorganic phosphate and potassium ions on the activity of phosphofructokinases from muscle and nervous tissues of vertebrates and invertebrates. *Biochem. J.* **150,** 113–122.

Swanson, R. A., Morton, M. M., Sagar, S. M., and Sharp, F. R. (1992). Sensory stimulation induces local cerebral glycogenolysis: Demonstration by autoradiography. *Neuroscience* **51,** 451–461.

Sykova, E. (1983). Extracellular K^+ accumulation in the central nervous system. *Progr. Biophysiol.* **42,** 135–189.

Takagaki, G. (1972). Control of aerobic glycolysis in guinea-pig cerebral cortex slices. *J. Neurochem.* **19,** 1737–1751.

Takahashi, S., Driscoll, B. F., Law, M. J., and Sokoloff, L. (1995). Role of sodium and potassium ions in regulation of glucose metabolism in cultured astroglia. *Proc. Natl. Acad. Sci. USA* **92,** 4616–4620.

Takahashi, S., Shibata, M., and Fukuuchi, Y. (1997). Effects of increased extracellular potassium on influx of sodium ions in cultured rat astroglia and neurons. *Dev. Brain Res.* **104,** 111–117.

Takita, M., Mikuni, M., and Takahashi, K. (1992). Habituation of lactate release responding to stressful stimuli in rat prefrontal cortex in vivo. *Am. J. Physiol.* **263,** R722–R727.

Tansey, F. A., Farooq, M., and Cammer, W. (1991). Glutamine synthetase in oligodendrocytes and astrocytes: new biochemical and immunocytochemical evidence. *J. Neurochem.* **56,** 266–272.

Tas, P. W. L., Massa, P. T., Kress, H. G., and Koschel, K. (1987). Characterization of an Na^+, K^+, Cl^- co-transport in primary cultures of rat astrocytes. *Biochim. Biophys. Acta* **903,** 411–416.

Taylor, D. L., Richards, D. A., Obrenovitch, T. P., and Symon, L. (1994). Time course of changes in extracellular lactate evoked by transient K^+- induced depolarization in the rat striatum. *J. Neurochem.* **62,** 2368–2374.

Thorlin, T., Eriksson, P. S., Ronnback, L., and Hansson, E. (1998). Receptor-activated Ca^{2+} increases in vibrodissociated cortical astrocytes: a nonenzymatic method for acute isolation of astrocytes. *J. Neurosci. Res.* **54,** 390–401.

Tildon, J. T., and Roeder, L. M. (1984). Glutamine oxidation by dissociated cells and homogenates of rat brain: Kinetics and inhibitor studies. *J. Neurochem.* **42,** 1060–1076.

Tower, D. B. (1980). Opening remarks: Some historical perspectives. *In* "Cerebral Metabolism and Neural Function" (J. V. Passonneau, R. A. Hawkins, W. D. Lust, and F. A. Welsh, eds.), pp. 1–8. Williams & Wilkins, Baltimore.

Turton, W. E., Ciriello, J., and Calaresu, F. R. (1986). Changes in forebrain hexokinase activity after aortic baroreceptor denervation. *Am. J. Physiol.* **251,** R274–R281.

van Erp, H. E., Roholl, P. J., Rijksen, G., Sprengers, E. D., Van Weelen, C. W., and Staal, G. E. (1988). Production and characterization of monoclonal antibodies against human type K pyruvate kinase. *Eur. J. Cell. Biol.* **47,** 388–394.

Varon, S., and McIlwain, H. (1961). Fluid contents and compartments in isolated cerebral tissues. *J. Neurochem.* **8,** 262–275.

Vannucci, S. J., Clark, R. R., Koehler-Stec, E., Li, K., Smith, C. B., Davies, P., Maher, P., and Simpson, I. A. (1998). Glucose transporter expression in brain: relationship to cerebral glucose utilization. *Dev. Neurosci.* **20,** 369–379.

Veech, R. L. (1980). Freeze-blowing of brain and the interpretation of the meaning of certain metabolite levels. *In* "Cerebral Metabolism and Neural Function" (J. V. Passonneau, R. A. Hawkins, W. D. Lust, and F. A. Welsh, eds.), pp. 34–41. Williams & Wilkins, Baltimore.

Waagepetersen, H. S., Qu, H., Hertz, L., Sonnewald, U., and Schousboe, A. (2002). Demonstration of pyruvate recycling in primary cultures of neocortical astrocytes but not in neurons. *Neurochem. Res.,* in press.

Wallgren, H. (1963). Ouabain-induced depression of the respiration of electrically stimulated brain slices in presence and absence of ethanol. *Ann. Med. Exp. Fenn.* **41,** 166–173.

Walz, W., and Hertz, L. (1982). Ouabain-sensitive and ouabain-resistant net uptake of potassium into astrocytes and neurons in primary cultures. *J. Neurochem.* **39,** 70–77.

Walz, W., and Hertz, L. (1983). Functional interactions between neurons and astrocytes II. Potassium homeostasis at the cellular level. *Prog. Neurobiol.* **20,** 133–183.

Walz, W., and Hertz, L. (1984). Intense furosemide-sensitive potassium accumulation into astrocytes in the presence of pathologically high extracellular potassium levels. *J. Cerebr. Blood Flow Metab.* **4,** 301–304.

Walz, W., and Wuttke, W. A. (1999). Independent mechanisms of potassium clearance by astrocytes in gliotic tissue. *J. Neurosci. Res.* **56,** 595–603.

Waniewski, R. A., and Martin, D. L. (1998). Preferential utilization of acetate by astrocytes is attributable to transport. *J. Neurosci.* **18,** 5225–5233.

Weller, R. O., Kida, S., and Zhang, E. T. (1992). Pathways of fluid drainage from the brain-morphological aspects and immunological significance in rat and man. *Brain Pathol.* **2,** 277–284.

Westergaard, N., Varming, T., Peng, L., Sonnewald, U., Hertz, L., and Schousboe, A. (1993). Uptake, release and metabolism of alanine in neurons and astrocytes in primary cultures. *J. Neurosci. Res.* **35,** 540–545.

Westergaard, N., Drejer, J., Schousboe, A., and Sonnewald, U. (1996). Evaluation of the importance of transamination versus deamination in astrocytic metabolism of [U-^{13}C]glutamate. *Glia* **17,** 160–168.

Wiesinger, H., Hamprecht, B., and Dringen, R. (1997). Metabolic pathways for glucose in astrocytes. *Glia* **21,** 22–34.

Wilkin, G. P., and Wilson, J. E. (1977). Localization of hexokinase in neural tissue: Light microscopic studies with immunofluorescence and histochemical procedures. *J. Neurochem.* **29,** 1039–1051.

Wilson, J. E. (1995). Hexokinases. *Rev. Physiol. Biochem. Pharmacol.* **126,** 65–198.

Wittendorp-Rechenmann, E., Lam, C. D., and Nehlig, A. (2001). First in vivo demonstration of the uptake of [^{14}C]deoxyglucose by astrocytes and neurons: A microautoradiographic study. *J. Cereb. Blood Flow Metab.* **21,** Suppl. 1, S321.

Wong-Riley, M., Anderson, B., Liebl, W., and Huang, Z. (1998). Neurochemical organization of the macaque striate cortex: correlation of cytochrome oxidase with $Na^{+}K^{+}$ATPase, NADPH-diaphorase, nitric oxide synthase and *N*-methyl-D-aspartate receptor subunit 1. *Neuroscience* **83,** 1025–1045.

Wree, A., Schleicher, A., Zilles, K., and Beck, T. (1988). Local cerebral glucose utilization in the Ammon's horn and dentate gyrus of the rat brain. *Hisotchemistry* **88,** 415–426.

Xiong, Z. Q., and Stringer, J. L. (1999). Astrocytic regulation of the recovery of extracellular potassium after seizures in vivo. *Eur. J. Neurosci.* **11,** 1677–1684.

Xiong, Z. Q., and Stringer, J. L. (2000). Sodium pump activity, not glial spatial buffering, clears potassium after epileptiform activity induced in the dentate gyrus. *J. Neurophysiol.* **83,** 1443–1451.

Yamamoto, T., Ochalski, A., Hertzberg, E. L., and Nagy, J. L. (1990). On the organization of astrocytic gap junctions in rat brain as suggested by LM and EM immunohistochemistry of connexin 43 expression. *J. Comp. Neurol.* **302,** 853–883.

Yarowsky, P., Boyne, A. F., Wierwille, R., and Brookes, N. (1986). Effect of monensin on deoxyglucose uptake in cultured astrocytes: energy metabolism is coupled to sodium entry. *J. Neurosci.* **6,** 859–866.

Yu, A. C. H., and Hertz, L. (1983). Metabolic sources of energy in astrocytes. *In* "Glutamine, Glutamate and GABA in the Central Nervous System" (L. Hertz, E. Kvamme, E. G. McGeer, and A. Schousboe, eds.), pp. 431–439. Alan R. Liss, New York.

Yu, A. C. H., Schousboe, A., and Hertz, L. (1982). Metabolic fate of (14C)-labelled glutamate in astrocytes. *J. Neurochem.* **39,** 954–966.

Yu, A. C. H., Drejer, J., Hertz, L., and Schousboe, A. (1983). Pyruvate carboxylase activity in primary cultures of astrocytes and neurons. *J. Neurochem.* **41,** 1484–1487.

Yu, S., and Ding, W. G. (1998). The 45 kDa form of glucose transporter 1 (GLUT 1) is localized in oligodendrocytes and astrocytes but not in microglia in the rat brain. *Brain Res.* **797,** 65–72.

Yudkoff, M. (1997). Brain metabolism of branched-chain amino acids. *Glia* **21,** 92–98.

Zeitschel, U., Bigl, M., Eschrich, K., and Bigl, V. (1996). Cellular distribution of 6-phosphofructo-1-kinase in brain. *J. Neurochem.* **67,** 2573–2580.

Zeller, K., Rahner-Welsch, S., and Kuschinsky, W. (1997). Distribution of Glut1 glucose transporters in different brain structures compared to glucose utilization and capillary density of adult rat brains. *J. Cereb. Blood Flow Metab.* **17,** 204–209.

Zhao, Z. (1992). Effects of alpha2 agonists and benzodiazepines on cytosolic calcium concentration in primary cultures of astrocytes and cerebellar granule cells. M. Sc. Thesis. University of Saskatchewan, Saskatoon, Saskatchewan, Canada.

Zhao, Z., Code, W. E., and Hertz, L. (1992). Dexmedetomidine, a potent and highly specific alpha2 agonist, evokes cytosolic calcium surge in astrocytes but not in neurons. *Neuropharmacology* **31,** 1077–1079.

Zhao, Z., Hertz, L., and Code, W. E. (1996). Effects of benzodiazepines on potassium induced increase in free cytosolic calcium concentration in astrocytes and neurons in primary cultures. *Can. J. Physiol. Pharmacol.* **74,** 273–277.

Zielke, H. R., Collins, R. M., Jr., Baab, P. J., Huang, Y., Zielke, C. L., and Tildon, W. T. (1998). Compartmentation of [^{14}C] glutamate and [^{14}C] glutamine oxidative metabolism in the rat hippocampus as determined by microdialysis. *J. Neurochem.* **71,** 1315–1320.

THE CEREBRAL GLUCOSE-FATTY ACID CYCLE: EVOLUTIONARY ROOTS, REGULATION, AND (PATHO)PHYSIOLOGICAL IMPORTANCE

Kurt Heininger

Department of Neurology, Heinrich Heine University, D-40597 Düsseldorf, Germany

I. Introduction

Metabolism under changing fuel supply is controlled by the glucose-fatty acid cycle. Its evolutionary prototypes were programmed to provide optimal resilience in the "feast and famine" lifestyle of microorganisms. The counterregulatory processes, activated during nutrient deprivation, include hypometabolism and oxidative stress and aim to preserve glucose for anabolic needs and promote the oxidative utilization of fatty acids and ketone bodies. Conserved from bacteria to man, the program ensures the survival of a deprived organism and controls such divergent processes as sporulation, hibernation, and starvation. The cycle is controlled by the antagonistic insulin/glucocorticoid couple. In the brain, the "feast"-side is supplemented by gonadal hormones, insulin-like growth factors (IGFs), dehydroepiandrosterone (DHEA), neuropeptide Y (NPY), neurotrophins, thyroid hormones, and somatostatin, together with acetylcholine, glutamate, serotonin and noradrenaline. Protagonist effectors of the dystrophic leg are glucocorticoids, galanin, GABA and amyloid-β protein (Aβ). Both systems are

0074-7742/02 $35.00

intricately interlocked by a network of regulatory crosstalks and shared signaling pathways.

The pattern of hormone/neurotransmitter changes in a variety of CNS ontogenetic and accidental/degenerative events is surprisingly uniform: desensitization/degeneration of proglucolytic agents and activation of prolipolytic hormones/neurotransmitters paralleled by hypometabolism and downregulation of glucose utilization. Thus, probing various brain pathologies (e.g., stroke, traumatic injury, epilepsy, Alzheimer disease) and ontogenetic events (maturation, aging) for metabolic stress signatures reveals their shared heritage as metabolic deprivation syndromes. Neurotransmitters are dual conveyors of intercellular communication (enabling affective and cognitive processes) and energy homeostasis, establishing a psychosomatic continuum. In this holistic approach, psychosocial stress, depression, and psychosis can be understood as heterogeneous phenotypes of psychosocial/metabolic stress pathologies.

II. Energy Homeostasis and the Glucose-Fatty Acid Cycle

Energy is the motor of life. Those organisms had a huge advantage which "learned" to store energy as chemical fuels during times of relative abundance to use it for energy production during times of fuel shortage. Organisms as early as at the unicellular level have succeeded in taking advantage of the chemical properties of fuel substrates and particularly the most important ones, glucose and fatty acids. Glucose is relatively rich in potential energy (although already partly oxidized), highly soluble in water and hence when energy demand suddenly increases can easily be released from its stores and even yields energy when sufficient oxygen is not available (by anaerobic glycolysis). Fatty acids (FA) are a particularly energy-dense storage fuel with, relative to glucose, a more than twice as high energy yield upon complete oxidation. However, because of their bad solubility in water and relative inertness of the C—C bonds, FA are not as readily mobilized and rapidly oxidizable. Thus, glucose is predestined as abundance fuel, allowing an active life style while FA are the optimal storage fuel which is mobilized during deprivation. Given the central role of fuel combustion for energy homeostasis, the metazoan cell has evolved an intricate homeostatic system to sense and adapt to varying fuel supplies. This system has been described as the reciprocal glucose-fatty acid cycle almost 40 years ago (reviewed by Randle, 1998; Ruderman *et al.*, 1999). Glucose provision promotes glucose oxidation, glucose and lipid storage, and inhibits FA oxidation. Supply of FA enhances FA oxidation, lipid and glucose storage, and inhibits glucose oxidation. In addition to supply, perceived or anticipated conditions of increased energy demand due to stressful events activate the regulatory agents of the glucose-FA cycle, hormones, and neurotransmitters. These counterregulatory events are exerted and intricately tuned at a variety of levels with the main regulation at the level of carnitine palmitoyl transferase I (CPT-I) and the pyruvate dehydrogenase complex (PDH). CPT-I controls the oxidation of fatty acids by regulating their transfer into mitochondria. PDH catalyzes the irreversible oxidative decarboxylation of pyruvate to acetyl-CoA. Glucose oxidation increases acetyl-CoA and citrate which activate acetyl-CoA carboxylase to form malonyl-CoA, an intermediate of the FA biosynthetic pathway and allosteric inhibitor of CPT-I. On the other hand, FA oxidation lowers PDH activity mediated by increased PDH kinase activity (Kelley *et al.*, 1993; Randle, 1998) followed by insulin resistance (Kim *et al.*, 1996). The latter control is exerted both at the systemic and Langerhans islet cell level following for instance high fat feeding, obesity, and starvation (Zhou *et al.*, 1996). AMP-activated protein kinase and malonyl CoA may be integrating interfaces of these metabolic pathways (Ruderman *et al.*, 1997, 1999; Winder and Hardie, 1999). Thus, both nutrients and a variety of stressors modulate the glucose-FA cycle (Chatelain *et al.*, 1996).

III. The Evolutionary Roots of the Glucose-FA Cycle

Too small to accumulate substantial energy depots during phases of nutrient surfeit, microbial populations enter stationary phases when facing nutrient shortage. The characteristic pattern of metabolic adaptations to deprivation was established as early as in stationary phase bacteria and protozoa: hypometabolism and downregulation of oxidative respiration, upregulation of glycogen synthesis, and storage and oxidation of lipids from endogenous polyhydroxybutyrate/alkanoate reserves, membranes, and exogenous sources (Heininger, 2001). Thus, the cell takes advantage of the higher energy yield from FA β-oxidation. Glucose, on the other hand, the precursor of amino acids and nucleotides, is saved for anabolic purposes (Heininger, 2000b). These adaptations subserve the dual purpose to reduce the generation of toxic oxygen species and to minimize the utilization of endogenous reserves. Dependent on fuel supply, these adaptive responses are subject to repression/derepression cycles. Catabolite repression is a common feature of prokaryote and protist metabolic regulation and deprivation responses and is highly coordinated by regulons (Heininger, 2001). The somatic storage of fuels during phases of food abundance represented a metazoan evolutionary breakthrough. Multicellular organisms evolved cells which are specialized to store energy-rich fuels, particularly triglycerides, during phases of nutrient excess which can be mobilized and catabolized during shortage, boosting the organisms' energetic stress-related fitness. The glucose-FA cycle (Randle, 1998) can be regarded as a metazoan elaboration of the unicellular catabolite repression/derepression regulation (Heininger, 2001).

Recently, I defined a deprivation syndrome which is a universal response pattern to environmental stress and has been programmed as prototypic response of organisms to metabolic stress (Heininger, 2001). These processes evolved in unicellular organisms and serve to either resist or mutagenically adapt to environmental stressors and in a consecutive step were developed as key mechanisms of differentiation, apoptosis, and reproductive behavior. These mechanisms have been fixed and programmed as life- and cell-cycle events in multicellular organisms and constitute the core of sexual reproduction and ontogenetic and oncogenetic events. For instance, differentiation events were programmed as responses, e.g., to nutrient shortage, and served to ensure the survival of extremely resilient germ-like cells as spores. The building blocks and nutrients to fuel this complex metamorphosis, however, are depleted in the environment and hence, have to be provided by dead siblings (Heininger, 2001). Thus, the differentiating spores inflict death upon their kin ensuring the provision of resources in the deprived habitat. For instance in *Dictyostelium,* a social amoeba, the prespores secrete so-called morphogens that kill prestalk cells. These processes were programmatically fixed as life- and cell-cycle events but retained their deprivation response phenotypes. Particularly, ontogenetic events, including aging, unwind under a variety of metabolic features which suggest that induced by hormones and mediated by stress transduction pathways the cells experience hypometabolic or energetic stress-like states (Heininger, 2001, 2002).

Hormones are conveyors of energetic/metabolic informations thus replacing in the metazoan organism the environmental signals to which microorganisms respond (Tomkins, 1975, Heininger, 1999a). The control of energy balance at the systemic level and the cellular adaptation to the changing fuel supply are primarily mediated by the reciprocal insulin/glucocorticosteroid (GC) systems (Dallman *et al.,* 1995; Schwartz *et al.,* 1999; Saltiel and Kahn, 2001) and modulated by the thyroid system (Ingenbleek and Bernstein, 1999). This hormonal regulation is complemented by the neurotransmitter system, particularly the dual acetylcholine/catecholamine system (Bahnsen *et al.,* 1984; Heininger, 2000b). In principle, the eutrophic, proglucolytic insulin/acetylcholine is opposed by the dystrophic, prolipolytic GC/catecholamine system. Both animals and plants resort to FA oxidation following glucose starvation (Heininger, 2000b). In addition, as fixed in microorganisms, the adaptive mechanism to a variety of environmental stressors is hypometabolism like in hibernation, estivation, and anaerobiosis of amphibians, insects, reptiles, and mammals (Heininger, 2000b). Animals living in a wide variety of deprived habitats, e.g., deserts or caves, exhibit a reduced metabolic rate (Parsons, 1993), as is also observed under caloric restriction (Heininger, 2002). The deprivation response is part of a phylogenetically primordial system which aims at energy conservation and stress resistance in adverse conditions, preparing the organism for sporulation (e.g., in bacteria and fungi), cold adaptation (e.g., in fish), and hibernation (e.g., in reptiles and mammals) (Heininger, 2000b, 2001).

IV. The Regulation of Brain Energy Homeostasis

The brain is the metabolically most active organ and is therefore highly dependent on a continuous supply of its fuel. To meet its very high energy demands, the brain (around 1/40 of the body weight) possesses a relatively high blood flow and glucose consumption equal in amount to about one-fifth of the entire resting requirements of the body. In mammals, the regulation of fuel metabolism is regulated principally to serve the needs of brain and muscle, the major consumers of fuel. The adult mammalian brain relies almost completely on glucose as energy source while ketone bodies (KB) are preferentially directed toward lipid synthesis (Roeder *et al.*, 1982; Yeh, 1984). The KB consist of acetoacetate, 3-hydroxybutyrate, and acetone. In nonruminant mammals, the liver is the only significant site of KB formation through fatty acid β-oxidation. Cultured astrocytes, however, may produce KB at rates similar to those of hepatocytes and like hepatocytes appear to be ketogenic cells (Blazquez *et al.*, 1998; Guzman and Blazquez, 2001). After entering the blood, KB are oxidized in extrahepatic tissues, under particular circumstances also the brain, by mitochondrial enzymes to form acetyl-CoA, the substrate of the tricarboxylic acid (TCA) cycle. Glycolysis not only meets the brain's constant need for fuel but also provides the substrate for anabolic processes, namely pyruvate which via the TCA cycle and anaplerotic pathways is the source for a variety of amino acids and neurotransmitters such as GABA and glutamate, and for acetyl-CoA which is used for lipid and acetylcholine (ACh) synthesis.

Evidence is accumulating that brain metabolism is compartmentalized (reviewed by Magistretti *et al.*, 1995; Pellerin *et al.*, 1998b). Glucose is predominantly taken up by specialized processes, the astrocytic end feet which surround brain capillaries. Other astrocytic processes ensheath synapses and express receptors and transporters for various neurotransmitters to sense synaptic activity. Neuronal activity increases the demand for energy.

Glutamate, released by neuronal activity, is taken up by and stimulates glycolysis and glycogenolysis in astrocytes (Pellerin and Magistretti, 1994; Fray *et al.*, 1996; Hu and Wilson, 1997; Pellerin *et al.*, 1998b). There, glucose is metabolized to lactate which, together with glucose, is then released into the extracellular space and avidly taken up by neurons to be used as substrate for oxidative phosphorylation (Dringen *et al.*, 1993; McKenna *et al.*, 1993). When glucose becomes scarce, astrocytes appear to supply *in situ* generated ketone bodies (Guzman and Blazquez, 2001). The higher reliance of neurons on oxidative combustion that, on one hand, yields much more ATP, but on the other hand, carries the hazard of increased oxidative stress (Brand and Hermfisse, 1997), may account for the enhanced vulnerability of neurons (Almeida *et al.*, 2001).

With some abstraction, two largely interlocked systems regulate the metabolic homeostasis of the brain: an abundance system, active in conditions of abundant glucose supply and a deprivation system aiming to minimize the sequelae of glucose deprivation. The former is set up by the neurotrophins, thyroid hormones, estrogens, insulin, IGFs, DHEA, vasoactive intestinal peptide (VIP) (Heininger, 1999a), ACh, noradrenaline (NA), serotonin (5-HT), and glutamate, the latter depends on GC, galanin, and GABA (Heininger, 2000b). In principle, the agents synergize within their system and antagonize the agents of the other system on a multitude of levels.

A. Neurotransmitters

Neurotransmitters evolved at least 1 billion years ago, before the evolution of multicellular organisms, and hence all the classical neurotransmitters occur in virtually all phyla (Walker *et al.*, 1996). The archaic role of neurotransmitters was to regulate cellular energy metabolism. Neurotransmitters are either amino acids or amino acid metabolites and their trophic role may have evolved from their capacity to act in primitive cells as signals for environmental nutrient availability (Tomkins, 1975). Very early, probably at least 800 million years ago, transmitters developed the ability to regulate a range of ion channels via their receptors (Venter *et al.*, 1988; Walker *et al.*, 1996). Recent evidence even suggests that this evolutionary achievement may have occurred at the level of prokaryotes, at least 1.5–1.8 billion years ago (Chen *et al.*, 1999; Miller, 2000). With the advent of multicellular organisms some 700 million years ago, this system enabled intercellular communication. Neurotransmitters were the primordial growth regulatory signals, controlling mitosis, proliferation, differentiation, cell motility, and metamorphosis (Lauder, 1993; Turlejski, 1996; Fanburg and Lee, 1997). This is still their principal function in the entire metazoan body via the parasympathetic, sympathetic, and serotonergic systems providing short-term adaptations in adjunct to long-term hormonal control. Thus, long before evolution could elaborate their sophisticated roles in perception, affect and cognition, neurotransmitters, by controlling the ion homeostasis of cells, regulated the energy homeostasis of cells and controlled the flow of fuels in an integrated, interactive network. For instance, the simple nervous system of the nematode *Caenorhabditis elegans* already exhibits highly complex transmitter circuits encompassing all classical neurotransmitters (identified in 166 of the 302 neurons), of which, however, only ACh is vital (reviewed by Brownlee and Fairweather, 1999). Intriguingly, cholinergic agonists mediate recovery from diapause (a state of metabolic dormancy in response to harsh environmental conditions) to the reproductive stage in nematodes by metabotropic signaling via the insulin-like transduction pathway (Tissenbaum *et al.*, 2000). Likewise, serotonin acts, at least in part, via insulin-like pathways to regulate the sensation and ingestion of food and in the control of metabolism (Sze *et al.*, 2000). Since each signal transmission is associated with the consumption of energy, it would make sense from an evolutionary point of view as a type of preadaptation in the Darwinian sense, that it was their nutrient-signaling and metabolism-controlling capacity which qualified these agents as neurotransmitters. This may be another evolutionary application of the principle that the agent which causes energy consumption should carry the signal that ensures that this energy is provided. This principle also rules mitochondrial stimulus-response-metabolism coupling (McCormack and Denton, 1986) by which Ca^{2+} as a signal adapts the supply of high-energy phosphates to the stimulus-evoked energy demand (Heininger, 1999a). It should be stressed that each neurotransmitter elicits or inhibits the movement of Ca^{2+} across biological membranes through direct or indirect modulation of Ca^{2+} channels. In the hierarchic system of homeostatic control, neurotransmitters and the later evolved hormones collectively can be regarded as a higher level of "environmental" signals regulating the primordial cellular messenger network consisting of phosphates, Ca^{2+} and the redox balance. In addition, by modulating both neuronal activity and vascular tone, neurotransmitters constitute the coupling mechanism between cerebral metabolism and blood flow (see Edvinsson *et al.*, 1993) thus balancing demand and supply. In fact, by regulating the cellular ion homeostasis neurotransmitters are the short-term life-sustaining system, controlling blood and fuel flow inside and outside the brain.

Here only general principles of the role of neurotransmitters in the regulation of cerebral blood flow (CBF) and metabolism can be delineated without dwelling in detail upon the complexity in terms of multiple receptor subclass-mediated and brain area-specific effects (for further reading the book by Edvinsson *et al.,* 1993 is recommended). A dense network of cholinergic, noradrenergic, dopaminergic, serotonergic, and GABAergic nerve fibers invests the cerebral blood vessels in all mammals. Together with perivascular astrocytes, nerve fibers and microvessels form a tripartite functional unit (Vaucher *et al.,* 1997) whose integrating signals, coupling supply to demand, are intercellular Ca^{2+} waves (Braet *et al.,* 2001). Under physiological conditions, CBF and glucose utilization (GU) are tightly coupled, neuronal activity and hence metabolic demand being the driving force (Edvinsson *et al.,* 1993; Leuchter *et al.,* 1999). This tight coupling may be loosened during activation when glucose uptake is more increased than oxygen uptake and CBF (Madsen *et al.,* 1998). Furthermore, with glucose deprivation the brain has to utilize alternative energy sources as discussed above. Maintaining the coupling under these circumstances would lead to detrimental consequences. Accordingly, CBF and glucose consumption have to be regulated, at least in part, independently. Neurotransmitters, neuropeptides, and hormones have a role in this regulation and may have independent effects on CBF and GU (Edvinsson *et al.,* 1993). However, the effort to outline these differential effects faces a wealth of conflicting findings obtained in awake and anesthetized animals (Edvinsson *et al.,* 1993). These discrepancies may be reconciled by taking into account that during anaesthesia resting brain glucose metabolism is reduced by pharmacological means (Sokoloff *et al.,* 1977; Crane *et al.,* 1978; Shulman *et al.,* 1999), adding another dimension to the complexity of neurotransmitter-induced effects on CBF and GU. Thus, anesthesia may suppress coupled neurotransmitter-related effects on GU and CBF (e.g., of nicotine, Bedran de Castro *et al.,* 1984) while discordant effects on CBF and GU may be unmasked (e.g., of GABA, Edvinsson *et al.,* 1980). A coherent picture emerges by comparison of data obtained in awake and anesthetized animals.

The *cholinergic* system has profound effects on CBF (reviewed by Sato and Sato, 1992; Vaucher *et al.,* 1997) and GU. The morphological substrate is formed by intimate neurovascular associations of basalocortical cholinergic fibers (Vaucher *et al.,* 1997). Cortical CBF is increased by electrical stimulation of the substantia innominata (SI), i.e., the rodent equivalent of the nucleus basalis Meynert (Biesold *et al.,* 1989; Lacombe *et al.,* 1989; Dauphin *et al.,* 1991), or by microinjection of an ACh agonist into the SI (Barbelivien *et al.,* 1995). The effects are mediated by cholinergic pathways (Biesold *et al.,* 1989; Kurosawa *et al.,* 1989; Dauphin *et al.,* 1991) and are elicited by muscarinic agonists and blocked by muscarinic antagonists (reviewed by Edvinsson *et al.,* 1993). Increase of CBF following cortical neuronal activity is abolished by inhibition of cortical ACh synthesis (Fukuyama *et al.,* 1996) and a muscarinic antagonist (Tsukada *et al.,* 1997). The effects are achieved either directly by cholinergic relaxation of vessels (Kuschinsky *et al.*, 1974; Furchgott and Zawadzki, 1980; Marshall *et al.*, 1988) and/or via mediation of nitric oxide (NO)-containing neurons (Marshall *et al.*, 1988; Vaucher *et al.*, 1997). In humans, muscarinic inhibition attenuated memory-task-induced regional CBF increases (Grasby *et al.*, 1995).

GU is largely dependent on cholinergic transmission. Lesion of the basal forebrain impairs GU in the cerebral cortices (Orzi *et al.*, 1988; Ogawa *et al.*, 1996). Intriguingly, the initial decline of CBF is less pronounced, leading to an uncoupling of GU and CBF (Ogawa *et al.*, 1996). GU is increased by cholinesterase inhibitors in rodent brain slices (Peterson, 1990). In rats, activation of central muscarinergic neurotransmission also enhanced cerebral GU, glycogenolysis, the activity of glycogenolytic enzymes, and lactate production (von Schwarzenfeld *et al.*, 1975; Husain and Ansari, 1988). Lowest doses were effective to increase GU in the hippocampus and median raphe, while higher doses of a muscarinic agonist led to a generalized metabolic enhancement (Soncrant *et al.*, 1985). Electrical stimulation of the medial septum increased GU in the terminal fields of the septohippocampal pathway while ablation of this projection reduced hippocampal GU (Harrell and Davis, 1985). The proglycolytic actions of muscarinergic signaling pathways may be causally involved in their anti-apoptotic effects (Leloup *et al.*, 2000). Septohippocampal and hippocampal-hypothalamic cholinergic fibers are also involved in the CNS-mediated systemic glucoregulation. Injection of cholinergic agonists into the hippocampus, hypothalamus, and third ventricle induces systemic hyperglycemia and hepatic gluconeogenesis which is suppressed by muscarinic antagonists (Nonogaki and Iguchi, 1997). Hypothalamic cholinergic afferents also mediate the hyperglycemia elicited by neuronal glycopenia after 2-deoxyglucose (Takahashi *et al.*, 1996). Finally, muscarinic agonists promote β-cell insulin release and thus delivery of glucose to the tissues (Miller, 1981).

Cholinergic nicotinic receptors regulate brain CBF and glucose metabolism on a variety of levels. Regional CBF is increased by nicotine in conscious rats (Grünwald *et al.*, 1991). Since this finding was also obtained in anaesthetized animals, it indicates that this effect is independent of a concomitant effect on GU (Uchida *et al.*, 1997). In distinct brain areas, nicotinic agonists increase GU (Grünwald *et al.*, 1988; London *et al.*, 1988; Duelli *et al.*, 1998), cerebral glucose transporter densities (Duelli *et al.*, 1998), and expression of the glycolytic enzyme glyceraldehyde-3-phosphate dehydrogenase (Prendergast and Buccafusco, 1998). In humans, a nicotinic antagonist produces a temporoparietal perfusion deficit (Gitelman and Prohovnik, 1992). Nicotinic agonists induce via nicotinic autoreceptors the cortical and hippocampal release of ACh and thus may also act by potentiating muscarinic transmission (Beani *et al.*, 1985; De Sarno and Giacobini, 1989). Nicotine also enhances choroid plexus transthyretin secretion

(Li *et al.*, 2000) and thus the cerebral delivery of proglycolytic thyroid hormones (see Heininger, 1999a). Importantly, the nicotinic system inhibits lipolysis and reduces plasma KB concentrations (Mayor *et al.*, 1967; Ammon *et al.*, 1969).

Glucose and cholinergic neurotransmission establish a mutually augmentative cycle. Glucose enhances cognition and various cholinergic functions. Glucose attenuates the amnesia induced by insulin-induced hypoglycemia, scopolamine, a muscarinic receptor antagonist, or intraseptal morphine injections; it enhances choline uptake, ACh synthesis, and learning-induced ACh release (Lefresne *et al.*, 1973; Kopf and Baratti, 1995; Micheau *et al.*, 1995; Ragozzino *et al.*, 1996). The memory-improving actions of glucose are blocked by central muscarinic and nicotinic antagonists and facilitated by a cholinomimetic agent (Kopf and Baratti, 1996).

Both central and sympathetic *adrenergic* nerves innervate cerebral vessels. It appears that sympathetic fibers supply cerebral arterioles and that central adrenergic fibers are associated with microvessels (Edvinsson *et al.*, 1993). NA is the mediator of stress and attentional responses in the CNS (reviewed by Stratakis and Chrousos, 1995) and as such also regulates CBF and energy metabolism during stress and neuronal activation (Bryan, 1990). Although a plethora of studies has been done about the sympathetic regulation of CBF (reviewed by Edvinsson *et al.*, 1993) only little attention has been paid to its cerebral catecholaminergic regulation. Intracerebral noradrenergic pathways appear to have little or no effect on resting CBF (Dahlgren *et al.*, 1981; Edvinsson *et al.*, 1993). Intracerebral NA depletion, however, abolishes the increase of CBF following neuronal activity (Harik *et al.*, 1979; Tsubokawa *et al.*, 1980) and attenuates metabolic recovery (LaManna *et al.*, 1981). Lesion of the locus coeruleus (LC), the primary source of noradrenergic projections, has little effect on basal GU (Savaki *et al.*, 1984), but increases the area of stimulated GU during stress or neuronal activation (Craik *et al.*, 1987; Justice *et al.*, 1989) indicating that NA acts during stress to limit unnecessary activity and to focus on relevant stimuli and appropriate coping responses. Blockade of $\alpha 1$ and 2-adrenoceptors or LC lesion significantly reduces GU and oxidative metabolism in the active state in LC projection areas (LaManna *et al.*, 1981; Inoue *et al.*, 1991; French *et al.*, 1995). At the cellular level, NA stimulates glucose uptake by and glycogen metabolism in astrocytes and thus activates the glial-neuronal glucose and lactate shuttle (reviewed by Magistretti *et al.*, 1995; Fillenz *et al.*, 1999). Long-term effects on astrocyte glucose uptake may be mediated by NA-elicited arachidonic acid release (Yu *et al.*, 1993). At the hypothalamic level α_2-adrenoceptor agonists induce a systemic hyperglycemia (Smythe and Edwards, 1991).

Central ascending *serotonergic* (5-HT) pathways from the raphe nuclei invest pial vessels in various brain regions and 5-HT may also be released from adrenergic vasomotor fibers as "false" transmitter along with NA (Edvinsson *et al.,* 1993). *In vitro,* 5-HT induces concentration-dependent contractions of cerebral arteries (Edvinsson *et al.,* 1993; Cohen *et al.,* 1996). Stimulation of the dorsal raphe nucleus elicits a decrease of CBF in anesthetized rats (Bonvento *et al.,* 1989) and an increase of CBF in conscious rats (Cudennec *et al.,* 1993) suggesting that the effects on CBF are coupled to GU in a complex way. Central 5-HT projections have only minimal tonic influence on cerebral GU as evidenced by raphe nucleus lesions (Cudennec *et al.,* 1988a). In contrast, stimulation of the raphe nucleus elicits a widespread increase of GU in neocortical areas and the limbic and extrapyramidal system of conscious (Cudennec *et al.,* 1988b) and anesthetized rats (Bonvento *et al.,* 1991). A comparison of central 5-HT-mediated effects on CBF and GU revealed a general modification of the flow-metabolism coupling (Cudennec *et al.,* 1993). In humans, 5-HT activation appears to elicit hemispheric lateralized increases of GU in the left parietotemporal and prefrontal cortices (Mann *et al.,* 1996). At the cellular level, 5-HT activates astroglial glycogenolysis by mechanisms which may involve 5-HT_{1C} and 5-HT_{2A} receptors (Zhang *et al.,* 1993; Magistretti *et al.,* 1995; Poblete and Azmitia, 1995). The raphe nuclei also elicit a systemic hyperglycemia upon stimulation which is mediated by the sympathetic adrenal system (Lin and Shian, 1991).

Cerebral blood vessels are not innervated by *glutamatergic* nerve fibers. Glutamate, however, exerts its essential effects on GU at the cellular level. Glutamate released during neuronal activity is taken up by astrocytes and there activates glucose metabolism and lactate release (Pellerin and Magistretti, 1994; Fray *et al.,* 1996). Thus, cerebral GU is coupled very closely to glutamate neurotransmitter cycling between neurons and astrocytes and reflects synaptic glutamatergic activity (Sibson *et al.,* 1998). In a closed feedback loop, virtually all brain glutamate appears to be derived from glucose in healthy human brain (Gruetter *et al.,* 1994). Glutamate-associated Ca^{2+} influx activates neuronal NO synthase and subsequent NO release (Garthwaite *et al.,* 1988, 1989; Bredt and Snyder, 1989, 1990). Hence, the coupling of glutamate-elicited GU and CBF may be accomplished by NO, the principal regulator of blood vessel relaxation, maintainer of blood pressure, and mediator of activation-flow coupling (Ignarro *et al.,* 1987; Palmer *et al.,* 1987; Greenberg *et al.,* 1999).

The pattern of changes elicited by *GABA* agonists in conscious and anesthetized rats suggests a vasodilatatory, CBF-increasing, and glucose metabolism-depressant effect. *In vitro,* GABA relaxes arterial blood vessels via $GABA_A$ receptors (Fujiwara *et al.,* 1975; Edvinsson and Krause, 1979; Anwar and Mason, 1982) and presumably presynaptically via $GABA_B$ receptors on sympathetic fibers (Anwar and Mason, 1982). In anesthesia, $GABA_A$ agonists decrease cerebral glucose uptake (Gjedde and Rasmussen, 1980) and metabolism (Crane *et al.*, 1978) and increase local blood flow (Edvinsson *et al.*, 1980). In awake animals and man, however, they also reduce GU which, due to the coupling, leads to a decreased CBF (Kelly and McCulloch, 1983; Roland and Friberg, 1988; Kelly *et al.*, 1989). Thus, GABAergic agonists, e.g., diazepam and phenobarbital, reduce cerebral GU, an effect which is antagonized by a GABA antagonist (Ableitner *et al.*, 1985; Hodes *et al.*, 1985; Laurie and Pratt, 1989). Intraseptal muscimol, a GABAergic agonist, induces memory impairments which could be attenuated by hippocampal but not by intraseptal glucose injections (Parent *et al.*, 1997). Human PET data demonstrate a decrease of GU during barbiturate medication and an increase of GU after drug withdrawal (Theodore *et al.*, 1986). Data on the effect of GABA on KB utilization is scarce. Chronic GABA agonists, e.g., phenobarbital or diazepam shift the cerebral energy metabolism in favor of KB in the developing rat while decreasing brain GU (Pereira de Vasconcelos *et al.*, 1987, 1990; Schroeder *et al.*, 1991, 1994, 1995). Similar to the metabolic links between glucose and ACh (see above), the KB-GABA couple also sets up a mutual metabolic/regulatory loop (Erecinska *et al.*, 1996; Daikhin and Yudkoff, 1998). This synergistic feedback loop may operate at the hypothalamic-pituitary level as well (Elias *et al.*, 2000). These features predispose GABA as the main mediator of metabolic depression in adverse conditions (Nilsson and Lutz, 1993). *Gammahydroxybutyrate,* a GABA catabolite, may assist the GABAergic system in the depression of energy utilization (Mamelak, 1989; Cash, 1994). Moreover, *kynurenate,* an endogenous broad-spectrum glutamate receptor antagonist (reviewed by Moroni, 1999; Sun and Cheng, 1999), is regulated by fuel supply and energy demand (Hodgkins and Schwarcz, 1998). Reduction of energy demand increases kynurenic acid production while increased energy metabolism reduces kynurenate production. As excitatory amino acid receptor antagonist, kynurenate decreases CBF and GU in conditions of metabolic stress (Hovda *et al.*, 1990; Kawamata *et al.*, 1992; Iadecola *et al.*, 1996) and thus exerts its neuroprotective effects (Moroni, 1999; Sun and Cheng, 1999). Region- and receptor subtype, specifically *dopamine,* may control GU in the limbic system (Meguro *et al.*, 1997; Levant *et al.*, 1998; Conti *et al.*, 2001).

Importantly, there is a profound cross-talk between GU-enhancing neurotransmitters which form a complex, mutually reinforcing network. Impact of these cooperations on behavior, neurotransmitter balances, and neuronal activity have been investigated (Dringenberg, 2000). Their consequences for GU and CBF, however, have not yet been addressed experimentally but due to the activity-metabolism coupling concurrent effects can be expected. So far, only bilateral interactions have been studied. Multilateral relations are expected to add another dimension of complexity to these interactions. Overall, the basically reinforcing nature of the interactions is evidenced by the additive or potentiated impairment of memory performance elicited by dual inhibitions. Although, inhibitory interactions have been reported as well which may serve to protect against overstimulation. This cooperative network is antagonized by GABA and kynurenate on multiple levels. A detailed presentation of this network goes beyond the scope of this paper but is given elsewhere (Heininger, 2000b). The role of neurotransmitters in the regulation of CNS glucose supply is also reflected by the effects of central neural mechanism on hepatic glucose metabolism and disposal. While activation of central cholinergic, noradrenergic, and serotonergic neurons rapidly increases hepatic output via the sympathetic nervous system, the GABAergic system is inhibitory on these pathways (reviewed by Nonogaki and Iguchi, 1997).

B. Hormones

In the last 15 years our knowledge about both neuropeptides and neurohormones has been greatly expanded. It has emerged that these powerful substances by their autocrine, paracrine, and endocrine actions are involved in the regulation and modulation of neuronal development, trophism, and repair (Heininger, 1999a). Often coexisting with classical neurotransmitters, they participate in signaling at synaptic and nonsynaptic sites and particularly play an important role in intercellular communication as volume transmission signals (Fuxe *et al.*, 1994). Preferentially conferring a slow, long-lasting, syndromic message (Fuxe *et al.*, 1994), they modulate neuronal energy homeostasis and excitability, regulate neurotransmitter balances, and drive circadian rhythms.

Both the control of energy balance at the *systemic* level and the cellular adaptation to the changing fuel supply and utilization is primarily achieved by the insulin/GC systems (reviewed by Dallman *et al.*, 1993, 1995; Schwartz *et al.*, 1999; Saltiel and Kahn, 2001) and modulated by the thyroid system (Ingenbleek and Bernstein, 1999). In principle, the eutrophic insulin is opposed by the metabolic stress-induced, dystrophic GC system. In the *brain,* while the general principles also apply, the control of intermediary metabolism is even more complex, particularly in the regulation of the eutrophic system. Cellular GU and mitochondrial efficiency is ensured by an integrated, albeit locally heterogeneous, trophic system consisting of insulin/IGFs, NPY, thyroid hormones (TH), neurotrophins, estrogens (E), DHEA, and somatostatin (Heininger, 1999a). In short, insulin and IGF-1 upregulate brain glucose transporters through independent pathways at the transcriptional and posttranscriptional level (Werner *et al.*, 1989). Insulin and IGF-1 enhance astrocytic glycolysis while neurons are stimulated to release factors which upregulate astrocytic lactate production (Sonnewald *et al.,* 1996). IGFs appear to reduce brain blood flow resistance (Gillespie *et al.,* 1997). Treatment with insulin and IGF-1 protects and even rescues neurons against damage induced by ischemia (Gluckman *et al.,* 1993; Zhu and Auer, 1994; Johnston *et al.,* 1996), glucose deprivation, and impairment of energy production (Mattson and Cheng, 1993). The developmental role of IGF-1 appears to depend on an insulin-like role in brain glucose metabolism (Cheng *et al.,* 2000). NPY improves the economy of energy homeostasis, it increases the respiratory quotient, reduces energy expenditure, and stimulates glucose turnover (Marks and Waite, 1997). DHEA has multiple effects on cellular energy metabolism. It was shown to stimulate glucose uptake and expression of glucose transporter (Nakashima *et al.,* 1995), improve efficiency of oxygen utilization at the tissue level (Schauer *et al.,* 1990) and elevate mitochondrial respiration (Mohan and Cleary, 1991; McIntosh *et al.,* 1993). The cellular energy metabolism is augmented by estrogens via a stabilization of mitochondrial transmembrane potential and function (Mattson *et al.,* 1997), induction of cytochrome c oxidase (Bettini and Maggi, 1992) and increase of glucose transport, uptake, and utilization through induction of key glycolytic enzymes (Nehlig *et al.,* 1985; Kostanyan and Nazaryan, 1992; Bishop and Simpkins, 1994). Neurotrophins enhance glucose uptake and utilization, increase mRNA and biochemical activity of enzymes of the glycolytic pathway, and increase cellular content of energy-rich phosphates (Liuzzi *et al.,* 1968; Skaper and Varon, 1979; Morelli *et al.,* 1986). Somatostatin enhances cortical and hippocampal glucose uptake (Shibata *et al.,* 1993). In development and maturity, mitochondrial respiration and oxidative phosphorylation is stimulated by TH in a variety of organs including the brain (Popovici *et al.,* 1980; Nelson, 1990; Soboll, 1993; Katyare *et al.,* 1994). Long-term influences of TH on mitochondrial activity include the expression of both nuclear and mitochondrial-encoded respiratory genes (Nelson, 1990; Vega-Nunez *et al.,* 1995; Pillar and Seitz, 1997), increase of the activity of TCA cycle enzymes (Diez-Guerra *et al.,* 1981) and stimulation of mitochondrial oxygen consumption (Horst *et al.,* 1989). In the brain, TH increases the activity of glycolytic enzymes (Sabell *et al.,* 1985; Srivastava and Baquer, 1985) and modulate the expression of the cerebral glucose transporter (Mooradian *et al.,* 1997).

Proglycolytic hormones and neuropetides establish a trophic cerebral milieu via an extensive regulatory network. Our insights into the cooperative control circuits, however, is poor and at best bilateral interactions have been investigated. For instance, synergistic interactions and reciprocal regulations have been described for neurotrophins on one hand and TH, insulin/IGFs, estrogens, NPY, and somatostatin on the other, for estrogen and TH, insulin, and IGF, and for TH and insulin/IGF-1 (Heininger, 1999a).

There appears to be intensive co-localization, cross-talk, cross-coupling, substitution, and feedback control between signal transduction systems (Toran-Allerand *et al.*, 1992; Dikic *et al.*, 1994; Brewster *et al.*, 1995; Azcoitia *et al.*, 1999). For instance, insulin and NGF elicit differential responses using the same cell signaling system. Overexpression of the insulin receptor leads to a more pronounced activation of its transduction system and allows insulin to mimic phenotypically the cellular actions of NGF (Dikic *et al.*, 1994). Similarly, estradiol can reverse the behavioral and hypometabolic effects of intracerebroventricular streptozotocin and thus, at least in part, substitute insulin (Lannert *et al.*, 1998). Proglycolytic hormones and neurotransmitters interact as well. In a feedback cycle, proglycolytic growth factors, e.g., estrogens, nerve growth factor, IGFs, and TH promote the survival and tropism of cholinergic basal forebrain neurons (Heininger, 1999a; Silva *et al.*, 2000). Insulin, on the other hand, may exert its memory-impairing actions through a hypoglycemia-induced decrement of cholinergic transmission (Kopf and Baratti, 1995, 1999).

Protagonists of the cerebral dystrophic system are GC and galanin. With unlimited glucose supply, GC increase blood glucose and insulin, but not KB (Schade *et al.*, 1980; Johnston *et al.*, 1982), responding to the increased, stress-related demand for fuel. In contrast, in conditions of glucose deprivation or insulin deficiency such as fasting and diabetes, GC induce a reduced systemic and cerebral GU and increased FA/KB concentration and utilization (Owen and Cahill, 1973; Schade and Eaton, 1980; Johnston *et al.*, 1982). Compatible with these findings, the drive for the hypoglycemic stress-related HPA activation is largely attenuated by KB infusion (Amiel *et al.*, 1991). Conversely, suppression of fat store mobilization doubled GC levels during fasting (Fery *et al.*, 1996). The pattern of GC-induced fuel preferences is also detectable at the cellular level. GC mediate the control of the glucose-FA cycle by means of the FA-regulated PDH flux (Cipres *et al.*, 1994). Remarkably, GC upregulate KB metabolizing enzymes in glial cells (Poduslo, 1989). GC disrupt neuronal and glial energetics by inhibition of glucose transport and utilization (Sapolsky, 1985; Virgin *et al.*, 1991; Doyle *et al.*, 1994). Mitochondrial respiratory rates are decreased and oxidative phosphorylation is uncoupled resulting in decreased ATP synthesis (Jani *et al.*, 1991; Martens *et al.*, 1991). GC impair cellular GU rendering neurons vulnerable to stressors (Heininger, 1999a). Supporting the metabolic rationale of GC actions, the endangerment elicited by GC can be abolished by either glucose or hydroxybutyrate, the latter being more effective (Sapolsky, 1986). Interactions with pro- and antiglycolytic neurotransmitters contribute to the dystrophic actions of GC. There are multiple, reciprocal relationships between the cholinergic system and the hypothalamo-pituitary-adrenal (HPA) axis. Hippocampal injection of a muscarinic antagonist enhances the HPA axis responsiveness to stress while depletion of hypothalamic ACh resulted in an impaired adaptation to chronic intermittent stress (Ramade and Bayle, 1989; Bhatnagar *et al.,* 1997). Chronic GC leads to the degeneration of cholinergic neurons in the medial septal area (Tizabi *et al.,* 1989). Furthermore, chronic GC administration decreases nicotine sensitivity and brain nicotinic receptor binding in mice, the hippocampal, hypothalamic, and frontal cortical regions being particularly sensitive (Pauly *et al.,* 1990; Pauly and Collins, 1993). GC also render cholinergic neurons more susceptible to other neurotoxic agents (Hortnagl *et al.,* 1993). The glutamatergic system, known to be activated by GC, may participate in the mediation of the degenerative effects (Michel and Agid, 1995). Serotonergic transmission is also inhibited on a multitude of regulatory levels including receptor, particularly 5-HT_{1A}, sensitivity and downregulation (Lopez *et al.,* 1998; Chaouloff, 2000; Wissink *et al.,* 2000). On the other hand, GC stimulate GABAergic function (Losada, 1988; Mishunina and Kononenko, 1990). In addition, HPA axis hormones activate dopamine release and metabolism (Walker and Diforio, 1997; Posener *et al.,* 1999). Overall these actions are compatible with an antiglucolytic, proketonolytic pattern of neurotransmitter regulation.

The neurochemical profile of galanin and its effects on fuel homeostasis (reviewed by Bartfai, 1995; Wang *et al.,* 1998) match the pattern of an agent that is actively involved in the adaptations to conditions of glucose deprivation and anoxia (Ben-Ari and Lazdunski, 1989). Galanin coexists with ACh, NA and 5-HT (reviewed by Bartfai *et al.,* 1993; Robinson and Crawley, 1993) and blocks the action of these proglycolytic neurotransmitters. Thus, galanin inhibits cholinergic, noradrenergic, serotonergic, and glutamatergic, but not GABAergic, transmission and neuronal activity. Galanin is upregulated by the lesional or electrophysiological de-efferentation of septal neurons from their hippocampal targets, in the dorsal raphe nucleus after decortication (Cortes *et al.,* 1990; Unger and Schmidt, 1993; Agoston *et al.,* 1994) and in the cortex following ischemia (Dubal *et al.,* 1999). This upregulation appears to have a neuroprotective rationale (Hökfelt *et al.,* 1999; O'Meara *et al.,* 2000) as also evidenced by the damage-attenuating effects of galanin pretreatment of ischemia, traumatic brain injury (TBI) and seizures (Ben-Ari and Lazdunski, 1989; Liu *et al.,* 1994; Mazarati *et al.,* 1998). The cholinergic vasodilatatory action and upregulation of CBF is antagonized by galanin (Barbelivien *et al.,* 1998). At the behavioral level, these actions result in the compromise of learning and memory (reviewed by Robinson and Crawley, 1993; Bartfai, 1995) which can be reversed by the intraseptal administration of glucose (Stefani and Gold, 1998). Metabolically, galanin induces a reduction in energy expenditure (Menendez *et al.,* 1992) and regulates fat intake, metabolism, and storage (Leibowitz, 1998; Wang *et al.,* 1998) putatively in a cooperative feedback loop with GC (Malendowicz *et al.,* 1994; Leibowitz, 1998). By its central and peripheral actions, galanin has profound stimulatory effects on the intake of fat and inhibits glucose-induced insulin release and GU (reviewed by Leibowitz, 1998).

Neurotrophic/neuroprotective and neurodystrophic agents inter- and counteract via multiple regulatory and effector mechanisms. Thereby, in a complex interplay second messenger pathways converge in the regulation of effector functions, constantly integrating synergistic and antagonistic influences (Jehan *et al.,* 1993). For instance, GC and neurotrophins are antagonistically linked (Heininger, 1999a). DHEA has antiglucocorticoid effects in that it blocks GC synthesis and antagonizes a variety of glucocorticoid actions, particularly on learning and memory like fear conditioning (Kalimi *et al.,* 1994; Fleshner *et al.,* 1997). The suppression of GC synthesis might be regulated via the GABAergic action of DHEA on dorsomedial and paraventricular hypothalamic neurons (Bartanusz *et al.,* 1995). However, in stress or serious illness, pregnenolone metabolism is shifted away from DHEA to GC. Gonadal hormones and GC show multiple, mostly antagonistic interactions. Estrogens and androgens interfere with GC receptor function by reducing GC receptor message and impairing its feedback control (Burgess and Handa, 1992; Turner, 1997). Oral contraceptive users show blunted free cortisol responses to psychological and physical stress (Kirschbaum *et al.,* 1996) and, similarly, testosterone inhibited HPA-responses to stress (Viau and Meaney, 1996). Stress and GC induce neuronal death in the hippocampus of castrated rats which can be blocked by substitution with testosterone (Mizoguchi *et al.,* 1992). Somatostatin secretions are inhibited by GC (Liu and Patel, 1995), and stimulated by testosterone, while somatostatin infusions blunt the insulin-induced hypoglycemia-stimulated elevation of cortisol (Rubinow *et al.,* 1995).

C. Signal Transduction Pathways and Effectors

In response to hormones and by integrating stress sensing, reporting, and responding, the adaptive responses of the glucose-FA cycle are regulated by a network of signal transduction pathways. The insulin signaling pathways, interconnected by a complex network involving multiple substrates and feedback loops, play a vital role in cellular development and homeostasis (Folli *et al.,* 1996; Virkamäki *et al.,* 1999; Saltiel and Kahn, 2001). Importantly, a variety of other growth factors such as IGFs, neurotrophins, platelet derived-growth factor (PDGF), epidermal growth factor (EGF), and fibroblast growth factor (FGF) (Schlessinger and Ullrich, 1992; Williams and Roberts, 1994; Feldman *et al.,* 1997; Scrimgeour *et al.,* 1997) share, and network via these signaling pathways. This signaling actuates glucose uptake, glycogen and protein synthesis, and involves Ca^{2+}/calmodulin (Ashkenazy-Shahar *et al.,* 1998) and a multitude of phosphorylation cascades

(Srivastava and Pandey, 1998) via the phospholipase C, phophatidylinositol, protein kinase B (PKB), protein kinase C (PKC), and MAP kinase pathways (Heininger, 2000b). The yeast SNF1/mammalian AMP-activated protein kinase (AMPK) may be the metabolic master switch of intermediary metabolism (Hardie and Carling, 1997; Johnston, 1999). SNF1/AMPK are homologous mediators of cellular stress responses (Hardie, 1999). The SNF1/AMPK pathway is activated under conditions of an elevated AMP-ATP ratio and constitutes a cellular protection system which switches off ATP consuming processes and coordinates adaptations of the glucose-FA cycle (Hardie and Carling, 1997; Kemp *et al.*, 1999). SNF1 controls the derepression of many glucose-repressible genes (Trumbly, 1992; Gancedo, 1998), which suggests the catabolite repression pathway as primitive precursor of the metazoan glucose-FA cycle. cAMP can be regarded as a universal signal for carbon source deprivation and mediates the mobilization of fuel stores in response to a variety of hormones and neurotransmitters such as glucagon or adrenaline (Tomkins, 1975). cAMP-dependent PK signal transduction is a mediator of glucose transport inhibition (Prapong *et al.*, 2001). FA β-oxidation is stimulated by a variety of stress signal transduction systems such as cAMP (Blazquez *et al.*, 1998; McGarry and Brown, 1997), AMP (Hardie and Carling, 1997; Johnston, 1999), and ceramide (Blazquez *et al.*, 1999). In the brain, highly plastic astroglial cells (Vernadakis, 1996) when activated present immature metabolic features including β-oxidation of their favorite fuel, FA (Auestad *et al.*, 1991; Edmond, 1992; Cullingford *et al.*, 1998), to provide KB for the neurons. In fact, the cAMP signal transduction pathway stimulates astroglial ketogenesis from FA (Poduslo, 1989) by decreasing malonyl CoA levels through inhibition of acetyl CoA carboxylase activity and hence by elevating carnitine acyltransferase I activity (Blazquez *et al.*, 1998). The regulatory interfaces of the glucose-FA cycle also control cell cycle decisions (Heininger, 2001). The regulatory input of cell fuel metabolism is exemplified by the key functions of pleiotropic glycogen synthase kinase (Frame and Cohen, 2001; Harwood, 2001) and glyceraldehyde-3-phosphate dehydrogenase (Heininger, 2000b, 2001) during cell cycle events.

Regulatory protagonists of the glucose-FA cycle are GC and Aβ, a metabolite of amyloid precursor protein that in appreciation of its pleiotropic actions in this adaptation has been proposed to be called deprivin (Heininger, 2000b). GCs and Aβ are interlocked in a synergistic feedback loop with mutual regulation of production as one of its features. These agents also orchestrate the differentiation/apoptosis balance (Heininger, 2000b, 2001) enabling a smooth and coordinated transition from mild (metabolic switch as seen in sleep and exercise) to severe deprivation responses (ontogenetic events and tissue damage). Both GC and deprivin establish a link between the endocrine control of stress responses and the cellular redox regulation and DNA repair system (Heininger, 2000b).

Intriguingly, the cytosolic Ca^{2+}-redox balance, a target of Aβ signaling (Heininger, 2000b), is the final integrator of signals for fuel utilization. Glucose metabolism regulates Ca^{2+} homeostasis presumably by mediation of Na^+/Ca^{2+} exchange (Nijjar and Belgrave, 1997). On the other hand, glucose metabolism and the switch to alternative fuels are controlled by Ca^{2+}. When cytosolic Ca^{2+} levels are raised by physiological stimuli, glucose transport is activated (Bihler, 1988). Small Ca^{2+} increases inhibit FA oxidation (Otto and Ontko, 1978). Conversely, high cytosolic Ca^{2+} levels inhibit insulin-stimulated glucose transport in a variety of insulin target cells (Begum *et al.*, 1993) while substantial Ca^{2+} increases enhance the oxidation of FA and elevate the NADH/NAD$^+$ ratio in liver mitochondria (Otto and Ontko, 1978; Ontko and Westbrook, 1983; Okuda *et al.*, 1992). Thus, adult rat brain utilizes KB for oxidative purposes in conditions of uncoupling of oxidative phosphorylation (Lopes-Cardozo and Klein, 1982) as can be accomplished by Ca^{2+} increases (see Heininger, 1999a). Low ATP, i.e., sufficient energetic resources, inhibit this metabolic pathway (Lopes-Cardozo and Klein, 1982). Jointly with Ca^{2+}, the cellular redox balance drives fuel utilization and is targeted by fuel availability (Heininger, 1999a, 2001, 2002).

V. (Patho)physiological Importance of the Glucose-FA Cycle

The first line of stress defense engages a "metazoan" type of response taking advantage of stored carbon fuels. Thus, the graded response is initiated by an adaptation of the glucose-FA cycle mediated by an inhibition of the eutrophic acetylcholine/insulin system and an activation of the dystrophic catecholaminergic/GC stress system. More vital endangerments during cellular injury and repair evoke a more primitive "unicellular" program. This entails a recapitulation of cell cycle events leading to de-/redifferentiation and apoptosis decisions depending on the status of the Ca^{2+}-energy-redox homeostasis (Heininger, 2000b, 2001). Importantly, the entire cascade of events is regulated and coordinated by a unique signaling/effector system which integrates metazoan and unicellular types of organismal and cellular behavior. The mechanisms actuating the differentiation/apoptosis deprivation response are not unidirectional. It should be borne in mind that differentiation initially served to generate dormant, resistant life forms, e.g., spores that, under favorable environmental conditions, could dedifferentiate and, e.g., in the case of *Dictyostelium,* yield fully vital, vegetative amebae. This process is, at least in part, conserved in metazoans. During tissue injury and repair, a program recapitulates developmental processes in reverse order and allows populations of differentiated cells to reenter the cell cycle

(Hass, 1994; Heininger, 2000b). Importantly, retrodifferentiation is associated with the loss of specific function of differentiated tissues (Renault *et al.*, 1998; Heininger, 2000b, 2001) and together with apoptosis aggravates the damage-related initial dysfunction.

A. Starvation

During starvation, glucose is too precious to serve as fuel (Randle *et al.*, 1978). Equally importantly, satisfying the organism's need for glucose would necessitate the eventually complete consumption of protein as the primary substrate for gluconeogenesis (Owen *et al.*, 1967). Lipid fuel attenuates the glucose dependence and effectively leads to protein sparing (Palaiologos and Felig, 1976; Lowell and Goodman, 1987; Thompson and Wu, 1991; Fery *et al.*, 1996). Therefore, the oxidative metabolism is, at least in part, switched to other fuels, FA and KB, paralleled by a decline of glucose oxidation and development of insulin resistance (Henriksson, 1990; Mansell and Macdonald, 1990; Romijn *et al.*, 1990). The brain is the last organ which is converted to substantial KB metabolism after prolonged periods of food deprivation (Owen *et al.*, 1967; Soengas *et al.*, 1998) while muscle KB clearance declines (Balasse and Fery, 1989) to aid the redirection of fuel. Since the enzymatic activities of KB metabolism are present at much higher activities than necessary for measured rates of utilization (Hawkins *et al.*, 1971; Ruderman *et al.*, 1974), the animal and human brain can metabolize KB acutely (Kammula, 1976; Amiel *et al.*, 1991). However, the brain's KB uptake is limited by permeability (Hawkins and Biebuyck, 1979; Blomqvist *et al.*, 1995) and hence is a function of KB blood concentration (Hawkins *et al.*, 1971; Kammula, 1976; Blomqvist *et al.*, 1995). Tissue pH also affects astrocytic KB transport and conditions with brain acidosis may favor KB utilization (Tildon *et al.*, 1994). Thus, after only 12–16 h of fasting a modest brain KB uptake can be detected (Gottstein *et al.*, 1971) and after 2 days of starvation glucose oxidation is decreased by 12% in the rat brain (Mans *et al.*, 1987). Human brain KB levels increase substantially after fasting for 2 days (Pan *et al.*, 2000) and after 3 days its energy requirements are covered by approximately 25% by β-hydroxybutyrate (HB) (Hasselbalch *et al.*, 1994). After 1–2 weeks of starvation, adult brain GU is decreased by approx 50% (Redies *et al.*, 1989), and KB transport and uptake are increased significantly (Owen *et al.*, 1967; Gjedde and Crone, 1975; Pollay and Stevens, 1980). Brain HB dehydrogenase may be upregulated (Smith *et al.*, 1969; but see Sokoloff, 1973) and the fuel mix of the brain consists of about equal amounts of glucose and KB, while regional oxygen utilization, blood flow, and blood volume are unchanged (Redies *et al.*, 1989). Adding HB and acetoacetate

(AC) as fuel in conditions of glucose deprivation (the physiological ratio HB/AC is 3–6) has profound effects on cerebral energy homeostasis. This switch (1) reduces brain glucose extraction, (2) increases the mitochondrial metabolites of the TCA cycle, (3) impairs glycolysis through inhibition of PDH and citrate-mediated inhibition of 6-phosphofructokinase, (4) elevates the redox energy by reduction of the free [NAD^+/NADH] ratio, (5) increases the oxidation of the coenzyme Q couple, (6) attenuates the Ca^{2+}-dependence of oxidative phosphorylation, and (7) improves mitochondrial energy transduction, energy charge, and reserve (Heininger, 2000b). These effects counteract the metabolic stress-related deterioration of the redox ratio, acidosis, uncoupling of oxidative phosphorylation, and ATP depletion and thereby protect the cell from oxidative stress and energy shortage (Hillered *et al.,* 1984; Corbisier and Remacle, 1993; Humphries *et al.,* 1998). Particularly the modulation of the [NAD^+/NADH] ratio appears to have a key role in the KB-induced benefit by inhibiting glucose consumption (Gibson and Blass, 1976) and PDH (Pettit *et al.,* 1975; Randle *et al.,* 1978), preventing and reversing the mitochondrial Ca^{2+} release (Lehninger *et al.,* 1978; Harris *et al.,* 1982; Vercesi, 1984), preventing the loss of mitochondrial sulfhydryl groups (Bindoli *et al.,* 1997), improving mitochondrial resistance to the induction of permeability transition (Laughlin and Heineman, 1994; Beyer *et al.,* 1996; Humphries *et al.,* 1998), and both protecting against DNA damage and promoting DNA repair (Zhang *et al.,* 1998). In addition, the oxidation of the [Q/QH_2] couple, by decreasing the semiquinone form of reduced coenzyme Q, the major source of mitochondrial reactive oxygen species (Chance *et al.,* 1979), can be expected to attenuate free radical production (Sato *et al.,* 1995).

KB increase and maintain a high level of GABA by regulating the enzymes of the GABA shunt and diminish aspartate levels, an excitotoxin (Erecinska *et al.,* 1996; Daikhin and Yudkoff, 1998). These metabolic features confer KB a neuroprotective potential in models of hypoglycemia, hypoxia, ischemia, and brain trauma (Johnson and Weiner, 1978; D'Alecy *et al.,* 1990; Marie *et al.,* 1990; Hiraide *et al.,* 1991; Veneman *et al.,* 1994; Yager, 1994). They also render the brain metabolically more stable, result in a reduced cerebral excitability and elevated electroconvulsive threshold (Hori *et al.,* 1997; Stafstrom *et al.,* 1999), and make ketogenic diet a valuable therapy in childhood and adulthood epilepsy (Huttenlocher, 1976; Ross *et al.,* 1985; Hori *et al.,* 1997; Prasad and Stafstrom, 1998). Remarkably, acetone appears to be the predominant KB after ketogenic diet, at least in brain of children (Seymour *et al.,* 1999). The hormonal and metabolic changes induced by a isocaloric ketogenic diet or lipid infusion mimic those observed after a few days of total fasting, emphasizing the role of carbohydrate restriction for the adaptations (Fery *et al.,* 1982; Klein and Wolfe, 1992). The hormonal/metabolic adaptive changes in the feedback loop of the glucose-FA cycle, but not the KB levels (Bough *et al.*, 2000; Likhodii *et al.*, 2000), may constitute the underlying therapeutic principle. This conclusion is supported by the finding that caloric restriction also exerts a seizure-protective effect (Bough *et al.*, 1999). The beneficial effects of KB in optimizing the cellular energy homeostasis may be regarded as a means to achieve the primary goal, to ensure the cellular survival under adverse conditions. The character of KB as an austerity fuel is highlighted by their effects on pyrimidine biosynthesis. KB can impair cell proliferation by inhibiting *de novo* pyrimidine biosynthesis by 30% which may lead to lower brain weights in the newborn following ketonemic states in late pregnancy (Bhasin and Shambaugh, 1982). The brain may not be totally dependent on hepatic KB supply. Although neurons are not able to use FA, astroglial cells can degrade FA by β-oxidation to provide KB to the neurons (Edmond, 1992). In fact, the starving brain also extracts FA from the circulation (Padmini and Rao, 1991; Ishiwata *et al.*, 1996). In addition, lactate may serve as fuel in the glucoprivic brain (Veneman *et al.*, 1994; King *et al.*, 1997).

B. Brain Ontogenesis

Cellular life history events like differentiation and apoptosis evolved in unicellular organisms as reponses to environmental challenges, particularly nutrient deprivation (Heininger, 2001). Activation of lipolytic metabolism and inhibition of glycolysis as metabolic hallmarks of deprivation syndromes also control the differentiation/apoptosis balance (Heininger, 2001). Thus, in ontogenetic events like brain maturation and aging systemic metabolic conditions carry the signature of evolutionarily programmed deprivation responses (Heininger, 2001, 2002).

In nonprecocial species like human and rat which are born neurologically incompetent the enzyme machinery necessary to metabolize glucose only develops during the late suckling period (Booth *et al.*, 1980; Clark *et al.*, 1993). In the early neonatal period, lactate serves as major fuel (Mayor and Cuezva, 1985; Dombrowski *et al.*, 1989). Thereafter, milk provides FA as primary carbon source (Edmond *et al.*, 1985) and the brain uses KB and FA both as an energy source and lipid precursors for myelinogenesis (reviewed by Mayor and Cuezva, 1985; Nehlig, 1999). During this period, lactate is the major product of glucose metabolism (Lopes-Cardozo *et al.*, 1986). In addition, neonatal brain astrocytes readily β-oxidize FA to provide KB to the neurons (Auestad *et al.*, 1991; Edmond, 1992), particularly in conditions of nutritional stress (Padmini and Rao, 1991). Thus in human neonates and infants, KB concentration, cerebral uptake and turnover are markedly elevated (Settergren *et al.*, 1980; Bougneres *et al.*, 1986). In contrast, the increase of GU correlates temporally and topically with maturation of neuronal functions as evidenced in the rat brain (Nehlig, 1999). Notably, cholinergic innervation of the hippocampus and cortex only develop postnatally (Perry *et al.*, 1993), in parallel with the maturation of glycolytic metabolism. Even then, glucose is readily replaced by lactate as cerebral metabolic fuel in metabolically stressed weanling mice (Thurston and Hauhart, 1989).

Compatible with the pattern of fetal fuel use, *in utero* maternal GC play an important role for fetal organogenesis and maturation (Waddell, 1993) and are particularly necessary for brain development (Trejo *et al.*, 1998). GC regulate fuel utilization during development when the glycolytic pathway is immature (Thurston *et al.*, 1980; Poduslo *et al.*, 1990; Bhargava *et al.*, 1991). In the immature rat brain, high levels of 11β-hydroxysteroid dehydrogenase, converting inactive GC into active GC, are expressed (Moisan *et al.*, 1992; Diaz *et al.*, 1998). Cortisone-cortisol interconversion may also play a role in human fetal development (Murphy, 1981; Seckl, 1997). Lactating rats exhibit a high brain corticosterone level which may serve to promote the utilization of KB and give rise to the neonatal suppression of HPA axis responses to stress (Kakihana *et al.*, 1980). In nursing mice, GC also increase brain glucose and glycogen, presumably due to the shift of oxidative metabolism to KB (Thurston *et al.*, 1980) and lactate (Thurston and Hauhart, 1989).

Aging is a genetically programmed deprivation syndrome in which the soma, stressed by the degeneration of the trophic milieu, actively pursues a survival pathway characterized by hypometabolism, adjustments of the glucose-FA cycle, oxidative stress, and modulation of DNA repair (Heininger, 1999a, 2001, 2002). Overall, as evidence of a metabolic stress response aging is associated with an increased HPA axis activity exerting particularly CNS-targeted effects. Human aging is associated with increased CSF, particularly ventricular, cortisol levels (Swaab *et al.*, 1994; Murakami *et al.*, 1999). Due to an at best mild concomitant increase of serum cortisol levels the cortisol CSF/serum ratio is elevated (Murakami *et al.*, 1999) suggesting an intrathecal accumulation of GC. Conversion of inactive precursors into active agents, i.e., cortisone into cortisol by the action of 11β-hydroxysteroid dehydrogenase, at the level of target structures appears to be the underlying principle (Heininger, 2000b). The adaptive response of the glucose-FA cycle is set in motion during normal brain aging resulting in the reduction of CBF and GU (Heininger, 1999a), the dissociation of oxygen utilization and GU (Hoyer *et al.*, 1988, 1991), and increased KB utilization (Dickinson, 1996). Brain insulin receptor densities and signal transduction is decreasing during normal aging (Frölich *et al.*, 1999). Particularly during increased neuronal activity, increase of regional GU but not CBF is attenuated in aged compared to young individuals (Pietrini *et al.*, 1999) indicating that part of the increased energetic demand is covered by combustion of alternative fuels, possibly KB and lactate. The temporal sequence of predominant fuel utilization in the neonatal brain (which is lactate, KB and glucose) is recapitulated in reverse order by the aging brain (Heininger, 2000b).

During CNS aging, specifically proglycolytic neurotransmitters ACh, NA, 5-HT, and glutamate degenerate (Heininger, 1999a). The cholinergic effects on CBF and GU are attenuated with aging presumably due to the aging-related cholinergic decline (De Micheli and Soncrant, 1992; Lacombe *et al.*, 1997; Uchida *et al.*, 1997). While the muscarinergic transmission seems to be unaffected (Soncrant *et al.*, 1989), the nicotinergic mediation of CBF is profoundly impaired (Uchida *et al.*, 1997). In contrast, GABA is little affected (Heininger, 1999a) and kynurenate metabolism and tissue levels are increased (Moroni *et al.*, 1988; Gramsbergen *et al.*, 1992). In conjunction with its late maturation (Perry *et al.*, 1993), the cholinergic system follows a "last in-first out" behavior while the GABAergic system exhibits a "first in-last out" pattern (Davies *et al.*, 1998; Nägga *et al.*, 1999) in accordance with the, at least in part, changing fuel use.

There is a clear age-dependent propensity to accomplish and benefit from the starvation-related metabolic switch. Fasting-induced KB levels rise more readily in young than in aged individuals and activities of enzymes of KB metabolism are higher in suckling than adult rat brain (Klee and Sokoloff, 1967; Dierks-Ventling and Cone, 1971; Bilger and Nehlig, 1992). Thus, in prolonged fasting KB levels in aged individuals remain higher (London *et al.*, 1986) possibly reflecting reduced clearance. Moreover, monocarboxylate transporter densities are markedly higher in vessel and astrocytic end feet membranes of suckling than adult rats (Pellerin *et al.*, 1998a; Leino *et al.*, 1999). Hence, with increasing age brain uptake, utilization, and mitochondrial oxidative capacity of KB and cellular ATP decline (Kraus *et al.*, 1974; Patel, 1977) paralleled by the progressive failure of KB to sustain synaptic function in glucoprivic brain (Arakawa *et al.*, 1991; Wada *et al.*, 1997; Izumi *et al.*, 1998). The deterioration of synaptic function may even occur in hippocampal slices when sufficient KB are provided to maintain ATP levels (Arakawa *et al.*, 1991). Consequently, the capacity of KB to overcome respiration uncoupling is impaired in aged mitochondria (Corbisier and Remacle, 1993) and the antiepileptic potency of a ketogenic diet decreases (Bough *et al.*, 1999).

In accordance with this concept, GC, probably by enhancing cerebral KB supply and utilization (Dardzinski *et al.*, 2000), are neuroprotective in neonatal animals when given in advance of cerebral ischemia (reviewed by Tuor, 1997) and increase high-energy phosphates (Dardzinski *et al.*, 2000), a condition characteristic for KB oxidative metabolism. Compatible with the relative ease of the metabolic switch in neonate but not adult rats is the finding that this protective action is no longer detectable in one-month-old rats (Tuor *et al.*, 1995) which parallels the time course of loss of KB-related maintenance of synaptic function (Izumi *et al.*, 1998). In adult and aged brain, lactate can support the recovery of energy metabolism and synaptic activity after ischemia, hypoglycemia, and hypoxia (Schurr *et al.*, 1988, 1997; Bock *et al.*, 1993) and appears to be oxidized as replacement of glucose in hypercapnia (Miller and Corddry, 1981) and following TBI (Chen *et al.*, 2000).

C. Neurological Disorders

The brain has adaptive systems which allow it to survive short-term energy deprivations (Lutz, 1992; Reis *et al.*, 1997). In a variety of acutely developing conditions which are associated with persistent ($\geq$2 min) metabolic stress and particularly a lack of glucose supply, e.g., ischemia, TBI with secondary ischemia, hypoxia, and hypoglycemia, a different kind of syndrome develops. It is characterized by the phylogenetically fixed program aiming at curbing metabolism and saving glucose by both hypometabolism and the delayed degeneration of the proglycolytic cholinergic system. After an initial activation of the cholinergic system (Gorman *et al.*, 1989; Kumagae and Matsui, 1991) which can be detected before any energetic compromise takes place (Park *et al.*, 1987), a delayed degeneration of basal forebrain cholinergic neurons and loss of cholinergic innervation of the cortex and hippocampus ensues (Liberini *et al.*, 1994; Ni *et al.*, 1995) which cannot be prevented by reperfusion (Kumagae and Matsui, 1991; Ishimaru *et al.*, 1995). Following both permanent vessel occlusion and reperfusion, a syndrome develops with reduced CBF and regional dysautoregulation presumably due to an impaired vascular response to ACh (Ott *et al.*, 1975; Rosenblum, 1997). In a variety of energetic stress model systems, the cholinergic degeneration correlates with the fall of GU, appears to be mediated by Ca^{2+} uptake (Gibson and Mykytyn, 1988; Gibson *et al.*, 1989) and oxidative stress (Meyer *et al.*, 1994), and is characterized by presynaptic inhibition of neurotransmission in the basal forebrain, cortical decline of both ChAT activity, stimulated ACh release and muscarinic binding sites, loss of cholinergic receptor-G protein coupling, and increased tissue choline and ACh esterase levels (Heininger, 2000b). For instance, in human traumatized brain, markers of cholinergic neurotransmission are lost (Murdoch *et al.*, 1998). The loss of cholinergic forebrain neurons after TBI (Leonard *et al.*, 1994; Schmidt and Grady, 1995) leads to a disruption of the septohippocampal pathway (DeAngelis *et al.*,

1994; Leonard *et al.*, 1997) and an enhanced vulnerability to anticholinergic agents (Dixon *et al.*, 1995). Likewise, acute hypoglycemia diminishes cortical and striatal ACh concentration and synthesis (Ghajar *et al.*, 1985) resulting in impaired memory performance (Kopf and Baratti, 1995). The adaptive inhibition of the cholinergic system can also be elicited by inhibition of PDH (Frölich *et al.*, 1990). Generally, the time course of the degeneration appears to depend on the severity of the metabolic stress, is reversible in 14 days' time after 5 min of ischemia (Haba *et al.*, 1991) and chronic progredient after permanent vessel occlusion (Ni *et al.*, 1995). In permanent occlusion models, the adaptive changes of the cholinergic system are paralleled by a gradual decrease of regional energy metabolism (Nowicki *et al.*, 1988) and ensuing neuronal degeneration.

Key to the understanding of the deprivation syndrome is the realization that the tissue endangerments due to reperfusion injury and hyperglycemia and the neuroprotective effects of ischemic tolerance and induced hypometabolism highlight the flip-flop sides of the same pathophysiological process. An integrative interpretation of these phenomena (Heininger, 2000b) suggests that ischemic damage can be attenuated by hypometabolic (antiglycolytic) and aggravated by hypermetabolic (proglycolytic) manipulations. The downregulation of the proglycolytic cholinergic system fulfills a neuroprotective purpose as exemplified by the effects of pharmacologic manipulations. Muscarinergic agonists aggravate and muscarinic antagonists attenuate ischemia and TBI-induced deficits (Lyeth *et al.*, 1988; Robinson *et al.*, 1990; Shibata *et al.*, 1992), while a cholinergic deafferentation protects hippocampal pyramidal neurons from ischemia-induced degeneration (Kinoshita *et al.*, 1992). Likewise, noradrenergic inhibition attenuates the posttraumatic metabolic depression (Inoue *et al.*, 1991). However, the adaptive response needs time to develop. The evolutionary masterplan of the response only becomes evident with a free interval of hours after a preconditioning stimulus. Intriguingly, the adaptive response can be frustrated by an intervening hypermetabolism. Thus, a reinstitution of glycolytic metabolism (reperfusion) can have detrimental effects in a tissue actually adapting for hypometabolism.

In conditions of metabolic stress such as hypoxia, anoxia, ischemia and hypoglycemia, the brain fuel utilization, and mitochondrial redox state reflect the utilization of KB (Kirsch and D'Alecy, 1984; Vannucci and Brucklacher, 1994). KB or ketosis-inducing agents and diets are neuroprotective in models of hypoglycemia, hypoxia, ischemia, and brain trauma (Johnson and Weiner, 1978; D'Alecy *et al.*, 1990; Marie *et al.*, 1990; Hiraide *et al.*, 1991) and may aid the postischemic recovery of oxidative metabolism (Varela *et al.*, 1997; Brooks *et al.*, 1998). Hypercortisolism and galanin upregulation appear to regulate these metabolic changes in ischemia (Lebedev,

1978; Sapolsky, 1986; de Lacalle *et al.*, 1997) like they do in hibernation (Saboureau *et al.*, 1980; Dauphin-Villemant *et al.*, 1990; Gonzalez-Nicolini *et al.*, 1998).

Mammals go through attenuated cycles of caloric excess and deprivation with every meal followed by fasting and during sleep (Dallman *et al.*, 1993; Dinneen *et al.*, 1993). The same program is activated in times of increased need, e.g., prolonged physical activity and trauma, conditions which are characterized by ketosis. Since ketosis itself has an antiglycolytic effect (see Section V.A.) minor adaptations do not have to be necessarily associated with long-term changes of neurotransmitter balances. Prerequisite for these regulatory and even degenerative changes are severe, persistent energy deprivations. For instance, no long-term adaptations occur after an ischemia lasting up to 2 min (Kato *et al.*, 1992). Apparently, there is a severity-duration continuum (Dixon *et al.*, 1995), with less severe but more persistent changes, like in Alzheimer disease being able to cause adaptations as well (see Heininger, 1999b, 2000b). These adaptive responses carry the neurochemical and metabolic signature of a prototypic deprivation response (Heininger, 2000b).

D. Psychiatric Disorders

Due to the dual role of neurotransmitters as conveyors of intercellular communication (enabling affective and cognitive processes) and energy homeostasis, psychosocial and affective stressors feed into the somatic and cellular metabolic stress response system, establishing a psychosomatic continuum. Neurotransmitters provide the link between psychosocial stressors and somatic energetic responses giving rise to and representing the pathophysiological substrate of a whole array of psychosomatic diseases. Thus, in its last consequence and as its defining signature both biotic and abiotic stressors, mediated by neurotransmitters, elicit metabolic stress and trigger an adaptive metabolic response.

Contemporary stress research emphasizes the importance of the individual appraisal of the stressor. The degree to which the individual can cope with or defend against the psychosocial stressor determines the individual stress hormone profile and development of stress pathologies (Virgin and Sapolsky, 1997; Bookwala and Schulz, 1998; Olff, 1999). The individual stress vulnerability is modulated very early in life as a result of a gene-environment interaction (Plotsky and Meaney, 1993; Bouchard, 1994; Plomin *et al.*, 1994; Heim *et al.*, 1997). In animals, prenatal and postnatal stress increase HPA-axis responsivity and prolong stress-induced GC secretion in later life

(Maccari *et al.*, 1995; Vallee *et al.*, 1999). On the other hand, postnatal maternal care and neonatal handling reduce HPA reactivity, improve cognitive performance, and protect the hippocampus from age-related dysfunction and neuronal loss (Mohammed *et al.*, 1993; Sapolsky 1993; Liu *et al.*, 1997). In adult life, exposure to acute and chronic psychosocial stressors worsens cognitive performance in laboratory animals (McEwen and Sapolsky, 1995; Blanchard *et al.*, 1995; deQuervain *et al.*, 1998), deteriorates hippocampal blood flow (Endo *et al.*, 1999), increases GC levels, and impairs hippocampal function and dendritic morphology (McEwen *et al.*, 1992; Sapolsky, 1994; McEwen and Margarinos, 1997). Like with other stressors (Heininger, 2001), oxidative stress may be the final effector of emotional stress signaling pathways (Liu and Mori, 1994, 1999).

Acute stress activates the cholinergic system (Gilad, 1987; Kaufer *et al.*, 1998a) but may also attenuate choline acetyltransferase (ChAT) activity (Wahba and Soliman, 1992). Intermediate-term stress elicits mixed responses characterized by decreased choline uptake but an upregulation of muscarinic binding sites (Finkelstein *et al.*, 1985; Gonzalez and Pazos, 1992). In contrast, chronic stress induces cholinergic degeneration evidenced by enhanced sensitivity to muscarinic antagonists (Kaufer *et al.*, 1998b). The neurobiological effects are paralleled by behavioral changes. Ten days of stress induce a cholinergic hypersensitivity and resistance to scopolamine-induced amnesia, while 21–30 days of stress result in cholinergic hyposensitivity and learning deficits (Zerbib and Laborit, 1990; Sunanda *et al.*, 2000). An increased stress responsivity is associated with a premature degeneration of the cholinergic septohippocampal pathway and shorter life span (Gilad *et al.*, 1987). Chronic stress, like GC, decreases aminergic (5-HT, NA) hippocampal and cortical neurotransmission (Lopez *et al.*, 1998; Sunanda *et al.*, 2000; Blanchard *et al.*, 2001). Chronic stress also interacts antagonistically with hormonal activities of the hypothalamic-pituitary-gonadal, -thyroid, and -somatic axes and the neurotrophic agents (Heininger, 2000a).

Depression and chronic stress share pathophysiological pathways (Checkley, 1996; Wheatley, 1997; Lopez *et al.*, 1998; Nesse, 1999) causally linked to GC at the crossroads of the neuroendocrine network (Holsboer, 2001). Depression is associated with a variety of neuroendocrinological abnormalities (reviewed by Nemeroff and Krishnan, 1992; Holsboer, 1995; Plotsky *et al.*, 1995). The pattern of elevated antiglucolytic and prolipolytic and decreased proglucolytic hormonal and neurotransmitter activity suggest depression as a metabolic stress syndrome. Animal models indicate an increased galaninergic tone (Kask *et al.*, 1997; Bellido *et al.*, 2002). The HPA axis is dysregulated with elevated plasma cortisol levels which normalize after remission (Steckler *et al.*, 1999). The dexamethasone suppression test indicates an impaired HPA-axis feedback. Likewise, the hypothalmic-pituitary-thyroid axis is hypofunctional (reviewed by Prange *et al.*, 1987; Holsboer, 1995). The TSH response to TRH is often blunted, the nocturnal TSH rise absent, transthyretin is reduced in CSF (Hatterer *et al.*, 1993), and euthyroid sick syndrome may be present. Importantly, the peripheral thyroid stimulating hormone (TSH) levels are inversely correlated to the global and regional CBF and glucose metabolism (Marangell *et al.*, 1997). Systemic insulin resistence is a routine finding (Winokur *et al.*, 1988) and, like in the aging-associated metabolic syndrome (Heininger, 2002), visceral fat deposition is increased (Thakore *et al.*, 1997), arguing for systemic metabolic stress. This notion is supported by systemic findings of defective antioxidant defenses and oxidative stress (Peet *et al.*, 1998; Maes *et al.*, 2000) and disrupted Ca^{2+} homeostasis (Heininger *et al.*, 1998). The systemic alterations may mirror a CNS metabolic syndrome (Frölich *et al.*, 1999; Holden, 1999) that is also indicated by metabolic stress markers found in temporal cortex tissue of depressed suicide victims (Bown *et al.*, 2000). Imaging studies detected a reduced perfusion and glucose metabolism of regions in the limbic system and frontal lobes (Biver *et al.*, 1994; Nobler *et al.*, 1999; Videbech, 2000) in depression, particularly of old age. Successful therapy may be associated with increased GU (Kennedy *et al.*, 2001).

In a gene-environment interaction, stress is thought to play a role in the manifestation of schizophrenia (Parnas, 1999; Gispen-de Wied, 2000; Tsuang, 2000). Notably, schizophrenics display an impaired stress response system (Walker and Diforio, 1997; Muck-Seler *et al.*, 1999; Jansen *et al.*, 2000), including exaggerated responses to metabolic stress (Elman *et al.*, 1998). Stress, GC, and dopamine may be causally involved in behavioral sensitization and impaired behavioral control (Walker and Diforio, 1997; Lyons *et al.*, 2000). Neurochemically, cholinergic, particularly nicotinergic, neurotransmission is impaired (Kaufer and Soreq, 1999). Dopaminergic alterations (Walker and Diforio, 1997) may also be relevant for energetic homeostasis, particularly of the limbic system. The most consistent finding is a reduced CBF and GU in the frontal cortex and thalamic nuclei (Buchsbaum and Hazlett, 1998; Hazlett *et al.*, 1999; Clark *et al.*, 2001). Systemic and cerebral markers of oxidative stress exposure (Mahadik *et al.*, 2001; Yao *et al.*, 2001) indicate a metabolic stress syndrome (Heininger, 2001).

VI. Conclusion

The fingerprints of evolutionary pressures can be recognized upon a variety of features of metabolic stress responses. With regard to time course and fuel availability two main stress reactions can be distinguished: the acute fight/flight reaction and the chronic starvation response. Fight/flight as active behavior requires the mobilization of glucose to meet the suddenly increased demand for readily metabolizable fuel. In contrast, chronic starvation responses ask for economization of fuel use characterized by hypometabolism and combustion of energy-rich FA. Consistently, neurochemical responses to stress, both metabolic and psychosocial, mimic this pattern: after an acute phase with activation of proglycolytic neurotransmitters, e.g., ACh, 5-HT and NA, a chronic phase ensues in which these neurotransmitters are downregulated.

An intriguing finding is the high diversity of agents regulating eutrophic conditions, but the limited number of agents engaged in dystrophic responses. At the brink of life, selection pressures increase and safety margins become narrow. This limits the leeway of "mutagenetic experimentation" and the viability of its outcomes. Efficiency in the "low tolerance zone" is ensured by a limited set of highly efficient, pleiotropic agents such as Aβ. Similar laws appear to rule the diversity of organisms: diversity is highest in resourceful environments while it becomes low under harsher conditions.

The uniformity of the evolutionarily programmed metabolic stress response contrasts with the great variety of clinical pictures of CNS disorders. Adaptive entity of the phylogenetically old metabolic stress response is the individual cell. Largely independent on the location of these cells, they activate their archaic cellular survival responses; and, if a threshold number of these responders is affected, they induce the also very old adaptive responses of the limbic-hypothalamic-pituitary axes. Due to the structural organization of the brain, the location of the cells determines the manifold clinical pictures elicited by their malfunction and compensatory capacity. Thus, the brain intertwines in its response pattern two behaviors with widely differing evolutionary histories: the most mundane stress survival strategies programmed by egoistic microorganisms and the most advanced technology, highly plastic parallel computing of millions of cells.

References

Ableitner, A., Wuster, M., and Herz, A. (1985). Specific changes in local cerebral glucose utilization in the rat brain induced by acute and chronic diazepam. *Brain Res.* **359,** 49–56.

Agoston, D. V., Komoly, S., and Palkovits, M. (1994). Selective up-regulation of neuropeptide synthesis by blocking the neuronal activity: galanin expression in septohippocampal neurons. *Exp. Neurol.* **126,** 247–255.

Almeida, A., Almeida, J., Bolanos, J. P., and Moncada, S. (2001). Different responses of astrocytes and neurons to nitric oxide: The role of glycolytically generated ATP in astrocyte protection. *Proc. Natl. Acad. Sci. USA* **98,** 15294–15299.

Amiel, S. A., Archibald, H. R., Chusney, G., Williams, A. J., and Gale, E. A. (1991). Ketone infusion lowers hormonal responses to hypoglycaemia: evidence for acute cerebral utilization of a non-glucose fuel. *Clin. Sci.* **81,** 189–194.

Ammon, H. P., Estler, C. J., Heim, F., and Okoronkwo, B. (1969). Different effects of inhibition of lipolysis by nicotinic acid on the rate of glycolytic carbohydrate breakdown in brain and skeletal muscle. *Life Sci.* **8,** 213–221.

Anwar, N., and Mason, D. F. J. (1982). Two actions of (γ-aminobutyric acid on the responses of the isolated basilar artery from the rabbit. *Br. J. Pharmacol.* **75,** 177–181.

Arakawa, T., Goto, T., and Okada, Y. (1991). Effect of ketone body (D-3-hydroxybutyrate) on neural activity and energy metabolism in hippocampal slices of the adult guinea pig. *Neurosci. Lett.* **130,** 53–56.

Ashkenazy-Shahar, M., Ben-Porat, H., and Beitner, R. (1998). Insulin stimulates binding of phosphofructokinase to cytoskeleton and increases glucose 1,6-bisphosphate levels in NIH-3T3 fibroblasts, which is prevented by calmodulin antagonists. *Mol. Genet. Metab.* **65,** 213–219.

Auestad, N., Korsak, R. A., Morrow, J. W., and Edmond, J. (1991). Fatty acid oxidation and ketogenesis by astrocytes in primary culture. *J. Neurochem.* **56,** 1376–1386.

Azcoitia, I., Sierra, A., and Garcia-Segura, L. M. (1999). Neuroprotective effects of estradiol in the adult rat hippocampus: interaction with insulin-like growth factor-I signalling. *J. Neurosci. Res.* **58,** 815–822.

Bahnsen, M., Burrin, J. M., Johnston, D. G., Pernet, A., Walker, M., and Alberti, K. G. (1984). Mechanisms of catecholamine effects on ketogenesis. *Am. J. Physiol.* **247,** E173–E180.

Balasse, E. O., and Fery, F. (1989). Ketone body production and disposal: effects of fasting, diabetes, and exercise. *Diabetes Metab. Rev.* **5,** 247–270.

Barbelivien, A., MacKenzie, E. T., and Dauphin, F. (1995). Regional cerebral blood flow responses to neurochemical stimulation of the substantia innominata in the anaesthetized rat. *Neurosci. Lett.* **190,** 81–84.

Barbelivien, A., MacKenzie, E. T., and Dauphin, F. (1998). Autoradiographic mapping of cerebral blood flow responses to cholinergic stimulation of the rat substantia innominata: modulatory effect of galanin. *Brain Res.* **789,** 92–100.

Bartanusz, V., Muller, D., Gaillard, R. C., Sreit, P., and Kiss, J. Z. (1995). GABA and glutamatergic local networks control CRF neuron activity in organotypic cultures of the hypothalamic paraventricular nucleus. *Soc. Neurosci. Abstr.* **21,** 1386.

Bartfai, T. (1995). Galanin. A neuropeptide with important central nervous system actions. *In* "Psychopharmacology: The Fourth Generation of Progress" (F. E. Bloom and D. J. Kupfer, eds.), pp. 563–571. Raven Press, New York.

Bartfai, T., Hökfelt, T., and Langel, U. (1993). Galanin—a neuroendocrine peptide. *Crit. Rev. Neurobiol.* **7,** 229–274.

Beani, L., Bianchi, C., Nilsson, L., Nordberg, A., Romanelli, L., and Sivilotti, L. (1985). The effect of nicotine and cytisine on ^{3}H-acetylcholine release from cortical slices of guinea-pig brain. *Naunyn-Schmiedebergs Arch. Pharmakol.* **331,** 293–296.

Bedran de Castro, M. T. B., Crystal, G. J., Downey, H. F., and Bashour, F. A. (1984). Regional blood flow in canine brain during nicotine infusion: Pentobarbital vs. chloralose anesthesia. *Stroke* **15,** 690–694.

Begum, N., Leitner, W., Reusch, J. E. -B., Sussman, K. E., and Draznin, B. (1993). GLUT-4 phosphorylation and its intrinsic activity. Mechanism of Ca^{2+}-induced inhibition of insulin-stimulated glucose transport. *J. Biol. Chem.* **268,** 3352–3356.

Bellido, I., Diaz-Cabiale, Z., Jimenez-Vasquez, P. A., Andbjer, B., Mathe, A. A., and Fuxe, K. (2002). Increased density of galanin binding sites in the dorsal raphe in a genetic rat model of depression. *Neurosci. Lett.* **317,** 101–105.

Ben-Ari, Y., and Lazdunski, M. (1989). Galanin protects hippocampal neurons from the functional effects of anoxia. *Eur. J. Pharmacol.* **165,** 331–332.

Bettini, E., and Maggi, A. (1992). Estrogen induction of cytochrome c oxidase subunit III in rat hippocampus. *J. Neurochem.* **58,** 1923–1929.

Beyer, R. E., Segura-Aguilar, J., Di Bernardo, S., Cavazzoni, M., Fato, R., Fiorentini, D., Galli, M. C., Setti, M., Landi, L., and Lenaz, G. (1996). The role of DT-diaphorase in the maintenance of the reduced antioxidant form of coenzyme Q in membrane systems. *Proc. Natl. Acad. Sci. USA* **93,** 2528–2532.

Bhargava, H. K., Lalitha, T., and Telang, S. D. (1991). Corticosterone administration and lipid metabolism in brain regions during development. *Ind. J. Biochem. Biophys.* **28,** 214–218.

Bhasin, S., and Shambaugh, G. E. (1982). Fetal fuels. V. Ketone bodies inhibit pyrimidine biosynthesis in fetal rat brain. *Am. J. Physiol.* **243,** E234–E239.

Bhatnagar, S., Costall, B., and Smythe, J. W. (1997). Hippocampal cholinergic blockade enhances hypothalamic-pituitary-adrenal responses to stress. *Brain Res.* **766,** 244–248.

Biesold, D., Inanami, O., Sato, A., and Sato, Y. (1989). Stimulation of the nucleus basalis of Meynert increases cerebral cortical blood flow in rats. *Neurosci. Lett.* **8,** 39–44.

Bihler, I. (1988). The 1987 Upjohn award lecture. The role of membrane transport in the control of glucose metabolism and its coupling to cellular function. *Can. J. Physiol. Pharmacol.* **66,** 549–560.

Bilger, A., and Nehlig, A. (1992). Quantitative histochemical changes in enzymes involved in energy metabolism in the rat brain during postnatal development. II. Glucose-6-phosphatase dehydrogenase and β-hydroxybutyrate dehydrogenase. *Int. J. Dev. Neurosci.* **10,** 143–152.

Bindoli, A., Callegaro, M. T., Barzon, E., Benetti, M., and Rigobello, M. P. (1997). Influence of the redox state of pyridine nucleotides on mitochondrial sulfhydryl groups and permeability transition. *Arch. Biochem. Biophys.* **342,** 22–28.

Bishop, J., and Simpkins, J. W. (1994). Estradiol enhances brain glucose uptake in ovariectomized rats. *Brain Res. Bull.* **36,** 315–320.

Biver, F., Goldman, S., Delvenne, V., Luxen, A., De Maertelaer, V., Hubain, P., Mendlewicz, J., and Lotstra, F. (1994). Frontal and parietal metabolic disturbances in unipolar depression. *Biol. Psychiatry* **36,** 381–388.

Blanchard, D. C., Spencer, R. L., Weiss, S. M., Blanchard, R. J., McEwen, B., and Sakai, R. R. (1995). Visible burrow system as a model of chronic social stress: behavioral and neuroendocrine correlates. *Psychoneuroendocrinology* **20,** 117–134.

Blanchard, R. J., McKittrick, C. R., and Blanchard, D. C. (2001). Animal models of social stress: Effects on behavior and brain neurochemical systems. *Physiol. Behav.* **73,** 261–271.

Blazquez, C., Sanchez, C., Velasco, G., and Guzman, M. (1998). Role of carnitine palmitoyltransferase I in the control of ketogenesis in primary cultures of rat astrocytes. *J. Neurochem.* **71,** 1597–1606.

Blazquez, C., Sanchez, C., Daza, A., Galve-Roperh, I., and Guzman, M. (1999). The stimulation of ketogenesis by cannabinoids in cultured astrocytes defines carnitine palmitoyltransferase I as a new ceramide-activated enzyme. *J. Neurochem.* **72,** 1759–1768.

Blomqvist, G., Thorell, J. O., Ingvar, M., Grill, V., Widen, L., and Stone-Elander, S. (1995). Use of R-β-[1-^{11}C]hydroxybutyrate in PET studies of regional cerebral uptake of ketone bodies in humans. *Am. J. Physiol.* **269,** E948–E959.

Bock, A., Tegtmeier, F., Hansen, A. J., and Holler, M. (1993). Lactate and postischemic recovery of energy metabolism and electrical activity in the isolated perfused rat brain. *J. Neurosurg. Anesthesiol.* **5,** 94–103.

Bonvento, G., Lacombe, P., and Seylaz, J. (1989). Effects of electrical stimulation of the dorsal raphe nucleus on local cerebral blood flow in the rat. *J. Cereb. Blood Flow Metab.* **9,** 251–255.

Bonvento, G., Lacombe, P., MacKenzie, E. T., and Seylaz, J. (1991). Effects of dorsal raphe stimulation on cerebral glucose utilization in the anaesthetized rat. *Brain Res.* **567,** 325–327.

Bookwala, J., and Schulz, R. (1998). The role of neuroticism and mastery in spouse caregivers' assessment of and response to a contextual stressor. *J. Gerontol. B. Psychol. Sci. Social Sci.* **53,** P155–P164.

Booth, R. F., Patel, T. B., and Clark, J. B. (1980). The development of enzymes of energy metabolism in the brain of a precocial (guinea pig) and non-precocial (rat) species. *J. Neurochem.* **34,** 17–25.

Bouchard, T. J. (1994). Genes, environment and personality. *Science* **264,** 1700–1701.

Bough, K. J., Valiyil, R., Han, F. T., and Eagles, D. A. (1999). Seizure resistance is dependent upon age and calorie restriction in rats fed a ketogenic diet. *Epilepsy Res.* **35,** 21–28.

Bough, K. J., Yao, S. G., and Eagles, D. A. (2000). Higher ketogenic diet ratios confer protection from seizures without neurotoxicity. *Epilepsy Res.* **38,** 15–25.

Bougneres, P. F., Lemmel, C., Ferre, P., and Bier, D. M. (1986). Ketone body transport in the human neonate and infant. *J. Clin. Invest.* **77,** 42–48.

Bown, C., Wang, J. F., MacQueen, G., and Young, L. T. (2000). Increased temporal cortex ER stress proteins in depressed subjects who died by suicide. *Neuropsychopharmacology* **22,** 327–332.

Braet, K., Paemeleire, K., D'Herde, K., Sanderson, M. J., and Leybaert, L. (2001). Astrocyte-endothelial cell calcium signals conveyed by two signalling pathways. *Eur. J. Neurosci.* **13,** 79–81.

Brand, K. A., and Hermfisse, U. (1997). Aerobic glycolysis by proliferating cells: A protective strategy against reactive oxygen species. *FASEB J.* **11,** 388–395.

Bredt, D. S., and Snyder, S. H. (1989). Nitric oxide mediates glutamate-linked enhancement of cGMP levels in the cerebellum. *Proc. Natl. Acad. Sci. USA* **86,** 9030–9033.

Bredt, D. S., and Snyder, S. H. (1990). Isolation of nitric oxide synthetase, a calmodulin-requiring enzyme. *Proc. Natl. Acad. Sci. USA* **87,** 682–685.

Brewster, W. J., Fernyhough, P., Diemel, L. T., Mohiuddin, L., and Tomlinson, D. R. (1995). Changes in nerve growth factor and preprotachykinin messenger RNA levels in the iris and trigeminal ganglion in diabetic rats; effects of treatment with insulin or nerve growth factor. *Brain Res. Mol. Brain Res.* **29,** 131–139.

Brooks, K. J., Clark, J. B., and Bates, T. E. (1998). 3-Hydroxybutyrate aids the recovery of the energy state from aglycemic hypoxia of adult but not neonatal rat brain slices. *J. Neurochem.* **70,** 1986–1990.

Brownlee, D. J. A., and Fairweather, I. (1999). Exploring the neurotransmitter labyrinth in nematodes. *Trends Neurosci.* **22,** 16–24.

Bryan, R. M., Jr. (1990). Cerebral blood flow and energy metabolism during stress. *Am. J. Physiol.* **259,** H269–H280.

Buchsbaum, M. S., and Hazlett, E. A. (1998). Positron emission tomography studies of abnormal glucose metabolism in schizophrenia. *Schizophrenia Bull.* **24,** 343–364.

Burgess, L. H., and Handa, R. J. (1992). Chronic estrogen-induced alterations in adrenocorticotropin and corticosterone secretion, and glucocorticoid recetor-mediated functions in female rats. *Endocrinology* **131,** 1261–1269.

Cash, C. D. (1994). (γ-hydroxybutyrate: an overview of the pros and cons for it being a neurotransmitter and/or a useful therapeutic agent. *Neurosci. Behav. Rev.* **18,** 291–304.

Chance, B., Sies, H., and Boveris, A. (1979). Hydroperoxide metabolism in mammalian organs. *Physiol. Rev.* **59,** 527–605.

Chaouloff, F. (2000). Serotonin, stress and corticoids. *J. Psychopharmacol.* **14,** 139–151.

Checkley, S. (1996). The neuroendocrinology of depression and chronic stress. *Br. Med. Bull.* **52,** 597–617.

Chatelain, F., Kohl, C., Esser, V., McGarry, J. D., Girard, J., and Pegorier, J. P. (1996). Cyclic AMP and fatty acids increase carnitine palmitoyltransferase-I gene transcription in cultured fetal rat hepatocytes. *Eur. J. Biochem.* **235,** 789–798.

Chen, G. Q., Cui, C. H., Mayer, M. L., and Gouaux, E. (1999). Functional characterization of a potassium-selective prokaryotic glutamate receptor. *Nature* **402,** 817–821.

Chen, T., Qian, Y. Z., Rice, A., Zhu, J. P., Di, X., and Bullock, R. (2000). Brain lactate uptake increases at the site of impact after traumatic brain injury. *Brain Res.* **861,** 281–287.

Cheng, C. M., Reinhardt, R. R., Lee, W. H., Joncas, G., Patel, S. C., and Bondy, C. A. (2000). Insulin-like growth factor 1 regulates developing brain glucose metabolism. *Proc. Natl. Acad. Sci. USA* **97,** 10236–10241.

Cipres, G., Urcelay, E., Butta, N., Ayuso, M. S., Parrilla, R., and Martin-Requero, A. (1994). Loss of fatty acid control of gluconeogenesis and PDH complex flux in adrenalectomized rats. *Am. J. Physiol.* **30,** E528–E536.

Clark, C., Kopala, L., Li, D. K., and Hurwitz, T. (2001). Regional cerebral glucose metabolism in never-medicated patients with schizophrenia. *Can. J. Psychiatry* **46,** 340–345.

Clark, J. B., Bates, T. E., Cullingford, T., and Land, J. M. (1993). Development of enzymes of energy metabolism in the neonatal mammalian brain. *Dev. Neurosci.* **15,** 174–180.

Cohen, Z., Bonvento, G., Lacombe, P., and Hamel, E. (1996). Serotonin in the regulation of brain microcirculation. *Prog. Neurobiol.* **50,** 335–362.

Conti, G., Blandini, F., Tassorelli, C., Giubilei, F., Fornai, F., Zocchi, A., and Orzi, F. (2001). Intrastriatal injection of D-1 or D-2 dopamine agonists affects glucose utilization in both the direct and indirect pathways of the rat basal ganglia. *Neurosci. Lett.* **309,** 161–164.

Corbisier, P., and Remacle, J. (1993). Influence of the energetic pattern of mitochondria in cell aging. *Mech. Aging Dev.* **71,** 47–58.

Cortes, R., Villar, M. J., Verhofstad, A., and Hökfelt, T. (1990). Effects of central nervous system lesions on the expression of galanin: a comparative in situ hybridization and immunohistochemical study. *Proc. Natl. Acad. Sci. USA* **87,** 7742–7746.

Craik, R. L., Hand, P. J., and Levin, B. E. (1987). Locus coeruleus input affects glucose metabolism in activated rat barrel cortex. *Brain Res. Bull.* **19,** 495–499.

Crane, P. D., Braun, L. D., Cornford, E. M., Cremer, J. E., Glass, J. M., and Oldendorf, W. H. (1978). Dose dependent reduction of glucose utilization by pentobarbital in rat brain. *Stroke* **9,** 12–18.

Cudennec, A., Duverger, D., Nishikawa, T., McRae-Degueurce, A., MacKenzie, E. T., and Scatton, B. (1988a). Influence of ascending serotonergic pathways on glucose use in the conscious rat brain. I. Effects of electrolytic or neurotoxic lesions of the dorsal and/or median raphe nucleus. *Brain Res.* **444,** 214–226.

Cudennec, A., Duverger, D., Serrano, A., Scatton, B., and MacKenzie, E. T. (1988b). Influence of ascending serotonergic pathways on glucose use in the conscious rat brain. II. Effects of electrical stimulation of the rostral raphe nuclei. *Brain Res.* **444,** 227–246.

Cudennec, A., Bonvento, G., Duverger, D., Lacombe, P., Seylaz, J., and MacKenzie, E. T. (1993). Effects of dorsal raphe nucleus stimulation on cerebral blood flow and flow-metabolism coupling in the conscious rat. *Neuroscience* **55,** 395–401.

Cullingford, T. E., Dolphin, C. T., Bhakoo, K. K., Peuchen, S., Canevari, L., and Clark, J. B. (1998). Molecular cloning of rat mitochondrial 3-hydroxy-3-methylglutaryl-CoA lyase and detection of the corresponding mRNA and of those encoding the remaining enzymes comprising the ketogenic 3-hydroxy-3-methylglutaryl-CoA cycle in central nervous system of suckling rats. *Biochem. J.* **329,** 373–381.

Dahlgren, N., Lindvall, O., Sakabe, T., Stenevi, U., and Siesjö, B. K. (1981). Cerebral blood flow and oxygen consumption in the rat brain after lesions of the noradrenergic locus coeruleus system. *Brain Res.* **209,** 11–23.

Daikhin, Y., and Yudkoff, M. (1998). Ketone bodies and brain glutamate and GABA metabolism. *Dev. Neurosci.* **20,** 358–364.

D'Alecy, L. G., Lundy, E. F., Kluger, M. J., Harker, C. T., LeMay, D. R., and Shlafer, M. (1990). Beta-hydroxybutyrate and response to hypoxia in the ground squirrel, *Spermophilus tridecimlineatus. Comp. Biochem. Physiol.* **96,** 189–193.

Dallman, M. F., Strack, A. M., Akana, S. F., Bradbury, M. J., Hanson, E. S., Scribner, K. A., and Smith, M. (1993). Feast and famine: Critical role of glucocorticoids with insulin in daily energy flow. *Front. Neuroendocrinol.* **14,** 303–347.

Dallman, M. F., Akana, S. F., Strack, A. M., Hanson, E. S., and Sebastian, R. J. (1995). The neural network that regulates energy balance is responsive to glucocorticoids and insulin and also regulates HPA axis responsivity at a site proximal to CRF neurons. *Ann. N.Y. Acad. Sci.* **771,** 730–742.

Dardzinski, B. J., Smith, S. L., Towfighi, J., Williams, G. D., Vannucci, R. C., and Smith, M. B. (2000). Increased plasma β-hydroxybutyrate, preserved cerebral energy metabolism, and amelioration of brain damage during neonatal hypoxia ischemia with dexamethasone pretreatment. *Pediatr. Res.* **48,** 248–255.

Dauphin, F., Lacombe, P., Sercombe, R., Hamel, E., and Seylaz, J. (1991). Hypercapnia and stimulation of the substantia innominata increase rat frontal cortical blood flow by different cholinergic mechanisms. *Brain Res.* **553,** 75–83.

Dauphin-Villemant, C., Leboulenger, F., and Vaudry, H. (1990). Adrenal actvity in the female lizard *Lacerty vivipara Jacquin* during artificial hibernation. *Gen. Comp. Endocrinol.* **79,** 201–214.

Davies, P., Anderton, B., Kirsch, J., Konnerth, A., Nitsch, R., and Sheetz, M. (1998). First one in, last one out: The role of GABAergic transmission in generation and degeneration. *Prog. Neurobiol.* **55,** 651–658.

DeAngelis, M. M., Hayes, R. L., and Lyeth, B. G. (1994). Traumatic brain injury causes a decrease in M2 muscarinic cholinergic receptor binding in rat brain. *Brain Res.* **653,** 39–44.

de Lacalle, S., Kulkarni, S., and Mufson, E. J. (1997). Plasticity of galaninergic fibers following neurotoxic damage within the rat basal forebrain: Initial observations. *Exp. Neurol.* **146,** 361–366.

De Micheli, E., and Soncrant, T. T. (1992). Age-dependent cerebral metabolic effects of unilateral nucleus basalis magnocellularis ablation in rats. *Neurobiol. Aging* **13,** 687–695.

deQuervain, D. J. F., Roozendaal, B., and McCaugh, J. L. (1998). Stress and glucocorticoids impair retrieval of long-term spatial memory. *Nature* **394,** 787–790.

De Sarno, P., and Giacobini, E. (1989). Modulation of acetylcholine release by nicotinic receptors in the rat brain. *J. Neurosci. Res.* **22,** 194–200.

Diaz, R., Brown, R. W., and Seckl, J. R. (1998). Distinct ontogeny of glucocorticoid and mineralocorticoid receptor and 11β-hydroxysteroid dehydrogenase types I and II mRNAs in the fetal rat brain suggest a complex control of glucocorticoid actions. *J. Neurosci.* **18,** 2570–2580.

Dickinson, C. J. (1996). Cerebral oxidative metabolism in hypertension. *Clin. Sci.* **91,** 539–550.

Dierks-Ventling, C., and Cone, A. L. (1971). Acetoacetyl-coenzyme A thiolase in brain, liver, and kidney during maturation of the rat. *Science* **172,** 380–382.

Diez-Guerra, J., Aragon, M. C., Gimenez, C., and Valdivieso, F. (1981). Effect of thyroid hormones on the malic enzyme activity in rat brain during development. *Dev. Neurosci.* **4,** 130–133.

Dikic, I., Schlessinger, J., and Lax, I. (1994). PC12 cells overexpressing the insulin receptor undergo insulin-dependent neuronal differentiation. *Curr. Biol.* **4,** 702–708.

Dinneen, S., Alzaid, A., Miles, J., and Rizza, R. (1993). Metabolic effects of the nocturnal rise in cortisol on carbohydrate metabolism in normal humans. *J. Clin. Invest.* **92,** 2283–2290.

Dixon, C. E., Liu, S. J., Jenkins, L. W., Bhattachargee, M., Whitson, J. S., Yang, K. Y., and Hayes, R. L. (1995). Time course of increased vulnerability of cholinergic neurotransmission following traumatic brain injury in the rat. *Behav. Brain Res.* **70,** 125–131.

Dombrowski, G. J., Jr., Swiatek, K. R., and Chao, K. L. (1989). Lactate, 3-hydroxybutyrate, and glucose as substrates for the early postnatal rat brain. *Neurochem. Res.* **14,** 667–675.

Doyle, P., Guillaume-Gentile, C., Rohner-Jeanrenaud, F., and Jeanrenaud, B. (1994). Effects of corticosterone administration on local cerebral glucose utilization of rats. *Brain Res.* **645,** 225–230.

Dringen, R., Wiesinger, H., and Hamprecht, B. (1993). Uptake of L-lactate by cultured rat brain neurons. *Neurosci. Lett.* **163,** 5–7.

Dringenberg, H. C. (2000). Alzheimer's disease: more than a "cholinergic disorder"—evidence that cholinergic-monoaminergic interactions contribute to EEG slowing and dementia. *Behav. Brain Res.* **115,** 235–249.

Dubal, D. B., Wilson, M. E., Shughrue, P. J., Merchenthaler, I., and Wise, P. M. (1999). Induction of galanin gene expression in estradiol-mediated neuroprotection against cerebral ischemia. *Soc. Neurosci. Abstr.* **25,** 1449.

Duelli, R., Staudt, R., Grünwald, F., and Kuschinsky, W. (1998). Increase of glucose transporter densities (GLUT1 and GLUT3) during chronic administration of nicotine in rat brain. *Brain Res.* **782,** 36–42.

Edmond, J. (1992). Energy metabolism in developing brain cells. *Can. J. Physiol. Pharmacol.* **70** (Suppl.), S118–S129.

Edmond, J., Auestad, N., Robbins, R. A., and Bergstrom, J. D. (1985). Ketone body metabolism in the neonate: Development and the effect of diet. *Fed. Proc.* **44,** 2359–2364.

Edvinsson, L., and Krause, D. N. (1979). Pharmacological characterization of GABA receptors mediating vasodilatation of cerebral arteries in vitro. *Brain Res.* **173,** 89–97.

Edvinsson, L., Larsson, B., and Skärby, T. (1980). Effects of the GABA receptor agonist muscimol on regional cerebral blood flow in the rat. *Brain Res.* **185,** 445–458.

Edvinsson, L., MacKenzie, E. T., and McCulloch, J. (1993). "Cerebral Blood Flow and Metabolism." Raven Press, New York.

Elias, A. N., Guich, S., and Wilson, A. F. (2000). Ketosis with enhanced GABAergic tone promotes physiological changes in transcendental meditation. *Med. Hypotheses* **54,** 660–662.

Elman, I., Adler, C. M., Malhotra, A. K., Bir, C., Pickar, D., and Breier, A. (1998). Effect of acute metabolic stress on pituitary-adrenal axis activation in patients with schizophrenia. *Am. J. Psychiatry* **155,** 979–981.

Endo, Y., Nishimura, J. I., Kobayashi, S., and Kimura, F. (1999). Chronic stress exposure influences local cerebral blood flow in the rat hippocampus. *Neuroscience* **93,** 551–555.

Erecinska, M., Nelson, D., Daikhin, Y., and Yudkoff, M. (1996). Regulation of GABA level in rat brain synaptosomes: fluxes through enzymes of the GABA shunt and effects of glutamate, calcium, and ketone bodies. *J. Neurochem.* **67,** 2325–2334.

Fanburg, B. L., and Lee, S. L. (1997). A new role for an old molecule: Serotonin as a mitogen. *Am. J. Physiol.* **272,** L795–L806.

Feldman, E. L., Sullivan, K. A., Kim, B., and Russell, J. W. (1997). Insulin-like growth factors regulate neuronal differentiation and survival. *Neurobiol. Dis.* **4,** 201–214.

Fery, F., Bourdoux, P., Christophe, J., and Balasse, E. O. (1982). Hormonal and metabolic changes induced by an isocaloric isoproteinic ketogenic diet in healthy subjects. *Diabetes Metab.* **8,** 299–305.

Fery, F., Plat, L., Melot, C., and Balasse, E. O. (1996). Role of fat-derived substrates in the regulation of gluconeogenesis during fasting. *Am. J. Physiol.* **270,** E822–E830.

Fillenz, M., Lowry, J. P., Boutelle, M. G., and Fray, A. E. (1999). The role of astrocytes and noradrenaline in neuronal glucose metabolism. *Acta Physiol. Scand.* **167,** 275–284.

Finkelstein, Y., Koffler, B., Rabey, J. M., and Gilad, G. M. (1985). Dynamics of cholinergic synaptic mechanisms in rat hippocampus after stress. *Brain Res.* **343,** 314–319.

Fleshner, M., Pugh, C. R., Tremblay, D., and Rudy, J. W. (1997). DHEA-S selectively impairs contextual fear conditioning: Support for the antiglucocorticoid hypothesis. *Behav. Neurosci.* **111,** 512–517.

Folli, F., Ghidella, S., Bonfanti, L., Kahn, C. R., and Merighi, A. (1996). The early intracellular signaling pathway for the insulin/insulin-like growth factor receptor family in the mammalian central nervous system. *Mol. Neurobiol.* **13,** 155–183.

Frame, S., and Cohen, P. (2001). GSK3 takes centre stage more than 20 years after its discovery. *Biochem. J.* **359,** 1–16.

Fray, A. E., Forsyth, R. J., Boutelle, M. G., and Fillenz, M. (1996). The mechanism controlling physiologically stimulated changes in rat brain glucose and lactate: a microdialysis study. *J. Physiol.* (*Lond.*) **496,** 49–57.

French, N., Lalies, M. D., Nutt, D. J., and Pratt, J. A. (1995). Idazoxan-induced reductions in cortical glucose use are accompanied by an increase in noradrenaline release: complementary [^{14}C]2-deoxyglucose and microdialysis studies. *Neuropharmacology* **34,** 605–613.

Frölich, L., Strauss, M., Kornhuber, J., Hoyer, S., Sorbi, S., Riederer, P., and Amaducci, L. (1990). Changes in pyruvate dehydrogenase complex (PDHc) activity and [^{3}H]QNB-receptor binding in rat brain subsequent to intracerebroventricular injection of bromopyruvate. *J. Neural. Transm. Park. Dis. Dement. Sect.* **2,** 169–178.

Frölich, L., Blum-Degen, D., Riederer, P., and Hoyer, S. (1999). A disturbance in the neuronal insulin receptor signal transduction in sporadic Alzheimer's disease. *Ann. N.Y. Acad. Sci.* **893,** 290–293.

Fujiwara, M., Muramatsu, I., and Shibata, S. (1975). Gamma-aminobutyric acid receptors on vascular smooth muscle of dog cerebral arteries. *Br. J. Pharmacol.* **55,** 561–562.

Fukuyama, H., Ouchi, Y., Matsuzaki, S., Ogawa, M., Yamauchi, H., Nagahama, Y., Kimura, J., Yonekura, Y., Shibasaki, H., and Tsukada, H. (1996). Focal cortical blood flow activation is regulated by intrinsic cortical cholinergic neurons. *Neuroimage* **3,** 195–201.

Furchgott, R. F., and Zawadzki, J. V. (1980). The obligatory role of endothelial cells in the relaxation of arterial smooth muscle by acetylcholine. *Nature* **288,** 373–376.

Fuxe, K., Li, X.-M., Bjelke, B., Hedlund, P. B., Biagini, G., and Agnati, L. F. (1994). Possible mechanisms for the powerful actions of neuropeptides. *Ann. N.Y. Acad. Sci.* **739,** 42–59.

Gancedo, J. M. (1998). Yeast carbon catabolite repression. Microbiol. *Mol. Biol. Rev.* **62,** 334–361.

Garthwaite, J., Charles, S. L., and Chess-Williams, R. (1988). Endothelium-derived relaxing factor release on activation of NMDA receptors suggests role as intercellular messenger in the brain. *Nature* **336,** 385–388.

Garthwaite, J., Garthwaite, G., Palmer, R. M. J., and Moncada, S. (1989). NMDA receptor activation induces nitric oxide synthesis from arginine in rat brain slices. *Eur. J. Pharmacol.* **172,** 413–416.

Ghajar, J. B., Gibson, G. E., and Duffy, T. E. (1985). Regional acetylcholine metabolism in brain during acute hypoglycemia and recovery. *J. Neurochem.* **44,** 94–98.

Gibson, G. E., and Blass, J. P. (1976). A relation between (NAD^{+})/(NADH) potentials and glucose utilization in rat brain slices. *J. Biol. Chem.* **251,** 4127–4130.

Gibson, G. E., Barclay, L., and Blass, J. (1982). The role of the cholinergic system in thiamin deficiency. *Ann. N.Y. Acad. Sci.* **378,** 382–403.

Gibson, G. E., and Mykytyn, V. (1988). An in vitro model of anoxic-induced damage in mouse brain. *Neurochem. Res.* **13,** 9–20.

Gibson, G. E., Manger, T., Toral-Barza, L., and Freeman, G. (1989). Cytosolic-free calcium and neurotransmitter release with decreased availability of glucose and oxygen. *Neurochem. Res.* **14,** 437–443.

Gilad, G. M. (1987). The stress-induced response of the septo-hippocampal cholinergic system. A vectorial outcome of psychoneuroendocrinological interactions. *Psychoneuroendocrinology* **12,** 167–184.

Gilad, G. M., Rabey, J. M., Tizabi, Y., and Gilad, V. H. (1987). Age-dependent loss and compensatory changes of septohippocampal cholinergic neurons in two rat strains differing in longevity and response to stress. *Brain Res.* **436,** 311–322.

Gillespie, C. M., Merkel, A. L., and Martin, A. A. (1997). Effects of insulin-like growth factor-I and LR(3)IGF-I on regional blood flow in normal rats. *J. Endocrinol.* **155,** 351–358.

Gispen-de Wied, C. C. (2000). Stress in schizophrenia: An integrative view. *Eur. J. Pharmacol.* **405,** 375–384.

Gitelman, D. R., and Prohovnik, I. (1992). Muscarinic and nicotinic contributions to cognitive function and cortical blood flow. *Neurobiol. Aging* **13,** 313–318.

Gjedde, A., and Crone, C. (1975). Induction processes in blood-brain transfer of ketone bodies during starvation. *Am. J. Physiol.* **229,** 1165–1169.

Gjedde, A., and Rasmussen, M. (1980). Pentobarbital anesthesia reduces blood-brain glucose transfer in the rat. *J. Neurochem.* **35,** 1382–1387.

Gluckman, P. D., Guan, J., Beilharz, E. J., Klempt, N. D., Klempt, M., Miller, O., Sirimanne, E., Dragunow, M., and Williams, C. E. (1993). The role of the insulin-like growth factor system in neuronal rescue. *Ann. N.Y. Acad. Sci.* **692,** 138–148.

Gonzalez, A. M., and Pazos, A. (1992). Affinity changes in muscarinic acetylcholine receptors in the rat brain following acute immobilization stress: an autoradiographic study. *Eur. J. Pharmacol.* **214,** 261–268.

Gonzalez-Nicolini, M. V., Orezzoli, A. A., and Villar, M. J. (1998). An immunohistochemical study of temperature-related changes in galanin and nitric oxide synthase immunoreactivity in the hypothalamus of the toad. *Gen. Comp. Endocrinol.* **110,** 175–181.

Gorman, L. K., Fu, K., Hovda, D. A., Becker, D. P., and Katayama, Y. (1989). Analysis of acetylcholine release following concussive brain injury in the rat. *J. Neurotrauma* **6,** 203.

Gottstein, U., Müller, W., Berghoff, W., Gärtner, H., and Held, K. (1971). Utilization of non-esterified fatty acids and ketone bodies in human brain. *Klein. Wschr.* **49,** 406–411.

Gramsbergen, J. B. P., Schmidt, W., Turski, W. A., and Schwarcz, R. (1992). Age-related changes in kynurenic acid production in rat brain. *Brain Res.* **588,** 1–5.

Grasby, P. M., Frith, C. D., Paulesu, E., Friston, K. J., Frackowiak, R. S., and Dolan, R. J. (1995). The effect of the muscarinic antagonist scopolamine on regional cerebral blood flow during the performance of a memory task. *Exp. Brain Res.* **104,** 337–348.

Greenberg, J. H., Greenberg, M. Z., Takahashi, K., and Reivich, M. (1999). Nitric oxide plays a role in activation-flow coupling in both awake and anesthetized rats. *Soc. Neurosci. Abstr.* **25,** 900.

Gruetter, R., Novotny, E. J., Boulware, S. D., Mason, G. F., Rothman, D. L., Shulman, G. I., Prichard, J. W., and Shulman, R. G. (1994). Localized ^{13}C NMR spectroscopy in the human brain of amino acid labeling from D-[1-^{13}C]glucose. *J. Neurochem.* **63,** 1377–1385.

Grünwald, F., Schröck, H., Theilen, H., Biber, A., and Kuschinsky, W. (1988). Local cerebral glucose utilization of the awake rat during chronic administration of nicotine. *Brain Res.* **456,** 350–356.

Grünwald, F., Schröck, H., and Kuschinsky, W. (1991). The influence of nicotine on local cerebral blood flow in rats. *Neurosci. Lett.* **124,** 108–110.

Guzman, M., and Blazquez, C. (2001). Is there an astrocyte-neuron ketone body shuttle? *Trends Endocrinol Metab.* **12,** 169–173.

Haba, K., Ogawa, N., Mizukawa, K., and Mori, A. (1991). Time course of changes in lipid peroxidation, pre- and postsynaptic cholinergic indices, NMDA receptor binding and neuronal death in the gerbil hippocampus following transient ischemia. *Brain Res.* **540,** 116–122.

Hardie, D. G. (1999). Roles of the AMP-activated/SNF1 protein kinase family in the response to cellular stress. *Biochem. Soc. Symp.* **64,** 13–27.

Hardie, D. G., and Carling, D. (1997). The AMP-activated protein kinase—fuel gauge of the mammalian cell? *Eur. J. Biochem.* **246,** 259–273.

Harik, S. I., LaManna, J. C., Light, A. I., and Rosenthal, M. (1979). Cerebral norepinephrine: Influence on cortical oxidative metabolism in situ. *Science* **206,** 69–71.

Harrell, L. E., and Davis, J. N. (1985). Cholinergic influences on hippocampal glucose metabolism. *Neuroscience* **15,** 359–369.

Harris, E. J., Booth, R., and Cooper, M. B. (1982). The effect of superoxide generation on the ability of mitochondria to take up and retain Ca^{2+}. *FEBS Lett.* **146,** 267–272.

Harwood, A. J. (2001). Regulation of GSK-3: A cellular multiprocessor. *Cell* **105,** 821–824.

Hass, R. (1994). Retrodifferentiation and cell death. *Crit. Rev. Oncog.* **5,** 359–371.

Hasselbalch, S. G., Knudsen, G. M., Jakobsen, J., Hageman, L. P., Holm, S., and Paulson, O. B. (1994). Brain metabolism during short-term starvation in humans. *J. Cereb. Blood Flow Metab.* **14,** 125–131.

Hatterer, J. A., Herbert, J., Hidaka, C., Roose, S. P., and Gorman, J. M. (1993). CSF transthyretin in patients with depression. *Am. J. Psychiatry* **150,** 813–815.

Hawkins, R. A., Williamson, D. H., and Krebs, H. A. (1971). Ketone-body utilization by adult and suckling rat brain in vivo. *Biochem. J.* **122,** 13–18.

Hawkins, R. A., and Biebuyck, J. F. (1979). Ketone bodies are selectively used by individual brain regions. *Science* **205,** 325–327.

Hawkins, R. A., Mans, A. M., and Davis, D. W. (1986). Regional ketone body utilization by rat brain in starvation and diabetes. *Am. J. Physiol.* **250,** E169–E178.

Hazlett, E. A., Buchsbaum, M. S., Byne, W., Wei, T. C., Spiegel-Cohen, J., Geneve, C., Kinderlehrer, R., Haznedar, M. M., Shihabuddin, L., and Siever, L. J. (1999). Three-dimensional analysis with MRI and PET of the size, shape, and function of the thalamus in the schizophrenia spectrum. *Am. J. Psychiatry* **156,** 1190–1199.

Heim, C., Owens, M. J., Plotsky, P. M., and Nemeroff, C. B. (1997). Persistent changes in corticotropin-releasing factor systems due to early life stress: relationship to the pathophysiology of major depression and post-traumatic stress disorder. *Psychopharmacol. Bull.* **33,** 185–192.

Heininger, K. (1999a). A unifying hypothesis of Alzheimer's disease. I. Ageing sets the stage. *Hum. Psychopharmacol. Clin. Exp.* **14,** 363–414.

Heininger, K. (1999b). A unifying hypothesis of Alzheimer's disease. II. Pathophysiological processes. *Hum. Psychopharmacol. Clin. Exp.* **14,** 525–581.

Heininger, K. (2000a). A unifying hypothesis of Alzheimer's disease. III. Risk factors. *Hum. Psychopharmacol. Clin. Exp.* **15,** 1–70.

Heininger, K. (2000b). A unifying hypothesis of Alzheimer's disease. IV. Causation and sequence of events. *Rev. Neurosci.* **11,** 213–328.

Heininger, K. (2001). The deprivation syndrome is the driving force of phylogeny, ontogeny and oncogeny. *Rev. Neurosci.* **12,** 217–287.

Heininger, K. (2002). Ageing is a deprivation syndrome driven by a germ-soma conflict. *Aging Res. Rev.* **1,** 481–536.

Heininger, K., Hindmarch, I., and deVry, J. (1998). Psychotropic potential of calcium antagonists. *In* "Calcium Ion Modulators. The New Wave of Psychotropic Drugs" (K. Inoue and Y. Watanabe, eds.), pp. 147–160. Harwood Academic Publishers, Amsterdam.

Henriksson, J. (1990). The possible role of skeletal muscle in the adaptation to periods of energy deficiency. *Eur. J. Clin. Nutr.* **44** Suppl. 1, 55–64.

Hillered, L., Ernster, L., and Siesjö, B. K. (1984). Influence of in vitro lactic acidosis and hypercapnia on respiratory activity of isolated rat brain mitochondria *J. Cereb. Blood Flow Metab.* **4,** 430–437.

Hiraide, A., Katayama, M., Sugimoto, H., Yoshioka, T., and Sugimoto, T. (1991). Effect of 3-hydroxybutyrate on posttraumatic metabolism in man. *Surgery* **109,** 176–181.

Hodes, J. E., Soncrant, T. T., Larson, D. M., Carlson, S. G., and Rapoport, S. I. (1985). Selective changes in local cerebral glucose utilization induced by phenobarbital in the rat. *Anesthesiology* **63,** 633–639.

Hodgkins, P. S., and Schwarcz, R. (1998). Metabolic control of kynurenic acid formation in the rat brain. *Dev. Neurosci.* **20,** 408–416.

Hökfelt, T., Broberger, C., Diez, M., Xu, Z. Q., Shi, T., Kopp, J., Zhang, X., Holmberg, K., Landry, M., and Koistinaho, J. (1999). Galanin and NPY, two peptides with multiple putative roles in the nervous system. *Horm. Metab. Res.* **31,** 330–334.

Holden, R. J. (1999). The role of brain insulin in the neurophysiology of serious mental disorders: Review. *Med. Hypotheses* **52,** 193–200.

Holsboer, F. (1995). Neuroendocrinology of mood disorders. *In* "Psychopharmacology: The Fourth Generation of Progress" (F. E. Bloom and D. J. Kupfer, eds.), pp. 957–969. Raven Press, New York.

Holsboer, F. (2001). Stress, hypercortisolism and corticosteroid receptors in depression: Implications for therapy. *J. Affect. Disord.* **62,** 77–91.

Hori, A., Tandon, P., Holmes, G. L., and Stafstrom, C. E. (1997). Ketogenic diet: Effects on expression of kindled seizures and behavior in adult rats. *Epilepsia* **38,** 750–758.

Horst, C., Rokos, H., and Seitz, H. J. (1989). Rapid stimulation of hepatic oxygen consumption by 3,5-di-iodo-L-thyronine. *Biochem. J.* **261,** 945–950.

Hortnagl, H., Berger, M. L., Havelec, L., and Hornykiewicz, O. (1993). Role of glucocorticoids in the cholinergic degeneration in rat hippocampus induced by ethylcholine aziridinium (AF64A). *J. Neurosci.* **13,** 2939–2945.

Hovda, D. A., Yoshino, A., Kawamata, T., Katayama, Y., Fineman, I., and Becker, D. P. (1990). The increase in local cerebral glucose utilization following fluid percussion brain injury is prevented with kynurenic acid and is associated with an increase in calcium. *Acta Neurochir. Suppl. (Wien)* **51,** 331–333.

Hoyer, S., Oesterreich, K., and Wagner, O. (1988). Glucose metabolism as the site of the primary abnormality in early-onset dementia of Alzheimer type? *J. Neurol.* **235,** 143–148.

Hoyer, S., Nitsch, R., and Oesterreich, K. (1991). Predominant abnormality in cerebral glucose utilization in late-onset dementia of the Alzheimer type: a cross-sectional comparison against advanced late-onset and incipient early-onset cases. *J. Neural Transm. Park. Dis. Dement. Sect.* **3,** 1–14.

Hu, Y., and Wilson, G. S. (1997). A temporary local energy pool coupled to neuronal activity: fluctuations of extracellular lactate levels in rat brain monitored with rapid-response enzyme-based sensor. *J. Neurochem.* **69,** 1484–1490.

Humphries, K. M., Yoo, Y., and Szweda, L. I. (1998). Inhibition of NADH-linked mitochondrial respiration by 4-hydroxy-2-nonenal. *Biochemistry* **37,** 552–557.

Husain, K., and Ansari, R. A. (1988). Influence of cholinergic and adrenergic blocking drugs on hyperglycemia and brain glycogenolysis in diazinon-treated animals. *Can. J. Physiol. Pharmacol.* **66,** 1144–1147.

Huttenlocher, P. R. (1976). Ketonemia and seizures: metabolic and anticonvulsant effects of two ketogenic diets in childhood epilepsy. *Pediatr. Res.* **10,** 536–540.

Iadecola, C., Li, J., Xu, S., and Yang, G. (1996). Neural mechanisms of blood flow regulation during synaptic activity in cerebellar cortex. *J. Neurophysiol.* **75,** 940–950.

Ignarro, L. J., Buga, G. M., Wood, K. S., Byrns, R. E., and Chaudhuri, G. (1987). Endothelium-derived relaxing factor produced and released from artery and vein is nitric oxide. *Proc. Natl. Acad. Sci. USA* **84,** 9265–9269.

Ingenbleek, Y., and Bernstein, L. (1999). The stressful condition as a nutritionally dependent adaptive dichotomy. *Nutrition* **15,** 305–320.

Inoue, M., McHugh, M., and Pappius, H. M. (1991). The effect of α-adrenergic receptor blockers prazosin and yohimbine on cerebral metabolism and biogenic amine content of traumatized brain. *J. Cereb. Blood Flow Metab.* **11,** 242–252.

Ishimaru, H., Takahashi, A., Ikarashi, Y., and Maruyama, Y. (1995). Pentobarbital protects against CA1 pyramidal cell death but not dysfunction of hippocampal cholinergic neurons following transient ischemia. *Brain Res.* **673,** 112–118.

Ishiwata, K., Ishii, K., Ogawa, K., Nozaki, T., and Senda, M. (1996). A brain uptake study of [1-^{11}C]hexanoate in the mouse: the effect of hypoxia, starvation and substrate competition. *Ann. Nucl. Med.* **10,** 265–270.

Izumi, Y., Ishii, K., Katsuki, H., Benz, A. M., and Zorumski, C. F. (1998). β-hydroxybutyrate fuels synaptic function during development. Histological and physiological evidence in rat hippocampal slices. *J. Clin. Invest.* **101,** 1121–1132.

Jani, M. S., Telang, S. D., and Katyare, S. S. (1991). Effect of corticosterone treatment on energy metabolism in rat liver mitochondria. *J. Steroid Biochem. Mol. Biol.* **38,** 587–591.

Jansen, L. M., Gispen-de Wied, C. C., and Kahn, R. S. (2000). Selective impairments in the stress response in schizophrenic patients. *Psychopharmacology (Berlin)* **149,** 319–325.

Jehan, F., Neveu, I., Naveilhan, P., Brachet, P., and Wion, D. (1993). Complex interactions among second messenger pathways, steroid hormones, and protooncogenes of the Fos and Jun families converge in the regulation of the nerve growth factor gene. *J. Neurochem.* **60,** 1843–1853.

Johnson, J. K., and Berman, N. E. (1996). A transient phase of cell death in the developing medial forebrain of the perinatal ferret. *Brain Res. Dev. Brain Res.* **94,** 159–165.

Johnson, W. A., and Weiner, M. W. (1978). Protective effects of ketogenic diets on signs of hypoglycemia. *Diabetes* **27,** 1087–1091.

Johnston, B. M., Mallard, E. C., Williams, C. E., and Gluckman, P. D. (1996). Insulin-like growth factor-I is a potent neuronal rescue agent after hypoxic-ischemic injury in fetal lambs. *J. Clin. Invest.* **97,** 300–308.

Johnston, D. G., Gill, A., Orskov, H., Batstone, G. F., and Alberti, K. G. (1982). Metabolic effects of cortisol in man—studies with somatostatin. *Metabolism* **31,** 312–317.

Johnston, M. (1999). Feasting, fasting and fermenting. *Trends Genet.* **15,** 29–33.

Justice, A., Feldman, S. M., and Brown, L. L. (1989). The nucleus locus coeruleus modulates local cerebral glucose utilization during noise stress in rats. *Brain Res.* **490,** 73–84.

Kakihana, R., Moore, J. A., and Butte, J. C. (1980). Brain corticosterone in lactating rats: possible relation to the attenuation of the pituitary-adrenocortical response to stress. *Res. Commun. Chem. Pathol. Pharmacol.* **27,** 105–117.

Kalimi, M., Shafagoj, Y., Loria, R., Padgett, D., and Regelson, W. (1994). Anti-glucocorticoid effects of dehydroepiandrosterone (DHEA). *Mol. Cell. Biochem.* **131,** 99–104.

Kammula, R. G. (1976). Metabolism of ketone bodies by ovine brain in vivo. *Am. J. Physiol.* **231,** 1490–1494.

Kask, K., Berthold, M., and Bartfai, T. (1997). Galanin receptors: involvement in feeding, pain, depression and Alzheimer's disease. *Life Sci.* **60,** 1523–1533.

Kato, H., Araki, T., Murase, K., and Kogure, K. (1992). Induction of tolerance to ischemia: alterations in second-messenger systems in the gerbil hippocampus. *Brain Res. Bull.* **29,** 559–565.

Katyare, S. S., Bangur, C. S., and Howland, J. L. (1994). Is respiratory activity in the brain mitochondria responsive to thyroid hormone action? A critical re-evaluation. *Biochem. J.* **302,** 857–860.

Kaufer, D., and Soreq, H. (1999). Tracking cholinergic pathways from psychological and chemical stressors to variable neurodeterioration paradigms. *Curr. Opin. Neurol.* **12,** 739–743.

Kaufer, D., Friedman, A., Seidman, S., and Soreq, H. (1998a). Acute stress facilitates long-lasting changes in cholinergic gene expression. *Nature* **393,** 373–377.

Kaufer, D., Friedman, A., Pavlovsky, L., and Soreq, H. (1998b). Stress-induced transcriptional modulation elicits weeks-long changes in cholinergic transmission. *Soc. Neurosci. Abstr.* **24,** 1434.

Kawamata, T., Katayama, Y., Hovda, D. A., Yoshino, A., and Becker, D. P. (1992). Administration of excitatory amino acid antagonists via microdialysis attenuates the increase in glucose utilization seen following concussive brain injury. *J. Cereb. Blood Flow Metab.* **12,** 12–24.

Kelley, D. E., Mokan, M., Simoneau, J. A., and Mandarino, L. J. (1993). Interaction between glucose and free fatty acid metabolism in human skeletal muscle. *J. Clin. Invest.* **92,** 91–98.

Kelly, P. A. T., and McCulloch, J. (1983). The effect of the GABA agonist muscimol upon the relationship between local cerebral blood flow and glucose utilization. *Brain Res.* **258,** 338–342.

Kelly, P. A. T., Faulkner, A. J., and Burro, A. P. (1989). The effects of the GABA agonist muscimol upon blood flow in different vascular territories of the rat cortex. *J. Cereb. Blood Flow Metab.* **9,** 754–758.

Kemp, B. E., Mitchelhill, K. I., Stapleton, D., Michell, B. J., Chen, Z. P., and Witters, L. A. (1999). Dealing with energy demand: the AMP-activated protein kinase. *Trends Biochem. Sci.* **24,** 22–25.

Kennedy, S. H., Evans, K. R., Kruger, S., Mayberg, H. S., Meyer, J. H., McCann, S., Arifuzzman, A. I., Houle, S., and Vaccarino, F. J. (2001). Changes in regional brain glucose metabolism measured with positron emission tomography after paroxetine treatment of major depression. *Am. J. Psychiatry* **158,** 899–905.

Kim, J. K., Wi, J. K., and Youn, J. H. (1996). Metabolic impairment precedes insulin resistance in skeletal muscle during high-fat feeding in rats. *Diabetes* **45,** 651–658.

King, P., Parkin, H., MacDonald, I. A., Barber, C., and Tattersall, R. B. (1997). The effect of intravenous lactate on cerebral function during hypoglycaemia. *Diabetes Med.* **14,** 19–28.

Kinoshita, A., Yamada, K., Mushiroi, T., and Hayakawa, T. (1992). Cholinergic deafferentation prevents delayed neuronal death of the hippocampal CA1 pyramidal neurons after transient forebrain ischemia. *Neurol. Res.* **14,** 340–344.

Kirsch, J. R., and D'Alecy, L. G. (1984). Hypoxia induced preferential ketone utilization by rat brain slices. *Stroke* **15,** 319–323.

Kirschbaum, C., Platte, P., Pirke, K.-M., and Hellhammer, D. H. (1995). Adrenocortical activation following stressful exercise: further evidence for attenuated free cortisol responses in women using oral contraceptives. *Stress Med.* **12,** 137–143.

Klee, C. B., and Sokoloff, L. (1967). Changes in D^{-}-β-hydroxybutyric dehydrogenase activity during brain maturation in the rat. *J. Biol. Chem.* **242,** 3880–3883.

Klein, S., and Wolfe, R. R. (1992). Carbohydrate restriction regulates the adaptive response to fasting. *Am. J. Physiol.* **262,** E631–E636.

Kopf, S. R., and Baratti, C. M. (1995). The impairment of retention induced by insulin in mice may be mediated by a reduction in central cholinergic activity. *Neurobiol. Learn. Mem.* **63,** 220–228.

Kopf, S. R., and Baratti, C. M. (1996). Effects of posttraining administration of glucose on retention of a habituation response in mice: Participation of a central cholinergic mechanism. *Neurobiol. Learn. Mem.* **65,** 253–260.

Kopf, S. R., and Baratti, C. M. (1999). Effects of posttraining administration of insulin on retention of a habituation response in mice: Participation of a central cholinergic mechanism. *Neurobiol. Learn. Mem.* **71,** 50–61.

Kostanyan, A., and Nazaryan, K. (1992). Rat brain glycolysis regulation by estradiol-17β. *Biochim. Biophys. Acta* **1133,** 301–306.

Kraus, H., Schlenker, S., and Schwedesky, D. (1974). Developmental changes of cerebral ketone body utilization in human infants. *Hoppe Seylers Z. Physiol. Chem.* **355,** 164–170.

Kumagae, Y., and Matsui, Y. (1991). Output, tissue levels, and synthesis of acetylcholine during and after transient forebrain ischemia in the rat. *J. Neurochem.* **56,** 1169–1173.

Kurosawa, M., Sato, A., and Sato, Y. (1989). Stimulation of the nucleus basalis of Meynert increases acetylcholine release in the cerebral cortex in rats. *Neurosci. Lett.* **98,** 45–50.

Kuschinsky, W., Wahl, M., and Neiss, A. (1974). Evidence for cholinergic dilatatory receptors in pial arteries of cats. *Pflügers Arch.* **347,** 199–208.

Lacombe, P., Sercombe, R., Verrecchia, C., Philipson, V., MacKenzie, E. T., and Seylaz, J. (1989). Cortical blood flow increases induced by stimulation of the substantia innominata in the unanesthetized rat. *Brain Res.* **491,** 1–14.

Lacombe, P., Sercombe, R., Vaucher, E., and Seylaz, J. (1997). Reduced cortical vasodilatory response to stimulation of the nucleus basalis of Meynert in the aged rat and evidence for a control of the cerebral circulation. *Ann. N.Y. Acad. Sci.* **826,** 410–415.

LaManna, J. C., Harik, S. I., Light, A. I., and Rosenthal, M. (1981). Norepinephrine depletion alters cerebral oxidative metabolism in the "active" state. *Brain Res.* **204,** 87–101.

Lannert, H., Wirtz, P., Schuhmann, V., and Galmbacher, R. (1988). Effects of estradiol-17β on learning, memory and cerebral energy metabolism in male rats after intracerebroventricular administration of streptozotocin. *J. Neural Transm.* **105,** 1045–1063.

Lauder, J. M. (1993). Neurotransmitters as growth regulatory signals: role of receptors and second messengers. *Trends Neurosci.* **16,** 233–240.

Laughlin, M. R., and Heineman, F. W. (1994). The relationship between phosphorylation potential and redox state in the isolated working rabbit heart. *J. Mol. Cell. Cardiol.* **26,** 1525–1536.

Laurie, D. J., and Pratt, J. A. (1989). Local cerebral glucose utilization following subacute and chronic diazepam pretreatment: differential tolerance. *Brain Res.* **504,** 101–111.

Lebedev, E. D. (1978). 11-hydroxycorticosteroid content in the peripheral blood plasma and in the cerebrospinal fluid in acute disorders of the cerebral circulation. *Zh. Nevropatol. Psikhiatr.* **78,** 66–72.

Lefresne, P., Guyenet, P., and Glowinski, J. (1973). Acetylcholine synthesis from (2-^{14}C) pyruvate in rat striatal slices. *J. Neurochem.* **20,** 1083–1097.

Lehninger, A. L., Vercesi, A., and Bababunmi, E. A. (1978). Regulation of Ca^{2+} release from mitochondria by the oxidation-reduction state of pyridine nucleotides. *Proc. Natl. Acad. Sci. USA* **75,** 1690–1694.

Leibowitz, S. F. (1998). Differential functions of hypothalamic galanin cell groups in the regulation of eating and body weight. *Ann. N.Y. Acad. Sci.* **863,** 206–220.

Leino, R. L., Gerhart, D. Z., and Drewes, L. R. (1999). Monocarboxylate transporter (MCT1) abundance in brains of suckling and adult rats: A quantitative electron microscopic immunogold study. *Dev. Brain Res.* **113,** 47–54.

Leloup, C., Michaelson, D. M., Fisher, A., Hartmann, T., Beyreuther, K., and Stein, R. (2000). M1 muscarinic receptors block caspase activation by phosphoinositide 3-kinase- and MAPK/ERK-independent pathways. *Cell Death Differ.* **7,** 825–833.

Leonard, J. R., Maris, D. O., and Grady, M. S. (1994). Fluid percussion injury causes loss of forebrain choline acetyltransferase and nerve growth factor receptor immunoreactive cells in the rat. *J. Neurotrauma* **11,** 379–392.

Leonard, J. R., Grady, M. S., Lee, M. E., Paz, J. C., and Westrum, L. E. (1997). Fluid percussion injury causes disruption of the septohippocampal pathway in the rat. *Exp. Neurol.* **143,** 177–187.

Leuchter, A. F., Uijtdehaage, S. H. J., Cook, I. A., O'Hara, R., and Mandelkern, M. (1999). Relationship between brain electrical activity and cortical perfusion in normal subjects. *Psychiatry Res.* **90,** 125–140.

Levant, B., Cross, R. S., and Pazdernik, T. L. (1998). Alterations in local cerebral glucose utilization produced by D-3 dopamine receptor-selective doses of 7-OH-DPAT and nafadotride. *Brain Res.* **812,** 193–199.

Li, M. D., Kane, J. K., Matta, S. G., Blaner, W. S., and Sharp, B. M. (2000). Nicotine enhances the biosynthesis and secretion of transthyretin from the choroid plexus in rats: implications for β-amyloid formation. *J. Neurosci.* **20,** 1318–1323.

Liberini, P., Pioro, E. P., Maysinger, D., and Cuello, A. C. (1994). Neocortical infarction in subhuman primates leads to restricted morphological damage of the cholinergic neurons in the nucleus basalis of Meynert. *Brain Res.* **648,** 1–8.

Lin, M. T., and Shian, L. R. (1991). Stimulation of 5-hydroxytryptamine nerve cells in dorsal and median raphe nuclei elevates blood glucose in rats. *Pflügers Arch.* **417,** 441–445.

Liu, D., Diorio, J., Tannenbaum, B., Caldji, C., Francis, D., Freedman, A., Sharma, S., Pearson, D., Plotsky, P. M., and Meaney, M. J. (1997). Maternal care, hippocampal glucocorticoid receptors, and hypothalamic-pituitary-adrenal responses to stress. *Science* **277,** 1659–1662.

Liu, J., and Mori, A. (1994). Involvement of reactive oxygen species in emotional stress: A hypothesis based on the immobilization stress-induced oxidative damage and antioxidant defense changes in rat brain, and the effect of antioxidant treatment with reduced glutathione. *Int. J. Stress Manag.* **1,** 249–263.

Liu, J. K., and Mori, A. (1999). Stress, aging, and brain oxidative damage. *Neurochem. Res.* **24,** 1479–1497.

Liu, J. L., and Patel, Y. C. (1995). Glucocorticoids inhibit somatostatin gene expression through accelerated degradation of somatostatin messenger ribonucleic acid in human thyroid medullary carcinoma (TT) cells. *Endocrinology* **136,** 2389–2396.

Liu, S. L., Lyeth, B. G., and Hamm, R. J. (1994). Protective effect of galanin on behavioral deficits in experimental traumatic brain injury. *J. Neurotrauma* **11,** 73–82.

Liuzzi, A., Pocchiari, F., and Angeletti, P. U. (1968). Glucose metabolism in embryonic ganglia: effect of nerve growth factor [NGF] and insulin. *Brain Res.* **7,** 452–454.

London, E. D., Margolin, R. A., Duara, R., Holloway, H. W., Robertson-Tchabo, E. A., Cutler, N. R., and Rapoport, S. I. (1986). Effects of fasting on ketone body concentrations in healthy men of different ages. *J. Gerontol.* **41,** 599–604.

London, E. D., Connolly, R. J., Szikszay, M., Wamsley, J. K., and Dam, M. (1988). Effects of nicotine on local cerebral glucose utilization in the rat. *J. Neurosci.* **8,** 3920–3928.

Lopes-Cardozo, M., and Klein, W. (1982). Ketone-body utilization by homogenates of adult rat brain. *Neurochem. Res.* **7,** 687–703.

Lopes-Cardozo, M., Larsson, O. M., and Schousboe, A. (1986). Acetoacetate and glucose as lipid precursors and energy substrates in primary cultures of astrocytes and neurons from mouse cerebral cortex. *J. Neurochem.* **46,** 773–778.

Lopez, J. F., Chalmers, D. T., Little, K. Y., and Watson, S. J. (1998). Regulation of serotonin$_{1A}$, glucocorticoid, and mineralocorticoid receptor in rat and human hippocampus: implications for the neurobiology of depression. *Biol. Psychiatry* **43,** 547–573.

Losada, M. E. (1988). Changes in the central GABAergic system after acute treatment with corticosterone. *Naunyn-Schmiedebergs Arch. Pharmakol.* **337,** 669–674.

Lowell, B. B., and Goodman, M. N. (1987). Protein sparing in skeletal muscle during prolonged starvation. Dependence on lipid fuel availability. *Diabetes* **36,** 14–19.

Lutz, P. L. (1992). Mechanisms for anoxic survival in the vertebrate brain. *Annu. Rev. Physiol.* **54,** 601–618.

Lyeth, B. G., Dixon, C. E., Jenkins, L. W., Hamm, R. J., Alberico, A., Young, H. F., Stonnington, H. H., and Hayes, R. L. (1988). Effects of scopolamine treatment on long-term behavioral deficits following concussive brain injury to the rat. *Brain Res.* **452,** 39–48.

Lyons, D. M., Lopez, J. M., Yang, C., and Schatzberg, A. F. (2000). Stress-level cortisol treatment impairs inhibitory control of behavior in monkeys. *J. Neurosci.* **20,** 7816–7821.

Maccari, S., Piazza, P. V., Kabbaj, M., Barbazanges, A., Simon, H., and Lemoal, M. (1995). Adoption reverses the long-term impairment in glucocorticoid feedback induced by prenatal stress. *J. Neurosci.* **15,** 110–116.

Madsen, P. L., Linde, R., Hasselbalch, S. G., Paulson, O. B., and Lassen, N. A. (1998). Activation-induced resetting of cerebral oxygen and glucose uptake in the rat. *J. Cereb. Blood Flow Metab.* **18,** 742–748.

Maes, M., De Vos, N., Pioli, R., Demedts, P., Wauters, A., Neels, H., and Christophe, A. (2000). Lower serum vitamin estrogens concentrations in major depression. Another marker of lowered antioxidant defenses in that illness. *J. Affect. Disord.* **58,** 241–246.

Magistretti, P. J., Pellerin, L., and Martin, J.-L. (1995). Brain energy metabolism. An integrated perspective. *In* "Psychopharmacology: The Fourth Generation of Progress" (F. E. Bloom and D. J. Kupfer, eds.), pp. 657–670. Raven Press, New York.

Mahadik, S. P., Evans, D., and Lal, H. (2001). Oxidative stress and role of antioxidant and omega-3 essential fatty acid supplementation in schizophrenia. *Prog. Neuro-Psychopharmacol. Biol. Psychiatry* **25,** 463–493.

Malendowicz, L. K., Nussdorfer, G. G., Nowak, K. W., and Mazzocchi, G. (1994). The possible involvement of galanin in the modulation of the function of rat pituitary-adrenocortical axis under basal and stressful conditions. *Endocr. Res.* **20,** 307–317.

Mamelak, M. (1989). Gammahydroxybutyrate: an endogenous regulator of energy metabolism. *Neurosci. Biobehav. Rev.* **13,** 187–198.

Mann, J. J., Malone, K. M., Diehl, D. J., Perel, J., Nichols, T. E., and Mintun, M. A. (1996). Positron emission tomographic imaging of serotonin activation effects on prefrontal cortex in healthy volunteers. *J. Cereb. Blood Flow Metab.* **16,** 418–426.

Mans, A. M., Davis, D. W., and Hawkins, R. A. (1987). Regional brain glucose use in unstressed rats after two days of starvation. *Metab. Brain Dis.* **2,** 213–221.

Mansell, P. I., and Macdonald, I. A. (1990). The effect of starvation on insulin-induced glucose disposal and thermogenesis in humans. *Metabolism* **39,** 502–510.

Marangell, L. B., Ketter, T. A., George, M. S., Pazzaglia, P. J., Callahan, A. M., Parekh, P., Andreason, P. J., Horwitz, B., Herscovitch, P., and Post, R. M. (1997). Inverse relationship of peripheral thyrotropin-stimulating hormone levels to brain activity in mood disorders. *Am. J. Psychiatry* **154,** 224–230.

Marie, C., Bralet, A. M., Gueldry, S., and Bralet, J. (1990). Fasting prior to transient cerebral ischemia reduces delayed neuronal necrosis. *Metab. Brain Dis.* **5,** 65–75.

Marks, J. L., and Waite, K. (1997). Intracerebroventricular neuropeptide Y acutely influences glucose metabolism and insulin sensitivity in the rat. *J. Neuroendocrinol.* **9,** 99–103.

Marshall, J. J., Wie, E. P., and Kontos, H. A. (1988). Independent blockade of cerebral vasodilation from acetylcholine and nitric oxide. *Am. J. Physiol.* **255,** H847–H854.

Martens, M. E., Peterson, P. L., and Lee, C. P. (1991). In vitro effects of glucocorticoid on mitochondrial energy metabolism. *Biochim. Biophys. Acta* **1058,** 152–160.

Mattson, M. P., and Cheng, B. (1993). Growth factors protect neurons against excitotoxic/ischemic damage by stabilizing calcium homeostasis. *Stroke* **24,** I 136–I 140.

Mattson, M. P., Robinson, N., and Guo, Q. (1997). Estrogens stabilize mitochondrial function and protect neural cells against the pro-apoptotic action of mutant presenilin-1. *Neuroreport* **8,** 3817–3812.

Mayor, F., Veloso, D., and Williamson, D. H. (1967). Effects of nicotinic acid on the acetoacetate and 3-hydroxybutyrate concentrations of rat blood and liver. *Biochem. J.* **104,** 57P.

Mayor, F., and Cuezva, J. M. (1985). Hormonal and metabolic changes in the perinatal period. *Biol. Neonate* **48,** 185–196.

Mazarati, A. M., Liu, H. T., Soomets, U., Sankar, R., Shin, D., Katsumori, H., Langel, U., and Wasterlain, C. G. (1998). Galanin modulation of seizures and seizure modulation of hippocampal galanin in animal models of status epilepticus. *J. Neurosci.* **18,** 10070–10077.

McCormack, J. G., and Denton, R. M. (1986). Ca^{2+} as second messenger within mitochondria. *Trends Biol. Sci.* **11,** 258–262.

McEwen, B. S., Gould, E. A., and Sakai, R. R. (1992). The vulnerability of the hippocampus to protective and destructive effects of glucocorticoids in relation to stress. *Br. J. Psychiatry* **15** Suppl, 18–23.

McEwen, B. S., and Sapolsky, R. M. (1995). Stress and cognitive function. *Curr. Opin. Neurobiol.* **5,** 205–216.

McEwen, B. S., and Margarinos, A. M. (1997). Stress effects on morphology and function of the hippocampus. *Ann. N.Y. Acad. Sci.* **821,** 271–284.

McGarry, J. D., and Brown, N. F. (1997). The mitochondrial carnitine palmitoyltransferase system. From concept to molecular analysis. *Eur. J. Biochem.* **244,** 1–14.

McIntosh, M. K., Pan, J.-S., and Berdanier, C. D. (1993). In vitro studies on the effects of dehydroepiandrosterone and corticosterone on hepatic steroid receptor binding and mitochondrial respiration. *Comp. Biochem. Physiol.* **104A,** 147–153.

McKenna, M. C., Tildon, J. T., Stevenson, J. H., Boatright, R., and Huang, S. (1993). Regulation of energy metabolism in synaptic terminals and cultured rat brain astrocytes: differences revealed using aminooxyacetate. *Dev. Neurosci.* **15,** 320–329.

Meguro, K., Yamaguchi, S., Itoh, M., Fujiwara, T., and Yamadori, A. (1997). Striatal dopamine metabolism correlated with frontotemporal glucose utilization in Alzheimer's disease: a double-tracer PET study. *Neurology* **49,** 941–945.

Menendez, J. A., Atrens, D. M., and Leibowitz, S. F. (1992). Metabolic effects of galanin injections into the paraventricular nucleus of the hypothalamus. *Peptides* **13,** 323–327.

Meyer, E. M., Judkins, J. H., Momol, A. E., and Hardwick, E. O. (1994). Effects of peroxidation and aging on rat neocortical ACh-release and protein kinase C. *Neurobiol. Aging* **15,** 63–67.

Micheau, J., Messier, C., and Jaffard, R. (1995). Glucose enhancement of scopolamine-induced increase of hippocampal high-affinity uptake in mice: Relation to pasma glucose levels. *Brain Res.* **685,** 99–104.

Michel, P. P., and Agid, Y. (1995). Death of septal cholinergic neurons produced by chronic exposure to glutamate is prevented by the noncompetitive NMDA receptor/channel antagonist, MK-801: role of nerve growth factor and nitric oxide. *J. Neurosci. Res.* **40,** 764–775.

Miller, A. L., and Corddry, D. H. (1981). Brain carbohydrate metabolism in developing rats during hypercapnia. *J. Neurochem.* **36,** 1202–1210.

Miller, C. (2000). Ion channel surprises: prokaryotes do it again. *Neuron* **25,** 7–9.

Miller, R. E. (1981). Pancreatic neuroendocrinology: peripheral neural mechanisms in the regulation of the Islets of Langerhans. *Endocr. Rev.* **2,** 471–494.

Mishunina, T. M., and Kononenko, V. I. (1990). Hormonal control of metabolism of (γ-aminobutyric acid in the rat hypothalamus and hippocampus. *Ukr. Biokhim. Zh.* **62,** 71–79.

Mizoguchi, K., Kunishita, T., Chui, D. H., and Tabira, T. (1992). Stress induces neuronal death in the hippocampus of castrated rats. *Neurosci. Lett.* **138,** 157–160.

Mohammed, A. H., Henriksson, B. G., Söderström, S., Ebental, T., Olsson, T., and Seckl, J. R. (1993). Environmental influences on the central nervous system and their implications for the aging rat. *Behav. Brain Res.* **57,** 183–191.

Mohan, P. F., and Cleary, M. P. (1991). Short-term effects of dehydroepiandrosterone treatment in rats on mitochondrial respiration. *J. Nutr.* **121,** 240–250.

Moisan, M. P., Edwards, C. R., and Seckl, J. R. (1992). Ontogeny of 11β-hydroxysteroid dehydrogenase in rat brain and kidney. *Endocrinology* **130,** 400–404.

Mooradian, A. D., Girgis, W., and Shah, G. N. (1997). Thyroid hormone-induced GLUT-1 expression in rat cerebral tissue: effect of age. *Brain Res.* **747,** 144–146.

Morelli, A., Grasso, M., and Calissano, P. (1986). Effect of nerve growth factor on glucose utilization and nucleotide content of pheochromocytoma cells (clone PC12). *J. Neurochem.* **47,** 375–381.

Moroni, F. (1999). Tryptophan metabolism and brain function: focus on kynurenine and other indole metabolites. *Eur. J. Pharmacol.* **375,** 87–100.

Moroni, F., Russi, P., Lombardi, G., Beni, M., and Carla, V. (1988). Kynurenic acid in the mammalian brain increases during the aging process or after the administration of its precursors. *In* "Frontiers in Excitatory Amino Acid Research" (E. A. Cavalheiro, J. Lehmann, and L. Turski, eds.), pp. 629–636. AR Liss, New York.

Muck-Seler, D., Pivac, N., Jakovljevic, M., and Brzovic, Z. (1999). Platelet serotonin, plasma cortisol, and dexamethasone suppression test in schizophrenic patients. *Biol. Psychiatry* **45,** 1433–1439.

Murakami, K., Nakagawa, T., Shozu, M., Uchide, K., Koike, K., and Inoue, M. (1999). Changes with aging of steroidal levels in the cerebrospinal fluid of women. *Maturitas* **33,** 71–80.

Murdoch, I., Perry, E. K., Court, J. A., Graham, D. I., and Dewar, D. (1998). Cortical cholinergic dysfunction after human head injury. *J. Neurotrauma* **15,** 295–305.

Murphy, B. E. (1981). Ontogeny of cortisol-cortisone interconversion in human tissues: a role for cortisone in human fetal development. *J. Steroid Biochem.* **14,** 811–817.

Nägga, K., Bogdanovic, N., and Marcusson, J. (1999). GABA transporters (GAT-1) in Alzheimer's disease. *J. Neural Transm.* **106,** 1141–1149.

Nakashima, N., Haji, M., Sakai, Y., Ono, Y., Umeda, F., and Nawata, H. (1995). Effect of dehydroepiandrosterone on glucose uptake in cultured human fibroblasts. *Metabolism* **44,** 543–548.

Nehlig, A. (1999). Age-dependent pathways of brain energy metabolism: the suckling rat, a natural model of the ketogenic diet. *Epilepsy Res.* **37,** 211–221.

Nehlig, A., Porrino, L. J., Crane, A. M., and Sokoloff, L. (1985). Local cerebral glucose utilization in normal female rats: variations during the estrous cycle and comparison with males. *J. Cereb. Blood Flow Metab.* **5,** 393–400.

Nelson, B. D. (1990). Thyroid hormone regulation of mitochondrial function. Comments on the mechanism of signal transduction. *Biochim. Biophys. Acta* **1018,** 275–277.

Nemeroff, C. B., and Krishnan, K. R. R. (1992). Neuroendocrine alterations in psychiatric disorders. *In* "Neuroendocrinology" (C. B. Nemeroff, ed.), pp. 413–441. CRC Press, Boca Raton.

Nesse, R. M. (1999). Proximate and evolutionary studies of anxiety, stress and depression: synergy at the interface. *Neurosci. Biobehav. Rev.* **23,** 895–903.

Ni, J. W., Matsumoto, K., Li, H. B., Murakami, Y., and Watanabe, H. (1995). Neuronal damage and decrease of central acetylcholine level following permanent occlusion of bilateral common carotid arteries in rat. *Brain Res.* **673,** 290–296.

Nijjar, M., and Belgrave, R. L. (1997). Regulation of Ca^{2+} homeostasis by glucose metabolism in rat brain. *Mol. Cell. Biochem.* **176,** 317–326.

Nilsson, G. E., and Lutz, P. L. (1993). Role of GABA in hypoxia tolerance, metabolic depression and hibernation—possible links to neurotransmitter evolution. *Comp. Biochem. Physiol. C* **105,** 329–336.

Nobler, M. S., Pelton, G. H., and Sackeim, H. A. (1999). Cerebral blood flow and metabolism in late-life depression and dementia. *J. Geriat. Psychiatr. Neurol.* **12,** 118–127.

Nonogaki, K., and Iguchi, A. (1997). Role of central neural mechanisms in the regulation of hepatic glucose metabolism. *Life Sci.* **60,** 797–807.

Nowicki, J. P., Assumel-Lurdin, C., Duverger, D., and MacKenzie, E. T. (1988). Temporal evolution of regional energy metabolism following focal cerebral ischemia in the rat. *J. Cereb. Blood Flow Metab.* **8,** 462–473.

Ogawa, M., Fukuyama, H., Ouchi, Y., Yamauchi, H., Matsuzaki, S., Kimura, J., and Tsukada, H. (1996). Uncoupling between cortical glucose metabolism and blood flow after ibotenate lesion of the rat basal forebrain: a PET study. *Neurosci. Lett.* **204,** 193–196.

Okuda, Y., Kawai, K., Koide, Y., Kimura, S., and Yamashita, K. (1992). Evidence for regulation of hepatic ketogenesis in rats by Ca^{2+}-calmodulin messenger system. *Diabetes Nutr. Metab.* **5,** 3–8.

Olff, M. (1999). Stress, depression and immunity: the role of defense and coping styles. *Psychiatr. Res. Neuroimaging* **85,** 7–15.

O'Meara, G., Coumis, U., Ma, S. Y., Kehr, J., Mahoney, S., Bacon, A., Allen, S. J., Holmes, F., Kahl, U., Wang, F. H., Kearns, I. R., Ove-Ogren, S., Dawbarn, D., Mufson, E. J., Davies, C., Dawson, G., and Wynick, D. (2000). Galanin regulates the postnatal survival of a subset of basal forebrain cholinergic neurons. *Proc. Natl. Acad. Sci. USA* **97,** 11569–11574.

Ontko, J. A., and Westbrook, D. J. (1983). Dual effects of calcium on the oxidation of fatty acids to ketone bodies in liver mitochondria. *Biochem. Biophys. Res. Commun.* **116,** 173–179.

Orzi, F., Diana, G., Casamenti, F., Palombo, E., and Fieschi, C. (1988). Local cerebral glucose utilization following unilateral and bilateral lesions of the nucleus basalis magnocellularis in the rat. *Brain Res.* **462,** 99–103.

Ott, E.O., Abraham, J., Meyer, J. S., Achari, A. N., Chee, A. N., and Mathew, N. T. (1975). Disordered cholinergic neurotransmission and dysautoregulation after acute cerebral infarction. *Stroke* **6,** 172–180.

Otto, D. A., and Ontko, J. A. (1978). Activation of mitochondrial fatty acid oxidation by calcium. Conversion to the energized state. *J. Biol. Chem.* **253,** 789–799.

Owen, O. E., Morgan, A. P., Kemp, H. G., Sullivan, J. M., Herrera, M. G., and Cahill, G. F., Jr. (1967). Brain metabolism during fasting. *J. Clin. Invest.* **46,** 1589–1595.

Owen, O. E., and Cahill, G. F., Jr. (1973). Metabolic effects of exogenous glucocorticoids in fasted man. *J. Clin. Invest.* **52,** 2596–2605.

Padmini, S., and Rao, P. S. (1991). Enhanced β-oxidative utilization of [1-^{14}C]palmitate during active myelinogenesis in developing rat brain under nutritional stress. *Lipids* **26,** 83–85.

Palaiologos, G., and Felig, P. (1976). Effects of ketone bodies on amino acid metabolism in isolated rat diaphragm. *Biochem. J.* **154,** 709–716.

Palmer, R. M., Ferrige, A. G., and Moncada, S. (1987). Nitric oxide release accounts for the biological activity of endothelium-derived relaxing factor. *Nature* **327,** 524–526.

Pan, J. W., Rothman, D. L., Behar, K. L., Stein, D. T., and Hetherington, H. P. (2000). Human brain β-hydroxybutyrate and lactate increase in fasting-induced ketosis. *J. Cereb. Blood Flow Metab.* **20,** 1502–1507.

Parent, M. B., Laurey, P. T., Wilkniss, S., and Gold, P. E. (1997). Intraseptal infusions of muscimol impair spontaneous alternation performance: Infusions of glucose into the hippocampus, but not the medial septum, reverse the deficit. *Neurobiol. Learn. Mem.* **68,** 75–85.

Park, I. R., Thorn, M. B., and Bachelard, H. S. (1987). Threshold requirements for oxygen in the release of acetylcholine from, and in the maintenance of the energy state in, rat brain synaptosomes. *J. Neurochem.* **49,** 781–788.

Parnas, J. (1999). From predisposition to psychosis: progression of symptoms in schizophrenia. *Acta Psychiatr. Scand. Suppl.* **395,** 20–29.

Parsons, P. A. (1993). Stress, extinctions and evolutionary change: From living organisms to fossils. *Biol. Rev.* **68,** 313–333.

Patel, M. S. (1977). Age-dependent changes in the oxidative metabolism in rat brain. *J. Gerontol.* **32,** 643–646.

Pauly, J. R., Grun, E. U., and Collins, A. C. (1990). Chronic corticosterone administration modulates nicotine sensitivity and brain nicotinic receptor binding in C3H mice. *Psychopharmacology* **101,** 310–316.

Pauly, J. R., and Collins, A. C. (1993). An autoradiographic analysis of alterations in nicotinic cholinergic receptors following 1 week of corticosterone supplementation. *Neuroendocrinology* **57,** 262–271.

Peet, M., Murphy, B., Shay, J., and Horrobin, D. (1998). Depletion of omega-3 fatty acid levels in red blood cell membranes of depressive patients. *Biol. Psychiatry* **43,** 315–319.

Pellerin, L., and Magistretti, P. J. (1994). Glutamate uptake into astrocytes stimulates aerobic glycolysis: A mechanism coupling neuronal activity to glucose utilization. *Proc. Natl. Acad. Sci. USA* **91,** 10625–10629.

Pellerin, L., Pellegri, G., Martin, J. L., and Magistretti, P. J. (1998a). Expression of monocarboxylate transporter mRNAs in mouse brain: Support for a distinct role of lactate as an energy substrate for the neonatal vs. adult brain. *Proc. Natl. Acad. Sci. USA* **95,** 3990–3995.

Pellerin, L., Pellegri, G., Bittar, P. G., Charnay, Y., Bouras, C., Martin, J. L., Stella, N., and Magistretti, P. J. (1998b). Evidence supporting the existence of an activity-dependent astrocyte-neuron lactate shuttle. *Dev. Neurosci.* **20,** 291–299.

Pereira de Vasconcelos, A., Schroeder, H., and Nehlig, A. (1987). Effects of early chronic phenobarbital treatment on the maturation of energy metabolism in the developing rat brain. II. Incorporation of β-hydroxybutyrate into amino acids. *Brain Res.* **433,** 231–236.

Pereira de Vasconcelos, A., Boyet, S., and Nehlig, A. (1990). Consequences of chronic phenobarbital treatment on local cerebral glucose utilization in the developing rat. *Brain Res. Dev. Brain Res.* **53,** 168–178.

Perry, E. K., Piggott, M. A., Court, J. A., Johnson, M., and Perry, R. H. (1993). Transmitters in the developing and senescent human brain. *Ann. N.Y. Acad. Sci.* **695,** 69–72.

Peterson, C. (1990). Tetrahydroaminoacridine increases acetylcholine synthesis and glucose oxidation by mouse brain slices in vitro. *Neurosci. Lett.* **115,** 274–278.

Pettit, F. H., Pelley, J. W., and Reed, L. J. (1975). Regulation of pyruvate dehydrogenase kinase and phosphatase by acetyl-CoA/CoA and NADH/NAD ratios. *Biochem. Biophys. Res. Commun.* **65,** 575–582.

Pietrini, P., Ricciardi, E., Furey, M. L., Dani, A., Guazzelli, M., Schapiro, M. B., Rapoport, S. I., and Alexander, G. E. (1999). Effects of the healthy aging process on the relation of regional cerebral blood flow (rCBF) to glucose metabolism (rCMRglc) during rest and audiovisual stimulation in humans. *Soc. Neurosci. Abstr.* **25,** 1144.

Pillar, T. M., and Seitz, H. J. (1997). Thyroid hormone and gene expression in the regulation of mitochondrial respiratory function. Eur. *J. Endocrinol.* **136,** 231–239.

Plomin, R., Owen, M. J., and McGuffin, P. (1994). The genetic basis of complex human behaviors. *Science* **264,** 1733–1739.

Plotsky, P. M., and Meaney, M. J. (1993). Early, postnatal experience alters hypothalamic corticotropin-releasing factor (CRF) mRNA, median eminence CRF content and stress-induced release in adult rats. *Brain Res. Mol. Brain Res.* **18,** 195–200.

Plotsky, P. M., Owens, M. J., and Nemeroff, C. B. (1995). Neuropeptide alterations in mood disorders. *In* "Psychopharmacology: The Fourth Generation of Progress" (F. E. Bloom and D. J. Kupfer, eds.), pp. 971–981. Raven Press, New York.

Poblete, J. C., and Azmitia, E. C. (1995). Activation of glycogen phosphorylase by serotonin and 3,4-methylenedioxymethamphetamine in astroglial-rich primary cultures: involvement of the 5-HT_{2A} receptor. *Brain Res.* **680,** 9–15.

Poduslo, S. E. (1989). Induction of ketone body enzymes in glial cells. *Arch. Biochem. Biophys.* **272,** 318–322.

Poduslo, S. E., Pak, C. H., and Miller, K. (1990). Hydrocortisone induction during oligodendroglial differentiation. *Neurosci. Lett.* **113,** 84–88.

Pollay, M., and Stevens, F. A. (1980). Starvation-induced changes in transport of ketone bodies across the blood-brain barrier. *J. Neurosci. Res.* **5,** 163–172.

Popovici, D., Mihai, N., and Urbanavicius, V. (1980). Abnormalities of oxidative phosphorylation due to excess or deficiency of thyroid hormones. *Endocrinologie* **18,** 143–147.

Posener, J. A., Schatzberg, A. F., Williams, G. H., Samson, J. A., McHale, N. L., Bessette, M. P., and Schildkraut, J. J. (1999). Hypothalamic-pituitary-adrenal axis effects on plasma homovanillic acid in man. *Biol. Psychiatry* **45,** 222–228.

Prange, A. J. Jr., Garbutt, J. C., and Loosen, P. T. (1987). The hypothalamic-pituitary-thyroid axis in affective disorders. *In* "Psychopharmacology: The Third Generation of Progress" (H. Y. Meltzer, eds.), pp. 629–636. Raven Press, New York.

Prapong, T., Uemura, E., and Hsu, W. (2001). G protein and cAMP-dependent protein kinase mediate amyloid β-peptide inhibition of neuronal glucose uptake. *Exp. Neurol.* **167,** 59–64.

Prasad, A. N., and Stafstrom, C. E. (1998). Dietary therapy of epilepsy in the nineties: Renewed experience with the ketogenic diet. *Nutr. Res.* **18,** 403–416.

Prendergast, M. A., and Buccafusco, J. J. (1998). (−)-Nicotine increases mRNA encoding G3PDH and the vesicular acetylcholine transporter in vivo. *Neuroreport* **9,** 1385–1389.

Ragozzino, M. E., Unick, K. E., and Gold, P. E. (1996). Hippocampal acetylcholine release during memory testing in rats: Augmentation by glucose. *Proc. Natl. Acad. Sci. USA* **93,** 4693–4698.

Ramade, F., and Bayle, J. D. (1989). Adaptation of the adrenocorticotropic response to chronic intermittent stress was altered by intracerebroventricular infusion of hemicholinium-3. *Neuroendocrinology* **50,** 165–169.

Randle, P. J. (1998). Regulatory interactions between lipids and carbohydrates: the glucose fatty acid cycle after 35 years. *Diabetes Metab. Rev.* **14,** 263–283.

Randle, P. J., Sugden, P. H., Kerbey, A. L., Radcliffe, P. M., and Hutson, N. J. (1978). Regulation of pyruvate oxidation and the conservation of glucose. *Biochem. Soc. Symp.* **43,** 47–67.

Redies, C., Hoffer, L. J., Beil, C., Marliss, E. B., Evans, A. C., Lariviere, F., Marrett, S., Meyer, E., Diksic, M., Gjedde, A., and Hakim, A. M. (1989). Generalized decrease in brain glucose metabolism during fasting in humans studied by PET. *Am. J. Physiol.* **256,** E805–E810.

Reis, D. J., Golanov, E. V., Galea, E., and Feinstein, D. L. (1997). Central neurogenic neuroprotection: Central neural systems that protect the brain from hypoxia and ischemia. *Ann. N.Y. Acad. Sci.* **835,** 168–186.

Renault, E., Sarrazin, S., and Deschatrette, J. (1998). Evidence for interactions between rat hepatoma cell apoptosis and differentiation. *Biochem. Genet.* **36,** 1–13.

Robinson, J. K., and Crawley, J. N. (1993). The role of galanin in cholinergically-mediated memory processes. *Prog. Neuro-Psychopharmacol. Biol. Psychiatry* **17,** 71–85.

Robinson, S. E., Martin, R. M., Davis, T. R., Gyenes, C. A., Ryland, J. E., and Enters, E. K. (1990). The effect of acetylcholine depletion on behavior following traumatic brain injury. *Brain Res.* **509,** 41–46.

Roeder, L. M., Poduslo, S. E., and Tildon, J. T. (1982). Utilization of ketone bodies and glucose by established neural cell lines. *J. Neurosci. Res.* **8,** 671–682.

Roland, P. E., and Friberg, L. (1988). The effect of GABA-A agonist THIP on regional cortical blood flow in humans. A new test of hemispheric dominance. *J. Cereb. Blood Flow Metab.* **8,** 314–323.

Romijn, J. A., Godfried, M. H., Hommes, M. J., Endert, E., and Sauerwein, H. P. (1990). Decreased glucose oxidation during short-term starvation. *Metabolism* **39,** 525–530.

Rosenblum, W. I. (1997). Selective impairment of response to acetylcholine after ischemia/reperfusion in mice. *Stroke* **28,** 448–451.

Ross, D. L., Swaiman, K. F., Torres, F., and Hansen, J. (1985). Early biochemical and EEG correlates of the ketogenic diet in children with atypical absence epilepsy. *Pediatr. Neurol.* **1,** 104–108.

Rubinow, D. R., Davis, C. L., and Post, R. M. (1995). Somatostatin in the central nervous system. *In* "Psychopharmacology: The Fourth Generation of Progress" (F. E. Bloom and D. J. Kupfer, eds.), pp. 553–562. Raven Press, New York.

Ruderman, N. B., Ross, P. S., Berger, M., and Goodman, M. W. (1974). Regulation of glucose and ketone-body metabolism in brain of anaesthetized rats. *Biochem. J.* **138,** 1–10.

Ruderman, N. B., Saha, A. K., Vavvas, D., Heydrick, S. J., and Kurowski, T. G (1997). Lipid abnormalities in muscle of insulin-resistant rodents. The malonyl CoA hypothesis. *Ann. N.Y. Acad. Sci.* **827,** 221–230.

Ruderman, N. B., Saha, A. K., Vavvas, D., and Witters, L. A. (1999). Malonyl-CoA, fuel sensing, and insulin resistance. *Am. J. Physiol.* **276,** E1–E18.

Sabell, I., Morata, P., Quesada, J., and Morell, M. (1985). Effect of thyroid hormones on the glycolytic enzyme activity in brain areas of the rat. *Enzyme* **34,** 27–32.

Saboureau, M., Bobet, J. P., and Boissin, J. (1980). Cyclic activity of adrenal function and seasonal variations of cortisol peripheral metabolism in a hibernating mammal, the hedgehog. *J. Physiol. (Paris)* **76,** 617–629.

Saltiel, A. R., and Kahn, C. R. (2001). Insulin signalling and the regulation of glucose and lipid metabolism. *Nature* **414,** 799–806.

Sapolsky, R. M. (1985). A mechanism for glucocorticoid toxicity in the hippocampus: Increased vulnerability to metabolic insults. *J. Neurosci.* **5,** 1228–1232.

Sapolsky, R. M. (1986). Glucocorticoid toxicity in the hippocampus: reversal by supplementation with brain fuels. *J. Neurosci.* **6,** 2240–2244.

Sapolsky, R. M. (1993). Potential behavioral modification of glucocorticoid damage to the hippocampus. *Behav. Brain Res.* **57,** 175–182.

Sapolsky, R. M. (1994). The physiological relevance of glucocorticoid endangerment of the hippocampus. *Ann. N.Y. Acad. Sci.* **746,** 294–307.

Sato, A., and Sato, Y. (1992). Regulation of regional cerebral blood flow by cholinergic fibers originating in the basal forebrain. *Neurosci. Res.* **14,** 242–274.

Sato, K., Kashiwaya, Y., Keon, C. A., Tsuchiya, N., King, M. T., Radda, G. K., Chance, B., Clarke, K., and Veech, R. L. (1995). Insulin, ketone bodies, and mitochondrial energy transduction. *FASEB J.* **9,** 651–658.

Savaki, H. E., Graham, D. I., Grome, J. J., and McCulloch, J. (1984). Functional consequences of unilateral lesion of the locus coeruleus: a quantitative ^{14}C-2-deoxyglucose investigation. *Brain Res.* **292,** 357–366.

Schade, D. S., Eaton, R. P., and Peake, G. T. (1980). The ketotic effects of glucocorticoid and growth hormone in man. *Acta Diabetol. Lat.* **17,** 161–169.

Schade, D. S., and Eaton, R. P. (1980). The temporal relationship betweeen endogenously secreted stress hormones and metabolic decompensation in diabetic man. *J. Clin. Endocrinol. Metab.* **50,** 131–136.

Schauer, J. E., Schelin, A., Hanson, P., and Stratman, F. W. (1990). Dehydroepiandrosterone and a beta-agonist, energy transducers, alter antioxidant enzyme systems: influence of chronic training and acute exercise in rats. *Arch. Biochem. Biophys.* **283,** 503–511.

Schlessinger, J., and Ullrich, A. (1992). Growth factor signaling by receptor tyrosine kinases. *Neuron* **9,** 383–391.

Schmidt, R. H., and Grady, M. S. (1995). Loss of forebrain cholinergic neurons following fluid-percussion injury: implications for cognitive impairment in closed head injury. *J. Neurosurg.* **83,** 496–502.

Schroeder, H., Bomont, L., and Nehlig, A. (1991). Influence of early chronic phenobarbital treatment on cerebral arteriovenous differences of glucose and ketone bodies in the developing rat. *Int. J. Dev. Neurosci.* **9,** 453–461.

Schroeder, H., Collignon, A., Uttscheid, L., Pereira de Vasconcelos, A., and Nehlig, A. (1994). Effects of early chronic diazepam treatment on incorporation of glucose and β-hydroxybutyrate into cerebral amino acids: Relation to undernutrition. *Int. J. Dev. Neurosci.* **12,** 471–484.

Schroeder, H., Humbert, A. C., Koziel, V., Desor, D., and Nehlig, A. (1995). Behavioral and metabolic consequences of neonatal exposure to diazepam in rat pups. *Exp. Neurol.* **131,** 53–63.

Schurr, A., West, C. A., and Rigor, B. M. (1988). Lactate-supported synaptic function in the rat hippocampal slice preparation. *Science* **240,** 1326–1328.

Schurr, A., Payne, R. S., Miller, J. J., and Rigor, B. M. (1997). Brain lactate, not glucose, fuels recovery of synaptic function from hypoxia upon reoxygenation: an in vitro study. *Brain Res.* **744,** 105–111.

Schwartz, M. W., Baskin, D. G., Kaiyala, K. J., and Woods, S. C. (1999). Model for the regulation of energy balance and adiposity by the central nervous system. *Am. J. Clin. Nutr.* **69,** 584–596.

Scrimgeour, A. G., Blakesley, V. A., Stannard, B. S., and LeRoith, D. (1997). Mitogen-activated protein kinase and phosphatidylinositol 3-kinase pathways are not sufficient for insulin-like growth factor I-induced mitogenesis and tumorigenesis. *Endocrinology* **138,** 2552–2558.

Seckl, J. R. (1997). Glucocorticoids, feto-placental 11β-hydroxysteroid dehydrogenase type 2, and the early life origins of adult disease. *Steroids* **62,** 89–94.

Settergren, G., Lindblad, B. S., and Persson, B. (1980). Cerebral blood flow and exchange of oxygen, glucose, ketone bodies, lactate, pyruvate and amino acids in anesthetized children. *Acta Paedriatr. Scand.* **69,** 457–465.

Seymour, K. J., Bluml, S., Sutherling, J., Sutherling, W., and Ross, B. D. (1999). Identification of cerebral acetone by ^{1}H-MRS in patients with epilepsy controlled by ketogenic diet. *MAGMA* **8,** 33–42.

Shibata, S., Kodama, K., Tominaga, K., Tanaka, T., and Watanabe, S. (1992). Effect of muscarinic cholinergic drugs on ischemia-induced decreases in glucose uptake and CA1 field potentials in rat hippocampus slices. *Eur. J. Pharmacol.* **221,** 113–119.

Shibata, S., Koga, Y., Hamada, T., and Watanabe, S. (1993). Facilitation of 2-deoxyglucose uptake in rat cortex and hippocampus slices by somatostatin is independent of cholinergic activity. *Eur. J. Pharmacol.* **231,** 381–388.

Shulman, R. G., Rothman, D. L., and Hyder, F. (1999). Stimulated changes in localized cerebral energy consumption under anesthesia. *Proc. Natl. Acad. Sci. USA* **96,** 3245–3250.

Sibson, N. R., Dhankhar, A., Mason, G. F., Rothman, D. L., Behar, K. L., and Shulman, R. G. (1998). Stoichiometric coupling of brain glucose metabolism and glutamatergic neuronal activity. *Proc. Natl. Acad. Sci. USA* **95,** 316–321.

Silva, A., Montague, J. R., Lopez, T. F., and Mudd, L. M. (2000). Growth factor effects on survival and development of calbindin immunopositive cultured septal neurons. *Brain Res. Bull.* **51,** 35–42.

Skaper, S. D., and Varon, S. (1979). Nerve growth factor action on 2-deoxy-D-glucose transport in dorsal root ganglionic dissociates from chick embryo. *Brain Res.* **163,** 89–100.

Smith, A. L., Satterthwaite, H. S., and Sokoloff, L. (1969). Induction of brain D(−)-β-hydroxybutyrate dehydrogenase activity by fasting. *Science* **163,** 79–81.

Smythe, G. A., and Edwards, S. R. (1991). A role for central postsynaptic α_2-adrenoceptors in glucoregulation. *Brain Res.* **562,** 225–229.

Soboll, S. (1993). Long-term and short-term changes in mitochondrial parameters by thyroid hormones. *Biochem. Soc. Trans.* **21,** 799–803.

Soengas, J. L., Strong, E. F., and Andres, M. D. (1998). Glucose, lactate, and β-hydroxybutyrate utilization by rainbow trout brain: changes during food deprivation. *Physiol. Zool.* **71,** 285–293.

Sokoloff, L. (1973). Metabolism of ketone bodies by the brain. *Annu. Rev. Med.* **24,** 271–280.

Sokoloff, L., Reivich, M., Kennedy, C., Des Rosiers, M. H., Patlak, C. S., Pettigrew, K. D., Sakurada, O., and Shinohara, M. (1977). The [^{14}C]-deoxyglucose method for the measurement of local cerebral glucose utilization: Theory, procedure, and normal values in the conscious and anesthetized albino rat. *J. Neurochem.* **28,** 897–916.

Soncrant, T. T., Holloway, H. W., and Rapoport, S. I. (1985). Arecoline-induced elevations of regional cerebral metabolism in the conscious rat. *Brain Res.* **347,** 205–216.

Soncrant, T. T., Holloway, H. W., Greig, N. H., and Rapoport, S. I. (1989). Regional brain metabolic responsivity to the muscarinic cholinergic agonist arecoline is similar in young and aged Fischer-344 rats. *Brain Res.* **487,** 255–266.

Sonnewald, U., Wang, A. Y., Schousboe, A., Erikson, R., and Skottner, A. (1996). New aspects of lactate metabolism: IGF-I and insulin regulate mitochondrial function in cultured brain cells during normoxia and hypoxia. *Dev. Neurosci.* **18,** 443–448.

Srivastava, A. K., and Pandey, S. K. (1998). Potential mechanism(s) involved in the regulation of glycogen synthesis by insulin. *Mol. Cell. Biochem.* **182,** 135–141.

Srivastava, L. K., and Baquer, N. Z. (1985). Changes in phosphofructokinase and pyruvate kinase in rat brain following thyroidectomy. *Biochem. Int.* **11,** 491–500.

Stafstrom, C. E., Wang, C., and Jensen, F. E. (1999). Electrophysiological observations in hippocampal slices from rats treated with the ketogenic diet. *Dev. Neurosci.* **21,** 393–399.

Steckler, T., Holsboer, F., and Reul, J. M. H. M. (1999). Glucocorticoids and depression. *Best Pract. Res. Clin. Endocrinol. Metab.* **13,** 597–614.

Stefani, M. R., and Gold, P. E. (1998). Intra-septal injections of glucose and glibenclamide attenuate galanin-induced spontaneous alternation performance deficits in the rat. *Brain Res.* **813,** 50–56.

Stratakis, C. A., and Chrousos, G. P. (1995). Neuroendocrinology and pathophysiology of the stress system. *Ann. N.Y. Acad. Sci.* **771,** 1–18.

Sun, A., and Cheng, J. (1999). Novel targets for therapeutic intervention against ischemic brain injury. *Clin. Neuropharmacol.* **22,** 164–171.

Sunanda Rao, B. S. S., and Raju, T. R. (2000). Restraint stress-induced alterations in the levels of biogenic amines, amino acids, and AChE activity in the hippocampus. *Neurochem. Res.* **25,** 1547–1552.

Swaab, D. F., Raadsheer, F. C., Endert, E., Hofman, M. A., Kamphorst, W., and Ravid, R. (1994). Increased cortisol levels in aging and Alzheimer's disease in postmortem cerebrospinal fluid. *J. Neuroendocrinol.* **6,** 681–687.

Sze, J. Y., Victor, M., Loer, C., Shi, Y., and Ruvkun, G. (2000). Food and metabolic signaling defects in a *Caenorhabditis elegans* serotonin-synthesis mutant. *Nature* **403,** 560–564.

Takahashi, A., Ishimaru, H., Ikarashi, Y., Kishi, E., and Maruyama, Y. (1996). Hypothalamic cholinergic activity associated with 2-deoxyglucose-induced hyperglycemia. *Brain Res.* **734,** 116–122.

Thakore, J. H., Richards, P. J., Reznek, R. H., Martin, A., and Dinan, T. G. (1997). Increased intra-abdominal fat deposition in patients with major depressive illness as measured by computed tomography. *Biol. Psychiatry* **41,** 1140–1142.

Theodore, W. H., DiChiro, G., Margolin, R., Fishbein, D., Porter, R. J., and Brooks, R. A. (1986). Barbiturates reduce human cerebral glucose metabolism. *Neurology* **36,** 60–64.

Thompson, J. R., and Wu, G. (1991). The effect of ketone bodies on nitrogen metabolism in skeletal muscle. *Comp. Biochem. Physiol.* **100,** 209–216.

Thurston, J. H., Hauhart, R. E., Dirgo, J. A., and Jones, E. M. (1980). Mechanism of increased brain glucose and glycogen after hydrocortisone: possible clinical significance. *Ann. Neurol.* **7,** 515–523.

Thurston, J. H., and Hauhart, R. E. (1989). Effect of momentary stress on brain energy metabolism in weanling mice: apparent use of lactate as cerebral metabolic fuel concomitant with a decrease in brain glucose metabolism. *Metab. Brain Dis.* **4,** 177–186.

Tildon, J. T., McKenna, M. C., and Stevenson, J. H. (1994). Transport of 3-hydroxybutyrate by cultured rat brain astrocytes. *Neurochem. Res.* **19,** 1237–1242.

Tissenbaum, H. A., Hawdon, J., Perregaux, M., Hotez, P., Guarente, L., and Ruvkun, G. (2000). A common muscarinic pathway for diapause recovery in the distantly related nematode species *Caenorhabditis elegans* and *Ancylostoma caninum. Proc. Natl. Acad. Sci. USA* **97,** 460–465.

Tizabi, Y., Gilad, V. H., and Gilad, G. M. (1989). Effects of chronic stressors or corticosterone treatment on the septohippocampal cholinergic system of the rat. *Neurosci. Lett.* **105,** 177–182.

Tomkins, G. M. (1975). The metabolic code. *Science* **189,** 760–763.

Toran-Allerand, C. D., Miranda, R. C., Bentham, W. D., Sohrabji, F., Brown, T. J., Hochberg, R. B., and MacLusky, N. J. (1992). Estrogen receptors colocalize with low-affinity nerve growth factor receptors in cholinergic neurons of the basal forebrain. *Proc. Natl. Acad. Sci. USA* **89,** 4668–4672.

Trejo, J. L., Rua, C., Cuchillo, I., and Machin, C. (1998). Calbindin-D28k- and astroglial protein-immunoreactivities, and ultrastructural differentiation in the prenatal rat cerebral cortex and hippocampus are affected by maternal adrenalectomy. *Brain Res. Dev. Brain Res.* **108,** 161–177.

Trumbly, R. J. (1992). Glucose repression in the yeast *Saccharomyces cerevisiae. Mol. Microbiol.* **6,** 15–21.

Tsuang, M. (2000). Schizophrenia: genes and environment. *Biol. Psychiatry* **47,** 210–220.

Tsubokawa, T., Katayama, Y., Kondo, T., Ueno, Y., Hayashi, N., and Moriyasu, N. (1980). Changes in local cerebral blood flow and neuronal activity during sensory stimulation in normal and sympathectomized cats. *Brain Res.* **190,** 51–65.

Tsukada, H., Kakiuchi, T., Ando, I., Shizuno, H., Nakanishi, S., and Ouchi, Y. (1997). Regulation of cerebral blood flow response to somatosensory stimulation through the cholinergic system: A positron emission tomography study in unanaesthetized monkeys. *Brain Res.* **749,** 10–17.

Tuor, U. I. (1997). Glucocorticoids and the prevention of hypoxic-ischemic brain damage. *Neurosci. Biobehav. Rev.* **21,** 175–179.

Tuor, U. I., Chumas, P. D., and Del Bigio, M. R. (1995). Prevention of hypoxic-ischemic damage with dexamethasone is dependent on age and not influenced by fasting. *Exp. Neurol.* **132,** 116–122.

Turlejski, K. (1996). Evolutionary ancient roles of serotonin: long-lasting regulation of activity and development. *Acta Neurobiol. Exp.* (*Warsz.*) **56,** 619–636.

Turner, B. B. (1997). Influence of gonadal steroids on brain corticosteroid receptors: a minireview. *Neurochem. Res.* **22,** 1375–1385.

Uchida, S., Kagitani, F., Nakayama, H., and Sato, A. (1997). Effect of stimulation of nicotinic cholinergic receptors on cortical cerebral blood flow and changes in the effect during aging in anesthetized rats. *Neurosci. Lett.* **228,** 203–206.

Unger, J.W., and Schmidt, Y. (1993). Galanin-immunoreactivity in the nucleus basalis of Meynert in the rat: Age-related changes and differential response to lesion-induced cholinergic cell loss. *Neurosci. Lett.* **153,** 140–143.

Vallee, M., MacCari, S., Dellu, F., Simon, H., Le Moal, M., and Mayo, W. (1999). Long-term effects of prenatal stress and postnatal handling on age-related glucocorticoid secretion and cognitive performance: a longitudinal study in the rat. *Eur. J. Neurosci.* **11,** 2906–2916.

Vannucci, R. C., and Brucklacher, R. M. (1994). Cerebral mitochondrial redox states during metabolic stress in the immature rat. *Brain Res.* **653,** 141–147.

Varela, A., Testoni, G., Carregal, M., Dalamon, V., and Savino, E. A. (1997). Effects of 3-hydroxybutyrate on the hypoxic and reoxygenated atria from fed and fasted rats. *J. Physiol. Biochem.* **53,** 377–381.

Vaucher, E., Lacombe, P., and Hamel, E. (1997). Cholinergic basal forebrain neurons as modulators of the cortical microcirculation. *Med. Sci.* **13,** 534–541.

Vega-Nunez, E., Menendez-Hurtado, A., Garesse, R., Santos, A., and Perez-Castillo, A. (1995). Thyroid hormone-regulated brain mitochondrial genes revealed by differential cDNA cloning. *J. Clin. Invest.* **96,** 893–899.

Veneman, T., Mitrakou, A., Mokan, M., Cryer, P., and Gerich, J. (1994). Effect of hyperketonemia and hyperlacticacidemia on symptoms, cognitive dysfunction, and counterregulatory hormone responses during hypoglycemia in normal humans. *Diabetes* **43,** 1311–1317.

Venter, J. C., di Porzio, U., Robinson, D. A., Shreeve, S. M., Lai, J., Kerlavage, A. R., Fracek, S. P., Jr., Lentes, K. U., and Fraser, C. M. (1988). Evolution of neurotransmitter receptor systems. *Prog. Neurobiol.* **30,** 105–169.

Vercesi, A. E. (1984). Dissociation of $NAD(P)^{+}$-stimulated mitochondrial Ca^{2+} efflux from swelling and membrane damage. *Arch. Biochem. Biophys.* **232,** 86–91.

Vernadakis, A. (1996). Glia-neuron intercommunications and synaptic plasticity. *Prog. Neurobiol.* **49,** 185–214.

Viau, V., and Meaney, M. J. (1996). The inhibitory effect of testosterone on hypothalamic-pituitary-adrenal responses to stress is mediated by the medial preoptic area. *J. Neurosci.* **16,** 1866–1876.

Videbech, P. (2000). PET measurements of brain glucose metabolism and blood flow in major depressive disorder: A critical review. *Acta Psychiatr. Scand.* **101,** 11–20.

Virgin, C. E., and Sapolsky, R. M. (1997). Styles of male social behavior and their endocrine correlates among low-ranking baboons. *Am. J. Primatol.* **42,** 25–39.

Virgin, C., Ha, T., Packan, D., Tombaugh, G., Yang, S., Horner, H., and Sapolsky, R. (1991). Glucocorticoids inhibit glucose transport and glutamate uptake in hippocampal astrocytes: Implications for glucocorticoid neurotoxicity. *J. Neurochem.* **57,** 1422–1428.

Virkamäki, A., Ueki, K., and Kahn, C. R. (1999). Protein-protein interaction in insulin signaling and the molecular mechanisms of insulin resistance. *J. Clin. Invest.* **103,** 931–943.

von Schwarzenfeld, I., Fischer, H. D., and Oelszner, W. (1975). Interaction between cellular metabolism and acetylcholine turnover of rat brain cortex slices in the presence of arecoline. *Acta Biol. Med. Ger.* **34,** 1525–1528.

Wada, H., Okada, Y., Nabetani, M., and Nakamura, H. (1997). The effects of lactate and β-hydroxybutyrate on the energy metabolism and neural activity of hippocampal slices from adult and immature rat. *Dev. Brain Res.* **101,** 1–7.

Waddell, B. J. (1993). The placenta as hypothalamus and pituitary: possible impact and fetal adrenal function. *Reprod. Fertil. Dev.* **5,** 479–497.

Wahba, Z. Z., and Soliman, K. F. (1992). Effect of stress on choline acetyltransferase activity of the brain and the adrenal of the rat. *Experientia* **48,** 265–268.

Walker, E. F., and Diforio, D. (1997). Schizophrenia: A neural diathesis-stress model. *Psychol. Rev.* **104,** 667–685.

Walker, R. J., Brooks, H. L., and Holden-Dye, L. (1996). Evolution and overview of classical transmitter molecules and their receptors. *Parasitology* **113** Suppl., S3–S33.

Wang, J., Akabayashi, A., Yu, H. J., Dourmashkin, J., Alexander, J. T., Silva, I., Lighter, J., and Leibowitz, S. F. (1998). Hypothalamic galanin: Control by signals of fat metabolism. *Brain Res.* **804,** 7–20.

Werner, H., Raizada, M. K., Mudd, L. M., Foyt, H. L., Simpson, I. A., Roberts, C. T., Jr., and LeRoith, D. (1989). Regulation of rat brain/HepG2 glucose transporter gene expression by insulin and insulin-like growth factor-I in primary cultures of neuronal and glial cells. *Endocrinology* **125,** 314–320.

Wheatley, D. (1997). Stress, anxiety and depression. *Stress Med.* **13,** 173–177.

Williams, N. G., and Roberts, T. M. (1994). Signal transduction pathways involving the Raf proto-oncogene. *Cancer Metastasis Rev.* **13,** 105–116.

Winder, W. W., and Hardie, D. G. (1999). AMP-activated protein kinase, a metabolic master switch: Possible roles in type 2 diabetes. *Am. J. Physiol.* **277,** E1–E10.

Winokur, A., Maislin, G., Phillips, J. L., and Amsterdam, J. D. (1988). Insulin resistance after oral glucose tolerance testing in patients with major depression. *Am. J. Psychiatry* **145,** 325–330.

Wissink, S., Meijer, O., Pearce, D., van der Burg, B., and van der Saag, P. T. (2000). Regulation of the rat serotonin-1A receptor gene by corticosteroids. *J. Biol. Chem.* **275,** 1321–1326.

Yager, J. Y. (1994). Protective effects of fasting and ketone body supplementation on hypoxic-ischemic brain damage in the immature rat. *Ann. Neurol.* **36,** 540.

Yao, J. K., Reddy, R. D., and van Kammen, D. P. (2001). Oxidative damage and schizophrenia: an overview of the evidence and its therapeutic implications. *CNS Drugs* **15,** 287–310.

Yeh, Y. Y. (1984). Biosynthesis of phospholipids and sphingolipids from acetoacetate and glucose in different regions of developing brain in vivo. *J. Neurosci. Res.* **11,** 383–394.

Yu, N., Martin, J. L., Stella, N., and Magistretti, P. J. (1993). Arachidonic acid stimulates glucose uptake in cerebral cortical astrocytes. *Proc. Natl. Acad. Sci. USA* **90,** 4042–4046.

Zerbib, R., and Laborit, H. (1990). Chronic stress and memory: implication of the central cholinergic system. *Pharmacol. Biochem. Behav.* **36,** 897–900.

Zhang, J. R., Vrecko, K., Nadlinger, K., Tomic, D. S., Birkmayer, G. D., and Reibnegger, G. (1998). The reduced coenzyme nicotinamide adenine dinucleotide (NADH) repairs DNA damage of PC12 cells induced by doxorubicin. *J. Tumor Marker Oncol.* **13,** 5–17.

Zhang, X. H., Peng, L., Chen, Y., and Hertz, L. (1993). Stimulation of glycogenolysis in astrocytes by fluoxetine, an antidepressant acting like 5-HT. *Neuroreport* **4,** 1235–1238.

Zhou, Y. P., Berggren, P. O., and Grill, V. (1996). A fatty acid-induced decrease in pyruvate dehydrogenase activity is an important determinant of beta-cell dysfunction in the obese diabetic *db/db* mouse. *Diabetes* **45,** 580–586.

Zhu, C. Z., and Auer, R. N. (1994). Intraventricular administration of insulin and IGF-1 in transient forebrain ischemia. *J. Cereb. Blood Flow Metab.* **14,** 237–242.

EXPRESSION, REGULATION, AND FUNCTIONAL ROLE OF GLUCOSE TRANSPORTERS (GLUTs) IN BRAIN

Donard S. Dwyer

Departments of Psychiatry and Pharmacology, LSU Health Sciences Center
Shreveport, Louisiana 71130

Susan J. Vannucci

Department of Pediatrics, Penn State University College of Medicine
Hershey, Pennsylvania17033

Ian A. Simpson

Department of Neuroscience and Anatomy, Penn State University College of Medicine
Hershey, Pennsylvania 17033

I. Introduction

Glucose is the primary source of energy for the mammalian brain. As in most other tissues, glucose transport in the CNS is mediated by facilitative glucose transporter proteins, GLUTs. However, to date a thorough

0074-7742/02 $35.00

understanding of the mechanisms regulating glucose transport in the brain lags behind what is known in the periphery. The objective of this review is to expand the working knowledge of brain glucose uptake and utilization in the context of recent advances in the transport field. As a starting point, current molecular models of the GLUT protein structure will be presented and compared with the known structures of other related transporters. The various structures may provide insights into the mechanisms by which glucose is transported across cell membranes. Data regarding the different GLUT isoforms found in brain and their regional distribution will then be summarized. Much of this information has been reviewed previously (Maher *et al.*, 1994; Vannucci *et al.*, 1977). However, there has been a virtual explosion within the past two years resulting in a doubling of the members of the family of facilitative glucose transporters (see Joost and Thorens, 2002). Thus, this review will attempt to incorporate the new members of this family into the picture of mammalian glucose transport as a whole and their specific neural locations and functions in the CNS, as presented in Fig. 1.

At present, the GLUT family includes thirteen members, GLUT1–GLUT12 and the H^+/myoinositiol co-transporter (HMIT1) (Mueckler, 1990; Baldwin, 1993; Gould and Holman, 1993; Doege *et al.*, 2000a,b; Ibberson *et al.*, 2000; Phay *et al.*, 2000; McVie-Wylie *et al.*, 2001; Joost and Thorens, 2002). In general, GLUTs are comprised of approximately 500 amino acids and include a single asparagine-linked glycosylation site. The molecular weights of the mature proteins are in the range of 40–55 kDa, although larger species (presumed to be dimers) are also observed (Hebert and Carruthers, 1992). On the basis of hydropathy plots and secondary structure prediction algorithms, it has been proposed that GLUTs have 12 transmembrane (TM) helices that form a pore structure (Mueckler *et al.*, 1985). Studies with circular dichroism and Fourier transform infrared spectroscopy have established that greater than 70% of the protein is α-helical, which is consistent with these structural predictions (Alvarez *et al.*, 1987; Chin *et al.*, 1987). The N terminus, the C terminus and a large loop segment that connects TM6 and TM7 extend into the cytoplasm. On the other hand, the loops between TM1 and TM2 and between TM9 and TM10 are located on an extracellular face of the protein and include the glycosylation sites in GLUT1–5 (Baldwin, 1993; Gould and Holman, 1993) and GLUT6/GLUT8 (Doege *et al.*, 2000a,b; Ibberson *et al.*, 2000), respectively. Intracellular segments of the GLUTs are presumed to be important for the reversible associations with other proteins that are involved in trafficking (see Czech *et al.*, 1992; Verhey and Birnbaum, 1994; Jung, 1998).

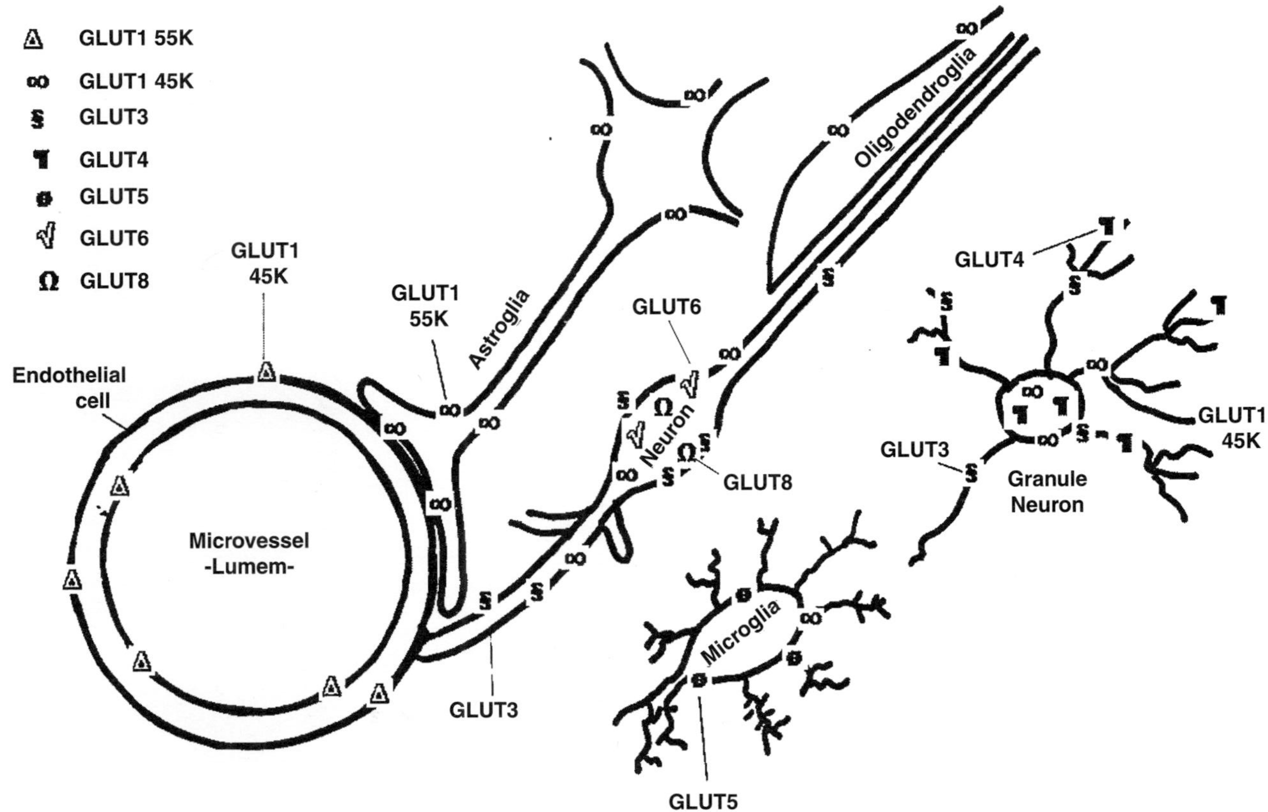

FIG. 1. Localization of GLUTs in the brain.

II. Structure of GLUTs

Presently, a high-resolution structure of a GLUT protein, or that of a closely related transporter molecule is not available. However, the X-ray crystal structures of several distantly related transporters and ion channel proteins have recently been determined and this information may be instructive. High-resolution maps of the glycerol transporter (Fu *et al.*, 2000), aquaporin (a water channel) (Murata *et al.*, 2000), and a mechanosensitive ion channel (MscL) (Chang *et al.*, 1998) reveal that each protein is comprised of multiple (7–10) membrane-spanning helices in right-hand barrel structures. Ribbon diagrams of these proteins are shown in Fig. 2A-C.

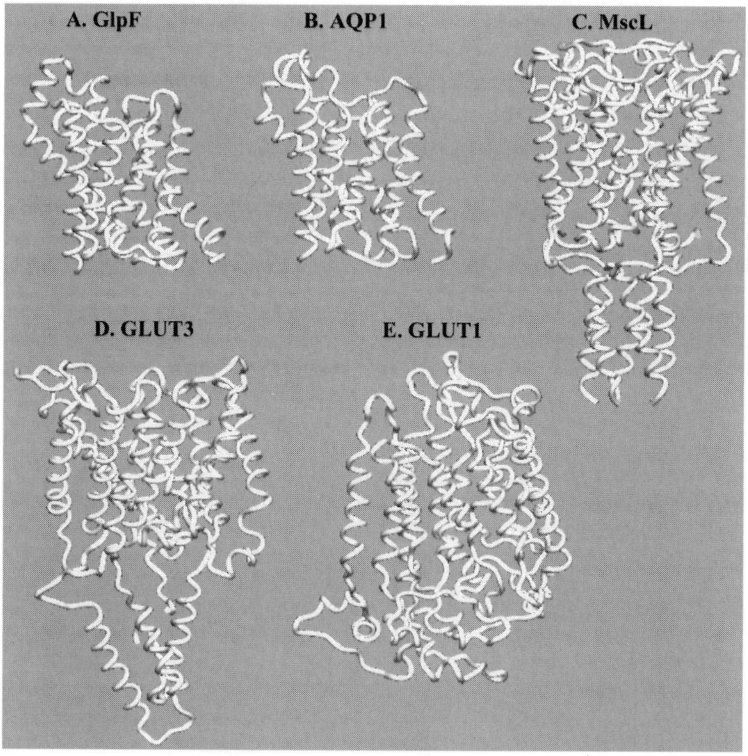

FIG. 2. Crystal structures of transporter proteins and ion channels, including the glycerol transporter (GlpF) (A), aquaporin-1 (AQP1) (B), MscL (C), and molecular models of GLUT3 (D) and GLUT1 (E).

Generally, the arrangement of the TM segments that form the pore is highly symmetrical. Overall, the structure of the transporters (GlpF and AQP1) is fairly similar to that of the ion channel proteins, represented here by MscL. It is worth noting that AQP1 (the water channel) and GlpF (the glycerol transporter) have 7 TM segments, whereas GLUTs have 12 TM helices. Thus, there may be a relationship between the number of TM segments in a transporter protein and the molecular radius of its substrate. Pore size and substrate selectivity are determined in each case by amino acid residues that are highly conserved within a protein family. A somewhat different picture has emerged from electron cryo-microscopy studies of the bacterial Na^+/H^+ antiporter (NhA). A low-resolution (7 Å) structure of this transport protein reveals 12 TM helices arranged in asymmetrical bundles (Williams, 2000). It is not clear which arrangement (symmetrical pore versus asymmetrical bundle) is present in the GLUT protein.

Several groups have devised molecular models of GLUTs to gain insight into possible structure-function relationships. Zeng *et al.* (1996) proposed that the 12 TM helices of GLUT1 are organized into several four-helix bundles. The pore is located at the interface of the bundles and a subset of five TM segments actually forms the opening in their model. The structure proposed in this paper bears a resemblance to the low-resolution image of the NhaA antiporter (Williams, 2000). By contrast, Dwyer (2001) proposed a molecular model of GLUT3 that is based on the MscL ion channel protein as the template (Fig. 2D). The model depicts a symmetrical arrangement of TM segments and an intracellular helix bundle, which may constitute an attractive site for protein-protein interactions involved in trafficking and turnover of GLUTs. Molecular dynamics studies suggest that, during transport of glucose, the TM helices change their tilt and secondary structure (reversible folding/unraveling) to accommodate the movement of substrate through the pore (Dwyer, 2001). This type of mechanism may allow for communication across the membrane bilayer (i.e., between inward-facing versus outward-facing substrate sites) and for the kinetic differences between GLUT isoforms. Fischbarg and colleagues (Zuniga *et al.*, 2001) recently proposed another model for the structure of GLUT1. In this case, the arrangement of the TM helices was based on a variety of observations, including insight from other transporter proteins, results of mutagenesis studies, and various physical constraints (i.e., acceptable ϕ, ψ angles). As seen in Fig. 2E, GLUT1 assumes a compact structure with the interconnecting loops packed against the TM segments. Although there are common features of the various molecular models, it is entirely possible that the actual three-dimensional structure of a GLUT protein, established by crystallographic or other methods, may reveal a different perspective.

III. Survey of GLUT Distribution in the Brain

The literature concerning GLUT 1–5 expression in brain has been extensively reviewed by us previously (Vannucci *et al.*, 1997). Therefore, the older data will be briefly summarized here and the major focus will concern more recent data and the new GLUTs that have been found in brain.

A. GLUT1

GLUT1 is ubiquitously expressed in the brain, as it is in the periphery. A major site of expression is at the blood-brain barrier (BBB)—in the endothelial cells of the cerebral microvasculature (Dick *et al.*, 1984; Gerhart *et al.*, 1989; Boado and Pardridge, 1990). On Western blot analysis of these cells, BBB GLUT1 is detected as a heavily glycosylated, high molecular weight (55 kDa) isoform (Dick *et al.*, 1984; Sivitz *et al.*, 1989; Maher *et al.*, 1992). High levels of the normally glycosylated, 45-kDa, GLUT1 protein have also been observed in the choroid plexus, and the ependymal lining of the cerebral ventricles, with a more diffuse expression in glial elements (Harik *et al.*, 1990; Devaskar *et al.*, 1992; Vannucci, 1994; Morgello *et al.*, 1995). Several studies support regional variations in GLUT1 (both 55 and 45 kDa) in rodent and human brain (Yano *et al.*, 1991; Maher *et al.*, 1992; Brant *et al.*, 1993; Rayner *et al.*, 1994; Simpson *et al.*, 1994; Morgello *et al.*, 1995). Immunohistochemical regional analysis of GLUT1 expression in brain (Zeller *et al.*, 1997), suggests that there is a strong correlation between the density of GLUT1, the density of BBB capillaries, and local cerebral glucose utilization.

B. GLUT2

Of all the family members, it would appear that GLUT2 has the lowest and most restricted expression in brain. By polymerase chain reaction (PCR) methods, mRNA for GLUT2 has been found in the nucleus tractus solaris, the motor nucleus of the vagus, the paraventricular nucleus and lateral area of the hypothalamus, the arcuate nucleus, and olfactory bulbs (Leloup *et al.*, 1994). This pattern of expression was generally confirmed by immunochemistry (Leloup *et al.*, 1994), but not by standard Western blotting (Rayner *et al.*, 1994). Leloup *et al.* (1994) reported that the expression of GLUT2 appeared to be restricted to a subpopulation of astrocytes. Brant *et al.* (1993) used enhanced chemiluminescence methods to detect GLUT2 protein in a number of brain areas, including the pituitary, the optic chiasma, the thalamus, hypothalamus, and cortex. However, the protein levels were very low. In human fetal tissue, GLUT2 immunoreactivity was detected in the granular layer of the cerebellum at 21 weeks, but not at earlier time points (Nualart *et al.,* 1999).

In the β-cells of the pancreatic islets, GLUT2 is coexpressed with glucokinase, and together they are involved in a glucose-sensing mechanism that controls insulin secretion (Sweet and Matschinsky, 1997). A recent study by Maekawa *et al.* (2000) located GLUT2, together with glucokinase, in ependymocytes. A functional role for GLUT2 in sugar transport in the brain remains to be established.

C. GLUT3

GLUT3 is the predominant neuronal glucose transporter (Maher *et al.,* 1992; Nagamatsu *et al.,* 1993; Maher and Simpson, 1994a; McCall *et al.,* 1994; Leino *et al.,* 1997). Regional differences in GLUT3 mRNA as detected by *in situ* hybridization, likely reflect neuronal cell density as well as activity (Vannucci *et al.,* 1998a). However, regional analysis of GLUT3 protein by Western blot suggests a more homogeneous expression in both rodent and human brain (Maher *et al.,* 1992; Simpson *et al.,* 1994). Immunohistochemistry studies generally detect the most intense staining of GLUT3 in the neuropil, reflecting high concentrations in axons and dendrites, (McCall *et al.,* 1994; Leino *et al.,* 1997; Vannucci *et al.,* 1997) consistent with high rates of regional glucose utilization ($1CMR_{glu}$) (Sokoloff, 1976). However, wider variations in observed $1CMR_{glu}$ than levels of immunoreactive GLUT3 have also been reported (Zeller *et al.,* 1995).

D. GLUT4

GLUT4 mRNA and protein expression has been detected by *in situ* hybridization and immunohistochemistry in discrete populations of neurons, and the ependymal cells of the cerebral ventricles (Kobayashi *et al.,* 1996; Vannucci *et al.,* 1998a; Apelt *et al.,* 1999; Maekawa *et al.,* 2000). Neuronal GLUT4 expression is most prominent in cerebellum, where it is detected in both Purkinje cells and granule cells, and in granule cells of the olfactory bulb and hippocampus, with less expression in cortical regions (Kobayashi *et al.,* 1996; Leloup *et al.,* 1996; Vannucci *et al.,* 1998b; El Massari *et al.,* 1998; Apelt *et al.,* 1999; Vannucci *et al.,* 2000). This regional localization has been confirmed by Western blot analysis (Vannucci *et al.,* 2000). Recently, GLUT4 immunoreactivity has been specifically localized to cholinergic cells in basal forebrain of rat, where it is co-expressed with GLUT3

(Apelt *et al.,* 1999). The physiological role of GLUT4 expression in the CNS is still unknown. GLUT4 is the major insulin-responsive transporter in peripheral tissues, such as skeletal muscle, heart, and brown and white adipose tissues. In these tissues, GLUT4 responds to insulin stimulation by translocation from an intracellular membrane compartment to the plasma membrane where it facilitates a rapid increase in glucose uptake. Major sites of expression of GLUT4 in brain are coincident with the expression of both insulin (Devaskar *et al.,* 1994) and insulin receptors (Hill *et al.,* 1986; Lesniak *et al.,* 1988), although cerebral concentrations of insulin are quite low. GLUT4 responds similarly to other stimuli, such as IGF-1 (Kohanski *et al.,* 1986; Bilan *et al.,* 1992), and to contraction and hypoxia, by an AMP-kinase mechanism (see section IV.B.2).

E. GLUT5

GLUT5, a putative fructose transporter expressed primarily in small intestine and sperm, is a very poor transporter of glucose (Burant *et al.,* 1992). The first detection of GLUT5 protein in brain was by Western blot analysis of human brain samples (Shepherd *et al.,* 1992). Immunohistochemical analyses of rat and human brain localized GLUT5 to microglial cells, the resident macrophages of the brain (Maher *et al.,* 1994; Payne *et al.,* 1997), consistent with expression in peripheral macrophages (Malide *et al.,* 1998). In humans, some staining with anti-GLUT5 antibodies has been detected in fetal brain in the cerebellum at about 21 weeks of gestation, but not at later stages of development (Nualart *et al.,* 1999).

F. GLUT6

Initially, the GLUT6 notation was assigned to a pseudogene that was not expressed. With the recent cloning of the new GLUT genes, what was initially referred to as GLUT9 (Doege *et al.,* 2000a), has been reassigned as GLUT6. In the periphery it is detected in the spleen, and in lymphocytes, and of the new GLUTs, it has the highest expression in brain (Doege *et al.,* 2000a). The cells expressing GLUT6 in brain have not been fully characterized, although preliminary studies in our laboratory suggest a primarily neuronal location.

G. GLUT8

GLUT8 represents first of the "new GLUTs," and was thus originally referred to as GLUTx1 (Ibberson *et al.,* 2000). GLUT8 mRNA has been detected in the hypothalamus, cerebellum, brain stem, and hippocampus (Ibberson *et al.,* 2000), where it has been localized to cell bodies and proximal dendrites of pyramidal neurons, granule neurons of the dentate gyrus, and nonprincipal neurons, in a primarily intracellular location (Reagan *et al.,* 2001). This pattern of cellular localization would be expected on the basis of the dileucine motif in the protein, similar to GLUT4 (Verhey and Birnbaum, 1994). This is especially interesting in that the cloning of GLUT8 (GLUTx or GLUTx1) was prompted by GLUT2-knockout (Ibberson *et al.,* 2000) and GLUT4-knockout mice (Katz *et al.,* 1995). Thus, it has been suggested that the expression of GLUT8 at the cell surface is regulated by hormones or other molecules, much like GLUT4 in muscle and adipose cells (Ibberson *et al.,* 2000; Doege *et al.,* 2000b). The potential for a recruitable glucose transporter, in addition to a constitutive neuronal transporter, i.e., GLUT3, could introduce a new level of complexity to cerebral glucose utilization. Clearly, it will be important to characterize the regulation of GLUT8 in neuronal cells. Of the new GLUTs, only GLUT6 and GLUT8 have been described in brain. The extent to which the other new family members will be detected in the CNS should soon be determined.

IV. Regulation of Glucose Transport in Brain

In view of the extensive array of GLUT family members that are expressed in the mammalian brain, an important question is raised: Why are so many different isoforms found in this tissue? Is this simply physiologic redundancy or is it purposeful? It is our hypothesis that the latter is the case and the most likely reasons for this are substrate specificity and activity, as well as different modes of regulation inherent in the different transporter isoforms.

A. Substrate Specificity, Activity, and Concentration

The major GLUT isoforms in mammalian brain are GLUT1 and GLUT3. Detailed quantitative studies using the photolabel, ATB-BMPA, suggest that despite their separate and distinct cellular localizations, they are present in rat brain in nearly identical concentrations (Maher *et al.,* 1994; Vannucci *et al.,* 1997). Relative to GLUT1 and GLUT3, even GLUT4 appears to have a minor level of expression, representing 5–10% of total transporter concentration in cerebellum, its site of highest expression (Vannucci *et al.,* 1997). GLUT1 and GLUT3 have low *Km* for glucose (1–3 m*M*) (Carruthers, 1990; Maher *et al.,* 1996), making these transporters ideally suited for transport at interstitial glucose concentrations in the brain; what distinguishes them is the rate at which they can transport glucose. GLUT3 has a higher turnover number, or catalytic center activity, than GLUT1, 853 sec^{-1} versus 123 sec^{-1} (Carruthers, 1990; Maher *et al.*, 1996; Vannucci *et al.*, 1997), which means that at the prevailing interstitial glucose concentrations the neurons have the capacity to transport seven more times glucose than glial cells expressing an equal number of transporters on the plasma membrane. GLUT2 has the highest *K*m (lowest affinity) for glucose and is thus not readily saturated even at elevated glucose levels enabling it to serve as a mediator of glucose homeostasis in the liver and a glucose-sensor in the β-cell of the pancreas. These well-described functions of GLUT2 in the periphery have provided the basis for proposing a glucose-sensing function for GLUT2 in the brain. GLUT4 has a *K*m of 5 mM and GLUT5 is a fructose, not a glucose, transporter. The *K*m for GLUT6 and GLUT8 remain to be determined.

The expression of multiple GLUTs in the brain may be also be related to the substrates transported in is tissue. Glucose is the major energy source of adult brain and is likely the main substrate in most cells. However, the presence of GLUT5 in microglial cells suggests transport of other substrates/molecules in these cells. The possibility that the newer GLUTs, such as GLUT6 and GLUT8, might transport other substrates besides glucose should be explored. Dehydroascorbic acid is transported by GLUTs 1, 3, and 4 (Vera *et al.*, 1993; Rumsey *et al.*, 1997), but not GLUT2 and GLUT5 (Rumsey *et al.*, 1997). Ascorbic acid plays an important neuromodulatory role in the CNS (Rebec and Pierce, 1994) and GLUTs 1, 3, and 4 may be involved in this function. It will be worthwhile to determine if additional substrates are transported by particular GLUT isoforms in the brain.

B. Modes of Regulation

This section will consider regulation of glucose transport by several different mechanisms: modulation of transporter activity, recruitment of GLUTs from intracellular pools to the plasma membrane, and control of gene/protein expression.

1. *Modulation of GLUT Activity*

The brain is quite heterogeneous relative to its rates of activity, metabolism, and thus glucose utilization. Local glucose metabolic demand is likely met by the regional variations in cerebral blood flow, capillary density, and concentrations of GLUTs, as discussed in Sections III.A and C for GLUT1 and GLUT3. During certain conditions, such as seizures, regional rates of glucose uptake and utilization become acutely increased. The classic response is a corresponding increase in blood flow, and it has been suggested that there may also be a redistribution of GLUT1 transporters between the lumenal and ablumenal membranes of the BBB (Farrell and Pardridge, 1991), as discussed in more detail below. In the light of recent observations, there may be the potential for additional contributions from specific modulations in intrinsic transport activity. It has been reported that palmitoylation (Pouliot and Beliveau, 1995) and conjugation of GLUTs with lipid peroxidation products (Reagan *et al.*, 2000) inhibit glucose transport activity. GLUT1 and GLUT4 share an 11-amino-acid sequence in their C-terminal tails, which can be conjugated with sentrin, a ubiquitin-like protein, by the enzyme mUbc9 (Giorgino *et al.*, 2000). Overexpression of mUbc9 in L6 myoblasts decreased the levels of GLUT1, but increased the levels of GLUT4. Interestingly, although GLUT3 has a very similar target sequence to GLUT1 and GLUT4, the transporter is not modified by mUbc9, suggesting that other recognition sites might exist (Giorgino *et al.*, 2000). Finally, it is possible that noncovalent interactions with proteins or small molecules may also modulate the activity of GLUT proteins (Jung, 1998; Bunn *et al.*, 1999).

Neuromodulators have been reported to enhance glucose transport and GLUT expression in brain. Boado and colleagues (1998, 1999) showed that Cerebrolysin, a peptide-enriched extract from pig brain, increased the expression of GLUT1 protein in endothelial cells of the BBB (Boado, 1998; Boado *et al.*, 1999). These changes in glucose transport capacity may explain some of the neuroprotective effects of Cerebrolysin in patients with Alzheimer disease (Windisch, 2000; Molloy and Standish, 2000). Insulin-like growth factor-1 (IGF-1) is another small polypeptide with important neurotrophic effects in developing brain, including the up-regulation of glucose transport and GLUT expression (Cheng *et al.*, 2000). The effects of IGF-1 on brain glucose metabolism have been reviewed in the chapter by Bondy and Cheng (Bondy, 2002). Nicotine, a cholinergic agonist, has been reported to increase the levels of both GLUT1 and GLUT3 in widespread areas of the brain, including the cortex, amygdala, and hypothalamus (Duelli *et al.*, 1998) in accordance with regional rates of local cerebral glucose utilization.

2. *Recruitment of GLUTs*

Translocation of glucose transporters from intracellular membrane compartments to the cell surface, with the consequent increase in the cellular uptake of glucose, is the primary mode of regulation of glucose transport in the periphery. It is likely that this mechanism operates in the CNS as well, although this has not yet been definitively demonstrated. Translocation in response to insulin was first demonstrated for GLUT1 in adipocytes (Lienhard *et al.*, 1982; Wheeler *et al.*, 1982). Acute transport stimulation, by growth factors, with or without PKC activation, was subsequently shown to promote GLUT1 translocation in a variety of cell types, including glia (Gibbs *et al.*, 1988; Werner *et al.*, 1989; Nishimura and Simpson, 1994; Klip *et al.*, 1994; Malide *et al.*, 1998). Significant pools of intracellular GLUT1 have been identified in endothelial cells of cerebral microvessels, astrocytes, and neurons *in vitro* (Farrell *et al.*, 1992; Maher *et al.*, 1992; Maher, 1995) supporting the potential for acute regulation of transport in these cells.

Insulin-stimulated translocation of GLUT4 has been extensively studied in muscle and adipose tissue, although it is unlikely that insulin is the major regulator of GLUT4 in brain, despite the co-localization with insulin receptors. Potential candidates certainly include both IGF-1 and IGF-2 which, like insulin, are able to stimulate GLUT4 translocation in adipose and muscle tissue (Kohanski *et al.*, 1986; Bilan *et al.*, 1992; Dardevet *et al.*, 1994). Both IGF-1 and IGF-2 are present at high levels in the developing embryo and neonatal animal and could act through either the insulin receptor or the IGF-1 receptor (Stylianopoulou *et al.*, 1988; Torres-Aleman, 1994). It is believed that insulin- or IGF-receptor stimulation initiates a phosphorylation cascade leading to activation of Akt and translocation of GLUT4. Akt is anti-apoptotic and may provide a mechanism for regulating the expression of GLUT4 in the developing embryo. In addition to Akt-mediated translocation, several other stimuli, e.g., contraction and hypoxia, can elicit GLUT4 translocation in muscle and heart. Common to both stimuli is a transient reduction in cellular ATP, with concomitant increase in AMP and stimulation of AMP kinase activity (Hayashi *et al.*, 1998; Kemp *et al.*, 1999; Abbud *et al.*, 2000; Mu *et al.*, 2001). In the CNS such transient reductions in ATP might accompany sustained bursts of neuronal activity. Central to all these mechanisms is the intracellular targeting of GLUT4 to specific vesicles in the basal state, which is believed to be mediated by the presence of dileucine motifs in the protein (Verhey and Birnbaum, 1994). GLUT6 and GLUT8 also have these dileucine motifs, suggesting trafficking and recruitment as modes of regulation for these isoforms. Although GLUT3 lacks this motif, rapid recruitment of GLUT3 to the cell surface in response to activation of PKC has been described in human platelets and neutrophils (Heijnen *et al.*, 1997; Sorbara *et al.*, 1997). Thus, the potential to modulate transport activity acutely and reversibly at the local level clearly exists and may well be involved in the regulation of cellular growth, survival, and differentiation in the CNS.

3. *Transcriptional/Translational Control of GLUT Expression*

GLUT expression is regulated by a number of known transcription factors. Upstream regulatory sites have been identified for the thyroid hormone nuclear receptor, myocyte enhancing factor 2 (MEF2), CCAAT/enhancer binding protein (C/EBP), hypoxia inducible factor-1α (HIF-1α), Sp1, and Sp3 (Ahn *et al.,* 1995; Ezaki, 1997; Iyer *et al.,* 1998). Other groups have reported that peroxisome proliferator-activated receptor (PPAR) (Long and Pekala, 1996), AP-1 (Kozlovsky *et al.,* 1997), and the cAMP responsive element (CRE) (Flores-Riveros *et al.,* 1993) are involved in the regulation of GLUT expression. Certain nuclear factors may control the transcription of a particular GLUT gene. For instance, MEF2 has been implicated in the regulation of GLUT4 expression in muscle cells (Ezaki, 1997) and C/EBP appears to be involved in the down-regulation of GLUT2 expression in cultured hepatocytes (Kim and Ahn, 1998). Moreover, Sp3 downmodulates the expression of GLUT1 during myogenesis, whereas Sp1 stimulates transcription of this gene (Fandos *et al.,* 1999). The reverse of this situation has been reported for the regulation of GLUT3 expression in neuroblasts (Rajakumar *et al.,* 1998). Thus, GLUT expression in neural cells may be regulated by the levels of activated transcription factors (Rajakumar *et al.,* 1998).

The hypoxia-inducible transcription factor, HIF1-α, is integral to the coordination of tissue oxygenation and the transport and metabolism of glucose. The HIF1-α response element was initially identified in the gene for vascular endothelial growth factor (VEGF) which stimulates angiogenesis in response to reduced cellular O_2 content (Semenza and Wang, 1992; Semenza *et al.,* 1996). Subsequently, a wide variety of cell types have been shown to produce adaptive responses to hypoxia, which involve HIF-1α-mediated transcriptional activation of genes encoding GLUT1 and GLUT3, and several glycolytic enzymes (Iyer *et al.,* 1998). Systemic hypoxia produces an increase in glucose transport at the BBB, together with an increase in the density of cerebral microvessels (Harik *et al.,* 1994) and increased expression of GLUT1 and GLUT3 (Harik *et al.,* 1996). Although hypoxia has been reported to increase the level of GLUT3 in neurons *in vivo* (Badr *et al.,* 1999), and *in vitro* (Bruckner *et al.,* 1999), the major physiologic response appears to involve BBB GLUT1 (Harik *et al.,* 1996; Badr *et al.,* 1999). The regulation of GLUT1 expression by hypoxia has recently been extensively reviewed by Zhang and colleagues (1999). They identified two phases in the GLUT1 response to hypoxia: an early phase, characterized by upregulation of GLUT1 activity via translocation of GLUT1 to the cell surface and/or activation of GLUTs already found there and a later response, where cellular levels of GLUT1 mRNA were dramatically elevated as a result of both an increase in transcription and enhanced stability of the GLUT1 transcripts (Zhang *et al.,* 1999). Bruckner *et al.* (1999) reported similar findings in primary cultures of both astrocytes and neurons subjected to hypoxia alone and hypoxia plus glucose deprivation, as the *in vitro* model for ischemia.

Transcriptional control of GLUT expression in the CNS *in vivo* is clearly evident in the response to ischemia. Various groups have shown that ischemia induces GLUT1 and GLUT3 mRNAs in adult (Lee and Bondy, 1993; Gerhart *et al.*, 1994; McCall *et al.*, 1995; Urabe *et al.*, 1996; Liu *et al.*, 1999) and immature rat brain (Vannucci *et al.*, 1996, 1998c). As for systemic hypoxia, the GLUT1 response is most prominent in the microvasculature, although reactive astrocytes also increase their expression of GLUT1 (Lee and Bondy, 1993; Vannucci *et al.*, 1997). Neurons do not normally express GLUT1, however, increased GLUT1 expression has been reported in hippocampal neurons very early in response to ischemia, and may represent a stress response (Lee and Bondy, 1993). The ability of neurons to increase their expression of the neuron-specific transporter, GLUT3, is less clear. Increased GLUT3 expression in response to cerebral ischemia and hypoxia-ischemia has been most commonly observed in penumbral regions, which also exhibit increased rates of glucose utilization (Ginsberg, 1990). Within the ischemic region, isolated "bursts" of GLUT3 mRNA expression have been observed in the immature rat brain (Vannucci *et al.*, 1998c). However, neuronal death is prominent in severe insults such as employed in all of the studies above, and any early increase in GLUT3 expression is ultimately lost. Studies of the precise mechanisms responsible for the upregulation of GLUT1 and GLUT3 expression are complicated by the fact that ischemia is a cessation of blood flow and as such involves a loss of oxygen as well as nutrients, i.e., glucose. Glucose deprivation is well known to affect both GLUT1 and GLUT3 expression in a variety of *in vitro* systems (Haspel *et al.*, 1986; Walker *et al.*, 1988; Wertheimer *et al.*, 1991; Boado and Pardridge, 1993; Klip *et al.*, 1994). Deleterious conditions, such as hypoxia and ischemia, are not the only stimuli that upregulate GLUT transcription. Several brain-derived peptides, such as TNFα and basic FGF, in addition to the phorbol ester TPA, increase BBB GLUT1 abundance both via transcriptional and mRNA stability mechanisms (Farrell *et al.*, 1992; Boado and Pardridge, 1993; Dwyer and Pardridge, 1993; Boado *et al.*, 1994; Boado *et al.*, 1996; Dwyer *et al.*, 1996). As mentioned previously, a commercially prepared brain-derived peptide preparation with neurotrophic properties, Cerebrolysin, induced GLUT1 mRNA in rat brain and cultured brain endothelial cells, largely as a result of enhanced stability of the transcript (Boado *et al.*, 1999; Gschanes *et al.*, 2000). The production of BBB GLUT1 is likely regulated by a variety of neurotrophic peptides and hormones, including circulating estrogen as demonstrated by Shi and colleagues (1997). Interestingly, the GLUT8 protein is subject to regulation by estrogen (Doege *et al.*, 2000b). Another transcription factor that merits further study in brain is PPARδ. A homolog of the consensus PPAR regulatory element is located upstream of the GLUT4 gene (Long and Pekala, 1996). Moreover, pioglitazone, which activates PPARγ, upregulates the expression of GLUT4 in obese Zucker rats (Hallakou *et al.*, 1997). The PPAR transcription factors bind to the retinoid X receptor (RXR) protein and activate gene transcription as a heterodimer. Recently, RXR has been reported to bind docosahexaenoic acid (DHA) (Mata de Urquiza *et al.*, 2000), which is highly enriched in fetal and neonatal brain tissue. Additional research is needed to fully characterize the transcriptional factors that regulate the expression of GLUTs in adult brain.

V. Conditions Associated with Modulation of GLUT Expression in Brain

Three situations where altered transporter expression has been extensively studied but the underlying transcriptional/translation basis has yet to be determined are development/aging, hypoglycemia/hyperglycemia, and neuronal stimulation/deprivation.

A. Development and Aging

Glucose is the predominant energy substrate for the adult brain; most *in vivo* investigations indicate that glucose is also the primary cerebral fuel for fetal and newborn animals—and presumably human infants—under physiologic conditions (Vannucci and Vannucci, 2002). During suckling in rodents, ketone bodies, i.e., β-hydroxybutyrate and acetoacetate, can supplement glucose to maintain normal cerebral metabolism (Hawkins *et al.*, 1971; Nehlig *et al.*, 1991). Human infants have relatively high levels of circulating ketone bodies, supporting the importance of this fuel in normal cerebral development (Stanley *et al.*, 1979). Lactate also contributes to cerebral energy metabolism in the immature brain during periods of glucose deficiency (Hellmann *et al.*, 1982; Vicario and Medina, 1992). Cerebral energy requirements of most mammals at birth are low, reflecting an immaturity of the CNS, and increase with brain growth and functional maturation. Studies from our own and other laboratories demonstrate that cerebral maturation of glucose transporter expression directly corresponds to levels of energy demand and available substrate. Thus GLUT1 levels in the BBB increase developmentally with the increased reliance on carbohydrate metabolism at the time of weaning; GLUT3 mRNA and protein increase in a temporal and regional manner coincident with neuronal maturation, synaptogenesis, and functional activity (Sivitz *et al.*, 1989; Devaskar *et al.*, 1991; Vannucci, 1994; Vannucci *et al.*, 1998a).

Although it was originally maintained that GLUT1 was the sole glucose transporter of fetal life, it is now appreciated that other isoforms are expressed embryonically. mRNAs for GLUT1 and GLUT3 have been detected in embryonic tissues of mouse and rat, including the neural tube, at 7–9 days of gestation (Smith and Gridley, 1992; Takao *et al.*, 1993) and protein is also present at this time. Increased levels of GLUT1 expression at E10–E12 (Smith and Gridley, 1992; Takao *et al.*, 1993; Vannucci *et al.*, 2000) coincide with an early increase in glucose utilization in the embryo (Tanimura and Shepard, 1970). GLUT1 mRNA is detected in the microvasculature of the developing nervous system as early as E13 in the rat (Bondy *et al.*, 1992; Vannucci *et al.*, 2000), coincident with the earliest expression of barrier properties in these vessels (Laterra *et al.*, 1992). GLUT3 mRNA is detected in the preimplantation embryo (Pantaleon *et al.*, 1997), and subsequently is expressed in early embryonic neurons, such as the large motor neurons of the dorsal horn of the spinal cord (Bondy *et al.*, 1992), and in the more primitive cerebral areas of the midbrain (Vannucci *et al.*, 2000). During fetal development (E14–E19) in the rat, the expression of GLUT3 in the brain is associated with the levels of HIF-1α and the thyroid hormone nuclear receptor subunit TRβ1 (Royer *et al.*, 2000). GLUT4 has a discrete embryonic expression, primarily in the ventricular cells, choroid plexus, and external granular cells of the developing cerebellum (Royer *et al.*, 2000; Vannucci *et al.*, 2000).

With normal aging, there appears to be a progressive decline in cerebral glucose metabolism in rodents (Patel, 1977; Leong *et al.*, 1981), although this is not a universal finding in humans (Horwitz *et al.*, 1986). Early studies in rats showed that one effect of aging was a decrease in the influx of glucose into the CNS apparently as a result of decreased transport (Daniel *et al.*, 1978). Vorbrodt *et al.* (1999) have reported reductions in GLUT1 in the cerebral vasculature with aging that varied depending upon the brain region. An age-related decrease in GLUT1 expression in brain has been confirmed by Gschanes *et al.* (2000). Finally, there is some evidence for age-related changes in GLUT3 expression especially in discrete areas of the hippocampus, including the dentate gyrus and the mossy fibers of the CA3 region (Fattoretti *et al.*, 2001). Again, this does not appear to reflect the human condition (Simpson *et al.*, 1994).

In the clinical setting, changes in cerebral glucose metabolism are more clearly associated with abnormal degenerative conditions, such as Alzheimer disease and Parkinson disease. It is now well-established that Alzheimer disease is associated with major metabolic deficiencies in the brain, in particular a decrease in glucose utilization (Friedland *et al.*, 1989; Hoyer, 1996; Ogawa *et al.*, 1996; Heininger, 2000). It has been suggested that the impairment in cerebral glucose metabolism is a cause rather than a result of neuronal degeneration in Alzheimer patients (Hoyer, 1996). There is widespread reduction in both GLUT1 (Kalaria and Harik, 1989) and GLUT3 in brain tissue obtained from patients that may underlie the deficits in glucose utilization (Simpson *et al.*, 1994). The loss of GLUT1 from cerebral vascular endothelium in Alzheimer disease has been confirmed by others (Horwood and Davies, 1994; Mooradian *et al.*, 1997). Although reductions in GLUT3 could readily be attributed to neuronal loss, absolute loss of GLUT3 in the hippocampus and temporal and parietal cortices of Alzheimer patients was still observed even after correcting for loss of synaptic density by levels of SNAP-25 protein (Simpson *et al.*, 1994). GLUT3 reductions have been confirmed in discrete brain regions such as the molecular layer of the dentate gyrus, which may reflect cell loss in the entorhinal cortex (Harr *et al.*, 1995).

B. Hypoglycemia/Hyperglycemia

Neurological symptoms of hypoglycemia are manifested when blood glucose concentrations fall below ~2 mM and include confusion and cognitive impairment. Coma and even death can result if the circulating concentration of glucose falls below 1 mM (Sokoloff, 1976). The brain adapts to hypoglycemia by increasing glucose transport and the expression of GLUT1 at the BBB. Early studies demonstrated that chronic hypoglycemia induced by insulin administration was accompanied by an increase in the permeability of brain capillaries to glucose (McCall *et al.*, 1986; Pelligrino *et al.*, 1990). McCall and colleagues (1986) attributed the increase in glucose uptake to enhanced expression of GLUTs at the BBB. Increased GLUT1 mRNA and protein expression in response to hypoglycemia has been confirmed by others (Kumagai *et al.*, 1995; Simpson *et al.*, 1999). The latter study demonstrated both an increase in total microvascular GLUT1 and a specific increase in the levels of GLUT1 in the lumen. However, no effect of hypoglycemia was observed on GLUT1 expression in astrocytes or GLUT3 in neurons.

In poorly controlled diabetes mellitus (both type 1 and type 2), hyperglycemia is a significant problem and it is associated with an increased risk of heart disease, stroke, peripheral neuropathy, and other serious complications. Early studies found a significant reduction in glucose transport at the BBB in rats with chronic hyperglycemia induced by treatment with streptozotocin (Gjedde and Crone, 1981; McCall *et al.*, 1982). Some studies found a commensurate decrease in GLUT1 in cerebral microvessels of diabetic rats (Pardridge *et al.*, 1990; Mooradian and Morin, 1991), whereas others found a very modest reduction of 7.5 % (Duelli *et al.*, 2000), or no change at all in GLUT1 expression (Simpson *et al.*, 1999). The level of GLUT3 in neuronal cell populations was unaltered in the hyperglycemic animals in the studies of Simpson *et al.* (1999) and Duelli *et al.* (2000), however, specific increases in GLUT3 mRNA and protein were observed in the hippocampus of the diabetic animal with further enhancement when animals were subjected to stress (Reagan *et al.*, 1999). In spontaneously diabetic mice (the *db/db* strain), there were no changes in GLUT1 or GLUT3 protein expression in the brain. However, the level of GLUT1 mRNA in microvessels in both *db/db* mice and the STZ diabetic rats was increased (Vannucci *et al.*, 1997; Simpson *et al.*, 1999).

C. Neuronal Stimulation/Deprivation

There is a close, but not perfect, relationship between the metabolic activity of the various brain regions and the levels of GLUT1 and GLUT3 expression (Zeller *et al.* 1995; Vannucci *et al.*, 1998a). A similar relationship between metabolic activity and the expression of GLUT1 and GLUT4 is seen in muscle, where both transporters and transport activity are increased in response to chronic electrical stimulation and exercise training (Ploug *et al.*, 1990; Etgen *et al.*, 1993a,b; Hofmann and Pette, 1994; Chilibeck *et al.*, 1999). Prolonged seizure activity represents an analogous situation in the CNS. Seizures induce a significant increase in local cerebral glucose utilization (Lothman and Collins, 1981), however, hypometabolism is also observed with sustained neuronal firing. Chronic seizures are associated with increased GLUT1 at the BBB (Gronlund *et al.*, 1996; Cornford *et al.*, 1998) and GLUT3 in many areas of the brain (Gronlund *et al.*, 1996). The rise in GLUT3 precedes that of GLUT1 suggesting that metabolic demand increases most acutely in the affected neurons. Early studies by Kadekaro *et al.* (1992) demonstrated that dehydration due to water deprivation, which stimulates the hypothalamic-neurohypophysial axis and the secretion of vasopressin, increased cerebral glucose utilization ($lCMR_{glu}$) in the neural lobe of the pituitary . These studies were subsequently confirmed by others (Vannucci *et al.*, 1994; Koehler-Stec *et al.*, 2000) who demonstrated that the increased neuronal activity in the neurohypophysis due to 3 days of water deprivation was accompanied by an increase in the expression of both GLUT1 and GLUT3 protein in this region. Moreover, $lCMR_{glu}$ and GLUT1 levels returned to normal within 24 hours of rehydration, whereas the levels of GLUT3 required 7 days for normalization. Similar observations were made in an *in vitro* paradigm in which cerebellar granule neurons were grown in the presence of depolarizing concentrations of potassium or glutamate. This continual stimulation resulted in an increase in glucose transport activity and in GLUT1 and GLUT3 expression (Maher and Simpson, 1994b).

In a related vein, there is a clear relationship between glucose transport activity and cell survival. This was demonstrated by Deckwerth and Johnson (1993) who found that a striking reduction in glucose uptake was the first detectable change in neuronal cells undergoing apoptosis in response to withdrawal of nerve growth factor. Along similar lines, several groups have reported that cells are more susceptible to harmful conditions when fuels other than glucose are used as the major energy source or when there is inefficient utilization of glucose (Kan, 1994; Goossens *et al.*, 1996; Dwyer *et al.*, 1999). Furthermore, it has been suggested that high levels of GLUT expression may provide protection against programmed cell death (Moley and Mueckler, 2000; Gupta *et al.*, 2001). Sapolsky's group (Lawrence *et al.*, 1995, 1996) has shown that increased GLUT expression in brain (with viral vectors coding for GLUT1) protected against injury from seizures and stroke. Conversely, neurons that express lower levels of GLUTs, perhaps as a consequence of decreased overall synaptic activity, may be more susceptible to injury and death at critical stages of development. Consistent with this notion, it has recently been shown that neuronal inactivity (chronic visual deprivation) is accompanied by a significant decrease in both glucose utilization and GLUT expression (Duelli *et al.*, 1998a). This may also be the case in Alzheimer disease where reductions in $lCMR_{glu}$, are associated with decreases in GLUT1 and GLUT3; the question then arises as to whether the affected neurons may have difficulty in maintaining high levels of firing due to limitations in their energy supply, which ultimately contributes to their demise.

In summary, there are a number of physiologic and pathologic conditions in which alterations in cerebral glucose metabolism are associated with alterations in GLUT expression and/or function. These *in vivo* situations, and appropriately designed *in vitro* studies, should provide the experimental paradigms to further investigate transcriptional control of GLUT expression in the brain, including potential involvement of the new GLUTs in addition to GLUTs 1, 3, and 4. Transcriptional regulation may represent an attractive target for drug development aimed at optimizing the expression of GLUTs in the brain in order to minimize cellular injury in devastating clinical disorders, such as stroke and neurodegenerative disease.

References

Abbud, W., Habinowski, S., Zhang, J. Z., Kendrew, J., Elkairi, F. S., Kemp, B. E., Witters, L. A., and Ismail-Beigi, F. (2000). Stimulation of AMP-activated protein kinase (AMPK) is associated with enhancement of Glut1-mediated glucose transport. *Arch. Biochem. Biophys.* **380,** 347–352.

Ahn, Y. H., Kim, J. W., Han, G. S., Lee, B. G., and Kim, Y. S. (1995). Cloning and characterization of rat pancreatic beta-cell/liver type glucose transporter gene: A unique exon/intron organization. *Arch. Biochem. Biophys.* **323,** 387–396.

Alvarez, J., Lee, D. C., Baldwin, S. A., and Chapman, D. (1987). Fourier transform infrared spectroscopic study of the structure and conformational changes of human erythrocyte glucose transporter. *J. Biol. Chem.* **262,** 3502–3509.

Apelt, J. A., Mehlhorn, G., and Schliebs, R. (1999). Insulin-sensitive GLUT4 glucose transporters are colocalized with GLUT3-expressing cells and demonstrate a chemically distinct neuron-specific localization in rat brain. *J. Neurosci. Res.* **57,** 693–705.

Badr, G. A., Zhang, J. Z., Tang, J., Kern, T. S., and Ismail-Beigi, F. (1999). Glut1 and Glut3 expression, but not capillary density, is increased by cobalt chloride in rat cerebrum and retina. *Brain Res. Mol. Brain Res.* **64,** 24–33.

Baldwin, S. A. (1993). Mammalian passive glucose transporters: Members of an ubiquitous family of active and passive transport proteins. *Biochim. Biophys. Acta* **1154,** 17–49.

Bilan, P. J., Mitsumoto, Y., Maher, F., Simpson, I. A., and Klip, A. (1992). Detection of the GLUT3 facilitative glucose transporter in rat L6 muscle cells: regulation by cellular differentiation, insulin and insulin-like growth factor-I. *Biochem. Biophys. Res. Commun.* **186,** 1129–1137.

Boado, R. J. (1998). Brain-derived peptides increase blood-brain barrier GLUT1 glucose transporter gene expression via mRNA stabilization. *Neurosci. Lett.* **255,** 147–150.

Boado, R. J., and Pardridge, W. M. (1990). The brain-type glucose transporter mRNA is specifically expressed at the blood-brain barrier. *Biochem. Biophys. Res. Commun.* **166,** 174–179.

Boado, R. J., and Pardridge, W. M. (1993). Glucose deprivation causes posttranscriptional enhancement of brain capillary endothelial glucose transporter gene expression via GLUT1 mRNA stabilization. *J. Neurochem.* **60,** 2290–2296.

Boado, R. J., Tsukamoto, H., and Pardridge, W. M. (1996). Evidence for translational control elements within the 5′-untranslated region of GLUT1 glucose transporter mRNA. *J. Neurochem.* **67,** 1335–1343.

Boado, R. J., Wang, L., and Pardridge, W. M. (1994). Enhanced expression of the blood-brain barrier GLUT1 glucose transporter gene by brain-derived factors. *Brain Res. Mol. Brain Res.* **22,** 259–267.

Boado, R. J., Wu, D., and Windisch, M. (1999). In vivo upregulation of the blood-brain barrier GLUT1 glucose transporter by brain-derived peptides. *Neurosci. Res.* **34,** 217–224.

Bondy, C. A., and Cheng, C. M. (2002). IGF-1 and brain glucose metabolism. *In* "International Review of Neurobiology" (D. Dwyer, ed.), Vol. 51. Academic Press, San Diego, CA.

Bondy, C. A., Lee, W., and Zhou, J. (1992). Ontogeny and cellular distribution of brain glucose transporter gene expression. *Mol. Cell. Neurosci.* **3,** 305–314.

Brant, A. M., Jess, T. J., Milligan, G., Brown, C. M., and Gould, G. W. (1993). Immunological analysis of glucose transporters expressed in different regions of the rat brain and central nervous system. *Biochem. Biophys. Res. Commun.* **192,** 1297–1302.

Bruckner, B. A., Ammini, C. V., Otal, M. P., Raizada, M. K., and Stacpoole, P. W. (1999). Regulation of brain glucose transporters by glucose and oxygen deprivation. *Metabolism* **48,** 422–431.

Bunn, R. C., Jensen, M. A., and Reed, B. C. (1999). Protein interactions with the glucose transporter binding protein GLUT1CBP that provide a link between GLUT1 and the cytoskeleton. *Mol. Biol. Cell* **10,** 819–832.

Burant, C. F., Takeda, J., Brot-Laroche, E., Bell, G. I., and Davidson, N. O. (1992). Fructose transporter in human spermatozoa and small intestine is GLUT5. *J. Biol. Chem.* **267,** 14523–14526.

Carruthers, A. (1990). Facilitated diffusion of glucose. *Physiol. Rev.* **70,** 1135–1176.

Chang, G., Spencer, R. H., Lee, A. T., Barclay, M. T., and Rees, D. C. (1998). Structure of the MscL homolog from *Mycobacterium tuberculosis:* A gated mechanosenstive ion channel. *Science* **282,** 2220–2226.

Cheng, C. M., Reinhardt, R. R., Lee, W.-H., Joncas, G., Patel, S. C., and Bondy, C. A. (2000). Insulin-like growth factor 1 regulates developing brain glucose metabolism. *Proc. Natl. Acad. Sci. USA* **97,** 10236–10241.

Chilibeck, P. D., Bell, G., Jeon, J., Weiss, C. B., Murdoch, G., MacLean, I., Ryan, E., and Burnham, R. (1999). Functional electrical stimulation exercise increases GLUT-1 and GLUT-4 in paralyzed skeletal muscle. *Metabolism* **48,** 1409–1413.

Chin, J. J., Jung, E. K. Y., and Jung, C. Y. (1987). Structural basis of human erythrocyte glucose transporter function in proteoliposome vesicles: circular dichroism measurements. *Proc. Natl. Acad. Sci. USA* **84,** 4113–4116.

Cornford, E. M., Hyman, S., Cornford, M. E., Landaw, E. M., and Delgada-Escueta, A. V. (1998). Interictal seizure resections show two configurations of endothelial GLUT1 glucose transporter in the human blood-brain barrier. *J. Cereb. Blood Flow Metab.* **18,** 26–42.

Czech, M. P., Clancy, B. M., Pessino, A., Woon, C.-W., and Harrison, S. A. (1992). Complex regulation of simple sugar transport in insulin-responsive cells. *Trends Biochem. Sci.* **17,** 197–201.

Daniel, P. M., Love, E. R., and Pratt, O. E. (1978). The effect of age upon the influx of glucose into the brain. *J. Physiol.* **274,** 141–148.

Dardevet, D., Sornet, C., Attaix, D., Baracos, V. E., and Grizard, J. (1994). Insulin-like growth factor-1 and insulin resistance in skeletal muscles of adult and old rats. *Endocrinology* **134,** 1475–1484.

Deckwerth, T. L., and Johnson, E. M., Jr. (1993). Temporal analysis of events associated with programmed cell death (apoptosis) of sympathetic neurons deprived of nerve growth factor. *J. Cell Biol.* **123,** 1207–1222.

Devaskar, S., Chunda, K., Zahm, D. S., Holtzclaw, L., and Holloran, K. (1992). The neonatal rabbit brain glucose transporter. *Dev. Brain Res.* **67,** 95–103.

Devaskar, S. U., Giddings, S. J., Rajakumar, P. A., Carnaghi, L. R., Menon, R. K., and Zahm, D. S. (1994). Insulin gene expression and insulin synthesis in mammalian neuronal cells. *J. Biol. Chem.* **269,** 8445–8454.

Devaskar, S., Zahm, D. S., Holtzclaw, L., Chundu, K., and Wadzinski, B. E. (1991). Developmental regulation of the distribution of rat brain insulin-insensitive (Glut1) glucose transporter. *Endocrinology* **129,** 1530–1540.

Dick, A. P., Harik, S. I., Klip, A., and Walker, D. M. (1984). Identification and characterization of the glucose transporter of the blood-brain barrier by cytochalasin B binding and immunological reactivity. *Proc. Natl. Acad. Sci. USA* **81,** 7233–7237.

Doege, H., Bocianski, A., Joost, H.-G., and Schuermann, A. (2000a). Activity and genomic organization of human glucose transporter 9 (GLUT9), a novel member of the family of sugar-transport facilitators predominantly expressed in brain and leucocytes. *Biochem. J.* **350,** 771–776.

Doege, H., Schuermann, A., Bahrenberg, G., Brauers, A., and Joost, H.-G. (2000b). GLUT8, a novel member of the sugar transport facilitator family with glucose transport activity. *J. Biol. Chem.* **275,** 16275–16280.

Duelli, R., Maurer, M. H., and Kuschinsky, W. (1998a). Decreased glucose transporter densities, rate constants and glucose utilization in visual structures of rat brain during chronic visual deprivation. *Neurosci. Lett.* **250,** 49–52.

Duelli, R., Maurer, M. H., Staudt, R., Heiland, S., Duembgen, L., and Kuschinsky, W. (2000). Increased cerebral glucose utilization and decreased glucose transporter Glut1 during chronic hyperglycemia in rat brain. *Brain Res.* **858,** 338–347.

Duelli, R., Staudt, R., Gruenwald, F., and Kuschinsky, W. (1998b). Increase of glucose transporter densities (Glut1 and Glut3) during chronic administration of nicotine in rat brain. *Brain Res.* **782,** 36–42.

Dwyer, D. S. (2001). Model of the 3-D structure of the GLUT3 glucose transporter and molecular dynamics simulation of glucose transport. *Proteins Struct. Func. Genet.* **42,** 531–541.

Dwyer, D. S., Liu, Y., and Bradley, R. J. (1999). An ethanol-sensitive variant of the PC12 neuronal cell line: sensitivity to ethanol is associated with increased cell adhesion and decreased glucose accumulation. *J. Cell. Physiol.* **178,** 93–101.

Dwyer, K. J., Boado, R. J., and Pardridge, W. M. (1996). Cis-element/cytoplasmic protein interaction within the 3′-untranslated region of the GLUT1 glucose transporter mRNA. *J. Neurochem.* **66,** 449–458.

Dwyer, K. J., and Pardridge, W. M. (1993). Developmental modulation of blood-brain barrier and choroid plexus GLUT1 glucose transporter messenger ribonucleic acid and immunoreactive protein in rabbits. *Endocrinology* **132,** 558–565.

El Messari, S., Leloup, C., Quignon, M., Brisorgueil, M.-J., Penicaud, L., and Arluison, M. (1998). Immunocytochemical localization of the insulin-responsive glucose transporter 4 (Glut4) in the rat central nervous system. *J. Comp. Neurol.* **399,** 492–512.

Etgen, G. J., Jr., Brozinick, J. T., Jr., Kang, H. Y., and Ivy, J. L. (1993a). Effects of exercise training on skeletal muscle glucose uptake and transport. *Am. J. Physiol.* **264,** C727–C733.

Etgen, G. J., Jr., Farrar, R. P., and Ivy, J. L. (1993b). Effect of chronic electrical stimulation on GLUT-4 protein content in fast-twitch muscle. *Am. J. Physiol.* **264,** R816–R819.

Ezaki, O. (1997). Regulatory elements in the insulin-responsive glucose transporter (GLUT4) gene. *Biochem. Biophys. Res. Commun.* **241,** 1–6.

Fandos, C., Sanchez-Feutrie, M., Santalucia, T., Vinals, F., Cadefau, J., Guma, A., Cusso, R., Kaliman, P., Canicio, J., Palacin, M., and Zorzano, A. (1999). GLUT1 glucose transporter gene transcription is repressed by Sp3. Evidence for a regulatory role of Sp3 during myogenesis. *J. Mol. Biol.* **294,** 103–119.

Farrell, C. R., Boado, R. J., and Pardridge, W. M. (1992). Enhanced GLUT1 glucose transporter and cytoskeleton gene expression in cultured bovine brain capillary endothelial cells after treatment with phorbol esters and serum. *Brain Res. Mol. Brain Res.* **15,** 221–226.

Farrell, C. L., and Pardridge, W. M. (1991). Blood-brain barrier glucose transporter is asymmetrically distributed on brain capillary endothelial lumenal and ablumenal membranes: An electron microscopic immunogold study. *Proc. Natl. Acad. Sci. USA* **88,** 5779–5783.

Farrell, C. L., Yang, J., and Pardridge, W. M. (1992). GLUT-1 glucose transporter is present within apical and basolateral membranes of brain epithelial interfaces and in microvascular endothelia with and without tight junctions. *J. Histochem. Cytochem.* **40,** 193–199.

Fattoretti, P., Bertoni-Freddari, C., Di Stefana, G., Casoli, T., Gracciotti, N., Solazzi, M., and Pompei, P. (2001). Quantitative immunohistochemnistry of glucose transport protein (Glut3) expression in the rat hippocampus during aging. *J. Histochem. Cytochem.* **49,** 671–672.

Flores-Riveros, J. R., Kaestner, K. H., Thompson, K. S., and Lane, M. D. (1993). Cyclic AMP-induced transcriptional repression of the insulin-responsive glucose transporter (GLUT4) gene: Identification of a promoter region required for down-regulation of transcription. *Biochem. Biophys. Res. Commun.* **194,** 1148–1154.

Friedland, R. P., Jagust, W. J., Huesman, R. H., Koss, E., Knittel, B., Mathis, C. A., Ober, B. A., Mazoyer, B. M., and Budinger, T. F. (1989). Regional cerebral glucose transport and utilization in Alzheimer's disease. *Neurology* **39,** 1427–1434.

Fu, D., Libson, A., Mierke, L. J., Weitzman, C., Nollert, P., Krucinski, J., and Stroud, R. M. (2000). Structure of a glycerol-conducting channel and the basis for its selectivity. *Science* **290,** 481–486.

Gerhart, D. Z., LeVasseur, R. J., Broderius, M. A., and Drewes, L. R. (1989). Glucose transporter localization in brain using light and electron immunocytochemistry. *J. Neurosci. Res.* **22,** 464–472.

Gerhart, D. Z., Leino, R. L., Taylor, W. E., Borson, N. D., and Drewes, L. R. (1994). GLUT1 and GLUT3 gene expression in gerbil brain following brief ischemia: An in situ hybridization study. *Brain Res. Mol. Brain Res.* **25,** 313–322.

Gibbs, E. M., Lienhard, G. E., and Gould, G. W. (1988). Insulin-induced translocation of glucose transporters to the plasma membrane precedes full stimulation of hexose transport. *Biochemistry* **27,** 6681–6685.

Ginsberg, M. D. (1990). Local metabolic responses to cerebral ischemia. *Cerebrovasc. Brain Metab. Rev.* **2,** 58–93.

Giorgino, F., de Robertis, O., Laviola, L., Montrone, C., Perrini, S., McCowen, K. C., and Smith, R. J. (2000). The sentrin-conjugating enzyme mUbc9 interacts with GLUT4 and GLUT1 glucose transporters and regulates transporter levels in skeletal muscle cells. *Proc. Natl. Acad. Sci. USA* **97,** 1125–1130.

Gjedde, A., and Crone, C. (1981). Blood-brain glucose transfer: Repression in chronic hyperglycemia. *Science* **214,** 456–457.

Goossens, V., Grooten, J., and Fiers, W. (1996). The oxidative metabolism of glutamine. *J. Biol. Chem.* **271,** 192–196.

Gould, G. W., and Holman, G. D. (1993). The glucose transporter family: Structure, function and tissue-specific expression. *Biochem. J.* **295,** 329–341.

Gronlund, K. M., Gerhart, D. Z., Leino, R. L., McCall, A. L., and Drewes, L. R. (1996). Chronic seizures increase glucose transporter abundance in rat brain. *J. Neuropathol. Exp. Neurol.* **55,** 832–840.

Gschanes, A., Boado, R., Sametz, W., and Windisch, M. (2000). The drug cerebrolysin and its peptide fraction E021 increase the abundance of the blood-brain barrier GLUT1 glucose transporter in brains of young and old rats. *Histochem. J.* **32,** 71–77.

Gupta, A., Ho, D. Y., Brooke, S., Franklin, L., Roy, M., McLaughlin, J., Fink, S. L., and Sapolsky, R. M. (2001). Neuroprotective effects of an adenoviral vector expressing the glucose transporter: a detailed description of the mediating cellular events. *Brain Res.* **908,** 49–57.

Hallakou, S., Doare, L., Foufelle, F., Kergoat, M., Guerre-Millo, M., Berthault, M.-F., Dugail, I., Morin, J., Auwerx, J., and Ferre, P. (1997). Pioglitazone induces in vivo adipocyte differentiation in the obese Zucker fa/fa rat. *Diabetes* **46,** 1393–1399.

Harik, S. I., Behmand, R. A., and LaManna, J. C. (1994). Hypoxia increases glucose transport at blood-brain barrier. *J. Appl. Physiol.* **77,** 896–901.

Harik, N., Harik, S. I., Kuo, N. T., Sakai, K., Przybylski, R. J., and LaManna, J. C. (1996). Time-course and reversibility of the hypoxia-induced alterations in cerebral vascularity and cerebral capillary glucose transporter density. *Brain Res.* **737,** 335–338.

Harik, S. I., Kalaria, R. N., Andersson, L., Lundahl, P., and Perry, G. (1990). Immunocytochemical localization of the erythroid glucose transporter: abundance in tissues with barrier functions. *J. Neurosci.* **10,** 3862–3872.

Harr, S. D., Simonian, N. A., and Hyman, B. T. (1995). Functional alterations in Alzheimer's disease: Decreased glucose transporter 3 immunoreactivity in the perforant pathway terminal zone. *J. Neuropathol. Exp. Neurol.* **54,** 38–41.

Haspel, H. C., Wilk, E. W., Birnbaum, M. J., Cushman, S. W., and Rosen, O. M. (1986). Glucose deprivation and hexose transporter polypeptides of murine fibroblasts. *J. Biol. Chem.* **261,** 6778–6789.

Hawkins, R. A., Williamson, D. H., and Krebs, H. A. (1971). Ketone-body utilization by adult and suckling rat brain in vivo. *Biochem. J.* **122,** 13–18.

Hayashi, T., Hirshman, M. F., Kurth, E. J., Winder, W. W., and Goodyear, L. J. (1998). Evidence for 5′ AMP-activated protein kinase mediation of the effect of muscle contraction on glucose transport. *Diabetes* **47,** 1369–1373.

Hebert, D. N., and Carruthers, A. (1992). Glucose transporter oligomeric structure determines transporter function. Reversible redox-dependent interconversions of tetrameric and dimeric GLUT1. *J. Biol. Chem.* **267,** 23829–23838.

Heijnen, H. F., Oorschot, V., Sixma, J. J., Slot, J. W., and James, D. E. (1997). Thrombin stimulates glucose transport in human platelets via the translocation of the glucose transporter GLUT-3 from alpha-granules to the cell surface. *J. Cell Biol.* **138,** 323–330.

Heininger, K. (2000). A unifying hypothesis of Alzheimer's disease. IV. Causation and sequence of events. *Rev. Neurosci.* **11,** 213–328.

Hellmann, J., Vannucci, R. C., and Nardis, E. E. (1982). Blood-brain barrier permeability to lactic acid in the newborn dog: lactate as a cerebral metabolic fuel. *Pediatr. Res.* **16,** 40–44.

Hill, J. M., Lesniak, M. A., Pert, C. B., and Roth, J. (1986). Autoradiographic localization of insulin receptors in rat brain: prominence in olfactory bulbs and limbic areas. *Neuroscience* **17,** 1127–1138.

Hofmann, S., and Pette, D. (1994). Low-frequency stimulation of rat fast-twitch muscle enhances the expression of hexokinase II and both the translocation and expression of glucose transporter-4 (GLUT-4). *Eur. J. Biochem.* **219,** 307–315.

Horwitz, B., Duara, R., and Rapoport, S. I. (1986). Age differences in intercorrelations between regional cerebral metabolic rates for glucose. *Ann. Neurol.* **19,** 60–67.

Horwood, N., and Davies, D. C. (1994). Immunolabelling of hippocampal microvessel glucose transporter protein is reduced in Alzheimer's disease. *Virchows Arch.* **425,** 69–72.

Hoyer, S. (1996). Oxidative metabolism deficiencies in brains of patients with Alzheimer's disease. *Arch. Neurol. Scand. Suppl.* **165,** 18–24.

Ibberson, M., Uldry, M., and Thorens, B. (2000). GLUTX1, a novel mammalian glucose transporter expressed in the central nervous system and insulin-sensitive tissues. *J. Biol. Chem.* **275,** 4607–4612.

Iyer, N. V., Kotch, L. E., Agani, F., Leung, S. W., Laughner, E., Wenger, R. H., Gassmann, M., Gearhart, J. D., Lawler, A. M., Yu, A. Y., and Semenza, G. L. (1998). Cellular and developmental control of O_2 homeostasis by hypoxia-inducible factor 1α. *Genes Dev.* **12,** 149–162.

Joost, H.-G., and Thorens, B. (2001). The extended GLUT-family of sugar/polyol transport facilitators—nomenclature, sequence characteristics, and potential function of its novel members. *Mol. Membr. Biol.* **18,** 247–256.

Jung, C. Y. (1998). Proteins that interact with facilitative glucose transporters: Implications for function. *Exp. Physiol.* **83,** 267–273.

Kadekaro, M., Summy-Long, J. Y., Freeman, S., Harris, J. S., Terrell, M. L., and Eisenberg, H. M. (1992). Cerebral metabolic responses and vasopressin and oxytocin secretions during progressive water deprivation in rats [published erratum appears in *Am. J. Physiol.* 1992 Jun 262(6 Pt 3):section R following table of contents]. *Am. J. Physiol.* **262,** R310–317.

Kalaria, R. N., and Harik, S. I. (1989). Reduced glucose transport at the blood-brain barrier and in cerebral cortex in Alzheimer disease. *J. Neurochem.* **53,** 1083–1088.

Kan, O., Baldwin, S., and Whetton, A. (1994). Apoptosis is regulated by the rate of glucose transport in an interleukin 3-dependent cell line. *J. Exp. Med.* **180,** 917–923.

Katz, E. B., Stenbit, A. E., Hatton, K., DePinho, R., and Charron, M. J. (1995). Cardiac and adipose tissue abnormalities but not diabetes in mice deficient in GLUT4. *Nature* **377,** 151–155.

Kemp, B. E., Mitchelhill, K. I., Stapleton, D., Michell, B. J., Chen, Z. P., and Witters, L. A. (1999). Dealing with energy demand: the AMP-activated protein kinase. *Trends Biochem. Sci.* **24,** 22–25.

Kim, J. W., and Ahn, Y. H. (1998). CCAAT/enhancer binding protein regulates the promoter activity of the rat GLUT2 glucose transporter gene in liver cells. *Biochem. J.* **336,** 83–90.

Klip, A., Tsakiridis, T., Marette, A., and Ortiz, P. A. (1994). Regulation of expression of glucose transporters by glucose: a review of studies in vivo and in cell cultures. *FASEB J.* **8,** 43–53.

Kobayashi, M., Nikami, H., Morimatsu, M., and Saito, M. (1996). Expression and localization of insulin-regulatable glucose transporter (GLUT4) in rat brain. *Neurosci. Lett.* **213,** 103–106.

Koehler-Stec, E. M., Li, K., Maher, F., Vannucci, S. J., Smith, C. B., and Simpson, I. A. (2000). Cerebral glucose utilization and glucose transporter expression: Response to water deprivation and restoration. *J. Cereb. Blood Flow Metab.* **20,** 192–200.

Kohanski, R. A., Frost, S. C., and Lane, M. D. (1986). Insulin-dependent phosphorylation of the insulin receptor-protein kinase and activation of glucose transport in 3T3-L1 adipocytes. *J. Biol. Chem.* **261,** 12272–12281.

Kozlovsky, N., Rudich, A., Potashnik, R., Ebina, Y., Murakami, T., and Bashan, N. (1997). Transcriptional activation of the Glut1 gene in response to oxidative stress in L6 myotubes. *J. Biol. Chem.* **272,** 33367–33372.

Kumagai, A. K., Kang, Y.-S., Boado, R. J., and Pardridge, W. M. (1995). Upregulation of blood-brain barrier GLUT1 glucose transporter protein and mRNA in experimental chronic hypoglycemia. *Diabetes* **44,** 1399–1404.

Laterra, J. J., Stewart, P. A., and Goldstein, G. W. (1992). Development of the blood-brain barrier. *In* "Fetal and Neonatal Physiology" (R. A. Polin and W. W. Fox, eds.), pp. 1525–1531. W. B. Saunders, Philadelphia.

Lawrence, M., Ho, D., Dash, R., and Sapolsky, R. (1995). Herpes simplex virus vectors overexpressing the glucose transporter gene protect against seizure-induced neuron loss. *Proc. Natl. Acad. Sci. USA* **92,** 7247–7251.

Lawrence, M., Sun, G., Kunis, T., Saydam, T., Dash, R., Ho, D., Sapolsky, R., and Steinberg, G. (1996). Overexpression of the glucose transporter gene with a herpes simplex viral vector protects striatal neurons against stroke. *J. Cereb. Blood Flow Metab.* **16,** 181–185.

Lee, W.-H., and Bondy, C. A. (1993). Ischemic injury induces brain glucose transporter gene expression. *Endocrinology* **133,** 2540–2544.

Leino, R. L., Gerhart, D. Z., van Bueren, A. M., McCall, A. L., and Drewes, L. R. (1997). Ultrastructural localization of GLUT1 and GLUT3 glucose transporters in rat brain. *J. Neurosci. Res.* **49,** 617–626.

Leloup, C., Arluison, M., Kassis, N., Lepetit, N., Cartier, N., Ferre, P., and Penicaud, L. (1996). Discrete brain areas express the insulin-responsive glucose transporter GLUT4. *Brain Res. Mol. Brain Res.* **38,** 45–53.

Leloup, C., Arluison, M., Lepetit, N., Cartier, N., Marfaing-Jallat, P., Ferre, P., and Penicaud, L. (1994). Glucose transporter 2 (GLUT2): Expression in specific brain nuclei. *Brain Res.* **638,** 221–226.

Leong, S. F., Lai, J. C., Lim, L., and Clark, J. B. (1981). Energy-metabolizing enzymes in brain regions of adult and aging rats. *J. Neurochem.* **37,** 1548–1556.

Lesniak, M. A., Hill, J. M., Kiess, W., Rojeski, M., Pert, C. B., and Roth, J. (1988). Receptors for insulin-like growth factors I and II: autoradiographic localization in rat brain and comparison to receptors for insulin. *Endocrinology* **123,** 2089–2099.

Lienhard, G. E., Kim, H. H., Ransome, K. J., and Gorga, J. C. (1982). Immunological identification of an insulin-responsive glucose transporter. *Biochem. Biophys. Res. Commun.* **105,** 1150–1156.

Liu, D., Smith, C. L., Barone, F. C., Ellison, J. A., Lysko, P. G., Li, K., and Simpson, I. A. (1999). Astrocyte demise precedes delayed neuronal death in focal ischemic rat brain. *Mol. Brain Res.* **68,** 29–41.

Long, S. D., and Pekala, P. H. (1996). Regulation of GLUT4 gene expression by arachidonic acid. *J. Biol. Chem.* **271,** 1138–1144.

Lothman, E. W., and Collins, R. C. (1981). Kainic acid induced limbic seizures: Metabolic, behavioral, electroencephalographic and neuropathological correlates. *Brain Res.* **218,** 299–318.

Maekawa, F., Toyoda, Y., Torii, N., Miwa, I., Thompson, R. C., Foster, D. L., Tsukahara, S., Tsukamura, H., and Maeda, K.-I. (2000). Localization of glucokinase-like immunoreactivity in the rat lower brain stem: For possible location of brain glucose-sensing mechanisms. *Endocrinology* **141,** 375–384.

Maher, F. (1995). Immunolocalization of GLUT1 and GLUT3 glucose transporters in primary cultured neurons and glia. *J. Neurosci. Res.* **42,** 459–469.

Maher, F., Davies-Hill, T. M., and Simpson, I. A. (1996). Substrate specificity and kinetic parameters of GLUT3 in rat cerebellar granule neurons. *Biochem. J.* **315,** 827–831.

Maher, F., and Simpson, I. A. (1994a). The GLUT3 glucose transporter is the predominant isoform in primary cultured neurons: assessment by biosynthetic and photoaffinity labelling. *Biochem. J.* **301,** 379–384.

Maher, F., and Simpson, I. A. (1994b). Modulation of expression of glucose transporters GLUT3 and GLUT1 by potassium and *N*-methyl-D-aspartate in cultured cerebellar granule neurons. *Mol. Cell. Neurosci.* **5,** 369–375.

Maher, F., Vannucci, S. J., and Simpson, I. A. (1994). Glucose transporter proteins in brain. *FASEB J.* **8,** 1003–1011.

Maher, F., Vannucci, S., Takeda, J., and Simpson, I. A. (1992). Expression of mouse-GLUT3 and human-GLUT3 glucose transporter proteins in brain. *Biochem. Biophys. Res. Commun.* **182,** 703–711.

Malide, D., Davies-Hill, T. M., Levine, M., and Simpson, I. A. (1998). Distinct localization of GLUT-1, -3, and -5 in human monocyte-derived macrophages: Effects of cell activation. *Am. J. Physiol.* **274,** E516–526.

Mata de Urquiza, A., Liu, S., Sjoberg, M., Zetterstroem, R. H., Griffiths, W., Sjovall, J., and Perlmann, T. (2000). Docosahexaenoic acid, a ligand for the retinoid X receptor in mouse brain. *Science* **290,** 2140–2144.

McCall, A. L., Fixman, L. B., Fleming, N., Tornheim, K., Chick, W., and Ruderman, N. B. (1986). Chronic hypoglycemia increases brain glucose transport. *Am. J. Physiol.* **251,** E442–E447.

McCall, A. L., Millington, W. R., and Wurtman, R. J. (1982). Metabolic fuel and amino acid transport into the brain in experimental diabetes mellitus. *Proc. Natl. Acad. Sci. USA* **79,** 5406–5410.

McCall, A. L., Moholt-Siebert, M., Van Bueren, A., Cherry, N. J., Lessov, N., Tiffany, N., Thompson, M., Downes, H., and Woodward, W. R. (1995). Progressive hippocampal loss of immunoreactive GLUT3, the neuron-specific glucose transporter, after global forebrain ischemia in the rat. *Brain Res.* **670,** 29–38.

McCall, A. L., Van Bueren, A. M., Moholt-Siebert, M., Cherry, N. J., and Woodward, W. R. (1994). Immunohistolocalization of the neuron specific glucose transporter (GLUT3) to neuropil in adult rat brain. *Brain Res.* **659,** 292–297.

McVie-Wylie, A. J., Lamson, D. R., and Chen, Y. T. (2001). Molecular cloning of a novel member of the Glut family of transporters, slc2a10 (Glut10), localized on chromosome 20q13.1: A candidate gene for NIDDM susceptibility. *Genomics* **72,** 113–117.

Moley, K. H., and Mueckler, M. M. (2000). Glucose transport and apoptosis. *Apoptosis* **5,** 99–105.

Molloy, D. W., and Standish, T. I. (2000). Clinical experience with cerebrolysin. *J. Neural Trans. Suppl.* **59,** 293–300.

Mooradian, A. D., Chung, H. C., and Shah, G. N. (1997). GLUT-1 expression in the cerebra of patients with Alzheimer's disease. *Neurobiol. Aging* **18,** 469–474.

Mooradian, A. D., and Morin, A. M. (1991). Brain uptake of glucose in diabetes mellitus: The role of glucose transporters. *Am. J. Med. Sci.* **301,** 173–177.

Morgello, S., Uson, R. R., Schwartz, E. J., and Haber, R. S. (1995). The human blood-brain barrier glucose transporter (GLUT1) is a glucose transporter of gray matter astrocytes. *Glia* **14,** 43–54.

Mu, J., Brozinick, J. T., Jr., Valladares, O., Bucan, M., and Birnbaum, M. J. (2001). A role for AMP-activated protein kinase in contraction- and hypoxia-regulated glucose transport in skeletal muscle. *Mol. Cell* **7,** 1085–1094.

Mueckler, M. (1990). Family of glucose-transporter genes. Implications for glucose homeostasis and diabetes. *Diabetes* **39,** 6–11.

Mueckler, M., Caruso, C., Baldwin, S. A., Panico, M., Blench, I., Morris, H. R., Allard, W. J., Lienhard, G. E., and Lodish, H. F. (1985). Sequence and structure of a human glucose transporter. *Science* **229,** 941–945.

Murata, K., Mitsuoka, K., Hirai, T., Walz, T., Agre, P., Heymann, J. B., Engel, A., and Fujiyoshi, Y. (2000). Structural determinants of water permeation through aquaporin-1. *Nature* **407,** 599–605.

Nagamatsu, S., Sawa, H., Kamada, K., Nakamichi, Y., Yoshimoto, K., and Hoshino, T. (1993). Neuron-specific glucose transporter (NSGT): CNS distribution of GLUT3 rat glucose transporter (RGT3) in rat central neurons. *FEBS Lett.* **334,** 289–295.

Nehlig, A., Boyet, S., and Pereira de Vasconcelos, A. (1991). Autoradiographic measurement of local cerebral β-hydroxybutyrate uptake in the rat during postnatal development. *Neuroscience* **40,** 871–878.

Nishimura, H., and Simpson, I. A. (1994). Staurosporine inhibits phorbol 12-myristate 13-acetate- and insulin-stimulated translocation of GLUT1 and GLUT4 glucose transporters in rat adipose cells. *Biochem. J.* **302,** 271–277.

Nualart, F., Godoy, A., and Reinicke, K. (1999). Expression of hexose transporters GLUT1 and GLUT2 during the early development of the human brain. *Brain Res.* **824,** 97–104.

Ogawa, M., Fukuyama, H., Ouchi, Y., Yamauchi, H., and Kimura, J. (1996). Altered energy metabolism in Alzheimer's disease. *J. Neurol. Sci.* **139,** 78–82.

Pantaleon, M., Harvey, M. B., Pascoe, W. S., James, D. E., and Kaye, P. L. (1997). Glucose transporter GLUT3: ontogeny, targeting, and role in the mouse blastocyst. *Proc. Natl. Acad. Sci. USA* **94,** 3795–3800.

Pardridge, W. M., Triguero, D., and Farrell, C. R. (1990). Downregulation of blood-brain barrier glucose transporter in experimental diabetes. *Diabetes* **39,** 1040–1044.

Patel, M. S. (1977). Age-dependent changes in the oxidative metabolism in rat brain. *J. Gerontol.* **32,** 643–646.

Payne, J., Maher, F., Simpson, I., Mattice, L., and Davies, P. (1997). Glucose transporter Glut5 expression in microglial cells. *Glia* **21,** 327–331.

Pelligrino, D. A., Segil, L. J., and Albrecht, R. F. (1990). Brain glucose utilization and transport and cortical function in chronic vs. acute hypoglycemia. *Am. J. Physiol.* **259,** E729–E735.

Phay, J. E., Hussain, H. B., and Moley, J. F. (2000). Cloning and expression analysis of a novel member of the facilitative glucose transporter family. *Genomics* **66,** 217–220.

Ploug, T., Stallknecht, B., Pedersen, O., Kahn, B. B., Ohkuwa, T., Vinten, J., and Galbo, H. (1990). Effect of endurance training on glucose transport capacity and glucose transporter expression in rat skeletal muscle. *Am. J. Physiol.* **259,** E778–E786.

Pouliot, J.-F., and Beliveau, R. (1995). Palmitoylation of the glucose transporter in blood-brain barrier capillaries. *Biochim. Biophys. Acta* **1234,** 191–196.

Rajakumar, R. A., Thamotharan, S., Menon, R. K., and Devaskar, S. U. (1998). Sp1 and Sp3 regulate transcriptional activity of the facilitative glucose transporter isoform-3 in mammalian neuroblasts and trophoblasts. *J. Biol. Chem.* **273,** 27474–27483.

Rayner, D. V., Thomas, M. E., and Trayhurn, P. (1994). Glucose transporters (GLUTs 1-4) and their mRNAs in regions of the rat brain: Insulin-sensitive transporter expression in the cerebellum. *Can. J. Physiol. Pharmacol.* **72,** 476–479.

Reagan, L. P., Gorovits, N., Hoskin, E. K., Alves, S. E., Katz, E. B., Grillo, C. A., Piroli, G. G., McEwen, B. S., and Charron, M. J. (2001). Localization and regulation of GLUTx1 glucose transporter in the hippocampus of streptozotocin diabteic rats. *Proc. Natl. Acad. Sci. USA* **98,** 2820–2825.

Reagan, L. P., Magarinos, A. M., Lucas, L. R., Van Bueren, A., McCall, A. L., and McEwen, B. S. (1999). Regulation of GLUT-3 glucose transporter in the hippocampus of diabetic rats subjected to stress. *Am. J. Physiol.* **276,** E879–E886.

Reagan, L. P., Magarinos, A. M., Yee, D. K., Swzeda, L. I., Van Bueren, A., McCall, A. L., and McEwen, B. S. (2000). Oxidative stress and HNE conjugation of GLUT3 are increased in the hippocampus of diabetic rats subjected to stress. *Brain Res.* **862,** 292–300.

Rebec, G. V., and Pierce, R. C. (1994). A vitamin as neuromodulator: ascorbate release into the extracellular fluid of the brain regulates dopaminergic and glutamatergic transmission. *Prog. Neurobiol.* **43,** 537–565.

Royer, C., Lachuer, J., Crouzoulon, G., Roux, J.-C., Peyronnet, J., Mamet, J., Pequignot, J.-M., and Dalmaz, Y. (2000). Effects of gestational hypoxia on mRNA levels of Glut3 and Glut4 transporters, hypoxia inducible factor-1 and thyroid hormone receptors in developing rat brain. *Brain Res.* **856,** 119–128.

Rumsey, S. C., Kwon, O., Xu, G. W., Burant, C. F., Simpson, I., and Levine, M. (1997). Glucose transporter isoforms GLUT1 and GLUT3 transport dehydroascorbic acid. *J. Biol. Chem.* **272,** 18982–18989.

Semenza, G. L., Jiang, B. H., Leung, S. W., Passantino, R., Concordet, J. P., Maire, P., and Giallongo, A. (1996). Hypoxia response elements in the aldolase A, enolase 1, and lactate dehydrogenase A gene promoters contain essential binding sites for hypoxia-inducible factor 1. *J. Biol. Chem.* **271,** 32529–32537.

Semenza, G. L., and Wang, G. L. (1992). A nuclear factor induced by hypoxia via de novo protein synthesis binds to the human erythropoietin gene enhancer at a site required for transcriptional activation. *Mol. Cell. Biol.* **12,** 5447–5454.

Shepherd, P. R., Gibbs, E. M., Wesslau, C., Gould, G. W., and Kahn, B. B. (1992). Human small intestine facilitative fructose/glucose transporter (GLUT5) is also present in insulin-responsive tissues and brain. *Diabetes* **41,** 1360–1365.

Shi, J., Zhang, Y. Q., and Simpkins, J. W. (1997). Effects of 17β-estradiol on glucose transporter 1 expression and endothelial cell survival following focal ischemia in the rats. *Exp. Brain Res.* **117,** 200–206.

Simpson, I. A., Chundu, K., Davies-Hill, T., Honer, W. G., and Davies, P. (1994). Decreased concentrations of GLUT1 and GLUT3 glucose transporters in the brains of patients with Alzheimer's disease. *Ann. Neurol.* **35,** 546–551.

Simpson, I. A., Appel, N. M., Hokari, M., Oki, J., Holman, G. D., Maher, F., Koehler-Stec, E. M., Vannucci, S. J., and Smith, Q. R. (1999). Blood-brain barrier glucose transporter: Effects of hypo- and hyperglycemia revisited. *J. Neurochem.* **72,** 238–247.

Sivitz, W., DeSautel, S., Walker, P. S., and Pessin, J. E. (1989). Regulation of the glucose transporter in developing rat brain. *Endocrinology* **124,** 1875–1880.

Smith, D. E., and Gridley, T. (1992). Differential screening of a PCR-generated mouse embryo cDNA library: Glucose transporters are differentially expressed in early postimplantation mouse embryos. *Development* **116,** 555–561.

Sokoloff, L. (1976). Circulation and energy metabolism in the brain. *In* "Basic Neurochemistry" (G. J. Siegel, R. W. Albers, R. Katzman, and B. W. Agranoff, eds.), pp. 388–413. Little, Brown and Company, Boston.

Sorbara, L. R., Davies-Hill, T. M., Koehler-Stec, E. M., Vannucci, S. J., Horne, M. K., and Simpson, I. A. (1997). Thrombin-induced translocation of GLUT3 glucose transporters in human platelets. *Biochem. J.* **328,** 511–516.

Stanley, C. A., Anday, E. K., Baker, L., and Delivoria-Papadopolous, M. (1979). Metabolic fuel and hormone responses to fasting in newborn infants. *Pediatrics* **64,** 613–9.

Stylianopoulou, F., Efstratiadis, A., Herbert, J., and Pintar, J. (1988). Pattern of the insulin-like growth factor II gene expression during rat embryogenesis. *Development* **103,** 497–506.

Sweet, I. R., and Matschinsky, F. M. (1997). Are there kinetic advantages of GLUT2 in pancreatic glucose sensing? *Diabetologia* **40,** 112–119.

Takao, Y., Akazawa, S., Matsumoto, K., Takino, H., Akazawa, M., Trocino, R. A., Maeda, Y., Okuno, S., Kawasaki, E., Uotani, S., Yokota, A., and Nagataki, S. (1993). Glucose transporter gene expression in rat conceptus during high glucose culture. *Diabetologia* **36,** 696–706.

Tanimura, T., and Shepard, T. H. (1970). Glucose metabolism by rat embryos in vitro. *Proc. Soc. Exp. Biol. Med.* **135,** 51–54.

Torres-Aleman, I., Pons, S., and Arevalo, M. A. (1994). The insulin-like growth factor I system in the rat cerebellum: developmental regulation and role in neuronal survival and differentiation. *J. Neurosci. Res.* **39,** 117–126.

Urabe, T., Hattori, N., Nagamatsu, S., Sawa, H., and Mizuno, Y. (1996). Expression of glucose transporters in rat brain following transient focal ischemic injury. *J. Neurochem.* **67,** 265–271.

Vannucci, S. J. (1994). Developmental expression of GLUT1 and GLUT3 glucose transporters in rat brain. *J. Neurochem.* **62,** 240–246.

Vannucci, R. C., Brucklacher, R. M., and Vannucci, S. J. (1996). The effect of hyperglycemia on cerebral metabolism during hypoxia-ischemia in the immature rat. *J. Cereb. Blood Flow Metab.* **16,** 1026–1033.

Vannucci, S. J., Clark, R. R., Koehler-Stec, E., Li, K., Smith, C. B., Davies, P., Maher, F., and Simpson, I. A. (1998a). Glucose transporter expression brain: Relationship to cerebral glucose utilization. *Dev. Neurosci.* **20,** 369–379.

Vannucci, S. J., Gibbs, E. M., and Simpson, I. A. (1997). Glucose utilization and glucose transporter proteins GLUT-1 and GLUT-3 in brains of diabetic (db/db) mice. *Am. J. Physiol.* **272,** E267–E274.

Vannucci, S. J., Koehler-Stec, E. M., Li, K., Reynolds, T. H., Clark, R., and Simpson, I. A. (1998b). GLUT4 glucose transporter expression in rodent brain: effect of diabetes. *Brain Res.* **797,** 1–11.

Vannucci, S. J., Maher, F., and Simpson, I. A. (1997). Glucose transporter protein in brain: delivery of glucose to neurons and glia. *Glia* **21,** 2–21.

Vannucci, S. J., Maher, F., Koehler, E., and Simpson, I. A. (1994). Altered expression of GLUT-1 and GLUT-3 glucose transporters in neurohypophysis of water-deprived or diabetic rats. *Am. J. Physiol.* **267,** E605–E611.

Vannucci, S. J., Reinhart, R., Maher, F., Bondy, C. A., Lee, W. H., Vannucci, R. C., and Simpson, I. A. (1998c). Alterations in GLUT1 and GLUT3 glucose transporter gene expression following unilateral hypoxia-ischemia in the immature rat brain. *Brain Res. Dev. Brain Res.* **107,** 255–264.

Vannucci, S. J., Rutherford, T., Wilkie, M. B., Simpson, I. A., and Lauder, J. M. (2000). Prenatal expression of the GLUT4 glucose transporter in the mouse. *Dev. Neurosci.* **22,** 274–282.

Vannucci, R. C., and Vannucci, S. J. (2002). Glucose, acidosis, and the developing brain. *In* "Birth Asphyxia and the Brain: Basic Science and Clinical Implications" (S. M. Donn and M. L. Chiswick, eds.), pp. 89–119. Futura Publishing Co., Armonk, New York.

Vera, J. C., Rivas, C. I., Fischbarg, J., and Golde, D. W. (1993). Mammalian facilitative hexose transporters mediate the transport of dehydroascorbic acid. *Nature* **364,** 79–82.

Verhey, K. J., and Birnbaum, M. J. (1994). A leu-leu sequence is essential for COOH-terminal targeting signal of GLUT4 glucose transporter in fibroblasts. *J. Biol. Chem.* **269,** 2353–2356.

Vicario, C., and Medina, J. (1992). Metabolism of lactate in the rat brain during the early neonatal period. *J. Neurochem.* **59,** 32–40.

Vorbrodt, A. W., Dobrogowska, D. H., Meeker, H. C., and Carp, R. I. (1999). Immunogold study of regional differences in the distribution of glucose transporter (GLUT-1) in mouse brain associated with physiological and accelerated aging and scrapie infection. *J. Neurocytol.* **28,** 711–719.

Walker, P. S., Donovan, J. A., Van Ness, B. G., Fellows, R. E., and Pessin, J. E. (1988). Glucose-dependent regulation of glucose transport activity, protein, and mRNA in primary cultures of rat brain glial cells. *J. Biol. Chem.* **263,** 15594–15601.

Werner, H., Raizada, M. K., Mudd, L. M., Foyt, H. L., Simpson, I. A., Roberts, C. T., Jr., and LeRoith, D. (1989). Regulation of rat brain/HepG2 glucose transporter gene expression by insulin and insulin-like growth factor-I in primary cultures of neuronal and glial cells. *Endocrinology* **125,** 314–320.

Wertheimer, E., Sasson, S., Cerasi, E., and Ben-Neriah, Y. (1991). The ubiquitous glucose transporter GLUT-1 belongs to the glucose-regulated protein family of stress-inducible proteins. *Proc. Natl. Acad. Sci. USA* **88,** 2525–2529.

Wheeler, T. J., Simpson, I. A., Sogin, D. C., Hinkle, P. C., and Cushman, S. W. (1982). Detection of the rat adipose cell glucose transporter with antibody against the human red cell glucose transporter. *Biochem. Biophys. Res. Commun.* **105,** 89–95.

Williams, K. A. (2000). Three-dimensional structure of the ion-coupled transport protein NhaA. *Nature* **403,** 112–115.

Windisch, M. (2000). Approach towards an integrative drug treatment of Alzheimer's disease. *J. Neural Trans. Suppl.* **59,** 301–313.

Yano, H., Seino, Y., Inagaki, N., Hinokio, Y., Yamamoto, T., Yasuda, K., Masuda, K., Someya, Y., and Imura, H. (1991). Tissue distribution and species difference of the brain type glucose transporter (GLUT3). *Biochem. Biophys. Res. Commun.* **174,** 470–477.

Zeller, K., Duelli, R., Vogel, J., Schroek, H., and Kuschinsky, W. (1995). Autoradiographic analysis of the regional distribution of Glut3 glucose transporters in the rat brain. *Brain Res.* **698,** 175–179.

Zeller, K., Rahner-Welsch, S., and Kuschinsky, W. (1997). Distribution of Glut1 glucose transporters in different brain structures compared to glucose utilization and capillary density of adult rat brains. *J. Cereb. Blood Flow Metab.* **17,** 204–209.

Zeng, H., Parthasarathy, R., Rampal, A. L., and Jung, C. Y. (1996). Proposed structure of putative glucose channel in GLUT1 facilitative glucose transporter. *Biophys. J.* **70,** 14–21.

Zhang, J.-Z., Behrooz, A., and Ismail-Beigi, F. (1999). Regulation of glucose transport by hypoxia. *Am. J. Kidney Dis.* **34,** 189–202.

Zuniga, F. A., Shi, G., Haller, J. F., Rubashkin, A., Flynn, D, R., Iserovich, P., and Fischbarg, J. (2001). A three-dimensional model of the human facilitative glucose transporter GLUT1. *J. Biol. Chem.* **276,** 44970–44975.

INSULIN-LIKE GROWTH FACTOR-1 PROMOTES NEURONAL GLUCOSE UTILIZATION DURING BRAIN DEVELOPMENT AND REPAIR PROCESSES

Carolyn A. Bondy and Clara M. Cheng

Developmental Endocrinology Branch, NICHD, NIH Bethesda, Maryland 20892

I. Introduction

The brain requires enormous supplies of fuel and substrate to support neuroglial growth and process formation during early postnatal development. Murine and human brains consume over half the energy available to the organism as a whole during this critical period characterized more by synapse formation than synaptic activity. Some neurons grow into gargantuan cells with surface areas exceeding all other cells in the body. How this anabolic feat is achieved when all brain cells are exposed to the same extracellular nutrient supply is unclear. Insulin preferentially enhances fuel and substrate utilization by specific peripheral tissues, but is not directly involved in the regulation of brain metabolism. This chapter reviews evidence from *in vivo* murine model systems indicating that insulin-like growth factor-1

0074-7742/02 $35.00

(IGF-1) acts in an autocrine fashion within the developing brain to promote preferential glucose utilization by IGF-1-expressing neurons, fueling prolific process growth and synaptogenesis. Limited data on IGF-system expression and function in the primate brain is also discussed briefly.

II. The Insulin/IGF System

Insulin and insulin-like growth factors-1 and -2 (IGFs 1 and 2) belong to a family of anabolic peptides sharing a common evolutionary origin (LeRoith *et al.*, 1986; LeRoith *et al.*, 1993; Reinecke and Collet, 1998; Navarro *et al.*, 1999). It appears that an ancestral gene encoding an insulin-like peptide gave rise to multiple genes encoding more specialized peptides about the time gastroenteric and central nervous systems differentiated in early chordate evolution (Reinecke and Collet, 1998). From that time insulin became progressively more specialized in terms of undergoing secondary processing (proteolytic excision of the C peptide and joining of the A and B peptides by disulfide bonds) and packaging in acidic secretory granules. Insulin expression is largely restricted to pancreatic beta cells, where its synthesis and secretion are tightly coupled to ingested substrates (Tager *et al.*, 1981). IGFs 1 and 2, in contrast, did not acquire such extensive posttranslational processing, and have continued to be widely expressed in many cell types demonstrating constitutive secretion (Clemmons, 1989; Sussenbach, 1989). The A and B domains in insulin, IGF-1 and IGF-2 share about 50% sequence identity and the tertiary structures of these peptides appear quite similar (Van den Brande *et al.*, 1986; LeRoith *et al.*, 1992).

The insulin and IGF-1 receptors also arose from a common ancestor, in parallel with the ligands evolution (LeRoith *et al.*, 1993; Reinecke and Collet, 1998; Navarro *et al.*, 1999). These receptors consist of ligand-binding α-subunits and membrane-spanning, tyrosine kinase β-subunits (Fig. 1) (Rechler and Nissley, 1986; Seta *et al.*, 1993). The α and β sequences are encoded by a single gene and derived from posttranslational cleavage followed by disulfide linking (Nissley *et al.*, 1985; Yarden and Ullrich, 1988). The receptors are inserted in the plasma membrane as disulfide linked dimers, which upon binding a single ligand molecule, undergo conformational changes resulting in kinase activation and autophosphorylation. The ligand binding domains of these receptors have diverged enough to confer relative selectivity for binding of the cognate ligands, but crossover, particularly for IGF-2 at the insulin receptor, is significant (Fig. 1). The IGF-1 receptor, however, mediates most of the biological effects of both IGF-1 and -2 (Seta *et al.*, 1993; Werner *et al.*, 1994). The IGF-2/mannose 6-phosphate receptor, which is structurally unrelated to the tyrosine kinase

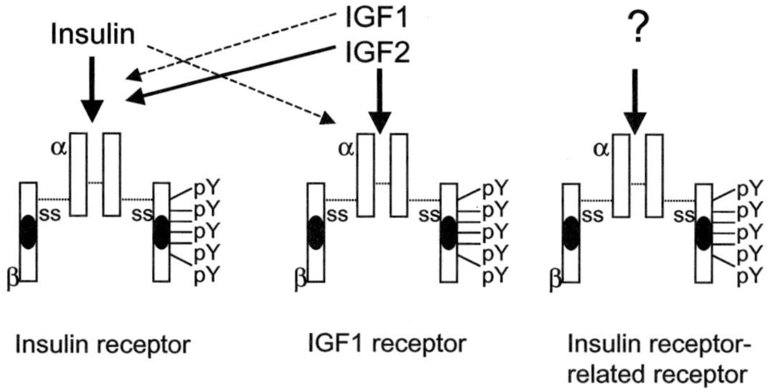

FIG. 1. Schematic diagram comparing insulin/IGF receptor family members. The ligand-binding α-subunits are entirely extracellular and are inserted in the plasma membrane as disulfide-linked dimers. The α-subunit is also disulfide bonded to the intracellular β-subunit, which has intrinsic tyrosine kinase activity and undergoes autophosphoryation upon ligand binding. Some of the phosphotyrosine sites are indicated (pY). Insulin selectively binds the insulin receptor with high affinity and the IGF-1 receptor with about 10-fold lower affinity. Likewise for IGF-1 and its cognate receptor. IGF-2, however, binds both the IGF-1 and insulin receptors with relatively high affinity. The ligand for the insulin receptor-related receptor (IRR) is unknown.

receptor family, promotes IGF-2 degradation and does not interact with IGF-1 (Kiess *et al.*, 1994; Braulke, 1999). This insulin/IGF receptor family includes a third member, termed the insulin receptor-related receptor (IRR), that subserves an unknown ligand (Seta *et al.*, 1993; Watt *et al.*, 1993).

The insulin, IGF-1 and IRR receptors, demonstrate ~85% conservation of the tyrosine kinase domain and interact with the same set of intracellular docking proteins (IRS 1–4, Shc, Gab1) and downstream signaling molecules, including phosphatidyl inositol 3 kinase (PI3K), protein tyrosine phosphatases, and Grb2/Sos (De Meyts *et al.*, 1994; Werner *et al.*, 1994; Chen *et al.*, 1997; Butler *et al.*, 1998). Despite the apparent convergence in signaling pathways, insulin and IGFs 1 and 2 have distinct physiological roles, with insulin responsible for maintaining glucose and lipid homeostasis in the circulation while IGF-1 and -2 are primarily responsible for promoting somatic growth. Much effort has been devoted to trying to delineate metabolic signaling pathways specific to the insulin receptor and growth or mitogenic pathways unique to the IGF-1 receptor (Adamo *et al.*, 1992; Siddle *et al.*, 2001). Mitogenic is probably not the most accurate term, since while IGFs promote progression through the mitotic cycle by proliferating cells, they don't stimulate entry into mitosis. In any case, both insulin and IGFs acting via the insulin and IGF-1 receptors, respectively, are able to promote mitosis (Ish-Shalom *et al.*, 1997) and both promote metabolic effects such as cell glucose utilization (Cascieri *et al.*, 1986; Shefi-Friedman *et al.*, 2001).

It seems likely that the functional outcomes of various *in vitro* assay systems reflect more on cell type, differentiation, culture, and experimental conditions than on intrinsic molecular properties of these ligand-receptor interactions. No obvious differences in hard-wired signaling pathways unique to the insulin or IGF-1 receptor have been identified, although subtle differences in the degree of kinase activation are discerned in some genetically manipulated *in vitro* systems (Nakae *et al.*, 2000; Kim *et al.*, 2001).

The different physiological roles served by insulin and IGFs may be explained more by differences in tissue- and developmental stage-specific bioavailability of the ligands than by intrinsic differences in ligand-receptor interactions. For example, IGF-2 is highly abundant in the fetal circulation (Sara and Carlsson-Skwirut, 1986; Owens, 1991) and is critical for normal fetal somatic growth (Baker *et al.*, 1993). IGF-1 is relatively scarce during embryogenesis but produced in great abundance by the liver and diverse peripheral tissues during postnatal development and is responsible primarily for postnatal somatic growth (Daughaday and Rotwein, 1989). In this latter capacity, IGF-1 is regulated by growth hormone (GH). IGF-1 expression in the brain, however, is independent of GH (Lupu *et al.*, 2001). High levels of these growth factors are maintained in the circulation and tissue compartments by more or less continuous, constitutive secretion and also by protection from clearance and/or proteolysis via tight binding to IGF binding proteins (Clemmons, 1998). This is in marked contrast to insulin availability. Insulin secretion is tightly coupled to ingested substrates; fasting insulin levels are normally quite low, then rise dramatically in response to a meal, followed by rapid clearance from the circulation (insulin does not bind to binding proteins). Thus peripheral tissues exposure to insulin and to IGFs is quite different. It seems likely that the IGF's growth promoting effects are due at the most fundamental level to insulin-like actions, directing the allocation of resources (e.g., glucose, lipids, amino acids) to cells and tissues with extraordinary anabolic needs during development and during tissue repair processes.

III. IGF/Insulin System Expression in the Brain

A. IGF-1 Receptor

The IGF-1 receptor is highly expressed from the time of neural tube formation (Bondy *et al.*, 1990) and Fig. 2. IGF-1 receptor mRNA is concentrated in undifferentiated neuroepithelial cells, especially in the ventral floor plate epithelium of the spinal cord and hind brain early in development (Bondy *et al.*, 1990). Interestingly, IGF effect has been implicated in neural and head

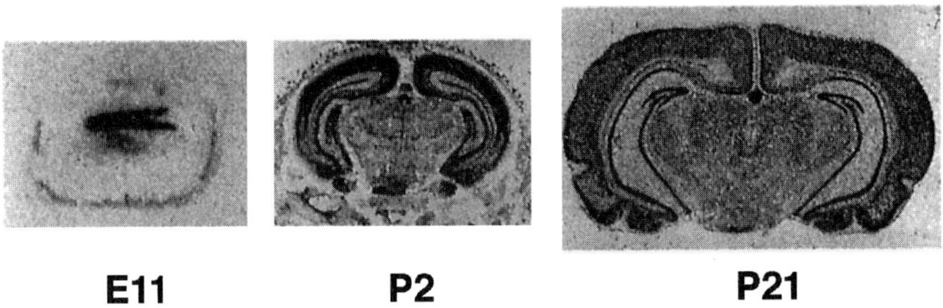

FIG. 2. IGF-1 receptor gene expression is abundant in the developing brain. *In situ* hybridization for the IGF-1 receptor is shown in film autoradiographs taken in the same hybridization from embryonic day 11 (E11) and postnatal days 2 and 21 (P2 and P21). IGF-1 receptor mRNA is abundant in the neural tube seen in the E11 section and is most abundant in the neuron dense areas of the postnatal brain sections. The apparent diminution in IGF-1 receptor signal in the more mature brain reflects the increasing numbers of glial cells populating the brain during postnatal development rather than a reduction in IGF-1 receptor expression by neurons.

induction in the Xenopus embryo (Pera *et al.,* 2001), although mammals with IGF or IGF-1 receptor deficiency or deletion do not exhibit major defects in brain or cranial development. As differentiated neurons emerge and mature brain structures form, IGF-1 receptor mRNA is preferentially concentrated in neurons as opposed to glial components of the central nervous system (CNS) (Bondy *et al.,* 1992b). Studies using brain homogenates to evaluate developmental patterns of expression suggested that IGF-1 receptor mRNA levels decrease with brain maturation (Werner *et al.,* 1989), but *in situ* analyses show that the mRNA level per neuron does not decrease, and actually increases in some neurons during the time of peak IGF-1 expression postnatally (Bondy *et al.,* 1992b). The evident decline in receptor expression noted in homogenized tissue is explained by the increasing abundance of glial cells, which express relatively less IGF-1 receptor mRNA, during the course of development. IGF-1 receptor expression is very widespread, in general reflecting neuronal density throughout the brain. The IGF-1 receptor, however, displays selective high-level expression in specific cell populations, including cerebellar and sensory system projection neurons (e.g., mitral and Purkinje cells, superior and inferior olives, red, cochlear, and vestibular, medial and lateral geniculate nuclei, inferior colliculus, and ventrobasilar complex) during the time of peak IGF-1 expression early in postnatal development (see Table I and the next section). IGF-1 binding sites generally coincide with IGF-1 receptor mRNA concentrations in the brain (Bohannon *et al.,* 1988; Lesniak *et al.,* 1988; Werther *et al.,* 1989; Devaskar *et al.,* 1991; Bondy *et al.,* 1992a) suggesting that the receptor is expressed on neuronal soma and dendrites.

TABLE I

IGF-1 GENE EXPRESSION IN SPECIFIC NEUROANATOMICAL LOCI OF THE DEVELOPING RAT CENTRAL NERVOUS SYSTEM

System	Component
Cerebellar	Purkinje cell
	Inferior olive
	Deep cerebellar n.
	Red n.
Somatosensory	Principal sensory
	Ventrobasilar n.
	Gelatinosa n.
	Intralaminar n.
Olfactory	Mitral cell
	Tufted cell
	Piriform cortex[a]
Visual	Retinal ganglion cells
	Superior colliculus
	Lateral geniculate n.
	Lateral posterior n.
	Anterior pretectal n.
Auditory-vestibular	Medial vestibular n.
	Superior vestibular n.
	Dorsal cochlear n.
	Ventral cochlear n.
	Superior olive
	Lateral lemniscal n.
	Medial geniculate n.
	Inferior colliculus
	Interstitial n.
Other	Nucleus basalis
	Diagonal band n.
	Lateral septal n.
	Medial habenula
	Ventral tegmental n.
	Dorsal tegmental n.
	Entopeduncular n.
	Parafascicular n.
	Retroflexus n.
	Parabrachial n.
	Substantia nigra, pars lateralis
	Hippocampus[b]
	Neocortex[c]
	Subventricular zone[d]

[a] IGF-1 mRNA is concentrated in pyramidal cells.

[b] IGF-1 mRNA is contained in nonpyramidal cells, localized in the strata oriens and radiatum.

[c] IGF-1 mRNA is localized in large, pale cells which are found in a nonlaminar distribution.

[d] From Bondy (1991), with modifications.

B. IGF-1

In contrast to the IGF-1 receptor, brain IGF-1 expression demonstrates strong developmental regulation. IGF-1 mRNA is relatively scarce in the murine brain prenatally and increases rather dramatically in the early postnatal period, followed by a decline several weeks after birth (Rotwein *et al.*, 1988; Ayer-Le Lievre *et al.*, 1991; Bach *et al.*, 1991; Bartlett *et al.*, 1991; Bondy, 1991). IGF-1 mRNA is most abundant in growing projection neurons belonging to the cerebellar and sensory relay systems mentioned in the previous section in connection with high level IGF-1 receptor expression (Bondy, 1991). The full spectrum of IGF-1 expression in postnatal brain is summarized in Table I. An example of the developmental course of IGF-1 gene expression in the somatosensory nuclei of the anterior thalamus is shown in Fig. 3.

The co-localization of IGF-1 receptor and IGF-1 mRNAs in projection neurons of the pirifom and cerebellar cortices is illustrated in Fig. 4. The prototypical IGF-1-expressing neuron is the Purkinje cell; this cell type has a long axon and huge soma and an incredibly branched, highly complex dendritic arbor, resulting in the largest surface area of any cell type in mammals. IGF-1 mRNA is first detected in Purkinje cells perinatally, after the postmitotic cells have migrated to the cerebellar plate (Bondy, 1991). IGF-1 mRNA levels increase until about postnatal day 10 (P10) and remain stable through approximately P20 after which there is a dramatic decline. These observations are consistent in both rat and mouse (Bondy, 1991; Bartlett *et al.*, 1991) which

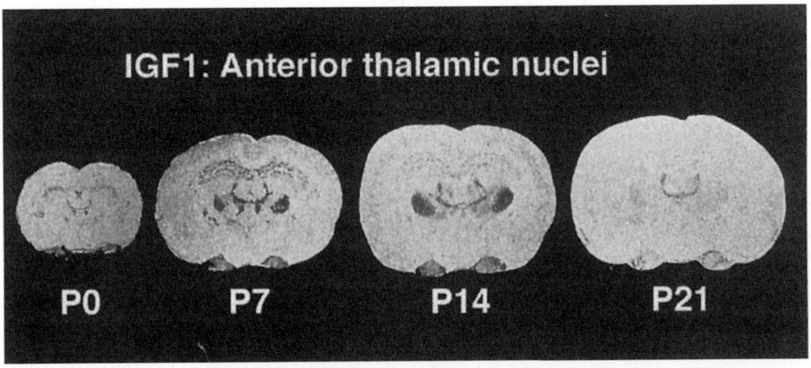

FIG. 3. IGF-1 gene expression in the anterior thalamus during postnatal brain development. Shown by film autoradiography of anatomically matched forebrain sections from postnatal (P) day 0, 7, 14, and 21 rats. IGF-1 mRNA is concentrated in the ventrobasilar, intralaminar, and parafascilular nuclei, with expression peaking between days 7 and 14 and receding by the beginning of the fourth week of life.

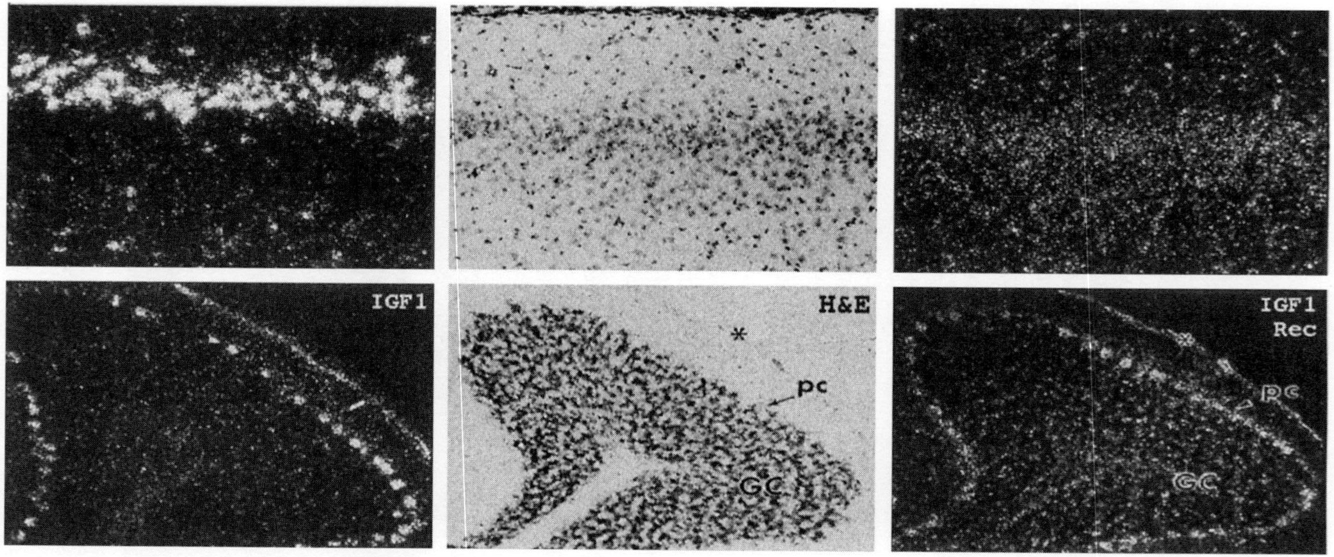

FIG. 4. Co-expression of IGF-1 and the IGF-1 receptor in projection neurons of the piriform (top panels) and cerebellar (lower panels) cortices. This figure shows dark field micrographs of emulsion-coated serial sections revealing IGF-1 mRNA hybrids on the left and IGF-1 receptor hybrids on the right side of the figure. The central panels show bright field views wherein the cell bodies can be seen. While IGF-1 mRNA is exclusively concentrated in projection neurons, i.e., Purkinje cells (PC) and pyramidal neurons, IGF-1 receptor mRNA is more widely expressed, e.g., also in cerebellar granule cells (GC).

exhibit quite similar time courses in brain maturation. Peak IGF-1 expression in Purkinje cells and other cerebellar and sensory projection neurons throughout the CNS coincides developmentally with somatic and dendritic growth and synaptogenesis (Bondy, 1991). IGF-1 immunoreactivity is localized in Purkinje perikarya and processes and is greatly enhanced shortly after blocking secretion with colchicine (Andersson *et al.*, 1988), suggesting that the peptide is constitutively released from processes relatively close to the perikarya. Of note, IGF-1 immunoreactivity is not detected in neurosecretory granules nor is it found concentrated in nerve terminals. The observations that IGF-1 binding and IGF-1 immunoreactivity are both concentrated in the immediate vicinity of the neurons synthesizing the receptor and peptide strongly suggests a local, autocrine, or paracrine mode of action for neuronal IGF-1. It is also possible that IGF-1 is transported down axons and released for action at distant sites. However, IGF-1 mRNA is abundant in *all* the relay neurons comprising these systems, e.g., IGF-1 mRNA is abundant in the inferior olive, Purkinje cells, deep cerebellar nuclei, and red nucleus (Bondy, 1991). Thus the transport of IGF-1 from inferior olive to Purkinje cell, or from Purkinje cell to deep cerebellar nuclei would seem somewhat redundant.

We have focused intently on the large neuron projection systems because their similar, well-defined morphological and functional phenotypes seemed likely to provide important insights into IGF-1's role in neuronal development. In fact, the functional correlation we have documented between IGF-1 expression and glucose utilization applies primarily to these projection systems (see Section IV). IGF-1 mRNA is detected in other sites in the developing murine brain that don't have such well-defined functions or connections, as noted in Table I under "Other."

C. IGF-2

IGF-2 is also synthesized within the CNS, for the most part by adventitial structures such as the leptomeninges and choroid plexii (Hynes *et al.*, 1988; Rotwein *et al.*, 1988; Stylianopoulou *et al.*, 1988; Bondy *et al.*, 1992b). Some investigators, however, have detected IGF-2 mRNA in cerebellar granule cells during early postnatal development (Hetts *et al.*, 1997). These authors have also noted that IGF-2 gene expression in the meninges and choroid plexus originates from both parental alleles, in contrast to the situation for most tissues, where IGF-2 expression is derived solely from the paternal allele. IGF-2's role in the CNS remains unclear, and there is no apparent effect of IGF-2 deletion on brain morphology, chemistry, or neurological function (our unpublished data).

D. Insulin and Insulin Receptor

Very little insulin is synthesized within the brain (Coker *et al.,* 1990) and much of the material earlier thought to be "brain" insulin, reviewed in Havrankova *et al.* (1981), may have represented the IGF-1 or -2 peptides which are synthesized in brain in abundance, or insulin derived from the circulation. *In situ* analysis has identified insulin mRNA in a few small periventricular cells in the anterior hypothalamus (Young, 1986) where insulin may play a role in the neuroendocrine systems regulating appetite, but seems quite unlikely to be involved in regulation of substrate homeostasis in diverse brain regions. The insulin receptor, however, is widely expressed in brain in a pattern that overlaps that of the IGF-1 receptor (Fig. 5). Both of these receptors are most highly expressed by neurons and demonstrate little expression in white matter areas where glial cells predominate. The IGF-1 receptor, however, is focally concentrated in select neuronal populations where IGF-1 expression is so abundant, while the insulin receptor appears more uniformly expressed, generally in parallel with neuronal density for each brain region (Hill *et al.,* 1986; Werther *et al.,* 1989; Bondy *et al.,* 1992a,b; Reinhardt and Bondy, 1994).

Given that little insulin is produced within the brain and relatively little crosses the blood-brain barrier (BBB), the significance of widespread insulin receptor expression in the brain remains unclear. A likely explanation is that the ancestral insulin-like peptide and its receptor were especially important for brain development (Smit *et al.,* 1998) and highly expressed in brain, and that the descendent insulin receptor gene retained regulatory regions specifying CNS expression while the ligand became more specialized for endocrine function in the periphery. Alternatively or additionally, it is possible that this receptor functions in hybrid form with the IGF-1 receptor, since these homologous subunits may readily combine when synthesized in the same cell (Bailyes *et al.,* 1997; Pandini *et al.,* 1999).

It should also be noted that both IGF-1 and insulin receptors are abundant in choroid plexii, meninges, and vascular structures of the brain, including the periventricular structures (Van Houten *et al.,* 1979; Bondy *et al.,* 1992b) thought to be important for transporting systemic molecules across the BBB. Before leaving the subject of expression patterns for insulin/IGF receptors in the brain, it is interesting to note that the enigmatic family member IRR is also expressed in the CNS, but in a much more focal or selective pattern than either of its cousins. The IRR is selectively expressed in TRK-expressing basal forebrain cholinergic neurons and in TRK-expressing sensory neurons in the developing sensory ganglia (Reinhardt *et al.,* 1993, 1994). As noted above, the ligand for this receptor is unknown, and its potential role in the nervous system likewise remains a mystery.

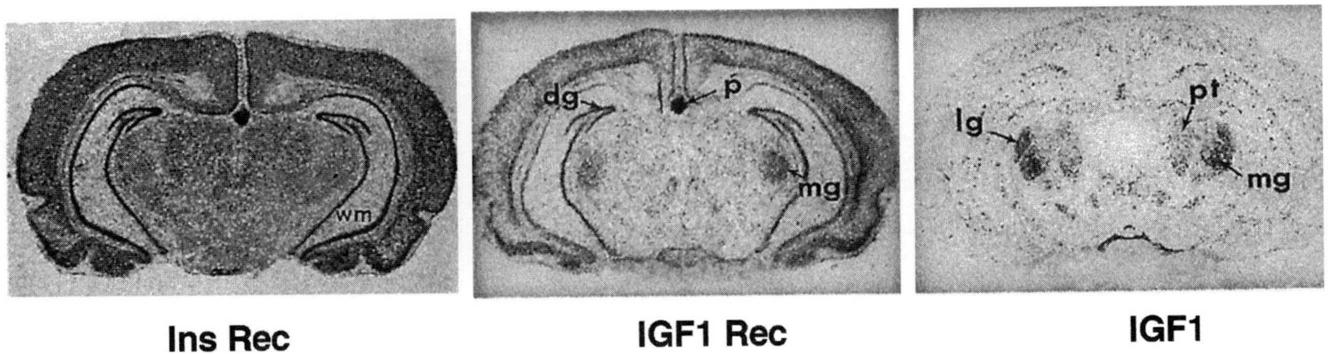

FIG. 5. Comparison of insulin receptor, IGF-1 receptor, and IGF-1 gene expression patterns in the developing brain. The insulin receptor hybridized sections were exposed to film for longer than the IGF-1 receptor sections, so the signal intensity should not be compared for the two probes. Note that both insulin and IGF-1 receptor mRNAs are widely expressed in gray matter with little signal in white matter (wm) areas. IGF-1 receptor displays some heterogeneity in distribution, correlating with local IGF-1 expression (e.g., in the medial geniculate nuclei [mg]).

It is thought that circulating insulin and IGF-1 may act as messengers of systemic nutritional status at the median eminence and other circumventricular structures outside the BBB (Schwartz *et al.*, 2000). IGF-1 receptor and insulin receptor mRNAs are both present in CNS capillary endothelium but while IGF-1 binding to brain capillaries is abundant, relatively little insulin binding is detected (Reinhardt and Bondy, 1994). We have compared transport of IGF-1 and insulin across the BBB using an *in vivo* carotid artery perfusion system and found that three-fold more radio-labeled IGF-1 accumulates in the cerebral cortex and almost 10 times more IGF-1 accumulates in the hypothalamic paraventricular nucleus compared with insulin (Reinhardt and Bondy, 1994). These findings are consistent with studies suggesting that augmentation of circulating IGF-1 levels promotes brain growth and repair processes (Laron and Klinger, 2000; Carro *et al.*, 2001; Guan *et al.*, 2001). A number of factors may explain IGF-1's relative facility in crossing the BBB. The co-expression of insulin and IGF-1 receptors in brain capillary endothelium may result in formation of hybrid receptors, which bind IGF-1 with substantially greater affinity than insulin (Soos *et al.*, 1993). In addition, IGF binding proteins are abundant in capillary endothelium, median eminence, and other circumventricular sites (Lee *et al.*, 1993), suggesting possible carrier-mediated IGF transport across the BBB.

E. IGF System Expression in CNS Injury

During normal brain development, IGF-1 gene expression is primarily or exclusively neuronal and diminishes after neuronal maturation, except in the olfactory system (Bondy, 1991). IGF-1 expression is, however, induced in mature brain astrocytes in response to a variety of insults, including ischemic, traumatic, and chemotoxic injury (Komoly *et al.*, 1992; Lee *et al.*, 1992; Lee and Bondy, 1993; Gehrmann *et al.*, 1994; Yao *et al.*, 1995; Walter *et al.*, 1997; Beilharz *et al.*, 1998; Li *et al.*, 1998). IGF-2 and a number of IGF binding proteins are also induced in these situations (Lee and Bondy, 1993; Walter *et al.*, 1999).

IV. IGF-1 and Brain Glucose Utilization

A. IGF-1/IGF-1 Receptor Expression Parallel Glucose Metabolism

Regional glucose utilization as measured by 14C-2-deoxyglucose uptake (2DGU) parallels IGF-1 and IGF-1 receptor gene expression in the developing murine brain (Fig. 6). High-level IGF-1 expression is seen in concert with intense 2DGU in maturing cerebellar, somatosensory, auditory-vestibular, olfactory, and visual system neurons. Generally speaking, IGF-1 expression is most abundant after birth, except for early maturing structures such as the olfactory bulb, where IGF-1 expression is abundant several days before birth, concordant with the precocious initiation of glucose utilization by this structure (Bondy, 1991) (Fig 6). It appears that IGF-1 promotes nutrient utilization during development, thus enhancing process growth and synaptogenesis in these structures, and that continued high local 2DGU after recession of IGF-1 expression reflects the extraordinary synaptic density at these sites. Alternatively or additionally, although IGF-1 expression is reduced by the fourth week of life compared with the first few weeks, it remains detectable and stable in these high glucose utilizing structures through maturity (Niblock *et al.*, 1998), suggesting a possible continuing role for IGF-1 in promoting glucose utilization by the mature brain.

As noted above, IGF-1 and other elements of the IGF system are strongly induced in response to diverse types of CNS injury. Interestingly, this injury-invoked IGF-1 expression is generally in astrocytes (Komoly *et al.*, 1992; Lee *et al.*, 1992; Gehrmann *et al.*, 1994; Yao *et al.*, 1995; Li *et al.*, 1998), where it is also strongly correlated with local 2DGU (Fig. 7). In this setting the reactive astroglia are involved in scar formation with high-level metabolic needs for phagocytosis and synthesis and secretion of matrix proteins. Exogenous IGF-1's ability to regulate brain glucose utilization in the normal adult brain is shown by IGF-1 microinjected into the cerebral cortex of normal adult animals producing a marked increase in local 2DGU, which was not seen with saline or with neuroactive peptides including nerve growth factor or fibroblast growth factor-2 (Cheng *et al.*, 2000). Another recent study found that IGF-1 injected into a lateral ventricle increased 2DGU in CA1 of the hippocampal formation, the cingulate gyrus, and the arcuate nucleus. This singular distribution of effects could derive from the mode of delivery or from the prolonged experimentation leading to downregulation of receptor or transporters in some regions (Lynch *et al.*, 2001).

B. Glucose Metabolism in the IGF-1 Deletion Brain

If endogenous IGF-1 normally promotes brain glucose metabolism, then IGF-1 deletion should result in impaired brain glucose utilization. We compared brain 2DGU in *IGF1–/–* and wild-type (WT) littermate mice at P10, a time of peak IGF-1 expression (Cheng *et al.*, 2000). Regional 2DGU patterns parallel *IGF1* expression in the developing WT mouse brain (Fig. 8) and, as predicted, 2DGU is significantly reduced in the *IGF1–/–* brain, most

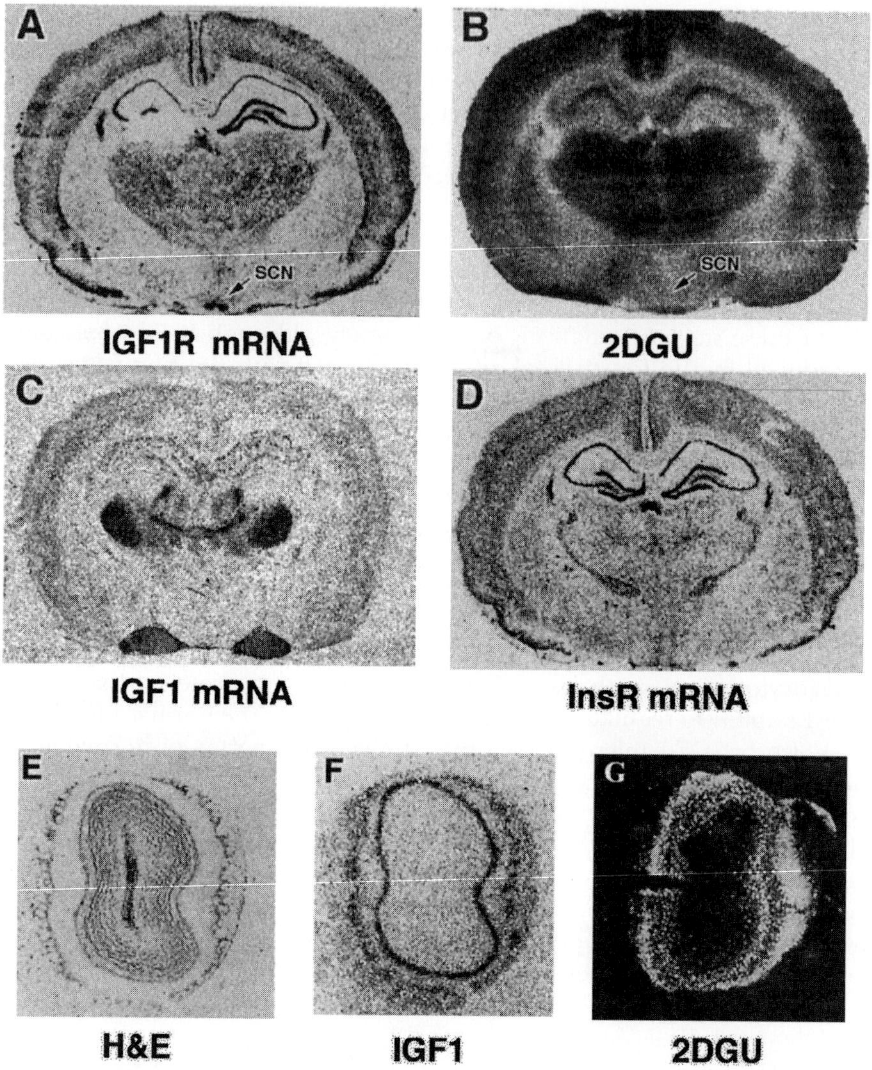

FIG. 6. 2-Deoxyglucose uptake (2DGU) correlates closely with IGF-1 and IGF-1 receptor but not so well with insulin receptor expression in the normal developing brain. Panels A–D are film autoradiographs from anatomically matched sections from the anterior thalamus from P14 rats. Panels E–G show the rat olfactory bulb in hematoxylin/eosin-stained view (E), with F showing the film autoradiograph of IGF-1 mRNA from the same section shown in E. Panel G shows a dark field micrograph of 14C-2DGU in an anatomically matched olfactory bulb section. IGF-1 mRNA is concentrated in the olfactory projection neurons (mitral and tufted cells) and 2DGU is concentrated in their dendritic synaptic fields.

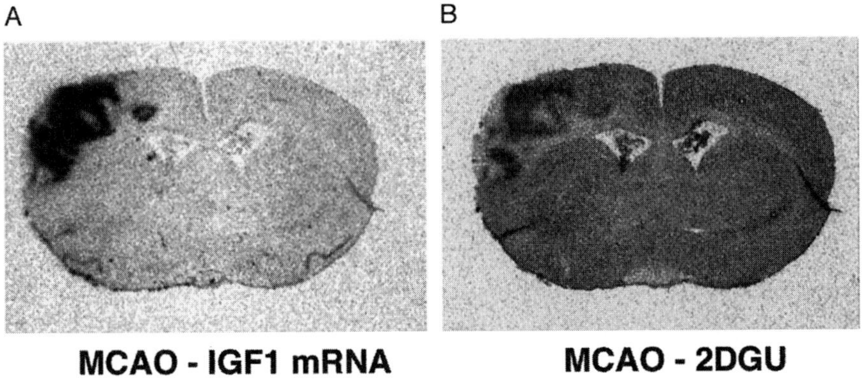

FIG. 7. IGF-1 gene expression (A) parallels 14C-2DGU (B) in the infarct zone several days after a middle cerebral artery occlusion (MCAO) in adult rat brain. For experimental protocol, see Lee *et al.* (1992).

profoundly in those structures where *IGF1* is highly expressed in the WT. For example, 2DGU is reduced by more than 50% in thalamic and brain stem sensory nuclei (Fig. 8). Glucose uptake was also investigated in isolated nerve terminals, or synaptosomes, prepared from *IGF-1–/–* and WT brains. Synaptosomes from *IGF1–/–* brains demonstrate an approximate 30% reduction ($p < 0.006$) in 2DGU *in vitro*, and addition of IGF-1 to the incubation medium normalizes the defect (Cheng *et al.*, 2000). Thus, the defective glucose utilization observed in the *IGF1–/–* brain *in vivo* is associated with reduced 2DGU at the nerve terminal level *in vitro*, which is completely reversed by IGF-1. These *in vitro* findings have important implications, suggesting that the defect in glucose utilization seen in the IGF-1 null brain *in vivo* is not due to reduced neural activity or reduced brain blood flow, neither of which affects the synaptosome preparation. Furthermore, the finding of reduced glucose uptake in isolated neuron terminals shows that IGF-1 normally promotes glucose uptake by terminals independent of glial effects, since glia are not present in the synaptosome fraction. Brain 2DGU levels were diminished in P40 *IGF1–/–* mice compared with WT, but the differences were not statistically significant at this stage.

C. Molecular Mechanisms

In peripheral tissues, insulin receptor activation triggers a kinase cascade leading through phosphatidylinositol 3-kinase to phosphorylation of protein kinase B/Akt (PKB/Akt) (Summers and Birnbaum, 1997). Activation of this kinase leads to translocation of facilitative glucose transporters

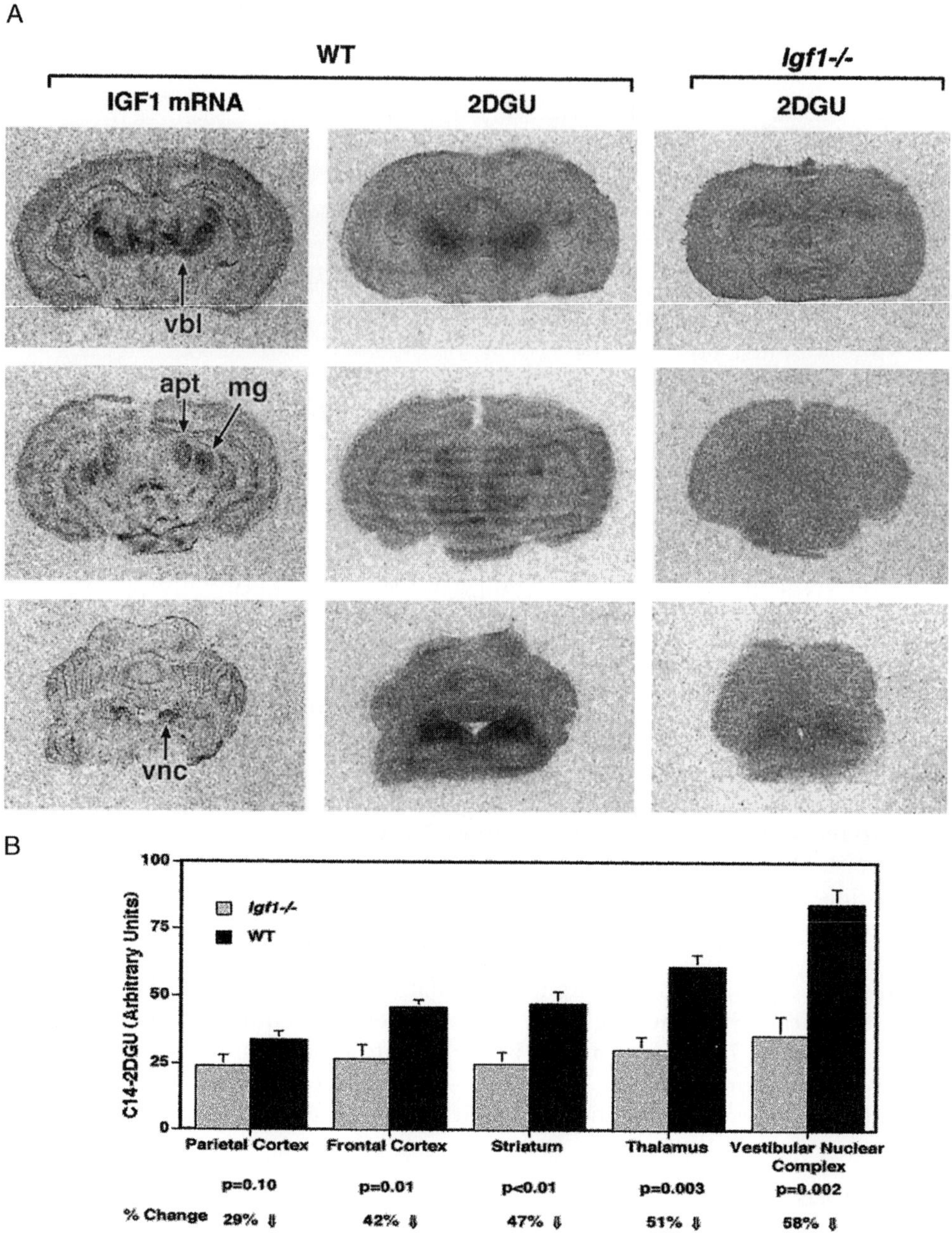

FIG. 8. Reduced glucose utilization in IGF-1 null brains (postnatal day 10). The first column in (A) shows IGF-1 mRNA expression in WT fore-, mid-, and hindbrain. The second column shows 2DGU in these same brain regions in WT mice, and the third column shows 2DGU in anatomically matched brain sections from IGF-1 null mice. Quantitative comparisons of 2DGU in select brain regions for the 2 groups are shown in (B). (Adapted from Cheng *et al.*, 2000, with modifications.)

(GLUTs), from intracellular pools to the plasma membrane, promoting glucose entry into cells (Kohn *et al.*, 1996; Summers and Birnbaum, 1997). The active, phosphorylated form of this kinase (phospho-serine^{473} and -threonine^{308} Akt) is concentrated in granular deposits in IGF-1-expressing neuronal processes in WT brains, but is barely detectable in these same neurons in *IGF1–/–* brains (Cheng *et al.*, 2000). In the WT brain, GLUT4 immunoreactivity is concentrated in the processes of IGF-1-expressing neurons in a pattern similar to phospho-Akt, suggesting a physical association of these proteins in translocation of GLUT4 to the plasma membrane. In *IGF1–/–* brains, however, GLUT4 immunoreactivity is reduced and largely confined to perikarya. GLUT4 mRNA is also decreased in *IGF1–/–* projection neurons (Cheng *et al.*, 2000). Thus, IGF-1-induced Akt phosphorylation appears linked to both production and translocation of neuronal GLUT4 from intracellular pools to nerve process membranes in the normal developing brain. Changes in GLUT1 and 3 mRNAs were not detected in IGF-1 null brains and investigations of differences in subcellular localization of these two transporters were inconclusive. Hexokinase activity is significantly reduced in IGF-1 null brains, suggesting a role for IGF-1 in regulation of brain hexokinase activity, which may contribute to the decreased glucose utilization in the IGF-1 null brain (Cheng *et al.*, 2000).

Another target of insulin/IGF signaling via Akt/PKB is glycogen synthase kinase 3β (GSK3β). Insulin/IGF-induced inhibitory phosphorylation of GSK3β on serine^{9} (Summers *et al.*, 1999) relieves GSK3β's inhibition of glycogen synthase and the translation initiation factor eIF2B, thus promoting glycogen and protein synthesis. This kinase has recently gained stature for its role in the Wnt signaling pathway, which is involved in the specification of cell fate in embryos and possibly maintaining stem cell pluripotency in adult tissues (Kim and Kimmel, 2000; Grimes and Jope, 2001). Data from the IGF-1 null mouse brain provide compelling evidence that GSK3β is crucially important in traditional anabolic pathways in brain development. Previous *in vitro* work has shown that both insulin and IGF-1 inhibit GSK3β in neural cells with the readout of tau-phosphorylation (Hong and Lee, 1997; Lesort and Johnson, 2000). Tau is a GSK3β substrate which is a microtubule-associated protein involved in neurofilament stabilization. When hyperphosphorylated, tau is prone to form intracelllular neurofibrillary tangles that contribute to neuronal degeneration.

Immunoreactivity specific for ser^{9}-phospho-GSK3β is selectively concentrated in the perikarya of large, IGF-1-expressing projection neurons in WT brain, associated with abundant glycogen accumulation in the same neurons (Fig. 9; see color insert) (Cheng *et al.*, 2000). In *IGF1–/–* brains, however, ser^{9}-phospho-GSK3β is barely detectable, and glycogen stores are profoundly reduced (Fig. 9). The co-localization of phospho-GSK3β with abundant glycogen stores specifically in IGF-1-expressing neurons suggests that IGF-1 acts in an autocrine manner to promote glucose uptake and storage as glycogen in developing projection neurons. Interestingly, while glycogen is associated with astrocytes in the mature brain, neuronal glycogen synthesis is abundant in postnatal development during the time of peak IGF-1 expression (Borke and Nau, 1984; Bondy, 1991). The significance of glycogen synthesis for growing neurons is uncertain; it could represent a mechanism to increase transport by effectively sequestering intracellular glucose-6-P.

Another important GSK3β target is the translation initiation factor eIF2B (Frame and Cohen, 2001). We did not evaluate eIF2B in this study, but predict that it too would be inhibited by GSK3β overactivity, contributing to the asthenia affecting IGF-1 null neurons. Tau phosphorylation, as expected, was increased in the IGF-1 null brain compared to WT littermates (Cheng and Bondy, manuscript submitted). The finding of tau hyperphosphorylation is another piece of evidence suggesting that IGF-1-induced inhibition of GSK3β is a central mechanism of IGF-1 action in brain. Interestingly, GSK3β appears to promote apoptosis in neurons (Crowder and Freeman, 2000; Hetman *et al.*, 2000; Li *et al.*, 2000; Tong *et al.*, 2001), possibly through hyperphosphorylation of tau and β-catenin (Lucas *et al.*, 2001). Thus, GSK3β hyperactivity in the IGF-1 null brain may contribute not only to hypoplastic neuronal development through reduced anabolic processes but also to increased neuronal loss, as seems to occur in the dentate gyrus (Cheng *et al.*, 2001a).

D. Functional Consequences of IGF-1 Deletion

IGF-1 deficiency results in decreased postnatal brain growth (Fig. 10; see color insert). This effect is clearly more profound in the nullizygous state, but even partial IGF-1 deficiency, as in *IGF1+/–* mice, results in significantly diminished brain growth. *IGF1–/+* brains are ~10% smaller than WT at P40 ($p < 0.0001$). Despite the reduction in size, *IGF1–/–* brain anatomy and cell numbers are for the most part normal, with the notable exceptions of the dentate gyrus, which is modestly reduced in size and cell number (Beck *et al.*, 1995; Cheng *et al.*, 2001a), and the olfactory bulb, which as seen in Fig. 10, is profoundly hypoplastic (Cheng *et al.*, 1998). Neuronal numbers in the cerebellum, thalamus, neocortex, and brain stem, however, are normal in the IGF-1 null brain. The cause of the reduced neuronal population in the hippocampal dentate gyrus appears due to increased cell death rather than decreased proliferation (Cheng *et al.*, 2001a).

The 30–40% reduction in brain size in adult IGF-1 null mice is not nearly accounted for by reduced numbers of dentate and olfactory bulb neurons, however. Cell density is significantly increased in the IGF-1 null brain (Beck *et al.,* 1995; Cheng *et al.,* 1998) suggesting decreased process growth, since the space between neurons, or neuropil, is normally occupied by extensively branched neuronal processes. The cross-sectional area of projection neurons in the IGF-1 null brain is reduced by ~25%, and nerve process length and branching are reduced by a similar amount (Fig. 11; see color insert). Thus it appears that IGF-1 deficiency impairs neuronal somatic growth and process formation, accounting in large part for the reduction in IGF-1 null brain size.

In terms of behavioral consequences, IGF-1 null mice show no obvious loss of motor skills or coordination and no evidence of myelinopathy or neurodegenerative disease (Cheng *et al.,* 1998). Learning ability in IGF-1 null dwarf mice has not yet been assessed due to technical limitations of conventional testing devices (mazes, etc.), that are designed for normal size animals. The decreased dendritic branching and synaptogenesis observed in IGF-1 null brains, however, certainly suggests that complex processing functions would be impaired. Supporting this view, the single human with homozygous IGF-1 gene deletions who has been examined has a small brain with normal myelination, but is profoundly mentally retarded (Woods *et al.,* 1996, 1997).

V. Effects of Transgenic IGF-1 Overexpression in Brain

Overexpression of an IGF-1 transgene in brain driven by a metallothionein promoter results in brain enlargement and increased myelin content (Carson *et al.,* 1993). DNA content and oligodendrocyte numbers are not increased, however, suggesting that the increased brain mass in this model is due primarily to increased cell size and process growth. Ye *et al.* (1995) showed that myelin sheath thickness is increased in proportion to the increased axonal diameter in these transgenic IGF-1-overexpressing brains. In the IGF-1 null brain, oligodendrocyte numbers are reduced in proportion to reduced neuronal numbers in the olfactory bulb, and whole brain myelin content is reduced in proportion to the reduction in brain size and protein content (Cheng *et al.,* 1998). Likewise, in the IGF-1 null peripheral nervous system, the thickness of myelin sheaths is reduced in proportion to reduced axonal diameter (Gao *et al.,* 1999) and IGF-1 null mice show no sign of myelinopathy (Cheng *et al.,* 1998; Gao *et al.,* 1999). These findings are consistent with current thinking that oligodendrocyte survival and function are regulated by neuronal fiber growth and activity (Stevens *et al.,* 1998; Barres and Raff, 1999). IGF-1 overexpression stimulates extraordinary process growth and the survival of more neurons, thus promoting additional oligodendrocyte biosynthetic activity and myelination. IGF-1 deficiency results in reduced process growth and neuronal survival, and there is a corresponding reduction in oligodendrocytes and myelination.

Interestingly, brain 2DGU is globally increased in IGF-1 overexpressing adult mice (Fig. 12; see color insert) (Gutierrez-Ospina *et al.*, 1997). It is not certain which cell types are responsible for the ectopic IGF-1 expression in these mice, but apparently IGF-1 is in excess through most of the brain for much of development. Thus, the generalized increase in 2DGU likely reflects local field potentials originating from more highly ramified dendritic arbors with greater synaptic density in IGF-1-overexpressing brains. The fact that pentobarbital anesthesia abolished the differences in glucose uptake between transgenic and WT mice in that study (Fig. 12) was thought by the authors to suggest that IGF-1 does not promote brain glucose utilization. However, pentobarbital interferes with glucose transporter function per se (Haspel *et al.*, 1999) and thus may obscure the study of the regulation of glucose transport. Furthermore, we have suggested that IGF-1, over a time frame of days to weeks, promotes glucose uptake and use by developing neurons, allowing them to grow larger and develop a more complex dendritic architecture. In maturity, these neurons will clearly be capable of more complex information-processing activity, associated with increased glucose utilization. We have not suggested that the rapid, transient alterations in glucose uptake associated with neural activity in the mature brain are due to direct effects by IGF-1 on glucose transport. These effects are related to rapid changes in cerebral blood flow and neurotransmitter metabolism, and involve glucose catabolism rather than the anabolic pathways associated with IGF-1 activity (i.e., glycolysis rather than glycogen synthesis).

VI. IGF-1 in the Primate Brain

All the studies described thus far were based on murine models and there is little data on IGF system expression or function in the brains of "higher" species. That said, there is ample evidence that fundamental anabolic pathways governed by IGF-1 are similar in mouse and man, with IGF-1 deficiency having very similar effects on somatic growth in all species (Daughaday, 1977). We have recently shown that IGF-1 gene expression is localized in cortical neurons in the adult rhesus monkey where it is co-expressed with GLUTs 3 and 4 (Cheng *et al.*, 2001b). Estrogen treatment increases IGF-1 and GLUT 3 and 4 expression in these neurons, raising the possibility that IGF-1 may regulate neuronal glucose transporter expression. Evidence about the relation between brain IGF-1 expression and brain function or pathology is difficult to obtain for human subjects, but evidence has accumulated in recent years that circulating IGF-1 may enhance brain growth and function (Sonntag *et al.*, 2000; Torres-Aleman, 2000; Laron, 2001). Finally, the finding provides compelling support for the view that IGF-1 is critically important for higher brain function in humans and that the single individual described with homozygous IGF-1 deletion has reduced brain size and severe mental retardation (Woods *et al.*, 1996, 1997).

VII. Summary

We have reviewed several lines of evidence suggesting that IGF-1 augments neuronal glucose utilization during brain development. To briefly recapitulate, brain glucose utilization parallels IGF-1 and IGF-1 receptor expression during brain development. In normal murine brain development, IGF-1 is produced in greatest abundance by growing cerebellar and sensory projection neurons during the time of dendrite elaboration and synaptogenesis. Glucose utilization is significantly reduced in the developing *IGF1–/–* brain, particularly in those sites where IGF-1 expression is normally most abundant. The defect in glucose utilization in *IGF1–/–* brains is demonstrable at the nerve terminal level *in vitro*, and is reversed by IGF-1. It appears that IGF-1 promotes glucose allocation to growing neurons in an autocrine manner, enabling the extraordinary elaboration of processes characterizing these complex information-processing systems. These sensory processing centers continue to exhibit high-level glucose utilization in the mature brain, after IGF-1 expression has receded, reflecting the extraordinary dendritic complexity and synaptic density achieved by these structures. IGF-1 is not likely to play a major role in the rapid, neural activity-based glucose utilization in the mature brain, as discussed in Section IV. It will be interesting to determine whether IGF-1 is involved in activity-induced synaptic remodeling, as a number of studies suggest (Torres-Aleman, 2000). IGF-1 is highly expressed by reactive (GFAP-positive) astrocytes in various injury models (Komoly *et al.*, 1992; Lee *et al.*, 1992; Lee and Bondy, 1993; Gehrmann *et al.*, 1994; Yao *et al.*, 1995; Walter *et al.*, 1997; Beilharz *et al.*, 1998; Li *et al.*, 1998). We have found that astrocytic IGF-1 expression several days after MCAO is closely correlated with increased glucose utilization in the injury site, as shown in Fig. 7 (Lee *et al.*, 1992). The significance of this local increase in glucose utilization is unclear, but it may be attributed to increased anabolic activity by astrocytes synthesizing and secreting collagen and other extracellular proteins involved in scar formation.

PKB/Akt and GSK3β appear to be central players in IGF signaling to the brain (Fig. 13). PKB/Akt phosphorylation in IGF-1-expressing neurons

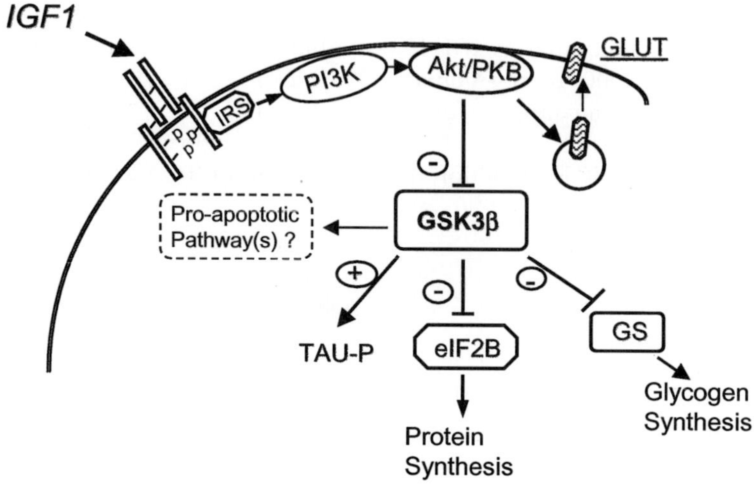

FIG. 13. Schematic of IGF-1 signaling pathways implicated in regulation of brain anabolic pathways and growth. IGF-1 binding to the IGF-1 receptor triggers receptor autophosphorylation and association with IRS docking proteins. Phosphatidylinositol 3 kinase (PI3K) is then activated and generates phospholipids which activate Akt/PKB. Akt/PKB promotes translocation of GLUTs (1, 3, and 4) from intracellular endosomal pools to the plasma membrane, thus augmenting glucose transport into the cell. Akt/PKB also serine phosphorylates GSK3β, causing inhibition. Since GSK3β normally inhibits glycogen synthase and eIF2b, inactivation of GSK3β promotes both glycogen and protein synthesis. The microtubule-associated protein tau is also a target for GSK3β and is hyperphosphorylated in the IGF-1 null brain, providing further evidence that IGF-1 normally inhibits brain GSK3β activity. Finally, GSK3β has been implicated as a proapoptotic factor in neurons, though there is no consensus on which cell death pathways are involved.

is associated with increased GLUT4 expression and translocation from intracellular compartments to the plasma membrane. IGF1's apparent link with this "insulin-sensitive" transporter may represent a specific anabolic pathway, distinct from the glucose transport pathways involving GLUTs 1 and 3. PKB/Akt phosphorylation also leads to the inhibition of GSK3β in IGF-1-expressing neurons (Figs. 11 and 13). This inhibition is expected to facilitate glycogen and protein synthesis, as GSK3β normally inhibits both glycogen synthase and eIF2B (Fig. 13). As a result, there is accumulation of glycogen in IGF-1-expressing neurons (Cheng *et al.*, 2000), which may serve to create a relative "sink" for G-6-P, promoting further glucose transport into the neuron.

GSK3β also phosphorylates the microtubule-associated protein tau. Tau is hyper-phosphorylated in the *IGF1* null brain, as expected if IGF-1 normally inhibits this multifunctional enzyme. Tau hyperphosphorylation is

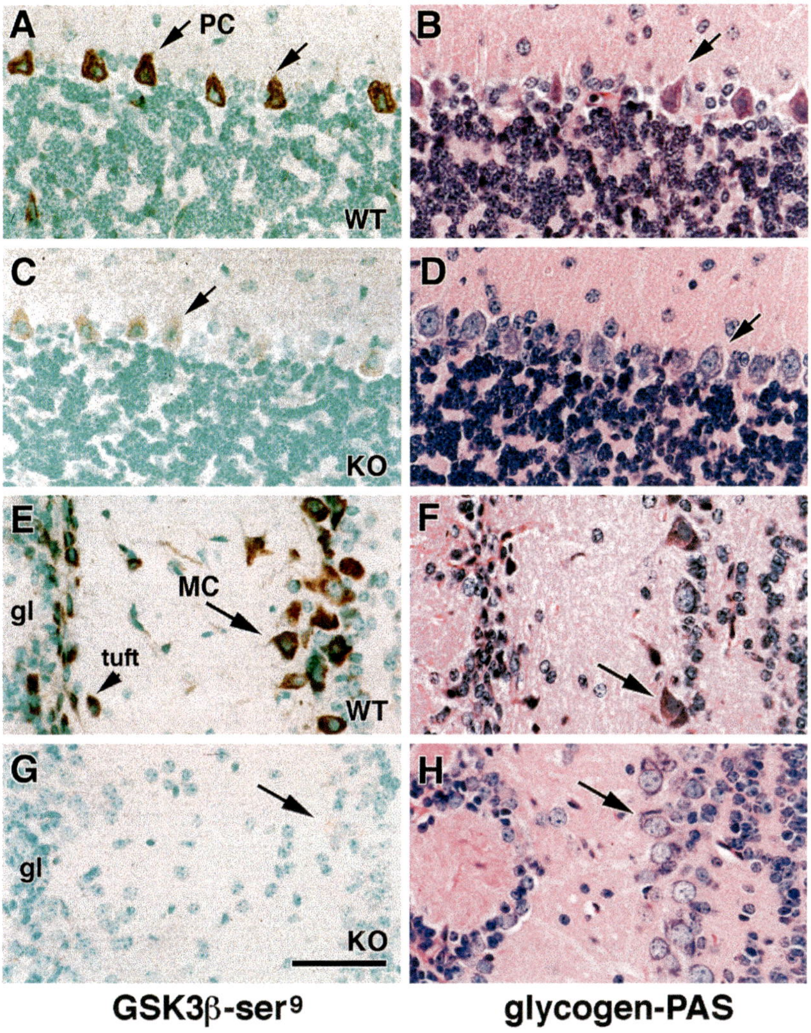

FIG. 4.9. Phosphorylation of glycogen synthase kinase 3β (GSK3β) and glycogen synthesis in WT and IGF-1 null cerebella (A–D) and olfactory bulbs (E–F). The antibody used in these sections recognizes only ser9-phosphorylated GSK3β, which is selectively concentrated in IGF-1-expressing projection neurons in WT brains. While GSK3β protein levels are not different in WT and IGF-1 null, the ser9-phospho-form of the kinase is barely detectable and glycogen synthesis (the purple PAS stain in B, D, F, and H) is correspondingly reduced in the IGF-1 null brain. (Adapted from Cheng, Reinhardt *et al.,* 2000, with modifications.)

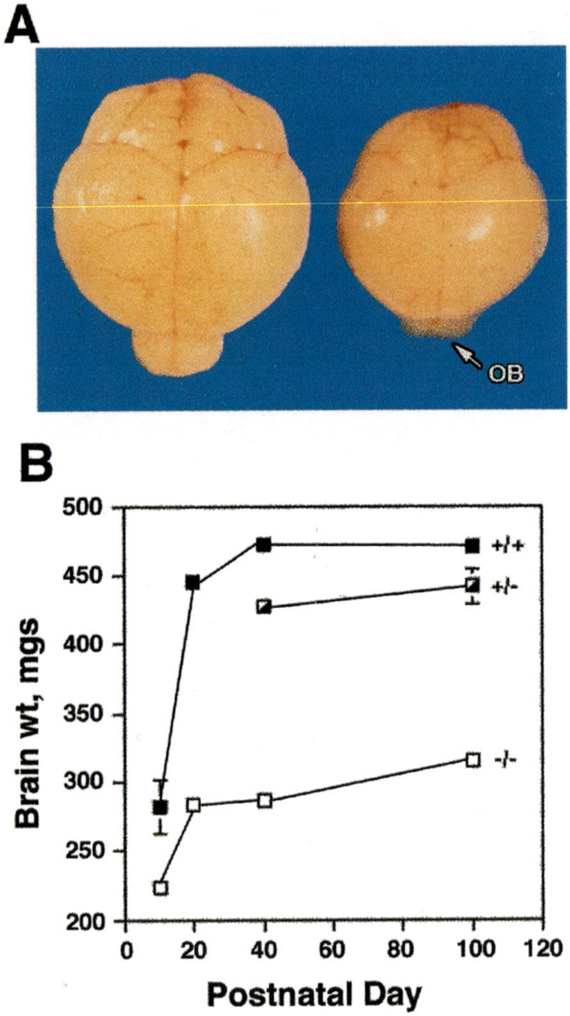

FIG. 4.10. Brain growth in WT and IGF-1 null mice. Panel A compares brains from representative WT and IGF-1 null mice at P40. Note the hypoplastic olfactory bulb in the IGF-1 null brain. Brain weights are compared graphically in panel B. The homozygous IGF-1 deletion results in significantly reduced brain size from P10 onward. The heterozygous IGF-1 deletion results in significant reduction in brain size at P40 and P100. (Adapted from Cheng, Cohen *et al.*, 2001, with modifications.)

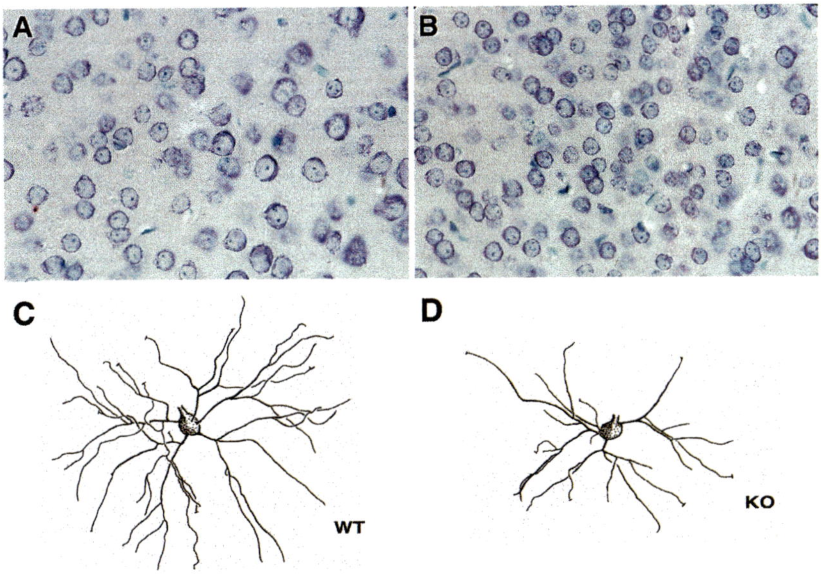

FIG. 4.11. Cortical neuron density and dendritic profiles in adult WT (A,C) and IGF-1 null mice (B and D). Representative toluidine-blue-stained sections from the frontoparietal cortex of P50 mice (A and B) show reduced somatic size and increased cell density in IGF-1 null mice. Camera lucida drawings of Golgi-stained cortical pyramids (layers II–III) reveal dramatically reduced dendritic profiles of the IGF-1 null neuron (C and D).

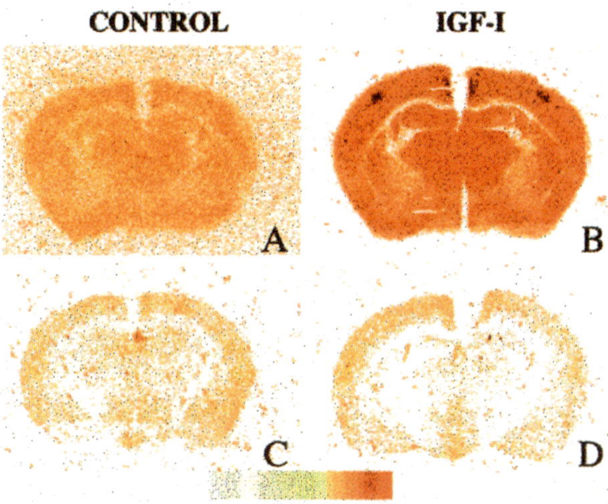

FIG. 4.12. Glucose utilization in wild-type (left panels) and IGF-1-overexpressing transgenic mice brains. The autoradiographic signal is color coded with red as the highest signal. 2-DGU is globally increased in the transgenic mouse brain (B). The lower panels compare 2-DGU in mice anesthetized with pentobarbital, which suppresses the signal from both brains. (From Gutierrez-Ospina, Saum *et al.*, Increased neural activity in transgenic mice with brain IGF-1 overexpression: a[^{3}H]2DG study, *NeuroReport*, **8**, 2907–2911, 1997, with modifications.)

associated with the formation of neurofibrillary tangles and neuronal degeneration, as in Alzheimer disease (AD). Several other aspects of AD could be related to IGF-1 deficiency. For example, glucose utilization is decreased and acetylcholine synthesis may be impaired secondary to glucose starvation. IGF-1 rescues neurons from many types of death *in vitro* (Dore *et al.*, 1997; Russell *et al.*, 1998), hence increased neuronal loss in AD might also be attributed to a relative IGF-1 deficiency. It is not known if IGF-1 prevents programmed neuron death in these experimental systems by inhibiting GSK3β, which has proapoptotic effects on neurons (Grimes and Jope, 2001). GSK3β proapoptotic effects are still not well defined, but as noted above, tau hyperphosphorylation is a clear result of GSKβ hyperactivity and may contribute to neurotoxicity. However, recent data suggest a coupling between increased glucose utilization, inhibition of GSK3β, and protection from cell death in endothelial cells—where tau presumably is not a player (Hall *et al.*, 2001).

The identification of GSK3β as a major target of brain IGF-1 signaling provides a unifying pathway for IGF1's well-established anabolic and antiapoptotic functions. Thus IGF-1-induced inhibition of GSK3β may have multifaceted anabolic and neuroprotective effects. IGF-1 deficiency during brain development results in smaller neurons with hypoplastic processes and decreased survival of dentate gyrus and olfactory neurons. In the human, global IGF-1 deficiency due to homozygous IGF-1 deletion results in a small brain and mental retardation, suggesting that these studies in mice are indeed relevant to human neurobiology.

VIII. Future Directions

IGF-1 clearly promotes neuronal survival, process growth, and synaptogenesis in normal brain development, making it a highly attractive candidate for therapies aimed at improving cognitive function and combating neurodegenerative disease. The challenge for future research is to learn to manipulate brain IGF-1 expression in a therapeutic manner that would enhance cognitive potential and neuronal longevity, without promoting potentially adverse consequences such as tumorigenesis. IGF-1 promotes anabolic functions and survival in most if not all cell types, including brain tumor cells, and IGF or IGF-1 receptor overexpression have been linked to brain tumorigenesis (Zumkeller and Schwab, 1999). Tumor development aside, misdirected or poorly timed effects by IGF-1 could result in overgrowth of inhibitory systems (e.g., inhibitory neurons or gliosis) which would impair rather than improve brain functioning. To this end, it is our goal to elucidate the mechanisms governing normal cell- and developmental stage-specific regulation of IGF-1 expression, and to attempt to restore normal patterns of expression in conditions such as fetal alcohol exposure and malnutrition, and possibly in other forms of mental retardation.

Acknowledgments

We are grateful to Wei-Hua Lee, Ph.D., and Rickey Reinhardt, M.D., Ph.D., for their contributions to these studies.

References

Adamo, M., Roberts, C. T., Jr., *et al.* (1992). How distinct are the insulin and insulin-like growth factor I signalling systems? *Biofactors* **3**(3), 151–157.

Andersson, I. K., Edwall, D., *et al.* (1988). Differing expression of insulin-like growth factor I in the developing rat cerebellum. *Acta. Physiol. Scand.* **132,** 167–173.

Ayer-Le Lievre, C., Stahlbom, P. A., *et al.* (1991). Expression of IGF-I and -II mRNA in the brain and craniofacial region of the rat fetus. *Development* **111,** 105–115.

Bach, M. A., Shen-Orr, Z., *et al.* (1991). Insulin-like growth factor I mRNA levels are developmentally regulated in specific regions of the rat brain. *Brain. Res. Mol. Brain. Res.* **10**(1), 43–48.

Bailyes, E. M., Nave, B. T., *et al.* (1997). Insulin receptor/IGF-I receptor hybrids are widely distributed in mammalian tissues: quantification of individual receptor species by selective immunoprecipitation and immunoblotting. *Biochem. J.* **327**(Pt 1), 209–215.

Baker, J., Liu, J. P., *et al.* (1993). Role of insulin-like growth factors in embryonic and postnatal growth. *Cell* **75**(1), 73–82.

Barres, B. A., and Raff, M. C. (1999). Axonal control of oligodendrocyte development. *J. Cell Biol.* **147**(6), 1123–1128.

Bartlett, W. P., Li, X. S., *et al.* (1991). Localization of insulin-like growth factor-1 mRNA in murine central nervous system during postnatal development. *Dev. Biol.* **147**(1), 239–250.

Beck, K. D., Powell-Braxton, L., *et al.* (1995). Igf1 gene disruption results in reduced brain size, CNS hypomyelination, and loss of hippocampal granule and striatal parvalbumin-containing neurons. *Neuron* **14**(4), 717–730.

Beilharz, E. J., Russo, V. C., *et al.* (1998). Co-ordinated and cellular specific induction of the components of the IGF/IGFBP axis in the rat brain following hypoxic-ischemic injury. *Brain Res. Mol. Brain Res.* **59**(2), 119–134.

Bohannon, N. J., Corp, E. S., *et al.* (1988). Localization of binding sites for insulin-like growth factor-I (IGF-I) in the rat brain by quantitative autoradiography. *Brain Res.* **444**(2), 205–213.

Bondy, C., Bach, M., *et al.* (1992a). Mapping of brain insulin and insulin-like growth factor receptor gene expression by in situ hybridization. *Neuroprotocols* **1,** 240–249.

Bondy, C., Werner, H., *et al.* (1992b). Cellular pattern of type-I insulin-like growth factor receptor gene expression during maturation of the rat brain: comparison with insulin-like growth factors I and II. *Neuroscience* **46**(4), 909–923.

Bondy, C. A. (1991). Transient IGF-I gene expression during the maturation of functionally related central projection neurons. *J. Neurosci.* **11**(11), 3442–3455.

Bondy, C. A., Werner, H., *et al.* (1990). Cellular pattern of insulin-like growth factor-I (IGF-I) and type I IGF receptor gene expression in early organogenesis: Comparison with IGF-II gene expression. *Mol. Endocrinol.* **4**(9), 1386–1398.

Borke, R. C., and Nau, M. E. (1984). Glycogen, its transient occurrence in neurons of the rat CNS during normal postnatal development. *Brain Res.* **318**(2), 277–284.

Braulke, T. (1999). Type-2 IGF receptor: A multi-ligand binding protein. *Horm. Metab. Res.* **31**(2–3), 242–246.

Butler, A. A., Yakar, S., *et al.* (1998). Insulin-like growth factor-I receptor signal transduction: At the interface between physiology and cell biology. *Comp. Biochem. Physiol. B Biochem. Mol. Biol.* **121**(1), 19–26.

Carro, E., J. Trejo, L., *et al.* (2001). Circulating insulin-like growth factor I mediates the protective effects of physical exercise against brain insults of different etiology and anatomy. *J. Neurosci.* **21**(15), 5678–56784.

Carson, M. J., Behringer, R. R., *et al.* (1993). Insulin-like growth factor I increases brain growth and central nervous system myelination in transgenic mice. *Neuron* **10**(4), 729–740.

Cascieri, M. A., Chicchi, G. G., *et al.* (1986). (Thr-59)-insulin-like growth factor I stimulates 2-deoxyglucose transport in BC3H1 myocytes through the insulin-like growth factor receptor, not the insulin receptor. *Biochim. Biophys. Acta.* **886**(3), 491–499.

Chen, H., Wertheimer, S. J., *et al.* (1997). Protein-tyrosine phosphatases PTP1B and syp are modulators of insulin-stimulated translocation of GLUT4 in transfected rat adipose cells. *J. Biol. Chem.* **272**(12), 8026–8031.

Cheng, C. M., Cohen, M., *et al.* (2001a). Endogenous IGF1 enhances cell survival in the postnatal dentate gyrus. *J. Neurosci. Res.* **64**(4), 341–347.

Cheng, C. M., Cohen, M., *et al.* (2001b). Estrogen augments glucose transporter and IGF1 expression in primate cerebral cortex. *FASEB J.* **15**(6), 907–915.

Cheng, C. M., Joncas, G., *et al.* (1998). Biochemical and morphometric analyses show that myelination in the insulin-like growth factor 1 null brain is proportionate to its neuronal composition. *J. Neurosci.* **18**(15), 5673–5681.

Cheng, C. M., Reinhardt, R. R., *et al.* (2000). Insulin-like growth factor 1 regulates developing brain glucose metabolism. *Proc. Natl. Acad. Sci. USA* **97**(18), 10236–10241.

Clemmons, D. R. (1989). Structural and functional analysis of insulin-like growth factors. *Br. Med. Bull.* **45**(2), 465–480.

Clemmons, D. R. (1998). Role of insulin-like growth factor binding proteins in controlling IGF actions. *Mol. Cell. Endocrinol.* **140**(1–2), 19–24.

Coker, G. T. 3rd, Studelska, D., *et al.* (1990). Analysis of tyrosine hydroxylase and insulin transcripts in human neuroendocrine tissues. *Brain Res. Mol. Brain Res.* **8**(2), 93–98.

Crowder, R. J., and Freeman, R. S. (2000). Glycogen synthase kinase-3 beta activity is critical for neuronal death caused by inhibiting phosphatidylinositol 3-kinase or Akt but not for death caused by nerve growth factor withdrawal. *J. Biol. Chem.* **275**(44), 34266–34271.

Daughaday, W. H. (1977). Hormonal regulation of growth by somatomedin and other tissue growth factors. *Clin. Endocrinol. Metab.* **6**(1), 117–135.

Daughaday, W. H., and Rotwein, P. (1989). Insulin-like growth factors I and II. Peptide, messenger ribonucleic acid and gene structures, serum, and tissue concentrations. *Endocrinol. Rev.* **10**(1), 68–91.

De Meyts, P., Wallach, B., *et al.* (1994). The insulin-like growth factor-I receptor structure, ligand-binding mechanism and signal transduction. *Horm. Res.* **42,** 152–169.

Devaskar, S. D., Zahm, S., *et al.* (1991). Developmental regulation of the distribution of rat brain insulin-insensitive (Glut 1) glucose transporter. *Endocrinology* **129**(3), 1530–1540.

Dore, S., Kar, S., *et al.* (1997). Insulin-like growth factor I protects and rescues hippocampal neurons against beta-amyloid- and human amylin-induced toxicity. *Proc. Natl. Acad. Sci. USA* **94**(9), 4772–4777.

Frame, S., and Cohen, P. (2001). GSK3 takes centre stage more than 20 years after its discovery. *Biochem. J.* **359**(Pt 1), 1–16.

Gao, W. Q., Shinsky, N., *et al.* (1999). IGF-I deficient mice show reduced peripheral nerve conduction velocities and decreased axonal diameters and respond to exogenous IGF-I treatment. *J. Neurobiol.* **39**(1), 142–152.

Gehrmann, J., Yao, D. L., *et al.* (1994). Expression of insulin-like growth factor-I and related peptides during motoneuron regeneration. *Exp. Neurol.* **128**(2), 202–210.

Grimes, C. A., and Jope, R. S. (2001). The multifaceted roles of glycogen synthase kinase 3beta in cellular signaling. *Prog. Neurobiol.* **65**(4), 391–426.

Guan, J., O. Miller, T., *et al.* (2001). Insulin-like growth factor-1 improves somatosensory function and reduces the extent of cortical infarction and ongoing neuronal loss after hypoxia-ischemia in rats. *Neuroscience* **105**(2), 299–306.

Gutierrez-Ospina, G., Saum, L., *et al.* (1997). Increased neural activity in transgenic mice with brain IGF-I overexpression: a [3H]2DG study. *Neuroreport* **8**(13), 2907–2911.

Hall, J. L., Chatham, J. C., *et al.* (2001). Upregulation of glucose metabolism during intimal lesion formation is coupled to the inhibition of vascular smooth muscle cell apoptosis. Role of GSK3beta. *Diabetes* **50**(5), 1171–1179.

Haspel, H. C., Stephenson, K. N., *et al.* (1999). Effects of barbiturates on facilitative glucose transporters are pharmacologically specific and isoform selective. *J. Membr. Biol.* **169**(1), 45–53.

Havrankova, J., Brownstein, M., *et al.* (1981). Insulin and insulin receptors in rodent brain. *Diabetologia* **20**(Suppl.), 268–273.

Hetman, M., Cavanaugh, J. E., *et al.* (2000). Role of glycogen synthase kinase-3beta in neuronal apoptosis induced by trophic withdrawal. *J. Neurosci.* **20**(7), 2567–2574.

Hetts, S. W., Rosen, K. M., *et al.* (1997). Expression and imprinting of the insulin-like growth factor II gene in neonatal mouse cerebellum. *J. Neurosci. Res.* **50**(6), 958–966.

Hill, J. M., Lesniak, M. A., *et al.* (1986). Autoradiographic localization of insulin receptors in rat brain: prominence in olfactory and limbic areas. *Neuroscience* **17**(4), 1127–1138.

Hong, M., and Lee, V. M. (1997). Insulin and insulin-like growth factor-1 regulate tau phosphorylation in cultured human neurons. *J. Biol. Chem.* **272**(31), 19547–19553.

Hynes, M. A., Brooks, P. J., *et al.* (1988). Insulin-like growth factor II messenger RNAs are synthesized in the choroid plexus of the rat brain. *Mol. Endocrinol.* **2,** 47–54.

Ish-Shalom, D., Christoffersen, C. T., *et al.* (1997). Mitogenic properties of insulin and insulin analogues mediated by the insulin receptor. *Diabetologia* **40**(Suppl. 2), S25–S31.

Kiess, W., Yang, Y., *et al.* (1994). Insulin-like growth factor II (IGF-II) and the IGF-II/mannose-6-phosphate receptor: the myth continues. *Horm. Res.* **41**(Suppl. 2), 66–73.

Kim, J. J., Park, B. C., *et al.* (2001). Mitogenic and metabolic effects of type I IGF receptor overexpression in insulin receptor-deficient hepatocytes. *Endocrinology* **142**(8), 3354–3360.

Kim, L., and Kimmel, A. R. (2000). GSK3, a master switch regulating cell-fate specification and tumorigenesis. *Curr. Opin. Genet. Dev.* **10**(5), 508–514.

Kohn, A. D., Summers, S. A., *et al.* (1996). Expression of a constitutively active Akt Ser/Thr kinase in 3T3-L1 adipocytes stimulates glucose uptake and glucose transporter 4 translocation. *J. Biol. Chem.* **271**(49), 31372–31378.

Komoly, S., Hudson, L. D., *et al.* (1992). Insulin-like growth factor I gene expression is induced in astrocytes during experimental demyelination. *Proc. Natl. Acad. Sci. USA* **89**(5), 1894–1898.

Laron, Z. (2001). Insulin-like growth factor 1 (IGF-1): A growth hormone. *Mol. Pathol.* **54**(5), 311–316.

Laron, Z., and Klinger, B. (2000). Comparison of the growth-promoting effects of insulin-like growth factor I and growth hormone in the early years of life. *Acta Paediatr.* **89**(1), 38–41.

Lee, W., Clemens, J., *et al.* (1992). Insulin-like growth factors in the response to cerebral ischemia. *Mol. Cell. Neurosci.* **3,** 36–43.

Lee, W. H., and Bondy, C. (1993). Insulin-like growth factors and cerebral ischemia. *Ann. N.Y. Acad. Sci.* **679,** 418–422.

Lee, W. H., Michels, K. M., *et al.* (1993). Localization of insulin-like growth factor binding protein-2 messenger RNA during postnatal brain development: correlation with insulin-like growth factors I and II. *Neuroscience* **53**(1), 251–265.

LeRoith, D., Delahunty, G., *et al.* (1986). Evolutionary aspects of the endocrine and nervous systems. *Recent Prog. Horm. Res.* **42,** 549–587.

LeRoith, D., Kavsan, V. M., *et al.* (1993). Phylogeny of the insulin-like growth factors (IGFs) and receptors: A molecular approach. *Mol. Reprod. Dev.* **35**(4), 332–336; discussion 337–338.

LeRoith, D., McGuinness, M., *et al.* (1992). Insulin-like growth factors. *Biol. Signals* **1**(4), 173–181.

Lesniak, M. A., Hill, J. M., *et al.* (1988). Receptors for insulin-like growth factors I and II: Autoradiographic localization in rat brain and comparison to receptors for insulin. *Endocrinology* **123**(4), 2089–2099.

Lesort, M., and Johnson, G. V. (2000). Insulin-like growth factor-1 and insulin mediate transient site-selective increases in tau phosphorylation in primary cortical neurons. *Neuroscience* **99**(2), 305–316.

Li, M., Wang, X., *et al.* (2000). Cyclic AMP promotes neuronal survival by phosphorylation of glycogen synthase kinase 3beta. *Mol. Cell. Biol.* **20**(24), 9356–9363.

Li, X. S., Williams, M., *et al.* (1998). Induction of IGF-1 mRNA expression following traumatic injury to the postnatal brain. *Brain Res. Mol. Brain Res.* **57**(1), 92–96.

Lucas, J. J., Hernandez, F., *et al.* (2001). Decreased nuclear beta-catenin, tau hyperphosphorylation and neurodegeneration in GSK-3beta conditional transgenic mice. *EMBO J.* **20**(1–2), 27–39.

Lupu, F., Terwilliger, J. D., *et al.* (2001). Roles of growth hormone and insulin-like growth factor 1 in mouse postnatal growth. *Dev. Biol.* **229**(1), 141–162.

Lynch, C. D., Lyons, D., *et al.* (2001). Insulin-like growth factor-1 selectively increases glucose utilization in brains of aged animals. *Endocrinology* **142**(1), 506–509.

Nakae, J., Barr, V., *et al.* (2000). Differential regulation of gene expression by insulin and IGF-1 receptors correlates with phosphorylation of a single amino acid residue in the forkhead transcription factor FKHR. *EMBO J.* **19**(5), 989–996.

Navarro, I., Leibush, B., *et al.* (1999). Insulin, insulin-like growth factor-I (IGF-I) and glucagon: The evolution of their receptors. *Comp. Biochem. Physiol. B Biochem. Mol. Biol.* **122**(2), 137–153.

Niblock, M. M., Brunso-Bechtold, J. K., *et al.* (1998). Distribution and levels of insulin-like growth factor I mRNA across the life span in the Brown Norway × Fischer 344 rat brain. *Brain Res.* **804**(1), 79–86.

Nissley, S. P., Haskell, J. F., *et al.* (1985). Insulin-like growth factor receptors. *J. Cell. Sci. Suppl.* **3,** 39–51.

Owens, J. A. (1991). Endocrine and substrate control of fetal growth: Placental and maternal influences and insulin-like growth factors. *Reprod. Fertil. Dev.* **3**(5), 501–517.

Pandini, G., Vigneri, R., *et al.* (1999). Insulin and insulin-like growth factor-I (IGF-I) receptor overexpression in breast cancers leads to insulin/IGF-I hybrid receptor overexpression: Evidence for a second mechanism of IGF-I signaling. *Clin. Cancer Res.* **5**(7), 1935–1944.

Pera, E. M., Wessely, O., *et al.* (2001). Neural and head induction by insulin-like growth factor signals. *Dev. Cell.* **1**(5), 655–665.

Rechler, M. M., and Nissley, S. P. (1986). Insulin-like growth factor (IGF)/somatomedin receptor subtypes: structure, function, and relationships to insulin receptors and IGF carrier proteins. *Horm. Res.* **24**(2–3), 152–159.

Reinecke, M., and Collet, C. (1998). The phylogeny of the insulin-like growth factors. *Int. Rev. Cytol.* **183,** 1–94.

Reinhardt, R. R., and Bondy, C. A. (1994). Insulin-like growth factors cross the blood-brain barrier. *Endocrinology* **135**(5), 1753–1761.

Reinhardt, R. R., Chin, E., *et al.* (1993). Insulin receptor-related receptor messenger ribonucleic acid is focally expressed in sympathetic and sensory neurons and renal distal tubule cells. *Endocrinology* **133**(1), 3–10.

Reinhardt, R. R., Chin, E., *et al.* (1994). Selective coexpression of insulin receptor-related receptor (IRR) and TRK in NGF-sensitive neurons. *J. Neurosci.* **14**(8), 4674–4683.

Rotwein, P., Burgess, S. K., *et al.* (1988). Differential expression of insulin-like growth factor genes in rat central nervous system. *Proc. Natl. Acad. Sci. USA* **85,** 265–269.

Russell, J. W., Windebank, A. J., *et al.* (1998). Insulin-like growth factor-I prevents apoptosis in neurons after nerve growth factor withdrawal. *J. Neurobiol.* **36**(4), 455–467.

Sara, V. R., and Carlsson-Skwirut, C. (1986). The biosynthesis of somatomedins and their role in the fetus. *Acta. Endocrinol. Suppl.* (*Copenh.*) **279,** 82–85.

Schwartz, M. W., Woods, S. C., *et al.* (2000). Central nervous system control of food intake. *Nature* **404**(6778), 661–671.

Seta, K. A., Kovacina, K. S., *et al.* (1993). The insulin receptor family. *Adv. Exp. Med. Biol.* **343,** 113–124.

Shefi-Friedman, L., Wertheimer, E., *et al.* (2001). Increased IGFR activity and glucose transport in cultured skeletal muscle from insulin receptor null mice. *Am. J. Physiol. Endocrinol. Metab.* **281**(1), E16–E24.

Siddle, K., Urso, B., *et al.* (2001). Specificity in ligand binding and intracellular signalling by insulin and insulin-like growth factor receptors. *Biochem. Soc. Trans.* **29**(4), 513–525.

Smit, A. B., van Kesteren, R. E., *et al.* (1998). Towards understanding the role of insulin in the brain: Lessons from insulin-related signaling systems in the invertebrate brain. *Prog. Neurobiol.* **54**(1), 35–54.

Sonntag, W. E., Lynch, C., *et al.* (2000). The effects of growth hormone and IGF-1 deficiency on cerebrovascular and brain ageing. *J. Anat.* **197**(Pt 4), 575–585.

Soos, M. A., Field, C. E., *et al.* (1993). Purified hybrid insulin/insulin-like growth factor-I receptors bind insulin-like growth factor-I, but not insulin, with high affinity. *Biochem. J.* **290**(Pt 2), 419–426.

Stevens, B., Tanner, S., *et al.* (1998). Control of myelination by specific patterns of neural impulses. *J. Neurosci.* **18**(22), 9303–11.

Stylianopoulou, F., Herbert, J., *et al.* (1988). Expression of the insulin-like growth factor II gene in the choroid plexus and the leptomeninges of the adult rat central nervous system. *Proc. Natl. Acad. Sci. USA* **85,** 141–145.

Summers, S. A., and Birnbaum, M. J. (1997). A role for the serine/threonine kinase, Akt, in insulin-stimulated glucose uptake. *Biochem. Soc. Trans.* **25**(3), 981–988.

Summers, S. A., Kao, A. W., *et al.* (1999). The role of glycogen synthase kinase 3beta in insulin-stimulated glucose metabolism. *J. Biol. Chem.* **274**(25), 17934–17940.

Sussenbach, J. S. (1989). The gene structure of the insulin-like growth factor family. *Prog. Growth Factor Res.* **1**(1), 33–48.

Tager, H. S., Steiner, D. F., *et al.* (1981). Biosynthesis of insulin and glucagon. *Methods Cell Biol.* **23,** 73–88.

Tong, N., Sanchez, J. F., *et al.* (2001). Activation of glycogen synthase kinase 3 beta (GSK-3beta) by platelet activating factor mediates migration and cell death in cerebellar granule neurons. *Eur. J. Neurosci.* **13**(10), 1913–1922.

Torres-Aleman, I. (2000). Serum growth factors and neuroprotective surveillance: Focus on IGF-1. *Mol. Neurobiol.* **21**(3), 153–160.

Van den Brande, J. L., Jansen, M., *et al.* (1986). Molecular aspects of the human somatomedins. *Horm. Res.* **24**(2–3), 131–140.

Van Houten, M., Posner, B. I., *et al.* (1979). Insulin binding in the rat brain: In vivo localization to the circumventricular organs by quantitative autoradiography. *Endocrinology* (105), 666.

Walter, H. J., Berry, M., *et al.* (1999). Distinct sites of insulin-like growth factor (IGF)-II expression and localization in lesioned rat brain: Possible roles of IGF binding proteins (IGFBPs) in the mediation of IGF-II activity. *Endocrinology* **140**(1), 520–532.

Walter, H. J., Berry, M., *et al.* (1997). Spatial and temporal changes in the insulin-like growth factor (IGF) axis indicate autocrine/paracrine actions of IGF-I within wounds of the rat brain. *Endocrinology* **138**(7), 3024–3034.

Watt, V. M., Shier, P., *et al.* (1993). IRR: A novel member of the insulin receptor family. *Adv. Exp. Med. Biol.* **343,** 125–132.

Werner, H., Adamo, M., *et al.* (1994). Molecular and cellular aspects of insulin-like growth factor action. *Vitam. Horm.* **48,** 1–58.

Werner, H., Woloschak, M., *et al.* (1989). Developmental regulation of the rat insulin-like growth factor I receptor gene. *Proc. Natl. Acad. Sci. USA* **86**(19), 7451–7455.

Werther, G. A., Hogg, A., *et al.* (1989). Localization and characterization of IGF-I receptors in rat brain and pituitary gland using in vitro autoradiography and computerized densitometry: Distinct distribution from insulin receptors. *J. Endocrinol.* **1,** 369–377.

Woods, K. A., Camacho-Hubner, C., *et al.* (1997). Insulin-like growth factor I gene deletion causing intrauterine growth retardation and severe short stature. *Acta. Paediatr. Suppl.* **423,** 39–45.

Woods, K. A., Camacho-Hubner, C., *et al.* (1996). Intrauterine growth retardation and postnatal growth failure associated with deletion of the insulin-like growth factor I gene. *N. Engl. J. Med.* **335**(18), 1363–1367.

Yao, D. L., West, N. R., *et al.* (1995). Cryogenic spinal cord injury induces astrocytic gene expression of insulin-like growth factor I and insulin-like growth factor binding protein 2 during myelin regeneration. *J. Neurosci. Res.* **40**(5), 647–659.

Yarden, Y., and Ullrich, A. (1988). Growth factor receptor tyrosine kinases. *Annu. Rev. Biochem.* **57,** 443–478.

Ye, P., Carson, J., *et al.* (1995). In vivo actions of insulin-like growth factor-I (IGF-I) on brain myelination: Studies of IGF-I and IGF binding protein-1 (IGFBP-1) transgenic mice. *J. Neurosci.* **15**(11), 7344–7356.

Young, W. S. (1986). Periventricular hypothalamic cells in rat brain contain insulin mRNA. *Neuropeptides* **8,** 93–97.

Zumkeller, W., and Schwab, M. (1999). Insulin-like growth factor system in neuroblastoma tumorigenesis and apoptosis: potential diagnostic and therapeutic perspectives. *Horm. Metab. Res.* **31**(2–3), 138–141.

CNS SENSING AND REGULATION OF PERIPHERAL GLUCOSE LEVELS

Barry E. Levin and Ambrose A. Dunn-Meynell

Neurology Service, VA Medical Center, East Orange, New Jersey 07018

Vanessa H. Routh

Department of Pharmacology and Physiology, New Jersey Medical School, Newark, New Jersey 07103

I. Introduction

A landmark year for the study of energy homeostasis was 1953. In that single year, two of the most influential and prescient hypotheses explaining how peripheral signals transmitted information about the metabolic status of the periphery to the brain were put forward. Mayer (1953) proposed that "glucoreceptors" in the hypothalamus could sense arteriovenous differences in glucose to alter ingestive behavior (and thereby plasma glucose levels). Kennedy (1953) postulated that adipose depots provided a signal that was sensed by the hypothalamus to act as a long-term regulator of body weight. Of the two, Kennedy's "lipostatic" hypothesis has been confirmed by the critical discovery of leptin. But we are still struggling to confirm Mayer's "glucostatic" hypothesis. In 1964, groups headed by Oomura (1964) and Anand (1964) showed that there were, in fact, neurons in the brain which altered their firing rate when plasma glucose levels changed. The idea that glucose could be used as a signaling molecule rather than simply a requisite substrate to fuel cellular metabolism was revolutionary and added important weight to Mayer's hypothesis. In fact, recent evidence suggests that these same neurons in the brain also sense a variety of other metabolic signals from the periphery. Thus, they are more properly thought of as metabolic sensors. Such metabolic sensing neurons also receive hard-wired neural inputs from the gastrointestinal tract and the primary senses, as well as a host of intrinsic neurotransmitter and neuropeptide signals from cognitive, reward, and memory systems of the brain. These signals are summated and integrated by their effect on the cell's membrane potential. This summated signal regulates either the firing rate and/or transmitter release of the neuron. Metabolic sensor neurons are highly integrated within other neuronal networks which provide efferent outputs to the autonomic, hypothalamo-pituitary, and motor systems governing energy homeostasis. All of these have direct or indirect influences on peripheral glucose levels.

At the most basic level, the brain must be able to sense and regulate the level of plasma glucose to maintain a constant source of glucose as a metabolic substrate for neural function (Sokoloff *et al.*, 1977). However, it is important to understand the glucosensing role of the brain as only a part of a totally integrated set of systems designed to maintain metabolic equilibrium in the body. While this review will focus on the glucosensing role of these neurons, we will make the case that they represent a distinct and unique type of neuron which has evolved specifically to sense and regulate energy homeostasis in the body. We shall address four questions relating to glucosensing neurons:

1. Where is glucose sensed?
2. How is glucose sensed?
3. Why is glucose sensed?
4. How is glucosensing altered in disease states?

II. Where Is Glucose Sensed?

The search for glucosensing neurons was carried out first in the hypothalamus because of ideas proposed by Anand and Brobeck (1951). They placed the hypothalamus at the center of the integratory centers controlling food intake. This gave rise to the "dual center" hypothesis of food intake in which the ventromedial hypothalamus is the "satiety center" and the lateral hypothalamus is the "feeding center" (Stellar, 1954). This hypothesis was based on studies showing that destruction of the ventromedial hypothalamus produced hyperphagia and obesity (Hetherington and Ranson, 1940), while destruction of the lateral hypothalamus produced aphagia and marked weight loss (Anand and Brobeck, 1951). As we have gained in sophistication, the idea of "centers" underlying specific neurologic functions has given way to the concept of integrated systems. Similarly, while there is still a strong hypothalamocentric emphasis in the field of energy homeostasis, it is clear that hypothalamic areas involved in this process are linked to extensive systems regulating biorhythms, arousal, emotions, memory, and reward pathways. Similarly, it is no longer possible to consider energy intake separately from the other components of energy homeostasis, i.e., energy expenditure and storage. The control of peripheral glucose levels is an integrated part of this overall regulatory system.

The first glucosensing neurons were described in the ventromedial hypothalamus and lateral hypothalamus (Anand *et al.,* 1964; Oomura *et al.,* 1964). Two general types of neurons were described and named by Oomura. He coined the term "glucose responsive" for those neurons that increase their firing rate as ambient glucose levels rise. He defined "glucose sensitive" neurons as those that are inhibited by rising glucose levels (Oomura *et al.,* 1964). While this terminology has been used for more than 40 years, we propose the more descriptive nomenclature of "glucose excited" (GE) for glucose responsive and "glucose inhibited" (GI) in place of glucose sensitive. Populations of GE and GI neurons are located in a variety of brain regions (Fig. 1). In addition, there are glucosensing elements in the portal vein (Matsuhisa *et al.,* 1997; Hevener *et al.,* 1997), gastrointestinal tract (Liu *et al.,* 1999), and carotid body (Alvarez-Buylla and de Alvarez-Buylla, 1988). The hypothalamus and the caudal brain stem are the two most studied brain areas because a large number of metabolic and visceral signals converge and are integrated at these sites (Fig. 1). In the hypothalamus, glucosensing neurons have been identified in the ventromedial (Anand *et al.,* 1964; Oomura *et al.,* 1964; Silver and Erecinska, 1998; Funahashi *et al.,* 1999), arcuate (ARC) (Muroya *et al.,* 1999; Funahashi *et al.,* 1999; Cowley *et al.,* 2001), paraventricular (Kow and Pfaff, 1989), and suprachaismatic nuclei (Nagai *et al.,* 1996; Hall *et al.,* 1997), as well as the lateral hypothalamus (Orsini *et al.,* 1990, 1991, 1992; Silver and Erecinska, 1998; Griffond *et al.,* 1999; Adachi *et al.,* 1984; Oomura *et al.,* 1969, 1975, 1985; Karadi *et al.,* 1990, 1992; Aou *et al.,* 1991). ARC neuropeptide Y (NPY) (Muroya *et al.,* 1999; Funahashi *et al.,* 1999) and pro-opiomelanocortin (POMC) (Cowley *et al.,* 2001) neurons are prototypic examples of neurons which respond to both glucose and a variety of other metabolic signals from the brain, periphery, and external environment. Other forebrain areas such as the septum (Shoji, 1992), amygdala (Nakano *et al.,* 1986), striatum (Lee *et al.,* 1998), and motor cortex (Lee *et al.,* 1996a) also contain glucosensing neurons. In the hindbrain, there are glucosensing neurons in the substantia nigra (Amoroso *et al.,* 1990; Roeper *et al.,* 1990; During *et al.,* 1995; Roper and Ashcroft, 1995; Liss *et al.,* 1999; Levin, 2000), locus coeruleus (Finta *et al.,* 1993), nucleus of the solitary tract (NTS) (Mizuno and Oomura, 1984; Adachi *et al.,* 1984; Dallaporta *et al.,* 1999, 2000; Yettefti *et al.,* 1997), dorsal vagal complex (Karschin *et al.,* 1998), and the area postrema (Adachi *et al.,* 1991). Most of these brain areas

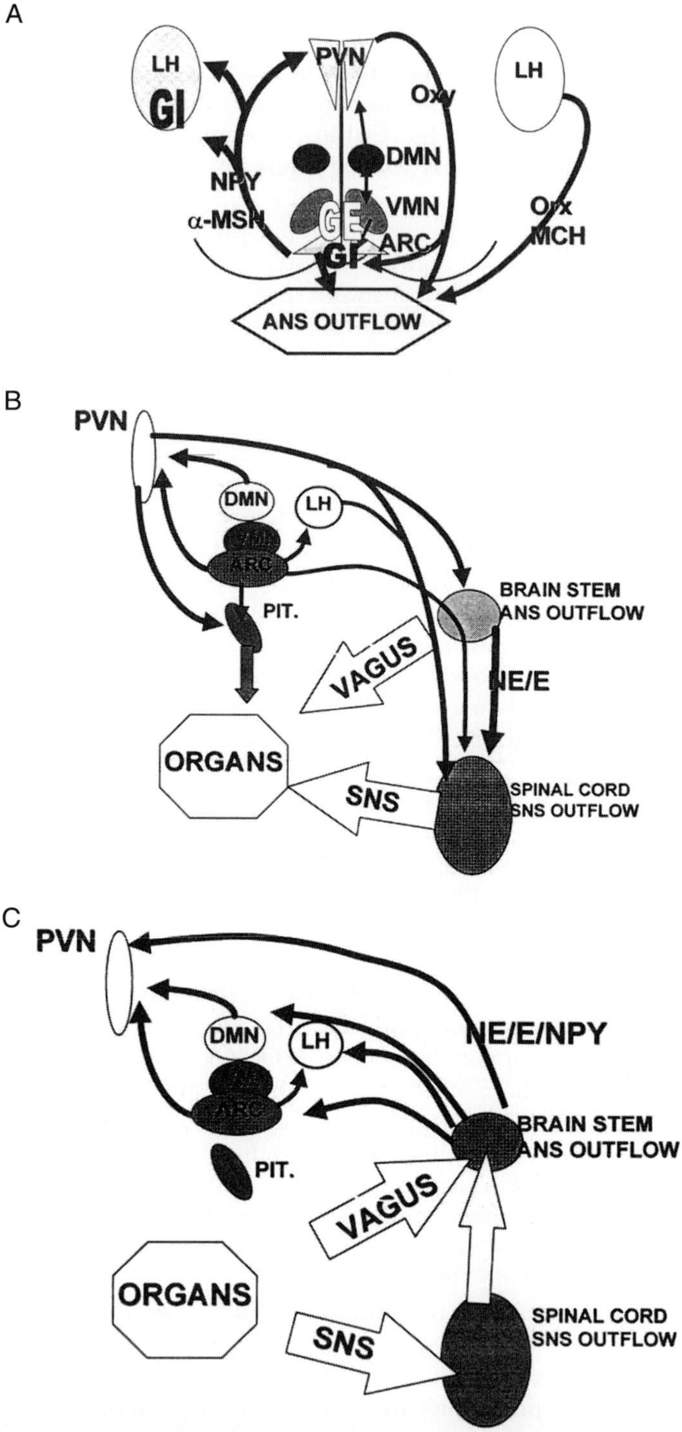
A
LH
GI
PVN
Oxy
LH
DMN
NPY
VMN
α-MSH
GE
ARC
Orx
MCH
GI
ANS OUTFLOW
B
PVN
DMN
LH
ARC
PIT.
BRAIN STEM
ANS OUTFLOW
VAGUS
NE/E
ORGANS
SNS
SPINAL CORD
SNS OUTFLOW
C
PVN
NE/E/NPY
DMN
LH
ARC
BRAIN STEM
ANS OUTFLOW
PIT.
VAGUS
ORGANS
SPINAL CORD
SNS OUTFLOW
SNS

containing glucosensing neurons are strategically placed to sense and integrate incoming signals and to pass these on to efferent pathways involved in the neuroendocrine and motor responses involved in energy homeostasis. While these areas have been identified using neurophysiologic techniques in experimental animals, functional imaging suggests that areas in the human hypothalamus are also responsive to changes in blood glucose (Matsuda *et al.*, 1999; Tataranni *et al.*, 1999; Mahankali *et al.*, 2000; Liu *et al.*, 2000). However, these studies do not show that glucose has a direct effect on glucosensing neurons in humans.

III. How Is Glucose Sensed?

Interstitial brain glucose levels are maintained at ~1–3 m*M* under physiologic conditions. Over the physiologic range of plasma glucose levels (~4–10 m*M*), extracellular brain glucose levels are about 30% of those in plasma when measured with glucosensing electrodes in anesthetized rats (Silver and Erecinska, 1994) or directly from freeze-clamped tissue (Paschen *et al.*, 1986). Estimates calculated using microdialysis show a lower ratio of about 20% (McNay and Gold, 1999, 2001; McNay *et al.*, 2001). There are also considerable regional differences (McNay and Gold, 1999, 2001;

FIG. 1. Location and anatomy of glucosensing neurons and their connections. (A) Model coronal section of the hypothalamus showing the location of glucose excited (GE) and glucose inhibited (GI) neurons. The arcuate (ARC), ventromedial (VMN), and dorsomedial (DMN) nuclei lie in the same plane and are interconnected. The ARC and DMN send projections to the paraventricular nucleus (PVN) which lies rostrally. Projections from the ARC to the PVN and lateral hypothalamic area (LH) contain neuropeptide Y (NPY) and α-melanocyte stimulating hormone (α-MSH). The ARC also sends direct projections to the autonomic nervous system (ANS) outflow in the spinal cord. The PVN sends oxytocin (Oxy) projections to both the spinal cord and brain stem autonomic outflow areas and LH orexin (Orx) and melanin concentrating hormone (MCH) neurons project to similar areas. (B) Sagittal view of the efferent projections of the hypothalamus, brain stem, and spinal cord to the peripheral organs. Both the PVN and ARC send efferents to the pituitary (PIT.) gland for release of hormones. The PVN, ARC, and LH project to brain stem parasympathetic (VAGUS) and/or spinal cord SNS outputs. Norepinephrine (NE) and epinephrine (E) neurons in the caudal brain stem project to the gray matter of the thoracic spinal cord which gives rise to sympathetic nervous system (SNS) outputs. These NE and E projections are required for the counterregulatory response to glucoprivation. (C) Sagittal view of the afferent projections carried by the vagus nerve and SNS fibers from peripheral organ brain centers regulating energy homeostasis. NE and E neurons in the caudal hindbrain both contain NPY and project to hypothalamic sites shown in part A. These NE and E projections are required for glucoprivic feeding.

McNay *et al.*, 2001; Paschen *et al.*, 1986). These regional differences become accentuated when plasma glucose levels fall. For example, local glucose concentrations fall precipitously in the neocortex, caudate-putamen, hippocampus, and cerebellum when plasma glucose levels drop below 4 m*M*, while they are better maintained in the thalamus, hypothalamus, and brain stems. Below 3 m*M* plasma glucose, the caudate-putamen becomes completely depleted of measurable glucose (Paschen *et al.*, 1986). These differences in regional glucose levels have important implications for cerebral metabolism, glucosensing, and selective vulnerability during states of hypoglycemia.

Regional differences in interstitial brain glucose levels represent the balance between transport from plasma across the blood-brain barrier into the brain, followed by uptake by neuronal and glial elements (Vannucci *et al.*, 1991). There is a family of glucose transporters (GLUT) containing more than eight members. GLUT1 is located at the blood-brain barrier in the endothelial cells of cerebral microvessels, as well as on astrocytes whose foot processes make up part of the barrier (Hu and Wilson, 1997; Simpson *et al.*, 2001). GLUT1 has a K_m of ~6–8 m*M* for transport of glucose across the blood-brain barrier (Vannucci *et al.*, 1991). Most neurons in the brain contain GLUT3 which has a K_m (~2–3 m*M*) within the physiologic range of brain glucose levels (Maher *et al.*, 1996). In addition, GLUT2 (the pancreatic β-cell transporter) (Jetton *et al.*, 1994; Leloup *et al.*, 1994; Vannucci *et al.*, 1991; Navarro *et al.*, 1996; Roncero *et al.*, 2000; Maekawa *et al.*, 2000), the insulin-sensitive GLUT4 (Leloup *et al.*, 1996; Rayner *et al.*, 1994; Vannucci *et al.*, 1991, 1998; El Messari *et al.*, 1998; Apelt *et al.*, 1999; Maekawa *et al.*, 2000; Kobayashi *et al.*, 1996), and GLUT8 (GLUTx1) (Ibberson *et al.*, 2000) are located on neuronal and glial elements in the brain. However, their role in cerebral glucose metabolism is largely unknown at present. Once glucose enters the neuron, it enters the pathway under the influence of hexokinase I, the predominant hexokinase in neurons (Jacobsson and Meister, 1994). Glucose is converted to glucose-6-phosphate (Liu *et al.*, 1991) which produces feedback inhibition of hexokinase I (German, 1993; Liu *et al.*, 1991). Pyruvate derived from this glycolytic process fuels oxidative production of ATP in the mitochondria.

In addition to glucose, neurons also utilize lactate produced by astrocytes (Fig. 2) (Magistretti *et al.*, 1993; Pellerin and Magistretti, 1994; Pellerin *et al.*, 1998a; Bittar *et al.*, 1996). Astrocytes take up glucose and store it as glycogen. This can serve as a short-term replacement for glucose under glucoprivic (low glucose availability) conditions (Ransom and Fern, 1997; Wender *et al.*, 2000). Lactate and ketone bodies can also enter the brain from the plasma under certain circumstances (Maran *et al.*, 1994; King *et al.*, 1998; De Bruin *et al.*, 1990; Thurston and Hauhart, 1989; Thurston *et al.*, 1983; Nemoto *et al.*, 1974; Veneman *et al.*, 1994). Transport of lactate and

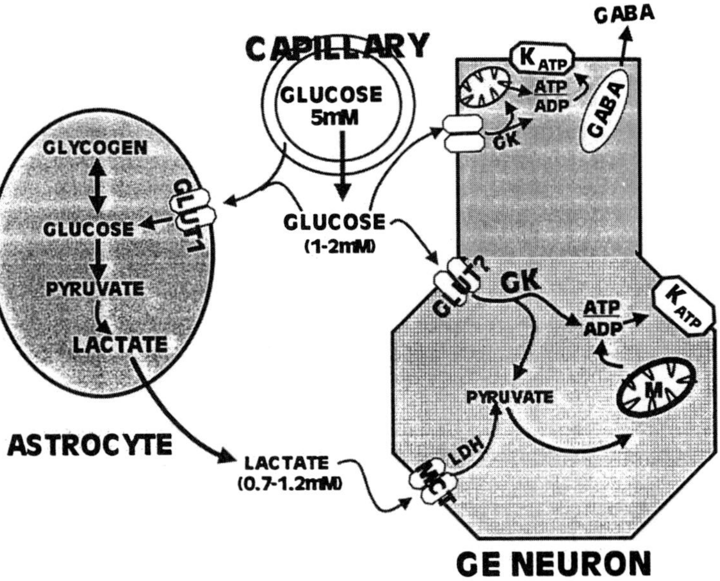

FIG. 2. Model of the metabolic relationship between astrocytes and glucose excited (GE) neurons. Glucose enters the brain by facilitated transport by GLUT1 glucose transporters. It enters GE neurons by a glucose transporter of unknown type (GLUT?) although nonglucosensing neurons have GLUT3 transporters. Glycolysis is regulated by glucokinase (GK) which increases the ATP/ADP ratio directly and by providing pyruvate to fuel oxidative metabolism in the mitochondria (M). ATP is bound to the ATP-sensitive K^+ channel (K_{ATP}) and inactivates (closes) it. This leads to firing of the neuron. Glucose can also be taken up in the nerve terminal (upper part of GE neuron) where inactivation of the K_{ATP} channel leads to neurotransmitter release (e.g., γ-aminobutyric acid [GABA]). Glucose can also be taken up by astrocytes containing GLUT1 glucose transporters. In astrocytes, glucose is stored as glycogen which can provide a short-term supply lactate when glucose supplies become limiting. Lactate is released into the extracellular space where it is taken up by neurons containing monocarboxylate transporters (MCT). Lactate is converted to pyruvate by lactate dehydrogenase (LDH) and can serve as an alternate source of ATP, particularly when glucose supplies are limiting. It is postulated that low levels of MCT and LDH expression make GK the "gatekeeper" of glucosensing in GE neurons through its regulation of glycolytic production of ATP.

ketone bodies is mediated by monocarboxylate transporters (MCT) which maintain extracellular brain lactate levels between $\sim$0.5–1.5mM (De Bruin *et al.,* 1990; Medina and Stavinoha, 1977; Medina *et al.,* 1975; Hu and Wilson, 1997; Fray *et al.,* 1996). Lactate and ketone bodies are a potential source of neuronal ATP (Pellerin *et al.,* 1998a; McKenna *et al.,* 1998; Deitmer, 2000; Hassel and Brathe, 2000; Schurr *et al.,* 1999). They are transported into neurons by MCT- 2, while glia and vascular endothelial cells at the blood-brain barrier transport them via the MCT-1 isoform (Pellerin *et al.,* 1998b; Broer *et al.*, 1997; Koehler-Stec *et al.*, 1998) (Fig. 2). Once lactate enters the neuron, it is converted to pyruvate by lactate dehydrogenase (LDH). Pyruvate then enters the mitochondria where it is oxidized to produce ATP. In human brain, hippocampal and occipital cortex neurons exclusively express LDH-1 (heart type) protein, while astrocytes express both LDH-1 and LDH-5 (muscle type) (Bittar *et al.*, 1996). The mRNA for LDH-1 is highly expressed in brain areas containing concentrated populations of neurons, while LDH-5 is more diffusely expressed (Laughton *et al.*, 2000; Bittar *et al.*, 1996; Tholey *et al.*, 1981; Pellerin *et al.*, 1998c). The LDH-1 isoform drives the reaction toward the production of pyruvate. The LDH-5 isoform has higher maximal velocity and is present in glycolytic tissues, favoring the formation of lactate from pyruvate. The specific kinetics of the MCTs and the isoforms of LDH on astrocytes versus neurons suggest that astrocytes are a lactate "source" while neurons are a lactate "sink" (Pellerin *et al.*, 1998a). Thus, in most neurons, astrocyte-derived lactate can be converted to pyruvate to provide a ready source of ATP production. This means that glucose can act as either a direct (through glycolysis in neurons) or indirect (through astrocyte metabolism) source of neuronal substrate to produce neuronal ATP (Fig. 2).

While the production of intracellular ATP production allows glucose to control the rate of cell firing and transmitter release in GE neurons, little is known about the mechanism of glucosensing in GI neurons. Potential candidates include the Na^+-K^+-ATP pump (Oomura *et al.*, 1974; Silver and Erecinska, 1998), an ATP-activated K^+ channel (Sellers *et al.*, 1992), and a chloride channel (Routh, unpublished observation). Much more is known about GE neurons because of their close homology to the pancreatic β-cell (Ashford *et al.*, 1990a; Yang *et al.*, 1999; Levin *et al.*, 1999) (Fig. 2). In both GE neurons (Miki *et al.*, 2001) and β-cells (Miki *et al.*, 1997, 1998), the ATP-sensitive K^+ (K_{ATP}) channel is required to confer glucosensing capability. The K_{ATP} channel responds to the intracellular ATP/ADP ratio (Trapp and Ashcroft, 1997; Detimary *et al.*, 1998; Newgard and McGarry, 1995). Low levels of extracellular glucose lead to a lowering of the ATP/ADP ratio. This activates (opens) the channel leading to egress of intracellular K^+ and hyperpolarization of the cell membrane. Phosphorylation of the channel acts to stabilize it in this active state (Routh *et al.*, 1997). When glucose levels increase, the ATP/ADP ratio rises and ATP is bound directly to the channel. This inactivates (closes) the channel and reduces egress of K^+ from the cell. Intracellular K^+ rises and the cell membrane depolarizes. This results in Ca^{2+} influx through a voltage-dependent Ca^{2+} channel. In the β-cell, this sequence results in insulin release (Trapp and Ashcroft, 1997). In GE neurons, inactivation of the channel on cell bodies leads to increased firing (Routh *et al.*, 1997), while channel inactivation on nerve terminals produces neurotransmitter release (Amoroso *et al.*, 1990; Lee *et al.*, 1995, 1996a; Levin, 2000).

The K_{ATP} channel is an octomeric functional unit composed of four pore-forming channels (Kir6.2) for K^+ and four binding sites for sulfonylureas, a class of insulin secretogogues (Trapp and Ashcroft, 1997; Aguilar-Bryan *et al.*, 1998; Babenko *et al.*, 2000). The Kir6.2 pore-forming unit is a member of the inwardly rectifying K^+ channel family (Inagaki *et al.*, 1995). The sulfonylurea receptor (SUR) is a member of the ATP-binding cassette family. As with binding of ATP to the channel, occupation of the SUR inactivates the K_{ATP} channel and stimulates insulin secretion in β-cells (Aguilar-Bryan *et al.*, 1995) and increased neuronal firing or transmitter release in GE neurons (Ashford *et al.*, 1990b; Roeper *et al.*, 1990; Lee *et al.*, 1995, 1996a; Roper and Ashcroft, 1995; Levin *et al.*, 1999b). Presumptive endogenous SUR ligands, α-endosulfine (Peyrollier *et al.*, 1996; Virsolvy-Vergine *et al.*, 1992) and related phosphoproteins (Horiuchi *et al.*, 1990; Girault *et al.*, 1990) have been identified in the brain. But little is known about their regulation or regional distribution. Thus, both high glucose levels and the presence of sulfonylureas (or endosulfines) lead to increased firing and/or transmitter release from GE neurons by their actions on the K_{ATP} channel. A class of drugs known as K^+ channel openers has the opposite effect of the sulfonylureas. They activate the K_{ATP} channel causing K^+ efflux, cellular hyperpolarization, and prevention of insulin secretion (Hausser *et al.*, 1991) or neuronal firing (During *et al.*, 1995; Maneuf *et al.*, 1996; Hausser *et al.*, 1991).

The K_{ATP} channel is present on GE neurons but not glia (Dunn-Meynell *et al.*, 1998). Kir6.2 (Inagaki *et al.*, 1995; Karschin *et al.*, 1997; Dunn-Meynell *et al.*, 1998) and both a high- (SUR1) and low-affinity (SUR2) form of the SUR have been cloned and identified in brain (Zini *et al.*, 1993; Levin *et al.*, 1996; Dunn-Meynell *et al.*, 1997; Inagaki *et al.*, 1996; Lee *et al.*, 1998, 1999a; Liss *et al.*, 1999). Kir6.2 and SUR1 are widely disseminated in the brain with closely overlapping distributions (Karschin *et al.*, 1997; Dunn-Meynell *et al.*, 1997, 1998; Schwanstecher and Panten, 1993; Levin *et al.*, 1996; Mourre *et al.*, 1990). SUR2 is more limited in its distribution to areas where GE neurons are concentrated (Levin *et al.*, 1996; Dunn-Meynell *et al.*, 1997). Binding studies, combined with neurotoxin lesions, suggest that the low affinity form (SUR2) is present on cell bodies, while the high-affinity form (SUR1) is likely to be on axon terminals of GE neurons (Dunn-Meynell *et al.*, 1997). A given ratio of Kir6.2 to either SUR1 or SUR2 determines the specific level of metabolic and sulfonylurea sensitivity of the K_{ATP} channel (Liss *et al.*, 1999). Finally, the K_{ATP} channel is an absolute requirement for glucosensing in GE neurons. Deletion of the Kir6.2 subunit of the K_{ATP} channel produces animals with no observable GE neurons in the ventromedial hypothalamus (Miki *et al.*, 2001).

While the GE neuronal cell body clearly has a major glucosensing function, both the K_{ATP} channel and a specific glucosensing mechanism are present on nerve terminals as well (Fig. 2) (Lee *et al.*, 1995, 1996a). This means that neurotransmitter release can occur solely by altering the ambient glucose concentration in the immediate vicinity of the nerve terminal without altering neuronal firing. This has particular importance in long projection neurons whose cell bodies lie in one area, such as the striatum (where glucose transport is particularly limited during hypoglycemia [Paschen *et al.*, 1986]), but whose axon terminals reside in areas with different glucose transport properties. Inhibitory γ-aminobutyric acid (GABA) neurons are an example of such neurons. They have their cell bodies in the striatum and axon terminals in the substantia nigra where glucose-mediated GABA release from their nerve terminals has a major impact on the firing of the glucosensing nigral dopamine neurons (During *et al.*, 1995; Amoroso *et al.*, 1990; Levin, 2000). The resultant firing rate of nigral dopamine neurons represents a complex interaction between glucose-induced GABA release from axon terminals and direct stimulation by glucose of the glucosensing dopamine neuron (Levin, 2000). Similarly, glucose releases glutamate from nerve terminals in the motor cortex independent of changes in cell body firing rate (Lee *et al.*, 1995, 1996a). On the other hand, norepinephrine neuron cell bodies in the locus coeruleus have glucosensing properties (Illes *et al.*, 1994), but their axon terminals in the hypothalamus do not (Dunn-Meynell *et al.*, 1997). Thus, the combination of regionally specific rates of glucose transport across the blood-brain barrier and the complement of glucosensing nerve cell bodies and/or terminals in a given area determines the resultant output of a particular neuronal system in response to changes in plasma glucose levels.

While the K_{ATP} channel is an absolute requirement for glucosensing in GE neurons, its near ubiquitous distribution makes it unlikely to be the defining feature of a GE neuron. In fact, neuroprotection is probably the major function of the K_{ATP} channel in the majority of nonglucosensing neurons (Zawar and Neumcke, 2000). When the supply of metabolic substrate (glucose or oxygen) becomes limited, intracellular ATP levels fall. This activates the K_{ATP} channel, hyperpolarizes the cell membrane, and interrupts cell firing. This counteracts the excitotoxin-induced damage associated with hypoglycemia and hypoxia (Lauritzen *et al.*, 1997). But a GE neuron must be able to sense glucose within the physiologic range of brain glucose levels and this requires an additional regulatory mechanism besides the K_{ATP} channel. There are several potential candidates (Fig. 2). For example, a glucosensing neuron might possess a glucose transporter other than GLUT3. Because of its high rate of glucose transport, GLUT3 is unlikely to be the rate-limiting or regulatory step in ATP production (Maher *et al.*, 1996). Both GLUT2 (Jetton *et al.*, 1994; Leloup *et al.*, 1994; Vannucci *et al.*, 1991; Navarro *et al.*, 1996; Roncero *et al.*, 2000; Maekawa *et al.*, 2000) and GLUT4 (Leloup *et al.*,

1996; Rayner *et al.*, 1994; Vannucci *et al.*, 1991, 1998; El Messari *et al.*, 1998; Apelt *et al.*, 1999; Maekawa *et al.*, 2000; Kobayashi *et al.*, 1996) are present on neurons and/or glia and both have characteristics that might provide regulatory control over glucose entry into neurons. However, neither has been localized definitively on GE neurons. Control of glycolysis is another process that might regulate the activity of the K_{ATP} channel in GE neurons. But neuronal hexokinase I is unlikely to play such a regulatory role because of the inhibitory feedback by its metabolic product, glucose-6-phosphate (German, 1993; Liu *et al.*, 1991b). On the other hand, glucokinase (GK; hexokinase IV) appears to a critical regulator of glycolytic production of ATP and K_{ATP} channel activity in the β-cell, (Matschinsky, 1996; German, 1993; De Vos *et al.*, 1995). GK is also present in the brain (Jetton *et al.*, 1994; Yang *et al.*, 1999; Roncero *et al.*, 2000; Maekawa *et al.*, 2000; Lynch *et al.*, 2000) and is highly localized in those areas where GE neurons are concentrated (Maekawa *et al.*, 2000; Lynch *et al.*, 2000). Further, GK mRNA is co-localized in both NPY and POMC neurons in the ARC (Lynch *et al.*, 2000). Thus, GK is very likely to be a defining feature of GE neurons (Yang *et al.*, 1999; Lynch *et al.*, 2000).

But lactate provided by glia is also a potential source of neuronal ATP (Pellerin *et al.*, 1998; McKenna *et al.*, 1998; Deitmer, 2000; Hassel and Brathe, 2000; Schurr *et al.*, 1999). Lactate utilization would bypass GK regulation of glycolysis and might overshadow its regulatory influences on the K_{ATP} channel. However, in the pancreatic β-cell, extracellular lactate appears to provide a minimal contribution to ATP production because both MCTs and LDH are present in low abundance (Ainscow *et al.*, 2000; Maechler and Wollheim, 2000). If GE neurons are comparable to the β-cell, they also should have low levels of MCT-2 and LDH-1 (or even LDH-5). This would make GK-mediated glycolysis the primary regulator of K_{ATP} channel activity in GE neurons under physiologic conditions. Under these conditions, lactate should serve as a signaling molecule only when glucose levels fall into the glucopenic range associated with hypoglycemia. Our preliminary data in both hypothalamic slices and dissociated hypothalamic neurons supports this hypothesis (unpublished data).

Even if we accept that GK is the primary regulatory step in the generation of ATP in GE neurons, both intracellular ATP and ADP levels are assumed to be highly buffered within the cell (Cesar and Wilson, 1998). This would make it difficult for relatively small changes (<1 m*M*) in extracellular glucose concentrations to alter the cytoplasmic ATP/ADP ratio sufficiently to regulate the activity of the plasma membrane-bound K_{ATP} channel. In fact, there actually appears to be quite distinct subcellular compartmentalization of ATP production, at least within isolated β-cells (Kennedy *et al.*, 1999). In these cells, the subplasmalemmal compartment is quite responsive to alterations in glucose availability. This creates a microenvironment to provide very localized ATP availability for regulation of the membrane-bound K_{ATP} channel activity. Such a mechanism could also operate in GE neurons as it does in β-cells.

The next issue is why there are so many GE neurons which differ in their firing characteristics, conductances, and sensitivities to glucose, sulfonylureas, and/or ATP (Sellers *et al.,* 1992; Silver and Erecinska, 1998). There are several apparent reasons for this observation. In a given population, individual GE neurons may contain different proportions of the SUR1, SUR2, and Kir6.2 units of the K_{ATP} channel. This confers upon them differing sensitivities to various metabolic signals (Lee *et al.,* 1998, 1999a; Liss *et al.,* 1999). Some GE neurons also show various subconductances of the K_{ATP} channel which may vary from moment to moment depending upon ambient conditions (Routh *et al.,* 1997). An additional factor is the interplay between pre- and postsynaptic inputs to a given neuron. Since both cell body firing rate and axon terminal transmitter release can be independently affected by ambient glucose concentrations (Lee *et al.,* 1995; Levin, 2000), the actual output of a given cell will be the sum of the direct effect of glucose on the cell body and the indirect effect of glucose-mediated transmitter release onto that cell from adjacent presynaptic terminals. Several types of neurons show this convergence of glucose-mediated interactions. These include dopamine neurons in the substantia nigra (Dunn-Meynell *et al.,* 1998; Roeper *et al.,* 1990; Roper and Ashcroft, 1995; Liss *et al.,* 1999), norepinephrine neurons in the locus coeruleus (Dunn-Meynell *et al.,* 1998), acetylcholine neurons in the striatum (Lee *et al.,* 1997), GABA neurons in the striatum (Dunn-Meynell *et al.,* 1998), and glutamate neurons in the motor cortex (Lee *et al.,* 1995).

In addition to glucose-mediated signals, these specialized sensing neurons also receive a host of other metabolic and neural signals. The ARC NPY and POMC neurons are especially good examples of this class of neurons (Stanley and Leibowitz, 1985; Schwartz *et al.,* 1997; Levin, 1999; Levin and Dunn-Meynell, 1997a). These neurons are specially equipped and located to play a key role in the regulation of energy homeostasis. ARC NPY neurons project to the hypothalamic paraventricular nucleus which is linked to both neuroendocrine and autonomic effector pathways (Bai *et al.,* 1985). Central NPY injections into such projection areas induce a host of anabolic events including increased food intake (Stanley and Leibowitz, 1985), insulin release (van Dijk *et al.,* 1994), decreased brown adipose thermogenesis (Giraudo *et al.,* 1994), and a shift from lipid to carbohydrate oxidation (Billington *et al.,* 1994). Both restriction of energy intake (Brady *et al.,* 1990a) and glucoprivation (Akabayashi *et al.,* 1993) increase NPY expression selectively in ARC NPY neurons. ARC POMC neurons have similar projection fields to

ARC NPY neurons (Elias *et al.*, 1999; Bagnol *et al.*, 1999). But these POMC neurons produce α-MSH which interacts with catabolic melanocortin receptors (Harrold *et al.*, 1999; Kistler-Heer *et al.*, 1998; Lu *et al.*, 2000). Occupancy of melanocortin-4 receptors decreases food intake and increases thermogenesis (Giraudo *et al.*, 1998; Haynes *et al.*, 1998; Grill *et al.*, 1998; Chen *et al.*, 2000). Both ARC NPY and POMC neurons express Kir6.2 (Dunn-Meynell *et al.*, 1998) and GK (Lynch *et al.*, 2000), as well as leptin (Hakansson *et al.*, 1998), and possibly insulin receptors (Marks *et al.*, 1990). This makes them likely to be GE neurons with the capability of "sensing" several metabolic signals from the periphery. ARC and other ventromedial hypothalamic GE neurons also respond to a host of neurotransmitters and peptides. These include monoamines (Kow and Pfaff, 1989; Baker *et al.*, 1996), GABA (Kow and Pfaff, 1989), vasopressin, and oxytocin (Kow and Pfaff, 1986). GI neurons have similar metabolic integrator properties (Dallaporta *et al.*, 1999; Funahashi *et al.*, 1999; Yettefti *et al.*, 1997; Karadi *et al.*, 1992; Kai *et al.*, 1988; Oomura *et al.*, 1975). Thus, numerous intrinsic and extrinsic signals related to energy homeostasis converge upon these glucosensing neurons. These converging signals are integrated and passed on to downstream effector areas responsible for energy intake, expenditure and storage.

If GK is the defining feature of GE neurons, it is the K_{ATP} channel which actually integrates many of metabolic and neural signals that converge on these metabolic sensor neurons (Fig. 3). Both leptin (Spanswick *et al.*, 1997) and insulin (Spanswick *et al.*, 2000) activate the K_{ATP} channel on GE neurons. Both leptin (Guan *et al.*, 1997; Baskin *et al.*, 1998; Fei *et al.*, 1998; Elmquist *et al.*, 1998a) and insulin receptors (Wilcox *et al.*, 1989; Marks *et al.*, 1990, 1991) are highly localized in ventromedial hypothalamic areas enriched in GE neurons. The long form of the leptin receptor (Ob-Rb) is a class I cytokine receptor linked to the JAK/STAT signaling pathway (Bjorbaek *et al.*, 1999; Banks *et al.*, 2000; Vaisse *et al.*, 1996; Ghilardi *et al.*, 1996). Receptor occupation can lead to either gene transcription or to more immediate effects mediated by activation of phosphoinositol-3-kinase (PI3-K) (Kim *et al.*, 2000; Harvey *et al.*, 2000a). PI3-K activation increases the formation of phosphoinositides (PIP_2, PIP_3) which promotes K_{ATP} channel activation and reduces cell firing by destabilizing channel-associated actin filaments (Harvey *et al.*, 2000b; Baukrowitz and Fakler, 2000). The insulin receptor is also linked to K_{ATP} channel function (Spanswick *et al.*, 2000). Insulin receptors in the ARC are co-localized with insulin receptor substrate-1 (Baskin *et al.*, 1993; Marks *et al.*, 1990) which amplifies the insulin signal (Myers, Jr. *et al.*, 1994). Insulin receptor occupancy (Myers, Jr. and White, 1996) causes tyrosine phosphorylation of insulin receptor substrate and activation of PI3-K (Harvey and Ashford, 1998; Myers, Jr. *et al.*, 1994). Thus, PI3-K activation serves as a final common pathway through which both insulin and leptin can affect K_{ATP}

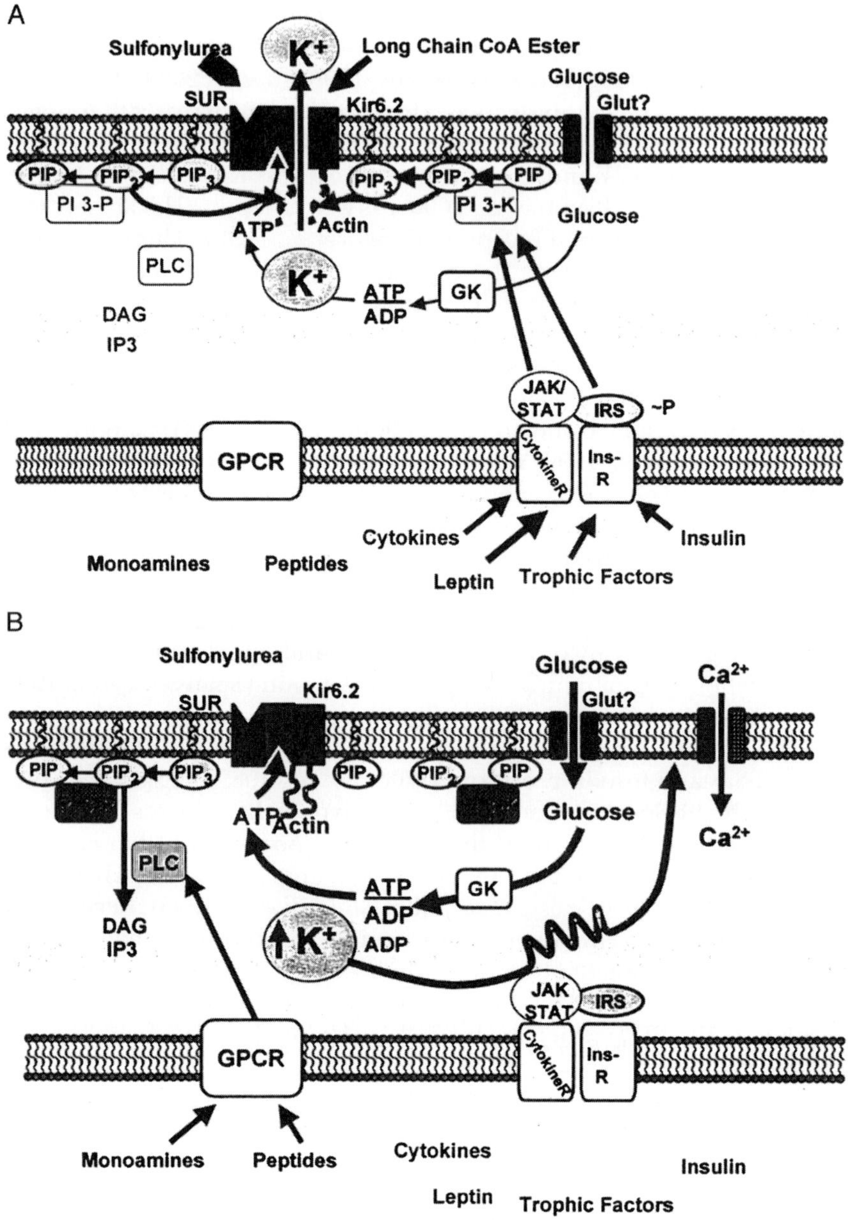

FIG. 3. Model of convergent metabolic and neural signals on the GE neuron. (A) Leptin and cytokines act through receptors linked to the JAK/STAT pathway. Insulin and trophic factors act at the insulin receptor or other tyrosine kinase receptors. Both pathways lead to the activation of phosphoinositide-3-kinase (PI3-K). The insulin receptor is linked to PI3-K activation

channel activity and neuronal firing. While both leptin and insulin are indirect signals informing the brain of the status of adipose stores, long chain fatty acids and their acyl-CoA esters act directly on the Kir6.2 pore-forming unit of the K_{ATP} channel to reduce its sensitivity to ATP and prolong the open state of the channel (Branstrom *et al.,* 1998; Gribble *et al.,* 1998; Liu *et al.,* 2001) (Fig. 3). Finally, neurotransmitters and neuropeptides act at G-protein coupled receptors to alter K_{ATP} channel activity (Xie *et al.,* 1999). Occupancy of stimulatory G-protein coupled receptors activates phospholipase C leading to degradation of phosphoinositides. This stabilizes actin filaments associated with the K_{ATP} channel and promotes inactivation of the channel (Fig. 3). Other agents related to monoamines also act on the K_{ATP} channel through mechanisms which are not yet determined (Lee *et al.,* 1996b). Of course, not all afferent signals impinging upon GE neurons need act through the K_{ATP} channel. It is sufficient for them to summate their effects on the membrane potential to promote or antagonize the effects of the various agents that alter membrane potential through their actions on the K_{ATP} channel.

GI neurons also act as integrators of various metabolic signals as well as being influenced by internal and external environmental cues. But there is no evidence that the K_{ATP} channel is involved in regulating the firing rate of these neurons. GI neurons in the lateral hypothalamus respond to signals transmitted from the liver (Shimizu *et al.,* 1983), as well as responding to pain, taste, and odor signals and the rewarding properties of food (Karadi *et al.,* 1990, 1992; Aou *et al.,* 1991; Sikdar and Oomura, 1985). GI neuronal firing rate is modulated by leptin (Funahashi *et al.,* 1999) and a variety of neurotransmitters including catecholamines (Karadi *et al.,* 1990, 1992; Nishino *et al.,* 1988; Miyahara and Oomura, 1982; Oomura *et al.,* 1975), serotonin (Kai *et al.,* 1988), opioids (Nishino *et al.,* 1988; Sikdar and

---

through phosphorylation (~P) of insulin receptor substrate (IRS). PI3-K activation increases the formation of phosphatidylinositol-4,5-bisphosphate (PIP_2) and phosphatidylinositol-3,4,5-triphosphate (PIP_3) from phosphatidyl-4-phosphate (PIP). This promotes dissociation of actin filaments associated with the K_{ATP} channel increasing the activated (open) state of the channel. Acyl-CoA esters of long chain fatty acids (long chain CoA ester) exert a direct effect on the channel to promote the open state of the channel. This leads to egress of intracellular K^+ through the pore-forming unit of the K_{ATP} channel (Kir6.2), hyperpolarization, and decreased firing of the neuron. These actions are opposed by intracellular metabolism of glucose or occupation of the sulfonylurea receptor (SUR). Occupation of G-protein-coupled receptors (GPCR) activates phospholipase C (PLC) which promotes formation of diacylglycerol (DAG) and inositol 3 phosphate (IP3). (B) As in part A., occupation of GPCR leads to activation of PLC which promotes formation of PIP from PIP_2 and PIP_3. This stabilizes actin filaments and leads to inactivation (closure) of the K_{ATP} channel. Glucose metabolism and occupation of the SUR also promote the closed state of the channel.

Oomura, 1985), and glutamate (Aou *et al.*, 1991). GI neurons in the NTS are well positioned to integrate a host of internal metabolic and neural signals (Fig. 1). Their firing rate is altered by afferent vagal signals from the liver (Mizuno and Oomura, 1984) and by catecholamines released by local neurons (Dallaporta *et al.*, 1999; Yettefti *et al.*, 1997). Thus, both GE and GI neurons integrate a host of signals from the internal and external environment. They generally reside in brain areas that are centrally positioned to receive and integrate signals from the periphery and interoceptive, reward, memory, and cognitive areas of the brain. In every way, glucosensing neurons truly represent a distinct subclass of metabolic sensor neurons.

IV. Why Is Glucose Sensed?

The obvious answer to this question is that glucose is sensed by the brain so that it can regulate peripheral glucose levels. The corollary to this is that the brain must be able to sense and regulate glucose levels to maintain a constant supply of glucose for its metabolic needs. Many attempts have been made to relate changes in plasma glucose to physiologic function. Unfortunately, we still have no conclusive idea about the true function of brain glucosensing under physiologic conditions. There are two major areas where glucosensing may well serve an important physiologic function. One is the modulation of ingestive behavior and the other is the regulation of hypothalamo-pituitary and autonomic activity.

Food intake is the most important source of glucose. This led Mayer to propose a feedback loop in which glucose oxidation was sensed and this information was used to regulate intake (Mayer, 1953). There are two major phases to ingestion, meal initiation and meal termination. In rats (Campfield *et al.*, 1985; Campfield and Smith, 1986; Louis-Sylvestre and Le Magnen, 1980; McBride and Sutin, 1984) and humans (Brady *et al.*, 1990b; McLean *et al.*, 1986; Melanson *et al.*, 1999), initiation of some meals is linked to a rise in plasma glucose levels which follows directly after a small (~10%), spontaneous dip in glucose levels. This amounts to only about a 0.5 m*M* (9 mg/dl) decrease in blood glucose. If this dip and rebound of glucose levels are a true signal for meal initiation, then brain (and peripheral) glucosensors must be capable of sensing such small changes. Since brain glucose levels are maintained at about 20–30% of plasma levels (Silver and Erecinska, 1994, 1998; McNay and Gold, 1999; McNay *et al.*, 2001), glucosensing neurons must be able to detect changes of as little as 0.1–0.15 m*M* to be able to sense 0.5-m*M* changes in plasma glucose levels. In fact, some hypothalamic neurons can respond within this range (Silver and Erecinska,

1998) and could theoretically be involved in meal initiation. However, transient dips in plasma glucose levels are clearly not the only stimulant of meal initiation since not all meals are preceded by such dips (Campfield and Smith, 1986; Brady *et al.*, 1990b; McLean *et al.*, 1986; Campfield *et al.*, 1985).

While a single laboratory has been able to demonstrate that manipulation of plasma glucose levels within the physiologic range can alter intake (Campfield and Smith, 1986; Smith *et al.*, 1988; Campfield *et al.*, 1985b; Smith and Campfield, 1993), the majority of studies have been carried out using glucose levels that fall considerably outside this range. The standard practice has been to use "glucoprivation" to study the effect of glucose on meal initiation (DiRocco and Grill, 1979; Ritter *et al.*, 1981, 2000; Smith *et al.*, 1972; Ritter and Slusser, 1980). Intracerebroventricular, intracarotid, intravenous, or intraportal vein glucose infusions have been used to study the effects of glucose on meal termination (Tordoff *et al.*, 1989; Gielkens *et al.*, 1998; Tsujii and Bray, 1990). Competitive inhibitors of glycolysis such as 2-deoxyglucose (2DG) or 5-thioglucose produce glucoprivic feeding by competitive inhibition of the glycolytic production of ATP from glucose. Because intracellular glucopenia produced by these agents stimulates both food intake and a counterregulatory neuroendocrine response (Scheurink and Ritter, 1993; Ritter and Slusser, 1980), it seems clear that glucoprivic feeding is not synonymous with normal diurnal ingestive behavior. Additionally, it is difficult to compare the degree to which such glucoprivic agents reduce intracellular ATP to those changes evoked by naturally occurring dips in plasma glucose that occur during the diurnal cycle. Alternatively, fasting and insulin-induced hypoglycemia have been used to stimulate feeding. Blood glucose does drop during nighttime fasting and rats will eat more rapidly when re-exposed to food after such a fast. The lower the level of glucose reached during fasting, the higher the rate of refeeding when food is made readily available (Larue-Achagiotis and Le Magnen, 1985). However, there are a number of other metabolic events that accompany fasting, such as a precipitous drop in plasma leptin (Weigle *et al.*, 1997; Ahima *et al.*, 1996), which are just as likely as low glucose levels to trigger meal initiation (Klein and Wolfe, 1992). In addition, the levels of plasma glucose needed to evoke glucoprivic meal initiation during insulin-induced hypoglycemia are far below those seen during normal diurnal cycles in plasma glucose. On the other hand, raising glucose in the brain (Tsujii and Bray, 1990), hepatic portal system (Tordoff *et al.*, 1989; Baird *et al.*, 1997), or systemically (Bergen *et al.*, 1996) has only variable and poorly reproducible effects on either meal initiation or termination.

Thus, the evidence that glucose serves as a physiologic regulator of ingestion remains equivocal. In fact, the pathways mediating glucoprivic feeding may well be separate from those modulating the physiologic control of ingestion. Under a variety of conditions in which glucoprivic feeding is rendered defective, feeding responses to other physiologic modulators of food intake such as NPY, leptin, and cholecystokinin are preserved (Miki *et al.,* 2001; Wiater and Ritter, 1999; Bergen *et al.,* 1996). All of these, and other transmitters and peptides which are intrinsic to the hypothalamus and other forebrain areas are considered to be physiologic regulators of ingestive behavior (Leibowitz *et al.,* 1982; Kyrkouli *et al.,* 1990; Kelly *et al.,* 1979, 1996; Stanley *et al.,* 1985, 1993; Currie and Wilson, 1992; Giraudo *et al.,* 1999), while glucoprivic feeding appears to be integrated and initiated primarily within hindbrain sties (DiRocco and Grill, 1979; Ritter *et al.,* 1981, 2000, 2001).

But neurons mediating glucoprivic feeding are undoubtedly integrated into systems involved in normal diurnal food intake. Selective destruction of hindbrain norepinephrine and epinephrine neurons that project to the hypothalamus (Fig. 1) obliterates glucoprivic feeding (Levitt and Moore, 1979; Ahlskog and Hoebel, 1973; Ritter *et al.,* 2001), while norepinephrine injection into hypothalamic targets of these neurons stimulates feeding (Leibowitz *et al.,* 1982). These same brain stem catecholamine neurons also co-localize and release NPY onto these hypothalamic targets (Harfstrand *et al.,* 1986; Benarroch, 1994; Jansen *et al.,* 1995; Gustafson and Moore, 1987). Despite this, when glucoprivic feeding is blocked by destruction of these brain stem catecholamine neurons (Ritter *et al.,* 2001) or by rendering GE neurons defective (Miki *et al.,* 2001), forebrain NPY injection still stimulated food intake. This suggests that there are separate but highly overlapping systems mediating glucoprivic and normal diurnal ingestive behavior.

Besides ingestive behavior, alterations in glucose availability also affect hypothalamo-pituitary and autonomic nervous system activity. The best example of this is the glucoprivic stimulation of the counterregulatory response. A graded counterregulatory response is triggered as plasma glucose levels decrease from the physiologic range to about 3.3–3.6 m*M* (Connolly *et al.,* 1992; Hamilton-Wessler *et al.,* 1994; Mitrakou *et al.,* 1991; Boyle *et al.,* 1994). Brain glucose uptake begins to decline at ~3.6 m*M* (Boyle *et al.,* 1994). As glucose levels fall, various counterregulatory systems are called into play to increase the supply of glucose to the brain. These include activation of the sympathoadrenal system and hypothalamo-pituitary axis (glucocorticoids, growth hormone, somatostatin) and pancreatic α-cells (glucagon) (Veneman *et al.,* 1993; Mitrakou *et al.,* 1991; Boyle *et al.,* 1994). The net result is a substantial mobilization of glucose from glycogen stores and an increase in hepatic glucose production. It is likely that both brain and hepatic portal glucosensors participate in this response (Hevener *et al.,* 2000; Cane *et al.,* 1986; Frizzell *et al.,* 1993; Hamilton-Wessler *et al.,* 1994; Donovan *et al.,* 1994). Caudal brain stem norepinephrine and epinephrine neurons that project to the spinal cord (Fig. 1) are necessary for the full expression of this response (Ritter *et al.,* 2001). However, forebrain sites can also modulate the counterregulatory response. Injection of 2DG into the ventromedial hypothalamus triggers the response (Borg *et al.,* 1995) and the response to systemic hypoglycemia can be attenuated by simultaneously raising ventromedial hypothalamic glucose levels (Borg *et al.,* 1997). Finally, lesions of the ventromedial hypothalamus significantly attenuate the counterregulatory response (Borg *et al.,* 1994). In the lateral hypothalamus, 2DG infusion stimulates activity in the adrenal nerve but inhibits sympathetic activation of brown adipose tissue (Egawa *et al.,* 1989). In the dorsomedial hypothalamus, 2DG infusion also inhibits adrenal nerve activity (Yoshimatsu *et al.,* 1991). While glucoprivation elicits a profound counterregulatory response, raising forebrain glucose availability produces less pronounced autonomic responses. Perfusion of the forebrain with glucose activates the sympathetic nervous system (Levin, 1991, 1992) and vagal nerve output to the pancreas (Seino *et al.,* 1997), while decreasing adrenal nerve activity (Niijima, 1975).

Finally, glucose has been implicated as a mediator of other physiologic functions unrelated to food intake and the autonomic activation. Glucose plays a role in facilitating memory function in humans (Manning *et al.,* 1998) and rodents (Ragozzino *et al.,* 1998), apparently by its action on the hippocampus. In addition, glucose has an interesting but unexplained effect on ligand binding to a variety of receptors. These include adrenergic (Angel *et al.,* 1990; Levin and Planas, 1993), morphine (Brase *et al.,* 1987), sulfonylurea (Levin and Dunn-Meynell, 1997b), and anorectic receptors (Angel *et al.,* 1985; Hauger *et al.,* 1986). These effects may be related to changes in brain neurotransmitter turnover produced by raising and lowering glucose levels (Di and Barth, 1991; Beverly *et al.,* 1995; Milusheva *et al.,* 1992; Gupta *et al.,* 1992; Specter *et al.,* 1996). However, the mechanism for this effect of glucose on receptor binding remains unknown.

V. How Is Glucosensing Altered in Disease States?

Major abnormalities of glucosensing are associated with both obesity and diabetes mellitus. In obese humans, functional magnetic resonance imaging demonstrates abnormal hypothalamic activation following oral glucose ingestion (Matsuda *et al.,* 1999). Obese humans also show a defect in growth hormone release in response to hypoglycemia (Bernini *et al.,* 1990). Rodent models of obesity also demonstrate several abnormalities of brain glucosensing. Neuronal activation in response to increased blood glucose is abnormal in hypothalamic areas involved in energy homeostasis and autonomic function (Levin *et al.,* 1998). This is associated with aberrant activation of the sympathetic nervous system in obesity-prone rats (Levin, 1992). Obesity-prone rats also have reduced low-affinity sulfonylurea binding (Levin *et al.,* 1996) and have abnormal glucose-mediated regulation of α_2-adrenoceptor binding in brain areas involved in energy homeostasis (Levin and Planas, 1993). Our preliminary data also show that the K_{ATP} channel on neurons in the ventromedial hypothalamic nucleus has reduced sensitivity to both ATP and sulfonylureas in obesity-prone rats (Routh *et al.,* 1998). A similar defect is seen in genetically obese Zucker rats (Spanswick *et al.,* 2000). Finally, neurotoxins such as gold thioglucose are thought to selectively damage glucosensing neurons. In mice, gold thioglucose produces hyperphagia and obesity associated with defects in central glucosensing (Bergen *et al.,* 1996).

Diabetes is also associated with abnormalities of glucosensing. While experimental diabetes produces some defects of brain glucosensing (Levin *et al.,* 1996; Levin and Dunn-Meynell, 1998), the most important abnormalities are those produced by hypoglycemia. The current standard for controlling the complications of type 1 diabetes calls for tight regulation of blood glucose levels within a relatively narrow range by intensive insulin therapy (The DCCT Research Group, 1993). But an unfortunate side effect of this tight control is an increasing incidence of severe hypoglycemic episodes. This treatment protocol carries a two-fold greater risk of developing hypoglycemia than conventional therapy (The DCCT Research Group, 1997). As mentioned above, hypoglycemia initiates the powerful neuroendocrine counterregulatory response which increases glucose availability. In addition, hypoglycemic patients usually become aware of their hypoglycemia because they experience symptoms of hunger, nervousness, tremor, palpitations, and sweating (Heller and Cryer, 1991). These symptoms become evident at plasma glucose levels of ~3.6–3.8 m*M* (Hvidberg *et al.,* 1996; Mitrakou *et al.,* 1991). Cognitive function begins to deteriorate at glucose levels of ~2.7–3.1 m*M* (Mitrakou *et al.,* 1991; Boyle *et al.,* 1994). However, even a single bout of hypoglycemia can significantly lower the threshold level of glucose required to initiate both hypoglycemia awareness and the counterregulatory response (Dagogo-Jack *et al.,* 1994; Boyle *et al.,* 1995; Cranston *et al.,* 1994). Moderate to severe hypoglycemia or glucoprivic stress both cause a profound downregulation of this counterregulatory response that can seriously impair the timely restoration of blood glucose levels (Dagogo-Jack *et al.,* 1993; Robinson *et al.,* 1995; Davis *et al.,* 1997; Tkacs *et al.,* 2000; Sanders and Ritter, 2000; Veneman *et al.,* 1993). Hypoglycemia awareness returns after a few days, while blunting of some elements of the counterregulatory response may persist for months after studious avoidance of hypoglycemia (Dagogo-Jack *et al.,* 1994; Powell *et al.,* 1993; Cranston *et al.,*

1994; Fanelli *et al.,* 1993, 1994). This means that the ability of the brain to sense and respond to lowered levels of plasma glucose is seriously impaired for variable lengths of time by prior bouts of hypoglycemia.

There is considerable uncertainty as to the pathogenesis of hypoglycemia unawareness and the blunted counterregulatory response caused by repeated bouts of hypoglycemia. In its most severe form, hypoglycemia can result in coma, seizures, severe brain damage, and even death (The DCCT Research Group, 1997; Fujioka *et al.,* 1997; Patrick and Campbell, 1990; Kiessling *et al.,* 1986; Perros *et al.,* 1997; Wredling *et al.,* 1990). During milder bouts, there is a reduction in cognitive function (Ovalle *et al.,* 1998; Maran *et al.,* 1995; Fanelli *et al.,* 1998a; King *et al.,* 1998) and slowed conduction of neural impulses within the brain (Jacob *et al.,* 1999; King *et al.,* 1998; Durrant *et al.,* 1991) as cerebral glucose uptake falls (Boyle *et al.,* 1994). In experimental animals, a single episode of moderate hypoglycemia that causes neither coma nor seizures can produce selective, delayed apoptosis in the ARC (Tkacs *et al.,* 2000). This is associated with reduced expression of ARC NPY and POMC mRNA and a significant attenuation of the counterregulatory response (Tkacs *et al.,* 2000). Impaired function of ARC NPY and POMC neurons might be partly responsible for the blunted counterregulatory response since these neurons project both rostrally and caudally to brain areas involved in the counterregulatory response (Elias *et al.,* 1998, 1999; Elmquist *et al.,* 1998b).

The apparent selective vulnerability of this population of ARC neurons may, in turn, be due to unique combination of their need to use glucose primarily as a signaling molecule and a strong complement of glutamatergic inputs to this area (Belousov and van den Pol, 1997; Diano *et al.,* 1998). If glucosensing neurons were similar to pancreatic β-cells (Ishihara *et al.,* 1999), they would have a limited capacity to use lactate provided by astrocytes as an alternate energy source when glucose supplies become limiting. This would make glucose the major source of intracellular ATP production and enable them to sense small changes in ambient glucose. But this increased sensitivity to glucose would carry a high price. Their reduced capacity to utilize lactate would impair their ability to mitigate the increased energy demands imposed by the excitotoxic levels of glutamate released during hypoglycemia (Wieloch, 1986; Maus *et al.,* 1999). While glucosensing neurons are found in many brain areas, only ARC neurons appear to exhibit such selective vulnerability to moderate levels of hypoglycemia (Tkacs *et al.,* 2000). These same neurons are selectively vulnerable to neonatal exposure to monosodium glutamate (Alessi *et al.,* 1988; Bergen *et al.,* 1998), suggesting a glutamatergic mechanism might underlie their selective vulnerability.

Interestingly, neonatal monosodium glutamate also destroys NTS neurons which do not show apoptosis following hypoglycemia in the adult

(Tkacs *et al.,* 2000). On the other hand, repeated bouts of hypoglycemia do reduce the activation of NTS and other caudal brain stem catecholamine neurons to subsequent bouts of hypoglycemia (Sanders and Ritter, 2000). These NTS neurons are a critical component of both glucoprivic feeding and the counterregulatory response (Ritter *et al.,* 2001) (Fig. 1). Apparent cell death in the ARC and reduced responsiveness of NTS neurons are certainly potentially serious results of hypoglycemia. However, this hypoglycemia-associated autonomic failure is rarely permanent (Dagogo-Jack *et al.,* 1994; Powell *et al.,* 1993; Cranston *et al.,* 1994; Fanelli *et al.,* 1993, 1994). If damage and/or impaired function of ARC and NTS neurons during hypoglycemia contribute to a reduced counterregulatory response, recovery probably occurs through a combination of neural plasticity and inherent redundancy of neural circuits involved in this response.

While brain damage may occur following hypoglycemic episodes, there is also somewhat controversial evidence that the brain can mount a protective response against recurrent hypoglycemic insults. For example, there is disagreement as to whether cognitive function may (Fanelli *et al.,* 1998a,b; Veneman *et al.,* 1993; Mitrakou *et al.,* 1991; Boyle *et al.,* 1994) or may not (Maran *et al.,* 1995) be preserved at lower glucose levels after prior bouts of hypoglycemia. (Segel *et al.,* 2001; Boyle *et al.,* 1994; Powers *et al.,* 1993). The expression of blood-brain barrier (GLUT1) and/or neuronal glucose transporters (GLUT3) may be variably increased by prior bouts of hypoglycemia (Kumagai *et al.,* 1995; Liu *et al.,* 1998; Vidal-Puig *et al.,* 1999; Simpson *et al.,* 1999). Glucose uptake into the brain may (Kumagai *et al.,* 1995; Simpson *et al.,* 1999; McCall *et al.,* 1986) or may not (Segel *et al.,* 2001; Boyle *et al.,* 1994; Powers *et al.,* 1993) be increased by antecedent bouts of hypoglycemia. In experimental animals, prior glucoprivation induces the expression of neuroprotective molecules like GRP78 and HSP70 which should help protect against subsequent bouts of glutamate toxicity and oxidative stress (Lee *et al.,* 1999b). While results vary as a function of the species examined and the experimental design, it seems likely that antecedent bouts of hypoglycemia do enhance the brain's ability to withstand future bouts of hypoglycemia.

VI. Summary and Conclusions

It is clear that the brain has evolved a mechanism for sensing levels of ambient glucose. Teleologically, this is likely to be a function of its requirement for glucose as a primary metabolic substrate. There is no question that the brain can sense and mount a counterregulatory response to restore very low levels of plasma and brain glucose. But it is less clear that the changes in glucose associated with normal diurnal rhythms and feeding cycles are sufficient to influence either ingestive behavior or the physiologic responses involved in regulating plasma glucose levels. Glucosensing neurons are clearly a distinct class of metabolic sensors with the capacity to respond to a variety of intero- and exteroceptive stimuli. This makes it likely that these glucosensing neurons do participate in physiologically relevant homeostatic mechanisms involving energy balance and the regulation of peripheral glucose levels. It is our challenge to identify the mechanisms by which these neurons sense and respond to these metabolic cues.

References

Adachi, A., Kobashi, M., Miyoshi, N., and Tsukamoto, G. (1991). Chemosensitive neurons in the area postrema of the rat and their possible functions. *Brain Res. Bull.* **26,** 137–140.

Adachi, A., Shimizu, N., Oomura, Y., and Kobashi, M. (1984). Convergence of heptoportal glucose-sensitive afferent signals to glucose-sensitive units within the nucleus of the solitary tract. *Neurosci. Lett.* **46,** 215–218.

Aguilar-Bryan, L., Clement IV, J. P., Gonzalez, G., Kunjilwar, K., Babenko, A., and Bryan, J. (1998). Toward understanding the assembly and structure of KATP channels. *Physiol. Rev.* **78,** 227–245.

Aguilar-Bryan, L., Nichols, C. G., Wechsler, S. W., Clement IV, J. P., Boyd, III, A. E., Gonzalez, G., Herrera-Sosa, H., Nguy, K., Bryan, J., and Nelson, D. A. (1995). Cloning of the β cell high-affinity sulfonylurea receptor: A regulator of insulin secretion. *Science* **268,** 423–426.

Ahima, R. S., Prabakaran, D., Mantzoros, C., Qu, D., Lowell, B., Maratos-Flier, E., and Flier, J. S. (1996). Role of leptin in the neuroendocrine response to fasting. *Nature* **382,** 250–252.

Ahlskog, J. E., and Hoebel, B. G. (1973). Overeating and obesity from damage to a noradrenergic system in the brain. *Science* **182,** 166–169.

Ainscow, E. K., Zhao, C., and Rutter, G. A. (2000). Acute overexpression of lactate dehydrogenase-A perturbs beta-cell mitochondrial metabolism and insulin secretion. *Diabetes* **49,** 1149–1155.

Akabayashi, A., Zaia, C. T., Silva, I., Chae, H. J., and Leibowitz, S. F. (1993). Neuropeptide Y in the arcuate nucleus is modulated by alterations in glucose utilization. *Brain Res.* **621,** 343–348.

Alessi, N. E., Quinlan, P., and Khachaturian, H. (1988). MSG effects on beta-endorphin and alpha-MSH in the hypothalamus and caudal medulla. *Peptides* **9,** 689–695.

Alvarez-Buylla, R., and de Alvarez-Buylla, E. R. (1988). Carotid sinus receptors participate in glucose homeostasis. *Respir. Physiol.* **72,** 347–360.

Amoroso, S., Schmid-Antomarchi, H., Fosset, M., and Lazdunski, M. (1990). Glucose, sulfonylureas, and neurotransmitter release: role of ATP-sensitive K^+ channels. *Science* **247,** 852–854.

Anand, B. K., and Brobeck, J. R. (1951). Hypothalamic control of food intake in rats and cats. *Yale J. Biol. Med.* **24,** 123–146.

Anand, B. K., Chhina, G. S., Sharma, K. N., Dua, S., and Singh, B. (1964). Activity of single neurons in the hypothalamus feeding centers: effect of glucose. *Am. J. Physiol.* **207,** 1146–1154.

Angel, I., Hauger, R. L., Luu, M. D., Giblin, B., Skolnick, P., and and Paul, S. M. (1985). Glucostatic regulation of (+)-[^{3}H]amphetamine binding in the hypothalamus: Correlation with Na^+,K^+-ATPase activity. *Proc. Natl. Acad. Sci. USA* **82,** 6320–6324.

Angel, I., Niddam, R., and Langer, S. Z. (1990). Involvement of *alpha*-2 adrenergic receptor subtypes in hyperglycemia. *J. Pharmacol. Exp. Ther.* **254,** 877–882.

Aou, S., Takaki, A., Karadi, Z., Hori, T., Nishino, H., and Oomura, Y. (1991). Functional heterogeneity of the monkey lateral hypothalamus in the control of feeding. *Brain Res. Bull.* **27,** 451–455.

Apelt, J., Mehlhorn, G., and Schliebs, R. (1999). Insulin-sensitive GLUT4 glucose transporters are colocalized with GLUT3-expressing cells and demonstrate a chemically distinct neuron-specific localization in rat brain. *J. Neurosci. Res.* **57,** 693–705.

Ashford, M. L. J., Boden, P. R., and Treherne, J. M. (1990a). Glucose-induced excitation of hypothalamic neurones is mediated by ATP-sensitive K^+ channels. *Pflugers Arch.* **415,** 479–483.

Ashford, M. L. J., Boden, P. R., and Treherne, J. M. (1990b). Tolbutamide excites rat glucoreceptive ventromedial hypothalamic neurones by indirect inhibition of ATP-K^+ channels. *Br. J. Pharmacol.* **101,** 531–540.

Babenko, A. P., Gonzalez, G., and Bryan, J. (2000). Pharmaco-topology of sulfonylurea receptors. Separate domains of the regulatory subunits of K(ATP) channel isoforms are required for selective interaction with K(+) channel openers. *J. Biol. Chem.* **275,** 717–720.

Bagnol, D., Lu, X.-Y., Kaelin, C. B., Day, H. E. W., Ollmann, M., Gantz, I., Akil, H., Barsh, G. S., and Watson, S. J. (1999). Anatomy of an endogenous antagonist: relationship between agouti-related protein and proopiomelanocortin in brain. *J. Neurosci.* **19,** 1–7.

Bai, F. L., Yamano, M., Shiotani, Y., Emson, P. C., Smith, A. D., Powell, J. F., and Tohyama, M. (1985). An arcuato-paraventricular and dorsomedial hypothalamic neuropeptide Y containing system which lacks norepinephrine in the rat. *Brain Res.* **331,** 172–175.

Baird, J. P., Grill, H. J., and Kaplan, J. M. (1997). Intake suppression after hepatic portal glucose infusion: all-or-none effect and its temporal threshold. *Am. J. Physiol.* **272,** R1454–R1460.

Baker, R. A., Herkenham, M., and Brady, L. S. (1996). Effects of long-term treatment with antidepressant drugs on proopiomelanocortin and neuropeptide Y mRNA expression in the hypothalamic arcuate nucleus in rats. *J. Neuroendocrinol.* **8,** 337–343.

Banks, A. S., Davis, S. M., Bates, S. H., and Myers, M. G. Jr. (2000). Activation of downstream signals by the long form of the leptin receptor. *J. Biol. Chem.* **275,** 14563–14572.

Baskin, D. G., Seeley, R. J., Kuijper, J. L., Lok, S., Weigle, D. S., Erickson, J. C., Palmiter, R. D., and Schwartz, M. W. (1998). Increased expression of mRNA for the long form of the leptin receptor in the hypothalamus is associated with leptin hypersensitivity and fasting. *Diabetes* **47,** 538–543.

Baskin, D. G., Sipols, A. J., Schwartz, M. W., and White, M. F. (1993). Immunocytochemical detection of insulin receptor substrate-1 (IRS-1) in rat brain: Colocalization with phosphotyrosine. *Regul. Pept.* **48,** 257–266.

Baukrowitz, T., and Fakler, B. (2000). K_{ATP} channels: Linker between phospholipid metabolism and excitability. *Biochem. Pharmacol.* **60,** 735–740.

Belousov, A. B., and van den Pol, A. N. (1997). Local synaptic release of glutamate from neurons in the rat hypothalamic arcuate nucleus. *J. Physiol.* **499,** 747–761.

Benarroch, E. E. (1994). Neuropeptides in the sympathetic system: presence, plasticity, modulation and implications. *Ann. Neurol.* **36,** 6–13.

Bergen, H. T., Mizuno, T. M., Taylor, J., and Mobbs, C. V. (1998). Hyperphagia and weight gain after gold-thioglucose: relation to hypothalamic neuropeptide Y and proopiomelanocortin. *Endocrinolology* **139,** 4483–4488.

Bergen, H. T., Monkman, N., and Mobbs, C. V. (1996). Injection with gold thioglucose impairs sensitivity to glucose: evidence that glucose-responsive neurons are important for long-term regulation of body weight. *Brain Res.* **734,** 332–336.

Bernini, G. P., Argenio, G. F., Vivaldi, M. S., Del Corso, C., Birindelli, R., Luisi, M., and Franchi, F. (1990). Impaired growth hormone response to insulin-induced hypoglycemia in obese patients: Restoration blocked by ritanserin after fenfluramine administration. *Clin. Endocrinol.* **32,** 453–459.

Beverly, J. L., Beverly, M. F., and Meguid, M. M. (1995). Alterations in extracellular GABA in the ventral hypothalamus of rats in response to acute glucoprivation. *Am. J. Physiol.* **269,** R1174–R1178.

Billington, C. J., Briggs, J. E., Harker, S., Grace, M., and Levine, A. S. (1994). Neuropeptide Y in hypothalamic paraventricular nucleus: a center coordinating energy metabolism. *Am. J. Physiol.* **266,** R1765–R1770.

Bittar, P. G., Charnay, Y., Pellerin, L., Bouras, C., and Magistretti, P. J. (1996). Selective distribution of lactate dehydrogenase isoenzymes in neurons and astrocytes of human brain. *J. Cereb. Blood Flow Metab.* **16,** 1079–1089.

Bjorbaek, C., El Haschimi, K., Frantz, J. D., and Flier, J. S. (1999). The role of SOCS-3 in leptin signaling and leptin resistance. *J. Biol. Chem.* **274,** 30059–30065.

Borg, M. A., Sherwin, R. S., Borg, W. P., Tamborlane, W. V., and Shulman, G. I. (1997). Local ventromedial hypothalamus glucose perfusion blocks counterregulation during systemic hypoglycemia in awake rats. *J. Clin. Invest.* **99,** 361–365.

Borg, W. P., During, M. J., Sherwin, R. S., Borg, M. A., Brines, M. L., and Shulman, G. I. (1994). Ventromedial hypothalamic lesions in rats suppress counterregulatory responses to hypoglycemia. *J. Clin. Invest.* **93,** 1677–1682.

Borg, W. P., Sherwin, R. S., During, M. J., Borg, M. A., and Shulman, G. I. (1995). Local ventromedial hypothalamic glucopenia triggers counterregulatory hormone release. *Diabetes* **44,** 180–184.

Boyle, P. J., Kempers, S. F., O'Conner, A. M., and Nagy, R. J. (1995). Brain glucose uptake and unawareness of hypoglycemia in patients with insulin-dependent diabetes mellitus. *N. Eng. J. Med.* **333,** 1726–1731.

Boyle, P. J., Nagy, R. J., O'Connor, A. M., Kempers, S. F., Yeo, R. A., and Qualls, C. (1994). Adaptation in brain glucose uptake following recurrent hypoglycemia. *Proc. Natl. Acad. Sci. USA* **91,** 9352–9356.

Brady, L. S., Smith, M. A., Gold, P. W., and Herkenham, M. (1990). Altered expression of hypothalamic neuropeptide mRNAs in food-restricted and food-deprived rats. *Neuroendocrinology* **52,** 441–447.

Branstrom, R., Leibiger, I. B., Leibiger, B., Corkey, B. E., Berggren, P. O., and Larsson, O. (1998). Long chain coenzyme A esters activate the pore-forming subunit (Kir6. 2) of the ATP-regulated potassium channel. *J. Biol. Chem.* **273,** 31395–31400.

Brase, D. A., Han, Y. -H., and Dewey, W. L. (1987). Effects of glucose and diabetes on binding of naloxone and dihydromorphine to opiate receptors in mouse brain. *Diabetes* **36,** 1173–1177.

Broer, S., Rahman, B., Pellegri, G., Pellerin, L., Martin, J. L., Verleysdonk, S., Hamprecht, B., and Magistretti, P. J. (1997). Comparison of lactate transport in astroglial cells and monocarboxylate transporter 1 (MCT 1) expressing *Xenopus laevis* oocytes. Expression of two different monocarboxylate transporters in astroglial cells and neurons. *J. Biol. Chem.* **272,** 30096–30102.

Campfield, A., Brandon, P., and Smith, F. J. (1985). On-line continuous measurement of blood glucose and meal pattern in free feeding rats: The role of glucose in meal initiation. *Brain Res. Bull.* **14,** 605–616.

Campfield, L. A., and Smith, F. J. (1986). Functional coupling between transient declines in blood glucose and feeding behavior: Temporal relationships. *Brain Res. Bull.* **17,** 427–433.

Cane, P., Artal, R., and Bergman, R. N. (1986). Putative hypothalamic glucoreceptors play no essential role in the response to moderate hypoglycemia. *Diabetes* **35,** 268–277.

Cesar, M. D. C., and Wilson, J. E. (1998). Further studies on the coupling of mitochondrially bound hexokinase to intramitochondrially compartmented ATP generated by oxidative phosphorylation. *Arch. Biochem. Biophys.* **350,** 109–117.

Chen, A. S., Metzger, J. M., Trumbauer, M. E., Guan, X. M., Yu, H., Frazier, E. G., Marsh, D. J., Forrest, M. J., Gopal-Truter, S., Fisher, J., Camacho, R. E., Strack, A. M., Mellin, T. N., MacIntyre, D. E., Chen, H. Y., and Van der Ploeg, L. H. (2000). Role of the melanocortin-4 receptor in metabolic rate and food intake in mice. *Transgenic Res.* **9,** 145–154.

Connolly, C. C., Adkins-Marshall, B. A., Neal, D. W., Pugh, W., Jaspan, J. B., and Cherrington, A. D. (1992). Relationship between decrements in glucose level and metabolic response to hypoglycemia in absence of counterregulatory hormones in the conscious dog. *Diabetes* **41,** 1308–1319.

Cowley, M. A., Smart, J. L., Rubinstein, M., Cerdan, M. G., Diano, S., Horvath, T. L., Cone, R. D., and Low, M. J. (2001). Leptin activates anorexigenic POMC neurons through a neural network in the arcuate nucleus. *Nature* **411,** 480–484.

Cranston, I., Lomas, J., Maran, A., Macdonald, I., and Amiel, S. A. (1994). Restoration of hypoglycaemia awareness in patients with long-duration insulin-dependent diabetes. *Lancet* **344,** 283–287.

Currie, P. J., and Wilson, L. M. (1992). Central injection of 5-hydroxytryptamine reduces food intake in obese and lean mice. *Neuroreport* **3,** 59–61.

Dagogo-Jack, S., Rattarasarn, C., and Cryer, P. E. (1994). Reversal of hypoglycemia unawareness, but not defective glucose counterregulation, in IDDM. *Diabetes* **43,** 1426–1434.

Dagogo-Jack, S. E., Craft, S., and Cryer, P. E. (1993). Hypoglycemia-associated autonomic failure in insulin-dependent diabetes mellitus. Recent antecedent hypoglycemia reduces autonomic responses to symptoms of, and defense against subsequent hypoglycemia. *J. Clin. Invest.* **91,** 819–828.

Dallaporta, M., Himmi, T., Perrin, J., and Orsini, J. C. (1999). Solitary tract nucleus sensitivity to moderate changes in glucose level. *Neuroreport* **10,** 2657–2660.

Dallaporta, M., Perrin, J., and Orsini, J. C. (2000). Involvement of adenosine triphosphate-sensitive K^+ channels in glucose-sensing in the rat solitary tract nucleus. *Neurosci. Lett.* **278,** 77–80.

Davis, S. N., Shavers, C., Mosqueda-Garcia, R., and Costa, F. (1997). Effects of differing antecedent hypoglycemia on subsequent counterregulation in normal humans. *Diabetes* **46,** 1328–1335.

De Bruin, L. A., Schasfoort, E. M., Steffens, A. B., and Korf, J. (1990). Effects of stress and exercise on rat hippocampus and striatum extracellular lactate. *Am. J. Physiol.* **259,** R773–R779.

De Vos, A., Heimberg, H., Quartier, E., Huypens, P., Bouwens, L., Pipeleers, D., and Schuit, F. (1995). Human and rat beta cells differ in glucose transporter but not in glucokinase gene expression. *J. Clin. Invest.* **96,** 2489–2495.

Deitmer, J. W. (2000). Glial strategy for metabolic shuttling and neuronal function. *Bioessays* **22,** 747–752.

Detimary, P., Gilon, P., and Henquin, J. C. (1998). Interplay between cytoplasmic Ca^{2+} and the ATP/ADP ratio: A feedback control mechanism in mouse pancreatic islets. *Biochem. J.* **333,** 269–274.

Di, S., and Barth, D. S. (1991). Topographic analysis of field potentials in rat vibrissa/barrel cortex. *Brain Res.* **546,** 106–112.

Diano, S., Naftolin, F., and Horvath, T. L. (1998). Kainate glutamate receptors (GluR5–7) in the rat arcuate nucleus: relationship to tanycytes, astrocytes, neurons and gonadal steroid receptors. *J. Endocrinol.* **10,** 239–247.

DiRocco, R. J., and Grill, H. J. (1979). The forebrain is not essential for sympathoadrenal hyperglycemic response to glucoprivation. *Science* **204,** 1112–1114.

Donovan, C. M., Hamilton-Wessler, M., Halter, J. B., and Bergman, R. N. (1994). Primacy of liver glucosensors in the sympathetic response to progressive hypoglycemia. *Proc. Natl. Acad. Sci. USA* **91,** 2863–2867.

Dunn-Meynell, A. A., Rawson, N. E., and Levin, B. E. (1998). Distribution and phenotype of neurons containing the ATP-sensitive K^+ channel in rat brain. *Brain Res.* **814,** 41–54.

Dunn-Meynell, A. A., Routh, V. H., McArdle, J. J., and Levin, B. E. (1997). Low affinity sulfonylurea binding sites reside on neuronal cell bodies in the brain. *Brain Res.* **745,** 1–9.

During, M. J., Leone, P., Davis, K. E., Kerr, D., and Sherwin, R. S. (1995). Glucose modulates rat substantia nigra GABA release in vivo via ATP-sensitive potassium channels. *J. Clin. Invest.* **95,** 2403–2408.

Durrant, J. D., Gerich, J. E., Mitrakou, A., Jenssen, T., and Hyre, R. J. (1991). Changes in BAEP under hypoglycemia: Temperature-related? Electroenceph. *Clin. Neurophysiol.* **80,** 547–550.

Egawa, M., Yoshimatsu, H., and Bray, G. A. (1989). Lateral hypothalamic injection of 2-deoxy-D-glucose suppresses sympathetic activity. *Am. J. Physiol.* **257**(Pt 2), R1386–92.

El Messari, S., Leloup, C., Quignon, M., Brisorgueil, M. J., Penicaud, L., and Arluison, M. (1998). Immunocytochemical localization of the insulin-responsive glucose transporter 4 (Glut4) in the rat central nervous system. *J. Comp. Neurol.* **399,** 492–512.

Elias, C. F., Aschkenasi, C., Lee, C., Kelly, J., Ahima, R. S., Bjorbaek, C., Flier, J. S., Saper, C. B., and Elmquist, J. K. (1999). Leptin differentially regulates NPY and POMC neurons projecting to the lateral hypothalamic area. *Neuron* **23,** 775–786.

Elias, C. F., Lee, C., Kelly, J., Aschkenasi, C., Ahima, R. S., Couceyro, P. R., Kuhar, M. J., Saper, C. B., and Elmquist, J. K. (1998). Leptin activates hypothalamic CART neurons projecting to the spinal cord. *Neuron* **21,** 1375–1385.

Elmquist, J. K., Ahima, R. S., Elias, C. F., Flier, J. S., and Saper, C. B. (1998b). Leptin activates distinct projections from the dorsomedial and ventromedial hypothalamic nuclei. *Proc. Natl. Acad. Sci. USA* **95,** 741–746.

Elmquist, J. K., Bjorbaek, C., Ahima, R. S., Flier, J. S., and Saper, C. B. (1998a). Distributions of leptin receptor mRNA isoforms in the rat brain. *J. Comp. Neurol.* **395,** 535–547.

Fanelli, C., Pampanelli, S., Epifano, L., Rambotti, A. M., Di Vincenzo, A., Modarelli, F., Ciofetta, M., Lepore, M., Annibale, B., and Torlone, E. (1994). Long-term recovery from unawareness, deficient counterregulation and lack of cognitive dysfunction during hypoglycaemia, following institution of rational, intensive insulin therapy in IDDM. *Diabetologia* **37,** 1265–1276.

Fanelli, C. G., Epifano, L., Rambotti, A. M., Pampanelli, S., Di Vincenzo, A., Modarelli, F., Lepore, M., Annibale, B., Ciofetta, M., and Bottini, P. (1993). Meticulous prevention of hypoglycemia normalizes the glycemic thresholds and magnitude of most of neuroendocrine responses to, symptoms of, and cognitive function during hypoglycemia in intensively treated patients with short-term IDDM. *Diabetes* **42,** 1683–1689.

Fanelli, C. G., Pampanelli, S., Porcellati, F., and Bolli, G. B. (1998b). Shift of glycaemic thresholds for cognitive function in hypoglycaemia unawareness in humans. *Diabetologia* **41,** 720–723.

Fanelli, C. G., Paramore, D. S., Hershey, T., Terkamp, C., Ovalle, F., Craft, S., and Cryer, P. E. (1998a). Impact of nocturnal hypoglycemia on hypoglycemic cognitive dysfunction in type 1 diabetes. *Diabetes* **47,** 1920–1927.

Fei, H., Okano, H. J., Li, C., Lee, G. -H., Zhao, C., Darnell, R., and Friedman, J. M. (1997). Anatomic localization of alternatively spliced leptin receptors (OB-R) in mouse brain and other tissues. *Proc. Natl. Acad. Sci. USA* **94,** 7001–7005.

Finta, E. P., Harms, L., Sevcik, J., Fischer, H.-D., and Illes, P. (1993). Effects of potassium channel openers and their antagonists on rat locus coeruleus neurones. *Br. J. Pharmacol.* **109,** 308–315.

Fray, A. E., Forsyth, R. J., Boutelle, M. G., and Fillenz, M. (1996). The mechanisms controlling physiologically stimulated changes in rat brain glucose and lactate: A microdialysis study. *J. Physiol.* (*Lond.*) **496,** 49–57.

Frizzell, R. T., Jones, E. M., Davis, S. N., Biggers, D. W., Myers, S. R., Connolly, C. C., Neal, D. W., Jaspan, J. B., and Cherrington, A. D. (1993). Counterregulation during hypoglycemia is directed by widespread brain regions. *Diabetes* **42,** 1253–1261.

Fujioka, M., Okuchi, K., Hiramatsu, K.-I., Sakaki, T., Sakaguchi, S., and Ishii, Y. (1997). Specific changes in human brain after hypoglycemic injury. *Stroke* **28,** 584–587.

Funahashi, H., Yada, T., Muroya, S., Takigawa, M., Ryushi, T., Horie, S., Nakai, Y., and Shioda, S. (1999). The effect of leptin on feeding-regulating neurons in the rat hypothalamus. *Neurosci. Lett.* **264,** 117–120.

German, M. S. (1993). Glucose sensing in pancreatic islet beta cells: The key role of glucokinase and the glycolytic intermediates. *Proc. Natl. Acad. Sci. USA* **90,** 1781–1785.

Ghilardi, N., Ziegler, S., Wiestner, A., Stoffel, R., Heim, M. H., and Skoda, R. C. (1996). Defective STAT signaling by the leptin receptor in diabetic mice. *Proc. Natl. Acad. Sci. USA* **93,** 6231–6235.

Gielkens, H. A. J., Verkijk, M., Lam, W. F., Lamers, C. B. H. W., and Masclee, A. A. (1998). Effects of hyperglycemia and hyperinsulinemia on satiety in humans. *Metabolism* **47,** 321–324.

Giraudo, S. Q., Billington, C. J., and and Levine, A. S. (1998). Feeding effects of hypothalamic injection of melanocortin 4 receptor ligands. *Brain Res.* **809,** 302–306.

Giraudo, S. Q., Grace, M. K., Billington, C. J., and Levine, A. S. (1999). Differential effects of neuropeptide Y and the m-agonist DAMGO on "palatability" vs. "energy." *Brain Res.* **834,** 160–163.

Giraudo, S. Q., Kotz, C. M., Grace, M. K., Levine, A. S., and Billington, C. J. (1994). Rat hypothalamic NPY mRNA and brown fat uncoupling protein mRNA after high-carbohydrate or high-fat diets. *Am. J. Physiol.* **266,** R1578–R1583.

Girault, J. A., Horiuchi, A., Gustafson, E. L., Rosen, N. L., and Greengard, P. (1990). Differential expression of ARPP-16 and ARPP-19, two highly related cAMP-regulated phosphoproteins, one of which is specifically associated with dopamine-innervated brain regions. *J. Neurosci.* **10,** 1124–1133.

Gribble, F. M., Proks, P., Corkey, B. E., and Ashcroft, F. M. (1998). Mechanism of cloned ATP-sensitive potassium channel activation by oleoyl-CoA. *J. Biol. Chem.* **273,** 26383–26387.

Griffond, B., Risold, P. Y., Jacquemard, C., Colard, C., and Fellmann, D. (1999). Insulin-induced hypoglycemia increases preprohypocretin (orexin) mRNA in the rat lateral hypothalamic area. *Neurosci. Lett.* **262,** 77–80.

Grill, H. J., Ginsberg, A. B., Seeley, R. J., and Kaplan, J. M. (1998). Brainstem application of melanocortin receptor ligands produces long-lasting effects on feeding and body weight. *J. Neurosci.* **18,** 10128–10135.

Guan, X.-M., Hess, J. F., Yu, H., Hey, P. J., and van der Ploeg, J. H. T. (1997). Differential expression of mRNA for leptin receptor isoforms in the rat brain. *Mol. Cell. Endocrinol.* **133,** 1–7.

Gupta, G., Azam, M., and Baquer, N. Z. (1992). Effect of experimental diabetes on the catecholamine metabolism in rat brain. *J. Neurochem.* **58,** 95–100.

Gustafson, E. L., and Moore, R. Y. (1987). Noradrenaline and neuropeptide Y innervation of the rat hypothalamus are differentially affected by 6-hydroxydopamine. *Neurosci. Lett.* **83,** 53–58.

Hakansson, M.-L., Brown, H., Ghilardi, N., Skoda, R. C., and Meister, B. (1998). Leptin receptor immunoreactivity in chemically defined target neurons of the hypothalamus. *J. Neurosci.* **18,** 559–572.

Hall, A. C., Hoffmaster, R. M., Stern, E. L., Harrington, M. E., and Bickar, D. (1997). Suprachiasmatic nucleus neurons are glucose sensitive. *J. Biol. Rhythms* **12,** 388–400.

Hamilton-Wessler, M., Bergman, R. N., Halter, J. B., Watanabe, R. M., and Donovan, C. M. (1994). The role of liver glucosensors in the integrated sympathetic response induced by deep hypoglycemia in dogs. *Diabetes* **43,** 1052–1060.

Harfstrand, A., Fuxe, K., Agnati, L. F., Eneroth, P., Zini, L., Zoli, M., Andersson, K., vonEuler, G., Terenius, L., Mutt, V., and Goldstein, M. (1986). Studies on neuropeptide Y-catecholamine interactions in the hypothalamus and in the forebrain of the male rat. Relationship to neuroendocrine function. *Neurochem. Int.* **8,** 355–376.

Harrold, J. A., Williams, G., and Widdowson, P. S. (1999). Changes in hypothalamic agouti-related protein (AGRP), but not a-MSH or pro-opiomelanocortin concentrations in dietary-obese and food-restricted rats. *Biochem. Biophys. Res. Commun.* **258,** 574–577.

Harvey, J., and Ashford, M. L. (1998). Role of tyrosine phosphorylation in leptin activation of ATP-sensitive K^+ channels in the rat insulinoma cell line CRI-G1. *J. Physiol.* (*Lond.*) **510,** 47–61.

Harvey, J., Hardy, S. C., Irving, A. J., and Ashford, M. L. J. (2000b). Leptin activation of ATP-sensitive K^+ (K_{ATP}) channels in rat CRI-G1 insulinoma cells involves disruption of the actin cytoskeleton. *J. Physiol.* (*London*) **527,** 95–107.

Harvey, J., McKay, N. G., Walker, K. S., Van der, K. J., Downes, C. P., and Ashford, M. L. J. (2000a). Essential role of phosphoinositide 3-kinase in leptin-induced K(ATP) channel activation in the rat CRI-G1 insulinoma cell line. *J. Biol. Chem.* **275,** 4660–4669.

Hassel, B., and Brathe, A. (2000). Cerebral metabolism of lactate in vivo: Evidence for neuronal pyruvate carboxylation. *J. Cereb. Blood Flow Metab.* **20,** 327–336.

Hauger, R., Hulihan-Giblin, B., Angel, I., Luu, M. D., Janowsky, A., Skolnick, P., and Paul, S. M. (1986). Glucose regulates [^{3}H](+)-amphetamine binding and Na^+K^+ ATPase activity in the hypothalamus: A proposed mechanism for the glucostatic control of feeding and satiety. *Brain Res. Bull.* **16,** 281–288.

Hausser, M. A., de Weille, J. R., and Lazdunski, M. (1991). Activation by cromakalim of pre- and post-synaptic ATP-sensitive K^+ channels in substantia nigra. *Biochem. Biophys. Res. Commun.* **174,** 909–914.

Haynes, W. G., Morgan, D. A., Djalali, A., and Mark, A. L. (1998). Activation of melanocortin-4 receptors enhances thermogenic sympathetic nerve activity to brown adipose tissue. *Hypertension* **32,** 620A.

Heller, S. R., and Cryer, P. E. (1991). Reduced neuroendocrine and symptomatic responses to subsequent hypoglycemia after 1 episode of hypoglycemia in nondiabetic humans. *Diabetes* **40,** 223–226.

Hetherington, A. W., and Ranson, S. W. (1940). Hypothalamic lesions and adiposity in the rat. *Anat. Rec.* **78,** 149–172.

Hevener, A. L., Bergman, R. N., and Donovan, C. M. (1997). Novel glucosensor for hypoglycemic detection localized to the portal vein. *Diabetes* **46,** 1521–1525.

Hevener, A. L., Bergman, R. N., and Donovan, C. M. (2000). Portal vein afferents are critical for the sympathoadrenal response to hypoglycemia. *Diabetes* **49,** 8–12.

Horiuchi, A., Williams, K. R., Kurihara, T., Nairn, A. C., and Greengard, P. (1990). Purification and cDNA cloning of ARPP-16, a cAMP-regulated phosphoprotein enriched in basal ganglia, and of a related phosphoprotein, ARPP-19. *J. Biol. Chem.* **265,** 9476–9484.

Hu, Y., and Wilson, G. S. (1997). A temporary local energy pool coupled to neuronal activity: Fluctuations of extracellular lactate levels in rat brain monitored with rapid-response enzyme-based sensor. *J. Neurochem.* **69,** 1484–1490.

Hvidberg, A., Fanelli, C. G., Hershey, T., Terkamp, C., Craft, S., and Cryer, P. E. (1996). Impact of recent antecedent hypoglycemia on hypoglycemic cognitive dysfunction in nondiabetic humans. *Diabetes* **45,** 1030–1036.

Ibberson, M., Uldry, M., and Thorens, B. (2000). GLUTX1, a novel mammalian glucose transporter expressed in the central nervous system and insulin-sensitive tissues. *J. Biol. Chem.* **275,** 4607–4612.

Illes, P., Sevcik, J., Finta, E. P., Frolich, R., Nieber, K., and Norenberg, W. (1994). Modulation of locus coeruleus neurons by extra- and intracellular adenosine 5′-triphosphate. *Brain Res. Bull.* **35,** 513–519.

Inagaki, N., Gonoi, T., Clement IV, J. P., Namba, N., Inazawa, J., Gonzalez, G., Aguilar-Bryan, L., Seino, S., and Bryan, J. (1995). Reconstitution of IKATP: An inward rectifier subunit plus the sulfonylurea receptor. *Science* **270,** 1166–1170.

Inagaki, N., Gonoi, T., Clement, J. P., Wang, C. Z., Aguilar-Bryan, L., Bryan, J., and Seino, S. (1996). A family of sulfonylurea receptors determines the pharmacological properties of ATP-sensitive K^+ channels. *Neuron* **16,** 1101–1107.

Ishihara, H., Wang, H., Drewes, L. R., and Wollheim, C. B. (1999). Overexpression of monocarboxylate transporter and lactate dehydrogenase alters insulin secretory responses to pyruvate and lactate in beta cells. *J. Clin. Invest.* **104,** 1621–1629.

Jacob, R. J., Dziura, J., Blumberg, M., Morgen, J. P., and Sherwin, R. S. (1999). Effects of recurrent hypoglycemia on brainstem function in diabetic BB rats. *Diabetes* **48,** 141–145.

Jacobsson, G., and Meister, B. (1994). Hexokinase I messenger rnRNA in the rat central nervous system. *Mol. Cell. Neurosci.* **5,** 658–677.

Jansen, A. S. P., Wessendorf, M. W., and Loewy, A. D. (1995). Transneuronal labeling of CNS neuropeptide and monoamine neurons after pseudorabies virus injections into stellate ganglion. *Brain Res.* **683,** 1–24.

Jetton, T. L., Liang, Y., Pettepher, C. C., Zimmerman, E. C., Cox, F. G., Horvath, K., Matschinsky, F. M., and Magnuson, M. A. (1994). Analysis of upstream glucokinase promoter activity in transgenic mice and identification of glucokinase in rare neuroendocrine cells in the brain and gut. *J. Biol. Chem.* **269,** 3641–3654.

Kai, Y., Oomura, Y., and Shimizu, N. (1988). Responses of rat lateral hypothalamic neuron activity to dorsal raphe nuclei stimulation. *J. Neurophysiol.* **60,** 524–535.

Karadi, Z., Oomura, Y., Nishino, H., Scott, T. R., Lenard, L., and Aou, S. (1990). Complex attributes of lateral hypothalamic neurons in the regulation of feeding of alert rhesus monkeys. *Brain Res. Bull.* **25,** 933–939.

Karadi, Z., Oomura, Y., Nishino, H., Scott, T. R., Lenard, L., and Aou, S. (1992). Responses of lateral hypothalamic glucose-sensitive and glucose-insensitive neurons to chemical stimuli in behaving rhesus monkeys. *J. Neurophysiol.* **67,** 389–400.

Karschin, A., Brockhaus, J., and Ballanyi, K. (1998). KATP channel formation by the sulphonylurea receptors SUR1 with Kir6.2 subunits in rat dorsal vagal neurons in situ. *J. Physiol.* **509,** 339–346.

Karschin, C., Ecke, C., Ashcroft, F. M., and Karschin, A. (1997). Overlapping distribution of K_{ATP} channel-forming unit Kir6.2 subunit and the sulfonylurea receptor SUR1 in rodent brain. *FEBS Lett.* **401,** 59–64.

Kelley, A. E., Bless, E. P., and Swanson, C. J. (1996). Investigation of the effects of opiate antagonists infused into the nucleus accumbens on feeding and sucrose drinking in rats. *J. Pharmacol. Exp. Ther.* **278,** 1499–1507.

Kelly, J., Rothstein, J., and Grossman, S. P. (1979). GABA and hypothalamic feeding systems. I. Topographic analysis of the effects of microinjections of muscimol. *Physiol. Behav.* **23,** 1123–1134.

Kennedy, G. C. (1953). The role of depot fat in the hypothalamic control of food intake in the rat. *Proc. R. Soc. Lond. B Biol. Sci.* **611,** 221–235.

Kennedy, H. J., Pouli, A. E., Ainscow, E. K., Jouaville, L. S., Rizzuto, R., and Rutter, G. A. (1999). Glucose generates sub-plasma membrane ATP microdomains in single islet beta-cells. Potential role for strategically located mitochondria. *J. Biol. Chem.* **274,** 13281–13291.

Kiessling, M., Auer, R. N., Kleihues, P., and Siesjo, B. K. (1986). Cerebral protein synthesis during long-term recovery from severe hypoglycemia. *J. Cereb. Blood Flow Metab.* **6,** 42–51.

Kim, Y. B., Uotani, S., Pierroz, D. D., Flier, J. S., and Kahn, B. B. (2000). In vivo administration of leptin activates signal transduction directly in insulin-sensitive tissues: Overlapping but distinct pathways from insulin. *Endocrinology* **141,** 2328–2339.

King, P., Kong, M. F., Parkin, H., Macdonald, I. A., Barber, C., and Tattersall, R. B. (1998). Intravenous lactate prevents cerebral dysfunction during hypoglycaemia in insulin-dependent diabetes mellitus. *Clin. Sci.* **94,** 157–163.

Kistler-Heer, V., Lauber, M. E., and Lichtensteiger, W. (1998). Different developmental patterns of melanocortin MC3 and MC4 receptor mRNA: predominance of Mc4 in fetal rat nervous system. *J. Neuroendocrinol.* **10,** 133–146.

Klein, S., and Wolfe, R. R. (1992). Carbohydrate restriction regulates the adaptive response to fasting. *Am. J. Phsyiol.* **262,** t–6.

Kobayashi, M., Nikami, H., Morimatsu, M., and Saito, M. (1996). Expression and localization of insulin-regulatable glucose transporter (GLUT4) in rat brain. *Neurosci. Lett.* **213,** 103–106.

Koehler-Stec, E. M., Simpson, I. A., Vannucci, S. J., Landschulz, K. T., and Landschulz, W. H. (1998). Monocarboxylate transporter expression in mouse brain. *Am. J. Physiol.* **275,** E516–E524.

Kow, L.-M., and Pfaff, D. W. (1986). Vasopressin excites ventromedial hypothalamic glucose-responsive neurons *in vitro*. *Physiol. Behav.* **37,** 153–158.

Kow, L.-M., and Pfaff, D. W. (1989). Responses of hypothalamic paraventricular neurons in vitro to norepinephrine and other feeding-relevant agents. *Physiol. Behav.* **46,** 265–271.

Kumagai, A. K., Kang, Y. -S., Boado, R. J., and Pardridge, W. M. (1995). Upregulation of blood-brain barrier GLUT1 glucose transporter protein and mRNA in experimental chronic hypoglycemia. *Diabetes* **44,** 1399–1404.

Kyrkouli, S. E., Stanley, B. G., Hutchinson, R., Seirafi, R. D., and Leibowitz, S. F. (1990). Peptide-amine interactions in the hypothalamic paraventricular nucleus: Analysis of galanin and neuropeptide Y in relation to feeding. *Brain Res.* **521,** 185–191.

Larue-Achagiotis, C., and Le Magnen, J. (1985). Feeding rate and responses to food deprivation as a function of fasting-induced hypoglycemia. *Behav. Neurosci.* **99,** 1176–1180.

Laughton, J. D., Charnay, Y., Belloir, B., Pellerin, L., Magistretti, P. J., and Bouras, C. (2000). Differential messenger RNA distribution of lactate dehydrogenase LDH-1 and LDH-5 isoforms in the rat brain. *Neuroscience* **96,** 619–625.

Lauritzen, I., de Weille, J. R., and Lazdunski, M. (1997). The potassium channel opener (−)–cromakalim prevents glutamate-induced cell death in hippocampal neurons. *J. Neurochem.* **69,** 1570–1579.

Lee, J., Bruce-Keller, A. J., Kruman, Y., Chan, S. L., and Mattson, M. P. (1999b). 2-deoxy-D-glucose protects hippocampal neurons against excitotoxic and oxidative injury: evidence for the involvement of stress proteins. *J. Neurosci. Res.* **57,** 48–61.

Lee, K., Brownhill, V., and Richardson, P. J. (1997). Antidiabetic sulphonylureas stimulate acetylcholine release from striatal cholinergic interneurons through inhibition of K_{ATP} channel activity. *J. Neurochem.* **69,** 1774–1776.

Lee, K., Dixon, A. K., Freeman, T. C., and Richardson, P. J. (1998). Identification of an ATP-sensitive potassium channel current in rat striatal cholinergic interneurones. *J. Physiol. (London)* **510,** 441–453.

Lee, K., Dixon, A. K., Richardson, P. J., and Pinnock, R. D. (1999a). Glucose-receptive neurones in the rat ventromedial hypothalamus express K_{ATP} channels composed of Kir6.1 and SUR 1 subunits.. *J. Physiol. (London)* **515,** 439–452.

Lee, K., Dixon, A. K., Rowe, I. C., Ashford, M. L., and Richardson, P. J. (1995). Direct demonstration of sulphonylurea-sensitive KATP channels on nerve terminals of the rat motor cortex. *Br. J. Pharmacol.* **115,** 385–387.

Lee, K., Dixon, A. K., Rowe, I. C. M., Ashford, M. L. J., and Richardson, P. J. (1996a). The high-affinity sulfonylurea receptor regulates K_{ATP} channels in nerve terminals of the rat motor cortex. *J. Neurochem.* **66,** 2562–2571.

Lee, K., Khan, R. N., Rowe, I. C., Ozanne, S. E., Hall, A. C., Papadakis, E., Hales, C. N., and Ashford, M. L. J. (1996b). Ciclazindol inhibits ATP-sensitive K^+ channels and stimulates insulin secretion in CR1-G1 insulin-secreting cells. *Mol. Pharmacol.* **49,** 715–720.

Leibowitz, S. F., Jhanwar-Uniyal, M., Dvorkin, B., and Makman, M. H. (1982). Distribution of α-adrenergic, β-adrenergic and dopminergic receptors in discrete hypothalamic areas of rat. *Brain Res.* **223,** 97–114.

Leloup, C., Arluison, M., Kassis, N., Lepetit, N., Cartier, N., Ferre, P., and Penicaud, L. (1996). Discrete brain areas express the insulin-responsive glucose transporter GLUT4. *Brain Res. Mol. Brain Res.* 45–53.

Leloup, C., Arluison, M., Lepetit, N., Cartier, N., Marfaing-Jallat, P., Ferre, P., and Penicaud, L. (1994). Glucose transporter 2 (GLUT 2): Expression in specific brain nuclei. *Brain Res.* **638,** 221–226.

Levin, B. E. (1991). Glucose increases rat plasma norepinephrine levels by direct action on the brain. *Am. J. Physiol. Regul. Integr. Comp. Physiol.* **261,** R1351–R1357.

Levin, B. E. (1992). Intracarotid glucose-induced norepinephrine response and the development of diet-induced obesity. *Int J Obesity* **16,** 451–457.

Levin, B. E. (1999). Arcuate NPY neurons and energy homeostasis in diet-induced obese and resistant rats. *Am. J. Physiol.* **276,** R382–R387.

Levin, B. E. (2000). Glucose-regulated dopamine release from substantia nigra neurons. *Brain Res.* **874,** 158–164.

Levin, B. E., Brown, K. L., and Dunn-Meynell, A. A. (1996). Differential effects of diet and obesity on high and low affinity sulfonylurea binding sites in the rat brain. *Brain Res.* **739,** 293–300.

Levin, B. E., and Dunn-Meynell, A. (1998). Effect of streptozotocin-induced diabetes on rat brain sulfonylurea biding sites. *Brain Res. Bull.* **46,** 513–518.

Levin, B. E., and Dunn-Meynell, A. A. (1997a). Dysregulation of arcuate nucleus preproneuropeptide Y mRNA in diet-induced obese rats. *Am. J. Physiol.* **272,** R1365–R1370.

Levin, B. E., and Dunn-Meynell, A. A. (1997b). In vivo and in vitro regulation of [^{3}H]glyburide binding to brain sulfonylurea receptors in obesity-prone and resistant rats by glucose. *Brain Res.* **776,** 146–153.

Levin, B. E., Dunn-Meynell, A. A., and Routh, V. H. (1999b). Brain glucose sensing and body energy homeostasis: Role in obesity and diabetes. *Am. J. Physiol.* **276,** R1223–R1231.

Levin, B. E., Govek, E. K., and Dunn-Meynell, A. A. (1998). Reduced glucose-induced neuronal activation in the hypothalamus of diet-induced obese rats. *Brain Res.* **808,** 317–319.

Levin, B. E., and Planas, B. (1993). Defective glucoregulation of brain α_2-adrenoceptors in obesity-prone rats. *Am. J. Physiol.* **264,** R305–R311.

Levin, B. E., Routh, V. H., and Dunn-Meynell, A. A. (1999a). Glucosensing neurons in the central nervous system. *In* "Neural and Metabolic Control of Macronutrient Intake" (H.-R. Berthoud and R. J. Seeley, eds.), pp. 325–337. CRC Press, Boca Raton, FL.

Levitt, P., and Moore, R. Y. (1979). Origin and organization of brainstem catecholamine innervation in the rat. *J. Comp. Neurol.* **186,** 505–528.

Liss, B., Bruns, R., and Roeper, J. (1999). Alternative sulfonylurea receptor expression defines metabolic sensitivity of K-ATP channels in dopaminergic midbrain neurons. *EMBO J.* **18,** 833–846.

Liu, F., Dong, Q., Myers, A. M., and Fromm, H. J. (1991). Expression of human brain hexokinase in *Escherichia coli:* Purification and characterization of the expressed enzyme. *Biochem. Biophys. Res. Commun.* **177,** 305–311.

Liu, G. X., Hanley, P. J., Ray, J., and Daut, J. (2001). Long-chain acyl-coenzyme A esters and fatty acids directly link metabolism to K(ATP) channels in the heart. *Circ. Res.* **88,** 918–924.

Liu, M., Seino, S., and Kirchgessner, A. L. (1999). Identification and characterization of glucoresponsive neurons in the enteric nervous system. *J. Neurosci.* **19,** 10305–10317.

Liu, Q., Bai, C., Chen, F., Wang, R., MacDonald, T., Gu, M., Zhang, Q., Morsy, M. A., and Caskey, C. T. (1998). Uncoupling protein-3: A muscle-specific gene upregulated by leptin in ob/ob mice. *Gene* **207,** 1–7.

Liu, Y., Gao, J.-H., and Fox, P. T. (2000). The temporal response of the brain after eating revealed by functional MRI. *Nature* **405,** 1058–1062.

Louis-Sylvestre, J., and Le Magnen, J. (1980). A fall in blood glucose level precedes meal onset in free-feeding rats. *Neurosci. Biobehav. Rev.* **4,** Suppl. 1, 13–15.

Lu, H., Buison, A., Jen, K. C., and Dunbar, J. C. (2000). Leptin resistance in obesity is characterized by decreased sensitivity to proopiomelanocortin products. *Peptides* **21,** 1479–1485.

Lynch, R. M., Tompkins, L. S., Brooks, H. L., Dunn-Meynell, A. A., and Levin, B. E. (2000). Localization of glucokinase gene expression in the rat brain. *Diabetes* **49,** 693–700.

Maechler, P., and Wollheim, C. B. (2000). Mitochondrial signals in glucose-stimulated insulin secretion in the beta cell. *J. Physiol.* **529**[1], 49–56.

Maekawa, F., Toyoda, Y., Torii, N., Miwa, I., Thompson, R. C., Foster, D. L., Tsukahara, S., Tsukamura, H., and Maeda, K. (2000). Localization of glucokinase-like immunoreactivity in the rat lower brain stem: For possible location of brain glucose-sensing mechanisms. *Endocrinolology* **141,** 375–384.

Magistretti, P. J., Sorg, O., Yu, N., Martin, J. -L., and Pellerin, L. (1993). Neurotransmitters regulate energy metabolism in astrocytes: Implications for the metabolic trafficking between neural cells. *Dev. Neurosci.* **15,** 306–312.

Mahankali, S., Liu, Y., Pu, Y., Wang, J., Chen, C. W., Fox, P. T., and Gao, J. H. (2000). In vivo fMRI demonstration of hypothalamic function following intraperitoneal glucose administration in a rat model. *Magn. Reson. Med.* **43,** 155–159.

Maher, F., Davies-Hill, T. M., and Simpson, I. A. (1996). Substrate specificity and kinetic parameters of glut3 in rat cerebellar granule neurons. *Biochem. J.* **315,** 827–831.

Maneuf, Y. P., Duty, S., Hille, C. J., Crossman, A. R., and Brotchie, J. M. (1996). Modulation of GABA transmission by diazoxide and cromakalim in the globus pallidus: Implications for the treatment of Parkinson's disease. *Exp. Neurol.* **139,** 12–16.

Manning, C. A., Stone, W. S., Korol, D. L., and Gold, P. E. (1998). Glucose enhancement of 24-h memory retrieval in healthy elderly humans. *Behav. Br. Res.* **93,** 71–76.

Maran, A., Cranston, I., Lomas, J., Macdonald, I., and Amiel, S. A. (1994). Protection by lactate of cerebral function during hypoglycaemia. *Lancet* **343,** 16–20.

Maran, A., Lomas, J., Macdonald, I. A., and Amiel, S. A. (1995). Lack of preservation of higher brain function during hypoglycaemia in patients with intensively-treated IDDM. *Diabetologia* **38,** 1412–1418.

Marks, J. L., King, M. G., and Baskin, D. G. (1991). Localization of insulin and type 1 IGF receptors in rat brain by in vitro autoradiography and in situ hybridization. *Adv. Exp. Med. Biol.* **293,** 459–70.

Marks, J. L., Porte, D., Jr., Stahl, W. L., and Baskin, D. G. (1990). Localization of insulin receptor mRNA in rat brain by in situ hybridization. *Endocrinology* **127,** 3234–3236.

Matschinsky, F. M. (1996). A lesson in metabolic regulation inspired by the glucokinase glucose sensor paradigm. *Diabetes* **45,** 223–241.

Matsuda, M., Liu, Y., Mahankali, S., Pu, Y., Mahankali, A., Wang, J., DeFronzo, R. A., Fox, P. T., and Gao, J. -H. (1999). Altered hypothalamic function in response to glucose ingestion in obese humans. *Diabetes* **48,** 1801–1806.

Matsuhisa, M., Morishima, T., Nakahara, I., Tomita, T., Shiba, Y., Kubota, M., Wada, M., Kanda, T., Kawamori, R., and Yamasaki, Y. (1997). Augmentation of hepatic glucose uptake by a positive glucose gradient between heptoportal and central nervous system. *Diabetes* **46,** 1101–1105.

Maus, M., Marin, P., Israel, M., Glowinski, J., and Premont, J. (1999). Pyruvate and lactate protect striatal neurons against *N*-methyl-D-aspartate-induced neurotoxicity. *Eur. J. Neurosci.* **11,** 3215–3224.

Mayer, J. (1953). Glucostatic mechanism of regulation of food intake. *N. Engl. J. Med.* **249,** 13–16.

McBride, R. L., and Sutin, J. (1984). Noradrenergic hyperinnervation of the trigeminal sensory nuclei. *Brain Res.* **324,** 211–221.

McCall, A. L., Fixman, L. B., Fleming, N., Tornheim, K., Chick, W., and Ruderman, N. B. (1986). Chronic hypoglycemia increases brain glucose transport. *Am. J. Physiol.* **251,** E442–E447.

McKenna, M. C., Tildon, J. T., Stevenson, J. H., Hopkins, I. B., Huang, X., and Couto, R. (1998). Lactate transport by cortical synaptosomes from adult rat brain: characterization of kinetics and inhibitor specificity *Dev. Neurosci.* **20,** 300–309.

McLean, S., Rothman, R. B., and Herkenham, M. (1986). Autoradiographic localization of mu- and delta-opiate receptors in the forebrain of the rat. *Brain Res.* **378,** 49–60.

McNay, E. C., and Gold, P. E. (1999). Extracellular glucose concentrations in the rat hippocampus measured by zero-net-flux: effects of microdialysis flow rate, strain, and age. *J. Neurochem.* **72,** 785–790.

McNay, E. C., and Gold, P. E. (2001). Age-related differences in hippocampal extracellular fluid glucose concentration during behavioral testing and following systemic glucose administration. *J. Gerontol.* **56A,** B66–B71.

McNay, E. C., McCarty, R. C., and Gold, P. E. (2001). Fluctuations in brain glucose concentration during behavioral testing: dissociations between brain areas and between brain and blood. *Neurobiol. Learn. Memory* **75,** 325–337.

Medina, M. A., Jones, D. J., Stavinoha, W. B., and Ross, D. H. (1975). The levels of labile intermediary metabolites in mouse brain following rapid tissue fixation with microwave irradiation. *J. Neurochem.* **24,** 223–227.

Medina, M. A., and Stavinoha, W. B. (1977). Labile intermediary metabolites in rat brain determined after tissue inactivation with microwave irradiation. *Brain Res.* **132,** 149–152.

Melanson, K. J., Westerterp-Plantenga, M. S., Saris, W. H., Smith, F. J., and Campfield, L. A. (1999). Blood glucose patterns and appetite in time-blinded humans: carbohydrate versus fat. *Am. J. Physiol.* **277,** R337–R345.

Miki, T., Liss, B., Minami, K., Shiuchi, T., Saraya, A., Kashima, Y., Horiuchi, M., Ashcroft, F. M., Minokoshi, Y., Roeper, J., and Seino, S. (2001). ATP-sensitive K^+ channels in the hypothalamus are essential for the maintenance of glucose homeostasis. *Nat. Neurosci.* **4,** 507–512.

Miki, T., Nagashima, K., Tashiro, F., Kotake, K., Yoshitomi, H., Tamamoto, A., Gonoi, T., Iwanaga, T., Miyazaki, J., and Seino, S. (1998). Defective insulin secretion and enhanced insulin action in KATP channel-deficient mice. *Proc. Natl. Acad. Sci. USA* **95,** 10402–10406.

Miki, T., Tashiro, F., Iwanaga, T., Nagashima, K., Yoshitomi, H., Aihara, H., Nitta, Y., Gonoi, T., Inagaki, N., Miyazaki, J., and Seino, S. (1997). Abnormalities of pancreatic islets by targeted expression of a dominant-negative KATP channel. *Proc. Natl. Acad. Sci. USA* **94,** 11969–11973.

Milusheva, E., Doda, M., Pasztor, E., Lajtha, A., Sershen, H., and Vizi, E. S. (1992). Regulatory interactions among axon terminals affecting the release of different transmitters from rat striatal slices under hypoxic and hypoglycemic conditions. *J. Neurochem.* **59,** 946–952.

Mitrakou, A., Ryan, C., Veneman, T., Mokan, M., Jenssen, T., Kiss, I., Durrant, J., Cryer, P., and Gerich, J. (1991). Hierarchy of glycemic thresholds for counterregulatory hormone secretion, symptoms, and cerebral dysfunction. *Am. J. Physiol.* **260,** E67–E74.

Miyahara, S., and Oomura, Y. (1982). Inhibitory action of the ventral noradrenergic bundle on the lateral hypothalamic neurons through alpha-noradrenergic mechanisms in the rat. *Brain Res.* **234,** 459–463.

Mizuno, Y., and Oomura, Y. (1984). Glucose responding neurons in the nucleus tractus solitarius of the rat, in vitro study. *Brain Res.* **307,** 109–116.

Mourre, C., Widmann, C., and Lazdunski, M. (1990). Sulfonylurea binding sites associated with ATP-regulated K^+ channels in the central nervous system: autoradiographic analysis of their distribution and ontogenesis, and of their localization in mutant mice cerebellum. *Brain Res.* **519,** 29–43.

Muroya, S., Yada, T., Shioda, S., and Takigawa, M. (1999). Glucose-sensitive neurons in the rat arcuate nucleus contain neuropeptide Y. *Neurosci. Lett.* **264,** 113–116.

Myers, M. G., Jr., Grammer, T. C., Wang, L. M., Sun, X. J., Pierce, J. H., Blenis, J., and White, M. F. (1994). Insulin receptor substrate-1 mediates phosphatidylinositol 3′-kinase and p70S6k signaling during insulin, insulin-like growth factor-1, and interleukin-4 stimulation. *J. Biol. Chem.* **269,** 28783–28789.

Myers, M. G., Jr., and White, M. F. (1996). Insulin signal transduction and the IRS proteins. *Ann. Rev. Pharmacol. Toxicol.* **36,** 615–658.

Nagai, K., Niijima, A., Nagai, N., Hiroshi, H., Chun, S.-J., Shimizu, K., and Nakagawa, H. (1996). Bilateral lesions of the hypothalamic suprachiasmatic nucleus eliminated sympathetic response to intracranial injection of 2-deoxy-D-glucose and VIP rescued this response. *Brain Res. Bull.* **39,** 293–297.

Nakano, Y., Oomura, Y., Lenard, L., Nishino, H., Aou, S., Yamamoto, T., and Aoyagi, K. (1986). Feeding-related activity of glucose- and morphine-sensitive neurons in the monkey amygdala. *Brain Res.* **399,** 167–172.

Navarro, M., Rodriquez, D. F., Alvarez, E., Chowen, J. A., Zueco, J. A., Gomez, R., Eng, J., and Blazquez, E. (1996). Colocalization of glucagon-like peptide-1 (GLP-1) receptors, glucose transporter GLUT-2, and glucokinase mRNAs in rat hypothalamic cells: Evidence for a role of GLP-1 receptor agonists as an inhibitory signal for food and water intake. *J. Neurochem.* **67,** 1982–1991.

Nemoto, E. M., Hoff, J. T., and Severinghaus, J. W. (1974). Lactate uptake and metabolism by brain during hyperlactatemia and hypoglycemia. *Stroke* **5,** 48–53.

Newgard, C. B., and McGarry, J. D. (1995). Metabolic coupling factors in pancreatic beta-cell signal transduction. *Ann. Rev. Biochem.* **64,** 689–719.

Niijima, A. (1975). Studies on the nervous regulatory mechanism of blood sugar levels. *Pharmacol. Biochem. Behav.* **3,** 139–143.

Nishino, H., Oomura, Y., Karadi, Z., Lenard, L., Kai, Y., Fukuda, A., Ito, C., Min, B. I., and Salaman, C. P. (1988). Internal and external information processing by lateral hypothalamic glucose-sensitive and insensitive neurons during bar press feeding in the monkey. *Brain Res. Bull.* **20,** 839–845.

Oomura, Y., Kimura, K., Ooyama, H., Maeo, T., Iki, M., and Kuniyoshi, N. (1964). Reciprocal activities of the ventromedial and lateral hypothalamic area of cats. *Science* **143,** 484–485.

Oomura, Y., Nakamura, T., and Manchanda, S. K. (1975). Excitatory and inhibitory effects of globus pallidus and substantia nigra on the lateral hypothalamic activity in the rat. *Pharmacol. Biochem. Behav.* **3,** 23–36.

Oomura, Y., Ono, T., Ooyama, H., and Wayner, M. J. (1969). Glucose and osmosensitive neurons of the rat hypothalamus. *Nature* **222,** 282–284.

Oomura, Y., Ooyama, H., Sugimori, M., Nakamura, T., and Yamada, Y. (1974). Glucose inhibition of the glucose-sensitive neurone in the rat lateral hypothalamus. *Nature* **247,** 284–286.

Orsini, J. C., Armstrong, D. L., and Wayner, M. J. (1992). Responses of lateral hypothalamic neurons recorded in vitro to moderate changes in glucose concentration. *Brain Res. Bull.* **29,** 503–505.

Orsini, J. C., Himmi, T., Wiser, A. K., and Perrin, J. (1990). Local versus indirect action of glucose on the lateral hypothalamic neurons sensitive to glycemic level. *Brain Res. Bull.* **25,** 49–53.

Orsini, J. C., Wiser, A. K., Himmi, T., Boyer, A., and Perrin, J. (1991). Sensitivity of lateral hypothalamic neurons to glycemic level: possible involvement of an indirect adrenergic mechanism. *Brain Res. Bull.* **26,** 473–478.

Ovalle, F., Fanelli, C. G., Paramore, D. S., Hershey, T., Craft, S., and Cryer, P. E. (1998). Brief twice-weekly episodes of hypoglycemia reduce detection of clinical hypoglycemia in type 1 diabetes mellitus. *Diabetes* **47,** 1472–1479.

Paschen, W., Siesjo, B. K., Ingvar, M., and Hossmann, K. A. (1986). Regional differences in brain glucose content in graded hypoglycemia. *Neurochem. Pathol.* **5,** 131–142.

Patrick, A. W., and Campbell, I. W. (1990). Fatal hypoglycaemia in insulin-treated diabetes mellitus: clinical features and neuropathological changes. *Diabetic Med.* **7,** 349–354.

Pellerin, L., and Magistretti, P. J. (1994). Glutamate uptake into astrocytes stimulates aerobic glycolysis: a mechanism coupling neuronal activity to glucose utilization. *Proc. Natl. Acad. Sci. USA* **91,** 10625–10629.

Pellerin, L., Pellegri, G., Bittar, P. G., Charnay, Y., Bouras, C., Martin, J. L., Stella, N., and Magistretti, P. J. (1998a). Evidence supporting the existence of an activity-dependent astrocyte-neuron lactate shuttle. *Dev. Neurosci.* **20,** 291–299.

Pellerin, L., Pellegri, G., Martin, J. L., and Magistretti, P. J. (1998b). Expression of monocarboxylate transporter mRNAs in mouse brain: support for a distinct role of lactate as an energy substrate for the neonatal vs. adult brain. *Proc. Natl. Acad. Sci. USA* **95,** 3990–3995.

Perros, P., Deary, I. J., Sellar, R. J., Best, J. J. K., and Frier, B. M. (1997). Brain abnormalities demonstrated by magnetic resonance imaging in adult IDDM patients with and without a history of recurrent severe hypoglycemia. *Diabetes Care* **20,** 1013–1018.

Peyrollier, K., Heron, L., Virsolvy-Vergine, A., Le Cam, A., and Bataille, D. (1996). Alpha endosulfine is a novel, molecule, structurally related to a family of phosphoproteins. *Biochem. Biophys. Res. Commun.* **223,** 583–586.

Powell, A. M., Sherwin, R. S., and Shulman, G. I. (1993). Impaired hormonal responses to hypoglycemia in spontaneously diabetic and recurrently hypoglycemic rats. Reversibility and stimulus specificity of the deficits. *J. Clin. Invest.* **92,** 2667–2674.

Powers, W. J., Boyle, P. J., Hirsch, I. B., and Cryer, P. E. (1993). Unaltered cerebral blood flow during hypoglycemic activation of the sympathochromaffin system in humans. *Am. J. Physiol.* **265,** R883–R887.

Ragozzino, M. E., Pal, S. N., Unick, K., Stefani, M. R., and Gold, P. E. (1998). Modulation of hippocampal acetylcholine release and spontaneous alternation scores by intrahippocampal glucose injections. *J. Neurosci.* **18,** 1595–1601.

Ransom, B. R., and Fern, R. (1997). Does astrocytic glycogen benefit axon function and survival in CNS white matter during glucose deprivation? *Glia* **21,** 134–141.

Rayner, D. V., Thomas, M. E. A., and Trayhurn, P. (1994). Glucose transporters (GLUTs 1–4) and their mRNAs in regions of the rat brain: insulin-sensitive transporter expression in the cerebellum. *Can. J. Physiol. Pharmacol.* **72,** 476–479.

Ritter, R. C., and Slusser, P. G. (1980). 5-Thio-D-glucose causes increased feeding and hyperglycemia in the rat. *Am. J. Physiol.* **238,** E141–E144.

Ritter, R. C., Slusser, P. G., and Stone, S. (1981). Glucoreceptors controlling feeding and blood glucose: location in the hindbrain. *Science* **213,** 451–452.

Ritter, S., Bugarith, K., and Dinh, T. T. (2001). Immunotoxic destruction of distinct catecholamine subgroups produces selective impairment of glucoregulatory responses and neuronal activation. *J. Comp. Neurol.* **432,** 197–216.

Ritter, S., Dinh, T. T., and Zhang, Y. (2000). Localization of hindbrain glucoreceptive sites controlling food intake and blood glucose. *Brain Res.* **856,** 37–47.

Robinson, A. M., Parkin, H. M., Macdonald, I. A., and Tattersall, R. B. (1995). Antecedent hypoglycemia in non-diabetic subjects reduces the adrenaline response for 6 days but does not affect the catecholamine response to other stimuli. *Clin. Sci.* **89,** 359–366.

Roeper, J., Hainsworth, A. H., and Ashcroft, F. M. (1990). Tolbutamide reverses membrane hyperpolarisation induced by activation of D_2 receptors and $GABA_B$ receptors in isolated substantia nigra neurones. *Pflugers Arch.* **416,** 473–475.

Roncero, I., Alvarez, E., Vazques, P., and Blazquez, E. (2000). Functional glucokinase isoforms are expressed in rat brain. *J. Neurochem.* **74,** 1848–1857.

Roper, J., and Ashcroft, F. M. (1995). Metabolic inhibition and low internal ATP activate K-atp channels in rat dopaminergic substantia nigra neurones. *Pflugers Arch.* **430,** 44–54.

Routh, V. H., Levin, B. E., and McArdle, J. J. (1998). Defective ATP-sensitive K^+(K_{ATP}) channel in ventromedial hypothalamic nucleus (VMN) of obesity-prone (DIO) rats. *FASEB J.* **12,** A864.

Routh, V. H., McArdle, J. J., and Levin, B. E. (1997). Phosphorylation modulates the activity of the ATP-sensitive K^+ channel in the ventromedial hypothalamic nucleus. *Brain Res.* **778,** 107–119.

Sanders, N. M., and Ritter, S. (2000). Repeated 2-deoxy-D-glucose-induced glucoprivation attenuates Fos expression and glucoregulatory responses during subsequent glucoprivation. *Diabetes* **49,** 1865–1874.

Scheurink, A., and Ritter, S. (1993). Sympathoadrenal responses to glucoprivation and lipoprivation in rats. *Physiol. Behav.* **53,** 995–1000.

Schurr, A., Miller, J. J., Payne, R. S., and Rigor, B. M. (1999). An increase in lactate output by brain tissue serves to meet the energy needs of glutamate-activated neurons. *J. Neurosci.* **19,** 34–39.

Schwanstecher, C., and Panten, U. (1993). Tolbutamide- and diazoxide-sensitive K^+ channel in neurons of substantia nigra pars reticulata. *Naunyn-Schmiedebergs Arch. Pharmacol.* **348,** 113–117.

Schwartz, M. W., Seeley, R. J., Woods, S. C., Weigle, D. S., Campfield, L. A., Burn, P., and Baskin, D. G. (1997). Leptin increases hypothalamic pro-opiomelanocortin mRNA expression in the rostral arcuate nucleus. *Diabetes* **46,** 2119–2123.

Segel, S. A., Fanelli, C. G., Dence, C. S., Markham, J., Videen, T. O., Paramore, D. S., Powers, W. J., and Cryer, P. E. (2001). Blood-to-brain glucose transport, cerebral glucose metabolism, and cerebral blood flow are not increased after hypoglycemia. *Diabetes* **50,** 1911–1917.

Seino, S., Inagaki, N., Namba, N., Wang, C. H., Kotake, K., Nagashima, K., Miki, T., Aguilar-Bryan, L., Bryan, J., and Gonoi, T. (1997). Molecular basis of functional diversity of ATP-sensitive K^+ channels. *J. J. Physiol.* **47,** Suppl. 1, S3–4.

Sellers, A. J., Boden, P. R., and Ashford, M. L. J. (1992). Lack of effect of potassium channel openers on ATP-modulated potassium channels recorded from rat ventromedial hypothalamic neurones. *Br. J. Pharmacol.* **107,** 1068–1074.

Shimizu, N., Oomura, Y., Novin, D., Grijava, C. V., and Cooper, P. H. (1983). Functional correlations between lateral hypothalamic glucose-sensitive neurons and hepatic portal glucose-sensitive units in rat. *Brain Res.* **265,** 49–54.

Shoji, S. (1992). Glucose regulation of synaptic transmission in the dorsolateral septal nucleus of the rat. *Synapse* **12,** 322–332.

Sikdar, S. K., and Oomura, Y. (1985). Selective inhibition of glucose-sensitive neurons in rat lateral hypothalamus by noxious stimuli and morphine. *J. Neurophysiol.* **53,** 17–31.

Silver, I. A., and Erecinska, M. (1994). Extracellular glucose concentrations in mammalian brain: continuous monitoring of changes during increased neuronal activity and upon limitation in oxygen supply in normo-, hypo-, and hyperglycemic animals. *J. Neurosci.* **14,** 5068–5076.

Silver, I. A., and Erecinska, M. (1998). Glucose-induced intracellular ion changes in sugar-sensitive hypothalamic neurons. *J. Neurophysiol.* **79,** 1733–1745.

Simpson, I. A., Appel, N. M., Hokari, M., Oki, J., Holman, G. D., Maher, F., Koehler-Stec, E. M., Vannucci, S. J., and Smith, Q. R. (1999). Blood-brain barrier glucose transporter: effects of hypo- and hyperglycemia revisited. *J. Neurochem.* **72,** 238–247.

Simpson, I. A., Vannucci, S. J., DeJoseph, M. R., and Hawkins, R. A. (2001). Glucose transporter asymmetries in the bovine blood-brain barrier. *J. Biol. Chem.* **276,** 12725–12729.

Smith, F. J., and Campfield, L. A. (1993). Meal initiation occurs after experimental induction of transient declines in blood glucose. *Am. J. Physiol.* **265,** R1423–R1429.

Smith, F. J., Driscoll, D. W., and Campfield, L. A. (1988). Short term effects of fructose on blood glucose dynamics and meal initiation. *Physiol. Behav.* **44,** 625–631.

Smith, G. P., Gibbs, J., Strohmayer, A. J., and Stokes, P. E. (1972). Threshold doses of 2-deoxy-D-glucose for hyperglycemia and feeding in rats and monkeys. *Am. J. Physiol.* **222,** 77–81.

Sokoloff, L., Reivich, M., Kennedy, C., DesRosiers, M. H., Patlak, C. S., Pettigrew, O., Sakaruda, O., and Shinohara, M. (1977). The [^{14}C]deoxyglucose method for the measurement of local cerebral glucose utilization: Theory, procedure, and normal values in the conscious and anesthetized albino rat. *J. Neurochem.* **23,** 897–916.

Spanswick, D., Smith, M. A., Groppi, V. E., Logan, S. D., and Ashford, M. L. (1997). Leptin inhibits hypothalamic neurons by activation of ATP-sensitive potassium channels. *Nature* **390,** 521–525.

Spanswick, D., Smith, M. A., Mirshamsi, S., Routh, V. H., and Ashford, M. L. (2000). Insulin activates ATP-sensitive K+ channels in hypothalamic neurons of lean, but not obese rats. *Nat. Neurosci.* **3,** 757–758.

Specter, S. E., Horwitz, B. A., and Beverly, J. L. (1996). Basal and glucoprivic-induced changes in extracellular GABA in the ventral hypothalamus of Zucker rats. *Am. J. Physiol.* **271,** R388–R392.

Stanley, B. G., Daniel, D. R., Chin, A. S., and Leibowitz, S. F. (1985). Paraventricular nucleus injections of peptide YY and neuropeptide Y preferentially enhance carbohydrate ingestion. *Peptides* **6,** 1205–1211.

Stanley, B. G., and Leibowitz, S. F. (1985). Neuropeptide Y injected in the paraventricular hypothalamus: a powerful stimulant of feeding behavior. *Proc. Natl. Acad. Sci. USA* **82,** 3940–3943.

Stanley, B. G., Magdalin, W., Seirafi, A., Thomas, W. J., and Leibowitz, S. F. (1993). The perifornical area: the major focus of (a) patchily distributed hypothalamic neuropeptide Y-sensitive feeding system(s). *Brain Res.* **604,** 304–317.

Stellar, E. (1954). The physiology of motivation. *Psychol. Rev.* **5,** 5–22.

Tataranni, P. A., Gautier, J. F., Chen, K., Uecker, A., Bandy, D., Salbe, A. D., Pratley, R. E., Lawson, M., Reiman, E. M., and Ravussin, E. (1999). Neuroanatomical correlates of hunger and satiation in humans using positron emission tomography. *Proc. Natl. Acad. Sci. USA* **96,** 4569–4574.

The DCCT Research, Group (1997). Hypoglycemia in diabetes control and complications trial. *Diabetes* **46,** 271–286.

The DCCT Research, Group (1993). The effect of intensive treatment of diabetes on the development and progression of long-term complications of insulin-dependent diabetes mellitus. *N. Eng. J. Med.* **329,** 977–986.

Tholey, G., Roth-Schechter, B. F., and Mandel, P. (1981). Activity and isoenzyme pattern of lactate dehydrogenase in neurons and astroblasts cultured from brains of chick embryos. *J. Neurochem.* **36,** 77–81.

Thurston, J. H., and Hauhart, R. E. (1989). Effect of momentary stress on brain energy metabolism in weanling mice: apparent use of lactate as cerebral metabolic fuel concomitant with a decrease in brain glucose utilization. *Metabol. Brain Dis.* **4,** 177–186.

Thurston, J. H., Hauhart, R. E., and Schiro, J. A. (1983). Lactate reverses insulin-induced hypoglycemic stupor in suckling-weanling mice: biochemical correlates in blood, liver, and brain. *J. Cereb. Blood Flow Metab.* **3,** 498–506.

Tkacs, N. C., Dunn-Meynell, A. A., and Levin, B. E. (2000). Presumed apoptosis and reduced arcuate nucleus neuropeptide Y and pro-opiomelanocortin mRNA in non-coma hypoglycemia. *Diabetes* **49,** 820–826.

Tordoff, M. G., Tluczek, J. P., and Friedman, M. I. (1989). Effect of hepatic portal glucose concentration on food intake and metabolism. *Am. J. Physiol.* **257,** R1474–R1480.

Trapp, S., and Ashcroft, F. M. (1997). A metabolic sensor in action: news from the ATP-sensitive K^+-channel. *News Physiol. Sci.* **12,** 255–263.

Tsujii, S., and Bray, G. A. (1990). Effects of glucose, 2-deoxyglucose, phlorizin, and insulin on food intake of lean and fatty rats. *Am. J. Physiol.* **258,** E476–E481.

Vaisse, C., Halaas, J. L., Horvath, C. M., Darnell, J. E., Jr., Stoffel, M., and Friedman, J. M. (1996). Leptin activation of Stat3 in the hypothalamus of wild-type and ob/ob mice but not db/db mice. *Nat. Genet.* **14,** 95–97.

van Dijk, G., Bottone, A. E., Strubbe, J. H., and Steffens, A. B. (1994). Hormonal and metabolic effects of hypothalamic administration of neuropeptide Y during rest and feeding. *Brain Res.* **660,** 96–103.

Vannucci, S. J., Koehler-Stec, E. M., Li, K., Reynolds, T. H., Clark, R., and Simpson, I. A. (1998). GLUT4 glucose transporter expression in rodent brain: effect of diabetes. *Brain Res.* **797,** 1–11.

Vannucci, S. J., Maher, F., and Simpson, I. A. (1991). Glucose transporter proteins in brain: Delivery of glucose to neurons and glia. *Glia* **21,** 2–21.

Veneman, T., Mitrakou, A., Mokan, M., Cryer, P., and Gerich, J. (1993). Induction of hypoglycemia unawareness by asymptomatic nocturnal hypoglycemia. *Diabetes* **42,** 1233–1237.

Veneman, T., Mitrakou, A., Mokan, M., Cryer, P., and Gerich, J. (1994). Effect of hyperketonemia and hyperlacticacidemia on symptoms, cognitive dysfunction, and counterregulatory hormone responses during hypoglycemia in normal humans. *Diabetes* **43,** 1311–1317.

Vidal-Puig, A., Rosenbaum, M., Considine, R. C., Leibel, R. L., Dohm, G. L., and Lowell, B. B. (1999). Effects of obesity and stable weight reduction on UCP2 and UCP3 gene expression in humans. *Obes. Res.* **7,** 133–140.

Virsolvy-Vergine, A., Leray, H., Kuroki, S., Lupo, B., Dufour, M., and Bataille, D. (1992). Endosulfine, an endogenous peptidic ligand for the sulfonylurea receptor: purification and partial characterization from ovine brain. *Proc. Natl. Acad. Sci. USA* **89,** 6629–6633.

Weigle, D. S., Duell, P. B., Connor, W. E., Steiner, R. A., Soules, M. R., and Kuijper, J. L. (1997). Effect of fasting, refeeding, and dietary fat restriction on plasma leptin levels. *J. Clin. Endocrinol. Metab.* **82,** 561–565.

Wender, R., Brown, A. M., Fern, R., Swanson, R. A., Farrell, K., and Ransom, B. R. (2000). Astrocytic glycogen influences axon function and survival during glucose deprivation in central with matter. *J. Neurosci.* **20,** 6804–6810.

Wiater, M. F., and Ritter, S. (1999). Leptin does not attenuate the hyperphagia induced by 2-deoxy-D-glucose. *Ann. N.Y. Acad. Sci.* **892,** 334–336.

Wieloch, T. (1986). Endogenous excitotoxins as possible mediators of ischemic and hypoglycemic brain damage. *Adv. Exp. Med. Biol.* **203,** 127–138.

Wilcox, B. J., Corp, E. S., Dorsa, D. M., Figlewicz, D. P., Greenwood, M. R. C., Woods, S. C., and Baskin, D. G. (1989). Insulin binding in the hypothalamus of lean and genetically obese Zucker rats. *Peptides* **10,** 1159–1164.

Wredling, R., Levander, S., Adamson, U., and Lins, P. E. (1990). Permanent neuropsychological impairment after recurrent episodes of severe hypoglycaemia in man. *Diabetologia* **33,** 152–157.

Xie, L. H., Horie, M., and Takano, M. (1999). Phospholipase C-linked receptors regulate the ATP-sensitive potassium channel by means of phosphatidylinositol 4,5-bisphosphate metabolism. *Proc. Natl. Acad. Sci. USA* **96,** 15292–15297.

Yang, X., Kow, L. -M., Funabashi, T., and Mobbs, C. V. (1999). Hypothalamic glucose sensor. Similarities to and differences from pancreatic β-cell mechanisms. *Diabetes* **48,** 1763–1772.

Yettefti, K., Orsini, J. C., and Perrin, J. (1997). Characteristics of glycemia-sensitive neurons in the nucleus tractus solitarii: possible involvement in nutritional regulation. *Physiol. Behav.* **61,** 93–100.

Yoshimatsu, H., Egawa, M., and Bray, G. A. (1991). Adrenal sympathetic nerve activity in response to hypothalamic injections of 2-deoxy-D-glucose. *Am. J. Physiol.* **261,** (Pt 2), R875–81.

Zawar, C., and Neumcke, B. (2000). Differential activation of ATP-sensitive potassium channels during energy depletion in CA1 pyramidal cells and interneurones of rat hippocampus. *Pflugers Arch. Eur. J. Physiol.* **439,** 256–262.

Zini, S., Zini, R., and Ben-Ari, Y. (1993). Nucleotides modulate the low affinity binding sites for [^{3}H]glibenclamide in the rat brain. *J. Pharmacol. Exp. Ther.* **264,** 701–708.

GLUCOSE TRANSPORTER PROTEIN SYNDROMES

Darryl C. De Vivo, Dong Wang, Juan M. Pascual, and Yuan Yuan Ho

Department of Neurology, Colleen Giblin Research Laboratories for Pediatric Neurology
Columbia University, New York 10032

I. Introduction

Disturbances of carbohydrate homeostasis have interested clinicians and scientists since the recognition of diabetes mellitus. A parallel line of study emerged in the study of patients with recurrent or chronic hypoglycemic syndromes. These studies eventually led to an understanding of fuel-hormonal interrelationships with insulin playing a prominent role (Cahill *et al.*, 1966; Owen *et al.*, 1967). However, this understanding has remained quite incomplete even following the discovery of several gene families that influence carbohydrate homeostasis. These families include the active co-transporters (SGLT) and the facilitative glucose transporters (GLUT). Recognition of these two gene families has expanded our understanding of carbohydrate metabolism, and introduced new insights into human disease states. Now we understand the molecular basis for old conditions such as the Fanconi-Bickel syndrome, and newer conditions such as the glucose-galactose malabsorption syndrome and the Glut1 deficiency syndrome. However, our understanding of disturbed carbohydrate homeostasis in the setting of diabetes mellitus remains incomplete, and it is expected that there will be further

0074-7742/02 $35.00

insight into the pathogenesis of this common human condition as we continue to explore the factors that control the movement of glucose across tissue membranes. This chapter will focus on the several clinical conditions that have been associated with documented mutations affecting the two glucose transporter gene families.

II. Review of Glucose Metabolism and Transport

Two major families of glucose transporters have been identified in humans: concentrative Na^+/glucose transporters (Sglts) and facilitative glucose transporters (Gluts) (Bell *et al.,* 1993; Carruthers, 1990; Longo and Elsas, 1998; Gould and Holman, 1993). The classification of glucose transporters is based on the sequence homology of amino acids among the transporters. Within each family, the members are numbered according to the chronological order of discovery. Capital letters are used to abbreviate the gene (e.g., GLUT1). Only the first letter of the abbreviation is capitalized when referring to the protein (e.g., Glut1).

The concentrative transporters are also known as active co-transporters or symporters. These symporters are energized by either H^+ or Na^+ gradients, which are generated and maintained by ATP-dependent ion pumps. Three types of concentrative Na^+/glucose transporters have been identified: Sglt1, Sglt2, and Sglt3. Secondary structure analysis suggests 14 transmembrane α-helices. Sglt1 is responsible for glucose absorption from the intestinal tract. Glucose absorption in the renal tubules is mediated by Sglt1 and Sglt2. The facilitative transporters permit D-glucose diffusion across cellular membranes by energy-independent, stereospecific mechanisms. This gene family consists of eight members (GLUT1–5, GLUT8–10). Originally GLUT6 was assigned to a pseudogene and GLUT7 was reported to encode a microsomal transporter that later was shown to be a cloning artifact (Burchell, 1998). Gluts are integral membrane proteins with predicted 12 transmembrane motifs. These proteins facilitate the bidirectional glucose movement across plasma membranes.

These two families of transporters are responsible for maintaining glucose homeostasis. Genetic mutations of these transporters have been associated with a variety of human diseases including familial glucose-galactose malabsorption (SGLT1), renal glycosuria (SGLT2), Glut1 deficiency syndrome (GLUT1), Fanconi-Bickel syndrome (GLUT2), and noninsulin-dependent diabetes mellitus (GLUT4).

A. Molecular Biology of Glucose Transporters

Each of the transporters is genetically distinct with a unique chromosomal assignment. The chromosomal locations and the molecular sizes of the three SGLT and eight GLUT genes and their products are summarized in Table I.

The GLUT1 structure is the most extensively studied (Mueckler *et al.*, 1985). The genomic sequence is approximately 35 kb, with 10 exons and 9 introns. The promoter region contains sequence elements for transcription factors, including a TATA box, two SpI sites, and a phorbol ester-responsive element. Two enhancer elements within the GLUT1 gene have been identified: the first is located between 3.3 and 2.7 kb upstream from the transcription initiation site, the second is located within the second intron between 16.7 and 18.0 kb downstream from the transcription initiation site (Murakami *et al.*, 1992). The second enhancer element contains two TRE-like sequences. These enhancer elements appear to be responsible for serum-induced GLUT1 expression.

Analysis of the amino acid sequences of the facilitative glucose transporters reveals a number of conserved regions ranging from single residues to sequence motifs. Examples of such sequence motifs include (GlyArgArgLys) located between transmembrane domains (TM) 2 and 3 and in a corresponding location between TM7 and 8, GluXXXXXXArg located between TM4 and 5 in the amino terminal half and between TM10 and 11

TABLE I
Molecular Details of Glucose Transporters

Gene	Chromosomal location	Transcript(s) (kb)	Transporter (kDa/amino acids)
SGLT1	22q13.1	2.2, 2.6, 4.8	75/664
SGLT2	16p11.2	2.4, 3.0, 3.5, 4.5	46/672
GLUT1	1p31.3–35	2.8	45, 55/492
GLUT2	3q26.1–26.3	2.8, 3.4, 5.4	58/524
GLUT3	12p13.3	2.7, 4.1	54/496
GLUT4	17p13	2.8, 3.5	55/509
GLUT5	1p36.2	2.0	50/501
GLUT8	9	2.1	38–47/478
GLUT9	9q34	2.6	48/N.A.
GLUT10	20q12–13.1	4.3	N.A./541

in the carboxyl terminal half. These motifs may be intimately involved in transporter function, and may be crucial to conformational stability of salt bridging between the helices. By applying mutagenic techniques to glucose transporters, selected conserved amino acids and domains have been altered, swapped, and deleted. The mutagenic data have provided insight into crucial locations within the glucose transporter sequence for substrate binding and for the conformational changes which result in D-glucose translocation.

B. Tissue Distribution and Protein Chemistry of Gluts

Each Glut has a unique pattern of tissue distribution and specific biochemical properties. These characteristics dictate specific functions in maintaining glucose homeostasis. Glut1 is highly expressed in brain and erythrocytes. It is a ubiquitously expressed isoform that facilitates basal glucose utilization by many tissues. Glut2, found predominately in liver, intestine, kidney, and pancreatic β-cells, is a low-affinity isoform acting as a high-capacity transporter to allow the uninhibited bidirectional flux of glucose in these tissues. Glut2 is also part of the glucose-sensing mechanism in pancreatic β-cells. Glut3 facilitates glucose movement across neuronal membranes. Glut4 is expressed exclusively in the insulin-sensitive tissues, fat, and striated muscle. It is responsible for increased postprandial glucose uptake in these tissues. Glut5 transports fructose in spermatozoa, apical

TABLE II
Tissue Distribution and Substrate Transport Kinetics of Gluts

Molecule	Tissue distribution	Substrate	*K*m
Glut1	Erythrocyte, fibroblast, adipocyte, brain endothelium, astrocyte, retina, placenta, kidney, colon, liver Kupffer cell	D-glucose DHA D-mannose	5–10 mM
Glut2	Liver (basolateral surface membrane), kidney, intestine, pancreatic β-cell	D-glucose	20–40 mM
Glut3	Brain neuron, fibroblast	D-glucose	1–4 mM
Glut4	Adipocyte, skeletal muscle, heart	D-glucose	1–4 mM
Glut5	Small intestine, testis, microglia	D-fructose	N.A.
Glut8	Testis, skeletal muscle, heart, small instine	D-glucose	N.A.
Glut9	Brain, spleen, leukocytes	D-glucose	N.A.
Glut10	Liver, pancreas, placenta, heart, lung, kidney, skeletal muscle	N.A.	N.A.

membrane of intestinal cells, and microglia. Glut8 is found predominately in testis and is regulated by insulin and gonadotropins (Doege *et al.*, 2000a; Carayannopoulos *et al.*, 2000). Glut9 is most abundant in brain and lymphoid tissues (Doege *et al.*, 2000b). Glut10 is highly expressed in liver and pancreas (McVie-Wylie *et al.*, 2001). The substrate specificity and transport kinetics of this protein remain to be characterized. In addition to glucose, DHA (dehydroascorbic acid), D-mannose, D-galactose, and D-fructose may also be transported by Gluts. Table II summarizes the tissue distribution and transporter kinetics of the Gluts.

III. Human Diseases Associated with Defects of Glucose Transport

Mutations affecting glucose transporters are associated with several human diseases (Table III).

A. Glucose-Galactose Malabsorption (OMIM 182380)

Genetic mutation of an intestinal monosaccharide transporter, SGLT1, causes glucose galactose malabsorption. This clinical entity was described almost simultaneously by Lindquist and Meeuwisse (1962) in Sweden and by LaPlane *et al.* (1962) in France. Both reports documented a familial entity of early infancy (with onset typically before the fourth day of life) characterized by chronic, watery acid diarrhea leading to life-threatening dehydration as long as glucose and galactose were maintained in the diet and that could be readily corrected by their substitution. The structure of the intestinal mucosa is remarkably preserved in this condition. Intestinal fermentation

TABLE III
Human Diseases Associated with Defects of Glucose Transport

Transporter	Disease
SGLT1	Glucose-galactose malabsorption
SGLT2	Renal glycosuria
GLUT1	Glut1 deficiency syndrome
GLUT2	Fanconi-Bickel syndrome
GLUT4(?)	Defective insulin-stimulated glucose transport (NIDDM)

of monosaccharide-enriched luminal contents leads to production of lactic acid and hydrogen, which can be detected in breath in excess of 20 parts per million after challenge with 2 g/kg of oral glucose or galactose, providing confirmatory diagnosis of the disease in the clinical setting (Wright, 1998). Substitution of the monosaccharides permits patients to experience catch-up grow and to develop normally, although sugar exposure later in life continues to render them symptomatic.

Autoradiography of duodenal biopsies exposed to ^{14}C-galactose revealed diminished intracellular accumulation of the radiolabel in the disease state, localizing the defect to the epithelial brush-border, the site of the glucose-galactose transporter (Schneider *et al.*, 1966), and kinetic analysis of transport in jejunal samples of a patient and her asymptomatic parents demonstrated an unsuspected partial impairment of absorption in the parents and a severely reduced transport capacity (Vmax) in the patient. A small degree of glycosuria is also usually detectable, owing to defective renal glucose reabsorption, most likely due to dysfunction of tubular transporters (Elsas *et al.*, 1970).

The pathophysiology of the disease is best understood in molecular terms. Dietary glucose and galactose released in the process of carbohydrate digestion are transported across the brush-border membrane by the Na^+/glucose cotransporter SGLT1. Transport of either monosaccharide is stoichiometrically coupled with and facilitated by the import of Na^+, which follows favorable electrical and concentration gradients. The transporter displays an apparent affinity for glucose of 0.4 mM and exhibits voltage-dependent apparent Na^+ affinity (3 mM at –150 mV). The ratio of transported Na/sugar is 2. Both sugars exit the enterocyte through the basolateral membrane via the facilitated transporter GLUT2 (Wright, 2001) and Na^+ is extruded via the Na^+/K^+-ATPase pump, which indirectly confers energy dependence to the transport cycle such that pump inhibitors prevent effective SGLT1 function.

The peptide sequence of mammalian SGLT1 was deduced in 1987 by expression cloning of its cDNA after unsuccessful efforts aimed to purify the ~75,000 d protein, which is present in very small proportion in the plasma membrane (Hediger *et al.*, 1987). Epithelial poly-$(A)^+$ RNA was assayed for potential transport-encoding activity by injection into *Xenopus* oocytes, which display a relatively low background activity. Size separation of the RNA in pools allowed for the isolation of an activity-enriched fraction which was subsequently divided into individual clones from which cDNA was synthesized and sequenced. The deduced peptide chain included 664 amino acid residues, hypothetically arranged in 14 transmembrane domains and exposing both amino and carboxyl termini to the extracellular solution. The SGLT1 gene is localized in chromosome 22q13.1, spanning 72 kb and containing 15 exons. To date, two other SGLTs of known tissue distribution (termed 2 and 3) exist in addition to several homologous genes of uncertain functionality (Wright, 2001). SGLT2 is present in kidney and, to a lesser degree, in ileum and SGLT3 is a low-affinity Na/glucose cotransporter (Wright, 2001). SGLT2 is localized to chromosome 16p11.2 and, interestingly, SGLT3 is located near SGLT1, at 22p12.1–12.3, suggesting duplication of an original locus. Some 60 SGLT1 homologs have been found in other species, emphasizing the fundamental mission of the transporter.

The first SGLT1 mutation (Asp28Asn) associated with glucose-galactose malabsorption was discovered in 1991 in two affected siblings, daughters of consanguineous parents and each bearing an heterozygous mutation (Turk *et al.,* 1991). The single nucleotide transition resulted in the loss of transport activity in *Xenopus* oocytes injected with cRNA. This finding was followed in subsequent years by the identification of more than three dozen mutations from a total of about 200 patients described to date. The many cases of parental consanguinity and the absence of symptomatology in carriers speak to the autosomal recessive character of the disease. Not unexpectedly, the mutations include all missense, nonsense, frameshift, and splice site transitions and result in either truncation of the protein, impaired trafficking to the plasma membrane, or production and correct targeting of a nonfunctional protein (Martin *et al.,* 1997). The result is a decrease in the glucose and galactose absorptive capacity of the epithelium with osmotic drag of water into the lumen and diarrhea. Associated phenomena such as dehydration and, if tolerated, growth failure ensue. Dietary avoidance of glucose and galactose circumvents the consequences of malabsorption and allows patients to develop normally.

B. Glut1 Deficiency Syndrome (OMIM 138140)

Glut1 deficiency syndrome (Glut1 DS) is a novel clinical disorder that was first described by De Vivo and colleagues (1991). The first patient, a 7-1/2-month-old infant boy, was evaluated for infantile-onset seizures since age 2 1/2 months, followed by delays in early neurological development. The CSF glucose was consistently low on four consecutive lumbar punctures, and the blood glucose concentrations were consistently normal. The hypoglycorrhachia also was associated with low CSF lactate concentrations. This combination of clinical and laboratory findings led De Vivo and colleagues to conclude that there was a defect in the transport of glucose from the blood, across the blood-brain barrier, into the brain. The second patient, a 27-month-old girl, also had suffered from infantile-onset seizures and delayed neurological development. The CSF glucose values also were low on repeated lumbar punctures. The seizures in each case failed to respond to antiepileptic drugs. The electroencephalograms initially were normal, and later showed ictal and interictal abnormalities, and neuroimaging studies were normal. These clinical and laboratory findings represented the diagnostic signature of the Glut1 DS, as it would later come to be known. It was postulated in1991 that there was a defect in the transport of glucose across the blood-brain barrier. Therefore, it was assumed that the patients would benefit from a ketogenic diet, providing the only alternative fuel source for brain metabolism. These speculations were rewarded by the immediate cessation of seizures in both patients after beginning the ketogenic diet. This response also reinforced the speculation that the patients were suffering from neuroglycopenia. Later, it would be shown that the first patient had a large-scale deletion involving one GLUT1 allele causing hemizygosity, and the second patient would be shown to have a heterozygous nonsense mutation (Y449X) (Seidner *et al.,* 1998).

In the intervening decade, additional patients were being identified, and as of the summer, 2001, approximately 60 patients have been identified clinically, and 20 patients have heterozygous or hemizygous mutations affecting GLUT1 (Seidner *et al.,* 1998; Klepper *et al.,* 1999b; Wang *et al.,* 2000).

1. *Clinical Features*

The phenotype for Glut1 DS is becoming more recognized. As a result, the prevalence of this clinical syndrome continues to increase throughout the world. The clinical signature is that of an infantile-onset epileptic encephalopathy associated with delayed neurological development, deceleration of head growth, and resulting microcephaly, ataxia, and spasticity. The single most important laboratory observation is hypoglycorrhachia. The CSF/blood glucose ratio, when measured in a postabsorptive preprandial state, is consistently decreased by 50%. Normally, the ratio is approximately 0.65. In patients with the Glut1 DS, the ratio is 0.33 (De Vivo *et al.,* 1995). This ratio may be perturbed by the blood glucose. Lower blood glucose values raise the ratio, and higher blood glucose values lower the ratio. Therefore, the ratio has limited value if specimens are not collected in a postabsorptive preprandial state. More important is the absolute CSF glucose value. In our experience, the CSF glucose values seldom, if ever, exceed 40 mg/dl. The corresponding CSF lactate value is low-normal or low reflecting the decreased availability of glucose for cerebral glycolysis.

Decreased erythrocyte glucose uptake is an added laboratory measure that supports the clinical diagnosis. As a group, these patients have a 50% reduction in glucose uptake (Klepper *et al.,* 1999a). The clinical features suggest sporadic mutations in most families. The loss of function suggests an autosomal dominant mechanism of transmission. In fact, three families now have been described (Brockman and Wang *et al.*, 2001; Klepper *et al.*, 2001; Ho *et al.*, 2001b) confirming this pattern of genetic transmission. These three families harbor novel heterozygous missense mutations. The pathogenicity of the haploinsufficient state is sufficiently muted to permit growth and development into adulthood and reproduction by affected family members. Fifty percent of the offspring are at risk as would be predicted by an autosomal dominant mode of inheritance.

Glucose is the principal fuel source for brain metabolism. Glut1 is expressed predominantly at the blood-brain barrier facilitating transport of glucose across the luminal and abluminal endothelial membranes of the cerebral microvessel. Glut1 also facilitates transport of glucose across the astroglial plasma membrane. Glut1 is dedicated to the transport of glucose, but also recognizes other substrates such as galactose, glycopeptides, water and DHA. Reduced transport of these substrates also may contribute to the pathophysiology of Glut1 DS.

Cerebral metabolic rate for glucose is low during fetal development and at the time of birth. The cerebral metabolic rate increases linearly after birth peaking around age 3 years (Chugani *et al.*, 1987). The increased metabolic rate continues for the remainder of the first decade of life, gradually receding during the second decade of life to values seen in early adulthood.

It, therefore, appears reasonable to anticipate that patients with Glut1 DS are at reduced clinical risk during fetal development and the newborn period. Such has been our experience. Patients appear normal at birth following an uneventful pregnancy and delivery. Birth weights and Apgar scores are normal. Seizures represent the first clinical indication of brain dysfunction, emerging usually in early infancy between 1 and 4 months of age (Fig. 1). Parents have described apneic episodes and episodic eye movement

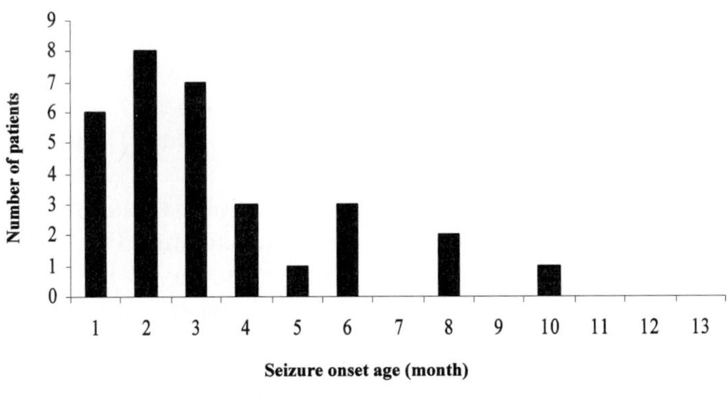

FIG. 1. Seizure onset time in 31 Glut1 DS patients is shown.

abnormalities simulating opsoclonus as the harbingers of early infantile seizures. The infantile seizures are clinically fragmented as is customarily the case, and the electroencephalographic correlate is that of multifocal spike discharges. With further brain maturation, the seizures become more synchronized and present clinically as generalized events associated with a classical 3-Hz spike and wave discharge electrically. Five seizure types have been described in these patients including generalized tonic or clonic seizures, myoclonic seizures, atypical absence, atonic, and unclassified seizures. The frequency of seizures has varied considerably from one patient to another. Some patients have suffered daily events; others have had only occasional seizures separated by days, weeks, or months; and one patient has never had a clinical seizure.

Other paroxysmal events also have been described in this patient population. It remains unclear whether these paroxysmal events are epileptic or nonepileptic nosologically. Paroxysmal events include intermittent ataxia, mental confusion, lethargy or somnolence, hemiparesis, abnormalities of movement or posture, total body paralysis, sleep disturbances, and recurrent headaches. These neurological symptoms generally fluctuate and may be influenced by environmental factors such as fasting or fatigue. All patients have degrees of speech and language impairment. Dysarthria is common and associated with dysfluency. Both receptive and expressive language skills are affected, but expressive language skills are affected disproportionately. Varying degrees of cognitive impairment have been described in this patient population ranging from learning disabilities to severe mental retardation. Social adaptive behavior is an exceptional strength in these patients. They are comfortable in the group setting and interact nicely with their peer group and with adults. Generally, they are very well accepted in the group setting and at school.

2. *Laboratory Findings*

Blood and urine studies generally are unremarkable in this patient population. Of note, the blood glucose values consistently are normal. The CSF abnormalities are distinctive with hypoglycorrhachia and decreased lactate concentrations. This profile is highly specific for Glut1 DS in the absence of hypoglycemia.

The electroencephalographic characteristics have been defined using continuous closed circuit television/EEG monitoring. The tracings have shown generalized spike or polyspike wave discharges in approximately 50% of patients, mild diffuse slowing in 33%, focal epileptiform discharges in 10–15%, and focal slowing in 5–10% (Fig. 2). In infancy, the clinical seizures often represent behavioral arrest and are correlated with focal temporal and posterior spike discharges. Later in childhood, there is a mixed motor seizure pattern with generalized spike-wave epileptiform discharges

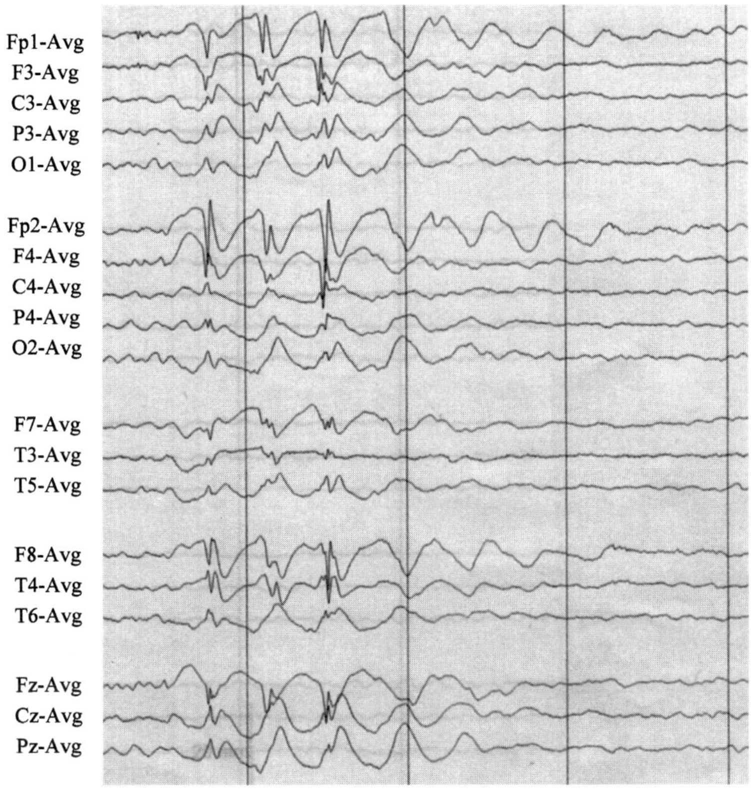

FIG. 2. This EEG demonstrates a 2-sec burst of generalized slow (2.5 Hz.) spike-wave discharges in a 6-year-old girl with Glut1 DS.

on EEG. Peculiar eye movement abnormalities simulating opsoclonus have been described in several infants, and often have represented the earliest manifestations of a paroxysmal disturbance. These abnormalities likely represent subtle seizures in early infancy.

Neuroimaging including computerized tomography and magnetic resonance imaging is usually normal. PET scan studies have shown a global reduction of glucose uptake in the brain, with more severe hypometabolism in the medial temporal lobes and the thalami. The thalamic hypometabolism accentuates the relative uptake of glucose in the basal ganglia (Fig. 3; see color insert) (Pascual *et al.*, 2002).

The erythrocyte glucose uptake study is a valuable screening test for this condition. Glut1, expressed in the erythrocyte membrane, is immunologically and chemically indistinguishable from the Glut1 isoform expressed in cerebral microvessels (Kalaria *et al.*, 1988). We developed the erythrocyte

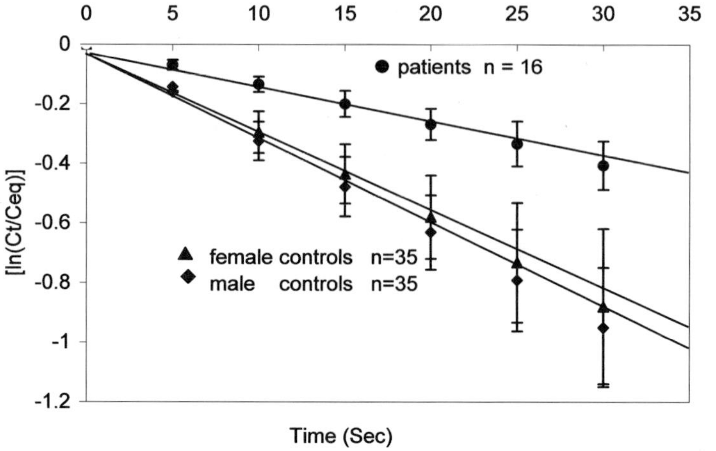

FIG. 4. 3-OMG uptake into erythrocytes from normal controls and Glut1 DS patients. The controls represent the mean value ± standard error of 35 separate experiments. The patients represent the mean value ± standard error of 16 separate experiments.

glucose uptake study using the reagent 3-O-methyl-D-glucose. A typical uptake curve from patients compared to their parents is shown in Fig. 4. Most, but not all patients with Glut1 DS, have a 50% reduction in erythrocyte glucose uptake. We have three patients who have Glut1 DS, proven by mutational analysis of the GLUT1 gene, who had normal erythrocyte glucose uptake studies. Therefore, the study is highly specific and highly sensitive, but false-negatives exist (Klepper *et al.*, 1999a). Also, false-positives exist for the Glut1 DS to the extent that hyperglycemic conditions cause downregulation of glucose uptake as we have shown recently in mitochondrial encephalopathy patients with the MELAS phenotype and associated diabetes mellitus (Fig. 5).

Additional laboratory testing can be performed to confirm Glut1 DS and the state of haploinsufficiency. Western blot analyses of erythrocyte membranes may confirm decreased gene product in patients who have heterozygous null alleles. Preparation of Western blots will permit visualization of the Glut1 band and a comparison with band 3, a protein contained within the erythrocyte membrane. The band 3 density can be used to normalize the Glut1 density. In our experience, we have found either normal expression of Glut1 in patient erythrocyte membranes or approximately 50% reduction in patients who have heterozygous null alleles (Seidner *et al.*, 1998).

Fluorescence *in situ* hybridization (FISH) studies can be performed on lymphocytes or cultured skin fibroblasts from affected patients using a fluorescence labeled p1 cloned DNA specific for the GLUT1 gene (Seidner *et al.*,

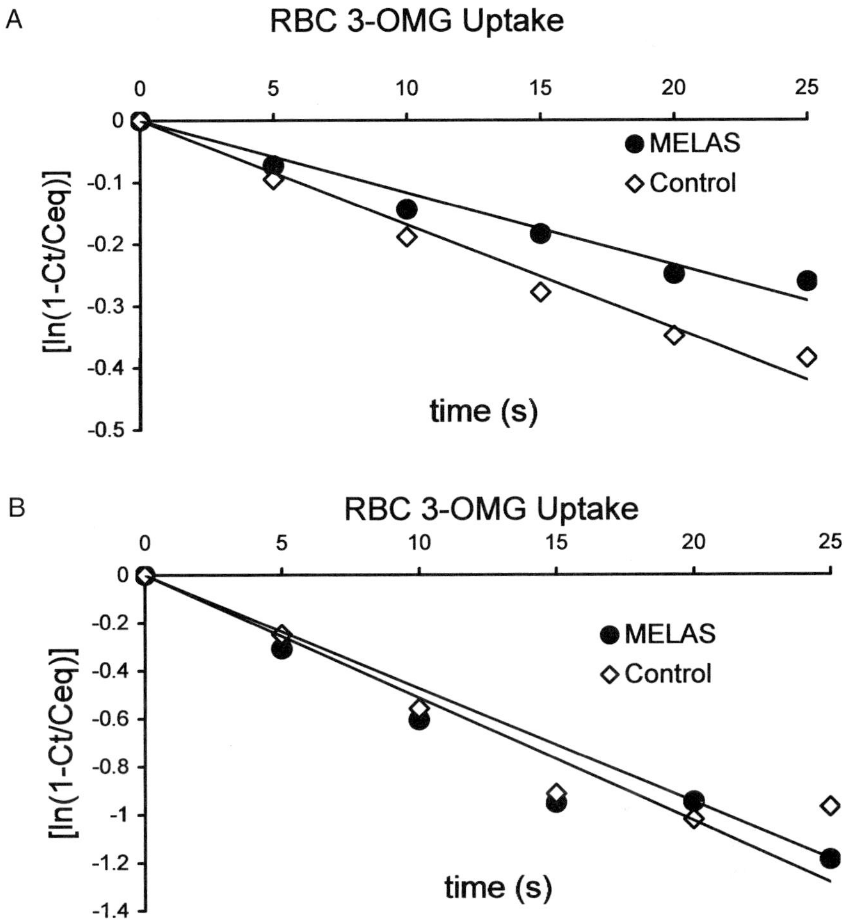

FIG. 5. 3-OMG uptake into erythrocytes from normal controls and MELAS patients. (A) Time course of the uptake of 3-OMG in the normal control and the MELAS patient with hemoglobin A1C 8.2%. The uptake value of the MELAS patient is 67% of the normal control. (B) Time course of the uptake of 3-OMG in the normal control and the MELAS patients with hemoglobin A1C 6.2%. The uptake value of the MELAS patient is 109% of the normal control.

1998). We have identified three patients with Glut1 DS who are hemizygous as determined by FISH analysis (Fig. 6; see color insert). This number represents approximately 10–15% of the patients who have been confirmed at the molecular level.

Glut1 genomic analysis is carried out in patients who are negative by FISH study. Genomic DNA can be extracted from the patient's blood or cultured

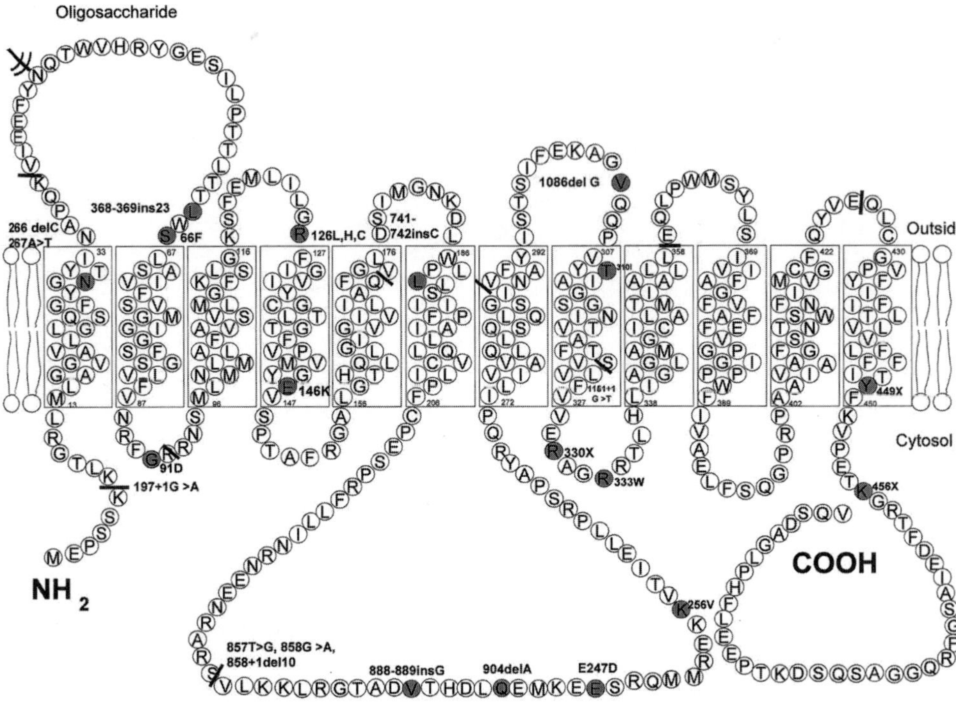

FIG. 7. Summary of GLUT1 mutations in Glut1 DS. All the identified mutations are shown in the predicted model of Glut1 in the plasma membrane; the hemizygous mutation is not represented. The 12 TM domains are shown in boxes. Solid bars indicate the introns in the GLUT1 gene.

skin fibroblasts or other available tissues. We have identified 22 heterozygous mutations including 10 missense mutations, 3 nonsense mutations, 3 insertions, 3 microdeletions, and 3 splice site mutations (Fig. 7 and Table IV).

3. *Familial Cases*

Three familial Glut1 DS cases also have been identified by Glut1 genomic DNA analysis. The first familial case was reported by Brockmann *et al.* (1999, 2001) and Wang *et al.* (2001c). Five patients in three generations were affected in this German family (Fig. 8A). The affected family members manifested mild to severe seizures, developmental delay, ataxia, hypoglycorrhachia in the three patients undergoing lumbar puncture, and decreased erythrocyte 3-O-methyl-D-glucose uptake. Several interesting caveats were identified clinically in the study of this family. First, the affected grandfather lived until age 78 years before suffering a fatal cerebral stroke. He had been gainfully employed and fathered four children. He also had

TABLE IV
SUMMARY OF THE MUTATIONS

Mutation number	Patient number	Nucleotide change	Location	Amino acid change	Mutation type
1	03	376C > T	Exon 3	S66F	Missense
2	40, 41, 42	G272A	Exon 3	G91D	Missense
3	13	556G > T	Exon 4	R126L	Missense
4	26, 27, 28	556G > A	Exon4	R126H	Missense
5	45, 46	555C > T	Exon 4	R126C	Missense
6	08	615G > A	Exon 4	E146K	Missense
7	22	920G > C	Exon 6	E247D	Missense
8	13	945A > G, 946A > T	Exon 6	K256V	Missense
9	16	1108C > T	Exon 7	T310I	Missense
10	17, 21, 39	1176C > G	Exon 8	R333W	Missense
11	07	1167C > T	Exon 7	R330X[a]	Nonsense
12	02	1526C > A	Exon 10	Y449X	Nonsense
13	09	1545A > T	Exon 10	K456X	Nonsense
14	04	266delC, 267A > T	Exon 2	FS[b] > STOP	Deletion
15	15	368-369ins23	Exon 3	FS > STOP	Insertion
16	23	741-742insC	Exon 5	FS > STOP	Insertion
17	20	888–889insG	Exon 6	FS > STOP	Insertion
18	10	904delA	Exon 6	FS > STOP	Deletion
19	12	1086delG	Exon 7	FS > STOP	Deletion
20	14	197 + 1G > A	Intron 1		Splice site
21	18	857T > G, 858G > A, 858 + 1del10	Intron 5		Splice site
22	11	1151 + 1G > T	Intron 7		Splice site
23	01, 06, 24	Large scale deletion			Hemizygosity

Mutation nomenclature is based on the guideline described by Antonarakis (1998). Nucleotide and amino acid positions are according to the published sequence. (From Mueckler *et al.*, 1985.)

[a] X specifies a termination codon.

[b] FS specifies a frame shift.

learned of a home remedy that would mitigate his neurological symptoms. With the anticipation of symptoms, he would take a dose of honey. Minutes later, he was feeling better. This home remedy was passed onto his two affected children, one man and one woman. Each had similar symptoms. Neither the grandfather nor his affected son was available for detailed neurological examination, and the son apparently lived a reclusive life as a logger. His daughter was more outgoing. She completed high school and learned to drive. She also recognized a peculiar sensitivity to coffee and to caffeine-containing beverages. As a result, she has avoided such beverages since childhood. She also appreciated clinical benefit from carbohydrates when symptomatic. She had three children as shown in Fig. 8A. Her

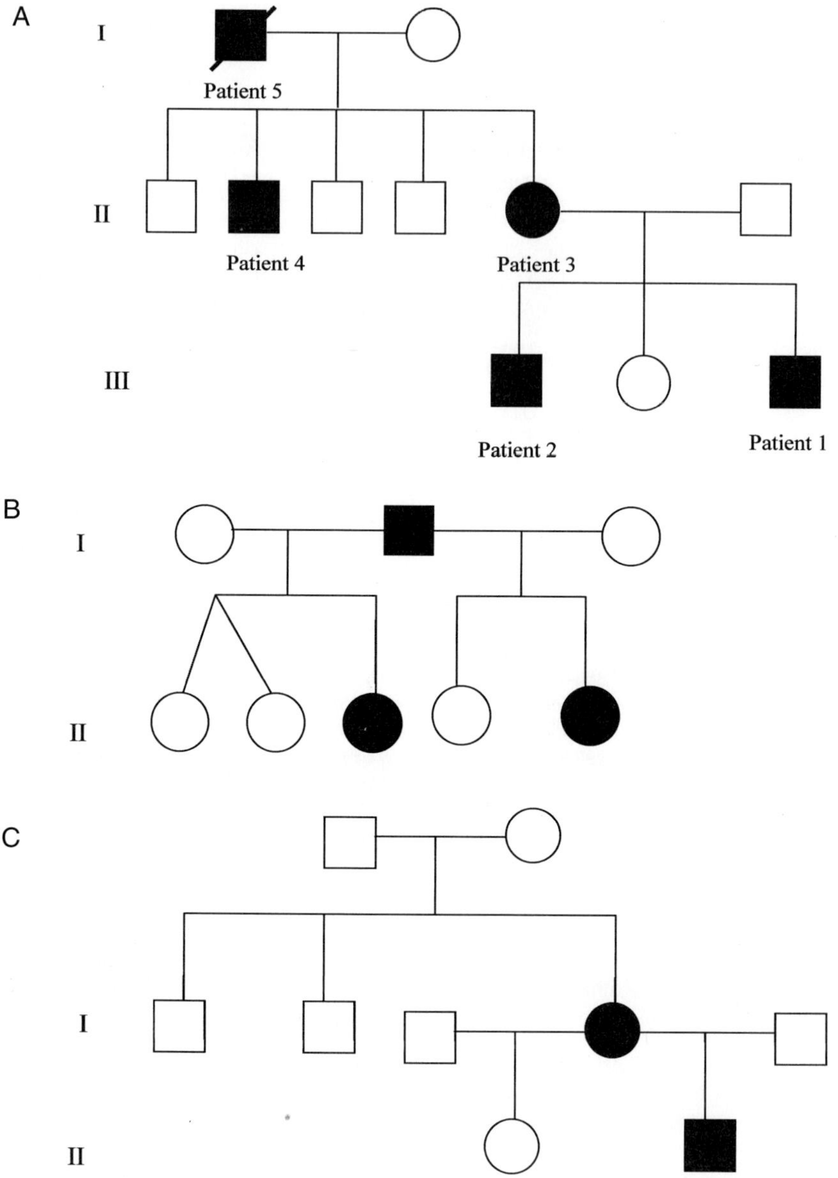

FIG. 8. Pedigree. (A) The proband and his son were not available for mutation screening. His daughter and his two grandsons were identified to carry a heterozygous R126H mutation (From Brockmann *et al.,* 2001, with modifications). (B) The pedigree of this second familial Glut1 DS case was adapted from Klepper *et al.* (Adapted from Klepper *et al.,* 2001, Fig. 1, p. 64, with modifications.) The proband and his two daughters carried the heterozygous mutation G91D. (C) The proband and her son were identified to carry the heterozygous mutation R126C (From Klepper *et al.,* 2001, Autosomal dominant transmission of GLUT1 deficiency, *Hum. Mol. Genet.,* **10,** 63–68, by permission of Oxford University Press.)

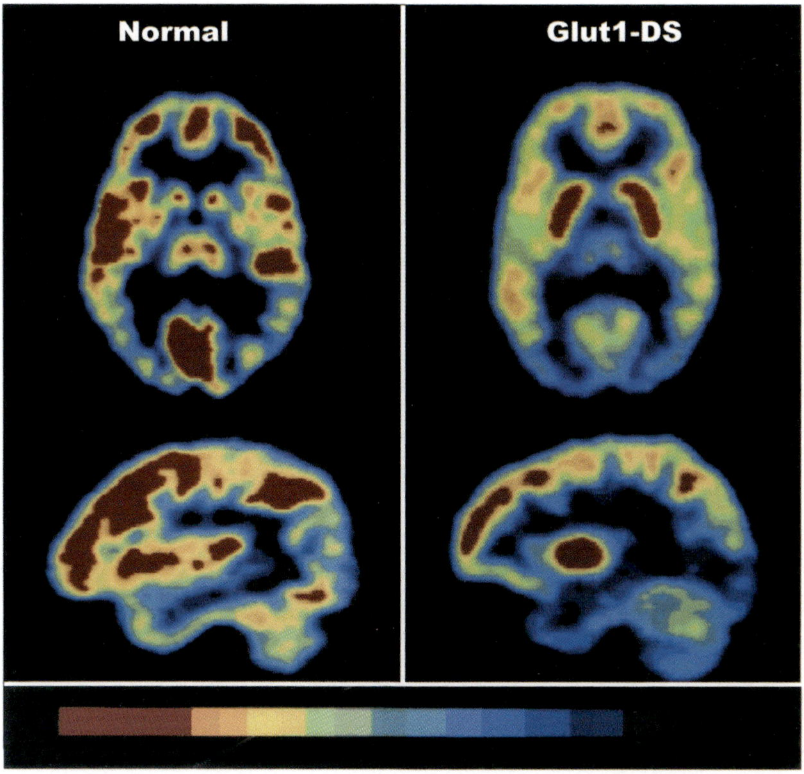

FIG. 6.3. PET scan, patients were fasted for at least 6 h prior to the injection of the radiopharmaceutical. Patient blood glucose levels were measured at the time of the study and found to be normal. Intravenous access was obtained at least 15 min prior to the radiopharmaceutical administration. The patients were then injected with 0.14 mCi/kg of 18F-2-deoxyglucose and scanned 30 min following injection. Studies were acquired on a Siemens ECAT EXACT HR (+), (Knoxville, TN) with FWH = 4 mm. Each study was acquired using a multiframe technique and auto attenuation correction in a dynamic scan mode (4 frames at 480 sec per frame) that used filter back projection. Reconstruction with auto attenuation correction used a Hann filter (cutoff 0.40 cycles/pixel). Post reconstruction transverse, oblique transverse, coronal, and sagittal plane images with a slice thickness of 0.50 cm were then produced and displayed using an inverted gray scale and rainbow (16 step) maps. PET scan studies in Glut1 DS patients have shown a global reduction of glucose uptake in the brain, with more severe hypometabolism in the medial temporal lobes and the thalami.

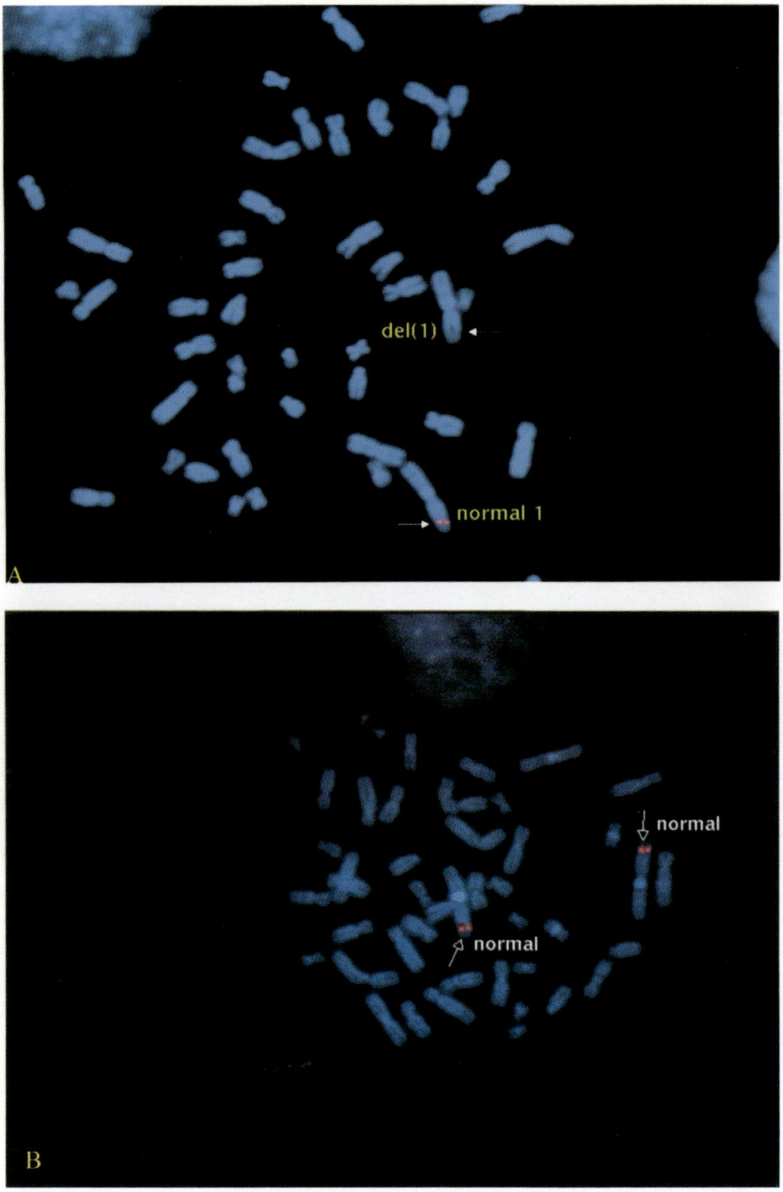

FIG. 6.6. Fluorescence *in situ* hybridization (FISH). Metaphase spreads of cultured skin fibroblasts from patients were probed with labeled p1 DNA specific for GLUT1. Hemizygosity has been found in 10–15% of Glut-1 DS patients. (A) A FISH-positive Glut1 DS patient who is hemizygous for GLUT1 is indicated by arrows with "normal" and "deletion." (B) A FISH-negative Glut1 DS patient who carries a heterozygous deletion (1186delG) is indicated by arrows with "normal."

daughter remains healthy and neurologically well. Her two sons are neurologically impaired and unable to carry out activities of daily living independently. The two boys and their mother underwent lumbar punctures, and all three had evidence of hypoglycorrhachia with CSF/blood glucose ratios of approximately 0.34. The erythrocyte glucose uptake studies were decreased, less so in the mother's cells than in her two boys' cells. This laboratory finding correlated with the severity of clinical symptoms. The pedigree also suggested a degree of anticipation through successive generations.

We identified an R126H missense mutation in the genomic DNA extracted from tissues of the mother and her two sons. Her daughter and her husband were negative for this pathogenic missense mutation. Also noted was a deletion (–173 del A) and heterozygous silent substitutions (GCT>GCC@224, TGC>TGT@578, CCG>CCA@767) in the younger brother, sister, mother, and father. These polymorphisms were absent in the more severely affected brother suggesting the possibility of a protective mechanism mitigating the pathogenicity of the primary R126H missense mutation.

The pathogenicity of this and other missense mutations can be determined by *in vitro* mutagenesis studies in *Xenopus laevis* oocytes. Western blot and confocal immunofluorescent microscopy also can determine the expression ratio of the mutant protein compared to the wild-type protein in the *Xenopus oocyte* membrane. Such studies were carried out in this pedigree and showed that the erythrocyte Western blot analysis was normal in the affected family members. The protein expression ratio also was similar in the *Xenopus oocyte* membrane preparation. Kinetic analysis demonstrated that the replacement of the arginine 126 by histidine in the mutant Glut1 caused a lower V_m. Glucose uptake at 5 mM by the *Xenopus* oocytes containing the mutant protein was only 8.9% compared to the oocytes containing wild-type protein.

A second familial case was reported by Klepper *et al.* (2001). In this pedigree, the father had two daughters by separate marriages (Fig. 8B). Each affected family member suffered seizures, mild to moderate mental retardation, hypoglycorrhachia, and decreased erythrocyte glucose uptake study. The two daughters also had ataxia. A heterozygous G91D missense mutation was present in genomic analysis of DNA from father and his two daughters. The Western blot analysis of the patients' erythrocytes was normal. *In vitro* mutagenesis studies in *Xenopus laevis* oocytes demonstrated a 40% decrease in the transport of 3-O-methyl-D-glucose. The expression ratio of mutant G91D Glut1 in the *Xenopus* oocyte membranes was similar to that of the wild-type preparation as determined by Western blot analysis and by confocal immunofluorescent microscopy.

The third family was identified in the United States (Ho *et al.,* 2001b). The mother and her son suffered apnea, seizures, hypoglycorrhachia, and nystagmus in infancy (Fig. 8C). The erythrocyte 3-O-methyl-D-glucose uptake studies were decreased in mother (54%) and son (65%). The apparent *K*m values were normal but the apparent Vmax values were decreased in the mother (38%) and her son (34%). Erythrocyte membrane immunoreactivity was normal as was true in the other two pedigrees. A heterozygous R126C missense mutation was identified by genomic analysis of DNA from mother and son. R126 is the last arginine in the second extracellular loop connecting transmembrane domains 3 and 4 (Fig. 7). This amino acid is conserved among all the known sugar transporters from different species such as rat, rabbit, bos mouse, chick, sheep, pig, and mouse.

We also have reported an example of Glut1 DS paternal mosaicism (Wang *et al.,* 2001a). In this family, the proband was an 11-year-old boy. He had seizures, hypotonia, delayed language development, hypoglycorrhachia, and decreased erythrocyte glucose uptake (40%) studies. He responded well to a ketogenic diet with control of seizures and improved neurobehavioral functioning. Direct sequencing of the PCR amplified genomic DNA identified a heterozygous R333W missense mutation in the patient but not in the asymptomatic parents. Restriction fragment length polymorphism (RFLP) analysis showed that the father was mosaic for the R333W mutation using DNA both from white blood cells and from cultured skin fibroblasts. Mutagenesis studies confirmed that the R333W missense mutation was pathogenic as discussed previously. Currently, we assume that the father's germ line cells also are mosaic, providing yet another mechanism for Glut1 DS inheritance as an autosomal dominant trait. This finding of mosaicism also might explain some sporadic patients who have variant phenotypes. Of greater importance is the need to consider parental mosaicism when families receive genetic counseling.

Finally, we have studied one severely affected patient with mixed seizures, acquired microcephaly, spastic quadriparesis, choreoathetosis, severe motor delay, and severely delayed expressive language development who had a CSF glucose value of 31 mg/dl (Wang *et al.,* 2001b). The erythrocyte glucose uptake studies also were decreased to 38% of control values. Previously, this patient was reported to carry heterozygous R126L and K256V missense mutations in trans producing the state of compound heterozygosity (Wang *et al.,* 2001b). The father is clinically well and has neither missense mutation. The patient's asymptomatic mother, on the other hand, carries the K256V heterozygous mutation. Mutagenesis studies in *Xenopus* oocytes showed that the R126L missense mutation was pathogenic with a glucose uptake study under zero trans influx conditions of 3%. The K256V missense mutation, under similar analysis, was 12% of control values. We have concluded from these observations that the asymptomatic mother has sufficient residual Glut1 activity to allow her to function normally. It remains unclear whether there is a synergistic interaction between the two missense mutations causing a more severe phenotype in the child with compound heterozygosity.

4. *Clinical Management*

The ketogenic diet was introduced in 1991 as a treatment for Glut1 DS, before it was confirmed that there was a decrease in glucose transport across tissue membranes. Ketone bodies are transported across tissue membranes by monocarboxylic transporters. As a result, this alternative brain fuel is available for uptake and metabolism by brain cells. Our findings over the past decade indicate that the ketogenic diet is highly effective in controlling the seizures, but the patients continue to have varying neurobehavioral deficits involving cognition and social adaptive behavior. Most, if not all patients, also suffer deceleration of head growth during infancy, and most develop acquired microcephaly with measurements that fall below the third percentile. It remains uncertain as to what brain structural elements are primarily at risk. Astroglial proliferation continues postnatally. Other postnatal events include myelination, dendritic arborization, and synaptogenesis. Any or all of these structural elements may be vulnerable in the postnatal period because of cerebral energy failure, and these vulnerabilities seemingly are only partially protected by the state of chronic ketosis (De Vivo *et al.*, 1995).

Transport of glucose across tissue barriers also may be influenced by other factors. It is known that a state of chronic hyperglycemia causes downregulation of glucose transporters. In contrast, the state of chronic hypoglycemia or acute hypoxia causes upregulation of Glut1 transporters. Alpha-lipoic acid (thioctic acid) also has been shown to facilitate glucose transport in cultured skeletal muscle cells that are Glut4 dependent (Klip *et al.*, 1994). Similar, although less convincing studies, have been carried out *in vitro* with Glut1 transport systems. For this reason, we have recommended, without any convincing clinical evidence, that patients take α-lipoic acid supplements (De Vivo *et al.*, 1996). The patient response has been modest, at best; but the dose taken by mouth may be inadequate to approximate the *in vitro* findings (Kulikova-Schupak *et al.*, 2001)

Other factors may influence glucose transport. Barbiturates are known to inhibit transport of glucose. We have noted that most patients with infantile-onset seizures are treated with phenobarbital as the most commonly used antiepileptic drug in this age group. Parents have reported anecdotally that phenobarbital did not improve seizure control, and worsened their child's clinical state. We have also shown *in vitro* evidence that barbiturates will aggravate the Glut1 transport defect in erythrocytes from patients with this clinical syndrome (Klepper *et al.*, 1999c). We have made similar observations with methylxanthines, also known to inhibit transport of glucose by Glut1

(Ho *et al.,* 2001a). Therefore, we strongly advise patients to avoid barbiturates, coffee, and other caffeine-containing beverages. Of note, one patient has reported to us that she was unduly sensitive to coffee and had avoided it since childhood (Brockmann *et al.,* 2001).

C. Fanconi-Bickel Syndrome

Originally named hepatorenal glycogenosis with renal Fanconi syndrome, Fanconi-Bickel syndrome was described in 1949 (Fanconi and Bickel, 1949). The index patient in their original report was subject to mutational analysis in 1997, culminating in the identification of a pathogenic mutation in GLUT2 (Santer *et al.,* 1998). Glut2 is the low-affinity facilitative transporter of hepatocytes, pancreatic β-cells, enterocytes, and renal tubular cells. The first patient identified, Claudio M., presented at 6 months of age with failure to thrive, polydipsia, and constipation. Through early childhood, osteopenia, short stature, hepatomegaly, and tubular nephropathy became noticeable. The latter was accompanied by hyperglycosuria, hyperphosphaturia, hypophosphatemia, and hyperuricemia, and hyperaminoaciduria with intermittent proteinuria (Santer *et al.,* 1998). The liver was enriched in glycogen and exhibited steatosis. Preprandial ketotic hypoglycemia was followed by postprandial hyperglycemia. At age 52, he was secluded in the mountains and able to sustain himself, with a stature of 150 cm and marked hepatomegaly and nephromegaly. Analysis of the primary pathophysiological anomalies, i.e., the high intracellular glucose and galactose concentration in liver and proximal tubule, the inhibition of glycogenolysis leading to accumulation of glycogen and the tubular dysfunction led to the postulate that facilitated transport into and out of the liver and proximal tubule was defective. The disease was also transiently known as glycogen storage disease type XI, when a defect of phosphoglucomutase was hypothesized (Hug *et al.,* 1987). This hypothesis was later disproved and therefore this term should not be utilized.

The reversibility of facilitated glucose transport becomes physiologically crucial in the liver, where postprandial uptake alternates with fasting efflux and where both gluconeogenesis and glycogenolysis are active. GLUT2 serves this dual role and also participates in glucose sensing by pancreatic β-cells. During the fed state, the hyperglycemia and hypergalactosemia can be explained on the basis of decreased uptake by the liver and hypoinsulinism due to offsetting of the glucose-sensing mechanism in β-cells. During fasting, there may be decreased monosaccharide release from the liver, accompanied by elevated intracellular accumulation; consequently inhibiting glycogenolysis and promoting excessive glycogen storage and hepatomegaly.

There also is decreased reabsorption of tubular glucose and significant urinary losses, similarly leading to renal accumulation of glycogen. A moderate degree of tubular dysfunction also is present, causing the further renal manifestations of the disease. A few patients also exhibit intestinal monosaccharide malabsorption and diarrhea, possibly due to limited enterocyte export and secondary glycogen accumulation (Santer *et al.*, 1998).

To date, approximately 85 patients have been described on clinical and laboratory grounds, but only about 20% have been studied and shown to carry mutations in GLUT2. Many cases (about two-thirds) are the product of a consanguineous conception, suggesting the autosomal recessive nature of the disease and several have been identified by newborn screening as positive for galactosemia. Nevertheless, the disorder is quite rare. In general, the disease is not thought to shorten life span, although postsurgical acidosis leading to death has been observed. Late in life the degree of dwarfism becomes more obvious and generalized osteoporosis may result in pathological fractures (Santer *et al.*, 1998). Interestingly, the severe hypoglycemia that has been accasionally documented (18 mg/dl) is not associated with neurological impairment, presumably because of the protective nature of ketosis.

The Glut2 gene was cloned in 1998 (Fukumoto *et al.*, 1988) and localized to chromosome 3q26.1–q26.3. In contains 11 exons and 10 introns, spanning about 30 kb. Most patients are homozygous for disease-related mutations, the remainder being compound heterozygtes (Santer *et al.*, 1998).

Treatment of the disease involves hydration and electrolyte replacement, plasma alkalinization, supplementation with vitamin D and phosphate, restriction of galactose, and a diabetic diet. Fructose provides a carbohydrate source whose metabolism is not altered by the disease and cornstarch affords slow release of glucose. However, these interventions may affect growth only marginally (Santer *et al.*, 1998).

D. Related Clinical Observations

Alterations of glucose transport have been investigated in other human disease states in an attempt to understand the pathophysiology of these conditions. Glut1 is the predominant glucose transporter in erythrocytes, and thus has served as an *in vitro* model for human conditions associated with disturbed glucose homeostasis. Reduction of erythrocyte glucose transport has been reported both in type 1 and in type 2 diabetes mellitus (Hu *et al.*, 2000; Yoa *et al.*, 1993). Both types of diabetes mellitus may coexist as a complication of mitochondrial DNA-associated conditions. The most common mitochondrial DNA point mutation, A3243G, involves the transfer RNA gene for leucine^{UUR}. This substitution is commonly referred to as the MELAS mutation. MELAS is the most common mitochondrial DNA-associated phenotype that is transmitted maternally as a non-Mendelian trait (De Vivo *et al.*, 1996). These patients classically develop normally in infancy and early childhood before developing neurological symptoms. The cardinal clinical features include a multifocal encephalopathy associated with lactic acidosis and stroke-like episodes. Gestational diabetes mellitus is common in women who are harboring the A3243G mitochondrial DNA point mutation, and diabetes mellitus often is a complication of MELAS. We have carried out erythrocyte studies on three patients with MELAS. One has normal carbohydrate homeostasis, the second has well-controlled insulin-dependent diabetes mellitus with a normal serum glycosylated hemoglobin A1C value of 6%, and the third patient has poorly controlled insulin-dependent diabetes mellitus and an elevated glycosylated hemoglobin A1C value of 8.2%. The third patient had a 67% reduction in erythrocyte glucose uptake studies whereas the first and second patients had normal erythrocyte glucose uptake studies (Fig. 5). These findings suggest the possibility that the neurological symptoms associated with MELAS and complicated by diabetes mellitus might be exacerbated by concomitant hyperglycemia and downregulation of brain Glut1 transporters.

Glucose transport also represents the rate-limiting step for glucose utilization in myocytes and in adipocytes. Noninsulin dependent diabetes mellitus (NIDDM) is associated with a reduction in the number of Glut4 transporters in cultured adipocytes (St. Denis and Cushman, 1998; Rice *et al.*, 1996). This reduction results from pretranslational suppression of Glut4 mRNA levels (Rice *et al.*, 1996). In humans, linkage studies and mutational analyses have indicated that Glut4 is not strongly associated with NIDDM. Studies involving several families transmitting NIDDM as an autosomal dominant trait have excluded an association with the GLUT1 gene. A Glut2 polymorphism resulting in the change of a valine at nucleotide position 197 to isoleucine has been reported in one patient with gestational diabetes mellitus (Mueckler *et al.*, 1994). However, other studies have failed to show meaningful correlations between Glut2 polymorphisms and NIDDM. Recently, a genome-wide screening effort has identified several susceptibility loci for NIDDM within the human genome. The newly cloned Glut10 gene has been mapped to chromosome 20q12–q13.1, one of the genomic loci associated with NIDDM. Therefore, the GLUT10 gene locus is considered to be a candidate for an NIDDM susceptibility gene (McVie-Wiley *et al.*, 2001). Further studies will be necessary to characterize the Glut10 transporter and the gene before a firm association can be established between the Glut10 transporter and diabetes mellitus. Current evidence suggests that diabetes mellitus is a multifactorial condition, and single gene variations alone may be insufficient to cause the clinical symptoms.

E. Relevant Animal Models

Several models have been investigated in laboratory animals to evaluate the pathophysiology of glucose transporter mutations. Homozygous GLUT2 gene disruption causes hyperglycemia and hypoinsulinemia in mice. These mice demonstrate abnormal glucose tolerance and loss of control of insulin gene expression by glucose (Guillam *et al.*, 1997). These research findings indicate that GLUT2 is required for maintaining normal glucose homeostasis and normal function and development of the endocrine pancreas. The clinical counterpart for the rodent GLUT2 knockout model is the Fanconi-Bickel syndrome, discussed earlier.

Female GLUT4 knockout mice exhibit normal plasma glucose levels in both the fed and the fasted states, and normal glucose tolerance. Male GLUT4 knockout mice show slightly elevated postprandial plasma glucose values and decreased fasting plasma glucose values compared to controls. There are no changes of other glucose transporter isoforms to compensate for the loss of GLUT4 expression in these knockout mice. Fed plasma insulin levels in both genders were elevated five- to sixfold indicating that GLUT4 knockout mice were insulin resistant. Male mice with heterozygous GLUT4 knockout genotype resemble human beings with NIDDM. The mice and the humans have reduced muscle glucose uptake, hypertension, and altered histopathology in heart and in liver (Stenbit *et al.*, 1997). GLUT4 knockout mice also have shown dramatic reduction in growth rates and shortened life span, possibly resulting from the reported cardiac hypertrophy.

IV. Concluding Remarks

The major clinical condition associated with disturbed carbohydrate metabolism is diabetes mellitus. Current wisdom suggests that diabetes is a polygenic disorder aggravated by numerous environmental factors. It stands to reason that a disturbance of glucose transport across tissue barriers would be fundamental to the pathophysiology of diabetes mellitus. However, proof of such a disturbance has been elusive. Several putative mutations in GLUT2 and GLUT4 genes have been described in diabetic patients, but none have withstood the test of time. Nevertheless, it remains likely that GLUT mutations may be contributory to or causative of the diabetic state, and the recent studies on GLUT10 and the linkage with chromosome 20q12–q13.1 seem promising. The animal models, thus far, have been equally tantalizing and at the same time frustrating. The GLUT2 knockout model appears to be more faithful to the Fanconi-Bickel syndrome than to type 1 or type 2 diabetes mellitus. The Fanconi-Bickel syndrome is well established as an autosomal recessive trait, a condition that affects liver and pancreatic function, and simulates a glycogen storage disease. It is the first example of a genetic disorder associated with mutations in a glucose transporter gene (GLUT2).

The Glut1 deficiency syndrome is a second human disorder resulting from mutations in another glucose transporter gene (GLUT1). This clinical condition is transmitted as an autosomal dominant loss of function trait. The clinical syndrome also can be likened to a state of persistent mild hypoglycemia to the extent that the brain glucose concentrations are presumably persistently low, a state of neuroglycopenia, if you will. Glut1 DS is the first example of a genetic disturbance of substrate transport across the blood-brain barrier. Also, it provides a human model for brain energy failure, and this condition can be used to address recent concepts of glial-neuronal metabolic homeostasis (Pellerin and Magistretti, 1994; Magistretti and Pellerin, 1997; Pellerin *et al.*, 1998). Magistretti and colleagues propose that glucose crosses the brain endothelial cell and the astroglial membrane by facilitated diffusion before being converted in the astrocyte to glycogen or being metabolized primarily to pyruvate and lactate. Glial-derived lactate then moves to the neuron for oxidative metabolism to carbon dioxide and water, yielding ATP. The Glut1 deficiency syndrome suggests that glucose transport is limited by the GLUT1 mutation, thereby decreasing glial lactate production, glial glycogen content, and robbing the neuron of its principal supply of fuel. If true, Glut1 deficiency syndrome ultimately causes neuronal dysfunction metabolically, and interferes with the reuptake of extracellular glutamate by the glial cell. Additional studies are necessary to pursue these speculations.

Other human diseases undoubtedly will be found to result from other GLUT mutations. One can only speculate, for example, about the phenotype associated with GLUT3 mutations. Would this also be an infantile epileptic encephalopathy, or does the neuron obtain its fuel mainly from the glial cell in the form of lactate? If so, a GLUT3 mutation might be clinically silent.

The SGLT mutations also have been informative as witnessed by the life-threatening osmotic diarrheal syndrome of infants associated with SGLT1 mutations. The renal glycosuria is a clinically minimal phenotype associated with SGLT2 mutations. To this extent, we may have seen the clinical spectrum associated with mutations of these glucose co-transporters.

V. Future Directions

Much has been learned over the past several decades regarding the transport of glucose across tissue membranes, and this information has helped us understand the biomolecular basis for several clinical syndromes. The glucose-galactose malabsorption syndrome was first described in 1962, and the Fanconi-Bickel syndrome was described in 1949. In 1991, the molecular basis for glucose-galactose malabsorption syndrome was determined to be a mutation involving the SGLT1 gene. Also in that same year, a new syndrome was described, and later named the Glut1 deficiency syndrome (Glut1 DS). In 1997, the molecular basis for the Fanconi-Bickel syndrome was attributed to a mutation involving the GLUT2 gene, and in 1998 the molecular basis for the Glut1 deficiency syndrome was attributed to the mutations involving the GLUT1 gene. In 2001, the newly described GLUT10 gene and the NIDDM susceptibility gene have been co-localized to chromosome 12q12–q13.1.

What can we expect from future studies? It is possible that this line of research ultimately will lead to a better understanding of the diabetic state and to further insights into molecular defects that are associated with NIDDM. Simultaneously, additional animal models will be developed for the various genes responsible for the transport of glucose across tissue membranes including the three SGLT genes and the eight GLUT genes. Our group continues to work on a homologous recombination knockout mouse model for GLUT1 expecting that the heterozygous animals will reflect the Glut1 DS phenotype. This model also will allow us to investigate the pathophysiology of this clinical syndrome and the cellular and structural alterations that cause deceleration of head growth postnatally and microcephaly. Parallel efforts will continue to explore more definitive treatments based on a gene therapy paradigm. The Glut1 deficiency syndrome represents an autosomal dominant loss of function condition, and pharmacological strategies are being pursued to facilitate upregulation of the existing wild-type allele. Our preliminary *in vitro* studies with α-lipoic acid are encouraging in this regard. If successful, this clinical condition could be cured, and it would provide a model for the treatment of other dominant loss of function syndromes. Currently, we know that thioctic acid effectively upregulates Glut1 activity in tissue culture and can correct glucose transport in cultured skin fibroblasts obtained from patients with Glut1 DS. This finding is promising and portends the development of future therapies that will be effective in treating these many human disorders.

Acknowledgments

The authors appreciate support from USPHS Grants NS37949, NS01698, RR00645; the Colleen Giblin Charitable Foundation, and the Will Foundation, and continued participation of patients and families who have taught us much about the clinical implications of these human conditions. We also acknowledge other colleages who are involved in our Glut1 DS studies including Veronica Hinton, Pamela Kranz-Eble, Hong Yang, Romana Kulikova-Schupak, Linda Leary, and Ronald van Heertum; and the many clinicians who have identified many patients and shared their cases with us.

References

Agus, D. B., Gambhir, S. S., Pardridge, W. M., Spielholz, C., Baselga, J., Vera, J. C., and Golde, D. W. (1997). Vitamin C crosses the blood-brain barrier in the oxidized form through the glucose transporters. *Clin. Invest.* **100,** 2842–2848.

Anon. (1996). The use of barbiturates in the control of intracranial hypertension. Brain Trauma Foundation [review]. *J. Neurotrauma* **13,** 711–714.

Antonarakis, S. E. (1998). Recommendation for a nonmenclature system for human gene mutations. *Hum. Mutat.* **11,** 1–3.

Bell, I. G., Burant, C. F., Takeda, J., and Gould, G. W. (1993). Structure and function of mammalian facilitative sugar transporters [review]. *J. Biol. Chem.* **268,** 19161–19164.

Brockman, K., Korenke, C. G., Moers, A. V., Weise, D., Klepper, J., De Vivo, D. C., and Hanefeld, F. (1999). Epilepsy with seizures after fasting and retardation—the first familial case of glucose transporter protein (GLUT1) deficiency. *Eur. J. Paed. Neurol.* **3,** A90–A91.

Brockman, K., Wang, D., Korenke, C. G., Moers, A. V., Ho, Y. Y., Pascual, J. M., Kuang, K. Y., Yang, H., Ma, L., Kranz-Eble, P., Fischbarg, J., Hanefeld, F., and De Vivo, D. C. (2001). Autosomal dominant Glut-1 deficiency syndrome and familial epilepsy. *Ann. Neurol.* **50,** 476–485.

Burchell, A. (1998). Re-evaluation of GLUT 7. *Biochem. J.* **331,** 973.

Cahill, G. F., Herrara, M. G., and Morgan, A. P. (1966). Hormone–fuel interelationships during fasting. *J. Clin. Ivest.* **45,** 1751–1769.

Carayannopoulos, M. O., Chi, M. M.-Y., Pingterhaus, J. M., Mcknight, R. A., Mueckler, M., Devaskar, S. U., and Moley, K. H. (2000). GLUT8 is a glucose transporter responsible for insulin-stimulated glucose uptake in the blastocyst, *Proc. Natl. Acad. Sci. USA* **97,** 7313–7318.

Carruthers, A. (1990). Facilitated diffusion of glucose. *Physiol. Rev.* **70,** 1135–1176.

Challiss, J. R., Taylor, L. P., and Holman, G. D. (1980). Sugar transport asymmetry in human erythrocytes—the effect of bulk haemoglobin removal and the addition of methylxanthines. *Biochem. Biophys. Acta* **602,** 155–166.

Chugani, H. T., Phelps, M. E., and Mazziotta, J. C. (1987). Positron emission tomography study of human brain functional development. *Ann. Neurol.* **22,** 487–497.

De Vivo, D. C. (1993). The expanding clinical spectrum of mitochondrial disease. *Brain Dev.* **15,** 1–22.

De Vivo, D. C., Garcia-Alvarez, M., Ronen, G. M., Behmand, R. A., and Trifiletti, R. R. (1995). Glucose transport protein deficiency: An emerging syndrome with therapeutic implications. *Int. Pediatr.* **10,** 51–56.

De Vivo, D. C., Garcia-Alvarez, M., and Tritschler, H. J. (1996). Deficiency of glucose transporter protein type I: Possible therapeutic role for α-lipoic acid (thioctic acid). *Diabetes Stoffwech.* **5,** 36–40.

De Vivo, D. C., Hirano, M., and DiMauro, S. (1996). Mitochondrial disorders. *In* "Handbook of Clinical Neurology" (P. J. Vinken and G. K. Bruyn, eds.), pp. 389–446. Elsevier Science BV, Amsterdam.

De Vivo, D. C., Trifiletti, R. R., Jacobson, R. I., Ronen, G. M., Behmand, R. A., and Harik, S. I. (1991). Defective glucose transport across the blood-brain barrier as a cause of persistent hypoglycorrhachia, seizures, and developmental delay. *N. Engl. J. Med.* **325,** 703–709.

Doege, H., Schurman, A., Bahrenberg, G., Brauers, A., and Joost, H.-G. (2000a). GLUT8, a novel member of the sugar transport facilitator family with glucose transport activity. *J. Biol. Chem.* **275,** 16275–16280.

Doege, H., Bocianski, A., Joost, H. G., and Schurmann, A. (2000b). Activity and genomic organization of human glucose transporter 9 (GLUT9), a novel member of the family of sugar-transport facilitators predominantly expressed in brain and leucocytes. *Biochem. J.* **350,** 771–776.

El-Barbary, A., Fenstermacher, J. D., and Haspel, H. C. (1996). Barbiturate inhibition of GLUT-1 mediated hexose transport in human erythrocytes exhibits substrate dependence for equilibrium exchange but not unidirectional sugar flux. *Biochemistry* **35,** 15222–15227.

Elsas, L. J., Hillman, R. E., Patterson, J. H., and Rosenberg, L. E. (1970). Renal and intestinal hexose transport in familial glucose-galactose malabsorption. *J. Clin. Invest.* **49,** 576–585.

Englard, S., and Seifter, S. (1986). The biochemical functions of ascorbic acid. *Ann. Rev. Nutr.* **6,** 365–406.

Fanconi, G., and Bickel, H. (1949). Die chronische Aminoacidurie (Aminosaeurediabetes oder nephrotisch-glukosurischer Zwergwuchs) bei der Glykogenose und der Cystinkrankheit. *Helv. Paediatr. Acta* **4,** 359–396.

Fischbarg, J., Kuang, K., Vera, J. C., Arant, S., Silverstein, S. C., Loike, J., and Rosen, O. M. (1990). Glucose transporters serve as water channels. *Proc. Natl. Acad. Sci. USA* **87,** 3244–3247.

Fukumoto, H., Seino, S., Imura, H., Seino, Y., Eddy, R. L., Fukushima, Y., Byers, M. G., Shows, T. B., and Bell, G. I. (1988). Sequence, tissue distribution, and chromosomal localization of mRNA encoding a human glucose transporter-like protein. *Proc. Natl. Acad. Sci. USA* **85,** 5434–5438.

Gjedde, A., and Rasmussen, M. (1980). Pentobarbital anesthesia reduces blood-brain glucose transfer in the rat. *J. Neurochem.* **35,** 1382–1387.

Gould, G. W., and Holman, G. D. (1993). The glucose transporter family: Structure, function and tissue-specific expresion [review]. *Biochem. J.* **295,** 329–341.

Guillam, M. T., Hummler, E., Schaerer, E., Yeh, J. I., Birnbaum, M. J., Beerman, F., Schmidt, A., Deriaz, N., Thorens, B., and Wu, J. Y. (1997). Early diabetes and abnormal postnatal pancreatic islet development in mice lacking Glut-2. *Nat. Genet.* **17,** 327–330.

Haslam, R. H. A. (1996). The nervous system. *In* "Nelson Textbook of Pediatrics" (R. E. Behrman, R. M. Kliegman, and A. M. Arvin, eds.), p. 1699. Saunders, Philadelphia, PA.

Hediger, M. A., Coady, M. J., Ikeda, T. S., and Wright, E. M. (1987). Expression cloning and cDNA sequencing of the Na^+/glucose co-transporter. *Nature* **330,** 379–381.

Ho, Y. Y., Yang, H., Klepper, J., Fischbarg, J., Wang, D., and De Vivo, D. C. (2001a). Glucose transporter type 1 deficiency syndrome (Glut-1DS): Methylxanthines potentiate GLUT1 haploinsufficiency in vitro. *Pediatr. Res.* **50,** 254–260.

Ho, Y. Y., Wang, D., Hinton, V., Yang, H., Vasilescu, A., Engelstad, K., Jhung, S., Hanson, K. K., Wolf, J. A., and De Vivo, D. C. (2001b). Glut-1 deficiency syndrome: Autosomal dominant transmission of the R126C missense mutation. *Ann. Neurol.* **50,** S125.

Honkanen, R. A., McBath, H., Kushmerick, C., Callender, G. E., Scarlata, S. F., Fenstermacher, J. D., and Haspel, H. C. (1995). Barbiturates inhibit hexose transport in cultured mammalian cells and human erythrocytes and interact directly with purified GLUT-1. *Biochemistry* **34,** 535–544.

Hornig, D. (1975). Distribution of ascorbic acid, metabolites and analogues in man and animals. *Ann. N.Y. Acad. Sci.* **258,** 103–118.

Hu, X. J., Peng, F., Zhou, H. Q., Zhang, Z. H., Cheng, W. Y., and Feng, H. F. (2000). The abnormality of glucose transporter in the erythrocyte membrane of chinese type 2 diabetic patients. *Biochem. Biophys. Acta.* **1466,** 306–314.

Hug, G. (1987). Defects in metabolism of carbohydrates. *In* "Nelson Textbook of Pediatrics" (R. E. Behrman, V. C. Vaughan, and W. E. Nelson, eds.), Saunders, Philadelphia, PA.

Ingwar, M., Abdul-Rahman, A., and Siesjo, B. K. (1980). Local cerebral glucose consumption in the artificially ventilated rat: influence of nitrous oxide analgesia and phenobarbital anesthesia. *Acta Physiol. Scand.* **109,** 177–185.

Kalaria, R. N., Gravina, S. A., Schmidley, J. W., Perry, G., and Harik, S. I. (1988). The glucose transporter of the human brain and blood-brain barrier. *Ann. Neurol.* **24,** 757–764.

Klepper, J., Vera, J. C., and De Vivo, D. C. (1998). Deficient transport of dehydroascorbic acid in the glucose transporter protein syndrome. *Ann. Neurol.* **44,** 286–287.

Klepper, J., Garcia-Alvarez, M., O'Driscoll, K. R., Parides, M. K., Wang, D., Ho, Y. Y., and De Vivo, D. C. (1999a). Erythrocyte 3-*O*-methyl-D-glucose uptake assay for diagnosis of glucose-transporter-protein syndrome. *Clin. Lab. Anal.* **13,** 116–121.

Klepper, J., Wang, D., Fischbarg, J., Vera, J. C., Jarjour, I. T., O'Driscoll, K. R., and De Vivo, D. C. (1999b). Defective glucose transport across brain tissue barriers: A newly recognized neurological syndrome. *Neurochem. Res.* **24,** 587–94.

Klepper, J., Fischbarg, J., Vera, J. C., Wang, D., and De Vivo, D. C. (1999c). GLUT1-deficiency: Barbiturates potentiate haploinsufficiency in vitro. *Pediatr. Res.* **46,** 677–683.

Klepper, J., Willemsen, M., Verrips, A., Guertsen, E., Herrmann, R., Kutzick, C., Florcken, A., and Voit, T. (2001). Autosomal dominant transmission of GLUT1 deficiency. *Hum. Mol. Genet.* **10,** 63–68.

Klip, A., Tsakiridis, T., Marette, A., and Ortiz, P. A. (1994). Regulation of expression of glucose transporters by glucose: a review of studies in vivo and in cell cultures. *FASEB J.* **8,** 43–53.

Kulikova-Schupak, R., Ho, Y. Y., Kranz-Eble, P., Yang, H., Wang, D., and De Vivo, D. C. (2001). Stimulation of GLUT-1 gene transcription by thioctic acid and its potential therapeutic value in Glut-1 deficiency syndrome (Glut-1 DS). *J. Inherit. Metab. Dis.* **24,** Suppl. 1, p. 106.

Laplane, R., Polonovski, C., Etienne, M., Debray, P., Lods, J.-C., and Pissarro, B. (1962). L'intolerance aux sucres a transfert intestinal actif. *Arch. Fr. Pediatr.* **19,** 895–944.

Lindquist, B., and Meeuwisse, G. W. (1962). Chronic diarrhoea caused by monosaccharide malabsorption. *Acta Paediatr.* **51,** 674–685.

Longo, N., and Elsas, L. J. (1998). Human glucose transporters. *Adv. Pediatr.* **45,** 293–313.

Magistretti, P. J., and Pellerin, L. (1997). Regulation by neurotransmitters of glial energy metabolism. *Adv. Exp. Med. Biol.* **429,** 137–143.

Martin, M. G., Lostao, M. P., Turk, E., Lam, J., Kreman, M., and Wright, E. M. (1997). Compound missense mutations in the sodium/D-glucose cotransporter result in trafficking defects. *Gastroenterology* **112,** 1206–1212.

McVie-Wylie, A. J., Lamson, D. R., and Chen, Y. T. (2001). Molecular cloning of a novel member of the GLUT family of transporters, SLC2a10 (GLUT10), localized on chromosome 20q13.1: A candidate gene for NIDDM susceptibility. *Genomics* **72,** 113–117.

Mueckler, M., Caruso, C., Baldwin, S. A., Panico, M., Blench, I., Morris, H. R, Allard, W. J., Lienhard, G. E., and Lodish, H. F. (1985). Sequence and structure of a human glucose transporter. *Science* **229,** 941–945.

Mueckler, M., Kruse, M., Stuube, M., Riggs, A. C., Chiu, K. C., and Permutt, M. A. (1994). A mutation in the Glut2 glucose transporter gene of a diabetic patient abolishes transport activity. *J. Biol. Hematol.* **269,** 17765–17767.

Murakami, T., Nishiyama, T., Shirotani, T., Shinohara, Y., Kan, M., Ishii, K., Kanai, F., Nakazuru, S., and Ebina, Y. (1992). Identification of two enhancer elements in the gene encoding the type 1 glucose transporter from the mouse which are responsive to serum, growth factor, and oncogenes. *J. Biol. Chem.* **267,** 9300–9306.

Nilsson, L., and Seisjo, B. K. (1975). The effect of phenobarbitone anesthesia on blood flow and oxygen consumption in the rat brain. *Acta Anaesthesiol. Scand.* **57,** 18–24.

Otsuka, T., Wei, L., Acuff, V. R., Shimizu, A., Pettigrew, K. D., Patlak, C. S., and Fenstermacher, J. (1991a). Variation in local cerebral blood flow response to high-dose pentobarbital sodium in the rat. *Am. J. Physiol.* **261,** H110–H120.

Otsuka, T., Wei, L., Bereczki, D., Acuff, V. R., Patlak, C. S., and Fenstermacher, J. (1991b). Pentobarbital produces dissimilar changes in glucose influx and utilization in brain. *Am. J. Physiol.* **261,** R265–R275.

Owens, O. E., Morgan, A. P., Kemp, H. G., Sullivan, J. M., Herrera, M. G., and Cahill, G. F., Jr. (1967). Brain metabolism during fasting. *J. Clin. Invest.* **46,** 1589–1595.

Padh, H. (1990). Cellular functions of ascorbic acid. *Biochem. Cell. Biol.* **68,** 1166–1173.

Pascual, J. M., van Heertum, R. L., Wang, D., Engelstad, K., and De Vivo, D. C. (2002). Regional deficits in cerebral glucose uptake caused by mutations of the Glut1 transporter. *Ann. Neurol.* (in press).

Pellerin, L., and Magistretti, P. J. (1994). Glutamate uptake into astrocytes stimulates aerobic glycolysis: A mechanism coupling neuronal activity to glucose utilization. *Proc. Natl. Acad. Sci. USA* **91,** 10625–10629.

Pellerin, L., Pellegri, G., Bittar, P. G., Charnay, Y., Bouras, C., Martin, J. L., Stella, N., and Magistretti, P. J. (1998). Evidence supporting the existence of an activity-dependent astrocyte-neuron lactate shuttle. *Dev. Neurosci.* **20,** 291–299.

Rice, J. E., Livingstone, C., and Gould, G. W. (1996). Trafficking, targeting and translocation of the insulin-responsive transporter, GLUT4, in adipocytes. *Biochem. Soc. Trans.* **24,** 540–546.

Saija, A., Princi, P., De Pasquale, R., and Costa, G. (1989). Modifications of the permeability of the blood-brain barrier and local cerebral metabolism in pentobarbital- and ketamine-anaesthetized rats. *Neuropharmacology* **28,** 997–1002.

Santer, R., Schneppenheim, R., Dombrowski, A., Gotze, H., Steinmann, B., and Schaub, J. (1997). Mutations in GLUT2, the gene for the liver-type glucose transporter, in patients with Fanconi-Bickel syndrome. *Nat. Genet.* **17,** 324–326.

Santer, R., Schneppenheim, R., Suter, D., Schaub, J., and Steinmann, B. (1998). Fanconi-Bickel syndrome—the original patient and his natural history, historical steps leading to the primary defect, and a review of the literature. *Eur. J. Pediatr.* **157,** 783–797.

Schneider, A. J., Kinter, W. B., and Stirling, C. E. (1966). Glucose-galactose malabsorption. Report of a case with autoradiographic studies of a mucosal biopsy. *N. Engl. J. Med.* **274,** 305–12.

Seidner, G., Alvarez, M. G., Yeh, J. I., O'Driscoll, K. R, Klepper, J., Stump, T. S., Wang, D., Spinner, N. B., Birnbaum, M. J., and De Vivo, D. C. (1998). GLUT-1 deficiency syndrome caused by haploinsufficiency of the blood-brain barrier hexose carrier. *Nat. Genet.* **18,** 188–191.

St. Denis, J. F., and Cushman, S. W. (1998). Role of SNARE's in the GLUT4 translocation response to insulin in adipose cells and muscle. *J. Basic Clin. Pharmacol.* **9,** 153–165.

Steinfelder, H. J., and Petho-Schramm, S. (1990). Methylxanthines inhibit glucose transport in rat adipocytes by two independent mechanisms. *Biochem. Pharmacol.* **40,** 1154–1157.

Stenbit, A. E., Tsao, T. S., Li, J., Burcelin, R., Geenen, D. L., Factor, S. M., Houseknecht, K., Katz, E. B., and Charron, M. J. (1997). GLUT4 heterozygous knockout mice develop muscle insulin resistance and diabetes. *Nat. Med.* **3,** 1096–1101.

Todd, M. M., and Warner, D. S. (1992). A comfortable hypothesis reevaluated: Cerebral metabolic depression and brain protection during ischemia. *Anesthesiology* **76,** 161–164.

Tsukaguchi, H., Tokui, T., Mackenzie, B., Berger, U. V., Chen, X. Z., Wang, Y., Brubaker, R. F., and Hediger, M. A. (1999). A family of mammalian Na^+-dependent L-ascorbic acid transporters. *Nature* **399,** 70–75.

Turk, E., Zabel, B., Mundlos, S., Dyer, J., and Wright, E. M. (1991). Glucose/galactose malabsorption caused by a defect in the Na^+/glucose cotransporter. *Nature* **350,** 354–356.

Wang, D., Kranz-Eble, P., and De Vivo, D. C. (2000). Mutational analysis of GLUT1 (SLC2A1) in Glut-1 Deficiency Syndrome. *Hum. Mutat.* **16,** 224–231.

Wang, D., Ho, Y. Y., Pascual, J., Hinton, V., Yang, H., Anolik, M., Kranz-Eble, P., Jhung, S., Engelstad, K., and De Vivo, D. C. (2001a). Glut-1 deficiency syndrome (Glut-1DS): R333W genotype and paternal mosaicism. *Ann. Neurol.* **50,** S124.

Wang, D., Pascual, J., Ho, Y. Y., Hinton, V., Yang, H., Kranz-Eble, P., Jhung, S., Engelstad, K., and De Vivo, D. C. (2001b). Glut-1 deficiency syndrome (Glut-1DS): A severe phenotype associated with compound heterozygosity in trans. *Ann. Neurol.* **50,** S125.

Wang, D., Brockman, K., Korenke, C. G., Von Moers, A., Ho, Y. Y., Pascual, J. M., Kuang, K. Y., Yang, H., Ma, L., Kranz-Eble, P., Fischbarg, J., Hanefeld, F., and De Vivo, D. C. (2001c). Glut-1 deficiency syndrome: Autosomal dominant transmission of the R126H missense mutation. *Ann. Neurol.* **50,** S125.

Wei, L., Otsuka, T., Acuff, V., Bereczki, D., Pettigrew, K., Patlak, C., and Fenstermacher, J. D. (1993). The velocities of red cell and plasma flows through parenchymal microvessels of rat brain are decreased by pentobarbital. *J. Cereb. Blood Flow Metab.* **13,** 487–497.

Wright, E. M. (1998). Genetic disorders of membrane transport. I. Glucose galactose malabsorption. *Am. J. Physiol.* **275,** G879–882.

Wright, E. M. (2001). Renal Na(+)-glucose cotransporters. *Am. J. Physiol. Renal Physiol.* **280,** F10–F18.

Yoa, R. G., Rapin, J. R., Wiernsperger, N. F., Martinand, A., and Belleville, I. (1993). Demonstration of defective glucose uptake and storage in erythrocytes from non-insulin dependent diabetic patients and effects of metformin. *Clin. Exp. Pharmacol. Physiol.* **20,** 563–567.

GLUCOSE, STRESS, AND HIPPOCAMPAL NEURONAL VULNERABILITY

Lawrence P. Reagan

Laboratory of Neuroendocrinology, The Rockefeller University
New York 10021

The hippocampus has been recognized to be an important integration center for learning and memory. A number of underlying neurochemical systems within the hippocampus contribute to cognitive function, as well as hippocampal glucoregulatory activities. Metabolic disorders that disrupt the interactions of these systems may compromise hippocampal neuronal homeostasis and contribute to cognitive impairments. Dysregulation of glucose utilization that is observed in diabetes mellitus or Alzheimer disease may represent an important component of the cognitive impairments observed in these disease states. The hippocampus is one of the major target areas for glucocorticoids (GCs) in the central nervous system. While basal levels of glucocorticoids are essential for neuronal maintenance, exposure to stress levels of GCs can reduce hippocampal glucose utilization and produce morphological changes that are accompanied by cognitive impairments. Diabetic subjects exhibit increased circulating levels of GCs, suggesting that neuronal allostatic load may be increased in the hippocampus

0074-7742/02 $35.00

by the combined actions of hyperglycemia and GCs. Accordingly, the aim of this chapter will be to review and update our understanding of the effects of stress and disease states that are associated with metabolic disturbances, such as diabetes mellitus, upon hippocampal glucoregulatory activities.

I. Introduction

A. Stress, GCs, and the Hippocampus

1. *Morphological Studies*

The adrenal gland produces two main classes of steroids, the corticosteroids and the androgens. Corticosteroids are further subdivided into the mineralocorticoids (MCs), such as aldosterone and the glucocorticoids such as cortisol and corticosterone. Both the production and release of the corticosteroids are regulated by various stimuli, including stress (Cade and Perenich, 1965). The physiological responses mediated by the corticosteroids in the mammalian central nervous system (CNS) may be loosely classified into four major groups: (1) developmental actions; (2) changes in behavior and adaptiveness to stress; (3) responsiveness to neuronal damage and mechanisms that promote repair; and (4) effects which potentiate neuronal damage and neuronal death (McEwen, 1989). Examination of the actions of adrenal steroids in the brain reveals an apparent paradox in that GCs and MCs are involved in central functions that are essential for neuronal plasticity, as well as neuronal death. As a result, GCs appear to have a biphasic effect upon neuronal development, function, and survival. The paradoxical actions of GCs are most evident in the hippocampus, a brain region proposed to serve as a critical integration center for memory formation and retrieval (McEwen and Sapolsky, 1995). In this regard, the absence of corticosteroids produces degeneration and neurogenesis of granule cells of the dentate gyrus (Gould and McEwen, 1993), while exposure to stress levels of GCs produces neuronal atrophy, or dendritic remodeling, of hippocampal pyramidal neurons. Specifically, 21-day administration of stress or stress levels of GCs produces selective atrophy of the apical dendrites of CA3 pyramidal neurons, morphological changes not observed in the basal CA3 dendrites, and CA1 pyramidal cells or granule cells of the dentate gyrus (Woolley *et al.*, 1990; Watanabe *et al.*, 1992c). We have previously speculated that the specificity of these stress-induced morphological changes may be due to the dense mossy fiber innervation from the dentate gyrus that terminates upon the apical dendrites of CA3 neurons (Reagan and McEwen,

1997). In support of this hypothesis, electron microscopic analysis revealed that repeated restraint stress produces vesicular rearrangement of mossy fiber terminals, and increased mitochondrial profile areas in mossy fiber terminals (Magariños *et al.*, 1997). Interestingly, atrophy of CA3 pyramidal neurons produced by exogenous application of GCs or repeated restraint stress is a reversible process, in that hippocampal neuronal architecture is restored when exposure to stress or stress levels of GCs is terminated.

A variety of neurotransmitter systems are involved in stress-mediated dendritic remodeling. For example, administration of NMDA receptor antagonists (Magariños and McEwen, 1995), as well as the antiepileptic drug phenytoin, inhibits stress-induced atrophy of CA3 pyramidal neurons (Watanabe *et al.*, 1992a). Tianeptine, a novel antidepressant drug that enhances serotonin uptake, also inhibits stress-induced remodeling (Watanabe *et al.*, 1992b). Lastly, administration of GABA receptor agonists inhibits stress-induced dendritic remodeling, suggesting that hippocampal interneurons also participate in stress-induced morphological changes (Magariños *et al.*, 1999). Collectively, these results suggest that a dynamic interplay exists between GCs and several neurotransmitter systems, including the glutamatergic, serotonergic, and GABAergic systems in the development of stress-induced neuronal remodeling.

In addition to the accumulated data from animal studies, more recent clinical studies have revealed that prolonged exposure to GCs or stress also produces hippocampal atrophy in humans (McEwen, 1997). For example, increases in basal levels of cortisol associated with aging reduce hippocampal volumes (Lupien *et al.*, 1998). Sheline and colleagues (1996) reported that left and right hippocampal volume is reduced in major depression patients, while total cerebral volumes are unchanged. Subsequent investigations confirmed and extended these findings in that reductions in hippocampal volume are observed in recurrent depressive illness patients (Sheline *et al.*, 1999). Since the hippocampus is a major site of GC action in the CNS, these results lead to the suggestion that stress may be responsible for hippocampal specific volume reductions observed in recurrent major depression. Specific decreases in hippocampal volume have also been observed in patients with posttraumatic stress disorder (PTSD) (Bremner *et al.*, 1995; Gurvits *et al.*, 1996) and schizophrenia (Bogerts *et al.*, 1993; Fukuzako *et al.*, 1996). Lastly, studies by Starkman, Schteingart, and co-workers revealed that patients with Cushing disease exhibit decreases in hippocampal formation volume (HFV) that are negatively correlated with urinary free cortisol concentrations (Starkman *et al.*, 1992). More recent studies by these same investigators revealed that decreases in HFV observed in Cushing patients are reversible upon restoration of normal circulating cortisol levels and that the magnitude of the increase in HFV is significantly correlated with the reduction in urinary free cortisol concentrations (Starkman *et al.*, 1999). These results, as well as studies performed in animal models of hippocampal dendritic remodeling, suggest that hippocampal volume reductions observed in affective disorders may be at least partially reversible and regulated in part by GCs.

2. *Hippocampal Glucose Utilization*

In contrast to reversible dendritic atrophy of CA3 pyramidal neurons, exposure to stress levels of GCs for prolonged periods, such as up to several months, produces neuronal atrophy as well as selective hippocampal neuronal loss of CA3 pyramidal cells in rodents and primates (Sapolsky *et al.*, 1985; Kerr *et al.*, 1991; Uno *et al.*, 1989; Sapolsky *et al.*, 1990). Glucocorticoids may also promote neuronal damage during neurotoxic events (Sapolsky, 1985). Glucocorticoid potentiation of ischemia and excitotoxicity-mediated neuronal damage can be prevented by energy supplementation (Sapolsky, 1986), suggesting that GCs create an energy compromised environment which makes hippocampal neurons more vulnerable to neurotoxic events (Reagan and McEwen, 1997). Indeed, GCs have been shown to decrease glucose utilization in both the clinical and the experimental settings. For example, elimination of endogenous GCs by adrenalectomy leads to a significant increase in hippocampal glucose utilization (Kadekaro *et al.*, 1988). Corticosterone and the synthetic GC dexamethasone decrease glucose utilization in hippocampal primary cultures in a dose-dependent fashion (Horner *et al.*, 1990; Virgin, Jr. *et al.*, 1991). Glucocorticoids also potentiate ATP depletion during metabolic insults in hippocampal primary neuronal cultures (Lawrence and Sapolsky, 1994), further supporting the hypothesis that GCs can impair hippocampal neuronal homeostasis. In the clinical setting, a single bolus administration of cortisol reduces hippocampal 2-deoxy-2-[^{18}F]fluoro-D-glucose uptake as determined by positron emission tomography (de Leon *et al.*, 1997). Collectively, these results suggest that GCs may increase neuronal vulnerability by decreasing hippocampal glucose utilization and also suggest that stress-induced dendritic remodeling may represent a form of neuronal plasticity that serves to protect vulnerable hippocampal populations.

3. *Consequences for Cognitive Function*

Morphological changes and neurochemical alterations such as decreases in glucose utilization ultimately contribute to GC-mediated impairments in cognitive function. For example, repeated restraint stress impairs hippocampal-dependent cognitive tests in rodents, including spatial performance in the Y maze and acquisition of spatial memory in the eight-arm radial maze

(Conrad *et al.*, 1996; Luine *et al.*, 1994). Similar to observations in dendritic remodeling, cognitive impairments in these behavioral tests were reversible (Luine *et al.*, 1994) and were blocked by tianeptine administration (Conrad *et al.*, 1996). More recently, stress or GC administration was shown to impair retrieval of spatial memory in the water maze test, an effect that was blocked by the GC synthesis inhibitor metyrapone (de Quervain *et al.*, 1998). In the clinical setting, dexamethasone (Newcomer *et al.*, 1994) or cortisol (Newcomer *et al.*, 1999) administration reversibly impairs declarative memory performance in healthy normal volunteers. Reductions in hippocampal volume resulting from age-related increases in basal levels of GCs are associated with cognitive impairments (Lupien *et al.*, 1998). Under pathophysiological conditions, cortisol hypersecretion (Rubinow *et al.*, 1984) and reductions in hippocampal volume (Sheline *et al.*, 1999) are associated with cognitive impairments in depressive illness. Similarly, reductions in hippocampal volumes observed in PTSD (Bremner *et al.*, 1995), and Cushing disease (Starkman *et al.*, 1992) are associated with memory impairments. Taken together, the results from clinical and preclinical studies demonstrate that GCs produce morphological changes that are associated with cognitive impairments. Fortunately, the current evidence suggests that stress-induced morphological changes and cognitive impairments are reversible. However, prolonged exposure to stress or stress levels of GCs produces irreversible neuronal damage. Identification of the factors that contribute to the transition from reversible dendritic remodeling to irreversible neuronal damage would have significant therapeutic applications in the prevention of neuronal damage following hypoxic/ischemic events, as well as the prevention of such complications in affective disorders. Accordingly, our laboratory has recently initiated experimental paradigms that attempt to accelerate the actions of GCs in the hippocampus as a way of identifying the mechanisms that contribute to GC-mediated plasticity and neuronal death.

II. Streptozotocin Diabetes as a Model of Increased Allostatic/Homeostatic Load

Diabetes mellitus is an endocrine disorder of carbohydrate metabolism resulting from inadequate insulin release (insulin-dependent diabetes mellitus, IDDM, or type 1 diabetes) or insulin insensitivity (noninsulin-dependent diabetes mellitus, NIDDM, or type 2 diabetes). According to current projections, over 200 million people worldwide will be diagnosed with diabetes by the year 2010 (Mandrup-Poulsen, 1998). While the symptomatology of diabetes can be controlled by insulin replacement, diet, and exercise, the long-term complications associated with diabetes can lead to cardiovascular disease, as well as renal and ocular disorders. In addition, diabetes leads to the development of peripheral neuropathies, including motor and sensory polyneuropathies, as well as autonomic neuropathy (Niakan *et al.,* 1986; Brown and Asbury, 1983). The neurological complications of diabetes are not limited to peripheral neuropathies (Biessels *et al.,* 1994; McCall, 1992; Mooradian, 1988) and as will be described in this chapter may include cognitive deficits, along with the morphological, cellular, and molecular alterations. Diabetes is also associated with neuroendocrine dysfunction, such as increased circulating levels of GCs, suggesting that increases in circulating GCs may contribute to the neurological complications associated with diabetes. These results also suggest that the effects of stress may be accelerated and increase hippocampal neuronal vulnerability in diabetic subjects.

In view of the similar metabolic demands that diabetes and corticosteroids exert upon neuronal homeostasis, we have recently initiated studies to examine the combined effects of stress and diabetes as a model of increased neuronal allostatic load. Allostasis may be defined as the ability to achieve stability, or homeostasis, in response to changes to changes in external or internal stimuli; allostatic load is the wear and tear associated with this process (McEwen, 1998). For our model of increased allostatic load, we have chosen the streptozotocin (STZ) diabetic rat. Streptozotocin is an *N*-nitroso derivative of glucosamine that possesses potent diabetogenic properties (Rakieten *et al.,* 1963). Administration of STZ leads to the selective destruction of insulin-producing β-cells of the pancreas, resulting in the hallmark features of diabetes: hyperglycemia, polyuria, polydipsia, and polyphagia (Junod *et al.,* 1969). Accordingly, STZ-induced hyperglycemia has been used extensively to examine the physiological and pathophysiological consequences of diabetes. Among these complications is neuroendocrine dysfunction, including greater sensitivity to both acute and chronic stressful events reflected by exaggerated rises in plasma GCs (Leedom *et al.,* 1987; Scribner *et al.,* 1991). Indeed, sustained elevations in basal corticosterone concentrations in diabetes (Oster *et al.,* 1988) have supported the use of the STZ-treated rat as a paradigm of chronic stress (Scribner *et al.,* 1993). In this model of increased allostatic load, the physiology, morphology, and neurochemistry of euglycemic, nonstressed control rats is compared with STZ-diabetic rats, as well as STZ-diabetic rats subjected to acute stress. For example, nonstressed diabetic rats gain less weight than nonstressed euglycemic control rats, while diabetic rats subjected to acute stress lose weight over the course of the STZ/stress paradigm (Reagan *et al.,* 1999a). Diabetic rats also exhibit increased basal levels of GCs, as well as impaired GC release in response to stressful stimuli (Magariños and McEwen, 2000). Therefore, the STZ-diabetic rat subjected to stress provides an excellent model of increased allostatic load in which to determine whether the actions of GCs may be accelerated in the hippocampus.

A. Accelerated Actions of GCs

To determine if increases in allostatic load would accelerate the actions of GCs in the hippocampus, we initiated studies in our laboratory to examine the morphological changes in diabetic rats and diabetic rats subjected to acute (seven days) restraint stress (Magariños and McEwen, 2000). These studies revealed that STZ-induced diabetes reduced the number of apical branch points, as well as total dendritic length of the apical dendrites of CA3 pyramidal neurons; diabetic rats subjected to acute restraint stress exhibited more pronounced dendritic remodeling. CA3 pyramidal neuron basal dendrites were unaffected by diabetes of the combination of diabetes and stress. These results demonstrate that increases in neuronal homeostatic load accelerate dendritic remodeling in the hippocampus and suggest diabetes may predispose hippocampal neurons to be more vulnerable to subsequent stressors.

The insulin-like growth factor (IGF) system represents another example of how the actions of GCs may be accelerated in diabetic subjects. The IGFs are proposed to mediate a variety of functions in the CNS to promote neuronal development, differentiation, and function (Jones and Clemmons, 1995). GCs represent an important component of the diabetes-mediated regulation of the IGF system. For example, diabetes-induced decreases in plasma IGF-I levels are partially reversed by adrenalectomy; GC replacement restores hyperglycemia-mediated decreases in IFG-I levels (Unterman *et al.*, 1993; Rodgers *et al.*, 1995). Moreover, GC modulation of IGF binding protein expression is most evident in insulin-deficient states (Unterman *et al.*, 1993; Rodgers *et al.*, 1995). In the hippocampus, diabetes or stress alone do not modulate IGF-I receptor (IGF-Ir) mRNA expression, while diabetic rats subjected to acute stress exhibit significant increases in IGF-Ir mRNA expression (Reagan *et al.*, 1999b). In addition, diabetic rats subjected to acute stress exhibit significant increases in hippocampal IGF-IIr mRNA expression compared with nonstressed diabetic rats (Reagan *et al.*, 1999b). Increased expression of IGF receptor mRNA may represent a compensatory mechanism to enhance the activity of the IGF system during periods of increased homeostatic load. These results provide another example of how the actions of stress may be potentiated in a model of increased allostatic load.

III. Diabetes, Stress, and Neuronal Glucose Homeostasis

A. Diabetes Modulation of Glucose Homeostasis

Examination of glucose transport and utilization in the brain during diabetes has produced equivocal results. For example, some investigators have reported that glucose utilization is unaffected by acute hyperglycemia (Duckrow, 1988a,b; Harik and LaManna, 1988) or chronic hyperglycemia (Harik and LaManna, 1988; Kainulainen *et al.*, 1993; Pelligrino *et al.*, 1990). Conversely, other studies have reported that brain glucose utilization is decreased by acute hyperglycemia (Ruderman *et al.*, 1974), as well as chronic hyperglycemia (Gjedde and Crone, 1981). McCall *et al.*, (1982) reported that glucose transport is decreased in several experimental models of diabetes while transport of basic and neutral amino acids remains unaffected, an observation supported by subsequent reports (Mooradian and Morin, 1991). It is possible that methodologies, site of administration, or time following STZ treatment may lead to these disparate observations. For example, a comprehensive study by Deulli and colleagues (1994) demonstrated that glucose utilization is decreased in response to icv streptozotocin administration in many brain regions, including the hippocampus. Interestingly, some regions exhibited little or no change in glucose utilization and some regions exhibited robust increases in glucose utilization. Subsequent studies by these same investigators revealed that CNS glucose utilization is not affected by acute hyperglycemia induced by peripheral STZ administration while chronic hyperglycemia increased glucose utilization in several brain regions, including the dentate gyrus (Duelli *et al.*, 2000). Therefore, the heterogeneity of responses induced by STZ administration in different neuronal populations may help to explain the discrepancies observed in studies examining glucose utilization and glucose uptake in experimentally induced diabetes.

B. Brain Glucose Transporters and Diabetes

The family of facilitative glucose transporter (GLUT) proteins is responsible for the entry of glucose into cells throughout the periphery and the brain (Vannucci *et al.*, 1997b; Maher *et al.*, 1994). The expression, regulation, and activity of glucose transporters plays an essential role in neuronal homeostasis, since glucose represents the primary energy source for the brain (Lund-Anderen, 1979; Pardridge, 1983). While many GLUT isoforms have been identified in the brain, the expression, regulation, and activity of four of these transporters has been extensively studied in the CNS. For example, GLUT5 is expressed in microglial cells, while GLUT1 is highly expressed in capillaries at the blood-brain barrier (BBB) and is responsible for the transport of circulating glucose into the CNS. GLUT4, the insulin-sensitive glucose transporter, is abundantly expressed in the cerebellum, as well as the hypothalamus (LeLoup *et al.*, 1996; Vannucci *et al.*, 1998; Messari *et al.*, 1998; Aplet *et al.*, 1999). More recently, GLUT4 was co-localized with GLUT1 in the BBB of the rat ventromedial hypothalamus (Ngarmukos *et al.*, 2001). In the hippocampus, GLUT4 is associated with pyramidal neuron cell bodies (Messari *et al.*, 1998; Aplet *et al.*, 1999) and is also localized to a select population of parvalbumin-positive hippocampal interneurons (Aplet *et al.*, 1999). GLUT3, the neuron-specific glucose transporter, is responsible for the essential function of glucose delivery into neurons. GLUT3 mRNA and protein exhibit a widespread distribution in the brain, including the hippocampus (Nagamatsu *et al.*, 1992; McCall *et al.*, 1994; Zeller *et al.*, 1995). GLUT3 has been localized to pre- and postsynaptic nerve terminals and small neuronal processes, but is relatively absent from neuronal cell bodies and apical dendrites (McCall *et al.*, 1994; Gerhart *et al.*, 1995; Leino *et al.*, 1997).

Recent cloning of two novel mammalian glucose transporters, known as GLUT8 (GLUTx1) and GLUT6 (formerly GLUT9), was stimulated by the unexpected phenotypes of GLUT2 and GLUT4 knockout mice (Katz *et al.*, 1995; Guillam *et al.*, 1997) and by the ability to search databases for sequence similarities with GLUTs 1–5 (Ibberson *et al.*, 2000). Northern blot analysis has revealed that GLUT8 is abundantly expressed in brain, including the hippocampus (Ibberson *et al.*, 2000). Studies from our laboratory have shown that GLUT8 mRNA is expressed in both principal and nonprincipal neurons of the rat hippocampus (Reagan *et al.*, 2001). Unlike GLUT3, which is localized to the neuropil, GLUT8 protein is localized to the cell bodies and the most proximal apical dendrites of hippocampal neurons (Reagan *et al.*, 2001). GLUT6 is expressed in human brain as determined by Northern blot analysis (Doege *et al.*, 2000). However, the regional and cellular distribution of GLUT6 in the CNS remains to be determined.

The regulation of glucose transporter expression in the brain during diabetes remains controversial. For example, some investigators have reported that GLUT1 mRNA expression is increased in experimental models of diabetes (Choi *et al.*, 1989; Lutz and Pardridge, 1993; Vannucci *et al.*, 1997a), while other studies reported that GLUT1 expression was not altered following chronic hyperglycemia (Kainulainen *et al.*, 1993; Nagamatsu *et al.*, 1994). GLUT1 protein expression, as measured by Western blot analysis, is decreased by acute hyperglycemia (Pardridge *et al.*, 1990) and chronic hyperglycemia (Duelli *et al.*, 2000) in STZ-diabetic rats. Investigations into the regulation of GLUT3 by diabetes have also failed to reach a consensus. In this regard, chronic hyperglycemia induced by STZ did not alter GLUT3 protein expression as determined by immunoblot analysis (Kainulainen *et al.*, 1993). GLUT3 mRNA expression exhibits region-specific regulation during development in *db/db* mice (Vannucci *et al.*, 1997a), but has been reported to be unaffected by STZ administration (Nagamatsu *et al.*, 1994). Our recent studies demonstrated that GLUT3 mRNA is increased in the hippocampus, but not the cortex of diabetic rats, while immunoreactive GLUT3 is not modulated by diabetes (Reagan *et al.*, 1999a). Similarly, chronic hyperglycemia did not modulate GLUT3 protein expression in the hippocampus (Duelli *et al.*, 2000). GLUT8 exhibits similar changes as those observed for GLUT3 in that steady-state mRNA levels are increased in the hippocampus of diabetic rats in the absence of increased GLUT8 protein expression (Reagan *et al.*, 2001). Immunoreactive levels of GLUT4 are decreased in the cerebellum of diabetic rats (Vannucci *et al.*, 1998) and GLUT4 mRNA levels are decreased in the lateral hypothalamic area of hyperglycemic-hyperinsulinemic rats (Alquier *et al.*, 2001). These results demonstrate that the regulation of glucose transporter mRNA and protein expression in the brain in response to diabetes remain equivocal. Some of the discrepancies in these studies may be a result of regionally specific changes that occur in the brain in response to diabetes, differences in experimental paradigms and/or duration of hyperglycemic states.

C. Stress and Diabetes Regulation of Glucose Transporter Expression

In our model of increased allostatic load, we have examined the expression of glucose transporters in the hippocampus of diabetic rats subjected to acute stress. The stress/diabetes modulation of glucose transporter expression and activity may have a significant impact upon hippocampal neuronal homeostasis, as well as hippocampal neuronal vulnerability. Our initial studies focused upon the stress/diabetes regulation of GLUT3 (Reagan *et al.*, 1999a). *In situ* hybridization histochemistry revealed that diabetes-induced increases in GLUT3 mRNA expression were not modulated by acute restraint stress in the hippocampus. Interestingly, while diabetes did not influence GLUT3 protein expression, diabetic rats subjected to acute stress exhibited significant increases in hippocampal GLUT3 protein expression when compared to vehicle-treated, nonstressed controls. These results suggested that STZ/stress-induced increases in GLUT3 mRNA and protein expression in the hippocampus represent a compensatory mechanism to increase glucose utilization during diabetes. However, since our subsequent studies revealed that GLUT3 serves as a target for oxidative stress in the hippocampus of diabetic rats subjected to stress (see Section III.D), the effect of increased GLUT3 protein upon glucose utilization remains to be determined.

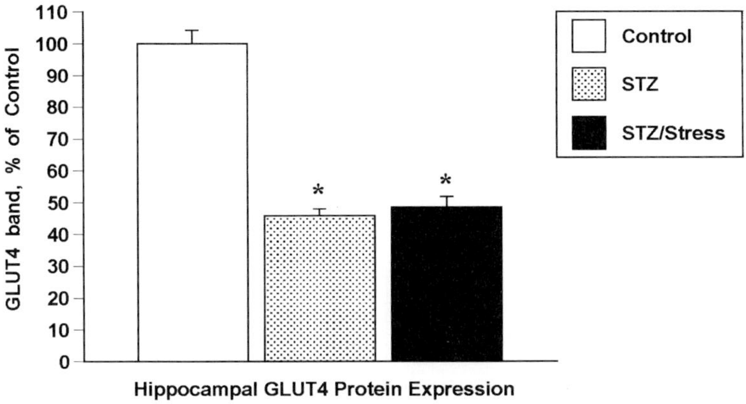

FIG. 1. Autoradiographic film analysis of the GLUT4 immunoreactive band in hippocampal synaptosomal fractions isolated from control, STZ-treated rats, and STZ rats subjected to acute stress. Diabetes and the combination of diabetes and stress significantly reduce membrane association of the insulin-sensitive glucose transporter GLUT4 in the rat hippocampus ($* = p < 0.01$).

As described above, GLUT4 protein expression is modulated in the cortex and cerebellum in experimental models of type 1 and type 2 diabetes (Vannucci *et al.,* 1998). In an attempt to expand upon these previous studies, we examined the regulation of GLUT4 in the hippocampus of diabetic rats subjected to stress. Western blot analysis was performed using synaptosomal membrane fractions isolated from diabetic rats, diabetic rats subjected to stress, and control rats. The results of these studies revealed that GLUT4 protein expression is significantly reduced in the hippocampus of diabetic rats, an effect that was not potentiated by acute restraint stress (Fig. 1). These results demonstrate that hippocampal GLUT4 responds to changes in insulin levels, as seen in the cerebellum and cortex. Moreover, decreased expression of GLUT4 in the hippocampus of diabetic rats and diabetic rats subjected to stress may compromise neuronal glucose availability and increase neuronal vulnerability in the hippocampus.

We recently described the localization and regulation of GLUT8 in the hippocampus of diabetic rats (Reagan *et al.,* 2001). As described above, GLUT8 mRNA levels are increased in the hippocampus of diabetic rats in the absence of concomitant increases in GLUT8 protein expression. More recently, we examined whether acute stress influences diabetes-induced increases in GLUT8 mRNA levels in the rat hippocampus. As shown in Fig. 2, acute restraint stress reverses STZ-induced increases in hippocampal GLUT8 mRNA expression. As described for the IGF system, it appears as if the actions of GCs upon GLUT8 mRNA expression are most evident during

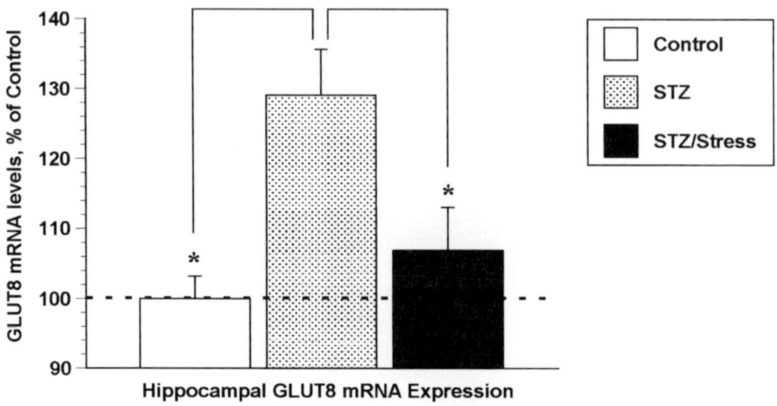

FIG. 2. Autoradiographic film analysis of hippocampal GLUT8 mRNA expression in control rats, diabetic rats, and diabetic rats subjected to acute stress. Statistical analysis revealed that STZ diabetes increases GLUT8 mRNA levels in the hippocampus compared to controls, while acute restraint stress attenuates diabetes-induced increases in GLUT8 mRNA expression. (Data expressed as percentage of control; $* = p < 0.01$.)

insulin deficient states. In order to determine whether diabetes/stress modulation of GLUT8 transcription would influence GLUT8 protein expression, we performed radioimmunocytochemical analysis using GLUT8 selective antisera and ^{35}S-labeled secondary antibodies. In agreement with our previous study, GLUT8 immunolabeling was localized to neuronal cell bodies (Fig. 3A). Autoradiographic image analysis revealed that GLUT8 protein expression was not regulated by STZ diabetes or the combination of diabetes plus acute stress (Fig. 3B). We previously hypothesized that GLUT8 mRNA levels are increased to compensate for increases in protein turnover in the hippocampus of diabetic rats (Reagan *et al.*, 2001). As such, the current results suggest that GCs may reverse diabetes-induced modulation of GLUT8 protein trafficking, eliminating the need for increased mRNA expression in the hippocampus of diabetic rats. Indeed, the stress modulation of glucose transporter expression in diabetic subjects is another example of compensation or modulation to increases in neuronal allostatic load.

D. Oxidative Stress as a Measure of Increased Neuronal Vulnerability

Oxidative stress, lipid peroxidation, and increased production of reactive oxygen species are increased in diabetes (Baynes, 1991; Wolff, 1993) and are proposed to contribute to the development of diabetic encephalopathy

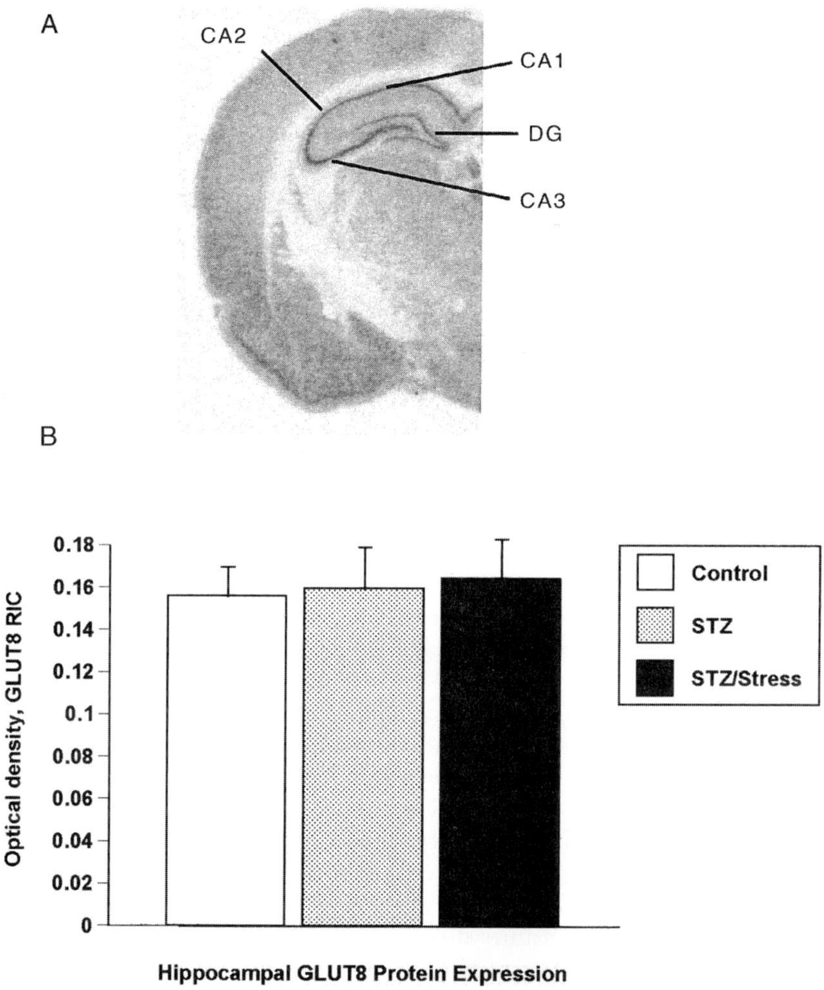

FIG. 3. Radioimmunocytochemical (RIC) analysis of the stress/diabetes regulation of GLUT8 expression. Panel A: representative autoradiograph of GLUT8 protein expression. In agreement with our previous studies, GLUT8 radioimmunolabeling is limited to neuronal cell bodies and the most proximal apical dendrites; adjacent sections incubated in the absence of primary antisera fail to detect GLUT8 protein (data not shown). Panel B: Autoradiographic analysis of GLUT8 protein expression revealed that control rats, diabetic rats, and diabetic rats subjected to acute stress exhibit similar levels of GLUT8 protein in the hippocampus.

(Gispen and Biessels, 2000). For instance, increases in superoxide production are observed in the serum of type 1 diabetic patients, increases that are reduced with improved glycemic control (Ceriello *et al.,* 1991). Moreover, oxidative damage in rat brain is increased by experimentally induced hyperglycemia (Aragno *et al.,* 1997). Lipid peroxidation products are also increased in the brains of type 2 diabetic mice, while the activity of antioxidant enzymes such as catalase and superoxide dismutase (SOD) are decreased (Kumar and Menon, 1993; Makar *et al.,* 1995b). Similar increases in oxidative stress and decreases in antioxidant activity have been reported in STZ-diabetic rats, hyperglycemia-induced alterations that were reversed by insulin replacement (Mooradian, 1995).

Increases in lipid peroxidation products, such as 4-hydroxynonenal (HNE) and malondialdehyde (MDA), reduce the activity of a variety of enzymes that are critical to neuronal homeostasis, including glucose transporters, glutamate transporters, and ion-motive ATPases (Mattson, 1998). HNE has been shown to mediate β-amyloid toxicity of hippocampal primary cultures (Mark *et al.,* 1997), as well as oxidative stress-induced apoptosis of hippocampal primary cultures (Kruman *et al.,* 1997). Exposure to HNE reduces respiration in intact cardiac mitochondria (Humphries *et al.,* 1998), possibly by reducing the activity of α-ketoglutarate dehydrogenase and pyruvate dehydrogenase (Humphries and Szweda, 1998). Similar decreases in pyruvate dehydrogenase activity are observed in the brains of type 2 diabetic mice, suggesting that mitochondrial enzymes may also be targets of oxidative stress in diabetes (Makar *et al.,* 1995a). In support of this hypothesis, HNE has been shown to impair mitochondrial function in brain synaptosomes (Keller *et al.,* 1997). Collectively, these results suggest that diabetes-mediated increases in oxidative stress, decreases in antioxidant enzyme activity, as well as diabetes-induced mitochondrial dysfunction, may make the hippocampus more vulnerable to subsequent stressors.

In an attempt to determine if stress would modulate diabetes-induced increases in oxidative stress, radioimmunocytochemical (RIC) analysis was performed using antisera selective for the lipid peroxidation products HNE and MDA and ^{35}S-labeled secondary antisera. Radioimmunocytochemical analysis performed for MDA revealed that diabetes produced small but nonsignificant increases in MDA levels. However, diabetic rats subjected to acute stress exhibited significant increases in MDA radioimmunoreactive levels compared with controls in all subregions of the hippocampus, including the CA1, CA2, and CA3 regions of Ammon's Horn and the dentate gyrus (Fig. 4). Such results provide another measure of increased allostatic load in the hippocampus of diabetic rats subjected to stress and support the findings from our previous studies that examined HNE protein conjugation. In particular, RIC revealed that HNE protein conjugation is increased in all subregions of the hippocampus of STZ-diabetic rats and STZ-diabetic rats subjected

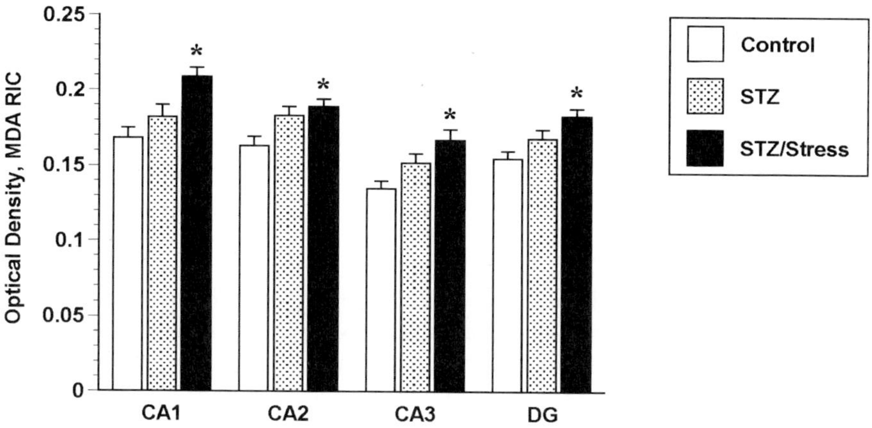

FIG. 4. Autoradiographic film analysis of radioimmunocytochemical labeling for MDA in the hippocampus of control rats, diabetic rats, and diabetic rats subjected to acute stress. Diabetic rats exhibited nonsignificant increases in MDA labeling in the hippocampus. Conversely, diabetic rats subjected to acute stress exhibited significant increases in MDA labeling in the hippocampus compared to control rats, providing another example of increases in allostatic load in the hippocampus of diabetic rats subjected to stress ($* = p < 0.01$).

to 7 days of stress, increases that were not observed in the cortex (Reagan *et al.*, 2000). Because increases in oxidative stress may disrupt hippocampal neuronal homeostasis, we examined whether GLUT3 serves as a target of HNE protein conjugation in the hippocampus of diabetic rats subjected to stress. GLUT3 immunoprecipitated from hippocampal membranes of diabetic rats subjected to stress exhibited significant increases in HNE protein conjugation compared to control rats. HNE protein conjugation of GLUT3 provides a potential mechanism for stress- and diabetes-mediated decreases in hippocampal neuronal glucose utilization. Moreover, these results demonstrate that diabetes and stress increase oxidative stress in the hippocampus and suggest that oxidative stress may contribute to the neurological complications associated with diabetes by disrupting hippocampal glucoregulatory activities.

IV. Stress, Diabetes, and Hippocampal Neuronal Activity: Behavioral Consequences

One of the earliest recognized neurological complications associated with diabetes was cognitive impairments (Miles and Root, 1922; Dashiell, 1930). More recently, the age of onset, but not the duration, of diabetes has been identified as the greatest risk factor in the development of cognitive impairments in children or adolescents with type 1 diabetes (Ryan, 1988; Ryan *et al.*, 1984, 1985; Rovet *et al.*, 1987). In adult type 1 patients, the degree of metabolic control has been identified as a possible risk factor in the development of neuropsychologic deficits (Ryan, 1988; Meuter *et al.*, 1980; Skenazy and Bigler, 1984; Holmes, 1986). Similar findings have been reported in elderly type 2 patients, in that the degree of metabolic control is a key determinant in the appearance of cognitive impairment (Reaven *et al.*, 1990; Gradman *et al.*, 1993; Meneilly *et al.*, 1993). In experimental models of diabetes, STZ administration produces impairments in learning and memory that are accompanied by cortical energy deficits (Lannert and Hoyer, 1998). STZ-diabetic rats also exhibit spatial learning deficits that are accompanied by decreases in hippocampal synaptic plasticity (Biessels *et al.*, 1996), impairments that are reversed by insulin replacement (Biessels *et al.*, 1998). A recent study by Gispen and co-workers demonstrated an interaction between aging and diabetes, in that aged diabetic rats performed more poorly in the water maze than young diabetic rats or nondiabetic aged rats (Kamal *et al.*, 2000).

Increases in oxidative stress have been proposed as contributing factors in the cognitive deficits observed in experimental models of type 1 diabetes (Gispen and Biessels, 2000). However, the relationship between oxidative stress and cognitive function is more firmly established in aging (Blass *et al.*, 2000). For example, age-related increases in oxidative stress are associated with cognitive impairments in senescent accelerated prone (SAMP8) mice and aged gerbils (Butterfield *et al.*, 1997; Carney *et al.*, 1991). Oxidative stress may also contribute to morphological changes in the hippocampus (Reagan *et al.*, 1999b). For example, oxidative stress reduces the number of dendritic spines in CA1 pyramidal neurons, morphological changes that are associated with deficits in passive avoidance behavior (Avila-Costa *et al.*, 1999). Collectively, these findings from clinical and experimental studies support the hypothesis that the neurological complications of diabetes include cognitive impairments and may be manifested by neurochemical and anatomical alterations. The following sections will examine and highlight some of the underlying neurochemical and neuroanatomical alterations that may contribute to or participate in cognitive impairments in diabetic subjects.

A. Neuroanatomical Changes in Diabetes: Relation to Cognition

While the development of peripheral neuropathies is well documented in diabetic patients (Niakan *et al.*, 1986; Brown and Asbury, 1983), the development of CNS complications in diabetes has previously been attributed to degenerative vascular disease (DeJong, 1977). However, seminal investigations by Reske-Nielsen and co-workers introduced the concept of diabetic encephalopathy independent of vascular disease (Reske-Nielsen and Lundbaek, 1963; Reske-Nielsen *et al.,* 1965). These investigators reported numerous morphologic abnormalities in type 1 diabetic patients, including neuronal degeneration in the cortex, cerebellum, and brain stem. Subsequent studies have confirmed and extended these initial findings (Olsson *et al.,* 1968). For example, type 1 diabetic patients exhibit significant increases in lesion size and lesion number when compared to an aged-matched control group (Dejgaard *et al.,* 1990). Cerebral atrophy has also been observed in diabetes patients, a morphological alteration that exhibits a significant interaction with aging (Pirttilä *et al.,* 1992; Perros *et al.,* 1997). For example, cerebral atrophy is significantly greater in diabetes patients in their 80s when compared with diabetic patients in their 60s (Araki *et al.,* 1994). In younger type 1 patients, cerebral atrophy has also been observed; these morphological changes could not be correlated with a number of diabetes-related complications, including disease duration, metabolic control, or the number of hypoglycemic episodes (Lunetta *et al.,* 1994). However, a number of studies have demonstrated that the frequency of hypoglycemic episodes contributes to morphological abnormalities and cognitive impairments in diabetic subjects (Perros and Frier, 1997). In this regard, cerebral atrophy associated with ventricular enlargement, neuronal loss, necrosis, and gliosis are among the morphological deficits observed in diabetic subjects with a history of recurrent severe hypoglycemia (Patrick and Campbell, 1990; Soininen *et al.,* 1992).

Neuroanatomical alterations are also observed in experimental models of type 1 diabetes. In particular, Bessitti and colleagues (1985, 1989) have reported that STZ-diabetic rats exhibit both functional and morphological alterations in the hypothalamus. Specifically, these investigators observed degenerative axons at both the light and electron microscopic levels that were associated with neurochemical and neuroendocrine deficits in the hypothalamus. Other investigators have reported that the long-term effects of STZ-induced diabetes give rise to reductions in brain weight, as well as reductions in neocortical volume and the density of neocortical neurons compared to age-matched controls (Jakobsen *et al.,* 1987). Electron microscopic analysis of the gracile nucleus in the STZ-diabetic rat has revealed morphological abnormalities, including degenerating dendritic and axonal terminals, mitochondrial swelling, and vesicular clustering (Tay and Wong, 1991). Our laboratory has examined the morphological changes that occur in the hippocampus of diabetic rats, including whether acute stress potentiates diabetes-induced morphological changes. As described above (see Section II.A), remodeling of the apical dendrites of CA3 pyramidal neurons is observed in STZ-diabetic rats (Magariños and McEwen, 2000). Moreover, acute stress potentiates neuronal remodeling, producing morphological changes that mimic those seen in euglycemic rats subjected to chronic stress. These results demonstrate that the neurological consequences of stress are accelerated in the hippocampus of diabetic rats, suggesting that hippocampal neuronal vulnerability is increased in diabetic rats subjected to stress. To more directly test this hypothesis, we recently examined the morphological alterations produced by chronic hyperglycemia. Similar to the changes observed by acute hyperglycemia, dendritic remodeling is observed following 3 weeks of STZ-induced diabetes (Magariños *et al.,* 2001). Such results suggest that chronic hyperglycemia or the combined actions of chronic hyperglycemia and chronic stress may produce neuronal atrophy as well as irreversible neuronal damage. To more directly address this possibility, Fluoro-Jade histochemistry, a histological stain that identifies damaged and degenerating neurons (Schmued *et al.,* 1997), was performed on hippocampal sections from diabetic rats subjected to chronic stress. As shown in Fig. 5, panel A, Fluoro-Jade histochemistry identifies damaged neurons in the rat hilus following pilocarpine-induced status epilecticus. Conversely, diabetic rats subjected to 21 days of daily restraint stress do not exhibit Fluoro-Jade staining in the hippocampus, including the hilus (Fig. 5B) and the CA3 region (Fig. 5C). These results suggest that even during periods of increased neuronal vulnerability, stress and diabetes do not produce irreversible neuronal damage at the time points examined. It is interesting to note that while clinical studies revealed degenerating neurons and neuronal loss in the brains of diabetic patients, our studies have not identified similar neuroanatomical features in the hippocampus of STZ-diabetic rats. Nonetheless, histological analysis performed at the light microscopic level cannot provide definitive support for this hypothesis. Indeed, our preliminary electron microscopic studies suggest that signs of irreversible neuronal damage following chronic hyperglycemia are observed in the hippocampus of STZ-diabetic rats (Magariños *et al.,* 2000). Collectively, the results from both the experimental and clinical settings reveal that diabetic encephalopathy is a complication associated with diabetes and suggests that such morphologic abnormalities may contribute to cognitive impairments, especially in elderly diabetic patients.

B. Cellular and Electrophysiological Markers of Cognitive Deficits

A neurological complication that may contribute to diabetes-related cognitive impairments is electrophysiological deficits. In this regard, STZ-diabetic rats exhibit reductions in hippocampal long-term potentiation

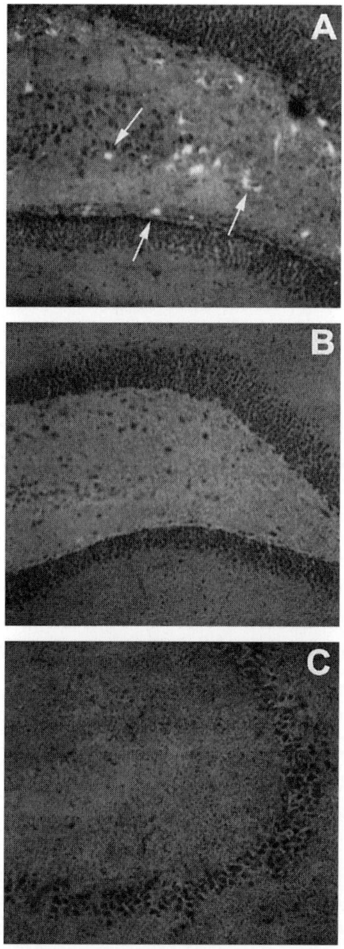

FIG. 5. Representative bright field micrographs of Fluoro-Jade histochemistry in the rat hippocampus. Panel A: Arrows identify degenerating neurons in the rat hilus following pilocarpine-induced status epilepticus, as detected by Fluoro-Jade histochemistry. Fluoro-Jade histochemistry does not detect degenerating neurons in the hilus (panel B) or the CA3 region (panel C) of diabetic rats subjected to chronic stress. These results suggest that during periods of increased allostatic load, irreversible neuronal damage cannot be detected at the light microscopic level in the hippocampus of diabetic rats subjected to chronic stress. (Panel A graciously provided by Daniel R. Rosell, The Rockefeller University, with permission.)

(LTP) (Kamal *et al.,* 1999; Biessels *et al.,* 1996, 1998). LTP, as a long-lasting increase in synaptic strength, has been identified as an electrophysiological marker of hippocampal-dependent learning (Bliss and Collingridge, 1993). In support of this hypothesis, inhibition of LTP impairs learning and memory (Morris *et al.,* 1986). Behavioral studies of STZ-diabetic rats lend further support to this hypothesis in that decreases in hippocampal synaptic plasticity observed in diabetic subjects are also associated with learning impairments. In particular, icv administration of STZ produces memory deficits in the holeboard task and passive avoidance tasks (Lannert and Hoyer, 1998). STZ-diabetic mice exhibit memory impairments in the T maze (Flood *et al.,* 1990), while impairments in water maze performance are observed in STZ-diabetic rats (Biessels *et al.,* 1996, 1998). As described above, the combination of diabetes and aging may produce more profound cognitive and electrophysiological impairments, an experimental paradigm that could be considered as another experimental model of increased allostatic load. In particular, LTP is reduced in aged diabetic rats compared with young diabetic rats or aged nondiabetic rats, electrophysiological deficits that are associated with greater cognitive impairments in these animals (Kamal *et al.,* 2000). These studies demonstrate that the electrophysiological properties of hippocampal neurons are altered by STZ diabetes and suggest that these impairments in LTP provide an electrophysiological marker of cognitive impairments in diabetic subjects.

Neuronal nitric oxide synthase (nNOS), which produces the short-lived second messenger nitric oxide (NO), has been identified as an enzyme that mediates the physiological activities of the glutamatergic system, including hippocampal dependent learning and LTP (Brown *et al.,* 1988; Zorumski and Izumi, 1993). For example, inhibition of NOS activity impairs hippocampal-dependent learning and LTP (Schuman and Madison, 1991; Bohme *et al.,* 1991, 1993; O'Dell *et al.,* 1991; Bon *et al.,* 1992; Ohno *et al.,* 1993). As such, modulation of nNOS expression may contribute to diabetes-induced impairments in these physiological and behavioral activities of the hippocampus. In an attempt to determine whether modulation of nNOS mRNA expression contributes to diabetes-induced cognitive impairments and alterations in hippocampal LTP, we examined the expression of nNOS mRNA in the hippocampus of diabetic rats by *in situ* hybridization histochemistry. Single cell emulsion autoradiography revealed that nNOS mRNA levels are decreased in hippocampal interneurons of diabetic rats and diabetic rats subjected to 7 days of restraint stress compared with vehicle-treated, nonstressed controls (Fig. 6). These results suggest that decreased expression of nNOS mRNA may account for decreases in hippocampal synaptic plasticity and ultimately to impaired hippocampal-dependent learning in diabetic rats.

Undoubtedly a variety of neurochemical alterations contribute to diabetes-induced cognitive deficits observed in the clinical and experimental

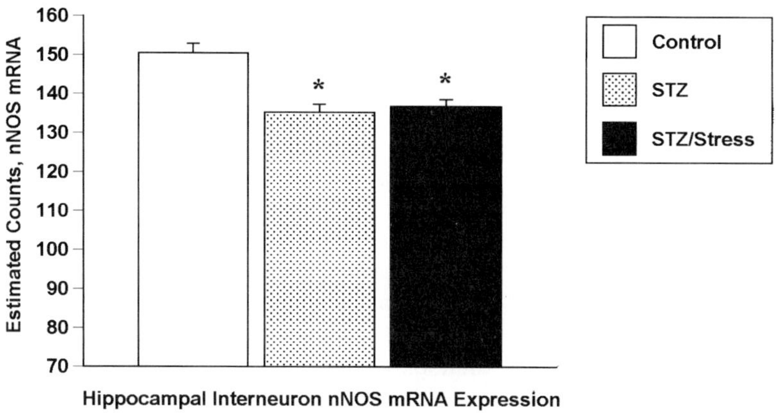

FIG. 6. Statistical analysis of emulsion autoradiography for nNOS mRNA expression in hippocampal interneurons. Diabetic rats and diabetic rats subjected to acute stress exhibit significant decreases in nNOS-reduced silver grains deposited over hippocampal interneurons compared with control rats ($* = p < 0.01$). Decreases in nNOS expression provide a cellular marker for decreases in hippocampal LTP and hippocampal dependent learning observed in diabetic rats.

settings (Biessels *et al.*, 1994; McCall, 1992; Mooradian, 1988). For example, ionotropic glutamate receptor occupation and subsequent activation of Ca^{2+}-dependent enzymes are critical for the induction and maintenance of hippocampal LTP, as well as hippocampal-dependent cognitive tasks (Bliss and Collingridge, 1993; Brown *et al.*, 1988). In this regard, modulation of AMPA receptor expression by Ca^{2+}-dependent processes has been suggested as a potential pharmacological mechanism for reductions in hippocampal LTP in STZ-diabetic rats (Chabot *et al.*, 1997; Gagne *et al.*, 1997). Transcriptional and posttranslational modification of the NMDA receptor may also contribute to diabetes-induced impairments in hippocampal synaptic plasticity (Di Luca *et al.*, 1999). Cerebrovascular complications and modulation of neurotransmitter activity and release, as well as disruption of glucose homeostasis may also contribute to cognitive impairments observed in diabetic patients (Mooradian, 1988). Since NOS inhibitors impair performance of hippocampal-dependent tasks, hyperglycemia-induced decreases in nNOS mRNA expression in hippocampal interneurons may represent another mechanism through which diabetes impairs hippocampal-dependent learning.

C. Hippocampal Glucocoregulation and the Cholinergic System

Based upon evidence from other studies, decreased activity of the cholinergic system may contribute to the cognitive deficits observed in diabetic rats. For example, hippocampal acetylcholine (ACh) release is correlated with cognitive performance in rats (Fadda *et al.,* 2000). Moreover, poor cognitive performance is associated with modulation of components of the cholinergic system, including ACh receptor expression (Smith *et al.,* 1995; Aubert *et al.,* 1995) and the high-affinity choline transporter (Slotkin *et al.,* 1994). Interestingly, transgenic mice that overexpress acetylcholinesterase, thereby decreasing the activity of the cholinergic system, exhibit dendritic atrophy of cortical neurons and increased expression of the high-affinity choline transporter, as well as cognitive impairments (Beeri *et al.,* 1997). In experimental models of type 1 diabetes, regulation of the cholinergic system is equivocal. Hippocampal ACh levels are similar in STZ-diabetic rats when compared with vehicle-treated control rats (Welsh and Wecker, 1991). Conversely, some studies suggest that ACh levels are increased in the hippocampus of diabetic rats (Szutowicz *et al.,* 1994). Central administration of STZ produces cognitive impairments that are associated with decreased activity of the ACh synthesizing enzyme choline aceyltransferase (ChAT). Therefore, while the role of the cholinergic system in cognition is well established, little is known about the modulation of the hippocampal cholinergic system in diabetes and how these changes may relate to the cognitive deficits observed in experimental models of type 1 diabetes.

Previous studies have established an important relationship between hippocampal glucoregulation, the cholinergic system, and cognitive function. For instance, neostigmine-induced hyperglycemia is inhibited by atropine, but not hexamethonium, suggesting that the activation of muscarinic ACh receptors mediate the cholinergic regulation of hippocampal glucoregulatory functions (Uemura *et al.,* 1989). Moreover, hippocampal ACh levels are increased during memory-associated tasks (Ragozzino *et al.,* 1998; Yamamuro *et al.,* 1995; Fadda *et al.,* 1996), an effect that is potentiated by glucose (Ragozzino *et al.,* 1996, 1998). Moreover, glucose administration improves spatial working memory in hippocampal-dependent tasks (Ragozzino *et al.,* 1998; McNay *et al.,* 2000). When compared with glucose utilization in the striatum, hippocampal glucose utilization is greater under basal conditions and is preferentially increased during hippocampal-dependent tasks (McNay *et al.,* 2000). Glucose administration also improves cognitive performance of hippocampal-dependent tasks in aged rats (Winocur and Gagnon, 1998) and in insulin-resistant states produced by high fat diets (Greenwood and Winocur, 2001). Conversely, destruction of cholinergic synapses by icv injection of cholinotoxic antisera specifically reduces hippocampal glucose utilization and produces spatial memory deficits in the water maze test (Browne *et al.,* 2001). Collectively, these studies and others (Messier *et al.,* 1990) suggest that glucose administration may improve cognitive performance by increasing the output of activated cholinergic neurons in the hippocampus.

In addition to morphological changes, GC modulation of glucose utilization and glucoregulatory functions of the hippocampus may contribute to cognitive impairments observed in clinical and experimental studies (Reagan and McEwen, 1997). In this regard, increased plasma levels of GCs observed in diabetic subjects may result in impairment of the glucoregulatory functions of the hippocampus and contribute to hyperglycemia-induced cognitive impairments. Indeed, the glucoregulatory activities of the hippocampus are attenuated in adrenalectomized rats (Uemura *et al.*, 1989), supporting the role of GCs in this hippocampal activity. Glucocorticoids decrease hippocampal glucose utilization in healthy elderly patients, an effect not observed in Alzheimer disease (AD) patients (de Leon *et al.*, 1997). Interestingly, glucose administration improves cognitive performance in AD patients (Manning *et al.*, 1993). Studies by Craft and co-workers (1993) revealed that hyperglycemia, as well as hyperinsulinemia improves cognitive function in AD patients. More recently, these same investigators reported that insulin is necessary for hyperglycemia-mediated improvements in cognitive performance (Craft *et al.*, 1999). Taken together, these results suggest that dysregulation of hippocampal glucoregulatory activities may contribute to the cognitive deficits observed in AD patients (Craft *et al.*, 1993) and perhaps diabetes patients. Indeed, the similarities between diabetes and AD has lead to the hypothesis that the life-long consequences of hyperglycemia may predispose diabetes patients to dementia of the Alzheimer type (Messier and Gagnon, 1996)

V. Future Perspectives

A variety of neurochemical and neuroanatomical abnormalities may contribute to the development of diabetic encephalopathy in the hippocampus of diabetic rats subjected to stress (Table I). These may include changes in glucose transporter expression, oxidative stress-mediated impairments in glucose homeostasis, and morphological abnormalities. The long-term consequences of hyperglycemia may ultimately contribute to cognitive impairments observed in diabetic patients. However, the plasticity of hippocampal neurons following exposure to stress levels of GCs suggests that diabetic encephalopathy may not be a permanent consequence of hyperglycemia. As described above, hippocampal neuronal architecture in animal models and hippocampal formation volumes in humans may be restored when GC levels are returned to basal levels. Such results suggest that the neurological

TABLE I
MARKERS OF INCREASED HIPPOCAMPAL ALLOSTATIC LOAD IN DIABETIC RATS SUBJECTED TO STRESS

	STZ	STZ + stress
I. Cellular/molecular markers		
1. HNE	↑	↑
2. MDA	↔	↑
3. nNOS mRNA	↓	↓
II. Morphology		
1. CA3 dendritic remodeling	↓	↓↓
III. Glucose Transporter expression		
1. GLUT3		
a. mRNA	↑	↑
b. Protein	↔	↑
2. GLUT4		
a. Protein	↓	↓
3. GLUT8		
a. mRNA	↑	↔
b. Protein	↔	↔
IV. Trophic Factors		
1. IGF-I receptor mRNA	↔	↑
2. IGF-II receptor mRNA	↓	↑

Neurological changes observed in the hippocampus of diabetic rats subjected to acute restraint stress. In our model of increased allostatic load, transcriptional, translational, and posttranslational changes, as well as morphological changes, may ultimately contribute to cognitive impairments in diabetic subjects. See text for details.

↔: no change; ↓: decrease; ↑: increase.

complications observed in diabetic subjects may be attenuated or reversed with improved glycemic control, a hypothesis supported by studies in the experimental and clinical settings. The challenge now facing investigators is to develop novel approaches to improve glycemic control while reducing the number of hypoglycemic episodes in diabetic subjects. One approach may include modulation of endocrine function in diabetic subjects, including circulating levels of GCs and dehydroepiandrosterone (DHEA). In this regard, diabetic patients in poor metabolic balance exhibit decreased levels of DHEA, as well as increased circulating levels of cortisol (Couch, 1992; Barret-Connor, 1992; Yamauchi *et al.,* 1996); type 1 patients in good metabolic control exhibit DHEA levels within the normal range (Ebeling *et al.,* 1995). In experimental models of diabetes, DHEA increases insulin sensitivity in type 2 diabetes (Buffington *et al.,* 1993), and increased insulin sensitivity and reduced serum glucose levels to within the normal range in *db/db* and *ob/ob* mice (Coleman *et al.,* 1982, 1984). However, the mechanism(s) of action of DHEA remains to be elucidated. Several studies have suggested that DHEA acts an antiglucocorticoid (Wright *et al.,* 1992; Browne *et al.,* 1992) and DHEA has also been proposed to serve as neurotrophic factor (Compagnone and Mellon, 1998). DHEA prevents oxidative stress-mediated damage induced by hyperglycemia in rat brain (Aragno *et al.,* 1997) and in hippocampal primary neuronal cultures (Bastianetto *et al.,* 1999). In addition, DHEA improves glucose uptake in isolated rat adipocytes (Tatsuo *et al.,* 1999). Such results have led to the proposal of using DHEA as a novel therapeutic agent in the treatment of diabetes (Assan *et al.,* 1996).

Pancreatic islet transplantation represents the most provocative approach in the treatment of type 1 diabetes. While diet, exercise, and insulin replacement improve glycemic control, these strategies may not eliminate all diabetes-related complications, such as hypoglycemic episodes. Recent advances in glucocorticoid-free immunosuppressive regimens provide hope that islet replacement strategies may achieve glycemic control in the absence of hyperglycemia-mediated complications in type 1 patients (Shapiro *et al.,* 2000). The success of the Edmonton Protocol in achieving insulin independence for type 1 patients has provided renewed optimism for islet transplantation. Islet transplantation has also provided provocative results in experimental models of diabetes (Keymeulen *et al.,* 1992; Hoffman *et al.,* 1983). Nonetheless, when examining electrophysiological parameters, islet transplantation is less effective in restoring diabetes-induced deficits in the CNS when compared with the peripheral nervous system (Morano *et al.,* 1996). However, our most recent studies revealed that pancreatic islet transplantation not only restores glycemic control, but also prevents dendritic remodeling of CA3 pyramidal neurons in the hippocampus of diabetic rats (Magariños *et al.,* 2001). These results suggest that islet transplantation may provide insulin independence for type 1 patients and may also prevent some of the neurological complications associated with diabetes. Moreover, these results from the clinical and experimental settings provide exciting new avenues of research to ameliorate or eliminate the neurological complications observed in diabetic patients.

Acknowledgments

The author's work is supported, in part, by a grant from The Juvenile Diabetes Foundational International. The author would like to acknowledge the efforts of his collaborators including: Bruce S. McEwen, Claudia Grillo, and Gerardo Piroli (The Rockefeller University); Anthony L. McCall (University of Virginia Health System); and Maureen J. Charron (Albert Einstein College of Medicine). The author would like to thank Randall R. Sakai for insightful comments and suggestions and Jocelyn Gnirrep for assistance in the final preparation of this chapter.

References

Alquier, T., LeLoup, C., Arnaud, E., Magnan, C., and Pénicaud, L. (2001). Altered Glut4 mRNA levels in specific brain areas of hyperglycemic-hyperinsulinemic rats. *Neurosci. Lett.* **308,** 75–78.

Aplet, J., Melhorn, G., and Schliebs, R. (1999). Insulin-sensitive GLUT4 glucose transporters are colocalized with GLUT3-expressing cells and demonstrate a chemically distinct neuron-specific localization in rat brain. *J. Neurosci. Res.* **57,** 693–705.

Aragno, M., Brignardello, E., Tamagno, E., Danni, O., and Boccuzzi, G. (1997). Dehydroepiandrosterone administration prevents the oxidative damage induced by acute hyperglycemia in rats. *J. Endocrinol.* **155,** 233–240.

Araki, Y., Nomura, M., Tanaka, H., Yamamoto, H., Yamamoto, T., Tsukaguchi, I., and Nakamura, H. (1994). MRI of the brain in diabetes mellitus. *Neuroradiology* **36,** 101–103.

Assan, R., Robel, P., and Baulieu, E.-E. (1996). Dehydroepiandrosterone (DHEA) for diabetic patients. *Eur. J. Endocrinol.* **135,** 37–38.

Aubert, I., Rowe, W., Meaney, M. J., Gauthier, S., and Quirion, R. (1995). Cholinergic markers in aged cognitively impaired Long-Evans rats. *Neuroscience* **67,** 277–292.

Avila-Costa, M. R., Colín-Barenque, L., Fortoul, T. I., Machado-Salas, J. P., Espinosa-Villanueva, J., Rugerio-Vargas, C., and Rivas-Arancibia, S. (1999). Memory deterioration in an oxidative stress model and its correlation with cytological changes on rat hippocampus CA1. *Neurosci. Lett.* **270,** 107–109.

Barret-Connor, E. (1992). Lower endogenous androgen levels and dyslipidemia in men with non-insulin dependent diabetes mellitus. *Ann. Intern. Med.* **117,** 807–811.

Bastianetto, S., Ramassamy, C., Poirer, J., and Quirion, R. (1999). Dehydroepiandrosterone (DHEA) protects hippocampal cells from oxidative stress-induced damage. *Mol. Brain Res.* **66,** 35–41.

Baynes, J. W. (1991). Role of oxidative stress in development of complications in diabetes. *Diabetes* **40,** 405–412.

Beeri, R., Le Novère, N., Mervis, R., Huberman, T., Grauer, E., Changeux, J. P., and Soreq, H. (1997). Enhanced hemicholinium binding and attenuated dendritic branching in cognitively impaired acetylcholinesterase-transgenic mice. *J. Neurochem.* **69,** 2441–2451.

Bestetti, G., Locatelli, V., Tirone, F., Rossi, G. L., and Müller, E. E. (1985). One month of streptozotocin-diabetes induces different neuroendocrine and morphological alterations in the pituitary axis of male and female rats. *Endocrinology* **117,** 208–216.

Bestetti, G. E., Reymond, M. J., Boujon, C. E., Lemarchand-Béraud, T., and Rossi, G. L. (1989). Functional and morphological aspects of impaired TRH release by mediobasal hypothalamus of STZ-induced diabetic rats. *Diabetes* **38,** 1351–1356.

Biessels, G.-J., Kappelle, A. C., Bravenboer, B., Erkelens, D. W., and Gispen, W. H. (1994). Cerebral function in diabetes mellitus. *Diabetologia* **37,** 643–650.

Biessels, G.-J., Kamal, A., Ramakers, G. M., Urban, I. J., Spruijt, B. M., Erkelens, D. W., and Gispen, W. H. (1996). Place learning and hippocampal synaptic plasticity in streptozotocin-induced diabteic rats. *Diabetes* **45,** 1259–1266.

Biessels, G.-J., Kamal, A., Urban, I. J., Spruijt, B. M., Erkelens, D. W., and Gispen, W. H. (1998). Water maze learning and hippocampal synaptic plasticity in streptozotocin-diabetic rats: Effects of insulin treatment. *Brain Res.* **800,** 125–135.

Blass, J. P., Sheu, R. K., and Gibson, G. E. (2000). Inherent abnormalities in energy metabolism in Alzheimer disease. Interaction with cerebrovascular compromise. *Ann. N.Y. Acad. Sci.* **903,** 204–221.

Bliss, T. V. P., and Collingridge, G. L. (1993). A synaptic model of memory: Long-term potentiation in the hippocampus. *Nature* **361,** 31–39.

Bogerts, B., Lieberman, J. A., Ashtari, M., Bilder, R. M., Degreef, G., Lerner, G., Johns, C., and Maiser, S. (1993). Hippocampus-amygdala volumes and psychopathology in chronic schizophrenia. *Biol. Psychiatry* **33,** 236–246.

Bohme, G. A., Bon, C., Stutzmann, J.-M., Doble, A., and Blanchard, J.-C. (1991). Possible involvement of nitric oxide in long-term potentiation. *Eur. J. Pharmacol.* **199,** 379–381.

Bohme, G. A., Bon, C., Lemaire, M., Reibaud, M., Piot, O., Stutzmann, J.-M., Doble, A., and Blanchard, J.-C. (1993). Altered synaptic plasticity and memory formation in nitric oxide synthase inhibitor-treated rats. *Proc. Natl. Acad. Sci. USA* **90,** 9191–9194.

Bon, C., Bohme, G. A., Doble, A., Stutzmann, J.-M., and Blanchard, J.-C. (1992). A role for nitric oxide in long-term potentiation. *Eur. J. Neurosci.* **4,** 420–424.

Bremner, J. D., Randall, P., Scott, T. M., Bronen, R. A., Seibyl, J. P., Southwick, S. M., Delaney, R. C., McCarthy, G., Charney, D. S., and Innis, R. B. (1995). MRI-based measurement of hippocampal volume of patients with combat-related posttraumatic stress disorder. *Am. J. Psychiatry* **152,** 973–981.

Brown, M. J., and Asbury, A. K. (1983). Diabetic neuropathy. *Ann. Neurol.* **15,** 2–12.

Brown, T. H., Chapman, P. F., Kairiss, E. W., and Keenan, C. L. (1988). Long-term synaptic potentiation. *Science* **242,** 724–728.

Browne, E. S., Wright, B. E., Porter, J. R., and Scev, F. (1992). Dehydroepiandrosterone: Antiglucocorticoid action in mice. *Am. J. Med. Sci.* **303,** 366–371.

Browne, S. E., Lin, L., Mattsson, A., Georgievska, B., and Isacson, O. (2001). Selective antibody-induced cholinergic cell and synapse loss produce sustained hippocampal and cortical hypometabolism with correlated cognitive deficits. *Exp. Neurol.* **170,** 36–47.

Buffington, C. K., Pourmotabbed, G., and Kitabchi, A. E. (1993). Case report: amelioration of insulin resistance in diabetics with dehdroepiandrosterone. *Am. J. Med. Sci.* **306,** 320–324.

Butterfield, D. A., Howard, B. J., Yatin, S., Allen, K. A., and Carney, J. M. (1997). Free radical oxidation of brain proteins in accelerated senescence and its modulation by *N-tert*-butyl-a-phenylnitrone. *Proc. Natl. Acad. Sci. USA* **94,** 674–678.

Cade, J., and Perenich, T. (1965). Secretion of aldosterone by rats. *Am. J. Physiol.* **208,** 1026–1030.

Carney, J. M., Starke-Reed, P. E., Oliver, C. N., Landum, R. W., Cheng, M. S., Wu, J. F., and Floyd, R. A. (1991). Reversal of age-related increase in brain protein oxidation, decrease in enzyme activity, and loss in temporal and spatial memory by chronic administration of the spin-trapping compound *N-tert*-butyl-a-phenylnitrone. *Proc. Natl. Acad. Sci. USA* **88,** 3633–3636.

Ceriello, A., Giugliano, D., Quatraro, A., Dello Russo, P., and Lefèbvre, P. J. (1991). Metabolic control may influence the increased superoxide generation in diabetic serum. *Diabetic Med.* **8,** 540–542.

Chabot, C., Massicotte, G., Milot, M., Trudeau, F., and Gagne, J. (1997). Impaired modulation of AMPA receptors by calcium-dependent processes in streptozotocin-induced diabetic rats. *Brain Res.* **768,** 249–256.

Choi, T. B., Boado, R. J., and Pardridge, W. M. (1989). Blood-brain barrier glucose transporter mRNA is increased in experimental diabetes mellitus. *Biochem. Biophys. Res. Commun.* **164,** 375–380.

Coleman, D. L., Leiter, E. H., and Schwizer, R. W. (1982). Therapeutic effects of dehydroepiandrosterone (DHEA) in diabetic mice. *Diabetes* **31,** 830–833.

Coleman, D. L., Schwizer, R. W., and Leiter, E. H. (1984). Effect of genetic background on the therapeutic effects of dehydroepiandrosterone (DHEA) in diabetes-obesity mutants and in aged normal mice. *Diabetes* **33,** 26–32.

Compagnone, N. A., and Mellon, S. H. (1998). Dehydroepiandrosterone: A potential signalling molecule for neocortical organization during development. *Proc. Natl. Acad. Sci. USA* **95,** 4678–4683.

Conrad, C. D., Galea, L. A., Kuroda, Y., and McEwen, B. S. (1996). Chronic stress impairs rat spatial memory on the Y maze, and this effect is blocked by tianeptine pretreatment. *Behav. Neurosci.* **110,** 1321–1334.

Couch, R. M. (1992). Dissociation of cortisol and adrenal androgen secretion in poorly controlled insulin-dependent diabetes mellitus. *Acta Endocrinol.* **127,** 115–117.

Craft, S., Dagogo-Jack, S. E., Wiethop, B. V., Murphy, C., Nevins, R. T., Fleischman, S., Rice, V., Newcomer, J. W., and Cryer, P. E. (1993). Effects of hyperglycemia on memory and hormone levels in dementia of the Alzheimer type: A longitudinal study. *Behav. Neurosci.* **6,** 926–940.

Craft, S., Asthana, S., Newcomer, J. W., Wilkinson, C. W., Matos, I. T., Baker, L. D., Cherrier, M., Lofgreen, C., Latendresse, S., Petrova, A., Plymate, S., Raskind, M., Grimwood, K., and Veith, R. C. (1999). Enhancement of memory in Alzheimer Disease with insulin and somatostatin, but not glucose. *Arch. Gen. Psychiatry* **56,** 1135–1140.

Dashiell, J. F. (1930). Variations in psycho-motor efficiency in a diabetic with changes in blood sugar level. *J. Comp. Psychol.* **10,** 187–189.

de Leon, M. J., McRae, T., Rusinek, H., Convit, A., De Santi, S., Tarshish, C., Golomb, J., Volkow, N., Daisley, K., Orentreich, N., and McEwen, B. S. (1997). Cortisol reduces hippocampal glucose metabolism in normal elderly, but not in Alzheimer's disease. *J. Clin. Endocrinol. Metab.* **82,** 3251–3259.

de Quervain, D. J.-F., Roozendaal, B., and McGaugh, J. L. (1998). Stress and glucocorticoids impair retrieval of long-term spatial memory. *Nature* **394,** 787–790.

Dejgaard, A., Gade, A., Larsson, H., Balle, V., Parving, A., and Parving, H.-H. (1990). Evidence for diabetic encephalopathy. *Diabetic Med.* **8,** 162–167.

DeJong, R. N. (1977). CNS manifestations of diabetes mellitus. *Postgrad. Med.* **61,** 101–107.

Di Luca, M., Ruts, L., Gardoni, F., Cattabeni, F., Biessels, G.-J., and Gispen, W. H. (1999). NMDA receptor subunits are modified transcriptionally and post-translationally in the brain of streptozotocin-diabetic rats. *Diabetologia* **42,** 693–701.

Doege, H., Bocianski, A., Joost, H. G., and Schurmann, A. (2000). Activity and genomic organization of human glucose transporter 9 (GLUT9), a novel member of the family of sugar-transport facilitators predominantly expressed in brain and leucocytes. *Biochem. J.* **350,** 771–776.

Duckrow, R. B. (1988a). Regional cerebral blood flow and glucose utilization during hyperinsulinemia. *Brain Res.* **462,** 363–366.

Duckrow, R. B. (1988b). Glucose transfer into rat brain during acute and chronic hyperglycemia. *Metab. Brain Dis.* **3,** 201–209.

Duelli, R., Schrock, H., Kuschinsky, W., and Hoyer, S. (1994). Intracerebroventricular injections of streptozotocin induces discrete local changes in cerebral glucose utilization in rats. *Int. J. Dev. Neurosci.* **12,** 737–743.

Duelli, R., Maurer, M. H., Staudt, R., Heiland, S., Duembgen, L., and Kuschinsky, W. (2000). Increased cerebral glucose utilization and decreased glucose transporter Glut1 during chronic hyperglycemia in rat brain. *Brain Res.* **858,** 338–347.

Ebeling, P., Stenman, U.-H., Seppälä, M., and Koivisto, V. A. (1995). Androgens and insulin resistance in type 1 diabetic men. *Clin. Endocrinol.* **43,** 601–607.

Fadda, F., Melis, F., and Stancampiano, R. (1996). Increased hippocampal acetylcholine release during a working memory task. *Eur. J. Pharmacol.* **307,** R1–R2.

Fadda, F., Cocco, S., and Stancampiano, R. (2000). Hippocampal acetylcholie release correlates with spatial learning performance in freely moving rats. *Neuroreport* **11,** 2265–2269.

Flood, J. F., Mooradian, A. D., and Morley, J. E. (1990). Characteristics of learning and memory in streptozotocin-induced diabetic mice. *Diabetes* **39,** 1391–1398.

Fukuzako, H., Fukuzako, T., Hashiguchi, T., Hokazono, Y., Takeuchi, K., Hirakawa, K., Ueyama, K., Takigawa, M., Kajiya, Y., Nakajo, M., and Fujimoto, T. (1996). Reduction in hippocampal formation volume is caused mainly by its shortening in chronic schizophrenia: Assessment by MRI. *Biol. Psychiatry* **39,** 938–945.

Gagne, J., Milot, M., Gelinas, S., Trudeau, F., Martinoli, M.-G., and Massicotte, G. (1997). Binding properties of glutamate receptors in streptozotocin-induced diabetes in rats. *Diabetes* **46,** 841–846.

Gerhart, D. Z., Leino, R. L., Borson, N. D., Taylor, W. E., Gronlund, K. M., McCall, A. L., and Drewes, L. R. (1995). Localization of glucose transporter GLUT 3 in brain: comparison of rodent and dog using species-specific carboxyl-terminal antisera. *Neuroscience* **66,** 237–246.

Gispen, W. H., and Biessels, G.-J. (2000). Cognition and synaptic plasticity in diabetes mellitus. *Trends Neurosci.* **23,** 542–549.

Gjedde, A., and Crone, C. (1981). Blood-brain glucose transfer: Repression in chronic hyperglycemia. *Science* **214,** 456–457.

Gould, E., and McEwen, B. S. (1993). Neuronal birth and death. *Curr. Opin. Neurobiol.* **3,** 676–682.

Gradman, T. J., Laws, A., Thompson, L. W., and Reaven, G. M. (1993). Verbal learning and/or memory improves with glycemic control in older subjects with non-insulin-dependent diabetes mellitus. *J. Am. Geriatr. Soc.* **41,** 1305–1312.

Greenwood, C. E., and Winocur, G. (2001). Glucose treatment reduces memory deficits in young adult rats fed high-fed diets. *Neurobiol. Learn. Mem.* 179–189.

Guillam, M. T., Hummler, E., Schaerer, E., Yeh, J. I., Birnbaum, M. J., Beermann, F., Schmidt, A., Deriaz, N., and Thorens, B. (1997). Early diabetes and abnormal postnatal pancreatic islet development in mice lacking Glut-2. *Nat. Genet.* **17,** 327–330.

Gurvits, T. V., Shenton, M. E., Hokama, H., Ohta, H., Lasko, N. B., Gilbertson, M. W., Orr, S. P., Kikinis, R., Jolesz, F. A., McCarley, R. W., and Pitman, R. K. (1996). Magnetic resonance imaging study of hippocampal volume in chronic, combat-related posttraumatic stress disorder. *Biol. Psychiatry* **40,** 1091–1099.

Harik, S. I., and LaManna, J. C. (1988). Vascular perfusion and blood-brain glucose transport in acute and chronic hyperglycemia. *J. Neurochem.* **51,** 1924–1929.

Hoffman, L., Mandel, T. E., Carter, W. M., Koulmanda, M., Martin, F. I., Campbell, D. G., and McMillan, N. (1983). A comparison between islet transplantation and parenteral insulin in the control of diabetes and the prevention of renal complications in mice. *Metabolism* **32,** 451–456.

Holmes, C. S. (1986). Neuropsychological profiles in men with insulin-dependent diabetes. *J. Consult. Clin. Psychol.* **54,** 386–389.

Horner, H. C., Packan, D. R., and Sapolsky, R. M. (1990). Glucocorticoids inhibit glucose transport in cultured hippocampal neurons and glia. *Neuroendocrinology* **52,** 57–64.

Humphries, K. M., and Szweda, L. I. (1998). Selective inactivation of α-ketoglutarate dehydrogenase and pyruvate dehydrogenase: reaction of lipoic acid with 4-hydroxy-2-nonenal. *Biochemistry* **37,** 15835–15841.

Humphries, K. M., Yoo, Y., and Szweda, L. I. (1998). Inhibition of NADH-linked mitochondrial respiration by 4-hydroxy-2-nonenal. *Biochemistry* **37,** 552–557.

Ibberson, M., Uldry, M., and Thorens, B. (2000). GLUT $\times$ 1, a novel mammalian glucose transporter expressed in the central nervous system and insulin-sensitive tissues. *J. Biol. Chem.* **275,** 4607–4612.

Jakobsen, J., Sidenius, P., Gundersen, H. J. G., and Osterby, R. (1987). Quantitative changes of cerebral neocortical structure in insulin-treated long-term streptozotocin-induced diabetes in rats. *Diabetes* **36,** 597–601.

Jones, J. I., and Clemmons, D. R. (1995). Insulin-like growth factors and their binding proteins: Biological actions. *Endocrinol. Rev.* **16,** 3–34.

Junod, A., Lambert, A. E., Stauffacher, W., and Renold, A. E. (1969). Diabetogenic action of streptozotocin: relationship of dose to metabolic response. *J. Clin. Invest.* **48,** 2129–2139.

Kadekaro, M., Ito, M., and Gross, P. M. (1988). Local cerebral glucose utilization is increased in acutely adrenalectomized rats. *Neuroendocrinology* **47,** 329–334.

Kainulainen, H., Schurmann, A., Vilja, P., and Joost, H. G. (1993). *In vivo* glucose uptake and glucose transporter proteins GLUT1 and GLUT3 in brain tissue from streptozotocin-diabetic rats. *Acta Physiol. Scand.* **149,** 221–225.

Kamal, A., Biessels, G.-J., Urban, I. J. A., and Gispen, W. H. (1999). Hippocampal synaptic plasticityin streptozotocin-diabetic rats: impairment of long-term potentiation and facilitation of long-term depression. *Neuroscience* **90,** 737–745.

Kamal, A., Biessels, G.-J., Duis, S. E. J., and Gispen, W. H. (2000). Learning and hippocampal synaptic plasticity in streptozotocin-diabetic rats: interaction of diabetes and ageing. *Diabetologia* **43,** 500–506.

Katz, E. B., Stenbit, A. E., Hatton, K., Depinho, R., and Charron, M. J. (1995). Cardiac and adipose tissue abnormalities but not diabetes in mice deficient in GLUT4. *Nature* **377,** 151–155.

Keller, J. N., Pang, Z., Geddes, J. W., Begley, J. G., Germeyer, A., Waeg, G., and Mattson, M. P. (1997). Impairment of glucose and glutamate transport and induction of mitochondrial oxidative stress and dysfunction in synaptosomes by amyloid β-peptide: Role of the lipid peroxidation product 4-hydroxynonenal. *J. Neurochem.* **69,** 273–284.

Kerr, D. S., Campbell, L. W., Applegate, M. D., Brodish, A., and Landfield, P. W. (1991). Chronic stress-induced acceleration of electrophysiologic and morphometric biomarkers of hippocampal aging. *J. Neurosci.* **11,** 1316–1324.

Keymeulen, B., Teng, H., Vetri, M., Gorus, F., In't Veld, P., and Pipeleers, D. G. (1992). Effect of donor islet mass on metabolic normalization in streptozotocin diabetic rats. *Diabetologia* **35,** 719–724.

Kruman, I., Bruce-Keller, A. J., Bredesen, D., Waeg, G., and Mattson, M. P. (1997). Evidence that 4-hydroxynonenal mediates oxidative stress-induced neuronal apoptosis. *J. Neurosci.* **17,** 5089–5100.

Kumar, J. S. S., and Menon, V. P. (1993). Effect of diabetes on levels of lipid peroxides and glycolipids in rat brain. *Metabolism* **42,** 1435–1439.

Lannert, H., and Hoyer, S. (1998). Intracerebroventricular administration of streptozotocin causes long-term diminutions in learning and memory abilities and in cerebral energy metabolism in adults rats. *Behav. Neurosci.* **112,** 1199–1208.

Lawrence, M. S., and Sapolsky, R. M. (1994). Glucorticoids accelerate ATP loss following metabolic insults in cultured hippocampal neurons. *Brain Res.* **646,** 303–306.

Leedom, L. J., Meehan, W. P., and Zeidler, A. (1987). Avoidance responding in mice with diabetes mellitus. *Physiol. Behav.* **40,** 447–451.

Leino, R. L., Gerhart, D. Z., Van Bueren, A. M., McCall, A. L., and Drewes, L. R. (1997). Ultrastructural localization of GLUT1 and GLUT3 glucose transporters in rat brain. *J. Neurosci. Res.* **49,** 617–626.

LeLoup, C., Arluison, M., Kassis, N., Lepetit, N., Cartier, N., Ferré, P., and Pénicaud, L. (1996). Discrete brain areas express the insulin-responsive glucose transporter GLUT4. *Mol. Brain Res.* **38,** 45–53.

Luine, V., Villegas, M., Martinez, C., and McEwen, B. S. (1994). Repeated stress causes reversible impairments of spatial memory impairments. *Brain Res.* **639,** 167–170.

Lund-Anderen, H. (1979). Transport of glucose from blood to brain. *Physiol. Rev.* **59,** 305–310.

Lunetta, M., Damanti, A. R., Fabbri, G., Lombardo, M., Di Mauro, M., and Mughini, L. (1994). Evidence by magnetic resonance imaging of cerebral alterations of atrophy type in young insulin-dependent diabetic patients. *J. Endocrinol. Invest.* **17,** 214–245.

Lupien, S. J., de Leon, M. J., De Santi, S., Convit, A., Tarshish, C., Nair, N. P. V., McEwen, B. S., Hauger, R. L., and Meaney, M. J. (1998). Cortisol levels during human aging predict hippocampal atrophy and memory deficits. *Nat. Neurosci.* **1,** 69–73.

Lutz, A. J., and Pardridge, W. M. (1993). Insulin therapy normalizes GLUT1 glucose transporter mRNA but not immunoreactive transporter protein in streptozotocin-diabetic rats. *Metabolism* **42,** 939–944.

Magariños, A. M., and McEwen, B. S. (1995). Stress-induced atrophy of apical dendrites of hippocampal CA3c neurons: involvement of glucocorticoid secretion and excitatory amino acid receptors. *Neuroscience* **69,** 89–98.

Magariños, A. M., and McEwen, B. S. (2000). Experimental diabetes in rats causes hippocampal dendritic and synaptic reorganization and increased glucocorticoid reactivity to stress. *Proc. Natl. Acad. Sci. USA* **97,** 11056–11061.

Magariños, A. M., Garcia Verdugo, J. M., and McEwen, B. S. (1997). Chronic stress alters synaptic terminal structure in the hippocampus. *Proc. Natl. Acad. Sci. USA* **94,** 14002–14008.

Magariños, A. M., Deslandes, A., and McEwen, B. S. (1999). Effects of antidepressants and benzodiazepine treatments on the dendritic structure of CA3 pyramidal neurons after chronic stress. *Eur. J. Pharmacol.* **371,** 113–122.

Magariños, A. M., Jain, K., Blount, E. D., Reagan, L., Smith, B. H., and McEwen, B. S. (2000). Implants of agarose macrobeads containing porcine islets prevent hyperglycemia and hippocampal morphological remodeling and damage observed in diabetic rats. *Soc. Neurosci. Abstr.* **26,** 1176.

Magariños, A. M., Jain, K., Blount, E. D., Reagan, L., Smith, B. H., and McEwen, B. S. (2001). Peritoneal implantation of microencapsulated porcine pancreatic islets in diabetic rats ameliorates severe hyperglycemia and prevents retraction and simplification of hippocampal dendrites. *Brain Res.* **902,** 282–287.

Maher, F., Vannucci, S. J., and Simpson, I. A. (1994). Glucose transporter proteins in brain. *FASEB J.* **8,** 1003–1011.

Makar, T. K., Hungund, B. L., Cook, G. A., Kashfi, K., and Cooper, A. J. L. (1995a). Lipid metabolism and membrane composition are altered in the brains of Type II diabetic mice. *J. Neurochem.* **64,** 2159–2168.

Makar, T. K., Rimpel-Lamhaouar, K., Abraham, D. G., Gokhale, V. S., and Cooper, A. J. L. (1995b). Antioxidant defense systems in the brains of Type II diabetic mice. *J. Neurochem.* **65,** 287–291.

Mandrup-Poulsen, T. (1998). Diabetes. *BMJ* **316,** 1221–1225.

Manning, C. A., Ragozzino, M. E., and Gold, P. E. (1993). Glucose enhancement of memory in patients with probable senile dementia of the Alzheimer's type. *Neurobiol. Aging* **14,** 523–528.

Mark, R. J., Lovell, M. A., Markesbery, W. R., Uchida, K., and Mattson, M. P. (1997). A role for 4-hydroxynonenal, an aldehyde product of lipid peroxidation, in disruption of ion homeostasis and neuronal death induced by amyloid β-peptide. *J. Neurochem.* **68,** 255–264.

Mattson, M. P. (1998). Modification of ion homeostasis by lipid peroxidation: roles in neuronal degeneration and adaptive plasticity. *Trends Neurosci.* **21,** 53–57.

McCall, A. L. (1992). The impact of diabetes on the CNS. *Diabetes* **41,** 557–570.

McCall, A. L., Millington, W. R., and Wurtman, R. J. (1982). Metabolic fuel and amino acid transport into the brain in experimental diabetes mellitus. *Proc. Natl. Acad. Sci. USA* **79,** 5406–5410.

McCall, A. L., Van Bueren, A. M., Moholt-Siebert, M., Cherry, N. J., and Woodward, W. R. (1994). Immunohistochemical localization of the neuron-specific glucose transporter (GLUT3) to neuropil in adult rat brain. *Brain Res.* **659,** 292–297.

McEwen, B. S. (1989). Endocrine effects on the brain and their relationship to behavior. *In* "Basic Neurochemistry". (G. J. Siegel, B. W. Agranoff, R. W. Albers, and P. B. Molinoff, eds.), pp. 893–913. Raven Press, New York.

McEwen, B. S. (1997). Possible mechanisms for atrophy of the human hippocampus. *Mol. Psychiatry* **2,** 255–262.

McEwen, B. S. (1998). Protective and damaging effects of stress mediators. *N. Engl. J. Med.* **338,** 171–179.

McEwen, B. S., and Sapolsky, R. M. (1995). Stress and cognitive function. *Curr. Opin. Neurobiol.* **5,** 205–216.

McNay, E. C., Fries, T. M., and Gold, P. E. (2000). Decreases in rat extracellular hippocampal glucose concentration associated with cognitive demand during a spatial task. *Proc. Natl. Acad. Sci. USA* **97,** 2881–2885.

Meneilly, G. S., Cheung, E., Tessier, D., Yakura, C., and Tuokko, H. (1993). The effect of improved glycemic control on cognitive functions in the elderly patient with diabetes. *J. Gerontol.* **48,** M117–M121.

Messari, S. E., LeLoup, C., Quignon, M., Brisorgueil, M.-J., Penicaud, L., and Arluison, M. (1998). Immunocytochemical localization of the insulin-responsive glucose transporter 4 (Glut4) in the rat central nervous system. *J. Comp. Neurol.* **399,** 492–512.

Messier, C., and Gagnon, M. (1996). Glucose regulation and cognitive functions: relation to Alzheimer's disease and diabetes. *Behav. Brain Res.* **75,** 1–11.

Messier, C., Durkin, T., Mrabet, O., and Destrade, C. (1990). Memory-improving action of glucose: indirect evidence for a facilitation of hippocampal acetylcholine synthesis. *Behav. Brain Res.* **39,** 135–143.

Meuter, F., Thomas, W., Grüneklee, D., Gries, F. A., and Lohmann, R. (1980). Psychometric evaluation of performance in diabetes mellitus. *Horm. Metab. Res. Suppl.* **9,** 9–17.

Miles, W. R., and Root, H. F. (1922). Psychologic tests applied to diabetic patients. *Arch. Int. Med.* **30,** 767–777.

Mooradian, A. D. (1988). Diabetic complications of the central nervous system. *Endocrinol. Rev.* **9,** 346–356.

Mooradian, A. D. (1995). The antioxidative potential of cerebral microvessels in experimental diabetes mellitus. *Brain Res.* **671,** 164–169.

Mooradian, A. D., and Morin, A. M. (1991). Brain uptake of glucose in diabetes mellitus: the role of glucose transporters. *Am. J. Med. Sci.* **301,** 173–177.

Morano, S., Sensi, M., Di Gregorio, S., Pozzessere, G., Petrucci, A. F. G., Valle, E., Pugliese, G., Caltabiano, V., Vetri, M., Di Mario, U., and Purrello, F. (1996). Peripheral, but not central, nervous system abnormalities are reversed by pancreatic islet transplantation in diabetic Lewis rats. *Eur. J. Neurosci.* **8,** 1117–1123.

Morris, R. G. M., Anderson, E., Lynch, G. S., and Baudry, M. (1986). Selective impairment of learning and blockade of long-term potentiation by an *N*-methyl-D-aspartate receptor antagonist, AP5. *Nature* **319,** 774–776.

Nagamatsu, S., Kornhauser, J. M., Burant, C. F., Seino, S., Mayo, K. E., and Bell, G. I. (1992). Glucose transporter expression in brain. *J. Biol. Chem.* **267,** 467–472.

Nagamatsu, S., Sawa, H., Inoue, N., Nakamichi, Y., Takeshima, H., and Hoshino, T. (1994). Gene expression of GLUT3 glucose transporter regulated by glucose *in vivo* in mouse brain and *in vitro* in neuronal cell cultures from rat embryos. *Biochem. J.* **300,** 125–131.

Newcomer, J. W., Craft, S., Hershey, T., Askins, K., and Bardgett, M. E. (1994). Glucocorticoid-induced impairmant in declarative memory performance in adult humans. *J. Neurosci.* **14,** 2047–2053.

Newcomer, J. W., Selke, G., Melson, A. K., Hershey, T., Craft, S., Richards, K., and Alderson, A. L. (1999). Decreased memory performance in healthy humans induced by stress-level cortisol treatment. *Arch. Gen. Psychiatry* **56,** 527–533.

Ngarmukos, C., Baur, E. L., and Kumagai, A. K. (2001). Co-localization of GLUT1 and GLUT4 in the blood-brain barrier of the rat ventromedial hypothalamus. *Brain Res.* **900,** 1–8.

Niakan, E., Harati, Y., and Comstock, J. P. (1986). Diabetic autonomic neuropathy. *Metabolism* **35,** 224–234.

O'Dell, T. J., Hawkins, R. D., Kandel, E. R., and Arancio, O. (1991). Tests of the roles of two diffusible substances in long-term potentiation: Evidence for nitric oxide as a possible early retrograde messenger. *Proc. Natl. Acad. Sci. USA* **88,** 11285–11289.

Ohno, M., Yamamoto, T., and Watanabe, S. (1993). Deficits in working memory following inhibition of hippocampal nitric oxide synthesis in the rat. *Brain Res.* **632,** 36–40.

Olsson, Y., Save-Soderbergh, J., Sourander, P., and Angervall, L. (1968). A patho-anatomical study of the central and peripheral nervous system in diabetes of early onset and long duration. *Pathol. Eur.* **3,** 62–79.

Oster, M. H., Castonguay, T. M., Keen, C. L., and Stern, J. S. (1988). Circadian rhythm of corticosterone in diabetic rats. *Life Sci.* **43,** 1643–1645.

Pardridge, W. M. (1983). Brain metabolism: a perspective from the blood-brain barrier. *Physiol. Rev.* **63,** 1481–1535.

Pardridge, W. M., Triguero, D., and Farrell, C. R. (1990). Downregulation of blood-brain barrier glucose transporter in experimental diabetes. *Diabetes* **39,** 1040–1044.

Patrick, A. W., and Campbell, I. A. (1990). Fatal hypoglycycaemia in insulin-treated diabetes mellitus: clinical features and neuropathological changes. *Diabetic Med.* **7,** 349–354.

Pelligrino, D. A., Lipa, M. D., and Albrecht, R. F. (1990). Regional blood-brain barrier glucose transfer and glucose utilization in chronically hyperglycemic, diabetic rats following acute glycemic normalization. *J. Cereb. Blood Flow Metab.* **10,** 774–780.

Perros, P., Deary, I. J., Sellar, R. J., Best, J. J. K., and Frier, B. M. (1997). Brain abnormalities demonstrated by magnetic resonance imaging in adult IDDM patients with and without a history of recurrent severe hypoglycemia. *Diabetes Care* **20,** 1013–1018.

Perros, P., and Frier, B. M. (1997). The long-term sequelae of severe hypoglycemia on the brain in insulin-dependent diabetes mellitus. *Horm. Metab. Res.* **29,** 197–202.

Pirttilä, T., Järvenpää, R., Laippala, P., and Frey, H. (1992). Brain atrophy on computerized axial tomography scans: interaction of age, diabetes and general morbidity. *Gerontology* **38,** 285–291.

Ragozzino, M. E., Unick, K., and Gold, P. E. (1996). Hippocampal acetylcholine release during memory testing in rats: augmentation by glucose. *Proc. Natl. Acad. Sci. USA* **93,** 4693–4698.

Ragozzino, M. E., Pal, S. N., Unick, K., Stefani, M. R., and Gold, P. E. (1998). Modulation of hippocampal acetylcholine release and spontaneous alteration scores by intrahippocampal glucose injections. *J. Neurosci.* **18,** 1595–1601.

Rakieten, N., Rakieten, M. L., and Nadkarni, M. V. (1963). Studies on the diabetogenic action of streptozotocin (NSC-37917). *Cancer Chemother. Rep.* **29,** 91–98.

Reagan, L. P., and McEwen, B. S. (1997). Controversies surrounding glucocorticoid-mediated cell death in the hippocampus. *J. Chem. Neuroanat.* **13,** 149–167.

Reagan, L. P., Magariños, A. M., Lucas, L. R., Van Bueren, A., McCall, A. L., and McEwen, B. S. (1999a). Regulation of GLUT3 glucose transporter in the hippocampus of diabetic rats subjected to stress. *Am. J. Physiol.* **276,** E879–E886.

Reagan, L. P., Magariños, A. M., and McEwen, B. S. (1999b). Neurological changes induced by stress in streptozotocin diabetic rats. *Ann. N.Y. Acad. Sci.* **893,** 126–137.

Reagan, L. P., Magariños, A. M., Yee, D. K., Szweda, L. I., Van Bueren, A., McCall, A. L., and McEwen, B. S. (2000). Oxidative stress and HNE conjugation of GLUT3 are increased in the hippocampus of diabetic rats subjected to stress. *Brain Res.* **862,** 292–300.

Reagan, L. P., Gorovits, N., Hoskin, E. K., Alves, S. E., Katz, E. B., Grillo, C., Pirilo, G., McEwen, B. S., and Charron, M. J. (2001). Localization and regulation of GLUTx1 glucose transporter in the hippocampus of streptozotocin diabetic rats. *Proc. Natl. Acad. Sci. USA* **98,** 2820–2825.

Reaven, G. M., Thompson, L. W., Nahum, D., and Haskins, E. (1990). Relationship between hyperglycemia and cognitive function in older NIDDM patients. *Diabetes Care* **13,** 16–21.

Reske-Nielsen, E., and Lundbaek, K. (1963). Diabetic encephalopathy: Diffuse and focal lesions of the brain in long-term diabetes. *Acta Neurol. Scand.* **39,** 273–290.

Reske-Nielsen, E., Lundbaek, K., and Rafaelsen, O. U. (1965). Pathological changes in the central and peripheral nervous systems of young long-term diabetics. 1. Diabetic encephalopathy. *Diabetologia* **1,** 233–241.

Rodgers, B. D., Strack, A. M., Dallman, M. F., Hwa, L., and Nicholl, C. S. (1995). Corticosterone regulation of insulin-like growth factor I, IGF-binding proteins, and growth in streptozotocin-induced diabetic rats. *Diabetes* **44,** 1420–1425.

Rovet, J. F., Ehrlich, R. M., and Hoppe, M. (1987). Intellectual deficits associated with early onset of insulin-dependent diabetes mellitus in children. *Diabetes Care* **10,** 510–515.

Rubinow, D. R., Post, R. M., Savard, R., and Gold, P. W. (1984). Cortisol hypersecretion and cognitive impairment in depression. *Arch. Gen. Psychiatry* **41,** 279–283.

Ruderman, N. B., Ross, P. S., Berger, M., and Goodman, M. N. (1974). Regulation of glucose and ketone-body metabolism in brain of anaesthetized rats. *Biochem. J.* **138,** 1–10.

Ryan, C. M. (1988). Neurobehavioral complications of Type I diabetes. Examination of possible risk factors. *Diabetes Care* **11,** 86–93.

Ryan, C., Vega, A., Longstreet, C., and Drash, A. (1984). Neuropsychological changes in adolescents with insulin-dependent diabetes. *J. Consult. Clin. Psychol.* **52,** 335–342.

Ryan, C., Vega, A., and Drash, A. (1985). Cognitive deficits in aldolescents who developed diabetes early in life. *Pediatrics* **75,** 921–927.

Sapolsky, R. M. (1985). A mechanism for glucocorticoid toxicity in the hippocampus: Increased neuronal vulnerability to metabolic insults. *J. Neurosci.* **5,** 1228–1232.

Sapolsky, R. M. (1986). Glucocorticoid toxicity in the hippocampus: reversal by supplementation with brain fuels. *J. Neurosci.* **6,** 2240–2244.

Sapolsky, R. M., Krey, L. C., and McEwen, B. S. (1985). Prolonged glucocorticoid exposure reduces hippocampal neuron number: implications for aging. *J. Neurosci.* **5,** 1222–1227.

Sapolsky, R. M., Uno, H., Rebert, C. S., and Finch, C. E. (1990). Hippocampal damage associated with prolonged glucocorticoid exposure in primates. *J. Neurosci.* **10,** 2897–2902.

Schmued, L. C., Albertson, C., and Slikker, Jr. (1997). Fluoro-Jade: A novel fluorochrome for the sensitive and reliable histochemical localization of neuronal degeneration. *Brain Res.* **751,** 37–46.

Schuman, E. M., and Madison, D. V. (1991). A requirement for the intracellular messenger nitric oxide in long-term potentiation. *Science* **254,** 1503–1506.

Scribner, K. A., Walker, C. D., Cascio, C. S., and Dallman, M. F. (1991). Chronic streptozotocin diabetes in rats facilitates the acute stress response without altering pituitary or adrenal responsiveness to secretagogues. *Endocrinology* **129,** 99–108.

Scribner, K. A., Akana, S. F., Walker, C. D., and Dallman, M. F. (1993). Streptozotocin-diabetic rats exhibit facilitated adrenocorticotropin responses to acute stress, but normal sensitivity to feedback by corticosteroids. *Endocrinology* **133,** 2667–2674.

Shapiro, A. M. J., Lakey, J. R. T., Ryan, E. A., Korbutt, G. S., Toth, E., Warnock, G. L., Kneteman, N. M., and Rajotte, R. V. (2000). Islet transplantation in seven patients with type 1 diabetes mellitus using a glucocorticoid-free immunosuppressive regimen. *New Eng. J. Med.* **343,** 230–238.

Sheline, Y. I., Wang, P. W., Gado, M. H., Csernansky, J. G., and Vannier, M. W. (1996). Hippocampal atrophy in recurrent major depression. *Proc. Natl. Acad. Sci. USA* **93,** 3908–3913.

Sheline, Y. I., Sanghavi, M., Mintun, M. A., and Gado, M. H. (1999). Depression duration but not age predicts hippocampal volume loss in medically healthy women with recurrent major depression. *J. Neurosci.* **19,** 5034–5043.

Skenazy, J. A., and Bigler, E. D. (1984). Neuropsychological findings in diabetes mellitus. *J. Clin. Psychol.* **40,** 246–258.

Slotkin, T. A., Nemeroff, C. B., Bissette, G., and Seidler, F. J. (1994). Overexpression of the high affinity choline transporter in cortical regions affected by Alzheimer's Disease. *J. Clin. Invest.* **94,** 696–702.

Smith, T. D., Gallagher, M., and Leslie, F. M. (1995). Cholinergic binding sites in rat brain: analysis by age and cognitive status. *Neurobiol. Aging* **16,** 161–173.

Soininen, H., Puranen, M., Helkala, E.-L., Laakso, M., and Riekkinen, P. J. (1992). Diabetes mellitus and brain atrophy: a computed tomography study in an elderly population. *Neurobiol. Aging* **13,** 717–721.

Starkman, M. N., Gebarski, S. S., Berent, S., and Schteingart, D. E. (1992). Hippocampal formation volume, memory dysfunction, and cortisol levels in patients with Cushing's syndrome. *Biol. Psychiatry* **32,** 756–765.

Starkman, M. N., Giordani, B., Gebarski, S. S., Berent, S., Schork, M. A., and Schteingart, D. E. (1999). Decrease in cortisol reverses human hippocampal atrophy following treatment of Cushing's Disease. *Biol. Psychiatry* **46,** 1595–1602.

Szutowicz, A., Tomaszewicz, M., Jankowska, A., and Kisielevski, Y. (1994). Acetylcholine synthesis in nerve terminals of diabetic rats. *Neuroreport* **5,** 2421–2424.

Tatsuo, I., Kajita, K., Miura, A., Ishizawa, M., Kanoh, Y., Itaya, S., Kimura, M., Muto, N., Mune, T., Morita, H., and Yasuda, K. (1999). DHEA improves glucose uptake via activations of protein kinase C and phosphatidylinositol 3-kinase. *Am. J. Physiol.* **276,** E196–E204.

Tay, S. S. W., and Wong, W. C. (1991). Gracile nucleus of streptozotocin-induced diabetic rats. *J. Neurocytol.* **20,** 356–364.

Uemura, K., Iguchi, A., Yatomi, A., Miura, H., Honmura, A., Yanase, M., and Sakamoto, N. (1989). Involvement of the hippocampus in central nervous system-mediated glucoregulation in rats. *Endocrinology* **124,** 2449–2455.

Uno, H., Tarara, R., Else, J. G., Suleman, M. A., and Sapolsky, R. M. (1989). Hippocampal damage associated with prolonged and fatal stress in primates. *J. Neurosci.* **9,** 1705–1711.

Unterman, T. G., Jentel, J. J., Oehler, D. T., Lacson, R. G., and Hofert, J. F. (1993). Effects of glucocorticoids on circulating levels and hepatic expression of insulin-like growth factor (IGF)-binding proteins and IGF-I in the adrenalectomized streptozotocin-diabetic rat. *Endocrinology* **133,** 2531–2539.

Vannucci, S. J., Gibbs, E. M., and Simpson, I. A. (1997a). Glucose utilization and glucose transporter proteins GLUT-1 and GLUT-3 in brains of diabetic (*db/db*) mice. *Am. J. Physiol.* **272,** E267–E274.

Vannucci, S. J., Maher, F., and Simpson, I. A. (1997b). Glucose transporter proteins in brain: Delivery of glucose to neurons and glia. *Glia* **21,** 2–21.

Vannucci, S. J., Koehler-Stec, E. M., Li, K., Reynolds, T. H., Clark, R., and Simpson, I. A. (1998). GLUT4 glucose transporter expression in rodent brain: effects of diabetes. *Brain Res.* **797,** 1–11.

Virgin, Jr., C. E., Ha, T. P.-T., Packan, D. R., Tombaugh, G. C., Yang, S. H., Horner, H. C., and Sapolsky, R. M. (1991). Glucocorticoids inhibit glucose transport and glutamate uptake in hippocampal astrocytes: implications for glucocorticoid neurotoxicity. *J. Neurochem.* **57,** 1422–1428.

Watanabe, Y., Gould, E., Cameron, H. A., Daniels, D. C., and McEwen, B. S. (1992a). Phenytoin prevents stress- and corticosterone-induced atrophy of CA3 pyramidal neurons. *Hippocampus* **2,** 431–436.

Watanabe, Y., Gould, E., Daniels, D. C., Cameron, H., and McEwen, B. S. (1992b). Tianeptine attenuates stress-induced morphological changes in the hippocampus. *Eur. J. Pharmacol.* **222,** 157–162.

Watanabe, Y., Gould, E., and McEwen, B. S. (1992c). Stress induces atrophy of apical dendrites of hippocampal CA3 pyramidal neurons. *Brain Res.* **588,** 341–345.

Welsh, B., and Wecker, L. (1991). Effects of streptozotocin-induced diabetes on acetylcholine metabolism in rat brain. *Neurochem. Res.* **16,** 453–460.

Winocur, G., and Gagnon, S. (1998). Glucose treatment attenuates spatial learning and memory deficts of aged rats on tests of hippocampal function. *Neurobiol. Aging* **19,** 233–241.

Wolff, S. P. (1993). Diabetes mellitus and free radicals. *Br. Med. Bull.* **49,** 642–652.

Woolley, C. S., Gould, E., and McEwen, B. S. (1990). Exposure to excess glucocorticoids alters dendritic morphology of adult hippocampal pyramidal neurons. *Brain Res.* **531,** 225–231.

Wright, B. E., Porter, J. R., Browne, E. S., and Svec, F. (1992). Antiglucocorticoid action of dehydroepiandrosterone in young obese Zucker rats. *Int. J. Obes. Relat. Metab. Disord.* **16,** 579–583.

Yamamuro, Y., Hori, K., Tanaka, J., Iwano, H., and Nomura, M. (1995). Septo-hippocampal cholinergic system under the discrimination learing task in the rat: a microdialysis study with the dual-probe approach. *Brain Res.* **684,** 1–7.

Yamauchi, A., Takei, I., Kasuga, A., Kitamura, Y., Ohashi, N., Nakano, S., Takayama, S., Nakamoto, S., Katsukawa, F., and Saruta, T. (1996). Depression of dehydroepiandrosterone in Japanese diabetic men-comparison between non-insulin-dependent diabetes mellitus and impaired glucose tolerance. *Eur. J. Endocrinol.* **135,** 101–104.

Zeller, K., Duelli, R., Vogel, J., Schrock, H., and Kuschinsky, W. (1995). Autoradiographic analysis of the regional distribution of Glut3 glucose transporters in the rat brain. *Brain Res.* **698,** 175–179.

Zorumski, C. F., and Izumi, Y. (1993). Nitric oxide and hippocampal synaptic plasticity. *Biochem. Pharmacol.* **46,** 777–785.

GLUCOSE/MITOCHONDRIA IN NEUROLOGICAL CONDITIONS

John P. Blass

Winifred Masterson Burke Professor of Neurology, Neuroscience and Medicine
Weill Medical College of Cornell University, Burke Medical Research Institute
White Plains, New York 10605

0074-7742/02 $35.00

I. Introduction

A. Historical Aspects

A relationship between impairments of glucose oxidation and deficiencies in neurological function was recognized even before the major metabolic pathways of glucose utilization had been worked out. In 1932, Judah Quastel wrote:

> The mental symptoms accompanying anoxemia (as, for instance, that following ascent to high altitudes) are well known. They include loss of judgment and memory, disorientation for time, irritability, and emotional instability. Abnormal mental symptoms accompany carbon monoxide poisoning, and there seems to be little question that anoxemia of the brain leads to irrational behavior. Anoxemia may not only be created by lack of oxygen, however, but by conditions set up which render the oxygen unavailable for oxidative purposes. Hence disturbances in the nervous system which result in diminished rates of oxidation will be as productive of mental disorder as lack of oxygen alone.

Scientists and clinicians promptly accepted this view in regard to secondary causes of impaired brain metabolism, such as those cited by Quastel (1932), but for over thirty years the existence of significant genetic deficiencies in oxidative/energy metabolism was disputed. The widely held assumption during those years was that genetic abnormalities severe enough to interfere with glucose oxidation would be fatal *in utero*. Consensus views in medicine have often been disproven by close study of a patient, and that occurred with genetic disorders of energy metabolism (Blass *et al.*, 1970). In fact, genetic variants that reduce but do not abolish glucose oxidation have proven not only to be compatible with life but to be common.

A common generalization is that the more severe a biological defect, the earlier the onset of the resulting illness and the worse the disability. For instance, a mutation that abolishes the activity of an enzyme is more likely to cause severe illness in infancy while a mutation that allows residual activity is more likely to lead to milder disease in adolescence or adult life. This view depends in part on an experimental tautology. When the activities of enzymes measured *in vitro* do not correlate with the severity of clinical illness, gifted investigators often devise new assays—if necessary even assays using intact cells—to find measurements that agree with the theory. In the literature on disorders of glucose metabolism that affect the nervous system, abnormalities that nearly abolish glucose oxidation have usually been identified in very sick infants. Milder abnormalities have tended to be associated with later onset and milder illness. Even milder abnormalities in the apparatus for the oxidation of glucose are now well documented in neurodegenerative diseases of aging, including such common one as Alzheimer (Blass, 2000) and Parkinson disease (Hu *et al.*, 1999, 2000).

Disorders of mitochondria can damage cells by leading to excessive levels of reactive oxygen species (ROS) and other free radicals (Floyd, 1999; Hu *et al.*, 1999). Our group has hypothesized that deficiencies in critical enzymes of oxidative metabolism in neurodegenerative diseases might arise from interaction between ROS and variations in the genes which render the encoded enzyme proteins more susceptible to damage by free radicals. Furthermore, the quenching of ROS requires reducing equivalents (electrons). Reducing equivalents are generated in the major pathways of glucose metabolism, i.e., as NADH and NADPH. Impairments in glucose metabolism are therefore likely to impair the physiological mechanisms that quench free radicals and thereby contribute to free radical damage. This idea has been formulated as "the mitochondrial spiral" (Fig. 1) (Blass, 2000).

B. Major Pathways of Glucose Metabolism

These are the same in the brain and other organs and in other species including plants. Textbooks of biochemistry describe these pathways and their regulation in detail. The pathways are *glycolysis* (the break down of glucose to pyruvate and lactate); the *pyruvate dehydrogenase* (PDHC) *complex* (the entry of pyruvate into oxidative metabolism); the Krebs *tricarboxylic acid* (TCA)

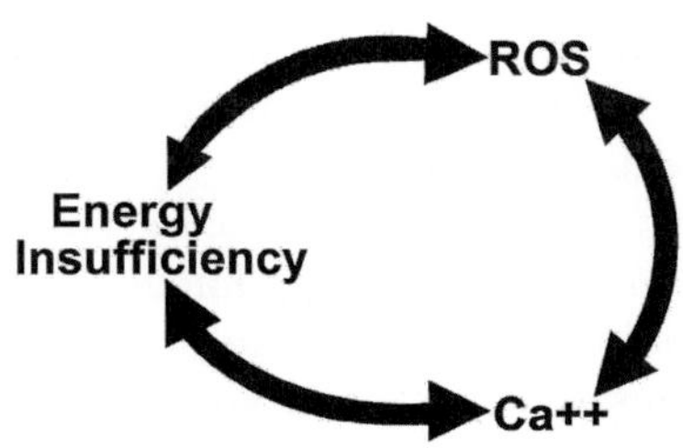

FIG. 1. The mitochondrial spiral. Impairments of energy metabolism, alterations in cellular calcium homeostasis and excess free radicals (ROS) interact with each other in mitochondria; inducing any one of them leads to abnormalities in the other two. The interaction can set up a deleterious, downward cycle. From Blass (2000), The mitochondrial spiral: An adequate cause of dementia in the Alzheimer syndrome, *Ann. N.Y. Acad. Sci.* **924,** 170–183. See Blass, 2000, for more detailed discussion.

cycle (the break down of pyruvate to CO_2 and reducing equivalents); and the *electron transport chain* (oxidation of reducing equivalents to water). Glycolysis occurs in the cytoplasm, and the enzymes involved are soluble when tissues are disrupted under conditions that leave intracellular organelles intact. PDHC, the TCA cycle, and electron transport occur in mitochondria, and the relevant enzymes travel with mitochondria when homogenates are fractionated. Transport of glucose and other substrates into the brain can also become rate-limiting (Pardridge, 1999).

The TCA cycle is important in amino acid as well as in glucose metabolism. The cycle oxidizes the products of amino acid metabolism. It also converts sugar metabolites into fragments that can be readily converted to amino acids by transamination. Glutamate is an important amino acid neurotransmitter that can be derived by the transamination of the TCA cycle intermediate α-ketoglutarate. In mammalian brain, the counts from [^{14}C]glucose accumulate more rapidly in glutamate than they do in CO_2, probably because the radioactive material is diluted by a large pool of unlabeled glutamate. Another important neurotransmitter that can be derived from glutamate and therefore from glucose is γ-aminobutyric acid (GABA), via catalysis by the enzyme glutamate decarboxylase (GAD) which is a marker for "GABAminergic" neurons.

Several other pathways are quantitatively much less important in brain than in other tissues. The oxidation of fats through the TCA cycle is normally almost negligible in brain *in vivo* although it is very important in many other tissues including the heart. In the very young, however, and in people undergoing starvation, the brain can efficiently oxidize ketone bodies derived from fat—namely acetoacetic acid and β-hydroxybutyrate. Anaplerosis (the conversion of three-carbon compounds to four-carbon compounds) and gluconeogenesis (the synthesis of new molecules of glucose) are quantitatively much more important in liver than they are in brain. The pentose phosphate pathway (an alternative pathway for the oxidation of glucose) can be detected in brain by sensitive methods. Brain contains some glycogen, which may supply fuel when metabolic demands outstrip the supply of glucose from the blood to the brain.

The levels of activity of enzymes of glucose metabolism vary markedly in mammalian brain. Table I lists reported activities for samples of human brain. These rates should be taken only as approximations for a tissue as large and heterogeneous as the brain. However, certain enzymes appear to have relatively low activities and appear to be potential control steps.

Abundant direct evidence indicates that the enzymes of glucose metabolism are not distributed uniformly in the brain. Many neurochemists are convinced that astroglia produce lactic acid that is taken up and used as a fuel by neurons, particularly in times of high metabolic activity. In other words, glucose metabolism probably shows cellular compartmentalization in the

TABLE I
ACTIVITIES OF ENZYMES OF ENERGY METABOLISM IN HUMAN BRAIN[a]

Glycolysis		
Hexokinase (soluble)	39	
Phosphohexose isomerase	581	
Phosphofructokinase	10	
Aldolase	35	
Triose phosphate isomerase	7742	
Glyceraldehyde-3-phosphate dehydrogenase	17	
Phosphoglyceromutase	41	
Pyruvate kinase	703	
Lactate dehydrogenase	662	
Pyruvate dehydrogenase (PDHC)	11	
Krebs TCA		
Citrate synthase	40	(120 in a biopsy sample)
Aconitase	9	
Isocitrate dehydrogenase	10	
Ketoglutarate dehydrogenase (KGDHC)	2	
Succinic dehydrogenase (Comple II)	98	
Fumarase	180	
Malate dehydrogenase (NAD-linked)	2491	
Electron Transport		
Complex I	18	
Complex II/III	98	
Complex IV	285	

[a] Values are nmol/min/mg whole brain protein. From Sheu and Blass (1999), The α-ketoglutarate dehydrogenase complex, *Ann. N.Y. Acad. Sci.* **893,** 61–78.

brain, with high rates of glycolysis in astroglia and high rates of PDHC, the TCA cycle, and electron transport in neurons. The number of mitochondria have been shown to differ among neurons (Ishihara *et al.,* 1997), and the composition of mitochondria to differ among different types of neurons (Aoki *et al.,* 1987; Milner *et al.,* 1987; Blass, 1993; Calingasan *et al.,* 1994) and among different types of astroglia (Aoki *et al.,* 1987). These variations are discussed in the following sections, including their relevance to selective vulnerability of brain structures in neurological diseases.

C. Substrates of Brain Metabolism

Quantitatively, the only significant substrate for brain oxidative metabolism is glucose, which accounts for essentially all of the oxygen the brain uses. Therefore, cerebral metabolic rate (CMR) for glucose (CMR_{glu}) and for oxygen (CMR_{O_2}) normally mirror each other. Cerebral blood flow (CBF) is usually tightly linked to CMR, as long as the mechanisms of autoregulation of the cerebral blood supply remain intact. Measurements of CBF have often been used as surrogates for technically more complex measurements of CMR. For instance, measurements of CBF by a variety of techniques have been used in neuropsychological studies to map the relationship between brain areas and functions. It is important to remember that these relationships hold within specific limits. They do not necessarily hold in disease. Glucose is transported into the brain from the blood by specific carriers, which can be altered in disease. The brain can outrun its supply of glucose and oxygen from the blood, for instance, in prolonged seizures (status epilepticus).

Compounds that the brain cannot synthesize but which it needs to function are *quantitatively* minor or negligible substrates of the brain *in vivo*. They include essential amino acids, vitamins, and a number of other compounds. Brain catabolism most often converts these compounds to intermediates that can be secreted, eventually into the blood, and then disposed of by the liver or kidney. Radioactive carbon from some of these compounds can often be detected as oxidized to CO_2, indicating some measure of oxidation by at least some cells in the brain.

Preparations of brain *in vitro* can efficiently oxidize more substrates than does mammalian brain *in vivo*. Destruction of the blood-brain barrier probably allows better access of those substrates to the tissue. The mechanisms by which materials other than glucose are transported into the brain *in vivo* have been described in detail (Pardridge, 1999), and are beyond the scope of this review.

D. Rate of Brain Metabolism and Metabolic Demand

Under physiological circumstances, the brain has the highest rate of oxidative metabolism of any organ of the body. Brain constitutes about 2% of body weight and utilizes about 20% of the oxygen. The high rate of cerebral oxidative metabolism compared to other tissues does not hold when tissues are studied *in vitro* under conditions that maximize metabolic rate (i.e., "forcing conditions"). These observations indicate that under physiological conditions, brain is oxidizing glucose at a rate closer to its maximum rate than do other tissues such as heart. Therefore, brain has *relatively little metabolic reserve* compared to other tissues.

The brain has a very close dependence on continuing glucose oxidation to maintain its function. A few seconds without cerebral glucose oxidation lead to unconsciousness. A few minutes lead to death of vulnerable brain cells. The first clinical signs of reduced blood oxygen or glucose are typically referable to the nervous system. Even partial impairment of glucose oxidation by the brain leads to abnormalities in function that present clinically as a brain disease. This point is discussed in Section VIII.A in relationship to delirium and dementia.

The rate at which the brain oxidizes glucose is proportional, within limits, to neuronal activity. CMR_{glu} and CMR_{O_2} classically increase with brain activity. This phenomenon underlies the use of positron emitting derivatives of glucose, i.e., $[^{15}F]$2-deoxyglucose, or of oxygen, i.e., $^{15}O^{16}O$, to measure brain activity by PET scanning. For instance, PET scanning using labeled glucose has indicated that simple visual stimuli activate primarily the calcarine cortex while complex, affectively loaded visual information activates much of the posterior third of the cortex. The tight coupling of nerve cell metabolism to function has also been elegantly demonstrated as well in simpler preparations including the isolated superior cervical ganglion.

E. Relation of Defect to Clinical Syndrome

For many years, it was believed that the exact site of the defect in a metabolic pathway would correlate with a specific clinical syndrome. This idea does not agree with a large body of newer data. High-throughput methods applied to the detection of mutations in relatively large populations typically show a relatively loose correspondence between mutation and syndrome. This issue is not more prominent for disorders of glucose metabolism than for other disorders. Consider, for instance, the lipidoses. Tay-Sachs disease (GM2 gangliosidosis) was described for decades as a single syndrome, "amaurotic idiocy." Extensive studies in the last three decades have documented unequivocally that this disease can be associated with a variety of clinical entities varying from severe infantile brain damage to typical hebephrenic schizophrenia (Suzuki *et al.*, 1994), *if* the disease is *defined* at the molecular level (as a defect in the *HEXA* gene) or at the protein level (deficiency of the α-subunit of hexosabminadese A).

Ascertainment error may have been a major reason for the now outdated view that mutations in a single gene led to a single clinical syndrome. Decades ago, when methods of analysis were more laborious, investigators were likely to test for a particular genetic/enzyme defect only in patients who fit the "classic" syndrome, and in their close relatives. Since patients were chosen for study because they fit a syndrome, essentially all the patients in whom the defects were found fit the syndrome.

Our understanding of disorders of glucose metabolism affecting the brain is following this pattern. For pathways whose role in brain disease has been relatively little studied, it is still possible to discuss the relationship of syndromes to deficiencies in specific metabolic components. This is the approach taken below for disorders of glucose transporters, glycolysis,

PDHC, the TCA cycle, and anaplerosis. For a pathway whose relationship to disease has been very intensively studied, such as electron transport, trying to relate the clinical syndromes to deficiencies in specific genes or proteins is no longer useful. In the discussion of electron transport in the next section, the components of the pathway are described and then the syndromes associated with defects in the pathway, without trying to associate specific defects with specific components of electron transport. Finally, there are disorders involving brain glucose metabolism that are most usefully discussed as syndromes without trying to attribute them to specific molecular deficits. The last part of this chapter discusses four such entities: cognitive disorders (delirium and dementia); psychoses (particularly "schizophrenia"); ischemic brain disease (e.g., stroke); and, very briefly, aging. Disorders of glucose/mitochondrial metabolism that do not affect the nervous system prominently including disorders affecting principally skeletal or cardiac muscle are not discussed, nor are brain tumors, since their metabolic characteristics are generally determined more by the nature and the malignancy of the tumor than by the nature of the tissue in which they occur.

II. Glucose Transport

A. Normal Transport of Glucose into Brain

D-glucose (but not L-glucose) is normally transported into the brain by well-studied mechanisms of facilitated transport. The Glut1 transporter is expressed primarily in glia and endothelial cells and the Glut3 transporter primarily in neurons. The capacity of this transport system is normally severalfold greater than the maximum rate of glucose metabolism in the brain. During convulsions, however, and particularly during continuous convulsions (status epilepticus), the energy demands of the brain can outrun the glucose supply (Pardridge, 1999).

B. Defects in Glucose Transport

There are genetic defects in the GLUT1 transporter. One example is a very rare syndrome of severe brain damage in infancy, with normal blood glucose but consistently low CSF glucose. This is associated with hemizygosity or nonsense mutations in the GLUT1 transporter (Seidner *et al.*, 1998). The GLUT1 transporter mRNA and protein both increase after global cerebral ischemia, in concert with heat shock protein (HSP) 70. The overexpression of GLUT1 may be an immediate early gene response (McCall *et al.*, 1996).

Upregulation of at least the Glut3 transporter has been reported to occur after diffuse brain trauma (Hamlin, 2001). Transport of glucose is reduced in Alzheimer disease (AD) in the areas of the brain in which glucose oxidation is reduced (Chandrasekaran *et al.,* 1996; Piert *et al.,* 1996a) (and see discussion of AD in the next section). These reductions alter the kinetic rate constants that determine entry (K1) and phosphorylation (K3) of glucose (Piert *et al.,* 1996a). Whether these alterations contribute to the deficiencies in glucose oxidation in AD brain or are adaptations to the reductions in CMR remains unclear. The reductions in glucose transport may relate to reductions in the activity of hexokinase, the enzyme that traps glucose inside cells by phosphorylating it (see in the next section). Abnormalities in glucose transport similar to those in AD have been described in Parkinson disease (PD) (Piert *et al.,* 1996b).

III. Glycolysis

Alterations in the activities of enzymes of glycolysis are hard to evaluate in brain diseases in which glial scarring occurs, because the activities of glycolytic enzymes in astroglia are high compared to those in neurons. Gliosis can reduce the magnitude or even obscure deficiencies in glycolysis in samples of whole brain (Bigl *et al.,* 1999). The catalytic capacities of most enzymes of glycolysis in human brain are present in large excess compared to the metabolic rate (Table I). The discussion here concentrates on enzymes that catalyze highly regulated steps in the pathway, i.e., potential "control-points."

A. Hexokinase

Hexokinase (HK) catalyzes the first reaction of glycolysis, in which glucose is phosphorylated to glucose-6-phosphate (G6P). This reaction is irreversible under physiological conditions. It has been proposed that this first irreversible reaction in glucose catabolism determines the rate of overall glucose oxidation. That assumption is usually not accurate, particularly in disease. For instance, in the PDHC deficiencies discussed further in this section, the rate of glucose catabolism is limited by the defective PDHC rather than by hexokinase.

Primary genetic defects of HK associated with brain disease are not known.

In AD, HK activity has been reported to be reduced in cerebral microvessels obtained at autopsy (Marcus and Freedman, 1997). This finding agrees with the results of studies on glucose transport into brain in AD discussed in the previous section. HK activity has been reported to be reduced in AD brain parenchyma by some (Liguri *et al.*, 1990) but not all (Bigl *et al.*, 1999) investigators, and to be reduced in cultured fibroblasts and leukocytes from some patients with familial AD (Sorbi *et al.*, 1990) but not in sporadic AD (Sorbi *et al.*, 1990; Antuono *et al.*, 1995).

Inhibitors of HK include copper ions (Lai and Blass, 1984). Copper inhibits many enzymes including HK. Copper inhibition of HK may be important in disorders of copper metabolism including Wilson disease and Menkes ("kinky-hair") disease.

B. Phosphofructokinase

Phosphofructokinase (PFK) catalyzes what is believed to be the rate-controlling step in glycolysis, the phosphorylation of fructose-6-phosphate (F6P) to fructose-1,6-bisphosphate (Table I).

Genetic deficiencies of PFK have been associated with muscle weakness (myopathy) rather than with brain disease (Servidei *et al.*, 1986; Tein, 1996), probably because of the selective involvement of a muscle isozyme of PFK.

In AD, studies of PFK have been relatively intensive. PFK activity has been reported to be reduced (Meier-Ruge and Bertoni-Freddari, 1997) and increased (Bigl *et al.*, 1999) in AD brain obtained at autopsy, to be normal in AD brain obtained by biopsy (Sims *et al.*, 1987), and to be normal in cultured AD skin fibroblasts (Sims and Blass, 1986; Sorbi *et al.*, 1990). Alzheimer amyloid can inactivate PFK, perhaps through a free radical mechanism (Bigl and Eschrich, 1995). Perhaps inactivation of PFK by amyloid has led to replicable but artifactual decreases of its activity in homogenized AD brain obtained at autopsies.

C. Glyceraldehyde-3-Phosphate Dehydrogenase (GAPDH)

This protein is a tetramer made up of 4 identical subunits of 38 kDa each. The tetrameric enzyme was originally identified as the catalyst for the first ATP-generating step in glycolysis (Table I). The 38-kDa monomer has turned out to have many other functions as well. These include endocytosis, translational control of gene expression, tRNA export from the nucleus, and nuclear DNA replication and repair (Sirover, 1997). The monomer also participates in apoptotic cascades (Kragten *et al.*, 1998). It is not surprising that this protein is affected in a variety of disease processes. More specific roles of GAPDH have been proposed in several neurological disorders. They are discussed in the following paragraphs.

Huntington disease (HD) is caused by a mutation that leads to a pathological expansion of a string of CAG triplet repeats in a specific, chromosome 4 gene. CAG encodes glutamine (Q), and the $(CAG)_n$ expansion leads to a pathological expansion of a string of Q residues in the encoded "huntingtin" protein. The protein containing the pathological Q_n expansion is widely believed to damage cells by a "deleterious gain of function." Proteins containing pathological Q_n expansions form complexes with GAPDH but do not by themselves reduce the enzymatic activity of the protein. Addition to the mixture of poly-Q protein and GAPDH of catalytic amounts of another enzyme, transglutaminase, does inactivate GAPDH (Cooper *et al.*, 1997a). Transglutaminase catalyzes the formation of a protein-protein bond, by attaching a glutamyl (Q) residue in one protein to a lysine (K) residue in another protein, forming a "γ-glutamyl-lysine" (GGL) bond. Available evidence supports the possibility that transglutaminase action mediates at least part of the "toxic gain of function" of the protein containing the pathological Q_n expansion (Cooper *et al.*, 1997b). However, two studies (Browne *et al.*, 1997; Tabrizi *et al.*, 1999) have found the activity of GAPDH in HD brain to be comparable to controls. A third study found an 18% decrease of GAPDH activity in the most selectively vulnerable area of HD brain, the caudate nucleus (Kish *et al.*, 1998).

In other $(CAG)_n/Q_n$ expansion disorders, namely spinocerebellar ataxias types 1, 2, and 3, the activity of brain GAPDH has been reported to be normal (Kish *et al.*, 1998).

In AD, the activity of GAPDH has been reported to be reduced in temporal lobe by –19% (Schulze *et al.*, 1993). Antibodies to the cytoplasmic domain of the Alzheimer amyloid precursor protein (Schulze *et al.*, 1993) and to amyloid plaques (Sunaga *et al.*, 1995) cross-react with GAPDH. Binding of the amyloid C fragment to GAPDH reduces the V_{max} of the enzyme slightly (Schulze *et al.*, 1993).

IV. The PDHC

The major product derived from glucose in glycolysis is lactic acid, a 3-carbon α-hydroxy acid. Lactic acid is in equilibrium with its oxidized, α-keto form, pyruvate acid. At physiological pH, both acids exist as their salts—lact*ate* and pyruv*ate*. Carbons from glucose enter the Krebs TCA cycle after the conversion of pyruvate to CO_2 and acetyl-coenzyme A (acetyl-CoA) with concomitant reduction of NAD to NADH. This physiologically irreversible reaction is catalyzed by PDHC. PDHC, like the enzymes of the TCA (Krebs) cycle, is located in the mitochondrial matrix, which is surrounded by the inner mitochondrial membrane. PDHC is a large and complex structure made up of an ordered array of six proteins: E1, an $\alpha_2\beta_2$ tetrammer which uses thiamin pyrophosphate as a co-factor; E2, which uses covalently bound lipoic acid as a co-factor; E3 (lipoamide dehydrogenase, LAD), a flavoprotein that interacts with NAD^+ and NADH; PDH kinase, which inactivates the E1 component by phosphorylating it and of which there are four isozymes; PDH phosphatase, which reactivates E1K by dephosphorylating it; and protein X, which also contains lipoic acid, is also involved in acetyl transfer, and appears to be vital for the binding of E3 to the core of the PDHC complex. The human genes for these proteins have been cloned and characterized. This complicated, highly regulated submitochondrial complex is one of the slow reactions of energy metabolism in the human brain (Table I). Maximal activity has been calculated to be less than double the activity required for brain function (Gibson and Blass, 1982). PDHC is found in higher concentrations in neurons than in glia and tends to be particularly high in cholinergic neurons (Milner *et al.,* 1987). The acetyl group of acetylcholine (ACh) is derived from pyruvate, and decreasing pyruvate oxidation leads to proportional reduction in ACh synthesis, even though the flux of pyruvate carbons to CO_2 is more than two orders of magnitude greater than that to the acetyl group of ACh (Gibson and Blass, 1982).

A. Inborn Errors of PDHC

These have been well documented over the last 30 years (Blass *et al.,* 1970; Robinson, 2001). As discussed below, different mutations cause different degrees of impairment in the activity of PDHC. Under the usual conditions of assay, there is a rough proportionality between the severity of the enzyme defect and the severity of the clinical syndrome. Very severe deficiencies in activity are usually associated with severe psychomotor defects in early infancy and milder deficiencies with neurodegenerative disease in adolescence. Relatively modest deficiencies have been described in degenerative diseases of late life including AD (see Section IV.D). The neuroanatomical abnormalities in children who survive long enough to come to neuropathology also relate to the severity of the enzyme defect. Severe early-onset PDHC deficiency is associated with dysmorphism in neurodevelopment including microcephaly. Dysgenesis of the corpus callosum is frequent; this might be related to an effective shortage of acetyl-CoA for the synthesis of fatty acids and other myelin lipids. In PDHC deficiencies, dysmorphisms of the brain tend to increase as the severity of the defect increases. In the clinically mildest forms of genetic PDHC deficiency, brain morphology may appear normal.

Deficiencies have been described in five of the six proteins of PDHC: E1p, E2p, E3, PDH-phosphatase, and the X protein. A number of mutations have been characterized at the DNA level (Lissens *et al.*, 1999, 2000; Robinson, 2001). The most common cause of PDHC deficiencies appears to be mutations in the αE1p subunit, which is encoded on the *PDHA1* gene on the X chromosome. Deficiencies in E3, encoded on the *DLD* gene on chromosome 7, lead to abnormalities not only of PDHC but also of the related ketoglutarate and branched chain dehydrogenase complexes, since the E3 protein (LAD) is a component of all of these three complexes. *DLD* mutations are discussed in the next section along with deficiencies of the ketoglutarate complex.

B. Infantile and Childhood Lactic Acidosis (Including Named Syndromes)

These can result from PDHC deficiency but are further discussed with disorders of electron transport. A single report raises the possibility that PDHC deficiency may be one of a variety of "inborn errors" associated with infantile or childhood autism (Coleman and Blass, 1985).

C. Hereditary Ataxic Syndromes

Milder forms of PDHC deficiency can present clinically as ataxic syndromes, sometimes intermittent but sometimes as "phenocopies" of known spinocerebellar degenerations such as Friedreich ataxia (FA). (See Section V.A.1, with deficiencies of α-ketoglutarate dehydrogenase.)

D. AD and Other Neurodegenerations

PDHC activity is also reduced in certain other neurodegenerative disorders in which mitochondria are involved, notably AD (see Section V.A.4) (Sorbi *et al.*, 1983; Gibson *et al.*, 1998b). PDHC activity has been reported to be diminished in HD (Sorbi *et al.*, 1983), but that observation has not been followed up vigorously. A number of environmental toxins, such as lead, inhibit PDHC effectively, and the inhibition of PDHC may contribute to their toxicity (Yun and Hoyer, 2000).

E. Therapies

Rational treatments for PDHC deficiency have had significant clinical success. A low carbohydrate, high-fat "ketogenic" diet provides ketone bodies that the brain can adapt to using as substrate, thereby bypassing the deficient PDHC-catalyzed reaction (Falk *et al.*, 1976). It has become a mainstay of therapy (Robinson, 2001). Intravenous treatment with a ketogenic emulsion was reported to reverse respiratory failure in a 38-year-old woman with partial pyruvate dehydrogenase deficiency and "adult Leigh disease" (Kumagai *et al.*, 1999). Dichloroacetate activates PDHC by inhibiting PDHC kinase and has been used to treat PDHC deficiency (Naito *et al.*, 1999; Robinson, 2001). It has also been reported to prolong the life span of transgenic "Huntington" mice (Andreasen *et al.*, 2001). A third treatment for PDHC deficiency is large doses of thiamin (vitamin B_1) (Naito *et al.*, 1999; Robinson, 2001). The E1p component of PDHC requires thiamin pyrophosphate (TPP) as a co-factor. The co-factor is not bound covalently to the enzyme protein and can be relatively easily dissociated from it. A small proportion of PDHC-deficient patients have a mutation in E1p that decreases the affinity of the enzyme for its TPP co-factor; raising the level of TPP in their cells can ameliorate their defect. This is an example of a "vitamin-dependency disorder," i.e., much larger doses of the vitamin are needed to maintain health in the patients than in normal people. Even in patients with other types of mutations in PDHC, treatment with large doses of thiamin has been proposed to stabilize the enzyme complex and to provide clinical benefits. Large doses of thiamin have not, unfortunately, proven to benefit AD patients (Nolan *et al.*, 1991). Other interventions to increase PDHC activity have not been extensively explored in AD.

V. The Krebs TCA Cycle

The TCA cycle (Krebs cycle) consists of a series of enzyme-catalyzed reactions in mitochondria. The series starts with the condensation of a four-carbon compound with a two-carbon compound. An ordered set of reactions then go on to convert the resulting six-carbon compound back into the original four-carbon compound, completing the cycle. The components of the TCA cycle are located in the inner mitochondrial matrix. The cycle and its control are discussed and diagrammed in standard textbooks of biochemistry.

A. Citrate Synthase

Citrate synthase carries out the condensation of oxaloacetic acid (a four-carbon acid) with acetyl-CoA to form citric acid (a six-carbon acid). Citrate synthase is often used as a marker of the number of mitochondria. Deficiencies in other mitochondrial components have been claimed to exist, based on the ratio of their activity to the activity of citrate synthase. The reason is "to correct for variation in the number of mitochondria." This manipulation obviously depends on the assumption that the amount of citrate synthase activity is constant among mitochondria even in disease. That assumption is experimentally convenient but not intuitively obvious. It makes it less likely that defects in citrate synthase activity will be discovered in neurological diseases.

Citrate synthase is one of several enzymes inhibited by "MPTP" (Villa *et al.*, 1994). That chemical induces a severe form of Parkinson disease, even in young people. The toxicity of MPTP was originally discovered because it contaminated a preparation of an illegal "designer drug," with disastrous consequences for the users. Its toxic derivatives are produced in brain mitochondria. MPTP toxicity is discussed in more detail with deficiencies of Complex I.

B. Aconitase

Aconitase converts citric acid to isocitric acid. Aconitase is one of the least active enzymes of glucose oxidation in human brain (Table I). The human gene (*ACO2*) has been cloned (Mirel *et al.*, 1998), but polymorphisms in this gene associated with brain disease have not been found, at least as yet.

Defects in aconitase have, however, been described in human brain in a number of neurodegenerative disorders and in transgenic mouse models. These are generally attributed to free radical toxicity. Aconitase is an iron-sulfur enzyme and appears to be very sensitive to the action of free radicals, in particular NO* (Schapira, 1999).

HD is associated with dramatic decreases of aconitase activity, in brain from patients (Schapira, 1999) and from the R6/2 transgenic "HD mouse" (Tabrizi *et al.*, 2000). The levels in histologically affected Huntington caudate can be as low as 8% of control (Schapira, 1999).

Friedreich ataxia (FA) is discussed in the next section. Aconitase activity is low in FA (Schapira, 1999), probably again because of excess action of free radicals in FA brain mitochondria.

Superoxide dismutase 2 mutant mice show low levels of aconitase, again presumably because of the action of free radicals on aconitase (Melov *et al.*, 1999).

Environmental toxins can inhibit aconitase, notably the metal manganese (Zheng *et al.*, 1998) and aluminum (Zatta *et al.*, 2000). Manganese toxicity can cause a syndrome similar to PD, but aconitase is normal in *idiopathic Parkinsonism* (Schapira, 1999). Aluminum is a neurotoxin when it gets into the brain. "Dialysis dementia" develops when renal dialysis fluid is contaminated with aluminum (Mach *et al.*, 1988). An etiologic role of aluminum in AD has been proposed but is not now widely accepted.

C. Isocitrate Dehydrogenase

Isocitrate dehydrogenase (ICDH) carries out the physiologically irreversible oxidation of isocitrate to α-ketoglutarate, with formation of CO_2 and reduction of NAD to NADH.

In Down syndrome (trisomy 21), ICDH activity in platelets has been reported to be about two-thirds of normal (Prince *et al.,* 1994).

D. α-Ketoglutarate Dehydrogenase Complex

The α-ketoglutarate dehydrogenase complex (KGDHC) oxidizes α-ketoglutarate to the coenzyme A derivative of succinic acid (succinyl-CoA) with the concomitant production of CO_2 and conversion of NAD to NADH. KGDHC resembles PDHC. It consists of three enzymes in an ordered array: E1k, which utilizes TPP as a co-factor; E2k, which utilizes lipoic acid covalently bound to the protein; and the same E3 enzyme as in PDHC. KGDHC may have the lowest overall activity of the enzymes of glucose metabolism in human brain (Table I). It is highly regulated. The distribution of KGDHC differs markedly among brain cells. Immunohistochemical studies in both rats (Calingasan, 1994) and humans (Blass, 1993; Ko *et al.,* 2001) have shown that it occurs at much higher concentrations in neurons than in glia, and in much higher concentrations in certain populations of neurons—notably cholinergic neurons—than in others.

Not surprisingly, impairments of the activity of this potentially rate-limiting enzyme complex have been associated with a number of diseases of the brain. Deficiencies of KGDHC, their effects on the nervous system, and their occurrence in diseases have recently been reviewed (Gibson *et al.,* 1998b, 1999; Sheu and Blass, 1999). Mutations have been described in the *DLD* gene which encodes the E3 protein that is also a constituent of PDHC. The mutations include *DLD* homozygotes (Robinson *et al.* 1989), compound heterozygotes (Hong *et al.,* 1996), and an insertion mutation in the mitochondrial leader sequence (Elpeleg *et al.,* 1997).

1. *Death in Utero*

Death *in utero* appears to be the result of the most severe defects in KGDHC, particularly those in the E3 enzyme which cause associated deficiencies in PDHC and branched chain ketoacid dehydrogenase (BCKDH, an enzyme of amino acid degradation) (Robinson *et al.,* 1989; Sheu and Blass, 1999). Knockout transgenic mice severely deficient in E3 die *in utero* in the perigastrulation stage (Johnson *et al.,* 1997).

2. *Infantile Lactic Acidosis*

Infantile lactic acidosis with severe psychomotor developmental disabilities can also result from deficiency of KGDHC as well as of other mitochondrial defects (see discussion is Section VII). Milder KGDHC deficiency has been associated with milder forms of psychomotor retardation in childhood (Robinson *et al.*, 1989; Sheu and Blass, 1999).

3. *Childhood Intermittent Encephalopathy*

Elpeleg and co-workers (Shaag *et al.*, 1999) have described a group of 13 patients with episodes of childhood intermittent encephalopathy accompanied by abdominal pain, vomiting, and occasionally with elevations of blood lactate and ketoacids and other manifestations. These 13 patients were from a single ethnic group, Ashkenazi Jews. They had normal IQs. Two patients who presented neonatally had two mutations in *DLD*, *G229C*, and also an insertion, 105insA (Y35X). These patients had residual neurological damage including attention deficit hyperactive disorder and motor problems. The other patients who had either the G229C mutation alone or the insertional mutation alone had no symptoms between attacks. The mutations in the genes reduced E3 activity to 8–21% of normal and E3 protein levels in muscle to 20–60% of the levels in controls.

4. *Hereditary Ataxias*

Deficiencies of KGDHC as well as of PDHC and other mitochondrial enzymes occur in patients with hereditary spinocerebellar ataxias (Blass *et al.*, 1976; Sorbi *et al.*, 1989; Schapira, 1999). The hereditary ataxias are a group of disorders in which motor incoordination, predominantly of a cerebellar type, is associated with a variety of other manifestations often including psychiatric disabilities (Davies, 1949; Blass *et al.*, 1976; Sorbi *et al.*, 1989; Kumagai *et al.*, 1999).

In FA the mutation is in a gene encoding a mitochondrial protein, frataxin. This protein is involved in mitochondrial iron metabolism and therefore with free radicals (Wong *et al.*, 1999). Aberrations in frataxin lead to deficiencies in aconitase (as noted in the beginning of Section V) (Schapira, 1999). Mutations in frataxin and resultant intramitochondrial oxidative stress may also explain the previously demonstrated abnormalities in KGDHC and PDHC in FA, perhaps by free radical damage to the E3 component they have in common (Blass *et al.*, 1976).

At least eight other forms of hereditary spinocerebellar ataxias have been recognized, each associated with the same type of mutation but each in a different gene. The mutations consist of expansions of CAG triplet repeats to a "pathological length," usually 40 or more repeats. CAG encodes glutamine

(Q), so these are CAG_n/Q_n repeat disorders. These disorders are also associated with deficiencies in mitochondrial enzymes, often including KGDHC and PDHC (Sheu *et al.*, 1988; Sorbi *et al.*, 1989). The mechanisms by which the CAG_n/Q_n expansions damage mitochondria are not known. KGDHC is a substrate for the enzyme transglutaminase (Cooper *et al.*, 1997a), as are proteins containing Q_n expansions (Cooper *et al.*, 1997a, 1999). The leader sequences for transporting proteins into mitochondria are rich in basic amino acids including K and seem likely substrates for transglutaminase. Perhaps components of KGDHC and of other nuclear-encoded mitochondrial constituents are trapped in the cytoplasm by transglutaminase-catalyzed covalent linkage to proteins containing Q_n expansions (Cooper *et al.*, 1997a, 1999).

Spinocerebellar ataxias resulting from CAG expansions are a particularly clear example of the principle that different mutations can lead to similar clinical manifestations and a single mutation can be associated with more than one clinically distinct syndrome. Attempts to classify the ataxias on clinical grounds alone created a forbidding forest of neurological erudition that has been clarified as well as simplified by the molecular genetic classification.

5. *Parkinson Disease*

This common condition is associated a defect in Complex I of the electron transport chain. More recent studies indicate that it is also associated with a deficiency in KGDHC (Mizuno *et al.*, 1994; Gibson and Blass; Kobayashi *et al.*, 1998). This was first reported on the basis of immunohistochemical studies (Mizuno *et al.*, 1994) and subsequently confirmed by direct measurement of enzyme activity in PD and control brains (Gibson and Blass). An association has been reported between PD and silent polymorphism in the *DLST* gene that encodes the core, E2k component of KGDHC (Kobayashi *et al.*, 1998). A compound that induces parkinsonism in humans, MPTP, inhibits KGDHC (Joffe *et al.*, 1998) as well as its well-known effect on Complex I. The effect on KGDHC may well be important in the mechanism of the disease, since the activity of this enzyme complex is so low compared to that of Complex I in human brain (Table I).

6. *Alzheimer Disease*

Deficiency of KGDHC in AD brain is extensively documented, and these findings have been extensively reviewed (Gibson *et al.*, 1999; Sheu and Blass, 1999; Ko *et al.*, 2001). At least four laboratories have found robust decreases in KGDHC activity in AD autopsy brain, and there are no contravening reports. (See Gibson *et al.*, 1998b, 1999; and Sheu and Blass, 1999) for original references.) The decreases in activity are to about 50% or less of control values and are more severe than the decreases in the amounts of the KGDHC proteins (Mastrogiacomo *et al.*, 1996). Deficiencies of KGDHC have also been found in lines of cultured skin fibroblasts from some but not all AD patients (Blass, 1993; Blass *et al.*, 1997; Gibson *et al.*, 1998b, 1999). The persistence of KGDHC deficiency in AD fibroblasts through serial culture suggests that the defect in KGDHC in AD is related, directly or indirectly, to the genetic information that predisposes an individual to develop AD, since only genetic information can be assumed to be maintained during the large dilutions that occur in serial culture (Blass and Gibson, 1993). Direct studies in AD of the genes encoding the components of KGDHC have given equivocal results. Neutral polymorphisms in the *DLST* gene that encodes the E2k core protein of KGDHC have been described in association with AD in at least three different research units on three different continents (Sheu *et al.*, 1997, 1999; Nakano *et al.*, 1997). The alleles associated with AD differ, however, among the different populations studied, and none lead to amino acid alterations in the encoded protein. They may, however, be a clue to the existence of potentially pathogenic mutations in this or in a nearby gene (Sheu *et al.*, 1997, 1999; Nakano *et al.*, 1997). KGDHC contains active sulfhydryl groups and is readily inactivated by free radicals (i.e., ROS and reactive nitrogen species such as NO*) (Park *et al.*, 1999). The decreases in KGDHC in AD (Park *et al.*, 1999) and in PD (Mizuno *et al.*, 1994; Gibson and Blass, ???) may be due at least in part to its inactivation by free radicals. Such an effect may explain the reduced activity of KGDHC in brains from patients who carry the "Swedish" amyloid mutation (Gibson *et al.*, 1998a), since AD amyloid exerts many of its effects through free radicals. Genetic and free radical mechanisms are not mutually exclusive. Genetic variations in the KGDHC genes might sensitize the proteins they encode to damage by free radicals.

The magnitude of the deficiency in KGDHC activity correlates well ($\rho \sim 0.7$) with the degree of clinical disability in AD, in patients with the variε4 allele of the *APOE* gene (Gibson *et al.*, 2000); this *APOE* allele is a major risk factor for AD. A decreased rate of brain metabolism (CMR) occurs in essentially all patients with AD. The extent of the decrease in CMR correlates with the degree of clinical disability in these patients (see Section VIII.A).

The brain cells that are selectively lost in temporal cortex in AD are normally enriched in KGDHC, suggesting that the abnormalities in KGDHC in AD brain may contribute to selective vulnerability in this syndrome (Ko *et al.*, 2001). The composition of mitochondria vary markedly among different cell types in the brain (Aoki *et al.*, 1987; Milner *et al.*, 1987; Calingasan *et al.*, 1994), and mitochondria have an important role in both apoptotic and necrotic cell death. My colleagues and I have proposed that variation among mitochondria in selectively vulnerable and selectively resistant cells may be one of the mechanisms that can lead to selective vulnerability (Blass, 1993, 1999; Ko *et al.*, 2001).

7. *Treatments*

Treatments for deficiencies in KGDHC have been developed. A patient with lactic acidosis due to E3 deficiency, and therefore deficiency of PDHC as well as KGDHC, reportedly responded to treatment with dicholoroacetate (a PDHC activator), thiamin (the source of a co-factor for the first enzymes of PDHC [E1p] and KGDHC [E1k]), and carnitine (which can be converted to acetylcarnitine). (Elpeleg *et al.,* 1995). This treatment is more rational for the defect in PDHC than in KGDHC. Failure of response to dichloroacetate in a patient with a mutation in E3 has also been reported (Craigen, 1996).

E. Thiaphorases

Thiaphorases catalyze the transfer of the coenzyme A moiety from succinyl-CoA to another acid with the production of succinate. No report of deficiency of thiaphorase in association with a brain disease was found during the preparation of this review.

F. Succinic Dehydrogenase

Succinic dehydrogenase removes two hydrogens from succinic acid to form fumaric acid. The enzyme is Complex II of the electron transport chain, and is discussed in Section VII.

G. Fumarase

Fumarase catalyzes the addition of water to fumaric acid to form malic acid. Severe fumarase deficiency has been described in two infants (Petrova-Benedict *et al.,* 1987; Gellera *et al.,* 1990). Both had severe, diffuse mental and motor deficiencies and died before their first birthdays. Both excreted excessive amounts of fumaric acid in their urine. One appeared to be deficient in both the cytoplasmic and mitochondrial enzymes (Gellera *et al.,* 1990) and the other only in the cytoplasmic form (Petrova-Benedict *et al.,* 1987), even though both forms appear to be coded on the same gene. Molecular genetic studies have not yet been reported.

H. Malic Dehydrogenase

Malic dehydrogenase catalyzes the removal of two hydrogens from malic acid to form oxaloacetic acid. Citrate sythetase then catalyzes the condensation of oxaloacetate (formed by malic enzyme) with acetyl-CoA (formed by PDHC) to form citrate, completing the TCA cycle.

In FA, a reported deficiency in malic enzyme (Stumpf, 1985) was not reproduced (Grossman *et al.,* 1987) and may have been secondary to inactivation of the enzyme because of the abnormalities of iron and free radical metabolism that are characteristic of the mitochondria of these patients (Wong *et al.,* 1999). The activity of malic dehydrogenase has been reported to be dramatically increased in the brains of patients with AD (Op den Velde and Stam, 1976).

VI. Anaplerosis

Anaplerosis is the process by which pyruvic acid, a three-carbon compound, is converted to a four-carbon dicarboxylic acid, oxaloacetate. The oxaloacetate can then be used to synthesize new molecules of glucose, i.e., to carry out *gluconeogenesis.* The four-carbon dicarboxylic acid is also needed to "prime" the Krebs TCA, as discussed in Section V. The two enzymes unique to this process are pyruvate carboxylase (PC) and phosphenolpyruvate carboxykinase (PCK). Pyruvate carboxylase is relatively unstable, and confident demonstrations of deficiencies of this enzyme have depended largely on measurements of activity in cultured skin fibroblasts from patients and controls. Pyruvate carboxylase is a mitochondrial enzyme. The human gene is on 11q13, and the cDNA has been sequenced (Freytag and Collier, 1984; Robinson, 2001). This biotin-containing enzyme is found in liver and kidney and only in low amounts in brain.

Clinically, deficiency of PC is associated with childhood lactic acidosis (De Vivo *et al.,* 1977; Gravel and Robinson, 1982; Robinson, 2001). Three main clinical forms have been recognized. In North American Indians who speak Algonquian languages, the disorder takes the form of impaired psychomotor development with lactic acidosis. In France and England, the syndrome is more complicated and deadly. Death typically occurs by 3 months of age, and the patients have excessive levels of ammonia, citrulline, and lysine in their blood as well as of lactic acid. In one girl, PC deficiency was associated with episodes of metabolic acidosis in infancy but no more than marginal clinical abnormalities by age 10. Her fibroblasts had 1–2% of normal PC activity (Van Coster *et al.,* 1991).

A single child has been convincingly reported to have a deficiency of the PCK enzyme, the cDNA for which has been cloned (Robinson, 2001). The neurological symptoms may be related to falls in blood glucose, i.e., episodes of hypoglycemia.

VII. Disorders of the Electron Transport Chain

The process carried out by the electron transport chain is also referred to as oxidative phosphorylation (OXPHOS). Brain disease associated with disordered electron transport has been studied very intensively (Wallace *et al.,* 1995, 2001; Beal, 1998; Schapira, 1998; Lemasters and Nieminen, 2001; Shoffner, 2001). Excellent reviews are available including those by Shoffner (2001) and Wallace *et al.,* (2001). This large and rapidly enlarging body of genetic and clinical information no longer fits attempts to associate specific clinical syndromes with deficiencies of specific components of this pathway. In a recent review, Shoffner (2001) wrote

> The last time this chapter was written, diseases were classified according to their molecular defect. However, this classification scheme is limited by the large amount of overlap in the clinical manifestations . . .

In the following discussion, the components of the electron transport chain are described, then special biological characteristics of electron transport are discussed including aspects of free radicals, and then well-known syndromes that have been associated with deficiencies of electron transport are reviewed.

A. Enzymology and Genetics of Electron Transport

The electron transport chain is a series of five mitochondrial complexes embedded in the inner mitochondrial membrane. Four complexes carry out the transfer of electrons (reducing equivalents) from NADH or from succinate to molecular oxygen, forming water. The fifth forms ATP from ADP and phosphate. Complex I transfers electrons from NADH to ubiquinone moieties incorporated into Complex III. It consists of approximately 42 polypeptides, of which 35 polypeptides are encoded on nDNA and 7 (MTNDA1, MTDNA2, MTDNA3, MTDNA 4, MTDNA 4L, MTDNA 5, and MTDNA6) on mitochondrial DNA (mtDNA; see the following section). Complex I contains three flavoproteins and seven iron-sulfur proteins (ISPs) and is susceptible to the action of a number of ROS. Complex II of the electron transport chain is also known as succinic dehydrogenase, and is a component of the TCA cycle. This complex removes two electrons from the Krebs cycle intermediate succinate, forming the cycle intermediate fumarate. The electrons are transferred to a ubiquinone in Complex III. Complex II consists of four proteins, each encoded on a different nuclear gene. Complex III accepts electrons from Complex I or Complex II onto its ubiquinone residues. It then transfers these electrons to cytochrome *c*, a protein containing an iron-heme group. The transfer of these electrons reduces ferricytochrome *c* to ferrocytochrome *c*. The energy released pumps two electrons across the mitochondrial membrane to synthesize ATP from ADP. Complex III consists of 11 proteins, 10 encoded on nDNA and 1 on mtDNA.

Complex IV utilizes molecular oxygen to oxidize cytochrome *c*. It is the only element of the electron transport chain that normally interacts with molecular oxygen. Its heme binding site for molecular oxygen also binds cyanide (CN^-) and carbon monoxide (CO) tightly, and the mechanism of brain toxicity of these materials is believed to be inhibition of Complex IV. Other names for Complex IV are "cytochrome oxidase" (COX), cytochrome *c* oxidase, and in older literature "cytochrome a/a_3 oxidase." Complex IV is composed of 13 polypeptides, 10 encoded on nDNA and 3 on mtDNA (namely *COI, COII, COIII*). The M_r of the whole complex is roughly 400 kDa. The reaction catalyzed by Complex IV takes electrons from reduced cytochrome *c* (ferrocytochrome *c*), transfers them to a copper site, then to cytochrome *a*, then to a combined copper-cytochrome *a* site, and then to molecular oxygen. It simultaneously uses the energy released from these steps to pump four electrons across the mitochondrial membrane. It has been proposed that the control of Complex IV depends on a cAMP-dependent, protein-kinase-A-catalyzed phosphorylation of subunits II (and/or III) and of component Vb (Bender and Kadenbach, 2000). After phosphorylation, the rate of Complex IV activity is controlled by the availability of ADP compared to the more phosphorylated ATP, i.e., by the ATP/ADP ratio. After dephosphorylation by a Ca^{2+}-stimulated phosphatase, the rate of the Complex IV reaction is substrate limited. Under these circumstances, the rate of oxygen utilization by the mitochondrion is presumably limited by the slowest steps of energy metabolism, such as KGDHC and aconitase (Table I).

The amount of Complex IV in a neuron appears to be a function of the rate of physiological activity of that neuron (Wong-Riley, 1999). This effect complicates interpretation of findings of low activity of Complex IV in brain diseases. Reduction in the level of this complex may reflect a reduced rate of activity of a diseased cell rather than damage to the complex itself. When found, mutations in a component of Complex IV combined with decreases in its activity and/or protein levels provide clearer evidence of a primary defect.

Complex V uses the electromotive force generated by earlier steps in the electron transport chain to synthesize ATP from ADP and inorganic phosphate (Pi). The resultant γ-phosphate bond in the resulting ATP has a high enough energy of hydrolysis to drive many biologically important processes, including ion transport and many synthetic reactions. Complex V is therefore the step in energy metabolism where energy is produced in a form widely useful to the cell. In human brain, the activity of Complex V is high compared to other complexes of the electron transport chain (Table I). The extremely interesting structural biology and enzyme mechanism of Complex V (Wallace *et al.*, 2001) is beyond the scope of this review.

B. The Mitochondrial Genome

Components of the electron transport chain are encoded on two genomes—the nuclear (nDNA) and the mitochondrial (mtDNA). Hundreds of other genes on nDNA are believed to influence the formation of the electron transport complexes even though they do not directly encode its components.

mtDNA is a circular double strand of DNA. It encodes only 13 components (~15%) of the electron transport chain and also the 22 tRNAs and the 12S and 16S mRNAs necessary for protein synthesis. mtDNA has been sequenced from several species including humans, facilitating its study in relation to diseases. Human mtDNA contains 16,569 nucleotide pairs. Each cell typically contains hundreds of mitochondria and thousands of strands of mtDNA (Wallace *et al.*, 1995).

mtDNA derives essentially completely from the mother. The number of mitochondria in the sperm head are negligible compared to the number in the egg, and what paternal mtDNA there is often eliminated during replication. Inherited disorders of mtDNA therefore typically show a pattern of *maternal inheritance,* i.e., diseases are passed through mothers to both sons and daughters but are not inherited from fathers.

The number of mtDNA molecules typically varies among mitochondria, and the composition of the mtDNA molecules often varies within an individual person, within an individual cell, and probably often within an individual mitochondrion (Wallace *et al.*, 1995, 2001). Individuals in whom the composition of mtDNA appears the same in all mitochondria are termed *homoplasmic;* individuals in whom they are not the same are *heteroplasmic.* Individuals can be homoplasmic for some parts of their mtDNA and still have variations (heteroplasmy) at other molecular sites.

During replication of mitochondria that contain strands of mtDNA with different sequences, transcription can favor one strand over another to the point where daughter mitochondria have only one type of DNA. The selective transcription can favor a "mutant" strand or a "wild-type" strand, depending on which strand is favored by the transcription mechanisms. A "mutant" strand may be favored because of being shorter or because feedback mechanisms (perhaps via NADH) favor the replication of mitochondria containing damaged mtDNA. *Mosaicism* refers to different predominant forms of mtDNA in different tissues or in different cells in the same tissue. This term can be confusing. In loose terminology, mosaicism has been used to refer to variability in mtDNA that could result from different forms of mtDNA within different mitochondria in a single cell or even in a single mitochondrion.

mtDNA has been claimed by many but not all investigators to accumulate proportionally more evidence of damage by ROS during life than does nDNA (Wallace *et al.,* 1995, 2001). Stated more formally, somatic mutations appear to be much more common in mtDNA than in nDNA. The greater accumulation of damage to mtDNA has been widely attributed to a less effective apparatus for repair of DNA in mitochondria than in the cell nucleus. Other factors that may contribute include the high rate of generation of reactive oxygen species in mitochondria and the potential for accumulation of damaged mtDNA strands within mitochondria. A practical consequence is that complicated genetic studies of families are often necessary to determine whether an abnormality in mtDNA in an individual patient is inherited or is an acquired “somatic mutation.”

Cybrids are cells in which the endogenous mitochondria have been depleted (ρ_O cells) and then fused with cells which lack nuclei but contain mitochondria and mtDNA. The ρ_O cells can be of a variety of types. The human cells most commonly used to replete their mtDNA are platelets. The persistence of abnormalities in cybrid cells in culture provides evidence of abnormalities in the mtDNA of the cells with which they were fused at the time the fusions were done. This powerful technique is, however, limited by the tendency for the abnormalities to disappear in serial culture. The assumption that the “healthy cells” with normal mtDNA overgrow the cells with damage mtDNA is hard to reconcile with evidence that selective multiplication of defective mtDNA is not rare, as discussed earlier in this section.

C. Free Radicals (ROS)

Perhaps 1% of the oxygen used by the brain is utilized by single electron transfer to form ROS. ROS can oxidize many materials thereby damaging them. However, it is important to remember that free radicals can also be critical signaling compounds. Life would probably be impossible without NO$^{\cdot}$. Evidence of free radical *action* does not of itself prove free radical *damage.* Minute amounts of damage to DNA or membranes might simply be a repairable consequence of increased cellular signaling.

Mitochondria are a major source of ROS since they are a major site of oxygen utilization. Complex III has been proposed to be an important site for generation of ROS. It can donate electrons directly to molecular oxygen to form superoxide ion (O_2^-). This reaction can become quantitatively important when adequate ferricytochrome *c* is not available. That can occur when ferrocytochrome *c* is not oxidized at an adequate rate by Complex IV. Superoxide ion reacts with two water molecules to form hydrogen peroxide (H_2O_2), which is itself an oxygen free radical (ROS). The enzyme catalase converts H_2O_2 to innocuous materials, water (H_2O), and oxygen. If, however, H_2O_2 comes into contact with an iron atom, it disproportionates by Fenton chemistry into a hydroxyl ion (OH^-) and an $OH^{\cdot}$ ion. The latter is a highly reactive species and therefore has a small capture radius. $OH^{\cdot}$ is very likely to damage molecules with which it comes in contact.

Free radicals play a major role in many forms of brain damage (Browne *et al.*, 1997; Beal, 1998; Floyd, 1999; Gibson *et al.*, 1999; Park *et al.*, 1999; Lemasters and Nieminen, 2001). A term often used to describe free radical action is "oxidative stress." Mechanisms of free radical generation, action, and removal are of great current interest in many areas of biology and even in the cosmetics industry. The literature on free radicals in the nervous system has been extensively reviewed (Beal, 1998; Lemasters and Nieminen, 2001). Mutations in a gene encoding an enzyme of mitochondrial free radical metabolism, *CuZn-SOD,* have been identified in motor neuron disease (amyotrophic lateral sclerosis; ALS) (Gaudette *et al.*, 2000). Over 500 reviews of free radicals in disease are listed on MEDLINE. Detailed discussion of this subject is beyond the scope of this review.

Therapy with "free radical scavengers" has been proposed for diseases in which oxidative stress occurs. Free radical scavengers are reducing agents that react with free radicals—and particularly with ROS—to form innocuous compounds. One of the most widely used free radical scavengers is vitamin E. Many other free radical scavengers have been and are being tested for diseases of other organs as well as for diseases of the brain.

D. Clinical Syndromes of Impaired Electron Transport

As discussed above, there is enormous overlap in the clinical manifestations of different deficiencies in electron transport. Excellent recent reviews provide details of these mutation-component-disease relationships, some of which have been found in only one patient (Wallace *et al.*, 1995, 2001). The following discussion describes some of the most common syndromes.

1. *Lactic Acidosis*

Severe impairments of the ability to oxidize substrates through any of the steps of PDHC, the TCA cycle, or electron transport are likely to lead to accumulation of lactate derived from pyruvate which is itself derived from glucose. Children with severe enough defects in mitochondrial function to give them lactic acidosis as infants or toddlers are likely to be very sick. They have an almost "pan-nervous system disorder," with poor mental development and then regression, spasticity relieved by hypotonia, decreased to absent reflexes, blindness, and incoordination. Alper disease is an eponymic disorder that falls within this group of conditions. Lethal infantile mitochondrial disease (LIMD) is a severe variant of infantile lactic acidosis typically leading to death with overwhelming lactic acidosis before age 6 months.

2. *Leigh Disease (Subacute Necrotizing Encephalomyopathy; SNE)*

This is another eponymic clinicopathological entity associated with lactic acidosis. Common causes of SNE include deficiencies in Complex I (19% of cases), Complex IV (18%), PDHC (10%), and in other mitochondrial components. The "classical" form of SNE develops in the first 2 years of life with death by age 5. Two clinical findings increase the index of suspicion of this syndrome of lactic acidosis. One is prominent respiratory difficulties. The second is marked exacerbation of the disorder, including the lactic acidemia, accompanying conditions that increase metabolic rate such as intercurrent infections particularly with fever. The neuropathology resembles Wernicke-Korsakoff syndrome, with periventricular hemorrhages as well as nerve cell loss. Milder juvenile and adult forms of SNE are also known. Serious impairment of vision due to pigmentary degeneration of the retina (retinitis pigmentosa) can coexist with SNE. A syndrome with related clinical manifestations is: neuropathy, ataxia, and retinitis pigmentosa (NARP).

3. *Mitochondrial Myopathy, Epilepsy, Lactic Acidosis, and Stroke-Like Episodes (MELAS)*

The classical clinical manifestations of MELAS are stroke-like episodes accompanied by lactic acidosis. The strokes can be visualized on CT or MRI, but can clear within hours or days. They are thought to reflect localized insufficiency of oxidative metabolism in both brain tissue and in the small blood vessels supplying the brain. A broad spectrum of other clinical manifestations can be seen in patients with MELAS mutations, including muscle disease (fatigability, pain, and myopathy), eye disease (retinitis pigmentosa), heart disease (cardiomyopathies and arrhythmias including ventricular arrhythmias), endocrine disease, kidney disease, and other types of damage to the nervous system (impaired eye movements, deafness, migraine, dementia), and formation of unusual autoantibodies. About 80% of patients with this condition have a heteroplasmic A3243G mutation in $tRNA^{Leucine\ (UUR)}$ (Snider *et al.*, 1999; Shoffner, 2001). This mutation alters the structure of the dihydrouridine loop of the tRNA and also interferes with the binding of a transcription factor (Shoffner, 2001). However, the most frequent clinical expression of the A3243G mutation is *not* MELAS but rather type II diabetes. The diabetes is "adult-onset type," with insulin resistance rather than a deficiency of insulin. The A3243G mutation may account for 1% of the cases of this very common disorder. The reason why some patients with this mtDNA mutation present clinically with diabetes and others with primary disease of the nervous system is not known. This observation illustrates yet again how misleading it can be to try postulate tight correlation between a single mutation and a single clinical syndrome.

4. *Myoclonic Epilepsy and Ragged Red Fiber Disease (MERFF)*

The classic clinical triad in this disorder is progressive myoclonic epilepsy, slowly progressive dementia, and a muscle disorder with "ragged red" muscle fibers evident on muscle biopsies. Age of onset varies from late childhood to adulthood. Ragged red fibers are degenerating muscle fibers in which excess numbers of mitochondria accumulate in bunches under the sarcolemma. These fibers stain red with Gomorri trichrome and can also be detected by histochemical staining for Complex II, i.e., for succinic dehydrogenase. Ragged red fibers are a rather nonspecific manifestation of mitochondrial disease.

Clinical manifestations associated with a "MERFF mutation" can vary even within the same kindred. For instance, in a thoroughly studied large kindred with a G8344A mutation in mtDNA, the proband had a complex syndrome that included myoclonic epilepsy, dementia, mitochondrial myopathy with ragged red fibers and paracrystalline inclusions, bilateral neurosensory hearing loss, cerebellar ataxia, hypoventilation with episodes of respiratory arrest, abnormal visual-evoked responses and an abnormal EEG, a mild cardiopathy, and lactic acidosis. (The clinical similarity in this patient to the description of Leigh disease is noticeable.) A spectrum of abnormalities was found in other family members: seven had mitochondrial myopathy and one had abnormalities only in visual-evoked responses.

About 80% of the patients with MERFF have a heteroplasmic G8344A mutation in the $tRNA^{Lysine}$ gene on mtDNA (Shoffner, 2001). Other mutations in this gene have also been described. The mutations lead to a variety of biochemical abnormalities. Combined deficits in Complex I and Complex IV are often found, but defects can also occur in Complex II, Complex III, Complex IV without deficiency of Complex I, and in coenzyme Q (a constituent of Complex III).

5. *Kearns-Sayre Syndrome (KSS) and Chronic Progressive External Ophthalmoplegia (CPEO)*

These are the syndromes most commonly associated with multiple deletions in mtDNA (Wallace *et al.*, 2001). By definition, if clinical onset is before age 20 the syndrome is KSS and if thereafter CPEO. The typical syndrome includes abnormal eye movements (ophthalmoplegia), atypical degeration of the retina (retinitis pigmentosa), and muscle weakness due to a ragged red myopathy, with at least one of the following: a cerebellar syndrome, a cardiac conduction defect, or elevated protein in the CSF (>1 mg/ml). In practice, a wide variety of clinical abnormalities are seen in patients with multiple deletions in mtDNA. They include various degrees of visual impairments, including optic atrophy as well as atypical retinitis pigmentosa; seizures; dementia; hearing loss; heart disease, including various types of arrhythmias; kidney failure of several types; and endocrine diseases, including diabetes and hypoparathyroidism. Some patients resemble those with MERFF and/or MELAS. Some patients who have clinical manifestations which mimic those of KSS start out with Pearson syndrome, another mtDNA-linked disease that first manifests itself as impairment of the formation of blood cells including red cells in infants.

6. *Leber's Hereditary Optic Atrophy (LHON)*

This disorder manifests itself as rather sudden loss of vision, first in one eye and then usually within a few years in the other eye (Novotny *et al.*, 1986; Wallace *et al.*, 2001). The most common age of onset is in early adult life, but sudden blindness can occur at any time during the life span depending in part on the specific mutation. Mutations that lead to LHON in some patients can lead to more complex neurological disease even in other members of a single kindred. For instance, in a five-generation pedigree with the MTND6*LDYT14459A mutation in Complex I and an early age of onset, 13 of 42 maternal relatives had dystonia and bilateral necrosis of the striatum, 8 had LHON, 1 had both, and 19 had neither (Wallace *et al.*, 2001). Again, the correlation of syndrome to mutation is hardly striking. The correlation between the severity of the clinical syndrome in a specific patient and the amount of enzyme activity or protein measured *in vitro* in LHON is also poor. Most studies of the MTND4*LHON11778A and MTND6*LHON14484 mutations in Complex I have found no decrease in overall Complex I activity although subtler abnormalities in enzyme kinetics have been variably demonstrated. At least 13 different mutations in mtDNA encoding components of Complex I have been described in LHON, as well as a mutation in a component of Complex V. Another 10 mutations associated with LHON appear not to play a primary role in causing this disorder. The 4 most common mutations account for about 90% of cases (Wallace *et al.*, 2001). A specific genetic background, the mtDNA haplogroup J, increases the likelihood that an LHON mutation will express itself clinically (Wallace *et al.*, 2001).

7. *Parkinson Disease*

This is a common disorder associated with mitochondrial abnormalities. Unlike the conditions discussed above, PD is largely although not entirely a disease of older people. Clinically, it is characterized by stiffness and rest tremor and neuropathologically by the degeneration of neurons in the pars compacta of the substantia nigra with accumulation of eosinophilic Lewy bodies that are rich in the protein synuclein. Although deficiency of other mitochondrial components—notably KGDHC—occurs in PD, the best known enzymatic defect in this disorder is in Complex I. The defect in Complex I is robust in the affected area of brain but has been only variably reported in other parts of the PD brain and body (Schapira *et al.*, 1999; Reichman and Janetzky, 2000). Two groups have reported a deficiency in Complex I activity in PD cybrids (Schapira *et al.*, 1998; Swerdlow *et al.*, 1998), but a third did not find this effect (Aomi *et al.*, 2001). Attempts to identify mutations in mtDNA components encoding polypeptides of Complex I have, in general, been unsuccessful when appropriate allowance is made for the spontaneous variation in mtDNA with age (Kosel *et al.*, 1998; Simon *et al.*, 2000).

Environmental factors, perhaps related to industrialization, have long been proposed to play a major role in the common forms of PD (Jenner, 2001). This disorder was not described before the industrial revolution. Its clinical characteristics are so striking that it seems unlikely that the skilled clinical observers of previous generations would have failed to recognize it. Evidence for hereditary factors causing the common forms of late-onset PD is slim, although specific genetic defects—notably in synuclein and in Parkin—have been described in rare, relatively early-onset, familial forms of this disorder. Environmental agents can cause syndromes clinically and pathologically indistinguishable from PD. A notable example is MPTP, which is taken up by dopaminergic neurons and oxidized to MPP^+ (Grunblatt *et al.*, 2001). The latter compound inactivates Complex I and other mitochondrial components (Grunblatt *et al.*, 2001) including KGDHC (Joffe *et al.*, 1998). Chronic treatment of rats with the lipophilic pesticide rotenone, a Complex I inhibitor, creates a model of PD that recapitulates many of the key clinical and pathological abnormalities of the human disorder (Betarbet *et al.*, 2000).

The normal metabolism of dopamine and similar compounds by the enzyme monoamine oxidase generates free radicals. It has been proposed that the generation of free radicals in dopaminergic neurons can, over time, decrease the activity of Complex I, aconitase, KGDHC, or other enzymes (Cohen, 2000; Lotharius and O'Malley, 2000; Shen *et al.*, 2000). Mutations that would increase the sensitivity of these or other components to ROS have not been identified in PD.

Alzheimer disease is a very common dementing disorder in which abnormalities of Complex IV have been found that will be discussed in Section VIII.A.

VIII. Impaired Mitochondrial Function in Other Disorders

Impairments in brain glucose metabolism occur in a number of conditions in which the fundamental mechanisms leading to the impairments have not been attributed to a single component of glucose metabolism. Some of these disorders have already been mentioned above. Four will be revieweded here: disorders of cognition, psychoses, cerebrovascular disease, and aging.

A. Disorders of Cognition: Delirium and Dementia

Neurobiologically, delirium and dementia have so much in common that they are considered together in this discussion. In internal medicine, however, it is now conventional to treat delirium and dementia as if they were different disorders, despite the frequent overlap of the two conditions in a single patient.

Both delirium and dementia are characterized by global cognitive impairment, i.e., cognitive impairment involving a number of domains (Engel and Romano, 1959). Both are also associated with variable and often disabling psychiatric manifestations. The difference between the two is that delirium is by definition transient and reversible while dementia is by definition long lasting and not spontaneously reversible. In clinical practice, particularly with older people, the distinction becomes blurred. Underlying dementia predisposes to delirium. Conditions that cause delirium, if prolonged, can also cause dementia (Blass, 1993, 2000; Gibson *et al.,* 1999; Blass and Gibson, 1999).

The characteristic neurobiological abnormality in both delirium and dementia is a reduced rate of brain metabolism ($\downarrow CMR_{glucose}$, $\downarrow CMR_{oxygen}$, $\downarrow$ CBF) (Blass, 1993, 2000; Gibson *et al.,* 1998b; Blass and Gibson, 1999). This generalization holds even though there are a variety of clinically distinct "nosological entities" that can cause delirium and/or dementia (Blass, 1993, 2000; Blass and Gibson, 1999). Brain metabolism is impaired even in AIDS dementia (Ernst *et al.,* 2000; Liow *et al.,* 2000). The cerebrometabolic abnormalities in the most common form of dementia, AD, are discussed in more detail later in this section. Even medications or toxins that induce delirium often interfere with brain metabolism, at least in localized regions of brain.

Neurologists often refer to delirium as "metabolic encephalopathy" (Plum and Posner, 1980). Classic causes of metabolic encephalopathy are hypoxia in its various forms, hypoglycemia, ammonia toxicity secondary to liver disease, and vitamin and nutritional deficiencies such as thiamin deficiency (Gibson and Blass, 1983; Blass, 1993, 2000; Blass and Gibson, 1999; Plum and Posner, 1980). Hypoxia and hypoglycemia interfere with brain metabolism by diminishing the supply of oxygen and of the brain's main substrate, glucose. Ammonia toxicity poisons brain metabolism, among other things by inhibiting KGDHC. Vitamin deficiencies impair the activity of the enzymes that depend on the vitamin-derived co-factors, for instance, impairment of KGDHC activity in thiamin deficiency. If relatively mild or transient, the metabolic encephalopathies are reversible when their causes are removed. For instance, pilots whose oxygen masks slip can regain their full wits when the mask is put back on or when they come to a lower altitude. Confusion in diabetics with transient hypoglycemia can end quickly when their blood sugar is restored to normal, for instance, after they drink a glass of sweetened orange juice. However, if the same metabolic insults are severe enough and prolonged enough then brain damage and impairment of cognition are permanent. Prolonged hypoxic and/or hypoglycemic brain damage are well-known causes of dementia. Thiamin deficiency leads first to the reversible confusion and neurological signs of Wernicke disease, but if not treated promptly it causes the irreversible confabulatory dementia of Korsakoff syndrome. The major mechanisms involved in brain damage due to metabolic encephalopathies include interference in neurotransmission (Gibson and Blass, 1983), excess action of free radicals (ROS) (Beal *et al.,* 1998; Heales *et al.,* 1999; Sims *et al.,* 2000; Lemasters and Nieminen, 2001), loss of cellular calcium homeostasis (Gibson and Peterson, 1987; Blass, 1993; Sims *et al.,* 2000), and eventually apoptotic or necrotic cell death (Gorman *et al.,* 2000).

Among neurotransmitter systems, cholinergic functions are exquisitely sensitive to impairment of mitochondrial function (Gibson and Blass, 1983). The acetyl group of acetylcholine is derived from pyruvate. Although the flux of pyruvate carbons to CO_2 is 200- to 300-fold that to the acetyl group of acetylcholine, the rate of acetylcholine synthesis is proportional to the flux to CO_2. This effect holds true even with minor decreases in pyruvate oxidation (Gibson and Blass, 1983). In contrast, interference with mitochondrial function increases the release of the potentially excitotoxic neurotransmitters glutamate and dopamine (Hirsch and Gibson, 1984; Gibson *et al.,* 1989). The effects on neurotransmission including cholinergic neurotransmission appear to be mediated by effects on calcium-dependent neurotransmitter release (Gibson *et al.,* 1989).

The alterations in neurotransmission can plausibly be linked to physiological and pathological effects of impairment of glucose/mitochondrial metabolism. Ascending cholinergic systems mediate arousal and attentional mechanisms, including activation of the EEG (Gibson and Blass, 1983). Impairment in these cholinergic mechanisms impairs learning and performance. The effects can be at least partially reversed by treatment with centrally acting cholinergic agonists such as the cholinesterase inhibitor physostigmine (Barclay *et al.*, 1981; Gibson and Blass, 1983). Glutamatergic neurotoxicity, often mediated via NMDA receptors, is believed to be a major mechanism of cell loss in hypoxia and related impairments of brain metabolism (Choi, 1990). NMDA receptors mediate the influx of calcium into neurons, and mitochondria are involved in reducing the toxic levels of Ca^{2+} and other materials that accumulate in the stimulated cells (Gibson and Peterson, 1987; Schousboe *et al.*, 1997; Blass, 2000; Sims *et al.*, 2000). Metabolic impairments can prevent neurons from maintaining healthy homeostasis even under normal circumstances of glutamatergic stimulation, when concentrations of glutamate in the synapse do not rise to levels that are normally neurotoxic. The effect of glutamate on neurons with such inherent metabolic impairments is described by the term "slow excitotoxicity" (Beal, 1992). Slow excitotoxicity may be an important mechanism of brain cell damage and death when glucose/mitochondrial metabolism is interfered with, including many of the conditions discussed in earlier parts of this review. As discussed in the previous section, free radicals appear to play a major role in many forms of brain damage in which mitochondrial functions are impaired.

Mitochondria have an important role in both necrotic and apoptotic cell death (Choi, 1990; Snider *et al.*, 1999; Zamzani and Kroemer, 2001). Brain cells are in a dynamic equilibrium that requires a continuous input of energy for them to maintain viability. Cutting off the supply of energy kills them by necrosis, often within minutes at normal body temperature, in part because they lose the ability to maintain ion homeostasis and therefore osmotic stability. Mitochondria are now recognized to play a critical role in apoptosis, i.e., in "programmed cell death." Mitochondria release specific factors that activate apoptosis, such as cytochrome *c* and "apoptosis inducing factor." An important mechanism in mitochondrial failure leading to apoptosis has been proposed to be the "mitochondrial permeability transition" (PT or MPT) (Bernardi, 1996). The PT is thought to involve the opening of a specific channel in the inner mitochondrial membrane that allows the free passage of materials up to M_r 1800 D. The adenine translocator is thought to be part of this pore. The usual method for opening the PT is to expose isolated liver mitochondria to unphysiologically high levels of Ca^{2+}. Other inducers such as products of the oxidation of catecholamines have also been described (Burke *et al.*, 1998). If the PT exists in brain mitochondria, its characteristics differ significantly from those of liver mitochondria (Kristal *et al.*, 2000). Cyclosporin blocks the PT (Bernardi, 1996; Burke *et al.*, 1998; Kristal *et al.*, 2000). Attempts have been made to use cyclosporin to implicate the PT in a variety of effects in intact cells including brain cells. Unfortunately, this drug can have so many other actions that reversal of an effect by cyclosporin is not, by itself, compelling proof that the PT was involved.

Alzheimer disease is, as mentioned previously, the most common dementing disorder known. It is defined by a clinicopathological correlation. However, the clinical syndrome can occur without the neuropathological findings and the full neuropathology can occur without clinical disability in significant numbers of older people (Snowden, 1997; Davis *et al.,* 1999; Blass, 2000; Price *et al.,* 2001). Amyloid plaques are the defining neuropathological abnormality in AD, and there is great interest in the hypothesis that amyloid dysmetabolism is the key pathogenetic abnormality in this disorder (the "amyloid cascade hypothesis"). However, it is hard to see how Alzheimer amyloid could be the immediate cause of the clinical disability, since the neuritic amyloid plaques which define the disease can occur in cognitively intact older people (Snowden, 1997; Davis *et al.,* 1999; Price *et al.,* 2001) as old as 101 years (Price *et al.,* 2001), and without loss of neurons at autopsy (Snowden, 1997; Davis *et al.,* 1999; Price *et al.,* 2001).

Association of AD with mitochondrial abnormalities in PDHC and KGDHC has been discussed in section IV with those mitochondrial complexes and reviewed elsewhere (Sorbi *et al.,* 1983; Blass and Gibson, 1993; Gibson *et al.,* 1998b; Gibson *et al.,* 1999; Sheu and Blass, 1999; Sheu *et al.,* 1999). Reduced activity of Complex IV has also been described (Kish *et al.,* 1999; Maurer *et al.,* 2000; Verwer *et al.,* 2000; Valla *et al.,* 2001). The abnormalities in KGDHC (Blass *et al.,* 1997) and Complex IV (Curti *et al.,* 1997) have been reported to persist in cultured skin fibroblasts from AD patients, indicating that they are not simply due to damage to brain cells. The decrease in activity of KGDHC in AD brain is proportional to the decrease in cognition, in patients who carry the variε4 allele of the *APOE* gene (Gibson *et al.,* 2000). The value of ρ was 0.7, suggesting that the decrease in KGDHC activity could account for about half of the variation in disability in these patients. The activity of Complex IV is decreased in AD brain including hippocampus and becomes more decreased as the disease progresses (Valla *et al.,* 2000; Verwer *et al.,* 2000). The amount of the mtDNA gene *CO1,* that encodes a component of Complex IV, has also been reported to be reduced in proportion to the disability in the brains of AD patients by simultaneous PCR of the mtDNA gene and nDNA pseudogenes (Brown *et al.,* 2001).

The reasons for the replicable decreases in the activities of PDHC, KG DHC, and Complex IV in AD brain are unclear. All three complexes contain active sulfhydryl groups and are sensitive to inactivation by free radicals. AD amyloid, including the 25–35 amino acid fragment of Aβ, has been reported to inactivate Complex IV (Canevari *et al.,* 1999; Parks *et al.,* 2001), presumably through a free radical mechanism, but the reduced Complex IV activity in AD brain is not associated with amyloid plaques (Hataanpa *et al.,*

1998). Molecular studies in AD of the genes encoding PDHC have not been published. Evidence for an abnormality in the *DLST* gene encoding the core protein of KGDHC is inconclusive (Nakano *et al.,* 1997; Sheu *et al.,* 1997, 1999). Evidence for a mutation in the *CO1* gene encoding a component of Complex IV was an artifact (Davis and Parker, 1998). A genetic variant in a gene for KGDHC or Complex IV might increase susceptibility to inactivation by a free radical but be physiologically silent until the cell is put under increased oxidative stress (Fig. 2).

Since the levels of Complex IV vary with neuronal activity (Wong-Riley, 1999), the significance of the deficits in Complex IV in AD brain is hard to know. These deficiencies are found in damaged but not undamaged parts of AD brain and become more marked with increased neuropathological damage (Valla *et al.,* 2001), and they may reflect reduced activity of the damaged neurons or some other relatively nonspecific disease process. "Genetic evidence supporting OXPHOS enzyme defects as a cause for AD is weak," according to a critical review by Shoffner (1997). However, reports continue to appear proposing mutations in genes encoding components of Complex IV in association with AD (Chagnon *et al.,* 1999). Persistence of

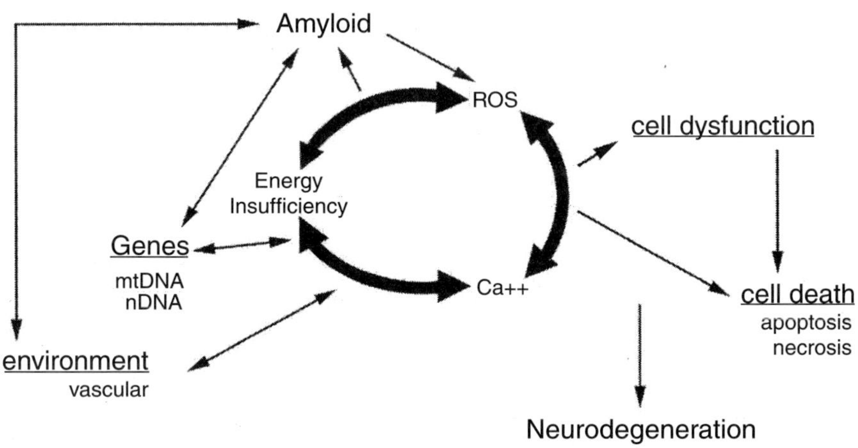

FIG. 2. The Mitochondrial Spiral in Alzheimer disease. Either genetic or nongenetic factors can lead to the mitochondrial spiral and to the accumulation of AD amyloid in the brain. Whether the same genes can lead directly to both abnormalities is not known. Vascular disease, which is the most common cause of death in the developed world, is a clinically prominent abnormality which can lead to increased expression of the AD amyloid precursor protein (APP) and to the mitochondrial spiral. AD amyloid appears to act on cells through free radical mechanisms, and impairment of oxidative/energy metabolism can lead to increased expression of APP. From Blass (2000), The mitochondrial spiral: An adequate cause of dementia in Alzheimer syndrome, *Ann. N.Y. Acad. Sci.* **924,** 170–183. See Blass, 2000, for more extensive discussion.

deficiency of Complex IV activity in cultured AD skin fibroblasts has been reported (Shoffner, 1997), but not independently confirmed or refuted. If replicated, this finding implies that the deficiency in Complex IV activity in AD is not due simply to brain damage but is somehow related to genetic information that is accurately transmitted in cultured cells through serial passage. Cybrid cells containing AD mtDNA and control nDNA have also been found to show abnormalities including abnormalities in Complex IV in some (Ghosh *et al.*, 1999) but not all studies (Ito *et al.*, 1999).

The impairment of brain glucose/mitochondrial metabolism in AD may well be the *proximate* cause of the clinical manifestations (Fig. 2) (Blass, 2001). The cerebrometabolic impairment always accompanies the clinical disabilities in AD, and is in fact proportional to the degree of clinical disability in studies of brain metabolism *in situ* (Jagust *et al.*, 1997; Frey *et al.*, 1998) and less tightly at the level of mitochondrial damage (Gibson *et al.*, 2000; Brown *et al.*, 2001). Inducing impairments in brain glucose oxidation induces mental impairment resembling that which occurs in AD, as discussed earlier (Blass and Gibson, 1979, 1999; Blass, 1993; Gibson *et al.*, 1998b; Blass, 2000). Inducing impairments in brain glucose oxidation in experimental animals leads to behavioral changes analogous to those that occur in AD (Blass and Gibson, 1979, 1999; Blass, 1993; Blass, 2000). Treatments directed toward improving brain glucose oxidation (glucose or a mixture of malate and glucose) have been reported to improve mental function in patients with AD (Craft *et al.*, 1993; Manning *et al.*, 1993; Boyt *et al.*, 2000; Blass, 2002).

B. Psychoses

People go mad for many reasons and in many different ways (Blass *et al.*, 1977; Blass, 1982). Attempts at finding a single discrete cause for madness—biological or environmental—have not been and do not seem likely to be fruitful in finding ways to help these unfortunate people. That is as true for abnormalities in brain glucose metabolism as for other abnormalities.

Early studies of the relationship between endogenous psychoses and brain glucose metabolism used the nonspecific techniques available in the 1930s and 1940s, gave nonspecific although reportedly positive results, and were "back-burnered." However, more recent studies by more specific modern techniques have again indicated that inherent abnormalities in oxidative metabolism—perhaps genetically determined—may contribute to susceptibility to madness in some subgroups of psychiatric patients. In patients with "schizophrenia," PET scanning has shown decreased CMR in frontal cortex and temporoparietal regions (Buchsbaum *et al.*, 1990). Neuropathological studies have shown a ~20% decrease in the number of mitochondria in schizophrenic striatum (Kung and Roberts, 1999). Oxidative stress is probably present in schizophrenic brain, according to several relatively preliminary studies (Reddy and Yao, 1996). Direct, confirmed measurements have shown decreases in the activity of Complex IV in schizophrenic caudate nucleus (Cavalier *et al.*, 1995; Prince *et al.*, 1999). Abnormal sensitivity of Complex IV to the inhibitor azide has been reported in schizophrenic brain, consistent with a structural abnormality of some kind in this complex (Whatley *et al.*, 1996). Decreases have also been described in the activity of Complex I in schizophrenic frontal cortex and of Complex II + III in schizophrenic temporal cortex (Maurer, 2001). Activities of enzymes of energy metabolism have also been reported to be increased in several other parts of schizophrenic brain (Maurer, 2001). At the mRNA level, elevated levels of five transcripts were found in schizophrenic brains, and all encoded mitochondrial genes—four mitochondrial rRNA and one the *CO2* gene (Whatley *et al.*, 1996). Abnormalities in mtDNA have tended to be more frequent in schizophrenics than in controls (Lindholm *et al.*, 1997), and the gene for the mitochondrial citrate transporter is in a region associated with schizophrenia as well as with other disorders (Stoffel *et al.*, 1996). Treatment with glucose has been reported to produce dose-dependent improvements in memory and attention in schizophrenia (Fucetola *et al.*, 1999). Accumulating evidence indicates that abnormalities in brain glucose metabolism also occur in subgroups of patients carrying other psychiatric diagnoses, including depression and mania (Henneman *et al.*, 1954; Kato *et al.*, 1997; Mendlewicz, 1997; Smythies, 1997; Kato and Kato, 2000; Ohara, 2001) and autism (Coleman and Blass, 1985; Chugani *et al.*, 1999). These include molecular genetic data (Mendlewicz, 1997; Ohara, 2001). The ability of intravenous infusion of lactate to induces panic attacks in susceptible people is well known and still unexplained.

The "specificity" and therefore the significance of the abnormalities in brain glucose metabolism in psychoses remains controversial largely because of diagnostic issues in psychiatry. The relation of psychiatric diagnoses to the biology of brain disease is a complex, evolving issue. Current problems have a historical base. The distinction between organic and functional psychoses arose to distinguish between madness for which an organic cause could be found and madness for which no organic cause was found, by the methods of the late 19th and early 20th centuries. "Organic psychoses" were (sensibly) classified biologically and "functional psychoses" behaviorally. In recent years, close attention has been paid to the details of the behavioral classification, in part because of the recognition after World War II that psychiatrists in London and New York were attaching different diagnostic labels to the same patients. It is now widely recognized that even detailed behavioral classification such as that in the DSM series cannot be assumed to differentiate biologically discrete groups. Nevertheless, biological psychiatry continues to search for biological factors that underlie specific behavioral patterns. Thus, in the paragraph above, the definitions of the "patient" populations studied in the cited references made it necessary to discuss separately abnormalities in people with "schizophrenia" and those with "depression" and "mania."

In fact, where the biological cause of madness is known, the form the insanity takes does not fall neatly into conventional behavioral categories. There are many examples. A metabolic encephalopathy such as hypoxia or hypoglycemia can induce a wide variety of psychotic manifestations. They include visual and/or auditory hallucinations, paranoia, and behavior that mimic mania or depression. Different psychiatric symptoms typically come and go in a single delirious patient. Not rarely, the same patient goes from apathetic to agitated delirium and back again. The psychiatric manifestations associated with discrete, well-characterized genetic abnormalities (i.e., "inborn errors of metabolism") can take the form of different "DSM" syndromes even within a single family (Blass *et al.*, 1977; Blass, 1982; Portin and Alanen, 1997; Zubenko, 2000). Modern molecular genetic studies demonstrate clearly that a single susceptibility locus can contribute to the development of different *behaviorally defined* mental disorders (Portin and Alanen, 1997; Zubenko, 2000).

The evidence for abnormalities of brain glucose metabolism in psychoses has at least two practical implications. Molecular geneticists should consider genes that encode components of brain glucose metabolism to be good candidates for psychiatric illness. Treatments to normalize brain glucose metabolism should be investigated as possible therapy for some psychotic people.

C. Cerebrovascular Disease

This is one of the two or three most common disorders in the U.S. (Small *et al.*, 1999). Exactly how common it is depends in part on how one interprets "minor" ischemic changes on MRI or CT in older people. In the elderly, neuroradiologists often interpret these as "normal for age" even in patients who show mental or motor symptoms. Quantitative criteria for how much cerebrovascular disease must be present to be "significant" have been proposed, on the basis of clinical-radiological correlations. Their accuracy in defining clinically meaningful pathology in individual patients is not certain.

Cerebrovascular disease including strokes damage the brain by reducing or cutting off the supply of O_2, glucose, and other materials. They can cause both necrotic and apoptotic cell death, depending on the precise location and severity of the ischemic lesion (Snider *et al.*, 1999). Excessive stimulation of NMDA receptors and impaired cellular Ca^{2+} homeostasis are believed to play an important role in the cell damage and death (Schousboe *et al.*, 1997; Snider *et al.*, 1999). Damage during reperfusion appears to be mediated primarily by free radical mechanisms (Kuroda and Siesjo, 1997; Huang *et al.*, 2001).

Although severe and prolonged cerebral ischemia causes generalized brain damage, lesser degrees of ischemia can cause brain damage in a pattern of exquisitely selective vulnerability (Blass and Gibson, 1979; Pulsinelli *et al.*, 1982; Saji *et al.*, 1994). A particularly clear experimental example is the four-vessel occlusion rat model of Pulsinelli *et al.* (1982). Controlled, transient global ischemia in this model leads to selective vulnerability of the striatum, followed by degeneration in the hippocampus and then in the substantia nigra. In the hippocampus, CA1 cells are exquisitely vulnerable but CA2 and CA3 are not (Pulsinelli *et al.*, 1982; Saji *et al.*, 1994). Three weeks after the injury neurons are lost in the dorsal substantia nigra, perhaps because of transneuronal degeneration (Saji *et al.*, 1994). Treatment with the antioxidant melatonin reduces the hippocampal cell loss (Cho *et al.*, 1997). The human hippocampus also appears to be selectively vulnerable to episodes of global cerebral ischemia (Volpe and Petito, 1985). Data on vitamin deficiencies (Gibson and Blass, 1985, 1998) and AD (Ko *et al.*, 2001) are also in accordance with the proposal that impairment of glucose oxidation can lead to selective vulnerability.

Cell death is, however, a relatively late consequence of cerebral hypoxia or ischemia (Gibson *et al.*, 1978; Blass and Gibson, 1979). The earliest known changes affect cell signaling, including cyclic nucleotides (Gibson *et al.*, 1978) and calcium (Gibson and Peterson, 1987). Neurotransmitters including acetylcholine are also affected early (Gibson and Blass, 1982; Milner *et al.*, 1987), probably because of changes in calcium-mediated release (Gibson and Peterson, 1987). In general, as the supply of energy decreases, cells including neurons decrease function to maintain structure, i.e., decrease anabolic and increase catabolic activities. Survival takes precedence over usefulness. If the hypoxic/ischemic insult is long enough and severe enough, the homeostatic mechanisms fail and apoptotic or necrotic mechanisms of cell death ensue (Snider *et al.*, 1999). The sequence of changes associated with graded hypoxia is likely to be relevant in other disorders of glucose/mitochondrial metabolism including genetic disorders (Blass and Gibson, 1979; Blass, 1997).

The mechanisms of brain damage in cerebrovascular disease have been the subject of intensive study for many years. The literature on this subject is too extensive to review in more detail here (Small *et al.*, 1999).

D. Aging

While by definition aging is not a disease, it is a risk factor for a number of common diseases that involve impaired brain glucose metabolism. They include strokes, AD, and PD. Many attempts have been made to try to identify fundamental processes of aging that contribute to susceptibility to these disorders. However, interpretation of the resulting data is difficult for several reasons.

There is no widely accepted biological marker for aging (Severion, 2000). Lacking a biological marker, investigators use chronological age which is recognized to be a poor surrogate. But people and other animals age biologically at different rates. The same person can "look ten years younger" after recovering from an illness or biological stress. The relation of aging to standard measures of time and to thermodynamics have been discussed elsewhere (Blass *et al.*, 1992).

A related problem is making the distinction between aging and disease. It is well documented that the rate of brain glucose utilization falls with age if one studies unselected populations of older people free of clinical brain or heart disease. It is equally well documented that one can cull from elderly populations groups of unusually healthy people in whom the rate of brain glucose utilization is similar to that of the average of younger people. Are the decreases found in most older people the result of "subclinical" brain disease? Is subclinical brain disease a manifestation of statistically normal aging? The "error-damage" theory of aging proposes that damage accumulates with age until a critical threshold is passed and the organism dies. Biological damage to younger people is typically defined as a disease. Is the error-damage theory of aging different from a "disease theory" of aging?

Excessive action of free radicals is now widely thought to play a role in aging. Accumulation of free radical damage has proven to be a relatively specific and testable damage theory of aging for which there is now abundant direct data (Mattson *et al.*, 2001; Wallace, 2001). These findings imply that reducing the burden of free radicals might modify unwanted age-related changes.

The one manipulation that has been robustly documented to expand mean and median life span is dietary restriction, i.e., the ingestion of fewer calories while still receiving adequate nutrition. This is a dose-related effect (Mattson *et al.*, 2001). Dietary restricted animal also show fewer signs of aging. Dietary restriction has been proposed to be a plausible way to reduce the development of AD and other degenerative disease (Mattson *et al.*, 2001).

IX. Summary and Conclusions

Impairments of glucose and mitochondrial function are important causes of brain dysfunction and therefore of brain disease. Abnormalities have been found in association with disease of the nervous system in most of the components of glucose/mitochondrial metabolism. In many, molecular genetic abnormalities have been defined. Brain glucose oxidation is abnormal in common diseases of the nervous system, including Alzheimer disease and other dementias, Parkinson disease, delirium, probably schizophrenia and other psychoses, and of course cerebrovascular disease. Defects in a single component and even a single mutation can be associated with different clinical phenotypes. The same clinical phenotype can result from different genotypes. The complex relationship between biological abnormality in brain glucose utilization and clinical disorder is similar to that in other disorders that have been intensively studied at the genetic level.

Genes for components of the pathways of brain glucose oxidation are good candidate genes for disease of the brain. Preliminary data support the proposal that treatments to normalize abnormalities in brain glucose oxidation may benefit many patients with common brain diseases.

References

Andreasen, O. A., Ferrante, R. J., Huang, H. M., Dedeoglu, A., Park, L., Ferrante, K. L., Kwon, J., Borchelt, D. R., Ross, C. A., Gibson, G. E., and Beal, M. F. (2001). Dichloroacetate exerts therapeutic effects in transgenic mouse models of Huntington's disease. *Ann. Neurol.* **50,** 112–116.

Antuono, P. G., Ravanelli-Meyer, J., Nicholson, K., and Bloom, A. S. (1995). Leukocyte hexokinase activity in aging and Alzheimer disease. *Dementia* **6,** 200–204.

Aoki, C., Milner, T. A., Sheu, K. F. R., Blass, J. P., and Pickel, V. M. (1987). Regional distribution of astrocytes with intense immunoreactivity for glutamate dehydrogenase in rat brain: implications for neuron-glia interactions in glutamate transmission. *J. Neurosci.* **7,** 2214–2231.

Aomi, Y., Chen, C. S., Nakada, K., Ito, S., Isobe, K., Murakami, H., Kuno, S. Y., Tawata, M., Matsuoka, R., and Hayashi, J. L. (2001). Cytoplasmic transfer of platelet mtDNA from elderly patients with Parkinson's disease to HeLa cells restores complete mitochondrial respiratory function. *Biochem. Biophys. Res. Commun.* **280,** 265–273.

Barclay, L. L., Gibson, G. E., and Blass, J. P. (1981). Impairment of behavior and acetylcholine metabolism in thia mine deficiency. *J. Pharmacol. Exp. Ther.* **217,** 537–543.

Beal, M. F. (1992). Mechanisms of excitotoxicity in neurologic diseases. *FASEB J.* **6,** 3338–44.

Beal, M. F. (1998). Mitochondrial dysfunction in neurodegenerative diseases. *Biochim. Biophys Acta.* **1366,** 211–223.

Bender, E., and Kadenbach, B. (2000). The allosteric ATP-inhibition of cytochrome *c* oxidase is reversibly switched on by cAMP-dependent phosphorylation. *FEBS Lett.* **456,** 130–134.

Bernardi, P. (1996). The permeability transition pore. Control points of a cyclosporin A-sensitive mitochondrial channel involved in cell death. *Biochim. Biophys. Acta.* **1275,** 5–9.

Betarbet, R., Sherer, T. B., MacKenzie, G., Garcia-Osuna, M., Panov, A. V., and Greenmayre, J. T. (2000). Chronic systemic pesticide exposure reproduces features of Parkinson's disease. *Nat. Neurosci.* **3,** 1301–1306.

Bigl, M., and Eschrich, K. (1995). Interaction of rat brain phosphofructokinase with Alzheimer's βA4 amyloid. *Neurochem. Int.* **26,** 69–75.

Bigl, M., Bruckner, M. K., Arendt, T., Bigl, V., and Eschrich, K. (1999). Activities of key glycolytic enzymes in the brains of patients with Alzheimer's disease. *J. Neural. Transm.* **106,** 499–511.

Blass, J. P. (1982). Inborn errors, inborn predispositions, and normal variations: some implications for psychiatric research. In I. Hanin and E. Usdin (Eds.), Biological Markers in Psychiatry and Neurology, Pergamon Press, Oxford, pp. 473–482.

Blass, J. P. (1993). Metabolic alterations common to neural and non-neural cells in Alzheimer's disease. *Hippocampus* **3,** 45–54.

Blass, J. P. (1997). Cerebral metabolic impairments. In Z. S. Khachaturian and T. S. Radebaugh (Eds.) Alzheimer's Disease: Cause(s), Diagnosis, Treatment, and Care, pp. 187–206. CRC Press, NY.

Blass, J. P. (1999). Conference summary: Mitochondria, neurodegenerative diseases, and selective neuronal vulnerability. *Ann. N.Y. Acad. Sci.* **893,** 434–439.

Blass, J. P. (2000). The mitochondrial spiral: An adequate cause of dementia in the Alzheimer syndrome. *Ann. N.Y. Acad. Sci.* **924,** 170–183.

Blass, J. P. (2001). Brain metabolism and brain disease: Is metabolic deficiency the proximate cause of Alzheimer dementia. *J. Dev. Neurosci.* (in press).

Blass, J. P. (2002). Metabolic enhancement for the treatment of Alzheimer disease. Research and Practice in Alzheimer's Disease and Cognitive Decline (in press).

Blass, J. P., and Gibson, G. E. (1979). Consequences of mild, graded hypoxia. *In* S. Fahn (Ed.), Advances in Neurology, pp. 229–250. Raven Press, New York.

Blass, J. P., and Gibson, G. E. (1993). Nonneural markers in Alzheimer Disease. *Alz. Dis. Assoc. Disord.* **6,** 205–224.

Blass, J. P., and Gibson, G. E. (1999). Cerebrometabolic aspects of delirium in relationship to dementia. *Dement. Geriatr. Cogn. Disord.* **10,** 335–338.

Blass, J. P., Avigan, J., and Uhlendorf, B. W. (1970). A defect in pyruvate decarboxylase in a child with an intermittent movement disorder. *J. Clin. Invest.* **49,** 423–432.

Blass, J. P., Kark, R. A. P., Menon, N., and Harris, S. H. (1976). Decreased activities of the pyruvate and α-ketoglutarate dehydrogenase complexes in fibroblasts from five patients with Fredreich's ataxia. *N. Engl. J. Med.* **295,** 62–66.

Blass, J. P., Milne, J. A., and Rodnight, R. (1977). Newer concepts of psychiatric diagnosis and biochemical research on mental illness. *Lancet* **1,** 738–740.

Blass, J. P., Cherniak, E. P., and Weksler, M. E. (1992). Theories of Aging. *In* Calkins, E., Ford, A. B., and Katz, P. R. (Eds.), *Practice of Geriatrics.* W. B. Saunders, Philadelphia, pp. 10–18.

Blass, J. P., Sheu, K.-F. R., Piacentini, S., and Sorbi, S. (1997). Inherent abnormalities in oxidative metabolism in AD: Interaction with vascular abnormalities. *Ann. N.Y. Acad. Sci.* **826,** 382–385.

Boyt, A. A., Taddei, T. K., Hallmayer, J., Helmerhorst, E., Gandy, S. E., Craft, R. S., and Martins, R. N. (2000). The effect of insulin and glucose on the plasma concentration of Alzheimer's amyloid precursor protein. *Neuroscience* **95,** 727–734.

Brown, A. M., Sheu, R. K.-F., Mohs, R., Haroutunian, V., and Blass, J. P. (2001). Correlation of the clinical severity of Alzheimer's Disease with an aberration in mitochondrial DNA (mtDNA). *J. Mol. Neurosci.* **16,** 41–48.

Browne, S. E., Bowling, A. C., MacGarvey, U., Balk, M. J., Berger, S. C., Muqit, M. M., Bird, E. D., and Beal, M. F. (1997). Oxidative damage and metabolic dysfunction in Huntington's disease: selective vulnerability of the basal ganglia. *Ann. Neurol.* **41,** 646–653.

Buchsbaum, M. S., Neuchterlein, K. H., Haier, R. J., Wu, J., Sicotte, N., Hazlett, E., Asarnow, R., Potkin, S., and Guich, S. (1990). Glucose metabolic rate in normals and schizophrenics during the continuous performance test assessed by positron emission tomography. *Br. J. Psychiatry* **156,** 216–227.

Burke, W. J., Kristal, B. S., Yu, B. P., Li, S. W., and Lin, T. S. (1998). Norepinephrine transmitter metabolite generates free radicals and activates mitochondrial permeability transition: a mechanism for DOPEGAL-induced apoptosis. *Brain. Res.* **787,** 328–332.

Calingasan, N. Y., Baker, H., Sheu, K. F., and Gibson, G. E. (1994). Distribution of the α-ketoglutarate dehydrogenase complex in rat brain. *J. Comp. Neurol.* **346,** 461–479.

Canevari, L., Clark, J. B., and Bates, T. E. (1999). β-Amyloid fragment 25-35 selectively decreases complex IV activity in isolated mitochondria. *FEBS Lett.* **457,** 131–134.

Cavelier, L., Jazin, E. E., Erikson, I., Prince, J., Bave, U., Oreland, L., and Gyllensten, U. (1995). Decreased cytochrome-*c* oxidase activity and lack of age-related accumulation of mitochondrial DNA deletions in the brains of schizophrenics. *Genomics* **29,** 217–224.

Chagnon, P., Gee, M., Filion, M., Robitaille, Y., Belouchi, M., and Gauvreau, D. (1999). Phylogenetic analysis of the mitochondrial genome indicates significant differences between patients with Alzheimer disease and controls in a French-Canadian founder population. *Am. J. Med. Genet.* **85,** 20–30.

Chandrasekaran, K., Hatanpaa, K., Brady, D. R., and Rapoport, S. I. (1996). Evidence for physiological down-regulation of brain oxidative phosphorylation in Alzheimer's disease. *Exp. Neurol.* **142,** 80–88.

Cho, S., Joh, T. H., Baik, H. H., Dibnis, C., and Volpe, T. (1997). Melatonin administration protects CA1 hippocampal neurons after transient forebrain ischemia in rats. *Brain Res.* **755,** 335–338.

Choi, D. W. (1990). Cerebral hypoxia: some new approaches and unanswered questions. *J. Neurosci.* **10,** 2493–501.

Chugani, D. C., Sundram, B. S., Behen, M., Lee, M. L., and Moore, G. J. (1999). Evidence of altered energy metabolism in autistic children. *Prog. Neuro-Psychopharm. Biol. Psychiat.* **23,** 635–641.

Cohen, G. (2000). Oxidative stress, mitochondrial respiration, and Parkinson's disease. *Ann. N.Y. Acad. Sci.* **899,** 112–120.

Coleman, M., and Blass, J. P. (1985). Autism and lactic acidosis. *J. Autism. Dev. Disord.* **15,** 1–8.

Cooper, A. J. L., Sheu, K.-F. R., Burke, J. R., Onodera, O., Strittmatter, W. J., Roses, A. D., and Blass, J. P. (1997a). Glutathione-S-transferase constructs containing medium to large polyglutamine inserts are substrates of guineapig liver transglutaminase: Does transglutaminase play a role in the pathogenesis of expanded CAG/Poly Q neurodegenerative diseases? *Proc. Natl. Acad. Sci. USA* **94,** 12604–12609.

Cooper, A. J. L., Sheu, K.-F. R., Burke, J. R., Onodera, O., Strittmatter, W. J., Roses, A. D., and Blass, J. P. (1997b). Polyglutamine domains are substrates of tissue transglutaminase: Does transglutaminase play a role in expanded CAG/poly-Q neurodegenerative diseases? *J. Neurochem.* **69,** 431–434.

Cooper, A. J. L., Sheu, K.-F. R., Burke, J. R., Strittmatter, W. J., Gentile, V., Peluso, G., and Blass, J. P. (1999). Pathogenesis of inclusion bodies in $(CAG)_n/Q_n$-expansion diseases with special reference to the role of tissue transglutaminase and to selective vulnerability. *J. Neurochem.* **72,** 889–899.

Craft, S., Dagago-Jack, S. E., Wiethop, B. V., Murphy, C., Nevins, R. T., Fleischman, S., Rice, V., Newcomer, J. W., and Cryer, P. E. (1993). Effects pf hyperglycemia on memory and hormone levels in dementia of the Alzheimer type—A longitudinal study. *Behav. Neurosci.* 926–940.

Craigen, W. J. (1996). Leigh disease with deficiency of lipoamide dehydrogenase: treatment failure with dichloroacetate. *Pediatr. Neurol.* **14,** 69–71.

Curti, D., Rognoni, F., Gasparini, L., Cattaneo, A., Paolillo, M., Racchi, M., Zanni, L., Bianchetti, A., Trabucci, A., Bergamaschi, S., and Govoni, S. (1997). Oxidative metabolism in cultured fibroblasts from sporadic Alzheimer's disease. *Neurosci. Lett.* **236,** 13–16.

Davies, D. L. (1949). Psychiatric changes associated with Friedreich's Ataxia. *J. Neurol. Neurosurg. Psychiat* **12,** 246–250.

Davis, J. N., 2nd, and Parker, W. D., Jr. (1998). Evidence that two reports of mtDNA cytochrome *c* oxidase "mutations" in Alzheimer's disease are based on nDNA pseudogenes of recent evolutionary origin. *Biochem. Biophys. Res. Commun.* **244,** 877–83.

Davis, D. G., Schmittm, F. A., Wekstein, D. R., and Markesbery, W. R. (1999). Alzheimer neuropathological alterations in cognitively normal subjects. *J. Neuropathol. Exp. Neurol.* **58,** 376–388.

DeVivo, D., Haymond, M. W., Leckie, M. P., Bussman, Y. L., McDougal, D. B., and Pagliara, A. S. (1977). Clinical and biochemical implications of pyruvate carboxylase deficiency. *J. Clin. Endocrinol. Metab.* **45,** 1281–1290.

Elpeleg, O. N., Ruitenbeek, W., Jakobs, C., Barash, V., DeVivo, V. C., and Amir, N. (1995). Congenital lacticacidemia caused by lipoamide dehydrogenase deficiency with favorable outcome. *J. Pediatr.* **126,** 72–74.

Elpeleg, O. N., Shaag, A., Glustein, J. Z., Anikster, Y., Joseph, A., and Saada, A. (1997). Ashkenazi Jews: an insertion mutation in the mitochondrial leader sequence. *Hum. Mutat.* **10,** 256–257.

Engel, G. L., and Romano, J. (1959). Delirium—A syndrome of cerebral insufficiency. *J. Chronic. Dis.* **9,** 260–267.

Ernst, T., Itti, E., Itti, L., and Chang, L. (2000). Changes in cerebral metabolism are detected prior to perfusion changes in early HIV-CMC: A coregistered (1) H MRS and SPECT study. *J. Magn. Res. Imaging* **12,** 859–865.

Falk, R. E., Cedarbaum, S. D., Blass, J. P., Pruss, R. J., and Carrel, R. E. (1976). Effects of a ketogenic diet in two brothers with pyruvate dehydrogenase deficiency. *Pediatrics* **58,** 713–721.

Floyd, R. A. (1999). Antioxidants, oxidative stress, and degenerative neurological disorders. *Proc. Soc. Exp. Biol. Med.* **222,** 236–245.

Frey, K. A., Minoshima, S., and Kuhl, D. F. (1998). Neurochemical imaging of Alzheimer's disease and other degenerative dementias. *Q. J. Nucl. Med.* **42,** 166–178.

Freytag, S. O., and Collier, K. J. (1984). Molecular cloning of a cDNA for human pyruvate carboxylase. *J. Biol. Chem.* **259,** 12831–12837.

Fucetola, R., Newcomer, J. W., Craft, S., and Melson, A. K. (1999). Age- and dose-dependent glucose-induced increases in memory and attention in schizophrenia. *Psychol. Res.* **88,** 1–13.

Gaudette, M., Hirano, M., and Siddique, T. (2000). Current status of SOD1 mutations in familial amyotrophic lateral sclerosis. *Amyotroph Lareal Scler Other Motor Neuron. Disord.* **1,** 83–89.

Gellera, C., Uziel, G., Rimoldi, M., Zeviani, M., Laverda, A., Carrara, F., and Di Donato, S. (1990). Fumarase deficiency is an autosomal recessive encephalopathy affecting both the mitochondrial and cystosolic enzymes. *Neurology* **40,** 495–499.

Ghosh, S. S., Swerdlow, R. H., Miller, S. W., Sheeman, B., Parker, W. D., and Davis, R. E. (1999). Use of cytoplasmic hybrid cell lines for elucidating the role of mitochondrial dysfunction in Alzheimer's disease and Parkinson's disease. *Ann. N.Y. Acad. Sci.* **893,** 176–191.

Kosel, S., Grasbon-Frodl, E. M., Mautsch, U., Egensperger, R., von Eitzen, U., Frishman, D., Hofmann, S., Mehraein, P., and Graeber, M. B. (1998). Novel mutations of mitochondrial complex I in pathologically proven Parkinson disease. *Neurogenetics* **1,** 197–204.

Kragten, E., Lalande, I., Zimmerman, K., Roggo, S., Schindler, P., Muller, D., van Ostrum, J., Waldmeier, P., and Furst, P. (1998). Glyceraldehyde-3-phosphate dehydrogenase, the putative target of antiapoptotic compounds CGP 3466 and R-(-)-deprenyl. *J. Biol. Chem.* **273,** 5821–5828.

Kristal, B. S., Staats, P. N., and Shestopalov, A. I. (2000). Biochemical characterization of the mitochondrial permeability transition in isolated forebrain mitochondria. *Dev. Neurosci.* **22,** 376–383.

Kumagai, R., Ichikawa, K., Yasui, T., Kageyama, Y., and Miyabashi, S. (1999). Adult Leigh syndrome: treatment with intravenous soybean oil for acute central respiratory failure. *Eur. J. Neurol.* **6,** 613–615.

Kung, L., and Roberts, R. C. (1999). Mitochondrial pathology in human schizophrenic striatum: a postmortem ultrastructural study. *Synapse* **31,** 67–75.

Kuroda, S., and Siesjo, B. K. (1997). Reperfusion damage following focal ischemia: pathophysiology and therapeutic windows. *Clin. Neurosci.* **4,** 199–212.

Lai, J. C., and Blass, J. P. (1984). Neurotoxic effects of copper: inhibition of glycolysis and glycolytic enzymes. *Neurochem. Res.* **9,** 1699–710.

Lemasters, J. J., and Nieminen, A.-L. eds. (2001). Mitochondria in Pathogenesis, pp. 1–529. Kluwer-Plenum, NY.

Liguri, G., Taddei, N., Nassi, P., Latorraca, S., Nediani, C., and Sorbi, S. (1990). Changes in Na^+,K^+-ATPase, Ca^{2+} ATPase and some soluble enzymes related to energy metabolism in brains of patients with Alzheimer's disease. *Neurosci. Lett.* **112,** 338–342.

Lindholm, E., Cavelier, L., Howell, W. M., Eriksson, I., Jalonen, P., Adolfsson, R., Blackwood, D. H., Muir, W. J., Brookes, A. J., Gyllensten, U., and Jazin, E. E. (1997). Mitochondrial sequence variants in patients with schizophrenia. *Eur. J. Hum. Genet.* **5,** 406–412.

Liow, J. S., Rehm, K., Strother, S. C., Anderson, J. R., Morch, N., Hansen, L. K., Schaper, K. A., and Rottenberg, D. (2000). Comparison of voxel- and volume-of-interest-based analyses in FDG PET scans of HIV positive and healthy individuals. *J. Nucl. Med.* **41,** 612–621.

Lissens, W., Vreken, P., Barth, P. G., Wijburg, F. A., Ruitenbeek, W., Wqanders, R. J., Seneca, S., Libaers, I., and De Meirleir, L. (1999). Cerebral palsy and pyruvate dehydrogenase deficiency: identification of two new mutations in the E1α gene. *Eur. J. Pediatr.* **158,** 853–857.

Lissens, W., De Meirleir, L., Seneca, S., Liebaers, I., Brown, G. K., Brown, R. M., Ito, M., Naito, E., Kuroda, Y., Kerr, D. S., Wexler, I. D., Patel, M. S., Robinson, B. H., and Seyda, A. (2000). Mutations in the X-linked pyruvate dehydrogenase (E1) alpha subunit gene (PDHA1) in patients with a pyruvate dehydrogenase complex deficiency. *Hum. Mutat.* **15,** 209–219.

Lotharius, J., and O'Malley, K. L. (2000). The Parkinsonism-inducing drug 1-methyl-4-phenylpyridium triggers intracellular dopamine oxidation. A novel mechanism of toxicity. *J. Biol. Chem.* **275,** 38581–38588.

Mach, J. R., Korchik, W. P., and Mahowald, M. W. (1988). Dialysis dementia. *Clin. Geriatr. Med.* **4,** 853–867.

Manning, C. A., Ragozzino, M. E., and Gold, P. E. (1993). Glucose enhancement of memory in patients with probable senile dementia of the Alzheimer type. *Neurobiol. Aging* **14,** 523–528.

Marcus, D. L., and Freedman, M. L. (1997). Decreased brain glucose metabolism in microvessels from patients with Alzheimer's disease. *Ann. N.Y. Acad. Sci.* **826,** 248–253.

Mastrogiacomo, F., Lindsay, J. G., Bettendorff, L., Rice, J., and Kish, S. J. (1996). Brain protein and α-ketoglutarate dehydrogenase complex activity in Alzheimer's disease. *Ann. Neurol.* **39,** 592–598.

Mattson, M. P., Duan, W., Lee, J., and Guo, Z. (2001). Suppression of brain aging and neurodegenerative disorders by dietary restriction and environmental enrichment: molecular mechanisms. *Mech. Ageing Dev.* **122,** 757–758.

Maurer, I., Sierz, S., and Moller, H. J. (2000). A selective defect of cytochrome oxidase *c* is present in brain of Alzheimer disease patients. *Neurobiol Aging* **21,** 455–462.

Maurer, I., Zierz, S., and Moller, H. (2001). Evidence for a mitochondrial oxidative phosphorylation defect in brains from patients with schizophrenia. *Schizophr. Res.* **48,** 125–136.

McCall, A. L., Van Buren, A. M., Nipper, V., Moholt-Siebert, M., Downes, H., and Lessov, N. (1996). Forebrain Ischemia increases GLUT1 protein in brain microvessels and parenchyma. *J. Cereb. Blood Flow Metab.* **16,** 69–76.

Meier-Ruge, W. A., and Bertoni-Freddari, C. (1997). Pathogenesis of decreased glucose turnover and oxidative phosphorylation in ischemic and trauma-induced dementia of the Alzheimer type. *Ann. N.Y. Acad. Sci.* **826,** 229–241.

Melov, S., Coskun, P., Patel, M., Tuinstra, R., Cottrel, B., Jun, A. S., Zastawny, T. H., Dizdaroglu, M., Goodman, S. I., Huang, T. T., Miziorko, H., Epstein, C. J., and Wallace, D. C. (1999). Mitochondrial disease in superoxide dismutase 2 mutant mice. *Proc. Natl. Acad. Sci. USA* **96,** 846–851.

Mendlewicz, J., Lindbald, K., Souery, D., Mahieu, B., Nylander, P. O., De Bruyn, A., Zander, C., Engström, C., Adolfsson, R., Van Broeckhoven, C., Schalling, M., and Lipp, O. (1997). Expanded trinucleotide CAG repeats in families with bipolar affective disorders. *Biol. Psychol.* **42,** 1115–1122.

Milner, T. A., Aoki, C., Sheu, K. F. R., Blass, J. P., and Pickel, V. M. (1987). Light microscopic immunocytochemical localization of pyruvate dehydrogenase in rat brain: topographical distribution and relation to cholinergic and catecholaminergic nuclei. *J. Neurosci.* **7,** 3171–3190.

Mirel, D. B., Marder, K., Graziano, J., Freyer, G., Shao, Q., Mayeux, R., and Wilhelmsen, K. C. (1998). Characterization of the human mitochondrial aconitase gene. *Gene* **21,** 205–218.

Mizuno, Y., Matuda, S., Yoshino, H., More, H., Hattori, N., and Ikebe, S.-I. (1994). An immunohistochemical study on α-ketoglutarate dehydrogenase complex in Parkinson's disease. *Ann. Neurol.* **35,** 204–210.

Naito, E., Ito, M., Yokoto, I., Saijo, T., Chen, S., Maehara, M., and Kuoda, Y. (1999). Concomitant administration of sodium dichloroacetate and thiamine in West syndrome caused by thiamine-responsive pyruvate dehydrogenase complex deficiency. *J. Neurol. Sci.* **171,** 56–59.

Nakano, K., Ohta, S., Nishimaki, K., Miki, T., and Matuda, S. (1997). Alzheimer's disease and DLST genotype. *Lancet* **350,** 1367–1368.

Nolan, K. A., Black, R. S., Sheu, K. F. R., Langberg, J., and Blass, J. P. (1991). A trial of thiamine in Alzheimer's disease. *Arch. Neurol.* **48,** 81–83.

Novotny, E. J., Singh, G., Wallace, D. C., Dorfman, L. J., Louis, A., Sogg, R. L., and Steinman, L. (1986). Leber's disease and dystonia: A mitochondrial disease. *Neurology* **36,** 1053–1059.

Ohara, K. (2001). Anticipation, imprinting, trinucleotide repeat expansions and psychoses. *Prog. Neuropsychopharmacol. Biol. Psych.* **25,** 167–192.

Op den Velde, W., and Stam, F. C. (1976). Some cerebral proteins and enzyme systems in Alzheimer's presenile and senile dementia. *J. Am. Geriatr. Soc.* **24,** 12–16.

Pardridge, W. M. (1999). Blood-brain barrier biology and methodology. *J. Neurovirol.* **5,** 556–569.

Park, L. C., Zhang, H., Sheu, K. F., Calingasan, N. Y., Kristal, B. S., Linsay, J. G., and Gibson, G. E. (1999). Metabolic impairment induces oxidative stress, compromises inflammatory responses, and inactivates a key mitochondrial enzyme in microglia. *J. Neurochem.* **72,** 1948–58.

Parks, J. K., Smith, T. S., Trimmer, P. A., Bennet, J. P. Jr., and Parker, W. D. Jr. (2001). Neurotoxic Aβ peptides increase oxidative stress in vivo through NMDA-receptor and synthetase mechanisms, and inhibit Complex IV activity and induce a mitochondrial permeability transition in vitro. *J. Neurochem.* **76,** 1050–1056.

Petrova-Benedict, R., Robinson, B. H., Stacey, T. E., Mistry, J., and Chalmers, R. A. (1987). Deficient fumarase activity in an infant with fumaric acidemia and its distribution between the different forms of the enzyme seen on isolelectric focusing. *Am. J. Hum. Genet.* **40,** 257–266.

Piert, M., Koeppe, R. A., Giordani, B., Berent, S., and Kuhl, D. E. (1996a). Diminished glucose transport and phosphorylation in Alzheimer's disease determined by dynamic FDG-PET. *J. Nucl. Med.* **37,** 201–208.

Piert, M., Koeppe, R. A., Giordani, B., Minoshima, S., and Kuhl, D. E. (1996b). Determination of regional rate constants from dynamic FDG-PET studies in Parkinson's disease. *J. Nucl. Med.* **37,** 1115–1122.

Plum, F., and Posner, J. B. (1980). The Diagnosis of Stupor and Coma. 3rd edition, pp. 1–373. Davis, Philadelphia.

Portin, P., and Alanen, Y. O. (1997). A critical review of genetic studies of schizophrenia. II. Molecular genetic studies. *Acta. Psychiatr. Scand.* **95,** 73–80.

Price, J. L., Ko, A. I., Wade, M. J., Tsou, S. K., McKeel, D. W., and Morris, J. C. (2001). Neuron number in entorhinal cortex and CA1 in preclinical Alzheimer disease. *Arch. Neurol.* **58,** 1395–1402.

Prince, J., Hia, S., Bave, U., Anneren, G., and Oreland, L. (1994). Mitochondrial enzyme deficiencies in Down's syndrome. *J. Neural Transm. Park Dis. Dement. Sect.* **8,** 171–181.

Prince, J. A., Blennow, K., Gottfries, C. G., Karlsson, I., and Oreland, L. (1999). Mitochondrial function is differentially latered in the basal ganglia of chronic schizophrenics. *Neuropsychopharmacology* **21,** 372–399.

Pulsinelli, W. A., Brierly, J. B., and Plum, F. (1982). Temporal profile of neuronal damage in a model of transient forebrain ischemia. *Ann. Neurol.* **11,** 491–498.

Quastel, J. H. (1932). Biochemistry and mental disorder. *Lancet* **2,** 1417–1425.

Reddy, R. D., and Yao, J. K. (1996). Free radical pathology in schizophrenia: a review. *Prostagland. Leukot. Essenti. Fatty Acids* **55,** 33–43.

Reichman, H., and Janetzky, B. (2000). Mitochondrial dysfunction—a pathogenetic factor in Parkinson's disease. *J. Neurol.* **247,** 63–68.

Robinson, B. H. (2001). Lactic academia: Disorders of pyruvate carboxylase and pyruvate dehydrogenase *In* "The Metabolic and Molecular Bases of Inherited Disease." (C. R. Scriver, A. L. Beaudet, W. S. Sly, and D. Valle, eds.), pp. 2275–2295. 8th edition. McGraw-Hill, New York.

Robinson, B. H., Chun, K., Mackay, N., Otulakowski, G., Petrova-Benedict, R., and Willard, J. (1989). Isolated and combined deficiencies of the α-ketoacid dehydrogenase complexes. *Ann. N.Y. Acad. Sci.* **573,** 337–346.

Saji, M., Cohen, M., Blau, A. D., Wessel, T. C., and Volpe, B. T. (1994). Transient forebrain ischemia induces delayed injury in the substantia nigra reticulata: GABA neurons, compensatory expression of GAD mRNA. *Brain. Res.* **643,** 234–244.

Schapira, A. H. (1998). Inborn and induced defects of mitochondria. *Arch. Neurol.* **51,** 1293–6.

Schapira, A. H., Gu, M., Taanman, J. W., Tabrizi, S. J., Seaton, T., Cleeter, M., and Cooper, M. J. (1998). Mitochondria in the etiology and pathogenesis of Parkinson's disease. *Ann. Neurol.* **44,** S89–S98.

Schapira, A. H. (1999). Mitochondrial involvement in Parkinson's disease, Huntington's disease, hereditary spastic paraplegia and Friedreich's Ataxia. *Biochim. Biophys. Acta* **1410,** 99–102.

Schousboe, A., Belhage, B., and Frandsen, A. (1997). Role of Ca^{2+} and other second messengers in excitatory amino acid receptor mediated neurodegeneration: clinical perspectives. *Clin. Neurosci.* **4,** 194–198.

Schulze, H., Schuler, A., Stuber, D., Dobeli, H., Langen, H., and Huber, G. (1993). Rat brain glyceraldehydes-3-phosphate dehydrogenase interacts with the recombinant cytoplasmic domain of Alzheimer's β-amyloid precursor protein. *J. Neurochem.* **60,** 1915–1922.

Seidner, G., Alvarez, M. G., Yeh, J. L., O'Driscoll, K. R., Klepper, J., Stump, T. S., Wang, D., Spinner, N. B., Birnbaum, M. J., and DeVivo, D. C. (1998). GLUT-1 deficiency syndrome caused by haploinsufficiency of the blood-brain barrier hexose carrier. *Nat. Genet.* **18,** 188–191.

Servidei, S., Bonilla, E., Diedrich, R. G., Kornfeld, M., Oates, J. D., Davidson, M., Vora, S., and DiMauro, S. (1986). Fatal infantile form of muscle phosphofructokinase deficiency. *Neurology* **36,** 1465–1470.

Severion, J., Allen, R. G., Balin, S., Balin, A., and Cristofalo, V. J. (2000). Is β-galactosidase staining a marker of senescence in vitro and in vivo? *Exp. Cell Res.* **257,** 162–167.

Shaag, A., Saada, A., Berger, I., Mandel, H., Joseph, A., Feigenbaum, A., and Elpeleg, O. N. (1999). Molecular basis of lipoamide dehydrogenase deficiency in Ashkenazi Jews. *Am. J. Med. Genet.* **82,** 177–182.

Shen, X. M., Li, H., and Dryhurst, G. (2000). Oxidative metabolites of 5-S-cysteinyldopamine inhibit the α-ketoglutarate dehydrogenase complex: possible relevance to the pathogenesis of Parkinson's disease. *J. Neural. Transm.* **107,** 959–978.

Sheu, K.-F. R., and Blass, J. P. (1999). The α-ketoglutarate dehydrogenase complex. *Ann. N.Y. Acad. Sci.* **893,** 61–78.

Sheu, R. K.-F., Blass, J. P., Cedarbaum, J. M., Kim, Y. T., Harding, B. K., and DeCicco, J. (1988). Mitochondrial enzymes in hereditary ataxias. *Metab. Brain Dis.* **3,** 151–160.

Sheu, K-F. R., Gibson, G. E., Lilius, L, Lannfelt, L., and Blass, J. P. (1997). KGDHC, the *DLST* gene and the metabolic lesion in Alzheimer's Disease. *Alzheimer's Rep.* **2**(suppl. 1), S55.

Sheu, K-F. R., Brown, A. M., Haroutunian, V., Kristal, B. S., Thaler, H., Lesser, M., Kalaria, R. N., Relkin, N. R., Mohs, R. C., Lilius, L., Lannfelt, L., and Blass, J. P. (1999). Modulation by *DLST* of the genetic risk of Alzheimer's disease in a very elderly population. *Ann. Neurol.* **45,** 48–53.

Shoffner, J. M. (1997). Oxidative phosphorylation defects and Alzheimer's disease. *Neurogenetics* **1,** 13–19.

Shoffner, J. M. (2001). Oxidative phosphorylation diseases In: "The Metabolic and Molecular Bases of Inherited Disease". (C. R. Scriver, A. L. Beaudet, W. S. Sly, and D. Valle, eds.), pp. 2367–2423. 8th edition, volume 2, McGraw-Hill, NY.

Simon, D. K., Mayeux, R., Marder, K., Kowall, N. W., Beal, M. F., and Johns, D. R. (2000). Mitochondrial DNA mutations in complex I and tRNA genes in Parkinson's disease. *Neurology* **54,** 703–709.

Sims, N. R., and Blass, J. P. (1986). Phosphofructokinase activity in fibroblasts from patients with Alzheimer's Disease and age- and sex-matched controls. *Metab. Brain Dis.* **1,** 83–90.

Sims, N. R., Blass, J. P., Murphy, C., Bowen, D. M., and Neary, D. (1987). Phosphofructokinase activity in the brain in Alzheimer's disease. *Ann. Neurol.* **21,** 509–510.

Sims, N. R., Anderson, M. F., Hobbs, L. M., Powell, J. A., and Zaidan, E. (2000). Brain mitochondrial responses to postischemic reperfusion: A role for calcium and hydrogen peroxide? *Dev. Neurosci.* **22,** 366–375.

Sirover, M. A. (1997). Role of the glycolytic protein, glyceraldehyde-3-phosphate dehydrogenase, in normal cell function and in cell pathology. *J. Cell Biochem.* **66,** 133–140.

Small, D. L., Morley, P., and Buchan, A. M. (1999). Biology of ischemic cerebral cell death. *Progr. Cardiovasc. Dis.* **42,** 185–207.

Smythies, J. R. (1997). Oxidative reactions and schizophrenia: a review-discussion. *Schizophr. Res.* **4,** 357–364.

Snider, B. J., Gottron, F. J., and Choi, D. W. (1999). Apoptosis and necrosis in cerebrovascular disease. *Ann. N.Y. Acad. Sci.* **893,** 243–253.

Snowden, D. A. (1997). Aging and Alzheimer's Disease: Lessons from the Nun study. *Gerontologist* **37,** 150–156.

Sorbi, S., Bird, E. D., and Blass, J. P. (1983). Decreased pyruvate dehydrogenase complex activity in Huntington and Alzheimer brain. *Ann. Neurol.* **13,** 72–78.

Sorbi, S., Piacentini, S., Fani, C., Tonini, S., Marini, P., and Amaducci, L. (1989). Abnormalities of mitochondrial enzymes in hyereditary ataxias. *Acta. Neurol. Scand.* **80,** 103–110.

Sorbi, S., Mortilla, M., Piacentini, S., Tonini, S., and Amaducci, L. (1990). Altered hexokinase activity in skin cultured fibroblasts and leukocytes from Alzheimer's disease patients. *Neurosci. Lett.* **117,** 165–168.

Stoffel, M., Karayiorgou, M., Espinosa, R. 3d, and Beau, M. M. (1996). The human mitochondrial citrate transporter gene (SLC20A3) maps to chromosome b within a region implicated in DeGeorge syndrome, velo-cardio-facial syndrome and schizophrenia. *Hum. Genet.* **98,** 113–115.

Stumpf, D. A. (1985). The inherited ataxias. *Neurol. Clin.* **3,** 47–57.

Sunaga, K., Takahashi, H., Chuang, D. M., and Ishitani, R. (1995). Glyceraldehyde-3-phosphate dehydrogenase is over-expressed during apoptotic death of neuronal cultures and is recognized by a monoclonal antibody against amyloid plaques from Alzheimer's brain. *Neurosci. Lett.* **200,** 133–136.

Suzuki, K., and Saul, R. (1994). Korey Lecture. Molecular genetics of Tay-Sachs and related disorders: a personal account. *J. Neuropathol. Exp. Neurol.* **53,** 344–50.

Swerdlow, R. H., Parks, J. K., Davis, J. N., 2nd Casarino, D. S., Trimmer, P. A., Currie, L. J., Dougherty, J., Bridges, W. S., Wooten, G. F., and Parker, W. D. (1998). Matrilineal inheritance of complex I dysfunction in a multigenerational Parkinson's disease family. *Ann. Neurol.* **44,** 873–881.

Tabrizi, S. J., Cleeter, M. W., Xuareb, J., Taanman, J. W., Coper, J. M., and Schapira, A. H. (1999). Biochemical abnormalities and excitotoxicity in Huntington's disease brain. *Ann. Neurol.* **45,** 25–32.

Tabrizi, S. J., Wrokman, J., Hart, P. E., Mangiarini, L., Mahal, A., Bates, G., Cooper, J. M., and Shapira, A. H. (2000). Mitochondrial dysfunction and free radical damage in the Huntington R6/2 transgenic mouse. *Ann. Neurol.* **47,** 80–86.

Tein, I. (1996). Metabolic myopathies. *Semin. Pediatr. Neurol.* **3,** 59–98.

Valla, J., Berndt, J. D., and Gonzalez-Lima, F. (2001). Energy hypometabolism in posterior cingulated cortex of Alzheimer's patients: superficial cytochrome oxidase associated with disease duration. *J. Neurosci.* **21,** 4923–4930.

Van Coster, R. N., Fernhoff, P. M., and DeVivo, D. C. (1991). Pyruvate carboxylase deficiency: A benign variant with normal development. *Pediatr. Res.* **30,** 1–10.

Verwer, R. W., Jansen, K. A., Sluiter, A. A., Pool, C. W., Kamphorst, W., and Swaab, D. F. (2000). Decreased hippocampal metabolic activity in Alzheimer patients is not reflected in the immunoreactivity of cytochrome oxidase subunits. *Exp. Neurol.* **163,** 440–451.

Villa, R. F., Arnaboldi, R., Ghigini, B., and Gorini, A. (1994). Parkinson-like disease by 1-methyl-4-phenyl-1,2,3,6-tetrahydropyridine (MPTP) toxicity in Macaca fascicularis: synaptosomal metabolism and action of dihydroergocriptine. *Neurochem. Res.* **19,** 229–236.

Volpe, B. T., and Petito, C. K. (1985). Dementia with bilateral medial temporal lobe ischemia. *Neurology* **35,** 1793–1797.

Wallace, D. C. (2001). A mitochondrial paradigm for degenerative diseases and ageing. *Novartis Found Symp.* **235,** 247–263.

Wallace, D. C., Shoffner, J. M., Trounce, I., Brown, M. D., Ballinger, S. W., Corral-Debrinski, M., Horton, T., Jun, A. S., and Lott, M. T. (1995). Mitochondrial DNA mutations in human degenerative diseases and aging. *Biochim. Biophys. Acta.* **1271,** 141–51.

Wallace, D. C., Lott, M. T., Brown, M. D., and Kerstann, K. (2001). Mitochondria and neuro-ophthalmologic diseases *In* "The Metabolic and Molecular Bases of Inherited Disease." (C. R. Scriver, A. L. Beaudet, W. S. Sly, and D. Valle, eds.), pp. 2425–2509. 8th edition, volume 2, McGraw-Hill, NY.

Whatley, S. A., Curti, D., and Marchbanks, R. M. (1996). Mitochondrial involvement in schizophrenia and other functional psychoses. *Neurochem. Res.* **21,** 995–1004.

Wong, A., Yang, J., Cavidini, P., Gellera, C., Lonnerdal, B., Taroni, F., and Cortopassi, G. (1999). The Friedreich's ataxia mutation confers cellular sensitivity to oxidant stress which is rescued by chelators of iron and calcium and inhibitors of apoptosis. *Hum. Mol. Genet.* **8,** 425–430.

Wong-Riley, M. T. (1999). Disproportionate regulation of nuclear- and mitochondrial-encoded cytochrome oxidase subunit proteins by functional activity in neurons. *Hum. Mol. Genet.* **8,** 2541–2549.

Yun, S. W., and Hoyer, S. (2000). Effects of low-level lead on glycolytic enzymes and pyruvate dehydrogenase of rat brain in vivo: relevance to sporadic Alzheimer's disease. *J. Neural Transm.* **107,** 355–368.

Zamzani, N., and Kroemer, G. (2001). The mitochondrion in apoptosis: how Pandora's box opens. *Nat. Rev. Mol. Cell. Biol.* **2,** 67–71.

Zatta, P., Lain, E., and Cagnolini, C. (2000). Effects of aluminum on activity of krebs [sic] cycle enzymes and glutamate dehydrogenase in rat brain homogenate. *Eur. J. Biochem.* **267,** 3049–3055.

Zheng, W., Ren, S., and Graziano, J. H. (1998). Manganese inhibits mitochondrial aconitase: a mechanism of manganese neurotoxicity. *Brain Res.* **799,** 334–342.

Zubenko, G. S. (2000). Do susceptibility loci contribute to the expression of more than one mental disorder? A genetics of Alzheimer's disease. *Mol. Psychiatry.* **5,** 131–136.

ENERGY UTILIZATION IN THE ISCHEMIC/REPERFUSED BRAIN

John W. Phillis

Department of Physiology, Wayne State University School of Medicine
Detroit, Michigan 48201

Michael H. O'Regan

Biomedical Sciences, University of Detroit Mercy, School of Dentistry
Detroit, Michigan 48219

I. Introduction

Stroke and cardiac arrest are among the leading contributors to mortality and morbidity. Elevated blood glucose levels, as in diabetes, are associated with an increase in ischemic brain injuries and a reduced likelihood of functional recovery. Pre-, but not post-, ischemic hyperglycemia exaggerates brain damage in animal models of cerebral ischemic injury. This outcome has been widely attributed to a rise in lactate production from anaerobic

0074-7742/02 $35.00

glycolysis, with a concomitant tissue acidosis. It has been assumed that acidosis is toxic to neurons, initiating the dying process. It was therefore disturbing to observe that hyperglycemia reduced the ischemia-evoked release of the excitotoxic amino acids, glutamate and aspartate, which is considered to be among the primary causes of cerebral ischemic injury. This result suggested that hyperglycemia might actually have a cerebroprotective action during the ischemic episode, possibly due to lactate formation in astrocytes, with lactate serving as a source of metabolic energy for neurons, and that the damaging effects of hyperglycemia occur during reperfusion, as the intracellular pH reverts to normal levels. Evidence in support of this suggestion came from the observation that there was a marked increase in brain extracellular free fatty acid levels during the initial minutes of reperfusion, indicative of a sudden increase in phospholipase activation and cell membrane degradation. Potentially damaging enzymes such as phospholipases, caspases, and endonucleases tend to have neutral or slightly alkaline pH optima and would be inhibited by acidosis. During reperfusion, such enzymes would be released from inhibition as the pH became more favorable. Acidosis may therefore represent a built-in protective mechanism, which though protecting cells during the ischemic episode, only delays its ultimate consequences until reperfusion. Since the degree of acidosis is increased by hyperglycemia, the gradient of its recovery would be steeper, allowing cells less time to readjust their intracellular ion and metabolite levels following the onset of reperfusion. The subsequent sections of this chapter will review recent findings on cerebral metabolism in normoxic and ischemic brains exposed to physiologic and elevated levels of glucose. The role of insulin, plasma levels of which are depleted in many diabetics, in brain metabolism is also discussed.

II. The Normoxic, Normoglycemic Brain

A. Oxidative Metabolism

In contrast to many other tissues, the normal brain depends almost entirely on glucose as its primary source of energy. In the normal conscious human brain the cerebral metabolic rate of glucose utilization (CMR Glu) is 30 μmol $100g^{-1}$ min^{-1} (5 mg $100g^{-1}$ min^{-1}) and CMR O_2 is 165 μmol $100g^{-1}$ min^{-1} (3.5 ml $100g^{-1}$ min^{-1}). Approximately one-third of the oxygen and one-tenth of the glucose delivered to the brain by the cerebral vasculature are used. The CMR O_2/CMR Glu molar ratio of 5.5 (rather than the 6 expected for complete oxidation of glucose) corresponds to the production of a small amount of lactate by glycolysis. In the resting state, the brain utilizes oxidative metabolism of glucose to furnish its energy requirements to fuel processes which may include extrusion of intracellular Na^+ (40–50%) and Ca^{2+} (3–7%); processing of neurotransmitters (10–20%); intracellular signaling systems (20–30%); axonal and dendritic transport (20–30%); and basic vegetative processes including protein synthesis (5–15%) (Ames, 2000).

Glycolysis via the Embden-Meyerhof pathway, the major route for utilization of glucose, uses oxygen if it is available (aerobic) but can also function in the absence of oxygen (anaerobic). Anaerobic glycolysis is important in that it allows tissues such as glial cells, which have energy reserves in the form of glycogen, to survive anoxic conditions for considerable periods of time. The distinction between aerobic and anaerobic glycolysis is somewhat misleading in that the chemical reactions are the same in the presence of oxygen as in its absence, except for the nature of the end products generated. Under aerobic conditions, pyruvate formed as the end product of glycolysis does not accumulate because it is further oxidized via the tricarboxylic acid (TCA) cycle to carbon dioxide and water. In the absence of oxygen, reoxidation of NADH formed from NAD^+ during glycolysis is impaired. Reduction of pyruvate to lactate by lactate dehydrogenase, coupled to the reoxidation of NADH to NAD^+, allows further glycolysis to proceed. When pyruvate is formed anaerobically from glucose the amount of energy liberated is limited to 2 mol of ATP per 1 mol of glucose in comparison to the 38 mol of ATP generated by aerobic glycolysis via the Embden-Meyerhof/TCA cycle pathways.

Conventionally it has been assumed that brain cells obtain virtually all of their energy requirements by the direct oxidation of glucose. However, recent evidence suggests that, in some circumstances, a substantial portion of the energy requirement of neurons may be derived from lactate generated in glial cells, with the formation of 36 mol of ATP per initial 1 mol of glucose (Magistretti *et al.*, 1999).

B. Glycolytic Energy Production Fuels Plasma Membrane Pumps

Studies on erythrocytes and smooth muscle have provided evidence of a close physical association between glycolytic enzymes and ion transport pumps in the plasma membrane, making glycolytically generated ATP readily available to the sodium and calcium pumps (Mercer and Dunham, 1981; Paul *et al.*, 1979, 1989; Campbell and Paul, 1992). Ion transport in cardiac muscle cells appears to be fueled by glycolytically generated ATP (Weiss and Lamp, 1989). In the CNS, Lipton and Robacker (1983) reported that $[K^+]_O$ activation of the Na^+/K^+-ATPase in hippocampal slices was dependent on glycolytically generated energy. Similar observations have been made by other investigators (Raffin *et al.*, 1992; Roberts, 1993; Silver *et al.*, 1997).

A preference for ATP produced by glycolysis, rather than by mitochondrial oxidative phosphorylation, suggests that there may be an advantage to locally generated energy supplies for the Na^+ pump. The association between some of the enzymes of the glycolytic pathway with the plasma membrane and its proteins (Knull, 1978) would provide an explanation for this finding. Proverbio and Hoffman (1977) used the term "micro-domain" to describe this juxtaposition of glyceraldehyde 3-phosphate and phosphoglycerate kinase, terminal enzymes in the glycolytic pathway for ATP generation, and the Na^+/K^+ pump apparatus.

C. Relationships between Energy Use and Neuronal Activity

Over the past several decades it has been generally accepted that there is a dynamic regional coupling between cerebral blood flow (CBF) and tissue metabolic rate in normal brain. Regional changes in neuronal electrical activity were assumed to evoke similar changes in local metabolic activity, which in turn modulated local blood flows. Blood flow has therefore been considered as a reliable, if indirect, index of cerebral metabolic rates. Also accepted has been the concept that functional activity in the brain is normally fueled by the obligatory oxidative metabolism of glucose.

The belief that CMR O_2 and CMR Glu remain obligatorily coupled during different states of functional activity was challenged when Fox and Raichle (1986), using position emission tomography (PET), reported that human somatosensory stimulation caused a regional uncoupling of CBF and CMR O_2 in the sensorimotor cortex, with stimulus-induced focal augmentation of CBF far exceeding the concomitant local increase in tissue metabolic rate. Visual stimulation raised CBF and CMR Glu with little change in CMR O_2 in the visual cortex (Fox *et al.,* 1988). The authors argued that for glucose metabolism to rise in excess of oxygen consumption, lactate production must have increased. An increase in lactate production was reported by Prichard *et al.* (1991) who used magnetic resonance spectroscopy (NMR) to demonstrate a similar effect in the rat visual cortex during optic nerve stimulation.

Sappey-Marinier *et al.* (1992), using magnetic resonance spectroscopy, also observed significant rises in human visual cortex lactate levels during photic stimulation and speculated on the mechanisms by which anaerobic glycolysis might be stimulated without a parallel increase in oxidative phosphorylation. One suggestion was that the plasma membrane ATPase is closely linked to rate-limiting enzymes of glycolysis, as already discussed in Section II.B. Further evidence of enhanced lactate production during functional activation of the brain has been reported by other investigators using the microdialysis technique (Kuhr and Korf, 1988; Fellows *et al.,* 1993) and enzyme-based microsensors. Using the latter technique, Hu and Wilson (1997) were able to demonstrate an immediate shift to significant use of lactate as an energy supply in response to acute neuronal activation. A local temporary fuel reservoir in the form of extracellular lactate was established behind the blood-brain barrier. This pool was rapidly depleted by massive acute neuronal stimulation and then replenished within 20 sec. Coincident with the increase in lactate levels was a decrease in mean extracellular glucose levels.

D. The "Glutamate-Glutamine Cycle" and the "Lactate Shuttle"

Glutamate is considered to be the principal excitatory neurotransmitter in the mammalian CNS (Orrego and Villaneuva, 1993; Hertz *et al.*, 2000). Van den Berg and Garfinkel (1971) originally proposed that a neuronal-astrocytic neurotransmitter cycle exists in the brain in which glutamate from the neuronal pool is released into the synaptic cleft, taken up by astrocytes, and converted to glutamine. Glutamine is subsequently returned to the neurons for subsequent hydrolysis to glutamate.

The significance of glutamatergic signaling for brain function and energy utilization has been studied by Sibson *et al.* (1998) using ^{13}C NMR spectroscopy. The stoichiometry between oxidative glucose metabolism and glutamate-neurotransmitter cycling in the cerebral cortex was determined to be close to 1:1, suggesting that glutamatergic signaling may account for the majority of cortical energy expenditure.

Astrocyte processes enclosing glutamatergic synapses contain numerous glutamate uptake sites with a higher uptake capacity than those on the neurons (Rothstein *et al.*, 1994; Lehre and Danbolt, 1998). This specialization allows astrocytes to rapidly clear the synaptic clefts of glutamate following its release from neurons (Clements *et al.*, 1992), thus enhancing the signal to noise ratio of glutamate synaptic transmission. Glutamate taken up by astrocytes can either be metabolized oxidatively or converted to glutamine in a reaction catalyzed by the glial enzyme, glutamine synthetase (Norenberg and Martinez-Hernandez, 1979; Tansey *et al.*, 1991). A substantial amount of astrocytic glutamine is then released into the extracellular space by the SNI or ASCT2 amino acid transporters, taken up by neurons via the SAT/ATA transporter (Broer and Brookes, 2001), and reconverted to glutamate by phosphate-activated glutaminase (Kvamme *et al.*, 1982; Hogstad *et al.*, 1988) in the glutamate-glutamine cycle (Hertz *et al.*, 2000).

Uptake of glutamate into astrocytes is mediated by two Na^+-dependent transporters: GLAST (EAAT1) and GLT-1 (EAAT2) (Danbolt, 2001). Uptake of each glutamate ion is coupled to the uptake of two or three Na^+ plus one H^+, and is also coupled to the outward transport of one K^+ (Bouvier *et al.*,

1992; Zerangue and Kavanaugh, 1996). Astrocytic Na^+/K^+ ATPase is activated by the increase in intracellular Na^+ concentration, with consumption of glucose by glycolysis and the production of lactate from pyruvate by lactate dehydrogenase (LDH-5) (Bittar *et al.*, 1998). Lactate is exported from astrocytes by a specific monocarboxylate transporter (MCT-1) and taken up into neurons by MCT-2 (Broer *et al.*, 1997, 1999). Lactate is subsequently converted to pyruvate by LDH-1. Pyruvate, via the formation of acetyl-CoA by pyruvate dehydrogenase enters the TCA, undergoing oxidative metabolism with ATP formation (Pellerin *et al.*, 1998; Magistretti *et al.*, 1999). This cycle of events has been termed the "astrocyte-neuron lactate" shuttle (Pellerin *et al.*, 1998).

There is evidence that lactate can serve as the primary source of energy to support CA1 neuronal function in hypoxic hippocampal slice preparations (Schurr *et al.*, 1997). Topical applications of lactate significantly reduced amino acid release from the ischemic/reperfused rat cerebral cortex (Fig. 1) (Phillis *et al.*, 1999b). In addition to its function as an energy substrate, the conversion of lactate to pyruvate, which is coupled to a reduction of NAD^+ to NADH, appears to provide neurons with a source of "reducing equivalents" which may be of critical importance for the reversal of oxidizing reactions in the ischemia/reperfused brain.

As proposed by Magistretti and colleagues (1999), glutamate uptake by astrocytes, following its release at activated synapses, would initiate a transient activation of aerobic glycolysis, resulting in lactate overproduction. However, glycolysis will also occur in the anaerobic conditions existing during cerebral ischemia if adequate amounts of glucose or glycogen are available in the astrocytes. When oxygen is in short supply, reoxidation of NADH formed from NAD^+ during glycolysis is impaired. Under these conditions, NADH can be reoxidized by coupling the reduction of pyruvate to lactate and the NAD^+ so formed will allow further glycolysis to proceed, providing ATP for plasma membrane Na^+/K^+-ATPase. Evidence in support of the relationship between glutamate transport and lactate release has been obtained with astrocytes preincubated with the transportable glutamate uptake inhibitor *threo-β*-hydroxyaspartate (THA). After such preexposure to THA, there was a significant inhibition of D-aspartate-stimulated glucose utilization which was paralleled by an inhibition of D-aspartate uptake (Debernardi *et al.*, 1999).

The proposal that glutamate uptake into astrocytes stimulates aerobic glycolysis has been questioned by Peng *et al.* (2001) on the basis of evidence that oxidative metabolism, mainly fueled by glutamate itself, provides the energy for active Na^+/K^+-ATPase catalyzed Na^+ extrusion following glutamate uptake in conjunction with Na^+. The reason for these differences in the observations of the two groups remains unresolved at this time but could be a consequence of differing experimental conditions (Hertz *et al.*, 1998).

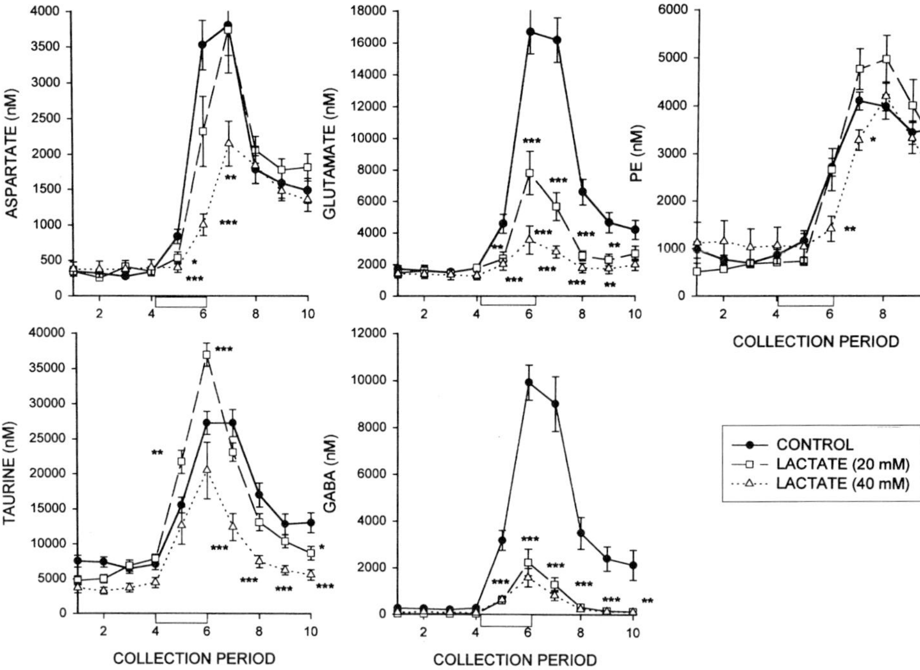

FIG. 1. The effect of topical lactate on ischemia/reperfusion-evoked releases of amino acids into rat cerebral cortical superfusates. Line plots show the time course of changes in amino acid concentrations (nM, mean ± SEM) before, during (collections 5 and 6), and after a 20-min period of four-vessel occlusion of CBF (□). Lactate (20 or 40 mM) was added to the artificial cerebrospinal fluid (aCSF) with removal of an equimolar amount of NaCl after the second basal collection and was present for the rest of the experiment. PE = phosphoethanolamine. Statistically significant differences between control and lactate treated animals were determined by ANOVA and Scheffe's test. $^{*}p < 0.05$; $^{**}p < 0.01$; $^{***}p < 0.001$. (Reprinted from *Neurosci. Lett.*, **272,** Phillis, J. W., Song, D., Guyot, L. L., and O'Regan, M. H., Lactate reduces amino acid release and fuels recovery of function in the ischemic brain, 195–198, copyright 1999, with permission from Elsevier Science.)

III. Metabolism in the Ischemic/Reperfused Brain

A. Energy Failure, Acidosis, Loss of Ion Homeostasis, and Edema

Ischemic brain injury can occur as a result of either global or focal ischemia. The former will be present during a cardiac arrest with a complete cessation of CBF, while focal ischemias or strokes, resulting from occlusion of cerebral blood vessels, lead to a partial reduction in blood flow. A third variant, which occurs in the laboratory, is forebrain ischemia which can be elicited in the rat by occlusion of the vertebral and common carotid arteries or in the gerbil by occlusion of the common carotid arteries. In these forebrain ischemia models the reduction in blood flow, as with stroke, is generally less than complete as a result of functional collateral arteries. These forebrain ischemia models provide the investigator with the convenience of being able to reperfuse the brain after a predetermined interval.

Because of its limited energy reserves, the brain's supply of high energy phosphate compounds is dependent on a continuous blood flow, with rapid depletion of ATP levels during ischemia (Phillis *et al.,* 1996), bioenergetic failure, and a loss of ion homeostasis. With energy depletion, the membrane potential is dissipated and neurons and glia depolarize. Depolarization activates somatodendritic and presynaptic terminal voltage-dependent calcium channels eliciting a release of glutamate into the extracellular space. Depolarization also results in an influx of Na^+ and Cl^- into neurons and glia, increasing the osmolarity of the intracellular fluid. Water follows this osmotic gradient into the cells, causing them to swell with ensuing brain edema.

Hypoxia and ischemia induce anerobic glycolysis, with the production of lactate which has frequently been considered to be the cause of ischemia-evoked acidosis (Rehncrona, 1985). The end product of glycolysis is lactate, not lactic acid, and the process does not generate H^+ unless accompanied by the hydrolysis of ATP (Krebs *et al.,* 1975; Hochachka and Mommsen, 1983; Busa and Nuccitelli, 1984; Dennis *et al.,* 1991). During aerobic glycolysis, the rapid hydrolysis and resynthesis of ATP and further metabolism of lactate maintain net production of H^+ at a low level without marked decreases in intracellular pH. Under anerobic conditions, the rates of ATP breakdown and H^+ production can exceed that of ATP resynthesis, with a large accumulation of protons and a decrease in intracellular pH (Erecinska and Silver, 1989). Lactate production during ischemia will therefore coincide with, but is not the direct cause of, intracellular acidification.

Cell swelling during ischemia occurs as a consequence of a loss of steady-state volume regulation by Na^+/K^+-ATPase-mediated Na^+ extrusion during metabolic failure. It can also be a result of glutamate-evoked depolarization due to the influx of sodium and chloride ions followed by water. Cell acidosis results in NaCl accumulation as a result of net uptake of Na^+ by activation of the Na^+/H^+ exchangers and of Cl^- by stimulation of Cl^-/HCO_3^- exchangers. Exposure to monocarboxylates, including lactate, causes intracellular acidosis with swelling as a consequence of their entry via the H^+-monocarboxylate cotransporter (Lomneth *et al.,* 1990).

There is considerable evidence in support of the hypothesis that the Na^+/H^+ exchanger contributes to cell swelling (Jakubowicz and Klip, 1989; Kempski *et al.,* 1988). Inhibition of the exchanger with amiloride inhibited both ischemia- and lactic acid-evoked swelling of glial cells (Jakubowicz *et al.,* 1987; Staub *et al.,* 1996). Loss of amino acids via swelling-activated membrane anion channels in a process called a regulatory volume decrease (RVD) is one mechanism by which swollen brain cells can regain their original volume during ischemia. Taurine and glutamate are two of the amino acids released into the extracellular space in this process with significant increases in their extracellular concentrations (Fig. 1) (Solis *et al.*, 1988; Phillis *et al.*, 1997, 1998a; Estevez *et al.*, 1999a). Confirmation of a link between ischemic swelling-evoked amino acid release from the rat cerebral cortex and activation of the Na^+/H^+ exchanger was obtained with the demonstration that 5-(*N*-ethyl-*N*-isopropyl)-amiloride (EIPA) inhibited the release of amino acids during ischemia/reperfusion (Fig. 2) (Phillis *et al.*, 1998a).

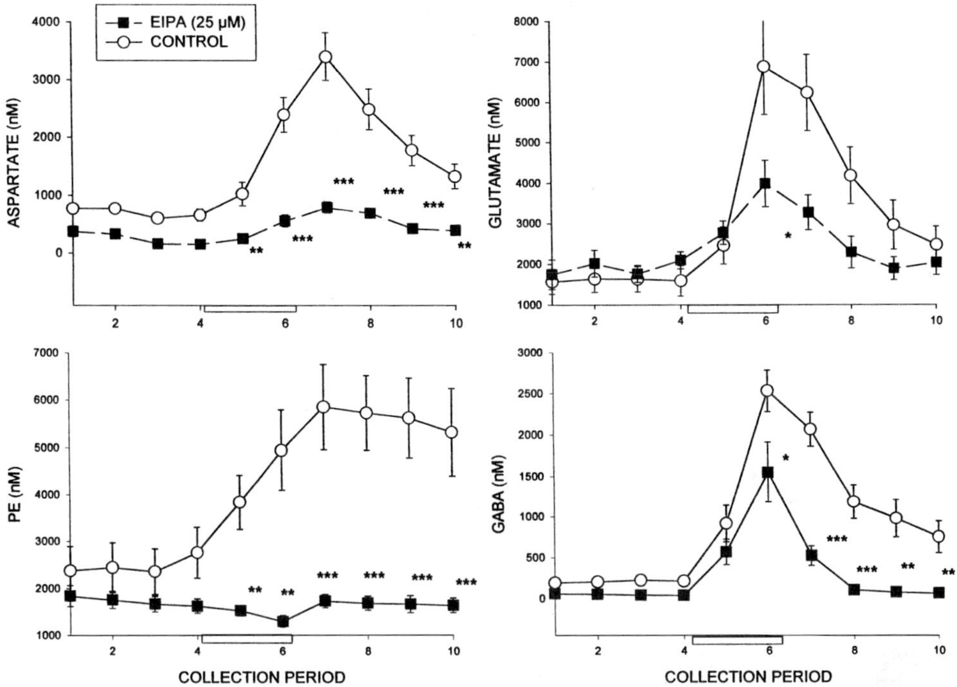

FIG. 2. Ischemia-evoked release of aspartate, glutamate, phosphoethanolamine (PE), and γ-aminobutyric acid (GABA) into rat cerebral cortical superfusates. EIPA (25 μM) was added to the aCSF after collection of the initial two basal samples, 15 min before start of collection 3, and was present in all subsequent collections. Four-vessel occlusion cerebral ischemia occurred during collections 5 and 6, followed by reperfusion. The EIPA data are compared with those from control animals. Data represent mean ± SEM. Statistical differences were analyzed with the Student's *t* test. *$p < 0.05$; **$p < 0.01$; ***$p < 0.001$. (Reprinted from *Brain Res.*, **812,** Phillis, J. W., O'Regan, M. H., and Song, D., 5-(*N*-ethyl-*N*-isopropyl)-amiloride inhibits amino acid release from the ischemic rat cerebral cortex: role of Na^+-H^+ exchange, 297–300, copyright 1998, with permission from Elsevier Science.)

B. Release of Excitotoxic Amino Acids, Regulatory Volume Decrease

Glutamate is considered to be an important contributor in the pathogenesis of neuronal injury and death induced by cerebral ischemia (Rothman and Olney, 1986) and enhanced amino acid release is considered as a reliable prognosticator of cerebral ischemia/reperfusion injury (Butcher *et al.*, 1990; Takagi *et al.*, 1993; Melani *et al.*, 1999). Extracellular brain concentrations of glutamate, and of the related excitatory amino acid aspartate, increase rapidly during ischemic episodes, contributing to neuronal depolarization and elevation of intracellular Ca^{2+} levels (Benveniste, 1991; Choi, 1992; Siesjo, 1992). The mechanisms underlying this increase in extracellular levels of the excitotoxic amino acids have been controversial. A depolarization induced, calcium-dependent, exocytotic release of glutamate undoubtedly contributes to the initial accumulation (Asai *et al.*, 2000). However, considering the rapid inactivation and desensitization of both voltage- and glutamate-receptor-gated channels resulting from ATP depletion and neuronal depolarization (Eckert and Chad, 1984; McBain and Mayer, 1994), this type of calcium-dependent release would be insufficient to account for the progressive, continuous increases in glutamate efflux which occur during lengthy periods of ischemia (Phillis *et al.*, 1996). Other studies have indicated that a Ca^{2+}-independent release of glutamate occurs during anoxia or ischemia (Ikeda *et al.*, 1989; Sanchez-Prieto and Gonzalez, 1988; Phillis *et al.*, 1994a; Lipton, 1999). Other proposed mechanisms to account for the increase in extracellular glutamate levels include a depolarization-induced reversal of the sodium and potassium coupled glutamate transporters in neuronal (EAAT3, EAAT4) and glial (EAAT1 [GLAST], EAAT2 [GLT]) plasma membranes (Danbolt, 2001). *In vivo* evidence in support of an ischemia-induced reversal of the glial EAAT2 transporter was obtained in studies using dihydrokainate (DHK), a selective, nontransported, inhibitor of this glutamate transporter, which caused a 44% decrease in extracellular glutamate levels during cerebral ischemia (Phillis *et al.*, 1994b; Seki *et al.*, 1999). Confirmation of this finding has come from experiments with DL-*threo*-β-benzyloxyaspartate (DL-TBOA), a newly developed, nontransported, competitive inhibitor of the EAAT1, 2, and 3 transporters (Shimamoto *et al.*, 1998). DL-TBOA application resulted in a 42% reduction in the ischemia-evoked increase in extracellular glutamate levels in cerebral cortical superfusates (Phillis *et al.*, 2000a). The similarity between the degrees of reduction in extracellular glutamate levels observed with DHK, a selective inhibitor of the EAAT2 transporter, and the less selective DL-TBOA strongly suggests that the glial EAAT2 transporter is the primary agent mediating reversal transport.

DL-TBOA dramatically reduced the increases in taurine release during the initial 10-min period of cerebral ischemia (Phillis *et al.,* 2000a). Taurine efflux is considered to be indicative of an RVD response to osmotically induced swelling (Law, 1995); thus it appears likely that DL-TBOA, by initially blocking glutamate uptake into glial cells together with Na^+ ions, actually delayed the onset of astrocytic swelling. DL-TBOA application also resulted in a sharp decline in extracellular glutamine levels in the cortical superfusates during ischemia. This reduction is consistent with current concepts (Section II.D) of astrocytic recycling of synaptically released glutamate. Extracellular glutamine levels rose during the ischemic interval in control rats, demonstrating that glutamate uptake was active with a continuing formation of glutamine, whereas inhibition of glutamate uptake by DL-TBOA curtailed glutamine formation.

The phenomenon of regulatory volume decrease has been extensively investigated in a variety of cell types. The majority of experiments on brain cells have, however, involved studies on astrocytic, C_6 glioma, or neuronal cultures and brain slices exposed to hyposmotic stresses (Jackson and Strange, 1993; Kimelberg *et al.,* 1990; Law, 1995; Pasantes-Morales, 1996) with few acute *in vivo* studies (Solis *et al.,* 1988; Barron *et al.,* 1988; Estevez *et al.,* 1999a,b). Osmolytes involved in RVD responses in neurons and glia include K^+ and Cl^-, a range of amino acids, and myoinositol. The efflux of amino acids occurs within minutes of the onset of hyposmotic stress and closely follows the time course of RVD.

Our interest in RVD as a potential contributor to ischemia-evoked amino acid release was stimulated by observation of a potent inhibitory action of arachidonic acid on glutamate and aspartate release from the ischemic rat cerebral cortex (O'Regan *et al.,* 1995a). An inhibitory effect of polyunsaturated fatty acids (PUFAs), including arachidonic acid, on RVD and amino acid efflux from cultured astrocytes had recently been reported by Sanchez-Olea *et al.* (1995). The authors proposed that arachidonic acid could be inhibiting the membrane Cl^- channels through which amino acids are presumed to pass in the RVD response (Pasantes-Morales, 1996). Subsequent studies confirmed that anion channel blockers, including 4-acetamido-4′-isothiocyanostilbene,2,2′-disulfonic acid (SITS), 5-nitro-2-(3-phenyl-propylamino)benzoic acid (NPPB), dipyridamole, tamoxifen, and 4,4′-dinitrostilbene 2,2′-disulfonic acid (DNDS) attenuated ischemia-evoked amino acid efflux in the brain (Phillis *et al.,* 1997, 1998b). A combination of DNDS and DHK reduced ischemia-evoked glutamate release by 83% (Seki *et al.,* 1999).

Yet another possible route of glutamate efflux could be diffusion across compromised plasma membranes. Intracellular levels of glutamate and taurine in the rat brain are of the order of 12.5 and 4.6 mM and, together with glutamine (5 mM), they constitute the bulk of the amino acid osmolarity (Perry, 1982). Extracellular levels of these three amino acids in cerebral cortical superfusates are of the order of 2, 4, and 32 μM, respectively. The transmembrane ratios for glutamate and taurine are therefore of the order 6000:1 and 1150:1, respectively. High transmembrane gradients also exist for aspartate, γ-aminobutyric acid, and phosphoethanolamine, whereas the concentration gradients for other amino acids, including asparagine, glutamine, valine, and phenylalanine are very low in comparison.

Recognition of the relationship between high transmembrane concentration gradients for certain amino acids and pronounced increases in their extracellular concentrations during cerebral ischemia suggested that compositional alterations of the plasma membrane during ischemia, with changes in its structural integrity, might allow these amino acids to diffuse down their concentration gradients. Evidence in support of this concept of an early disruption of the plasma membrane was furnished by experiments on the release of LDH into the interstitial spaces of the ischemic rat cerebral cortex. Increases in LDH levels in cortical superfusates became evident during a 30-min period of ischemia and the levels continued to rise during a subsequent 2-h reperfusion period (Phillis *et al.*, 1994b). In that LDH has a molecular weight of about 140,000, it is apparent that a 30-min period of ischemia must inflict considerable damage to the plasma membranes of at least some cells in the cerebral cortex. Amino acids with molecular weights in the 100–150 range would be expected to permeate membranes more readily.

Ultrastructural evidence of early ischemic damage to CA1 hippocampal neurons after 20 min of reperfusion following a 10-min period of complete cerebral ischemia included cell membrane breaks, which were mainly postsynaptic (Ekstrom von Lubitz and Diemer, 1982). Following injections of horse radish peroxidase (HRP); (MW 44,000) into the lateral ventricles 10–15 min after a 10-min period of ischemia, HRP could be observed in about 10% of the apical dendrites of CA1 neurons, whereas in nonischemic controls HRP was confined to the extracellular space (Diemer and Ekstrom von Lubitz, 1983). These authors concluded that there was evidence of early postischemic membrane damage occurring *in vivo* to the apical dendrites of CA1 pyramidal cells. Tanaka *et al.* (1999) described an early formation *in vitro* of small irregularities or blebs in ischemic rat CA1 pyramidal cell membranes, with swelling and the formation of larger blebs 1.5–2 min after the rapid anoxic depolarization. Dye studies suggested that micropores in the blebs provide a pathway for the diffusion of molecules with a molecular weight <520.

Hearse *et al.* (1973) have measured myocardial enzyme release from anoxic isolated, perfused, rat hearts. They were able to detect small releases of several enzymes, including creatine phosphokinase, myokinase, α-hydroxybutyrate dehydrogenase, and glyceraldehyde-3-phosphate dehydrogenase during a 30-min period of simulated ischemia (no glucose, anoxia). Reoxygenation resulted in an immediate and measurable release of all enzymes, an observation that may be relevant to phenomena observed during reperfusion of the ischemic brain.

Insight into the mechanisms whereby cerebral ischemia could rapidly modulate membrane permeability to amino acids was gained from a previous observation that the non-steroidal anti-inflammatory agent indomethacin inhibited ischemia-evoked glutamate and aspartate efflux from the rat cerebral cortex (Phillis *et al.,* 1994a). In addition to its well-known ability to block prostaglandin synthesis by cyclooxygenases, indomethacin also inhibits phospholipase A_2 (Franson *et al.,* 1980). Subsequent studies revealed that topical applications of phospholipases A_2 (PLA_2), and C (PLC) or melittin, the bee venom activator of endogenous PLA_2, enhanced amino acid release from the nonischemic rat cerebral cortex (O'Regan *et al.,* 1995b; Phillis *et al.,* 1999c). Inhibition of PLA_2 and PLC by a series of both nonselective and selective inhibitors reduced ischemia-evoked efflux of glutamate and aspartate from the rat cerebral cortex, confirming a role for these phospholipases in the release of amino acids from the ischemic brain (Phillis and O'Regan, 1996).

C. Elevation of Intracellular Calcium

One consequence of cerebral ischemia is a marked influx of calcium, which has been viewed as a major contributor to neuronal injury and death (Kristian and Siesjo, 1998). Activation of *N*-methyl-D-aspartate (NMDA) receptors by glutamate permits an influx of Ca^{2+}, and overstimulation of this type of receptor is one mechanism of calcium overload in neurons. In addition some variants of the α-amino-3-hydroxy-5-methyl-4-isoxazolepropionate (AMPA)-kainate receptors are coupled to Ca^{2+} channels and can contribute to the excessive calcium entry. Glutamate receptor-evoked membrane depolarization will also cause opening of voltage-dependent Ca^{2+} channels. Neuronal increases in Ca^{2+} can activate a series of enzymes such as phospholipases, proteases including the calpains, protein phosphatases, protein kinase C, endonucleases, and nitric oxide synthase. A coupling exists between the influx of calcium into cells and their production of reactive oxygen species such as $\cdot O_2^-$, $\cdot OH$, and H_2O_2. The cumulative effect of all these processes is to cause cellular disruption and ultimately death, which depending on its time course may be classified as necrotic (rapid) or apoptotic (slow).

D. Free Fatty Acid Formation, Lipid Peroxidation, and Membrane Degradation

Brain tissue has a high lipid content, mainly comprised by phospholipids which are the major structural component of cell membranes. The integrity of these membranes is a prerequisite for normal brain function. The effects of oxygen deprivation on brain lipid metabolism include a decomposition of membrane-bound phospholipids and release of free fatty acids (FFA) (Rehncrona *et al.,* 1981). Phospholipase A_2 can act on membrane phospholipids to alter membrane structure and release FFAs and lysophospholipids (Bazan, 1970; Farooqui *et al.,* 1997). Another pathway for FFA generation involves catalysis of phosphatidylinositol by phospholipase C to form inositol triphosphate and diacylglycerol. Diacyglycerol is further metabolized to FFA by lipases, including diacylglyceride lipase, which releases arachidonic acid. Umemura *et al.* (1992) have proposed that FFA release during the initial 2 min of a cerebral ischemia can be attributed mostly to the activation of PLC, which in turn influences the activity of PLA_2.

Calcium-dependent PLA_2 ($cPLA_2$) activity and protein levels were studied in cytosolic fractions of the rat cerebral cortex (Saluja *et al.,* 1999). Ten and 20 min of ischemia resulted in highly significant increases in $cPLA_2$ activity and protein levels, which then declined toward preischemia levels during 20 min of reperfusion. Extracellular levels of the FFAs in the normoxic rat cerebral cortex were in the low micromolar range. During the initial 10 min of a 20-min period of ischemia, there was a significant increase in cortical superfusate levels of arachidonic acid and nonsignificant increases in the levels of palmitic, myristic, linoleic, and oleic acids (Fig. 3). The levels declined during the second ischemic period and increased markedly during the initial two 10-min reperfusion periods, with a decline in the subsequent superfusate collections (Saluja *et al.,* 1999). This sequence of changes appears to indicate an initial ischemia-evoked enhancement of $cPLA_2$ activity, followed by a decline. Reperfusion was associated with a large increase in phospholipase activity, which declined toward baseline levels over the next 20–30 min.

There is evidence that the rate of lipolysis during ischemia is significantly greater in the selectively vulnerable areas than in other regions of the brain (Umemura, 1990). Evidence for a role of PLA_2 in cerebral ischemic injury has come from studies with "knockout" ($cPLA_2^{-/-}$) mice which had smaller infarcts and fewer neurological deficits than control mice following transient middle cerebral artery occlusion (Bonventre *et al.,* 1997). Further support for a role of phospholipases in cerebral ischemic injury was evident in experiments in which the phospholipase inhibitor quinacrine reduced hippocampal CA1 pyramidal cell loss in gerbils exposed to 5 min of

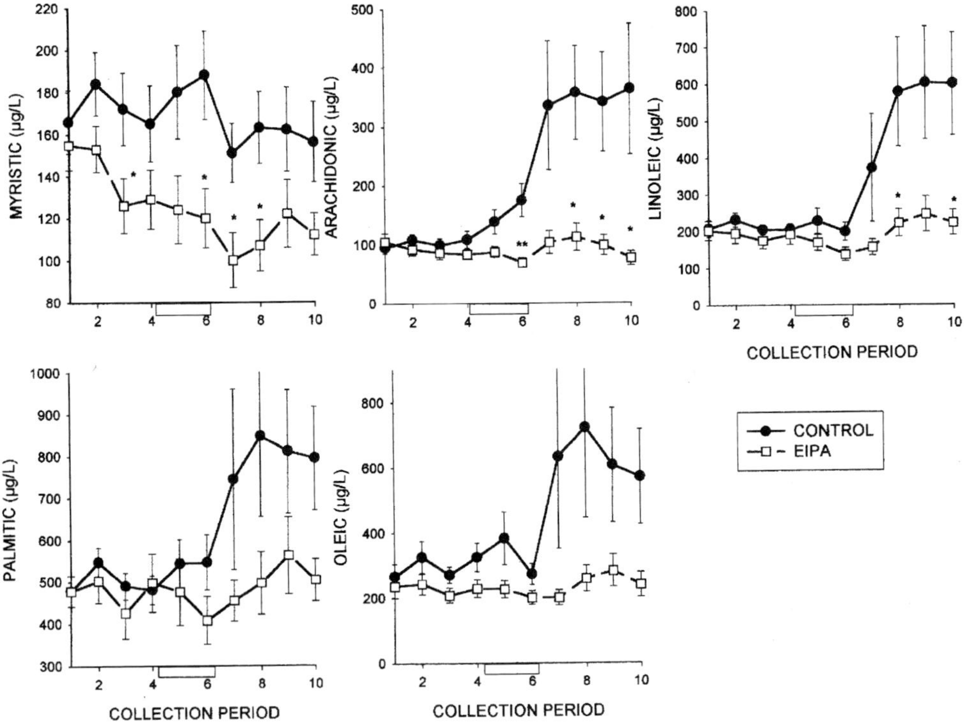

FIG. 3. Ischemia-evoked release of myristic, arachidonic, linoleic, palmitic, and oleic acids into rat cerebral cortical superfusates. The line plots show the changes in cortical superfusate concentrations of these FFAs before, during, and after a 20-min period of four-vessel occlusion cerebral ischemia (collections 5 and 6, open box). EIPA (25 μM) was added to the aCSF 15 min prior to the start of collection 3 and was present in all subsequent collections (line plots with open squares). See Fig. 2 legend for further details. (Reprinted from *Brain Res.*, **884,** Phillis, J. W., Ren, J., and O'Regan, M. H., Inhibition of Na^+/H^+ exchange by 5-(*N*-ethyl-*N*-isopropyl)-amiloride reduces free fatty acid efflux from the ischemic reperfused rat cerebral cortex, 155–162, copyright 2000, with permission from Elsevier Science.)

forebrain ischemia (Phillis, 1996) and corticostriatal injury in rats subjected to 2-h middle cerebral artery occlusion (Estevez and Phillis, 1997).

Lipid peroxidation is the primary mechanism by which free radicals damage cells. It occurs when free radicals abstract hydrogen from PUFAs in cell membranes and organelles to form fatty acid radicals. These lipid peroxy radicals can, in turn, react with other lipids, proteins, or nucleic acids, thus propagating the transfer of electrons and oxidation of substrates. Cell membranes contain large numbers of PUFAs, making them particularly susceptible to oxidative attacks, with resultant changes in membrane fluidity, permeability, and function. Oxygen-derived free radicals in the ischemic/reperfused brain can be generated by a number of pathways including: the electron transport process in mitochondrial membranes, metabolism of arachidonic acid by cyclooxygenases and lipoxygenases, purine catabolism when hypoxanthine is metabolized to xanthine and uric acid by the enzyme xanthine oxidase, the interaction between nitric oxide and superoxide anion radicals with the formation of peroxynitrite ($ONOO^-$), superoxide release by invading neutrophils during the respiratory burst, and as a consequence of the excitotoxic actions of glutamate and aspartate (Phillis, 1997). The interrelationships between free radical generation and excitatory amino acid release into the interstitial space has been emphasized by Pellegrini-Giampietro *et al.* (1990) who showed that inhibition of free radical formation in hippocampal slices reduced excitotoxic amino acid release under simulated ischemic conditions. Conversely incubation with enzymes and substrates known to cause the formation of free radicals increased the *in vitro* and *in vivo* release of glutamate and aspartate (Pellegrini-Giampietro *et al.*, 1990; O'Regan *et al.*, 1997).

E. Reperfusion, ATP Recovery, and the "pH Paradox" of Reperfusion Injury

The ability of brain tissues to recover their normal levels of high-energy phosphates during reperfusion has been used as a measure of mitochondrial functioning and tissue viability. Several early studies demonstrated a substantial recovery of rat brain adenine nucleotide levels following ischemic episodes (Nordstrom *et al.*, 1978a,b; Pulsinelli and Duffy, 1983). A more recent study (Phillis *et al.*, 1996), using microwave fixation to ensure minimal degradation of high-energy phosphates, evaluated the ability of cerebral cortical adenine nucleotides to recover following four-vessel occlusion ischemias of varying duration. Ischemias of 10-, 20-, and 40-min duration caused profound falls in ATP levels to 1–3% of control values. ADP levels were reduced to 20–30% of preischemic values with large increases in AMP and adenosine levels at all three times. The energy charge fell to around 10% of preischemic levels. After 10 min of reperfusion ATP levels had recovered to 74% of control values in the 10-min ischemia animals, 66% in the 20-min animals, and 41% in the 40-min animals. ADP levels also recovered to approximately the same extent, while AMP levels declined sharply, and adenosine levels continued to be elevated. After 45 min of reperfusion ATP levels were 84, 77, and 67% of control values in the three groups, respectively, and by 90 min of reperfusion ATP levels had recovered to 87 and 85% of control values in the 10- and 20-min ischemia groups, but had declined to 63% in the 40-min ischemia group. The apparent failure of full recovery after a 40-min period of cerebral ischemia is indicative of a more profound degree of ischemic injury with loss of cell viability. It is of some interest to compare the ATP recovery level of 87% in the *in vivo* cerebral cortex with those in hippocampal slices where ATP levels only recover to 45% of initial *in vivo* values (Lipton and Whittingham, 1984).

Given the initially substantial levels of recovery of energy metabolism following 10- and 20-min periods of ischemia, it would appear that cerebral cortical cells can actually survive ischemias of this duration and hypothetically emerge in reasonable condition. The critical problems may in fact be associated with reperfusion, the so-called "reperfusion injury." Insights into this problem have been gained from the studies of Lemasters and his colleagues (1996) who proposed that the reduction in tissue pH that develops during myocardial ischemia actually protects myocytes against cell death during the ischemic episode and that it is the return to normal pH during reperfusion that is responsible for their subsequent loss of viability. This worsening of injury, termed the "pH paradox," is mediated by the recovery of intracellular pH. The underlying hypothesis is that ischemia activates hydrolytic enzymes such as phospholipases and proteases, whose activity is inhibited at acidotic pHs. During reperfusion the recovery to normal pH releases this inhibition and hydrolytic injury ensues.

The role of acidosis as a primary cause of cerebral infarction has also been questioned (Tombaugh and Sapolsky, 1990), on the basis of mounting evidence that mild or moderate acidosis does not contribute to the selective neuronal necrosis that occurs after transient ischemia or in the penumbra of developing infarcts and could actually serve as a neuroprotective factor. By inhibiting cell excitability and blocking NMDA-induced membrane currents acidosis may reduce metabolic demand while at the same time restoring local blood flow via its potent vasodilatory effects. In that NMDA-activated receptors are considered to be among the primary mediators of excitotoxic ischemic injury (Choi, 1992), it is possible that acidosis might limit the extent of excitotoxic injury. Mild acidosis was subsequently shown to protect hippocampal and cortical neurons from ischemia-like conditions (Giffard *et al.*, 1990; Tombaugh and Sapolsky, 1990). Vornov *et al.* (1996) subsequently explored the possibility that recovery of pH during reperfusion was responsible for the ischemic injury by lowering the pH_e to 6.2 during the first hour of recovery of neuronal cultures from metabolic inhibition, demonstrating profound protective effects. Slowing the rate of pH_i recovery by incubation with blockers of Na^+/H^+ exchange also had protective effects. Measurements of pH_i confirmed that the blockers slowed recovery from intracellular acidosis and that less rapid recovery of pH_i was correlated with decreased injury as measured by LDH release.

F. Cerebroprotective Actions of Inhibition of Membrane Na^+/H^+ and Na^+/Ca^{2+} Exchangers

Lemasters and colleagues (Bond *et al.,* 1993; Lemasters *et al.,* 1996) used the term pH paradox to account for the protective action of intracellular acidosis on cardiac myocytes exposed to simulated ischemia, proposing instead that the injury actually occurs during the return to normal pH. Delays in the recovery of intracellular pH with inhibitors of Na^+/H^+ exchangers, dimethylamiloride or HOE 694, slowed the return to normal pH during reperfusion with protection of the myocytes. Acceleration of pH_i recovery with monensin, a Na^+/H^+ ionophore, accelerated the increase in intracellular pH and caused more rapid cell death. Loss of cell viability and PLA_2 activity (measured by arachidonic acid release) in rat hepatocytes increased in parallel during chemical hypoxia in a pH 7.4 medium (Harrison *et al.,* 1991). Acidosis (pH 6.5) or phospholipase inhibition with dibucaine and quinacrine delayed loss of cell viability and the release of arachidonic acid to a comparable extent. Staining of hypoxic rat hepatocytes with antibodies that recognize a pH-dependent group II phospholipase demonstrated a substantial increase in PLA_2 protein within 30 min of the initiation of hypoxic injury (Wang *et al.,* 1996). Treatment of hepatocytes with group II PLA_2-specific antisense DNA oligonucleotides abolished the hypoxia-evoked accumulation of PLA_2 protein, reduced arachidonic acid release, and significantly delayed cell death as indicated by propidium iodide uptake (Wang *et al.,* 1996). All of the above findings suggest that phospholipases play an important role in hypoxia/ischemia-evoked injury.

The pronounced increase in FFA levels in rat cerebral cortical superfusates at the onset of reperfusion following either 20- or 30-min periods of cerebral ischemia (Phillis *et al.,* 2000b) suggested that phospholipase activity, which was largely suppressed during ischemia, developed rapidly with the onset of reperfusion and pH normalization. This concept has been tested by slowing the recovery of pH with two inhibitors of the Na^+/H^+ exchanger EIPA (Fig. 3) and SM 20220. Both compounds significantly reduced FFA formation and release (Phillis *et al.,* 2000b; Pilitsis *et al.,* 2001a). EIPA has previously been observed to protect gerbil CA1 hippocampal pyramidal neurons from ischemic injury (Phillis *et al.,* 1999a) and similar effects of SM 20220 have been reported in the gerbil hippocampal model (Kuribayashi *et al.,* 2000) and in rat middle cerebral artery occlusion (MCAO) models (Kuribayashi *et al.,* 1999a,b).

During ischemia, ATP depletion leads to a loss of ion homeostasis, with increases in intracellular Na^+ and intracellular acidosis. Elevated levels of intracellular Na^+ can result in calcium entry via a reversed action of an Na^+/Ca^{2+} exchanger (NCX), which extrudes Na^+ ions in exchange for

Ca^{2+} (Egger and Niggli, 1999; Shigekawa and Iwamoto, 2001). The calcium overload can lead to a variety of deleterious effects including: mitochondrial accumulation of Ca^{2+}, activation of phospholipases and lipolysis, activation of proteases, and free radical production. Inhibition of the NCX with KB–R7943 was neuroprotective in an *in vitro* model of ischemia/reperfusion injury (Schroder *et al.*, 2001). When applied topically onto the rat cerebral cortex, KB–R7943 reduced the ischemia/reperfusion-evoked release of FFAs suggesting that it had decreased Ca^{2+} entry via the NCX with a reduction in activation of calcium-dependent phospholipases (Pilitsis *et al.*, 2001b). KB-R7943 also reduced the ischemia/reperfusion-elicited release of phosphoethanolamine into cortical perfusates. The latter's presence is considered to be indicative of the degradation of the plasma membrane phospholipids (Hagberg *et al.*, 1985). Another NCX, SEA 0400, attenuated reperfusion injury in a rat MCAO cerebral ischemia model (Matsuda *et al.*, 2001).

IV. Hyperglycemia, Diabetes, and Ischemic Injury

A. Effects on Extracellular Amino Acid and FFA Levels

It has been evident for over two decades that preischemic hyperglycemia can aggravate brain damage due to transient global or forebrain ischemia, as demonstrated by more extensive and rapidly developing neuronal necrosis with pannecrotic lesions (Ginsberg *et al.*, 1980; Pulsinelli *et al.*, 1982; Siesjo, 1988) and failure of ATP recovery during reperfusion (Katayama *et al.*, 1988). Exacerbation of brain lesions associated with stroke or middle cerebral artery occlusion in hyperglycemic subjects or experimental animals has also been observed (Pulsinelli *et al.*, 1983; Nedergaard, 1987), although a reduction in the extent of cerebral lesions was observed in rats subjected to permanent thrombolic neocortical infarction (Ginsberg *et al.*, 1987; Zaaslow *et al.*, 1989). The detrimental effects of glucose are observed when hyperglycemia is induced prior to, but not after, ischemia (Rehncrona *et al.*, 1981; Pulsinelli *et al.*, 1982). Another interesting observation has been that hyperglycemia increases infarct size in collaterally perfused, but not end-arterial, vascular territories (Prado *et al.*, 1988).

The mechanisms responsible for hyperglycemia-enhanced cerebral injury have remained uncertain. A rise in lactate production with concomitant tissue acidosis has been proposed as an explanation for the increase in brain damage (Smith *et al.*, 1986). Other investigators have questioned the relationship between lactate levels, acidosis, and ischemic brain injury

(Hoffman *et al.*, 1990). Additional causative factors may contribute to the glucose-induced increase in ischemic injury, including an attenuated cerebral production of adenosine, a neuroprotective compound (Phillis *et al.*, 1990; Hsu *et al.*, 1991), and a possible decrease of cerebral blood flow (Wagner *et al.*, 1992; Kawai *et al.*, 1998). The uncertainty as to whether hyperglycemia-enhanced ischemic injury was due to the resulting acidosis or an elevated level of glucose per se appears to have been resolved by the demonstration that an acidosis induced by hypercapnia was also effective in enhancing the ischemic injury (Katsura *et al.*, 1994).

The changes in intracellular pH during 15 min of ischemia followed by reperfusion in moderately hypoglycemic and hyperglycemic rats were measured by Smith *et al.* (1986). During hypoglycemic ischemia, pH_i declined to ~6.35, whereas in the hyperglycemic animals it was <6.00. Recovery was rapid in both groups of animals during the initial 5 min of reperfusion, with further normalization during the subsequent 25 min. pH_i in hyperglycemic animals then tended to rebound beyond the initial pH_i value of pH 7.15. Na^+/H^+ exchangers are inhibited by extracellular acidosis (Grinstein *et al.*, 1989) raising the possibility that intracellular lactate formation in hyperglycemic animals, with its efflux and subsequent dissociation into lactic acid in the extracellular space, could be a cause for the further drop in pH_i.

An initially attractive hypothesis to account for the effect of hyperglycemia on stroke injuries was presented by Choi *et al.* (1994), who rationalized that as glucose can be metabolized to glutamate via either the glycolytic pathway or the hexose monophosphate shunt, hyperglycemia could contribute to elevated levels of intracellular glutamate with an enhanced release of this amino acid during ischemia/reperfusion. Elevated extracellular glutamate levels could then account for the deleterious effects of hyperglycemia on the ischemic brain. However, a microdialysis study by the same authors, following glucose administration, revealed that glutamate levels in the dialysate were actually lower during global ischemia/reperfusion in comparison with normal controls. Comparable results were obtained in a study of amino acid efflux in the ischemic rat cerebral cortex following systemic glucose administration with reductions in the efflux of glutamate, GABA, and phosphoethanolamine in comparison with normoglycemic control ischemia rats (Fig. 4) (Phillis *et al.*, 1999d). The result of both studies suggest that hyperglycemia does not exert its detrimental effects by increasing the extracellular concentrations of glutamate. Rather glutamate release was reduced indicating that, at least during the ischemia itself, hyperglycemia might be cerebroprotective, allowing cells to sustain ATP production by anaerobic glycolysis. FFA efflux during ischemia/reperfusion was also depressed by hyperglycemia (Fig. 5), presumably as a consequence of the greater decrease in intracellular pH with a more pronounced inhibition of phospholipases (Phillis *et al.*, 1999d).

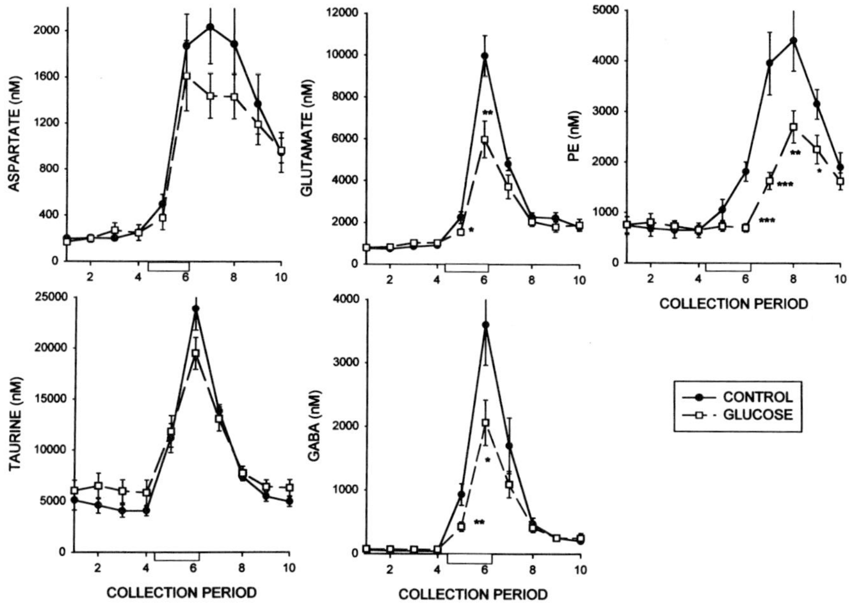

FIG. 4. Effects of systemic hyperglycemia (603 ± 144 mg/dl plasma glucose) on the ischemia-evoked release of amino acids into rat cerebral cortical superfusates. D-glucose (3.4 g/kg) was administered systemically after collection of the first two cortical superfusate samples, 10 min prior to the start of collection 3. Plasma glucose level in the control animals was 73.5 ± 4.0 mg/dl. Cerebral ischemia was induced by four-vessel occlusion during collections 5 and 6 followed by reperfusion. These data are presented in comparison with those from saline-injected normoglycemic, ischemic, animals. These data were recorded from the same animals as in Fig. 5. See Fig. 2 legend for further details. (Reprinted from *Brain Res.*, **837,** Phillis, J. W., Song, D., and O'Regan, M. H. (1999d). Effect of hyperglycemia on extracellular levels of amino acids and free fatty acids in the ischemic/reperfused rat cerebral cortex, 177–183, copyright 1999, with permission from Elsevier Science.)

In a subsequent study systemic administration of glucose was observed to enhance extracellular glutamate in the cerebral cortex, though not in the hippocampus (Li *et al.,* 2000), raising new questions regarding the effects of glucose on ischemia-evoked glutamate release. In order to avoid the potentially perturbing effects of systemically administered glucose, such as insulin release and systemic acidosis, further experiments were conducted with topically applied glucose (Guyot *et al.,* 2000a). Absence of glucose in the artificial CSF (aCSF) perfusing the rat cortex significantly enhanced ischemia-evoked glutamate, aspartate, GABA, and phosphoethanolamine in comparison to that during cortical perfusion with aCSF containing physiologic (66.5 mg/dl) amounts of glucose. Increasing glucose levels to 450 mg/dl did not significantly alter aspartate and glutamate efflux from that observed with 66.5 mg/dl. These experiments indicate that elevated

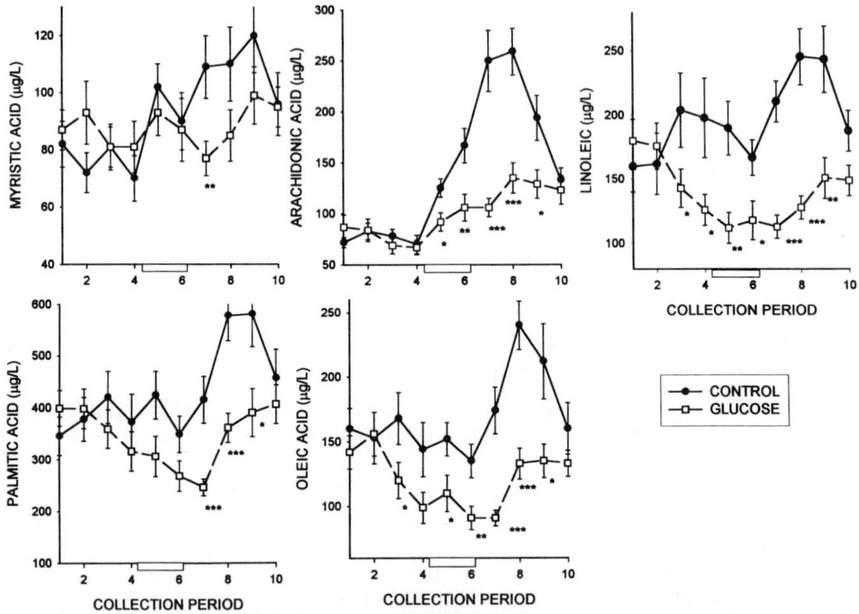

FIG. 5. Effects of systemic hyperglycemia (603 ± 144 mg/dl plasma glucose) on ischemia-evoked release of FFAs into cerebral cortical superfusates. Plasma glucose level in the control animals was 73.5 ± 4.0 mg/dl. Releases of myristic, arachidonic, palmitic, and oleic acids are compared in normoglycemic and hyperglycemic animals. D-glucose (3.4 g/kg) was administered 10 min prior to the start of collection 3 in the hyperglycemic animals. See Fig. 2 legend for further details. (Reprinted from *Brain Res.*, **837,** Phillis, J. W., Song, D., and O'Regan, M. H. Effect of hyperglycemia on extracellular levels of amino acids and free fatty acids in the ischemic/reperfused rat cerebral cortex, 177–183, copyright 1999, with permission from Elsevier Science.)

levels of topical glucose do not enhance ischemia/reperfusion-evoked release of the excitotoxic amino acids: rather it is the absence of glucose that results in the greatest accumulation of extracellular glutamate and aspartate.

A model of diabetes mellitus in rats can be created by intravenous streptozotocin (STZ) administration (Rakieten *et al.,* 1963). In comparison with normoglycemic controls, animals with hyperglycemia after STZ treatment (acute or chronic) sustain increased neurological damage after ischemia (Hoxworth *et al.,* 1999; Li *et al.,* 1998). The rats also showed no differences in the extent of histological damage following ischemia at 1 versus 4 weeks after STZ administration. A study on STZ-treated rats revealed that brain glucose utilization was initially elevated by 25%, but normalized after 4 weeks (Mans *et al.,* 1988) suggesting that brain energy metabolism may be affected during the course of STZ-induced diabetes. Cerebral blood flow studies failed to reveal any differences in the responses of acute or chronic streptozotocin-treated diabetic rats to hypoxia or hypercapnia (Simpson *et al.,* 1990; Pelligrino and Albrecht, 1991). However, increases in CBF in response to insulin-induced hypoglycemia were lost or attenuated in 4–6 months post-STZ rats (Pelligrino and Albrecht, 1991).

An examination of ischemia/reperfusion-evoked release of cerebral cortical amino acids in acute (1 week) and chronic (4–6 weeks) STZ-treated rats confirmed previous findings of hyperglycemia-evoked suppressions of glutamate release, which were most pronounced in the chronic animals (Guyot *et al.,* 2001). Plasma glucose levels in the three groups of rats were: controls, 72.4 mg/dl; acute STZ, 340.3 mg/dl; and chronic STZ, 344.3 mg/dl. There were no significant differences in mean arterial blood pressure and arterial blood pH. Another interesting finding was that cortical superfusate GABA levels during ischemia/reperfusion were significantly reduced from those in the control rats, with the most pronounced reduction in the chronic STZ group. Levels of other amino acids (alanine, aspartate, glutamine, glycine, phosphoethanolamine, serine, and taurine) were not affected. As a hyperpolarizing inhibitory neurotransmitter GABA is considered to be a potentially cerebroprotective agent. There is supporting evidence for this contention in that nipecotic acid, a GABA transport inhibitor, enhanced the rate of extracellular GABA accumulation in ischemia/reperfused rats (Phillis *et al.,* 1994b) and another GABA uptake inhibitor, CI–966, attenuated hippocampal injury in stroked gerbils (Phillis, 1995). Reductions in extracellular GABA levels after both short- and long-term hyperglycemia could be a contributing factor in the increased neuronal damage demonstrated after hyperglycemic strokes. A reduction in GABA levels after a hyperglycemic stroke, due to increased damage in the caudatoputamen, has been hypothesized as a reason for the increase in postischemic seizures in hyperglycemic rats (Smith *et al.,* 1988).

B. Can the pH Paradox Account for Hyperglycemia-Evoked Injury?

In the event of a stroke or cardiac arrest, the viability of brain cells is compromised by a number of factors including: glutamate receptor activation, a failure to regenerate ATP, with, in many instances, the collapse of plasma membrane barriers to ionic shifts, swelling, anoxic depolarization, Ca^{2+} entry, decreases in pH_i, and mitochondrial and DNA damage. Depending on the severity and duration of the ischemia, cells may either undergo a rapid (often as a result of excessive swelling) "necrotic" death due to lysis, whereas less susceptible cells or those in the penumbral area undergo a slower, more controlled, deterioration which culminates in "apoptotic" death. Over the past few years it has become fashionable to view necrosis and apoptosis as being the two extremes of a continuum, over which cells may display a variety of often overlapping degenerative characteristics (Roy and Sapolsky, 1999). It is very likely, however, that both processes are initiated by the same ischemia-induced insults, the outcome being dependent on cell susceptibility and the severity of the insult. Lemasters and colleagues, building on published reports of earlier investigators, were able to demonstrate that acidosis, which had previously been considered to be detrimental in ischemia, strongly protected hepatocytes and myocytes from ischemic injury (Gores *et al.*, 1988; Bond *et al.*, 1991). Thus arose the hypothesis that reperfusion injury involves a pH paradox whereby the return from acidotic to physiologic pH_i initiates the reperfusion injury (Lemasters *et al.*, 1996). Acidotic pH_i would inhibit the activity of pH-dependent enzymes, including phospholipases and proteases which would normally be activated by elevated calcium levels. Upon restitution of normal pH_i, these enzymes with neutral or alkaline pH optima would be released from inhibition, initiating cell injury. Na^+/H^+ exchange appears to require ATP (Orlowski and Grinstein, 1997) and would be depressed during the ischemia. Recovery of metabolic activity would promote the exteriorization of H^+ at the expense of an increase in intracellular Na^+. Intra- and extracellular acidification reduce the NCX rate (Egger and Niggli, 1999), enhancing the increase in intracellular calcium. During reperfusion the exchanger, operating in a reversed mode, would further augment intracellular Ca^{2+} levels whilst removing Na^+. Elevation of intracellular Ca^{2+} can lead to an accumulation of this ion by mitochondria with activation of a mitochondrial permeability transition (MPT), increased free radical production, and bioenergetic failure (Kristian and Siesjo, 1998). Cyclosporin A, which blocks the MPT pore, dramatically reduces ischemia-induced CA1 hippocampal neuronal loss (Uchino *et al.*, 1995).

When rat hepatocytes were exposed to anoxic buffer at pH 6.2, pH_i decreased to 6.3, mitochondria depolarized and ATP decreased to <1% of basal values, but MPT did not occur and viability remained at >90% (Qian *et al.*, 1997). After reperfusion at pH 7.4, pH_i recovered, the MPT occurred, and most hepatocytes lost viability. After reperfusion at pH 6.2, the MPT did not occur. When hepatocytes were reperfused with cyclosporin A at pH 7.4, the MPT was prevented and cell viability remained >80% although pH_i had recovered to 7.2, implying that opening of the MPT pore is a critical step in the initiation of necrotic-apoptotic cell death. Lemasters (1999) has suggested that cellular ATP levels during ischemia/reperfusion may determine whether death will be necrotic or apoptotic. With profound ATP depletion death will be necrotic, but if some mitochondria survive and continue to function apoptosis would result.

Our interest in reperfusion injury was stimulated by the observation of a marked rise in extracellular cerebral cortical FFA levels, which coincided with the onset of reperfusion following 20–30 min periods of ischemia, suggestive of a dramatic increase in phospholipase activity (Saluja *et al.*, 1999; Phillis *et al.*, 2000b). The increase in FFA efflux was significantly attenuated by prior exposure of the cerebral cortex to EIPA or SM 20220, Na^+/H^+ exchange inhibitors (Phillis *et al.*, 2000b; Pilitsis *et al.*, 2001a), and KB–R7943, an inhibitor of the NCX (Pilitsis *et al.*, 2001b). Ischemic injury to gerbil hippocampal CA1 pyramidal neurons was reduced by prior administration of mepacrine, a PLA_2 inhibitor (Phillis, 1996) and EIPA (Phillis *et al.*, 1999a). Mepacrine also reduced infarct size in rats after a 2-h middle cerebral artery occlusion (Estevez and Phillis, 1997). These observations, all of which are consistent with those of Lemaster's group, provide further support for the applicability of the pH paradox hypothesis to CNS tissues.

V. Insulin's Effects on Ischemic Injury

Insulin was first identified in the rat brain by using immunochemical staining with anti-insulin antibodies (Havrankova *et al.*, 1978). Insulin receptors were subsequently identified throughout the brain, with the highest concentrations in the olfactory bulb, hypothalamus, cerebral cortex, and hippocampus (Schulingkamp *et al.*, 2000). Insulin receptors have also been found on brain microvessels, indicating that it might affect the permeability of the blood-brain barrier (Frank and Pardridge, 1981).

Insulin's role in the brain still remains to be clearly defined. It has been shown to increase glucose transport across the blood-brain barrier (Hertz *et al.*, 1981), stimulate glycogen synthase to enhance glucose storage as glycogen in astrocytes (Hamai *et al.*, 1999), stimulate protein and RNA and DNA synthesis in the brain (Clarke *et al.*, 1985; Nakamura *et al.*, 1988), and inhibit spontaneous activity of pyramidal neurons in hippocampal slices (Pavolick *et al.*, 1984). The latter effect could be due to insulin's ability to recruit functional GABA receptors on neurons (Wan *et al.*, 1997).

Intraperitoneally administered insulin is neuroprotective when given prior to global (LeMay *et al.*, 1988; Warner *et al.*, 1992) or focal ischemia (Hamilton *et al.*, 1995) or during reperfusion following global (Fukuota *et al.*, 1989; Katz *et al.*, 1998) or focal ischemia (Izumi *et al.*, 1992). The neuroprotective effects of insulin appear to be independent of its hypoglycemic effects and may indeed be attenuated when blood glucose levels are reduced below 3.0 m*M* (Voll and Auer, 1991). Insulin increases the sensitivity of cultured cortical neurons to the excitotoxic effects of glutamate, possibly by increasing the expression of excitotoxic amino acid receptors (Schafer and Erdo, 1992).

Intravenously administered insulin (0.5 IU/kg) enhanced the ischemia/reperfusion-evoked release of aspartate, taurine, and phosphoethanolamine from the rat cerebral cortex in comparison with their releases in control rats (Guyot *et al.,* 2000b). When glucose was administered together with insulin during ischemia in order to maintain plasma glucose levels >80 mg/dl, increases in superfusate levels of glutamate and GABA, in addition to aspartate, taurine, and phosphoethanolamine were observed. During reperfusion, plasma glucose levels in the two insulin-injected groups of rats failed to rebound as in the control group. The observation that insulin administration increased aspartate but not glutamate release from the cortex, whereas insulin + glucose administration increased the release of both amino acids, finds certain parallels in experiments by Butterworth *et al.* (1982) in which insulin treatment of hypoglycemic rats, as compared to normoglycemic animals, evoked a significant increase in aspartate, but not of glutamate, levels in the cerebral cortex. The failure of glutamate efflux to increase may be a consequence of its utilization, via conversion to α-ketoglutarate, as an energy substrate in the mitochondrial TCA cycle with cycling through succinate, oxaloacetate, and ultimately aspartate formation. Glutamate utilization in this instance would be precipitated by the low glucose levels.

Topically applied insulin, in the absence of any changes in plasma glucose levels, also affected extracellular levels of the amino acids. At 1 mIU insulin/ml, insulin depressed ischemia/reperfusion-evoked releases of aspartate, glutamate, taurine, and phosphoethanolamine. At 100 mIU/ml, insulin increased the ischemia/reperfusion-evoked efflux of aspartate, glutamate, taurine, and GABA efflux, while depressing phosphoethanolamine formation (Guyot *et al.,* 2000c). The reductions in excitotoxic amino acid release in the presence of 1 mIU/ml of insulin may be a reflection of the ability of insulin to stimulate glycolysis and protein synthesis, thus enabling the glial cells to maintain energy sources and ion pumps during the ischemic episode with glycolytically formed lactate being available to the neurons. Such a scenario could account for the cerebroprotective actions of insulin. The detrimental effects of insulin 100 mIU/ml, with increasing extracellular excitotoxic amino acid levels, could be a consequence of insulin's upregulation of glutamate receptors, rendering cells more susceptible to the effects of glutamate, which include further release of this amino acid.

In a further study, the effect of topically applied insulin (1 and 100 mIU/ml) on ischemia/reperfusion-evoked amino acid efflux from the cerebral cortices of STZ-diabetic rats was evaluated (Guyot *et al.,* 2000d). In both acute (1 week) and chronic (4–6 weeks) STZ hyperglycemic rats, insulin at both concentrations enhanced glutamate, aspartate, and GABA release in comparison with that in acute or chronic STZ control rats. A potential explanation for the increases in amino acid efflux in the acute and chronic STZ groups with both doses of topical insulin may be alterations in the number of insulin receptors in the brain. Frank *et al.* (1986) have determined that brain capillary endothelial cells have a capacity to upregulate their insulin receptors in response to the low insulin environment in STZ-diabetic rats. A similar effect on neuronal receptors would explain why the lower dose of insulin, which depressed amino acid release in normoglycemic rats, now exerted a similar action to the higher dose, which increased ischemia/reperfusion elicited amino acid in control rats.

VI. Conclusions and Future Directions

During cerebral ischemia, ATP levels decline rapidly, accompanied by a marked reduction in tissue pH. Extracellular levels of amino acids, including the neurotransmitters glutamate, aspartate, γ-aminobutyric acid, and adenosine, are significantly elevated during the ischemic episode, and return toward preischemic levels during the initial 20–30 min of reperfusion, accompanied by a partial recovery of ATP and intracellular pH. Extracellular FFA levels rise slightly during ischemia, increase dramatically at the onset of reperfusion, and remain elevated during the subsequent 40–60 min. Hyperglycemia, following topical or systemic administration of glucose, depressed the ischemia-evoked release of both amino acids and FFA. Evidence for a role of phospholipases in amino acid and FFA effluxes was obtained by topical applications of melittin, a phospholipase activator, which enhanced release, and various phospholipase inhibitors, which attenuated release. Mepacrine, a phospholipase inhibitor, was cerebroprotective in both forebrain ischemia and middle cerebral artery occlusion models of ischemia/reperfusion.

Delaying the return of pH_i to normal levels with inhibitors of the Na^+/H^+ exchanger, such as EIPA and SM 20220, decreased FFA efflux during ischemia/reperfusion, and both agents were cerebroprotective in gerbil or rat ischemia models. KB–R7943, an inhibitor of Na^+/Ca^{2+} exchange, also inhibited ischemia/reperfusion-evoked FFA efflux, and this agent has been demonstrated to be cardioprotective. These findings are consistent with the hypothesis that acidosis may protect brain cells during an ischemic episode by inhibiting the activity of enzymes, such as phospholipases, which can degrade plasma and mitochondrial membranes. Injury occurs following their activation during reperfusion with the normalization of intracellular pH. Evidence that cyclosporin A, which prevents the opening of the mitochondrial permeability transition pore, has a cerebroprotective action indicates that increasing the pH_i during reperfusion may induce a mitochondrial permeability transition with mitochondrial depolarization and the failure of ATP regeneration. The ultimate fate of each cell may be largely dependent on the extent to which ATP is regenerated by surviving mitochondria with either lysis (necrosis), slow death (apoptosis), or life as the outcomes.

Future directions for research will involve a detailed examination of mitochondrial function in ischemia/reperfusion brains.

Acknowledgments

The studies reported herein were supported by funding from the USPHS Grant NS26912, the Wayne State University Department of Neurological Surgery, and the Morris Hood, Jr. Comprehensive Diabetes Grant. Thanks to Dr. J. G. Pilitsis for comments on the manuscript and Ms. L. McCraw for secretarial assistance.

References

Ames, A. (2000). CNS metabolism as related to function. *Brain Res. Rev.* **34,** 42–68.

Asai, S., Kunimatsu, T., Zhao, H., Nagata, T., Takahashi, Y., Ishii, Y., Kohno, T., and Ishikawa, K. (2000). Two distinct components of initial glutamate release synchronized with anoxic depolarization in rat global brain ischemia. *Neuroreport* **11,** 2947–2952.

Barron, K. D., Dentinger, M. P., Kimelberg, H. K., Nelson, L. R., Bourke, R. S., Keegan, S., Mankes, R., and Cragoe, E. J. (1988). Ultrastructural features of a brain injury model in cat. I. Vascular and neurological changes and the prevention of astroglial swelling by a fluorenyl (aryloxy) alkanoic acid derivative (L-644,711). *Acta Neuropathol.* (*Berl.*) **75,** 295–307.

Bazan, N. G. (1970). Effects of ischemia and electroconvulsive shock on fatty acid pool in brain. *Biochim. Biophys. Acta* **218,** 1–10.

Benveniste, H. (1991). The excitotoxin hypothesis in relation to cerebral ischemia. *Cerebrovasc. Brain Metab. Rev.* **3,** 213–245.

Bittar, P. G., Charnay, Y., Pellerin, L., Bouras, C., and Magistretti, P. J. (1998). Selective distribution of lactate dehydrogenase isoenzymes in neurons and astrocytes of human brain. *J. Cereb. Blood Flow Metab.* **16,** 1079–1089.

Bond, J. M., Herman, B., and Lemasters, J. J. (1991). Protection by acidotic pH against anoxia/reoxygenation injury to rat neonatal cardiac myocytes. *Biochem. Biophys. Res. Commun.* **179,** 798–803.

Bond, J. M., Chacon, E., Herman, B., and Lemasters, J. J. (1993). Intracellular pH and Ca^{2+} homeostasis in the pH paradox of reperfusion injury to neonatal rat cardiac myocytes. *Am. J. Physiol.* **265,** C129–C137.

Bonventre, J. V., Huang, Z., Taheri, M. R., O'Leary, E., Li, E., Moskowitz, M. A., and Sapirstein, M. (1997). Reduced fertility and postischaemic brain injury in mice deficient in cytosolic phospholipase A_2. *Nature* **390,** 622–625.

Bouvier, M., Szatkowski, M., Amato, A., and Attwell, D. (1992). The glial cell glutamate uptake carrier countertransports pH-changing anions. *Nature* **360,** 471–474.

Broer, S., and Brookes, N. (2001). Transfer of glutamine between astrocytes and neurons. *J. Neurochem.* **77,** 705–719.

Broer, S., Rahman, B., Pellegri, G., Pellerin, L., Martin, J. L., Verleysdonk, S., Hamprecht, B., and Magistretti, P. J. (1997). Comparison of lactate transport in astroglial cells and monocarboxylate transporter 1 (MCT 1) expressing Xenopus laevis oocytes. Expression of two different monocarboxylate transporters in astroglial cells and neurons. *J. Biol. Chem.* **272,** 30096–30102.

Broer, S., Broer, A., Schneider, H. P., Stegen, C., Halestrap, A. P., and Deitmer, J. W. (1999). Characterization of the high-affinity monocarboxylate transporter MCT 2 in Xenopus laevis oocytes. *Biochem. J.* **341,** 529–535.

Busa, W. B., and Nuccitelli, R. (1984). Metabolic regulation via intracellular pH. *Am. J. Physiol.* **246,** R409–R438.

Butcher, S. P., Bullock, R., Graham, D. I., and McCulloch, J. (1990). Correlation between amino acid release and neuropathologic outcome in rat brain following middle cerebral artery occlusion. *Stroke* **21,** 1727–1733.

Butterworth, R. F., Merkel, A. D., and Landreville, F. (1982). Regional amino acid distribution in relation to function in insulin hypoglycemia. *J. Neurochem.* **38,** 1483–1489.

Campbell, J. D., and Paul, R. J. (1992). The nature of fuel provision for the Na^+,K^+-ATPase in porcine vascular smooth muscle. *J. Physiol.* (*Lond.*) **447,** 67–82.

Choi, D. W. (1992). Excitotoxic cell death. *J. Neurobiol.* **23,** 1261–1276.

Choi, K. T., Illievich, U. M., Zornow, M. H., Scheller, M. S., and Strnat, M. A. P. (1994). Effect of hyperglycemia on peri-ischemia neurotransmitter levels in the rabbit hippocampus. *Brain Res.* **642,** 104–110.

Clarke, D. W., Boyd, F. T., Kappy, M. S., and Raizada, M. K. (1985). Insulin stimulates macromolecular synthesis in cultured glial cells from rat brain. *Am. J. Physiol.* **249,** C484–C489.

Clements, J. D., Lester, R. A. J., Tong, G., Jahr, C. E., and Westbrook, G. L. (1992). The time course of glutamate in the synaptic cleft. *Science* **258,** 1498–1501.

Danbolt, N. C. (2001). Glutamate uptake. *Progr. Neurobiol.* **65,** 1–105.

Debernardi, R., Magistretti, P. J., and Pellerin, L. (1999). Trans-inhibition of glutamate transport prevents excitatory amino acid-induced glycolysis in astrocytes. *Brain Res.* **850,** 39–46.

Dennis, S. C., Gevers, W., and Opie, L. H. (1991). Protons in ischemia: Where do they come from; where do they go to? *J. Mol. Cell. Cardiol.* **23,** 1077–1086.

Diemer, N. H., and Ekstrom von Lubitz, D. K. J. (1983). Cerebral ischemia in the rat: Increased permeability of postsynaptic membranes to horseradish peroxidase in the early post-ischaemic period. *Neuropathol. Appl. Neurobiol.* **9,** 403–414.

Eckert, R., and Chad, J. E. (1984). Inactivation of Ca channels. *Progr. Biophys. Mol. Biol.* **44,** 215–267.

Egger, M., and Niggli, E. (1999). Regulatory function of Na-Ca exchange in the heart: Milestones and outlook. *J. Memb. Biol.* **168,** 107–130.

Ekstrom von Lubitz, D. K. J., and Diemer, N. H. (1982). Complete cerebral ischemia in the rat: An ultrastructural and stereological analysis of the distal striatum radiatum in the hippocampal CA-1 region. *Neuropathol. Appl. Neurobiol.* **8,** 197–215.

Erecinska, M., and Silver, I. A. (1989). ATP and brain function. *J. Cereb. Blood Flow Metabol.* **9,** 2–19.

Estevez, A. Y., O'Regan, M. H., Song, D., and Phillis, J. W. (1999a). Effects of anion channel blockers on hyposmotically-induced amino acid release from the in vivo rat cerebral cortex. *Neurochem. Res.* **24,** 447–452.

Estevez, A. Y., O'Regan, M. H., Song, D., and Phillis, J. W. (1999b). Hyposmotically-induced amino acid release from the rat cerebral cortex: role of phospholipases and protein kinases. *Brain Res.* **844,** 1–9.

Estevez, A. Y., and Phillis, J. W. (1997). The phospholipase A_2 inhibitor, quinacrine, reduces infarct size in rats after transient middle cerebral artery occlusion. *Brain Res.* **752,** 203–208.

Farooqui, A. A., Yang, H.-C., Rosenberger, T. A., and Horrocks, L. A. (1997). Phospholipase A_2 and its role in brain tissue. *J. Neurochem.* **69,** 889–901.

Fellows, L. K., Boutelle, M. G., and Fillenz, M. (1993). Physiological stimulation increases nonoxidative glucose metabolism in the brain of the freely moving rat. *J. Neurochem.* **60,** 1258–1263.

Fox, P. T., and Raichle, M. E. (1986). Focal physiological uncoupling of cerebral blood flow and oxidative metabolism during somatosensory stimulation in human subjects. *Proc. Nat. Acad. Sci. USA* **83,** 1140–1144.

Fox, P. T., Raichle, M. E., Mintun, M. A., and Dence, C. (1988). Non-oxidataive glucose consumption during focal physiologic neural activity. *Science* **241,** 462–464.

Frank, H. J. L., and Pardridge, W. M. (1981). A direct *in vitro* demonstration of insulin binding to isolated brain microvessels. *Diabetes* **30,** 757–761.

Frank, H. J., Pardridge, W. M., Jankovic-Vokes, T., Vinters, H. V., and Morris, W. L. (1986). Insulin binding to the blood-brain barrier in the streptozotocin diabetic rat. *J. Neurochem.* **47,** 405–411.

Franson, R. C., Eisen, D., Jesse, R., and Lanni, C. (1994). Inhibition of highly purified mammalian phospholipases A_2 by non-steroidal anti-inflammatory agents. Modulation by calcium ions. *Biochem. J.* **186,** 633–636.

Fukuota, S., Yeh, H., Mandybur, T. I., and Tew, J. M. (1989). Effect of insulin on acute experimental cerebral ischemia in gerbils. *Stroke* **20,** 396–399.

Giffard, R. G., Monyer, H., Christine, C. W., and Choi, D. W. (1990). Acidosis reduces NMDA receptor activation, glutamate neurotoxicity, and oxygen-glucose deprivation neuronal injury in cortical cultures. *Brain Res.* **506,** 339–342.

Ginsberg, M. D., Welsh, F. A., and Budd, W. W. (1980). Deleterious effect of glucose pretreatment on recovery from diffuse cerebral ischemia in the cat. *Stroke* **11,** 347–354.

Ginsberg, M. D., Prado, R., Dietrich, W. D., Busto, R., and Watson, B. D. (1987). Hyperglycemia reduces the extent of cerebral infarction in rats. *Stroke* **18,** 570–574.

Gores, G. J., Fleishman, K. E., Dawson, T. E., Herman, B., Nieminem, A.-L., and Lemasters, J. J. (1988). Extracellular acidosis delays onset of cell death in ATP depleted hepatocytes. *Am. J. Physiol.* **255,** C315–C322.

Grinstein, S., Rotin, D., and Mason, M. J. (1989). Na^+/H^+ exchange and growth factor-induced cytosolic pH changes. Role in cellular proliferation. *Biochim. Biophys. Acta* **988,** 73–97.

Guyot, L. L., Diaz, F. G., O'Regan, M. H., Song, D., and Phillis, J. W. (2000a). Topical glucose and accumulation of excitotoxic and other amino acids in ischemic cerebral cortex. *Horm. Metab. Res.* **32,** 6–9.

Guyot, L. L., Diaz, F. G., O'Regan, M. H., Ren, J., and Phillis, J. W. (2000b). The effect of intravenous insulin on accumulation of excitotoxic and other amino acids in the ischemic rat cerebral cortex. *Neurosci. Lett.* **288,** 61–65.

Guyot, L. L., Diaz, F. G., O'Regan, M. H., Song, D., and Phillis, J. W. (2000c). Topical insulin and accumulation of excitotoxic and other amino acids in ischemic rat cerebral cortex. *Proc. Soc. Exp. Biol. Med.* **224,** 28–31.

Guyot, L. L., Diaz, F. G., O'Regan, M. H., Song, D., and Phillis, J. W. (2000d). The effect of topical insulin on the release of excitotoxic and other amino acids from the rat cerebral cortex during streptozotocin-induced hyperglycemic ischemia. *Brain Res.* **872,** 29–36.

Guyot, L. L., Diaz, F. G., O'Regan, M. H., Song, D., and Phillis, J. W. (2001). The effect of streptozotocin-induced diabetes on the release of excitotoxic and other amino acids from the ischemic rat cerebral cortex. *Neurosurgery* **48,** 385–390.

Hagberg, H., Lehman, S., Sandberg, M., Nystrom, B., Jacobson, I., and Hamberger, A. (1985). Ischemia-induced shift of inhibitory and excitatory amino acids from intra- to extracellular compartments. *J. Cereb. Blood Flow Metab.* **5,** 413–419.

Hamai, M., Minokoshi, Y., and Shimazu, T. (1999). L-glutamate and insulin enhance glycogen synthesis in cultured astrocytes from the rat brain through different intracellular mechanisms. *J. Neurochem.* **73,** 400–407.

Hamilton, M. G., Tranmer, B. I., and Auer, R. N. (1995). Insulin reduction of cerebral infarction due to transient focal ischemia. *J. Neurosurg.* **82,** 262–268.

Harrison, D. C., Lemasters, J. J., and Herman, B. (1991). A pH dependent phospholipase A_2 contributes to loss of plasma membrane integrity during chemical hypoxia in rat hepatocytes. *Biochem. Biophys. Res. Commun.* **174,** 654–659.

Havrankova, J., Schmechel, D., Roth, J., and Brownstein, M. (1978). Identification of insulin in rat brain. *Proc. Nat. Acad. Sci. USA* **75,** 5737–5741.

Hearse, D. J., Humphrey, S. M., and Chain, E. B. (1973). Abrupt reoxygenation of the anoxic potassium-arrested perfused rat heart: a study of myocardial enzyme release. *J. Mol. Cell. Cardiol.* **5,** 395–407.

Hertz, L., Swanson, R. A., Newman, G. C., Mariff, H., Juurlink, B. H. J., and Peng, L. (1998). Can experimental conditions explain the discrepancy whether or not glutamate stimulates aerobic glycolysis? *Dev. Neurosci.* **20,** 339–347.

Hertz, L., Yu, A. C. H., Kala, G., and Schousboe, A. (2000). Neuronal-astrocytic and cytosolic-mitochondrial metabolite trafficking during brain activation, hyperammonemia and energy deprivation. *Neurochem. Int.* **37,** 83–102.

Hertz, M. M., Paulson, O. B., Barry, D. I., Christiansen, J. S., and Svendsen, P. A. (1981). Insulin increases glucose transfer across the blood-brain barrier in man. *J. Clin. Invest.* **67,** 597–604.

Hochachka, P. W., and Mommsen, T. P. (1983). Protons and anaerobiosis. *Science* **219,** 1391–1397.

Hoffman, W. E., Braucher, E., Pelligrino, D. A., Thomas, C., Albrecht, R. A., and Miletich, D. J. (1990). Brain lactate and neurologic outcome following incomplete ischemia in fasted, non-fasted, and glucose-loaded rats. *Anesthesiology* **72,** 1045–1050.

Hogstad, S., Svenneby, G., Torgner, I. Aa., Kvamme, E., Hertz, L., and Schousboe, A. (1988). Glutamate in neurons and astrocytes cultured from mouse brain: Kinetic properties and effects of phosphate, glutamate and ammonia. *Neurochem. Res.* **13,** 383–388.

Hoxworth, J. M., Xu, K., Zhou, Y., Lust, W. D., and Lamanna, C. (1999). Cerebral metabolic profile, selective neuron loss, and survival of acute or chronic hyperglycemic rats following cardiac arrest and resuscitation. *Brain Res.* **821,** 467–479.

Hsu, S. F., Meno, J. R., Zhou, J.-G., Gordon, E. L., and Winn, H. R. (1991). Influence of hyperglycemia on cerebral adenosine production during ischemia and reperfusion. *Am. J. Physiol.* **261,** H398–H403.

Hu, Y., and Wilson, G. S. (1997). A temporary local energy pool coupled to neuronal activity: Fluctuations of extracellular lactate levels in rat brain monitored with rapid-response enzyme-based sensor. *J. Neurochem.* **69,** 1484–1490.

Ikeda, M., Nakazawa, T., Abe, K., Kaneko, T., and Yamatsu, K. (1989). Extracellular accumulation of glutamate in the hippocampus induced by ischemia is not calcium-dependent—*in vitro* and *in vivo* evidence. *Neurosci. Lett.* **96,** 202–206.

Izumi, Y., Pinard, E., Roussel, S., and Seylaz, J. (1992). Insulin protects brain tissue against focal ischemia in rats. *Neurosci. Lett.* **144,** 121–123.

Jackson, P. S., and Strange, K. (1993). Volume-sensitive anion channels mediate swelling-activated inositol and taurine efflux. *Am. J. Physiol.* **265,** C1489–C1500.

Jakubowicz, D. E., Grinstein, S., and Klip, A. (1987). Cell swelling following recovery from acidification in C_6 glioma cells: an in vitro model of post-ischemic brain edema. *Brain Res.* **435,** 138–146.

Jakubowicz, D. E., and Klip, A. (1989). Lactic acid-induced swelling in C_6 glial cells via Na^+/H^+ exchange. *Brain Res.* **485,** 215–224.

Katayama, Y., Fukuchi, T., McKee, A., and Terashi, A. (1988). Effect of hyperglycemia on pyruvate dehydrogenase activity and energy metabolites during ischemia and reperfusion in gerbil brain. *Brain Res.* **788,** 302–304.

Katsura, K., Kristian, T., Smith, M. L., and Siesjo, B. K. (1994). Acidosis induced by hypercapnia exaggerates ischemic brain damage. *J. Cereb. Blood Flow Metab.* **14,** 243–250.

Katz, L. M., Wang, Y., Ebmyer, U., Radovsky, A., and Safar, P. (1998). Glucose plus insulin infusion improves cerebral outcome after asphyxial cardiac arrest. *Neurochemistry* **9,** 3363–3367.

Kawai, N., Keep, R. F., Betz, A. L., and Nagao, S. (1998). Hyperglycemia induces progressive changes in the cerebral microvasculature and blood-brain barrier transport during focal cerebral ischemia. *Acta Neurochir.* (*Wien*) [Suppl.] **71,** 219–221.

Kempski, O., Staub, F., Jansen, M., Schödel, F., and Baethmann, A. (1988). Glial swelling during extracellular acidosis *in vitro. Stroke* **19,** 385–392.

Kimelberg, H. K., Goderie, S. K., Higman, S., Pang, S., and Waniewski, R. A. (1990). Swelling-induced release of glutamate, aspartate, and taurine from astrocyte cultures. *J. Neurosci.* **10,** 1583–1591.

Knull, H. R. (1978). Association of glycolytic enzymes with particulate fractions of nerve endings. *Biochim. Biophys. Acta* **522,** 1–9.

Krebs, H. A., Woods, H. F., and Alberti, K. G. M. M. (1975). Hyperlactatemia and lactic acidosis. *Essays Med. Biochem.* **1,** 81–103.

Kristian, T., and Siesjo, B. K. (1998). Calcium in ischemic cell death. *Stroke* **29,** 705–718.

Kuhr, W. G., and Korf, J. (1988). Extracellular lactic acid as an indicator of brain metabolism: Continuous on-line measurement in conscious, freely moving rats with intrastriatal dialysis. *J. Cereb. Blood Flow Metab.* **8,** 130–137.

Kuribayashi, Y., Horikawa, N., Itoh, N., Kitano, M., and Ohashi, N. (1999a). Delayed treatment of Na^+/H^+ exchange inhibitor SM-20220 reduces infarct size in both transient and permanent middle cerebral artery occlusion in rats. *Int. J. Tissue React.* **21,** 29–33.

Kuribayashi, Y., Itoh, N., Kitano, M., and Ohashi, N. (1999b). Cerebroprotective properties of SM-20220, a potent Na^+/H^+ exchange inhibitor, in transient cerebral ischemia in rats. *Eur. J. Pharmacol.* **383,** 163–168.

Kuribayashi, Y., Itoh, N., Horikawa, N., and Ohashi, N. (2000). SM-20220, a potent Na^+/H^+ exchange inhibitor, improves consciousness recovery and neurological outcome following transient cerebral ischemia in gerbils. *J. Pharm. Pharmacol.* **52,** 441–444.

Kvamme, E., Svenneby, G., Hertz, L., and Schousboe, A. (1982). Properties of phosphate activated glutaminase in astrocytes cultured from mouse brain. *Neurochem. Res.* **7,** 761–770.

Law, R. O. (1995). Taurine efflux and cell volume regulation in cerebral cortical slices during chronic hypernatraemia. *Neurosci. Lett.* **185,** 56–59.

Lehre, P., and Danbolt, N. C. (1998). The number of glutamate transproter subtype molecules at glutamatergic synapses: Chemical and sterological quantification in young adult rat brain. *J. Neurosci.* **18,** 8751–8757.

Lemasters, J. J. (1999). The mitochondrial permeability transition and the calcium, oxygen and pH paradoxes: One paradox after another. *Cardiovasc. Res.* **44,** 470–473.

Lemasters, J. J., Bond, J. M., Chacon, E., Harper, I. S., Kaplan, S. H., Obata, H., Trollinger, D. R., Herman, B., and Cascio, W. E. (1996). The pH paradox in ischemia-reperfusion injury to cardiac myocytes. *EXS* **76,** 99–114.

Lemasters, J. J., Quia, T., Bradham, C. A., Brenner, D. A., Cascio, W. E., Trost, L. C., Nishimura, Y., Nieminen, A. L., and Herman, B. (1999). Mitochondrial dysfunction in the pathogenesis of necrotic and apoptotic cell death. *J. Bioenerget. Biomemb.* **31,** 305–319.

Lemay, D. R., Gehua, L., Zelenock, G. B., and D'Alecy, L. G. (1988). Insulin administration protects neurologic function in cerebral ischemia in rats. *Stroke* **19,** 1411–1419.

Li, C., Li, P. A., He, Q. P., Ouyang, Y. B., and Siesjo, B. K. (1998). Effects of streptozotocin-induced hyperglycemia on brain damage following transient ischemia. *Neurobiol. Dis.* **5,** 117–128.

Li, P.-A., Shuaib, A., Miyashita, H., He, Q.-P., and Siesjo, B. K. (2000). Hyperglycemia enhances extracellular glutamate accumulation in rats subjected to forebrain ischemia. *Stroke* **31,** 183–192.

Lipton, P. (1999). Ischemic cell death in brain neurons. *Physiol. Rev.* **79,** 1431–1568.

Lipton, P., and Robacker, K. (1983). Glycolytic and brain functions: $[K^+]_O$ stimulation of protein synthesis and K^+ uptake requires glycolysis. *Fed. Proc.* **42,** 2875–2879.

Lipton, P., and Whittingham, T. S. (1984). Energy metabolism and brain slice function. *In* "Brain Slices" (R. Dingledine, ed.), pp. 113–153. Plenum Press, New York.

Lomneth, R., Medrano, S., and Gruenstein, E. I. (1990). The role of transmembrane pH gradients in the lactic acid induced swelling of astrocytes. *Brain Res.* **523,** 69–77.

Magistretti, P. J., Pellerin, L., Rothman, D. L., and Shulman, R. G. (1999). Energy on demand. *Science* **283,** 496–497.

Mans, A. M., DeJoseph, M. R., Davis, D. W., and Hawkins, R. A. (1988). Brain energy metabolism in streptozotocin-diabetes. *Biochem. J.* **249,** 57–62.

Matsuda, T., Arakawa, N., Takuma, K., Kishida, Y., Kawasaki, Y., Sakaue, M., Takahashi, T., Suzuki, T., Ota, T., Hamano-Takahashi, A., Onishi, M., Tanaka, Y., Kameo, K., and Baba, A. (2001). SEA 0400, a novel and selective inhibitor of the Na^+-Ca^{2+} exchanger, attenuates reperfusion injury in the *in vitro* and *in vivo* cerebral ischemic models. *J. Pharmacol. Exp. Ther.* **298,** 249–256.

McBain, C. J., and Mayer, M. L. (1994). *N*-Methyl-D-aspartic acid receptor structure and function. *Physiol. Rev.* **74,** 723–760.

Melani, A., Pantoni, L., Corsi, C., Bianchi, L., Monopoli, A., Bertorelli, R., Pepeu, G., and Pedata, F. (1999). Striatal outflow of adenosine, excitatory amino acids, γ-aminobutyric acid, and taurine in awake freely moving rats after middle cerebral artery occlusion. Correlations with neurological deficit and histopathological damage. *Stroke* **30,** 2448–2455.

Mercer, R. W., and Dunham, P. B. (1981). Membrane-bound ATP fuels the Na/K pump: Studies on membrane-bound glycolytic enzymes on inside-out vesicles from human red cell membranes. *J. Gen. Physiol.* **78,** 547–568.

Nakamura, A., Shitara, N., and Takakura, K. (1988). Insulin binds to specific receptors and stimulates macromolecular synthesis in C_6 glioma cells. *Acta Neurochir.* **93,** 10–12.

Nedergaard, M. (1987). Transient focal ischemia in hyperglycemic rats is associated with increased cerebral infarction. *Brain Res.* **408,** 79–85.

Nordstrom, C. H., Rehncrona, S., and Siesjo, B. K. (1978a). Effects of phenobarbital in cerebral ischemia. Part II. Restitution of cerebral energy state, as well as of glycolytic metabolites, citric acid cycle intermediates and associated amino acids after pronounced incomplete ischemia. *Stroke* **9,** 335–343.

Nordstrom, C. H., Rehncrona, S., and Siesjo, B. K. (1978b). Restitution of cerebral energy state, as well as glycolytic metabolites, citric acid cycle intermediates and associated amino acids after 30 minutes of complete ischemia in rats anaesthetized with nitrous oxide or phenobarbital. *J. Neurochem.* **30,** 479–486.

Norenberg, M. D., and Martinez-Hernandez, A. (1979). Fine structural localization of glutamine synthetase in astrocytes of rat brain. *Brain Res.* **161,** 303–310.

O'Regan, M. H., Perkins, L. M., and Phillis, J. W. (1995a). Arachidonic acid and lysophosphatidylcholine modulate excitatory transmitter amino acid release from the rat cerebral cortex. *Neurosci. Lett.* **193,** 85–88.

O'Regan, M. H., Smith-Barbour, M., Perkins, L. M., and Phillis, J. W. (1995b). A possible role for phospholipases in the release of neurotransmitter amino acids from ischemic rat cerebral cortex. *Neurosci. Lett.* **185,** 191–194.

O'Regan, M. H., Song, D., Vanderheide, S. J., and Phillis, J. W. (1997). Free radicals and the ischemia-evoked extracellular accumulation of amino acids in rat cerebral cortex. *Neurochem. Res.* **22,** 273–280.

Orlowski, J., and Grinstein, S. (1997). Na^+/H^+ exchangers of mammalian cells. *J. Biol. Chem.* **272,** 22373–22376.

Orrego, F., and Villaneuva, S. (1993). The chemical nature of the main central excitatory transmitter: A critical appraisal based upon release studies and synaptic vesicle localization. *Neuroscience* **56,** 539–555.

Pasantes-Morales, H. (1996). Volume regulation in brain cells: Cellular and molecular mechanisms. *Metabol. Brain Dis.* **11,** 187–204.

Paul, R. J., Bauer, M., and Pease, W. (1979). Vascular smooth muscle: Aerobic glycolysis linked to sodium and potassium transport processes. *Science* **206,** 1414–1416.

Paul, R. J., Hardin, D. C., Raeymaekers, L., Wuytack, F., and Casteels, R. (1989). Preferential support of Ca^{2+} uptake in smooth muscle plasma membrane vesicles by an endogenous glycolytic cascade. *FASEB J.* **3,** 2298–2301.

Pavolick, R. A., Phillips, M. I., Kappy, M. S., and Raizada, M. K. (1984). Insulin inhibits pyramidal neurons in hippocampal slices. *Brain Res.* **309,** 187–191.

Pellegrini-Giampietro, D. E., Cherici, G., Alesiani, M., Carla, V., and Moroni, F. (1990). Excitatory amino acid release and free radical formation may cooperate in the genesis of ischemia-induced neuronal damage. *J. Neurosci.* **10,** 1035–1041.

Pellerin, L., Pellegri, G., Bittar, P. G., Charnay, Y., Bouras, C., Martin, J. L., Stella, N., and Magistretti, P. J. (1998). Evidence supporting the existence of an activity-dependent astrocyte-neuron lactate shuttle. *Dev. Neurosci.* **20,** 291–299.

Pelligrino, D. A., and Albrecht, R. E. (1991). Chronic hyperglycemic diabetes in the rat is associated with a selective impairment of cerebral vasodilatory responses. *J. Cereb. Blood Flow Metab.* **11,** 667–677.

Peng, L., Swanson, R. A., and Hertz, L. (2001). Effects of L-glutamate, D-aspartate, and monensin on glycolytic and oxidative glucose metabolism in mouse astrocyte cultures: Further evidence that glutamate uptake is metabolically driven by oxidative metabolism. *Neurochem. Int.* **38,** 437–443.

Perry, T. L. (1982). Cerebral amino acid pools. *In* "Handbook of Neurochemistry" (A. Lajtha, ed.), Vol. 1, Chemical and Cellular Architecture, pp. 151–180. Plenum Press, New York.

Phillis, J. W. (1995). CI-966, a GABA uptake inhibitor, antagonizes ischemia-induced neuronal degeneration in the gerbil. *Gen. Pharmacol.* **26,** 1061–1064.

Phillis, J. W. (1996). Cerebroprotective action of the phospholipase inhibitor quinacrine in the ischemia/reperfused gerbil hippocampus. *Life Sci.* **58,** PL97–PL101.

Phillis, J. W. (1997). Oxygen free radical involvement in cerebral ischemia/reperfusion injury. *In* "Oxygen Radicals and the Disease Process" (C. E. Thomas and B. Kalyanaraman, eds.), pp. 15–39. Harwood Academic Publishers, Amsterdam.

Phillis, J. W., Estevez, A. Y., Guyot, L. L., and O'Regan, M. H. (1999a). 5-(*N*-ethyl-*N*-isopropyl)-amiloride, a Na^+-H^+ exchange inhibitor, protects gerbil hippocampal neurons from ischemic injury. *Brain Res.* **839,** 199–202.

Phillis, J. W., and O'Regan, M. H. (1996). Mechanisms of glutamate and aspartate release in the ischemic rat cerebral cortex. *Brain Res.* **730,** 150–164.

Phillis, J. W., O'Regan, M. H., Estevez, A. Y., Song, D., and Vanderheide, S. J. (1996). Cerebral energy metabolism during severe ischemia of varying duration and following reperfusion. *J. Neurochem.* **67,** 1525–1531.

Phillis, J. W., O'Regan, M. H., and Song, D. (1998a). 5-(*N*-ethyl-*N*-isopropyl)-amiloride inhibits amino acid release from the ischemic rat cerebral cortex: role of Na^+-H^+ exchange. *Brain Res.* **812,** 297–300.

Phillis, J. W., Ren, J., and O'Regan, M. H. (2000a). Transporter reversal as a mechanism of glutamate release from the ischemic rat cerebral cortex: Studies with DL-*threo*-β-benzyloxyaspartate. *Brain Res.* **868,** 105–112.

Phillis, J. W., Ren, J., and O'Regan, M. H. (2000b). Inhibition of Na^+/H^+ exchange by 5-(*N*-ethyl-*N*-isopropyl)-amiloride reduces free fatty acid efflux from the ischemic reperfused rat cerebral cortex. *Brain Res.* **884,** 155–162.

Phillis, J. W., Simpson, R. E., and Walter, G. A. (1990). The effect of hyperglycemia on extracellular levels of adenosine in the hypoxic rat cerebral cortex. *Brain Res.* **524,** 336–338.

Phillis, J. W., Smith-Barbour, M., and O'Regan, M. H. (1996). Changes in extracellular amino acid neurotransmitters and purines during and following ischemias of different durations in the rat cerebral cortex. *Neurochem. Int.* **29,** 115–120.

Phillis, J. W., Smith-Barbour, M., Perkins, L. M., and O'Regan, M. H. (1994a). Indomethacin modulates ischemia-evoked release of glutamate and adenosine from the rat cerebral cortex. *Brain Res.* **652,** 353–356.

Phillis, J. W., Smith-Barbour, M., Perkins, L. M., and O'Regan, M. H. (1994b). Characterization of glutamate, aspartate, and GABA release from the ischemic rat cerebral cortex. *Brain Res. Bull.* **34,** 457–466.

Phillis, J. W., Song, D., Guyot, L. L., and O'Regan, M. H. (1999b). Lactate reduces amino acid release and fuels recovery of function in the ischemic brain. *Neurosci. Lett.* **272,** 195–198.

Phillis, J. W., Song, D., and O'Regan, M. H. (1997). Inhibition by anion channel blockers of ischemia-evoked release of excitotoxic and other amino acids from rat cerebral cortex. *Brain Res.* **758,** 9–16.

Phillis, J. W., Song, D., and O'Regan, M. H. (1998b). Tamoxifen, a chloride channel blocker, reduces glutamate and aspartate release from the ischemic cerebral cortex. *Brain Res.* **780,** 352–355.

Phillis, J. W., Song, D., and O'Regan, M. H. (1999c). Melittin enhances amino acid and free fatty acid release from the *in vivo* cerebral cortex. *Brain Res.* **847,** 270–275.

Phillis, J. W., Song, D., and O'Regan, M. H. (1999d). Effect of hyperglycemia on extracellular levels of amino acids and free fatty acids in the ischemic/reperfused rat cerebral cortex. *Brain Res.* **837,** 177–183.

Pilitsis, J. G., Diaz, F. G., O'Regan, M. H., and Phillis, J. W. (2001a). Inhibition of Na^+/H^+ exchange by SM-20220 attenuates free fatty acid efflux in rat cerebral cortex during ischemia-reperfusion injury. *Brain Res.* **913,** 156–158.

Pilitsis, J. G., Diaz, F. G., O'Regan, M. H., and Phillis, J. W. (2001b). Inhibition of Na^+/Ca^{++} exchange by KB-R7943, a novel selective antagonist, attenuates phosphoethanolamine and free fatty acid efflux in rat cerebral cortex during ischemia-reperfusion injury. *Brain Res.* **916,** 192–198.

Prado, R., Ginsberg, M. D., Dietrich, W. D., Watson, B. D., and Busto, R. (1988). Hyperglycemia increases infarct size in collaterally perfused but not end-arterial vascular territories. *J. Cereb. Blood Flow Metab.* **8,** 186–192.

Prichard, J., Rothman, D., Novotny, E., Petroff, O., Kuwabara, T., Avison, M., Howseman, A., Hanstock, C. C., and Shulman, R. G. (1991). Lactate rise detected by ^1H NMR in human visual cortex during physiological stimulation. *Proc. Nat. Acad. Sci. USA* **88,** 5829–5831.

Proverbio, F., and Hoffman, J. F. (1977). Membrane compartmentalized ATP and its preferential use by the Na,K-ATPase of human red cell ghosts. *J. Gen. Physiol.* **69,** 605–632.

Pulsinelli, W. A., and Duffy, T. E. (1983). Regional energy balance in rat brain after transient forebrain ischemia. *J. Neurochem.* **40,** 1500–1503.

Pulsinelli, W. A., Levy, D. E., Sigsbee, B., Scherer, P., and Plum, F. (1983). Increased damage after ischemic stroke in patients with hyperglycemia with or without established diabetes mellitus. *Am. J. Med.* **74,** 540–544.

Pulsinelli, W. A., Waldman, S., Rawlinson, D., and Plum, F. (1982). Moderate hyperglycemia augments ischemic brain damage: a neuropathologic study in the rat. *Neurology* **32,** 1239–1246.

Qian, T., Nieminen, A.-L., Herman, B., and Lemasters, J. J. (1997). Mitochondrial permeability transition in pH-dependent reperfusion injury to rat hepatocytes. *Am. J. Physiol.* **273,** C1783–C1792.

Raffin, C. N., Rosenthal, M., Busto, R., and Sick, T. J. (1992). Glycolysis, oxidative metabolism and brain potassium ion clearance. *J. Cereb. Blood Flow Metab.* **12,** 34–42.

Rakieten, N., Rakieten, M. L., and Nadkarni, M. V. (1963). Studies on the diabetogenic action of streptozotocin (NSC-37917). *Cancer Chemother. Rep.* **29,** 91–98.

Rehncrona, S. (1985). Brain acidosis. *Ann. Emerg. Med.* **14,** 770–776.

Rehncrona, S., Rosen, I., and Siesjo, B. K. (1981). Brain lactic acidosis and ischemic cell damage: I. Biochemistry and neurophysiology. *J. Cereb. Blood Flow Metab.* **1,** 297–311.

Roberts, E. L. (1993). Glycolysis and recovery of potassium ion homeostasis and synaptic transmission in hippocampal slices after anoxia or stimulated potassium release. *Brain Res.* **620,** 251–258.

Rothman, S. M., and Olney, J. W. (1986). Glutamate and the pathophysiology of hypoxic/ischemic brain damage. *Ann. Neurol.* **19,** 105–111.

Rothstein, J. D., Martin, L., Levey, A. I., Dykes-Hoberg, M., Jin, L., Wu, D., Nash, N., and Kuncl, R. W. (1994). Localization of neuronal and glial glutamate transporters. *Neuron* **13,** 713–725.

Roy, M., and Sapolsky, R. (1999). Neuronal apoptosis in acute necrotic insults: Why is this subject such a mess? *Trends Neurosci.* **22,** 419–422.

Saluja, I., O'Regan, M. H., Song, D., and Phillis, J. W. (1999). Activation of $cPLA_2$, PKC and ERKs in the rat cerebral cortex during ischemia/reperfusion. *Neurochem. Res.* **24,** 669–677.

Sanchez-Olea, R., Morales-Mulia, M., Moran, J., and Pasantes-Morales, H. (1995). Inhibition by polyunsaturated fatty acids of cell volume regulation and osmolyte fluxes in astrocytes. *Am. J. Physiol.* **269,** C96–C102.

Sapolsky, R. M., Trafton, J., and Tombaugh, G. C. (1996). Excitotoxic neuron death, acidotic endangerment, and the paradox of acidotic protection. *In* "Cellular and Molecular Mechanisms of Ischemic Brain Damage" (B. K. Siesjo and T. Wieloch, ed.), pp. 237–244. Lippincott-Raven Publishers, Philadelphia.

Sappey-Marinier, D., Calabrese, G., Fein, G., Hugg, J. W., Biggins, C., and Weiner, M. W. (1992). Effect of photic stimulation on human visual cortex lactate and phosphates using ^1H and ^{31}P magnetic resonance spectroscopy. *J. Cereb. Blood Flow Metab.* **12,** 584–592.

Schafer, M., and Erdo, S. L. (1992). Insulin-specific sensitization of cultured cortical neurons to glutamate toxicity. *Brain Res.* **580,** 331–333.

Schroder, U. H., Breder, J., Sabelhaus, C. F., and Reymann, K. G. (2001). The novel Na^+/Ca^{2+} exchange inhibitor KB-R7943 protects CA1 neurons in rat hippocampus slices against hypoxic/hypoglycemic injury. *Neuropharmacology* **38,** 319–321.

Schulingkamp, R. J., Pagano, T. C., Hung, D., and Raffa, R. B. (2000). Insulin receptors and insulin action in the brain: review and clinical implications. *Neurosci. Biobehav. Rev.* **24,** 855–872.

Schurr, A., Payne, R. S., Miller, J. J., and Rigor, B. M. (1997). Brain lactate, not glucose, fuels the recovery of synaptic function from hypoxia upon reoxygenation: An *in vitro* study. *Brain Res.* **744,** 105–111.

Seki, Y., Feustel, P. J., Keller, R. W., Tranmer, B. I., and Kimelberg, H. K. (1999). Inhibition of ischemia-induced glutamate release in rat striatum by dihydrokainate and an anion channel blocker. *Stroke* **30,** 433–440.

Shigekawa, M., and Iwamoto, T. (2001). Cardiac Na^+-Ca^{2+} exchange: Molecular and pharmacological aspects. *Circ. Res.* **88,** 864–876.

Shimamoto, K., Lebrun, B., Yasuda-Katamani, Y., Sakaitani, M., Shigeri, Y., Yumoto, N., and Nakajima, T. (1998). DL-*threo*-β-Benzyloxyaspartate, a potent blocker of excitatory amino acid transporters. *Mol. Pharmacol.* **53,** 195–201.

Sibson, N. R., Dhankhar, A., Mason, G. F., Rothman, D.L., Behar, K. L., and Shulman, R. G. (1998). Stoichiometric coupling of brain glucose metabolism and glutamatergic neuronal activity. *Proc. Nat. Acad. Sci. USA* **95,** 316–321, 1998.

Siesjo, B. K. (1988). Acidosis and ischemic brain damage. *Neurochem. Pathol.* **9,** 31–88.

Siesjo, B. K. (1992). Pathophysiology and treatment of focal cerebral ischemia. Part I. Pathophysiology. *J. Neurosurg.* **77,** 169–184.

Silver, I. A., Deas, J., and Erecinska, M. (1997). Ion homeostasis in brain cells: Differences in intracellular ion responses to energy limitation between cultured neurons and glial cells. *Neuroscience* **78,** 589–601.

Simpson, R. E., Phillis, J. W., and Buchanan, J. (1990). A comparison of cerebral blood flow during basal, hypotensive, hypoxic and hypercapnic conditions between normal and streptozotocin diabetic rats. *Brain Res.* **531,** 136–142.

Smith, M. L., Von Hanwehr, R., and Siesjo, B. K. (1986). Changes in extra- and intracellular pH in the brain following ischemia in hyperglycemic and in moderately hypoglycemic rats. *J. Cereb. Blood Flow Metab.* **6,** 574–583.

Smith, M. L., Kalimp, H., Warner, D. S., and Siesjo, B. K. (1988). Morphological lesions in the brain preceding the development of postischemic seizures. *Acta Neuropathol.* (*Berl.*) **76,** 253–264.

Solis, J. M., Herranz, A. S., Herreras, O., Lerma, J., and Martin Del Rio, R. (1988). Does taurine act as an osmoregulatory substance in the rat brain? *Neurosci. Lett.* **91,** 53–58.

Staub, F., Winkler, A., Haberstok, J., Plesnila, N., Peters, J., Chang, R. C. C., Kempski, O., and Baethmann, A. (1996). Swelling, intracellular acidosis, and damage of glial cells. *Acta Neurochir.* [Suppl.] **66,** 56–62.

Takagi, K., Ginsberg, M. D., Globus, M. Y.-T., Dietrich, W. D., Martinez, E., Kraydieh, S., and Busto, R. (1993). Changes in amino acid neurotransmitters and cerebral blood flow in the ischemic penumbral region following middle cerebral artery occlusion in the rat: Correlation with histopathology. *J. Cereb. Blood Flow Metab.* **13,** 575–585.

Tanaka, E., Yamamoto, S., Inokuchi, H., Isagai, T., and Higashi, H. (1999). Membrane dysfunction induced by *in vitro* ischemia in rat hippocampal CA1 pyramidal neurons. *J. Neurophysiol.* **81,** 1872–1880.

Tansey, F. A., Farooq, M., and Cammer, W. (1991). Glutamine synthetase in oligodendrocytes and astrocytes: New biochemical and immunocytochemical evidence. *J. Neurochem.* **56,** 266–272.

Tombaugh, G. C., and Sapolsky, R. M. (1990). Mild acidosis protects hippocampal neurons from injury induced by oxygen and glucose deprivation. *Brain Res.* **506,** 343–345.

Uchino, H., Elmer, E., Uchino, K., Lindvall, O., and Siesjo, B. K. (1995). Cyclosporin A dramatically ameliorates CA1 hippocampal damage following transient forebrain ischemia in the rat. *Acta Physiol. Scand.* **155,** 469–471.

Umemura, A. (1990). Regional difference in free fatty acids release and the action of phospholipase during ischemia in rat brain. *No To Shinkei* **42,** 979–986.

Umemura, A., Mabe, H., Nagai, H., and Sugino, F. (1992). Action of phospholipases A_2 and C on free fatty acid release during complete ischemia in rat neocortex. Effect of phospholipase C inhibitor and *N*-methyl-D-aspartate antagonist. *J. Neurosurg.* **76,** 648–651.

Van Den Berg, C. J., and Garfinkel, D. (1971). A stimulation study of brain compartments. Metabolism of glutamate and related substances in mouse brain. *Biochem. J.* **123,** 211–218.

Voll, C. L., and Auer, R. N. (1991). Insulin attenuates ischemic brain damage independent of its hypoglycemic effect. *J. Cereb. Blood Flow Metab.* **11,** 1006–1014.

Voronov, J., Thomas, A., and Jo, D. (1996). Protective effects of extracellular acidosis and blockade of sodium/hydrogen exchange during recovery from metabolic inhibition in neuronal tissue culture. *J. Neurochem.* **67,** 2379–2389.

Wagner, K. R., Kleinholz, M., de Courten-Myers, G., and Myers, R. E. (1992). Hyperglycemia versus normoglycemic stroke: Topography of brain metabolites, intracellular pH, and infarct size. *J. Cereb. Blood Flow Metab.* **12,** 213–222.

Wan, Q., Xiong, Z. G., Man, H. Y., Ackerley, C. A., Braunton, J., Lu, W. Y., Becker, L. E., MacDonald, J. F., and Wang, Y. T. (1997). Recruitment of functional $GABA_A$ receptors on postsynaptic domains by insulin. *Nature* **388,** 636–640.

Wang, H., Harrison-Shostak, D. C., Lemasters, J. J., and Herman, B. (1996). Contribution of pH-dependent group II phospholipase A_2 to chemical hypoxic injury in rat hepatocytes. *FASEB J.* **10,** 1318–1325.

Warner, D. S., Gionet, T. X., Todd, M. M., and McAllister, A. M. (1992). Insulin-induced hypoglycemia improves ischemic outcome in hyperglycemic rats. *Stroke* **23,** 1775–1781.

Weiss, J. N., and Lamp, S. T. (1989). Cardiac ATP-sensitive K^+ channels. Evidence for preferential regulation by glycolysis. *J. Gen. Physiol.* **94,** 911–935.

Zaaslow, M. A., Pearl, R. G., Schuer, L. M., Steinberg, G. K., Lieberson, R. E., and Larson, C. P. (1989). Hyperglycemia decreases acute neuronal ischemic changes after middle cerebral artery occlusion in cats. *Stroke* **20,** 519–523.

Zerangue, N., and Kavanaugh, M. P. (1996). Flux coupling in a neuronal glutamate transporter. *Nature* **383,** 634–637.

DIABETES MELLITUS AND THE CENTRAL NERVOUS SYSTEM

Anthony L. McCall

James M. Moss Professor in Diabetes in Internal Medicine
University of Virginia School of Medicine, Division of Endocrinology & Metabolism
Department of Internal Medicine, UVA Health System
Charlottesville, Virginia 22908

0074-7742/02 $35.00

I. Introduction

Despite the traditional view that the brain is unaffected by insulin or its lack, diabetes mellitus (DM) disturbs central nervous system (CNS) function and may injure the brain (Mooradian, 1988, 1993, 1997, 1997a; McCall, 1992a,b, 1993, 1997, 1998; McCall and Figlewicz, 1997; Bhardwaj *et al.*, 1999; Biessels, 1999; Kumagai, 1999). An emerging body of literature has shown there are manifold biological effects of diabetes on the brain. Insulin action on the brain influences neurotransmission. Diabetes increases the risk of stroke and exacerbates stroke damage and mortality (Kuller, 1995; Bell, 1994). Stroke damage is worsened potentially both through the effects of hyperglycemia and lack of insulin upon brain biochemistry in the presence of ischemia.

The major treatment side effect of diabetes, which is hypoglycemia, may acutely impair brain function and, when severe and prolonged, may permanently injure the brain largely through excitotoxic mechanisms or cause death. Physiological adaptations of blood-brain barrier (BBB) function, brain metabolism, cerebral blood flow, and microvascular function may occur with either hyper- or hypoglycemia. One additional basis for altered brain function with diabetes or hypoglycemia is altered neurotransmitter metabolism. Diabetes may increase the risk of epilepsy because of altered metabolism or damage from several possible mechanisms. Intellectual function is adversely affected, albeit often subtly, in uncontrolled diabetes or marked hypoglycemia. Future research directions should delineate new areas and mechanisms of abnormalities in brain function in diabetes. Future research should explore the basis for CNS functional alterations in acute and chronic hyper- and hypoglycemia. It would be desirable to initiate clinical trials to limit the adverse CNS effects of diabetes on stroke. A better understanding of how hyper- and hypoglycemia impair brain function may help spare the brain from the harm of altered metabolism in diabetes.

This chapter focuses on clinically relevant pathophysiology. It deals with studies that relate altered insulin and glucose concentrations to both short- and long-term adverse consequences to the brain.

II. DM and Stroke

A. Clinical Issues: DM and Stroke Risk

Diabetes mellitus increases the risk of cerebrovascular disease (see Kuller, 1995; Bell, 1994 for reviews). Specifically, DM increases the risk of nonhemorrhagic stroke in most estimates between two- and sixfold. Both atherothrombotic brain infarction and lacunar stroke are increased by diabetes. Diabetes is the second leading cause of stroke after hypertension. Overall, 9.3% of surveyed diabetic patients over age 18 report a medical history of stroke with increased risk at all ages. Diabetes is thought to account for ~7% of stroke deaths. At autopsy, up to one quarter of patients with diabetes have evidence of cerebrovascular disease.

Stroke prevalence is higher both with recognized and unrecognized diabetes. The current prevalence of diabetes in the U.S. is about 16 million with the proportion of unrecognized cases about one-third of these. Epidemiological and clinical trial data indicate that diabetes is an important cause of stroke, particularly for the most common type, atherothrombotic brain infarction. For example, in the Multiple Risk Factor Intervention Trial (MRFIT), the relative risk (RR) for this type of stroke was increased 3.8-fold in men (Neaton *et al.,* 1993).

The risk of lacunar infarction, a type of stroke that is very small and not always clinically apparent, is also increased, especially when diabetes and hypertension are combined (Arboix *et al.,* 1990). In most studies, there is no increased risk for hemorrhagic stroke (Bell, 1994; Kuller, 1995). There may be, however, a slight increase in the risk of stroke with hemorrhagic transformation, which represents an indicator of more severe infarction. Nonetheless, there is probably no increase in subarachnoid hemorrhage because of diabetes, especially when the risk of hypertension itself, a common accompaniment of diabetes, is taken into account. Based on the Nurses Health Study, overall stroke risk may be even higher in women than in men with diabetes (4.1-fold overall) (Manson, 1991). Interestingly, in at least one rodent stroke study of the influence of diabetes, female mice were less susceptible to stroke damage than male mice, although diabetes increased stroke damage more than in control animals (Vannucci, 2001).

B. DM and Stroke Damage

Stroke damage and its associated morbidity and mortality are also increased in diabetes. For example, in the MRFIT study a 2.8-fold increased risk of death occurred among those with diabetes even after adjustment for other known risk factors (Neaton, 1993). The reasons why stroke damage is increased in diabetes are probably multiple, but appear to exist beyond simply the presence of multiple risk factors for vascular disease that commonly cluster in those with diabetes (see Table I).

TABLE I
POSSIBLE FACTORS INCREASING STROKE RISK AND PROGNOSIS IN DIABETES

Hyperglycemia	Hyperlipidemia
Insulin resistance	Inflammatory cytokines
Preexisting atherosclerosis	Procoagulant state
Proteinuria	Accelerated hypertension
Increased CNS vulnerability to damage	Autonomic neuropathy Ethnicity

C. Importance of Hyperglycemia

Hyperglycemia itself, even when relatively modest, such as in the high normal, i.e., nondiabetic range, is an important risk factor for stroke occurrence and for prognosis in acute stroke. Modestly elevated blood glucose after a glucose challenge has been shown to be a predictor of stroke (Bell, 1994). One reason for the predictive power of slight glucose elevations is related to this as an indicator of the metabolic syndrome of insulin resistance (Pyorala, 2000). Increased glycemia after oral glucose tolerance testing, usually termed impaired glucose tolerance (or IGT), typically precedes fasting hyperglycemia, and may be a more reliable predictor of vascular disease risk than fasting hyperglycemia (DECODE Study Group, 1999).

In patients undergoing an acute stroke, slight elevations of random glucose levels predict a larger size of cerebral infarction, a greater morbidity, an increased tendency to cerebral edema with acute infarction, and ultimately cause greater mortality (Berger and Hakim, 1986). Blood glucose values greater than 120 mg/dl (6.7 mM) have been reported to be associated with a markedly poorer prognosis in acute stroke. (Pulsinelli *et al.,* 1983). Hyperglycemia during stroke may simply be an index of severe metabolic and hormonal stress responses, which are extreme because of massive stroke damage. Thus, hyperglycemia might not necessarily be directly implicated in stroke morbidity mechanistically (Capes *et al.,* 2001). Thoughtful analysis and considerable work from animal models suggest, however, that hyperglycemia may directly or indirectly contribute to exacerbation of stroke damage (Helgason, 1988; Auer *et al.,* 1999).

D. How Hyperglycemia Relates to Stroke Damage

How does hyperglycemia relate to poor stroke prognosis? As indicated above, the first possibility is that this worse prognosis may indicate a stress response with stress hyperglycemia that is related to the size of the cerebral infarction. A large infarction causes greater hormonal stress responses that in turn exacerbate hyperglycemia (Woo *et al.,* 1988).

A second possibility is that hyperglycemia associated with poorer stroke prognosis may instead indicate that unrecognized DM exists with its accompanying multiple cardiovascular risk factors and greater extent of vascular disease (Riddle and Hart, 1982). The most practical way to assess whether diabetes is preexisting is to measure glycation of hemoglobin, typically hemoglobin A1 (which represents total glycated hemoglobin) or hemoglobin A1c (a subfraction). These glycated hemoglobins only become increased significantly when glucose levels have been elevated for weeks to months. The prevalence of unrecognized diabetes mellitus in various studies of first stroke or transient ischemic attacks (TIA) ranges between 7 and 40%. Thus, a search for unrecognized DM is important in cerebral ischemia, especially when there is also hyperglycemia.

A third possibility to explain the association of hyperglycemia and stroke damage is that there is a mechanistic association whereby hyperglycemia itself (or a closely allied metabolic factor) worsens stroke prognosis. An elevated blood glucose level may worsen stroke damage through direct effects, including through links to its associated metabolic abnormalities (e.g., acidotoxicity) and the ability to alter brain glucose metabolism in ischemia. High glucose levels per se may worsen metabolic abnormalities such as accumulation of lactate generated anaerobically within the ischemic brain. By doing so, they are associated with increased necrosis. Moreover, lack of insulin is also implicated as an important risk factor for stroke damage. Insulin acting through insulin-like growth factor (IGF-1) receptors may enhance survival of brain neurons in the face of certain types of cerebral ischemia (Auer *et al.,* 1999).

E. Hyperglycemia Effects Dependent Upon Type of Ischemia

Exacerbation of stroke damage by hyperglycemia probably differs depending upon the type and location of ischemia (Helgason, 1988). For example, in animal models of ischemia, which permit an examination of the effects of hyperglycemia absent confounding risk factors associated with human hyperglycemia (e.g., the metabolic syndrome of insulin resistance), global ischemia is severely worsened by either acute or chronic hyperglycemia (Auer *et al.,* 1999). Global ischemia is a type of stroke caused by widely reduced cerebral blood flow. It tends to manifest damage selectively in certain brain regions such as the hippocampus CA1 subfield. Histologically, selective neuronal necrosis with histological sparing of vascular and glial elements is usually observed. In humans, the most common causes of global ischemia are related to cardiac bypass surgery, severe hypotension, or cardiac arrest. In animal models, a combination of temporary clamping of arteries supplying the brain with the addition withdrawal of blood to produce systemic hypotension is a frequent model used to create global cerebral ischemia. Hyperglycemic exacerbation of stroke damage in global ischemia may be ameliorated significantly by systemic administration of insulin.

Work by Auer and colleagues (1999) suggests that the beneficial effect of insulin on stroke is partly due to correction of hyperglycemia. In addition, however, insulin's benefits may at least in significant measure be related to the effect of insulin per se and mediated through the IGF receptors in the brain (Voll and Auer, 1991).

In animal models of focal brain ischemia, such as middle cerebral artery infarction after ligation or embolization, which are similar to the common types of human atherothrombotic brain infarction, hyperglycemia also may exacerbate stroke damage. Insulin administration to reduce hyperglycemia has beneficial effects upon histological stroke damage. In contrast to global ischemia, Auer *et al.* (1999) suggests that the most important beneficial effect of insulin may be directly related to reversing the adverse effects of hyperglycemia, as neuroprotection is largely prevented by co-administration of glucose (Hamilton *et al.,* 1995). An important mitigating factor for the potential benefit for insulin administration and its glucose lowering in focal ischemia is that insulin overtreatment could reduce fuel to brain areas already compromised by partial ischemia. In this situation, which occurs particularly in the penumbral area of ischemia, marked hypoglycemia might exert a detrimental effect upon ischemic recovery.

The anaerobic metabolism of excess glucose (in hyperglycemic conditions) to lactic acid in the presence of partial blood flow to ischemic brain and tissue hypoxia is thought to be important for exacerbation of ischemic brain damage. This type of damage is sometimes called acidotoxicity. The accumulation of lactic acid both intra- and extracellularly in this kind of situation may contribute substantially to an extension of damage (pannecrosis) and lead to greater changes in both ionic dysequilibrium (K^+, Ca^{2+} and defective protein synthesis in the ischemic region (Auer and Siesjo, 1988).

Many possible mechanisms may exist to explain clinically why DM and hyperglycemia exacerbate stroke damage. Multiple cardiovascular risk factors associated with DM besides hyperglycemia and lack of insulin may be important factors. Coexisting hypertension, hypercoagulability, chronic changes in cytokines or serum protein patterns associated with chronic inflammation, proteinuria, hyperlipidemia, endothelial dysfunction, the existence of chronic diabetic complications such as neuropathy, retinopathy, and nephropathy may all play important roles.

In animal models of stroke with acute hyperglycemia one commonly observed abnormality is the increased prevalence of stroke-related generalized convulsions. Both stroke and seizures cause brain damage in part through excitotoxic amino acid release (i.e., glutamate, or aspartate release often with reduced energy-dependent synaptic reuptake). Presumably, since seizures are well known by themselves to increase brain metabolism, accelerating depletion of high-energy phosphate compounds and increasing the potentially damaging excitatory neurotransmitters, they may also directly contribute to brain damage. Even more likely one suspects that the combined insults of stroke and seizure exhibit cumulative or interactive contribution to excitotoxic damage.

The excitotoxic damage that may exist to the brain during stroke and hyperglycemia could be worsened by mechanisms that have yet to be tested involving clearance of excitatory amino acids (EAA) by specific transport proteins present in the brain. Thus, excessive release of EAA, reduced reuptake (that if normal would ordinarily detoxify the hyperstimulation of postsynaptic neurons), or some combination of such mechanisms is possible in brain ischemia during diabetes.

III. Hypoglycemia and the Brain

A. Glucose as the Primary Fuel for Brain

To promote normal function and energy metabolism, the brain is exquisitely dependent upon a continuing supply of glucose, which must transfer across the BBB to have access to neurons and glia (Clarke and Sokoloff, 1999). The primary use of high-energy phosphate compounds, such as creatine phosphate and adenosine triphosphate, in the brain is presumably to support synaptic transmission and its related, energy-dependent processes such as restoration of ionic equilibrium and neurotransmitter reuptake (axons, dendrites, and synapses). A significant proportion of fuel use by the brain occurs in the perikarya, however. Neuronal dysfunction in the form of abnormal latency of evoked potentials (e.g., P300) after visual or auditory stimuli occurs at values approximating 3 mmol/l (54 mg/dl) in nondiabetic adults (Mitrakou *et al.*, 1991; Jones *et al.*, 1990; Blackman *et al.*, 1990).

Glucose is sparingly lipid soluble and does not efficiently penetrate cellular barriers, such as the BBB, which is anatomically made up of the zonula occludens of the microvascular endothelial cells of the brain capillaries (Brightman *et al.*, 1970). Simple diffusion of glucose is not enough to supply adequate nutrition to the brain. To transfer adequate glucose for brain energy needs across the BBB and across the cellular barriers of brain parenchymal cells such as neurons and glia, facilitated diffusion via protein carriers is required. These carriers exist as a family of proteins with a number of isoforms of glucose transporter proteins (commonly called GLUTs) that facilitate transfer of glucose into the brain and into brain cells (Vannucci *et al.,* 1997). Recent work indicates that the superfamily of glucose transport proteins has a number of new and as yet incompletely characterized members, some of which are clearly expressed in brain (Joost and Thorens, 2001). GLUT8 appears to be susceptible to regulation by uncontrolled diabetes and stress as shown by Reagan and colleagues (2001).

B. Glucose Transporter Family

GLUT1 is the primary transport protein present in the cerebral microvasculature, which constitutes the BBB. It is also expressed widely in glial cells, especially astroglia. GLUT3 is thought to be the primary neuronal transporter (present in neuropil, but GLUT8 also is expressed in neurons, especially all cell bodies of principal neurons (Reagan *et al.,* 2001). Other transporters for glucose have been reported in the BBB as well, but their importance in normal situations and in situations of fuel deprivation to the brain is not clear at present. GLUT4 and GLUT5 are expressed in brain cells in animals (McCall *et al.,* 1997). The abundance of GLUT4 may be too low to be of physiological importance in brain endothelium and BBB function. The inefficiency as a glucose transporter for GLUT5, which appears to function more as a fructose transporter than a glucose transporter, makes this isoform an unlikely source of significant glucose availability to the brain (Shephard *et al.,* 1992). Moreover, GLUT5 is primarily expressed in brain microglial cells (Vannucci *et al.,* 1997). In addition to these well-characterized isoforms, there are several recently cloned isoforms. Their nomenclature has sometimes changed from that originally used. These include GLUTx1 (also called GLUT8), GLUT9 (now called GLUT6), GLUT10, GLUT11, and GLUT12. The role of GLUT8, which is a hormonally regulated neuronal transporter, is most clearly relevant to diabetes and the brain (Reagan *et al.,* 2001). Whether it will turn out to be an important provider of glucose to neurons as preliminary studies indicate will become clearer with additional research.

C. When Glucose Transport Limits Cerebral Energy Metabolism

Normally the transport of glucose into the brain is not the rate-limiting step in cerebral energy metabolism, despite the predominance of glucose as the normal energy source of the brain (Pardridge, 1983). Instead, type 1 hexokinase is probably more important because transport rates for glucose usually exceed rates of glucose metabolism by a factor of about 2. Under specialized circumstances, however, glucose transport may become the rate-limiting step in cerebral energy metabolism.

It is believed that BBB transport is likely to be rate-limiting step for brain energy metabolism in hypoglycemia. In addition, it could be important in other situations of decreased fuel supply, such as cerebral ischemia (whether global or focal). Transport would presumably be rate-limiting for metabolism only in regions of low (but not absent) cerebral blood flow (CBF), such as the ischemic penumbra. Moreover, there are other conditions in which there is brain hypermetabolism, such as with epileptic seizures, where the local energy metabolism of the brain is so accelerated that the supply of glucose across the BBB then could become rate-limiting for energy metabolism. The upregulation of GLUT1 and GLUT3 expression with seizures is consistent with this notion. Presumably, lack of adequate supply of energy is one of the more important reasons why seizures may also damage the brain.

A more controversial notion is that transport of glucose across cellular barriers within the brain, such as neuronal or glial membranes, could also be important for determining cellular fuel availability. This is a possibility that is raised by the association of hypometabolism of the brain and altered glucose transporter expression (Rapoport, 1997; Hoyer, 1994; Simpson *et al.,* 1994). The improvement in cognitive performance of demented subjects with glucose administration also is consistent with this possibility, especially as microdialysis studies from brain suggest that depletion of local hippocampal glucose concentrations may occur during cognitive testing (Korol and Gold, 1998; Gold, 1995). In dementing illness, it has been repeatedly been shown that glucose hypometabolism occurs in affected brain areas (Hoyer, 1994). Moreover, the expression of glucose transporters in human brain are reduced in such diseases as Alzheimer disease even after correction for a decreased synaptic density as shown by Simpson and Davies (Simpson *et al.,* 1994; Vannucci *et al.,* 1998). This issue remains controversial (Pardridge 1994), but nonetheless, it is intriguing to note that there is a significant parallel between the metabolic rate of areas of the brain and the capacity for glucose transport as measured by the density of the then known transporters, such as GLUT1 and GLUT3. Whether there will need to be further revisions to this correlative analysis with the more precise determination of transporter proteins for more recently recognized isoforms, such as GLUT8 (also called GLUTx1) and GLUT9 (now called GLUT6) or other isoforms cloned recently, based upon an analysis of conserved GLUT sequences, remains to be determined.

D. Chronic Hypoglycemia and Brain Glucose Transport

Chronic or recurrent hypoglycemia is an important issue related to the match of glucose transport and brain metabolism. There are both important physiological issues to be raised and the ramifications of these physiological issues affect the clinical treatment of DM.

The clinical interest in treatment of DM to lower blood glucose concentrations into the normal or near normal range has been responsible for much of the interest in the brain's adaptation to hypoglycemia. Nearly normal glucose levels clearly protect people with diabetes from the chronic complications of diabetes. These complications include retinopathy (a leading cause of blindness), nephropathy (the leading cause of kidney failure), and peripheral neuropathy that may lead to amputations because of loss of protective pain sensation.

E. Hypoglycemia Unawareness

Many patients with treated diabetes, particularly if diabetes is tightly controlled with insulin, do not undergo the typical sympathoadrenal (autonomic) discharge and related symptoms (shakiness, palpitations, sweating, etc.) with acute hypoglycemia (Cryer, 1999). This is called hypoglycemia unawareness; it may be defined as an inability to recognize hypoglycemia at an early stage that permits self-treatment. The existence of hypoglycemia unawareness is a major problem for the therapy of type 1 (previously called insulin-dependent) DM because perfect physiological replacement of insulin is impossible currently. When control is intensified to reduce chronic diabetes complications, frequent monitoring indicates that hypoglycemia is very frequent (Arias, 1985). In long-term studies of diabetes control in type 1 diabetes, a threefold increased risk of severe hypoglycemia occurs (DCCT Study Group, 1993, 1997). Hypoglycemia unawareness markedly increases the risk of severe hypoglycemia, a condition defined as requiring external help to recover and which sometimes severely damages the brain.

F. Defective Hypoglycemia Counterregulation

Often accompanying hypoglycemia unawareness is a second form of acquired autonomic neuropathy caused by hypoglycemia that is termed defective hypoglycemic counterregulation (White, 1983). Occasionally, defective counterregulation is due to severe autonomic neuropathy, a relatively fixed defect that prevents recognition and hormonal responses to acute hypoglycemia. More commonly, defective counterregulation is an acquired defect that occurs because of prior moderate, repeated hypoglycemia.

Normally, acute hypoglycemia causes an outpouring of stress hormone responses including the catecholamines, norepinephrine and epinephrine, cortisol, glucagon, and growth hormone. These hormone defenses, especially the catecholamines, generate autonomic symptoms that allow recognition and self-treatment. They also help to raise blood glucose levels back toward the normal range.

Hypoglycemia unawareness and defective hypoglycemia counterregulation are important predictors of severe hypoglycemia (Gold *et al.,* 1994; White, 1983). The brain is both the most vulnerable organ from hypoglycemia damage and is clearly responsible for organizing the signal of distress (through control of autonomic activation) to the rest of the body (McCall, 1997).

G. Sensing of Glucose Deprivation by the Brain

When acute hypoglycemia occurs, it is likely that a variety of areas of the brain specifically sense glucose lack (Frizell *et al.,* 1993). One particular region, the ventromedial nucleus of the hypothalamus, has come to be the focus of considerable interest. In a series of elegant studies, Shulman and colleagues have shown that this region of the brain is implicated in control of hormonal defenses against hypoglycemia. They found first that chemical toxin-induced lesions (ibotenic acid) in the ventromedial hypothalamus (VMH) markedly reduce the counterregulatory response to hypoglycemia (Borg *et al.,* 1994) when compared to lesions in other areas or sham operation. Next, these investigators instilled, via microdialysis catheters, 2-deoxy-D-glucose (to block local glucose utilization) into the VMH and were able to produce a counterregulation hormonal response (and increased plasma glucose) very similar to that seen with systemic hypoglycemia (Borg *et al.,* 1995). Finally, they showed that systemic hypoglycemia-induced counterregulation could be specifically blocked by the instillation of D-glucose but not L-glucose into the VMH area (Borg *et al.,* 1997). In seeming confirmation of the importance of the VMH area, Amiel and co-workers have recently identified in humans using PET scanning with fluoro-deoxyglucose technology a subthalamic region that exhibited significantly different behavior between the hypoglycemic aware and hypoglycemic unaware subjects (Cranston *et al.,* 2001). They believe that this region might be comparable to the observations on VMH in experimental animals and represent a crucial glucose sensor.

H. Clinical Studies

Studies of clinical physiology in humans have provided new insight into possibly important basic mechanisms by which sensing of lowered glucose occurs and the possible mechanisms underlying hypoglycemia unawareness. Taken together, these studies convincingly suggest that hypoglycemia unawareness and its companion autonomic failure partner, defective counterregulation, are due to brain-mediated, hormone-sensitive dysregulation of stress responses and that they are directly caused by prior hypoglycemia (McCall, 1998).

A single bout of moderately severe hypoglycemia (3.2 mM or 60 mg/dl), by itself even in nondiabetic individuals may blunt the symptom recognition and hormonal recovery from subsequent acute hypoglycemia (Heller and Cryer, 1991; Davis and Shamoon, 1991). The same acquisition of defective hypoglycemic counterregulation occurs in those with type 1 diabetes (Dagago-Jack *et al.*, 1994; Davis *et al.*, 1994). One interesting difference is that the duration of the abnormal counterregulation lasts less long in diabetic individuals. Nondiabetic individuals recover their normal counterregulation defenses after about 5 days while those with type 1 DM are recovered within 2 days (George *et al.*, 1995, 1997).

The acquired defect in counterregulation appears to be stimulus specific. Thus, although hormonal responses to hypoglycemia are commonly affected and important for defective early recovery of euglycemia, these hormones respond normally to such stimuli as amino acid infusion (growth hormone) and to exercise or standing (epinephrine) in diabetic subjects with defective counterregulation (Hirsch and Shamoon, 1987; Rattarasarn *et al.*, 1994). This implies that sensing of hypoglycemia by the brain rather than the effector arm of counterregulation stress hormone responses is impaired.

I. Insulin, Sex Steroids, and Counterregulation

Hormonal influences may modify hypoglycemia counterregulation. Evidence of the hormone sensitivity of counterregulation comes from both human and animal studies that have found that insulin itself may enhance hypoglycemic recovery and counterregulation. Although some debate has occurred about the dose-response characteristics, Davis and colleagues (1993a,b, 1994, 1995), have found in several studies that physiological levels of insulin influence counterregulation. Such data confirm that the brain is indeed insulin sensitive, as was also observed in studies on stroke outcome. The exact brain signaling mechanisms for such hormone sensitivity, however, are not yet clearly delineated.

It is likely that sex steroids also influence counterregulation and the maladaptive response that leads to hypoglycemia unawareness. Women do not acquire the defective counterregulation with recurrent hypoglycemia as completely as do men (Davis *et al.*, 2000a,b). CNS driven changes probably cause the differential gender effects. Moreover, women do not acquire defective counterregulation as commonly after exercise as do men (Galassetti *et al.*, 2001).

J. Recovery from Hypoglycemia Unawareness

Many, but not all hormone responses to hypoglycemia recover with strict avoidance of hypoglycemia (Fanelli *et al.*, 1993). Glucagon responses to hypoglycemia, however, do not appear to recover as do other hormone responses to hypoglycemia in type 1 DM patients. The reversibility of hypoglycemia-induced defects in counterregulation occurs more rapidly in type 1 (insulin-deficient) diabetic individuals than in nondiabetic persons (George *et al.*, 1995, 1997). This may imply that hyperglycemia or some factor associated with it also influences the brain's perception of or adaptation to hypoglycemia.

K. Glucose Transport and Transporters in Hypoglycemia

There is no evidence yet that a single bout of acute hypoglycemia alters glucose transport into the brain. Indeed this would be somewhat unexpected as a short-term adaptation given the brain's intolerance to acute hypoglycemia and the multiple, redundant defense mechanisms that characterize the response to acute hypoglycemia. There is, however, considerable evidence that repeated or chronic hypoglycemia alters glucose transport across the BBB into the brain (McCall *et al.*, 1986; Pelligrino *et al.*, 1990) and may alter the expression of brain glucose transporters and their mRNA (Kumagai *et al.*, 1995; Koranyi *et al.*, 1991; Uehara *et al.*, 1997; Simpson *et al.* 1999), especially BBB GLUT1. Some, but not all studies indicate that GLUT3, the primary neuronal transporter, is also increased in expression with recurrent hypoglycemia (Uehara, 1997; Nagamatsu, 1994).

In clinical studies in humans using arteriovenous differences, Boyle and colleagues (1994, 1995) have found that brain extraction of glucose is increased in chronic hypoglycemia in both nondiabetic and diabetic humans. Interestingly, it has not been possible so far using PET scanning methods to confirm this upregulation of transport or altered brain metabolism as a result of either chronic hyperglycemia or chronic hypoglycemia in human subjects (Fanelli *et al.*, 1998; Segal *et al.*, 2001).

There are three main hypotheses to explain hypoglycemia unawareness and defective counterregulation. First, Davis and colleagues (1996, 1997) have evidence in favor of a glucocorticoid-mediated adaptation, although the precise mechanisms are not entirely clear. Second, work described above showing altered BBB and brain transport mechanisms may be important in explaining altered distress signaling and defective counterregulation. The time course of these brain transport adaptations is relatively slow, however, and may explain adaptations only later in hypoglycemia unawareness. Third, based on reduced heart rate response to isoproterenol, there is a recent theory that altered beta-adrenergic sensitivity may be an important concomitant and possible explanation for altered hypoglycemia awareness with recurrent hypoglycemia (Fritsche *et al.,* 1998, 2000; Korytkowski, 1998). It is unclear, however, exactly how this observation explains defective counterregulation and whether it could account for reduced sympathetic nervous system activation with acute on chronic hypoglycemia.

L. Vicious Cycle of Hypoglycemia

It seems likely that hypoglycemia itself may produce conditions of physiological (mal)adaptation that are probably mediated largely through brain signaling mechanisms. Moreover, these adaptations may be hormone sensitive. Thus, a vicious cycle whereby hypoglycemia begets further and more severe hypoglycemia appears to exist in most cases of severe hypoglycemia.

How does such a vicious cycle come into existence? First, it appears that the ability to recognize and to mount a counter-insulin hormonal defense against hypoglycemia is modified by recurring moderate hypoglycemia. Elegant physiological studies by a number of clinical researchers in DM have shown that most patients with severe hypoglycemia do not have permanent defects in recognition and defense against hypoglycemia, i.e., severe diabetic autonomic neuropathy. Instead, what is observed is that tight control of diabetes causes mild hypoglycemia commonly. Because of mild hypoglycemia, hormone defenses are reduced with subsequent hypoglycemia, particularly the critical catecholamine responses (Heller and Cryer, 1991; Davis and Shamoon, 1991). Combined with an often preexisting impaired glucagon response to hypoglycemia, these defective hypoglycemia counterregulation defenses lead to serious hypoglycemia. The loss of catecholamine responses to hypoglycemia also reduces hypoglycemia awareness (Cryer, 1999).

IV. Brain Damage from Hypoglycemia

Severe hypoglycemia is typically defined clinically as that which requires help from an outside party to recover, that is, which cannot be self-treated.

Although the numbers of conditions that may cause serious hypoglycemia are large, in clinical practice by far the most common cause is insulin-treated DM. Treatment of DM with insulin or insulin secretagogues to achieve glycemia that is close to normal as evidenced by near normal hemoglobin A1c levels, substantially reduces the risk of the chronic complications of diabetes, including risk of visual loss from retinopathy, risk of renal failure from nephropathy, and risk of peripheral nerve damage from neuropathy. Type 1 DM, an autoimmune disorder associated with rapid destruction of insulin producing pancreatic β-cells requires treatment with insulin to permit survival. Type 2 DMs is nine times as common. Type 2 diabetes is thought to be noninsulin-dependent, but recent data clearly indicate that relative insulin deficiency is an integral part of the disorder and is apparently inexorably progressive; ultimately creating a condition of pronounced insulin deficiency as well that requires insulin treatment.

Physiological replacement of insulin therapy is complicated and many clinicians have insufficient time, energy, resources, or knowledge to do so in an optimum fashion. Moreover, many patients do not have the skills or motivation to achieve excellent glycemic control that may permit them to avoid or minimize the risk of long-term diabetes complications. As a result, clinicians dealing with insulin-treated diabetes frequently must deal with problems with hypoglycemia, often as well as with hyperglycemia.

Early experience with hypoglycemia made it clear that insulin therapy for treatment of DM may cause problems with normal brain function. Case reports of insulin hypoglycemia related brain dysfunction or damage punctuate the early literature but are often related to insulin shock therapy for psychiatric illness (see, for example, Wortis *et al.,* 1939). Therapeutic use of insulin-induced coma for treatment of psychiatric illness unfortunately contributed substantially to the literature illustrating that profound or prolonged hypoglycemia may damage the brain. Most estimates of risk of dying from hypoglycemia in type 1 diabetes (absolute insulin deficient) range from 2–4% mortality (Lins and Adamson, 1993).

Work on the neuropathology of hypoglycemia comes from three main sources. First are case reports or case series in which patients with diabetes have inadvertently or deliberately (suicide attempts) overdosed themselves with insulin (see, for example, Cooper, 1994; Perros *et al.,* 1994). Observations of brain damage in humans are helpful but are limited often by the individual circumstances and the lack of information about other factors (e.g., hypotension, respiratory depression, subsequent hyperglycemia, etc.) that may modify the neurological outcome. A second source of information on brain damage from hypoglycemia comes from animal studies either in primates or in rodents (Brierly, 1981; Auer and Siesjö, 1993).

Although there are similarities of hypoglycemic brain damage to ischemia in controlled animal studies, there are also notable differences. Selective neuronal necrosis in areas such as the hippocampus and neocortex characterizes the pathology of severe hypoglycemia. The biochemical mechanism appears to be a pure excitotoxic lesion. Aspartate and glutamate toxicity occurs without the energy-dependent reuptake required to detoxify their postsynaptic effects. The excitotoxicity may explain the common occurrence of seizures with hypoglycemia. If there is complicating hyperglycemia (delayed rebound), brain damage may extend to a more acidotoxic-like lesion, similar to cerebral ischemia in the face of hyperglycemia. In this latter lesion, glial and vascular necrotic lesions also may occur.

The importance of hypoglycemia has been brought to the forefront by clinical studies which attempt to achieve near normal glycemia levels to avoid chronic diabetes complications. The most important of these is the Diabetes Control and Complications Trial (DCCT) (DCCT Research Group 1995, 1997). In the DCCT, more than 1400 patients with type 1 diabetes reduced the risk of long-term diabetes complications by intensive insulin treatment. One of the primary prices to pay for this benefit was a threefold increased risk of severe hypoglycemia, including a threefold increased risk of coma and seizures because of severe hypoglycemia. No permanent brain damage resulted in this study. Moreover, neuropsychological testing was reassuring in finding no permanent subtle harm to the brain.

Despite the reassurance of such large clinical trials, clinical reports of hypoglycemic brain damage continue to appear in recent medical literature (Perros *et al.,* 1997a,b). In fact, the primary organ of the body that is at risk because of severe hypoglycemia is the brain. The risk factors for severe hypoglycemia have been examined in both humans and in animal models. The existence of mild to moderate hypoglycemia that is not sufficient to produce brain dysfunction and/or damage may nonetheless be implicated in the risk for later brain damage as they underlie the subsequent development of severe hypoglycemia.

V. Effects of Profound Hypoglycemia Upon Neurochemistry

Lack of glucose supply to the brain affects the concentrations of many metabolites of glucose. These include arachidonate, acetylcholine, aspartate, and glutamate, and of course, if sufficiently profound, lack of cerebral energy generation (McCall, 1993). Partial depletion of high-energy phosphate compounds (ATP and creatine phosphate) occurs, cerebral oxygen consumption declines (due to substrate lack), glucose metabolism declines to a greater degree, most levels of the intermediate compounds in the citric acid cycle are depressed, and lipolysis is accelerated leading to increases in polyunsaturated free fatty acids, especially arachidonic acid (Agardh *et al.*, 1980). There is a decrease in cerebral protein synthesis and reduced incorporation of amino acids into protein in a regionally specific manner. Areas most affected in the brain are those most damaged by hypoglycemia.

Based on extensive work in rat models of profound hypoglycemia, several other characteristics of hypoglycemia that are serious enough to produce an isoelectric EEG have emerged (Siesjö, 1988). These include a loss of ionic homeostasis: brain extracellular potassium rises (Wieloch *et al.*, 1984) while extracellular calcium falls (Harris *et al.*, 1984). Rises of intracellular calcium may activate a number of degradative enzymes including proteases and lipases whose action may lead to mitochondrial damage and swelling (Auer *et al.*, 1985a,b, Kalimo *et al.*, 1985). Extracellular pH rises initially and eventually intracellular pH rises with alkalosis itself helping to minimize the extent of damage (preventing pannecrosis) (Pelligrino *et al.*, 1981). Loss of high-energy adenine nucleotides and depletion of the nucleotide pools occurs (Agardh and Siesjo, 1981; Agardh *et al.*, 1980, 1981). Levels of amino acids and amino acid neurotransmitters are affected by profound hypoglycemia (Lewis *et al.*, 1974; Norberg and Siesjö, 1976). There are decreased concentrations of glutamate, glutamine, γ-aminobutyric acid (GABA), and alanine although intrasynaptic concentrations of glutamate may rise. Levels of aspartate and ammonia may rise. These excitotoxic amino acids (glutamate and aspartate) are thought to bind selectively to dendrites and perikarya of neurons and by doing so produce selective neuronal necrosis. Evidence of the excitotoxicity mediated through NMDA receptors is indicated by the ability of NMDA receptor blocking drugs to ameliorate brain damage from hypoglycemia (Wieloch, 1985). So far, unfortunately, such insights have not lead to a practical remedy to prevent severe hypoglycemic damage in susceptible individuals.

VI. Physiological Adaptations of the Brain to Hyper- and Hypoglycemia

A. CBF—Cerebrovascular Dysfunction in Diabetes

Mooradian and his colleagues have performed a great deal of work in this area and he has written a comprehensive review of brain microvascular dysfunction in diabetes. Marked hyperglycemia in animal models (e.g., streptozotocin-induced diabetes) is associated with a regionally specific decrease in CBF. For example, in the studies of Duckrow *et al.* (1987), hindbrain blood flow was more reduced than forebrain and hyperglycemia dependent reductions in CBF occurred. In such studies, an osmotic effect can be ruled out as isosmotic mannitol control experiments showed no CBF change. In contrast to hyperglycemia, characteristically in animal models, both acute and chronic hypoglycemia may increase CBF, presumably as a compensatory mechanism to maintain adequate delivery of glucose fuel to the brain (Pelligrino *et al.*, 1991).

Adaptations to chronic hyperglycemia and chronic hypoglycemia may occur with vasoreactive properties of cerebral blood vessels in animal models impaired by some unknown mechanism. For example, Pelligrino and colleagues (1990) found in chronically hyperglycemic rats, the increased CBF characteristic after acute hypoglycemia is blunted, as it is for several other stimuli, such as hypoxia or hypercarbia. Interestingly, chronic hypoglycemia in animal models produces less increase in CBF than does acute hypoglycemia, indicating some possible adaptation to the chronically hypoglycemic state. *In vitro* studies of cerebral blood vessel responses to elevated glucose concentrations show a number of abnormalities as well. Cipolla *et al.* (1997) have found that high glucose concentrations reduce the myogenic tone of cerebral blood vessels through an endothelial dependent mechanism and cause vasodilatation in isolated blood vessels.

Human studies of hyper- and hypoglycemia are less consistent than animal model studies. Reasons for this may be multiple and include that the extremes of hyper- and hypoglycemia that may have been tested in animal models are not safely reproduced in human experimentation. The interpretation of human experiments showing altered CBF in many studies could also be confounded by the presence of subtle vascular disease. Nonetheless, a variety of humans studies have shown that diabetes alters cerebrovascular reactivity and resistance function properties (Lass *et al.*, 1989; Tkac *et al.*, 2001) of blood vessels in the brain. Decreases in CBF (Wakisaka, 1990), no change in basal CBF but increases in CBF with hypoglycemia (Neil *et al.*, 1987), and increases in CBF (Grill *et al.*, 1990) of patients with diabetes have been described. Perhaps more consistent in human studies than an altered basal CBF (absent cerebrovascular disease) are abnormalities in vasoreactivity of a variety of types. Abnormalities include impaired autoregulation in type 1 DM (Kastrup, 1986), impaired microvascular reactivity (Kastrup, 1990), and abnormal autoregulation during cardiac bypass surgery (Croughwell, 1990).

B. Brain Metabolism

Overall brain energy metabolism (i.e., oxygen consumption) has been found to be depressed in classic studies of brain metabolism in extreme hyperglycemia associated with diabetic ketoacidosis and coma (Kety, 1948).

Some of these effects could be related to the depressed utilization of fuel in the comatose state. Animal studies have identified reductions in overall glucose metabolism or changes in regional glucose metabolism associated with poorly controlled diabetes (Ruderman *et al.*, 1974, Jakobsen *et al.*, 1990). When ketonemia is present in uncontrolled diabetes, it may displace glucose utilization as a source for energy metabolism in brain. Such findings are not mirrored, however, by findings in studies of brain metabolism in human diabetic subjects with less severe dysmetabolism (Fanelli *et al.*, 1998).

C. Mechanisms Involving BBB and Brain Cellular Transport

A number of studies have suggested that blood-brain transport of glucose may be downregulated in uncontrolled diabetes (see Mooradian, 1997a for review) and upregulated in chronic or recurrent hypoglycemia (see McCall, 1998). Reduced efficiency of transport in uncontrolled diabetes remains controversial (Simpson *et al.*, 1999), although a number of studies using differing techniques have found this. One wonders whether either subtle or as yet unrecognized differences in the models or in the techniques might play an important role in the discrepant findings of these studies. The findings of upregulation of glucose transport are less controversial, but questions are raised by the absence of confirmation in some recent human studies. The failure at times to confirm findings in animal studies when examined using less invasive techniques in human studies will require additional studies to yield more definitive answers.

D. Metabolism of the Brain Microvasculature in Diabetes and Hypoglycemia

The cerebral microvasculature appears to be less affected by diabetes than retinal vessels, but nonetheless may suffer chronic diabetic changes. Loss of pericytes can occur in brain capillaries (Mukai *et al.*, 1980). Pathological changes in the microvasculature of the brain, including thickening of basement membranes, occur in longstanding diabetes (Reske-Nielsen *et al.*, 1965; Johnson *et al.*, 1982).

Metabolic abnormalities of the cerebral microvessels *in vitro* exist in both uncontrolled diabetes and recurrent hypoglycemia in animal models (McCall, 1992b). Hingorani and Brecher (1987) observed a reciprocal relationship between glucose and fatty acid oxidation by isolated rabbit brain microvessels, reminiscent of the glucose fatty acid cycle. Moreover, in uncontrolled diabetes a depression of glucose oxidation with no effect upon microvascular fatty acid oxidation was found. In other studies, a greater than 50% reduction of glucose oxidation and reduction in glycolytic flux was observed in microvessels harvested from rats with uncontrolled diabetes (McCall *et al.*, 1984). Restoration of fuel metabolism was produced in the isolated microvessels either by insulin treatment or 48 h of starvation (which restored glycemia to near normal). In chronically hypoglycemic rats (rendered hypoglycemic by subcutaneous grafting of insulinoma producing tumors), the glucose metabolism of isolated brain microvessels shows a marked acceleration (McCall *et al.*, 1988). A reciprocal relationship between oxidation of glucose and that of β-hydroxybutyrate (ketone body; a form of short chain fatty acid) was observed in microvessels from brain of chronically hypoglycemic rats. In contrast, in microvessels from diabetic rats β-hydroxybutyrate oxidation was increased as glucose oxidation declined. Shorter term hypoglycemia produced by insulin pumps or insulin injection produced no consistent change in metabolism of isolated brain microvessels. No changes in the levels of ATP or ATP/ADP ratios were observed in microvessels from diabetic or insulin rats. The consequences of these modifications in energy metabolism and substrate utilization are not clear but could underlie an enhanced oxidative stress by the microvessels.

VII. Altered Neurotransmitter Metabolism in Diabetes and Hypoglycemia

Several neurotransmitter abnormalities have been described in uncontrolled diabetes and profound hypoglycemia. The monoamine neurotransmitters, serotonin, dopamine, and norepinephrine are affected by uncontrolled diabetes in a variety of studies. Serotonin synthesis in uncontrolled diabetes is depressed, possibly because of reduced brain availability of the neutral amino acid precursor L-tryptophan (relative to the entry of other more abundant amino acids due to the lack of insulin effect) as shown by Crandall and Fernstrom (1983) or possibly related to other mechanisms (Manjarrez-Gutierrez *et al.*, 2000). Insulin therapy restores serotonin synthesis to normal. Reduced concentrations of intrasynaptic serotonin could be an important contributor to depression in poorly controlled diabetes. Interestingly, according to one report, treatment with the SSRI drug fluoxetine can cause the acquisition of hypoglycemia unawareness, an anecdotal finding that might (with further research) prove useful for treatment of the risk of hypoglycemia (Sawka *et al.*, 2000).

Dopaminergic neurotransmission may be reduced in diabetes. ^{3}H-spiperone binding is increased in either alloxan or streptozotocin diabetes presumably because of reduced neurotransmission. Dopamine receptor binding is normalized by insulin treatment. Dopamine metabolites (dihydrophenylalanine and homovanillic acid) are reduced in the striatum of diabetic rats. Hyperglycemia may alter neuronal electrical activity in dopaminergic neurons. Altered sensitivity to dopamine agonists has been observed in patients with diabetes. One basis for altered dopamine neurotransmission is the effects of insulin on dopamine synaptic reuptake. As shown by Figlewicz *et al.* (1994) the mRNA of the transporter responsible for synaptic reuptake of dopamine (DAT) is upregulated by insulin in areas such as the substantia nigra and ventral tegmentum.

Studies by Ganzini *et al.* (1993) suggest that altered dopamine neurotransmission in the brain may underlie an increased risk of involuntary movement issues in those with diabetes. In contrast to DAT, Figlewicz and colleagues (1993) had previously found an opposite effect of insulin on the mRNA levels for the norepinephrine transporter (NET). In situ hybridization for NET showed a decrease in message levels in the locus coeruleus.

What are the consequences of these physiological observations for diabetes? In hypoinsulinemic diabetes, induced by the β-cell toxin streptozotocin, Figlewicz *et al.* (1996) found that NET mRNA was significantly elevated (159% of control) while DAT mRNA was nonsignificantly decreased (78% of control). Additionally, the rate-limiting enzymatic step for synthesis of these catecholamines, tyrosine hydroxylase (TH), was affected. TH mRNA was increased in the locus coeruleus (131% of control), but reduced in the ventral tegmentum and substantia nigra (79% of control). Chronic insulin administration via osmotic minipump to the brain has been found to alter startle responses to sound in a manner similar to noradrenergic blockade with drugs such as desipramine, presumably due to inhibition of norepinephrine uptake causing enhanced noradrenergic synaptic activity (Thompson *et al.*, 2000).

Complex effects on peptide neurotransmitters may be produced by uncontrolled DM. Sipols *et al.* (1995) have shown that uncontrolled diabetes increases the mRNA levels for neuropeptide Y (NPY; a stimulus to diabetic hyperphagia) more than 2.5-fold. Cholecystokinin mRNA increased by 50% while CRH *in situ* hybridization in the paraventricular nucleus was decreased by one-third. These latter peptides inhibit feeding. Insulin administration to diabetic animals substantially but not completely restores NPY mRNA, but has lesser effects on these other peptides. Havel *et al.* (2000) have shown that in untreated diabetic rats, hypothalamic pro-opiomelanocortin (POMC) mRNA expression (*in situ* hybridization) was reduced by 80%, whereas agouti related protein mRNA levels were increased by 60%, suggesting a marked decrease of hypothalamic melanocortin signaling. Similar to the findings of NPY studies, insulin administration to the CNS reverses the POMC changes but has little effect on agouti related protein.

A more comprehensive review of the many changes that uncontrolled diabetes or the central action of insulin exerts upon neurotransmitters is beyond the scope of this chapter. Nonetheless, work in this area is likely to yield significant new insights into how diabetes and insulin action affect the brain and behavior.

VIII. DM and Epilepsy

Uncontrolled DM and hypoglycemia may disturb energy and neurotransmitter metabolism as well as ionic equilibria within the brain. Extremes of glycemic control, such as hyperosmolar coma and severe hypoglycemia, are well recognized clinically as causes of epileptiform seizures. There is some evidence that diabetes and hypoglycemia may affect the brain by increasing the tendency to chronic seizures disorders. The possibility is raised but is not yet capable of a scientifically definite answer, whether diabetes and/or hypoglycemia predispose to chronic epilepsy.

Seizures and diabetes or hypoglycemia coexist in certain genetic syndromes. While the possibility exists that the genetic abnormality is an independent cause of the relationship to chronic epilepsy, the author views it as likely that there may be a causal or mechanistic association with diabetes and even more likely with recurrent hypoglycemia. Clearly, certain relatively uncommon variants of diabetes, for example, the mitochrondrial encephalopathy, lactic acidosis, and stroke (MELAS) syndrome, a mitochrondrial genetic abnormality that is associated with both diabetes and epileptic seizures, exemplifies a definite connection of this type. MELAS syndrome is typically characterized by normal early development and childhood-onset of recurrent neurologic deficits (stroke-like episodes), seizures, short stature, lactic acidosis, and ragged red fibers on muscle biopsy specimens. It is commonly, although not invariably, associated with the A3243G point mutation in the mitochondrial DNA tRNALeu (UUR) gene (Sue *et al.*, 1999).

Another relatively rare genetic syndrome, a partial deficiency of glucose transporter (GLUT1) in brain, also appears to mimic hypoglycemia and is associated with difficult to control seizures. De Vivo and colleagues (1991) had originally described two patients with infantile seizures, delayed development, and acquired microcephaly who have normal circulating blood glucose, low-to-normal cerebrospinal fluid (CSF) lactate, but persistent hypoglycorrhachia (low CSF glucose) and diminished transport of hexose into isolated red blood cells (RBC). These workers suggested the possibility of a defect in glucose transport across the BBB to explain this constellation of findings. They have reported two distinct classes of mutations as the molecular basis for the functional defect of glucose transport: the first, a hemizygosity of GLUT1, the second, nonsense mutations resulting in truncation of the GLUT1 protein (Scidncr *et al.*, 1998). These GLUT1 deficient children are being increasingly recognized. Their recognition is important because they respond to a ketogenic diet, one high in fats but low in carbohydrate. This diet may ameliorate seizures and potentially may prevent more severe neurologic manifestations, such as retardation or developmental delay (Nordli and De Vivo, 1997). Seizures manifest in these children most commonly as absence seizures. Diagnostic tests may include assays of 3-*O*-methylglucose glucose transport in red cells taken from subjects or family members (Klepper *et al.*, 1999a,b).

Another important, albeit again relatively rare (1 in 50,000 live births) syndrome is due to a genetic defect which is associated with persistent hyperinsulinemic hypoglycemia of infancy (PHHI) (Sharma *et al.*, 2000). This syndrome appears to be due to mutations in the ATP-sensitive, inwardly rectifying potassium channel (KIR 6.2), or in glucokinase, the sulfonylurea receptor (SUR), and glutamate dehydrogenase (Meissner *et al.*, 1999). Seizures are a major manifestation of this disorder and commonly manifest in the early perinatal period. Of interest, the characteristic seizure modality present in these children involves the developing neocortex, particularly the occipital and parietal lobes. Involvement of hypoglycemic damage to these areas of the brain may be detectable on imaging studies. The seizures in these children are often extremely difficult to treat. If therapies such as diazoxide or other drugs that minimize hyperinsulinemia fail, then pancreatic surgery may be required with the end result often being the presence of DM as a result of treatment.

One important aspect of this disorder is that it appears quite clear that it can show that hypoglycemic brain damage is capable of producing true epilepsy, as distinct from seizures that are only manifest in the face of hypoglycemia. This disorder also illustrates another important biologic issue, i.e., the developing brain is particularly susceptible to hypoglycemic brain damage. The reasons for such susceptibility are at least partly understood. They include the large head size of infants and young children relative to body size, which results in a greater proportion of glucose utilization by the younger brain relative to the overall body economy. In addition, metabolic rates of glucose consumption by the brain (CMRglu) are known to be higher in children relative to the CMRglu in the adult brain (Kinnala *et al.*, 1996). A final important issue probably relates to the different fate of glucose carbon within the developing brain. Glucose is not only used as fuel for brain energy metabolism during development; it may also be used in the production of macromolecules involved in structural development, a final special susceptibility of the developing brain to hypoglycemia. These considerations may be important when considering the issue of seizures because of insulin overdosage in treatment of diabetes.

More common than these relatively rare genetic syndromes and their associations with diabetes or brain glucose deprivation with epilepsy, however, is garden-variety insulin overdose depriving the brain of needed glucose. For example, tight glycemic control in type 1 diabetes risks severe hypoglycemia (DCCT Research Study Group, 1993). In the DCCT, tight glucose control predisposed to a threefold increased risk of coma or seizures despite considerable attempts to minimize such risks. It appears plausible and even likely that severe hypoglycemia bad enough to result in biochemical or anatomical brain injury may also produce an enduring form of epilepsy. On this point, however, scientific evidence is incomplete, and further study should be undertaken.

Diabetes itself may also be associated with EEG abnormalities even without hypoglycemia (Hauser *et al.*, 1995; Inui *et al.*, 1998; Cochin *et al.*, 1994; Soltesz and Acsadi, 1989). Whether neurotransmitter metabolic abnormalities that may be observed in diabetes influence EEG changes or the risk of seizures and chronic epilepsy is not known but certainly is a possibility. Both uncontrolled diabetes and hypoglycemia may represent conditions that predispose to epilepsy. The possible coincidence of epilepsy and diabetes, both common conditions, makes it more difficult in individual patients to say with certainty that diabetes or hypoglycemia per se are the major etiologic factors in the generation of epilepsy.

IX. Effects of Uncontrolled Diabetes and Hypoglycemia on Cognitive Performance

Both hyper- and hypoglycemia are unavoidable consequences experienced by those with treated diabetes, particularly marked with type 1 diabetes. There is evidence that both short-term hypoglycemia and hyperglycemia may have an adverse impact on cognitive function. Mechanisms for these changes are incompletely understood.

A. Short-Term Hypoglycemia

Hypoglycemia in the short term may cause a variety of negative effects on emotion and on cognitive and/or motor function (Gonder-Frederick *et al.*, 1997). Indeed, the earliest descriptions of the presentation of hypoglycemia described sensory and motor aphasia, loss of fine motor control in the hands, dysarthria, confusion, disorientation, and even delirium

(Fletcher and Campbell, 1922). Early controlled studies of mild to moderate hypoglycemia in those with type 1 diabetes showed slower reaction times to cognitive testing but testing accuracy was much less affected (Holmes *et al.*, 1983). Interestingly, hyperglycemia also slowed reaction time although to a lesser degree than hypoglycemia. Pramming and colleagues (1986) studied graded hypoglycemia in a series of tasks and found complex tasks more affected than simple psychomotor tasks such as finger tapping. Of some practical importance, prompting of individuals with hypoglycemia may be required during testing (Deary, 1993).

Visual-evoked potentials and auditory-evoked potentials may be significantly delayed in those with controlled experimental hypoglycemia. For example, P300 latency from visual or auditory stimuli become prolonged during moderate hypoglycemia (2.6 mM; 47 mg/dl) (Blackman *et al.*, 1990) and do not completely recover for 45–75 min after administration of glucose returns blood glucose concentration to normal. This observation suggests a metabolic "stunning" of brain function that may persist. In this study, another evoked potential (P140; a measure of sensory processing) was normal and the authors concluded from this study that neither sensory nor motor processing per se was abnormal, but instead that there was an abnormality in decision making. This obviously presents significant concern for patients who have hypoglycemia unawareness and defective hypoglycemic counterregulation if complex cognitive function and decision making are impaired by moderate hypoglycemia as these defects may alter the ability to self-treat and recover. The persistence of cognitive performance problems is greater with hypoglycemia experienced by those with hypoglycemia unawareness (Gold *et al.*, 1995).

B. Effects of Short-Term Hyperglycemia

Short duration hyperglycemia may also adversely affect cognitive function. Davis and colleagues (1996), examined the effects of transient hyperglycemia (20–30 mM; 360–540 mg/dl) in children with type 1 diabetes and found that a reduction in performance IQ with a mean of nearly a 10% decline in percentile score. On the other hand, Draelos *et al.* (1995) in a study of short duration type 1 diabetes in adults with a wide range of glycemic control and graded hypoglycemia to hyperglycemia found only with hypoglycemia did cognitive function worsen, and that glucose values of 21 mM did not impair cognitive performance. Pennebaker *et al.* (1981) found that acute hyperglycemia in 30 hospitalized diabetic patients was reliably associated with mood disturbances and/or physical symptoms more than half the time. Such an observation might have bearing on cognitive performance. Based on self-monitoring and cognitive assessment involving a handheld computer, Cox and colleagues have found that graded hyperglycemia is associated with deterioration in verbal fluency and speed of performing mental subtractions in treated insulin-dependent diabetic subjects (Kovatchev *et al.,* 2001).

C. Chronic Hyperglycemia and Cognition

Chronic, poorly controlled type 1 diabetes may result in reduced cognitive function and pathological changes similar to ischemia and degenerative dementing illness (Reske-Nielsen, 1965). Animal studies of diabetes have certainly identified both biochemical and structural abnormalities that could underlie serious cognitive deficits. Uncontrolled streptozotocin diabetes, an animal model similar to type 1 diabetes, is associated with cognitive abnormalities, particularly memory retention and retrieval, which are reversed by treatment with insulin (Flood *et al.,* 1990). Mooradian (1997a,b) has summarized a variety of reasons why the brain may be damaged by chronic hyperglycemia including altered vascular function, metabolic abnormalities, and comorbid conditions (such as hypertension, heart failure, etc.) that may occur. Within each of these categories there are expanding pathophysiological changes that might underlie cognitive defects.

In the elderly, even impaired glucose tolerance is associated with cognitive defects, although deficits are more pronounced in those with known DM (Kalmijn *et al.,* 1995). One reason for cumulative impairment of cognitive function in elderly diabetic subjects is cerebrovascular disease, even though often not clinically recognized. One MRI study found a very high prevalence of lacunar strokes and this predicted intellectual dysfunction in older subjects with type 2 diabetes (Fushimi *et al.,* 1994). When one compares MRI and auditory brain stem evoked potentials in longstanding vs. short duration type 1 diabetes, there is a high proportion of patients with longstanding diabetes that have abnormal latency of evoked potentials. In these subjects, there is also a high prevalence of MRI abnormalities in cortical and brain stem structures (Dejgaard *et al.,* 1991). Similarly, in another study evoked potential abnormalities were found to correlate with subtle cognitive dysfunction that might be missed on routine psychometric testing (Pozzessere *et al.,* 1991). Gispen and Biessels (2000), have suggested that one possible reason for the increased risk of dementia observed in patients with longstanding diabetes in addition to cerebrovascular disease and that of hyperglycemia per se might be related to lack of beneficial effects of insulin on the brain.

Numerous studies have found cognitive deficits in older type 2 diabetes subjects. Reaven and colleagues (1990) found that verbal learning, abstract reasoning, and complex psychomotor functioning were performed more poorly by diabetic 70-year-old subjects than matched nondiabetic subjects. Moreover, the degree of impairment on learning, reasoning and complex psychomotor tasks was more affected in those with poorer metabolic control. Mooradian *et al.* (1988a) showed diabetic subjects over age 60 had poorer visual retention and serial learning task performance than controls and that there were more commonly EGG abnormalities in older diabetic subjects (frequency band analysis revealed slowing and reduced alpha activity in parietal areas). Perlmuter and colleagues have found that older subjects with type 2 diabetes more frequently report cognitive difficulties, especially memory-related tasks and learning (Tun *et al.*, 1987, 1990). Poor glucose control and elevated blood triglyceride concentrations (as an independent variable) were found to predict cognitive testing abnormalities (Perlmuter *et al.*, 1984, 1988).

X. Future Research Directions

Some future directions for research are suggested below that are pertinent to several of the subsections of this chapter. In the area of cerebrovascular disease and diabetes, there is a need for studies to examine ways to manipulate either glucose transport or lactic acid transport that could potentially be important to provide new clues on how to minimize vascular brain damage. One paper has shown a slight benefit of adenoviral transfer of GLUT1 into an ischemic brain area in experimental stroke (nondiabetic conditions) (Lawrence *et al.*, 1996). The consistent beneficial effects of lowering hyperglycemia through insulin administration, however, may also offer a more immediate type of potential benefit if animal studies are applicable to human ischemia. Thus, it has been shown, for example, in cardiac ischemia that insulin administration to diabetic patients with acute myocardial infarction substantially reduces morbidity and mortality (30%) in five-year follow-up of the DIGAMI study (Diabetes Insulin Glucose Acute Myocardial Infarction; Malmberg *et al.*, 1997). An analogous type of study should be proposed and conducted for patients with DM and acute brain infarction and might be termed the DIGABI study. This study could show whether short-term (a few days to a few months) tight glycemic control with insulin has a beneficial effect on stroke outcome in nonhemorrhagic stroke in patients with diabetes.

In the area of hypoglycemia and brain, work on the new glucose transporters needs to be done. The fact that several newly described glucose transporters exist and that at least some of these are present within the brain (GLUT8, GLUT9, and others) raises the possibility that such transporters may act to serve as an emergency reserve of inducible carriers for additional glucose supply to the brain in situations such as chronic or repeated hypoglycemia. Indeed, the original description of GLUT8 suggested such a possibility and it is clearly a protein responsive to the hormonal milieu of diabetes. Much physiological study on their role in hypoglycemia is potentially possible given the existence of these newly described additional transporter isoforms.

Further work also needs to be done to characterize the time course of brain adaptations to hypoglycemia. Early as opposed to later mechanisms of adaptation may differ and may complement each other. One unexplored possibility is that brain transporter translocation, such as has been observed with seizures (Cornford *et al.*, 1993), might occur early on in hypoglycemia whereas altered GLUT expression could be a later phenomenon. On another mechanism of adaptation to hypoglycemia, it may be important to explore the role of the brain in altered beta-adrenergic sensitivity and hypoglycemia unawareness and defective counterregulation. It should be possible to determine aspects of beta receptor biochemistry in multiple tissues after hypoglycemia (acute versus chronic). It may also be of interest to examine hypoglycemia counterregulation in mutant animals with defective beta receptor kinase activity (involved in receptor downregulation) to gain new insights into adaptations to hypoglycemia.

Brain damage from hypoglycemia may be limited if effective and nontoxic agents could be found to block downstream pathways activated by excitotoxicity. Clearly, drugs interfering with the NMDA receptor activation are effective as shown by Wieloch (1985). The problem is that effective binding of these receptors by antagonists may interfere with crucial normal functions. Preventing hypoglycemic brain damage in diabetes may also in the future better exploit physiological observations than currently. For example, it has recently been found that the simple pattern of glycemia based upon home monitoring may successfully predict more than 50% of future serious hypoglycemic reactions (Kovatchev *et al.*, 2000). Similarly, it might be that a pattern of physiological stress response in overtreated diabetic patients could be more readily discerned and prevent many of the serious hypoglycemic episodes.

Further studies of signal transduction by the brain in acute and chronic hypoglycemia also seem warranted. The hope is that eventually an artificial pancreas or a cure for type 1 diabetes will be found.

The effects of hyperglycemia upon cognitive function need more investigation as to their importance and as to their mechanisms. One suspects that neurotransmitter related changes or electrophysiological effects of altered glycemia on neurons may be likely to explain the possibility of short duration of hyperglycemia impairing cognitive function. Other possibilities exist, however, and merit additional study.

In short, there is much more investigation, both basic and clinical, that can and should be done in the area of diabetes and the brain. Better understanding of the basic biology will, one hopes, lead to an improved clinical outcome and sparing of brain function for those with diabetes.

References

Agardh, C. D., Kalimo, H., Olsson, Y., and Siesjö, B. K. (1981). Hypoglycemic brain injury: metabolic and structural findings in rat cerebellar cortex during profound insulin-induced hypoglycemia and in the recovery period following glucose administration. *J. Cerebr Blood Flow Metab.* **1,** 71–84.

Agardh, C. D., and Siesjö, B. K. (1981). Hypoglycemic brain injury: Phospholipids, free fatty acids, and cyclic nucleotides in the cerebellum of the rat after 30 and 60 minutes of severe insulin-induced hypoglycemia. *J. Cereb. Blood Flow Metab.* **1,** 267–275.

Agardh, C. D., Westerberg, E., and Siesjö, B. K. (1980). Severe hypoglycemia leads to accumulation of arachidonic acid in brain tissue. *Acta Physiol. Scand.* **109,** 115–116.

Arboix, A., Marti Vilalta, J. L., and Garcia, J. H. (1990). Clinical study of 227 patients with lacunar infarcts. *Stroke* **21,** 842–847.

Arias, P., Kerner, W., Zier, H., Navascues, I., and Pfeiffer, E. F. (1985). Incidence of hypoglycemic episodes in diabetic patients under continuous subcutaneous insulin infusion and intensified conventional insulin treatment: Assessment by means of semiambulatory 24-hour continuous blood glucose monitoring. *Diabetes Care* **8,** 134–140.

Auer, R. N., and Siesjö, B. K. (1993). Hypoglycaemia: brain neurochemistry and neuropathology. *Baillieres Clin. Endocrinol. Metab.* **7**(3), 611–625.

Auer, R. N., Colbourne, F., and Sutherland, G. R. (1999). Insulin, blood glucose levels, and ischemic brain damage. [Review] [32 refs] Electron microscopic evidence against apoptosis as the mechanism of neuronal death in global ischemia. *Neurology* **19**(11), 4200–4210.

Auer, R. N., and Siesjö, B. K. (1988). Biological differences between ischemia, hypoglycemia, and epilepsy. *Ann. Neurol.* **24,** 699–707.

Auer, R. N. (1986). Progress review: Hypoglycemic brain damage. *Stroke* **17,** 699–708.

Auer, R. N., Kalimo, H., Olsson, Y., and Siesjö, B. K. (1985a). The temporal evolution of hypoglycemic brain damage. II. Light- and electron-microscopic findings in the hippocampal gyrus and subiculum of the rat. *Acta Neuropathol. (Berlin)* **67,** 25–36.

Auer, R. N., Kalimo, H., Olsson, Y., and Siesjö, B. K. (1985b). The temporal evolution of hypoglycemic brain damage. I. Light- and electron-microscopic findings in the rat cerebral cortex. *Acta Neuropathol. (Berlin)* **67,** 13–24.

Bell, D. S. (1994). Stroke in the diabetic patient. *Diabetes Care* **17,** 213–219.

Berger, L., and Hakim, A. M. (1986). The association of hyperglycemia with cerebral edema in stroke. *Stroke* **17,** 865–871.

Bhardwaj, S. K., Sandhu, S. K., Sharma, P., and Kaur, G. (1999). Impact of diabetes on CNS: role of signal transduction cascade. *Brain Res. Bull.* **49**(3), 155–162, 1999.

Biessels, G. J. (1999). Cerebral complications of diabetes: Clinical findings and pathogenetic mechanisms. *Netherlands J. Med.* **54**(2), 35–45.

Blackman, J. D., Towle, V. L., Lewis, G. F., Spire, J. P., and Polonsky, K. S. (1990). Hypoglycemic thresholds for cognitive dysfunction in humans. *Diabetes* **39,** 828–835.

Borg, M. A., Sherwin, R. S., Borg, W. P., Tamborlane, W. V., and Shulman, G. I. (1997). Local ventromedial hypothalamus glucose perfusion blocks counterregulation during systemic hypoglycemia in awake rats. *J. Clin. Invest.* **99**(2), 361–365.

Borg, W. P., Sherwin, R. S., During, M. J., Borg, M. A., and Shulman, G. I. (1995). Local ventromedial hypothalamus glucopenia triggers counterregulatory hormone release. *Diabetes* **44**(2), 180–184.

Borg, W. P., During, M. J., Sherwin, R. S., Borg, M. A., Brines, M. L., and Shulman, G. I. (1994). Ventromedial hypothalamic lesions in rats suppress counterregulatory responses to hypoglycemia. *J. Clin. Invest.* **93**(4), 1677–1682.

Boyle, P. J., Kempers, S. F., O'Connor, A. M., and Nagy, R. J. (1995). Brain glucose uptake and unawareness of hypoglycemia in patients with insulin-dependent diabetes mellitus. *N. Engl. J. Med.* **333**(26), 1726–1731.

Boyle, P. J., Nagy, R. J., O'Connor, A. M., Kempers, S. F., Yeo, R. A., and Qualls, C. (1994). Adaptation in brain glucose uptake following recurrent hypoglycemia. *Proc. Natl. Acad. Sci. USA* **91**(20), 9352–9356.

Brierley, J. B. (1981). Brain damage due to hypoglycaemia, *In* "Hypoglycaemia" (V. Marks and F. C. Rose, eds.), pp. 488–494. Blackwell Scientific Publications, Oxford.

Brightman, M. W., Reese, T. S., and Feder, N. (1970). Assessment with the electron microscope of the permeability to peroxidase of cerebral endothelium and epithelium in mice and sharks. *In* "Capillary Permeability" (C. Crone and N. A. Lassen, eds.). Academic Press, New York.

Bryan, R. M., Jr., and Pelligrino, D. A. (1988). Cerebral blood flow during chronic hypoglycemia in the rat. *Brain Res.* **475,** 397–400.

Capes, S. E., Hunt, D., Malmberg, K., Pathak, P., and Gerstein, H. C. (2001). Stress hyperglycemia and prognosis of stroke in nondiabetic and diabetic patients—A systematic overview. *Stroke* **32,** 2426–2432.

Chapman, A. G., Westerberg, E., and Siesjö, B. K. (1981). The metabolism of purine and pyrimidine nucleotides in rat cortex during insulin-induced hypoglycemia and recovery. *J. Neurochem.* **36,** 179–189.

Cipolla, M. J., Porter, J. M., and Osol, G. (1997). High glucose concentrations dilate cerebral arteries and diminish myogenic tone through an endothelial mechanism. *Stroke* **28**(2), 405–410.

Clarke, D. D., and Sokoloff, L. (1999). Circulation and energy metabolism of the brain, *In* "Basic Neurochemistry" (G. J. Siegel, B. W. Agranoff, R. W. Albers, S. K. M. P. B. Fisher, and M. D. Uhler, eds.), pp. 637–669. Lippincott-Raven, Philadelphia.

Cochin, J. P., Hannequin, D., Delangre, T., Guegan-Massardier, E., and Augustin, P. (1994). Continuous partial epilepsy disclosing diabetes mellitus. *Rev. Neurol. (Paris)* **150**(3), 239–241.

Cooper, A. J. (1994). Attempted suicide using insulin by a non diabetic: A case study demonstrating the acute and chronic consequences of profound hypoglycemia. *Can. J. Psychiatry,* **39**(2), 103–107.

Cornford, E. M., Hyman, S., and Pardridge, W. M. (1993). An electron microscopic immunogold analysis of developmental upregulation of the blood-brain barrier GLUT1 glucose transporter. *J. Cereb. Blood Flow Metab.* **13**(5), 841–854.

Crandall, E. A., and Fernstrom, J. D. (1983). Effect of experimental diabetes on the levels of aromatic and branched-chain amino acids in rat blood and brain. *Diabetes* **32,** 222–230.

Cranston, I., Reed, L. J., Marsden, P. K., and Amiel, S. A. (2001). Changes in regional brain (18)F-fluorodeoxyglucose uptake at hypoglycemia in type 1 diabetic men associated with hypoglycemia unawareness and counter-regulatory failure. *Diabetes* **50**(10), 2329–2336.

Croughwell, N., Lyth, M., Quill, T. J., Newman, M., Greeley, W. J., Smith, L. R., and Reves, J. G. (1990). Diabetic patients have abnormal cerebral autoregulation during cardiopulmonary bypass. *Circulation* **82,** IV407–IV412.

Cryer, P. E. (1999). Symptoms of hypoglycemia, thresholds for their occurrence, and hypoglycemia unawareness. [Review] [23 refs]. *Endocrinol. Metab. Clin. N. Am.* **28**(3), 495–500.

Dagogo-Jack, S., Rattarasarn, C., and Cryer, P. E. (1994). Reversal of hypoglycemia unawareness, but not defective glucose counterregulation, in IDDM. *Diabetes* **43**(12), 1426–1434.

Davis, M. R., and Shamoon, H. (1991). Counterregulatory adaptation to recurrent hypoglycemia in normal humans. *J. Clin. Endocrinol. Metab.* **73**(5), 995–1001.

Davis, E. A., Soong, S. A., Byrne, G. C., and Jones, T. W. (1996). Acute hyperglycaemia impairs cognitive function in children with IDDM. *J. Pediatr. Endocrinol. Metab.* **9**(4), 455–461.

Davis, S. N., Colburn, C., Dobbins, R., Nadeau, S., Neal, D., Williams, P., and Cherrington, A. D. (1995). Evidence that the brain of the conscious dog is insulin sensitive. *J. Clin. Invest.* **95**(2), 593–60.

Davis, M., Mellman, M., Friedman, S., Chang, C. J., and Shamoon, H. (1994a). Recovery of epinephrine response but not hypoglycemic symptom threshold after intensive therapy in type 1 diabetes. *Am. J. Med.* **97**(6), 535–542.

Davis, S. N., Shavers, C., and Costa, F. (2000a). Differential gender responses to hypoglycemia are due to alterations in CNS drive and not glycemic thresholds. *AJP Endocrinol. Metab.* **279**(5), E1054–E1063.

Davis, S. N., Shavers, C., and Costa, F. (2000b). Gender-related differences in counterregulatory responses to antecedent hypoglycemia in normal humans. *J. Clin. Endocrinol. Metab.* **85**(6), 2148–2157.

Davis, S. N., Shavers, C., Davis, B., and Costa, F. (1997). Prevention of an increase in plasma cortisol during hypoglycemia preserves subsequent counterregulatory responses. *J. Clin. Invest.* **100,** 429–438.

Davis, S. N., Shavers, C., Collins, L., Cherrington, A. D., Price, L., and Hedstrom, C. (1994b). Effects of physiological hyperinsulinemia on counterregulatory response to prolonged hypoglycemia in normal humans. *Am. J. Physiol.* **267**(3: Pt 1):Pt 1), E402–10.

Davis, S. N., Dobbins, R., Colburn, C., Tarumi, C., Jacobs, J., Neal, D., and Cherrington, A. D. (1993a). Effects of hyperinsulinemia on the subsequent hormonal response to hypoglycemia in conscious dogs. *Am. J. Physiol.* **264**(5:Pt 1), E748–55.

Davis, S. N., Goldstein, R. E., Jacobs, J., Price, L., Wolfe, R., and Cherrington, A. D. (1993b). The effects of differing insulin levels on the hormonal and metabolic response to equivalent hypoglycemia in normal humans. *Diabetes* **42**(2), 263–272.

De Vivo, D. C., Trifiletti, R. R., Jacobson, R. I., Ronen, G. M., Behmand, R. A., and Harik, S. I. (1991). Defective glucose transport across the blood-brain barrier as a cause of persistent hypoglycorrhachia, seizures, and developmental delay. *N. Engl. J. Med.* **325,** 703–709.

Deary, I. J. (1993). Effects of hypoglycaemia on cognitive function. *In* "Hypoglycaemia and Diabetes: Clinical. and Physiological Aspects" (B. Frier and M. Fisher, eds.), pp. 80–92. Edward Arnold, London.

The DECODE Study Group. European Diabetes Epidemiology Group. Diabetes Epidemiology: Collaborative Analysis Of Diagnostic criteria in Europe. (1999). Glucose tolerance and mortality: Comparison of WHO and American Diabetic Association diagnostic criteria. *Lancet* **354**(9179), 617–621.

Dejgaard, A., Gade, A., Larsson, H., Balle, V., Parving, A., and Parving, H. H. (1991). Evidence for diabetic encephalopathy. *Diabet. Med.* **8,** 162–167.

Di Giulio, A. M., Tenconi, B., La Croix, R., Mantegazza, P., Abbracchio, M. P., Cattabeni, F., and Gorio, A. (1989). Denervation and hyperinnervation in the nervous system of diabetic animals. II. Monoaminergic and peptidergic alterations in the diabetic encephalopathy. *J. Neurosci. Res.* **24,** 362–368.

The Diabetes Control and Complications Trial (DCCT) Research Group (1993). The effect of intensive treatment of diabetes on the development and progression of long-term complications in insulin-dependent diabetes mellitus. *N. Engl. J. Med.* **329,** 977–986.

The Diabetes Control and Complications Trial (DCCT) Research Group (1997). Hypoglycemia in the diabetes control and complications trial. *Diabetes* **46**(2), 271–286.

Draelos, M. T., Jacobson, A. M., Weinger, K., Widom, B., Ryan, C. M., Finkelstein, D. M., and Simonson, D. C. (1995). Cognitive function in patients with insulin-dependent diabetes mellitus during hyperglycemia and hypoglycemia. *Am. J. Med.* **98**(2), 135–144.

Duckrow, R. B., Beard, D. C., and Brennan, R. W. (1987). Regional cerebral blood flow decreases during chronic and acute hyperglycemia. *Stroke* **18,** 52–58.

Fanelli, C. G., Dence, C. S., Markham, J., Videen, T. O., Paramore, D. S., Cryer, P. E., and Powers, W. J. (1998). Blood-to-brain glucose transport and cerebral glucose metabolism are not reduced in poorly controlled type 1 diabetes. *Diabetes* **47**(9), 1444–1450.

Fanelli, C. G., Epifano, L., Rambotti, A. M., Pampanelli, S., Di Vincenzo, A., Modarelli, F., Lepore, M., Annibale, B., Ciofetta, M., Bottini, P., Porcellati, F., Scionti, L., Santeusanio, F., Brunetti, P., and Bolli, G. B. (1993). Meticulous prevention of hypoglycemia normalizes the glycemic thresholds and magnitude of most of neuroendocrine responses to, symptoms of, and cognitive function during hypoglycemia in intensively treated patients with short-term IDDM. *Diabetes* **42**(11), 1683–1689.

Fanelli, C. G., Paramore, D. S., Hershey, T., Terkamp, C., Ovalle, F., Craft, S., and Cryer, P. E. (1998). Impact of nocturnal hypoglycemia on hypoglycemic cognitive dysfunction in type 1 diabetes. *Diabetes* **47**(12), 1920–1927.

Figlewicz, D. P., Brot, M. D., McCall, A. L., and Szot, P. (1996). Differential gene regulation of CNS noradrenergic and dopaminergic neurons by diabetes. *Brain Res.* **736,** 54–60.

Figlewicz, D. P., Szot, P., Chavez, M., Woods, S. C., and Veith, R. C. (1994). Intraventricular insulin increases dopamine transporter mRNA in rat VTA/substantia nigra. *Brain Res.* **644**(2), 331–334.

Figlewicz, D. P., Szot, P., Israel, P. A., Payne, C., and Dorsa, D. M. (1993). Insulin reduces norepinephrine transporter mRNA in vivo in rat locus coeruleus. *Brain Res.* **602**(1), 161–164.

Fletcher, A. A., and Campbell, W. R. (1922). The blood sugar following insulin administration and the symptom complex-hypoglycaemia. *J. Metab. Res.* **(2),** 637–649.

Flood, J. F., Mooradian, A. D., and Morley, J. E. (1990). Characteristics of learning and memory in streptozocin-induced diabetic mice. *Diabetes* **39,** 1391–1398.

Franceschi, M., Cecchetto, R., Minicucci, F., Smizne, S., Baio, G., and Canal, N. (1984). Cognitive processes in insulin-dependent diabetes. *Diabetes Care* **7,** 228–231.

Fritsche, A., Stumvoll, M., Grub, M., Sieslack, S., Renn, W., Schmulling, R. M., Haring, H. U., and Gerich, J. E. (1998). Effect of hypoglycemia on beta-adrenergic sensitivity in normal and type 1 diabetic subjects. *Diabetes Care* **21**(9), 1505–1510.

Fritsche, A., Stumvoll, M., Haring, H. U., and Gerich, J. E. (2000). Reversal of hypoglycemia unawareness in a long-term type 1 diabetic patient by improvement of beta-adrenergic sensitivity after prevention of hypoglycemia. *J. Clin. Endocrinol. Metab.* **85**(2), 523–525.

Frizzell, R. T., Jones, E. M., Davis, S. N., Biggers, D. W., Myers, S. R., Connolly, C. C., Neal, D. W., Jaspan, J. B., and Cherrington, A. D. (1993). Counterregulation during hypoglycemia is directed by widespread brain regions. *Diabetes* **42,** 1253–1261.

Fushimi, H., Inoue, T., Yamada, Y., Udaka, F., and Kameyama, M. (1994). Asymptomatic lacunes and their relationship to intellectual disturbances. *Int. Med.* **33**(6), 317–320.

Galassetti, P., Neill, A. R., Tate, D., Ertl, A. C., Wasserman, D. H., and Davis, S. N. (2001). Sexual dimorphism in counterregulatory responses to hypoglycemia after antecedent exercise. *J. Clin. Endocrinol. Metab.* **86**(8), 3516–3524.

Ganzini, L., Casey, D. E., Hoffman, W. F., and McCall, A. L. (1993). The prevalence of metoclopramide-induced tardive dyskinesia and acute extrapyramidal movement disorders. *Arch. Intern. Med.* **153,** 1469–1475.

George, E, Marques, J. L., Harris, N. D., Macdonald, I. A., Hardisty, C. A., and Heller, S. R. (1997). Preservation of physiological responses to hypoglycemia 2 days after antecedent hypoglycemia in patients with IDDM. *Diabetes Care.* **20,** 1293–1298.

George, E., Harris, N., Bedford, C., Macdonald, I. A., Hardisty, C. A., and Heller, S. R. (1995). Prolonged but partial impairment of the hypoglycaemic physiological response following short-term hypoglycaemia in normal subjects. *Diabetologia* **38**(10), 1183–1190.

Gispen, W. H., and Biessels, G. J. (2000). Cognition and synaptic plasticity in diabetes mellitus. [Review] [76 refs]. *Trends Neurosci.* **23**(11), 542–549.

Gold, A. E., MacLeod, K. M., and Frier, B. M. (1994). Frequency of severe hypoglycemia in patients with type I diabetes with impaired awareness of hypoglycemia. *Diabetes Care* **17**(7), 697–703.

Gold, A. E., MacLeod, K. M., Deary, I. J., and Frier, B. M. (1995). Hypoglycemia-induced cognitive dysfunction in diabetes mellitus: effect of hypoglycemia unawareness. *Physiol. Behav.* **58**(3), 501–511.

Gold, P. E. (1995). Role of glucose in regulating the brain and cognition. [Review]. *Am. J. Clin. Nutr.* **61**(4:Suppl.), 987S–995S.

Gonder-Frederick, L. A., Clarke, W. L., and Cox, D. J. (1997). The emotional, social, and behavioral implications of insulin-induced hypoglycemia. *Semin. Clin. Neuropsychiatry* **2**(1), 57–65.

Grill, V., Gutniak, M., Bjorkman, O., Lindqvist, M., Stone-Elander, S., Seitz, R. J., Blomqvist, G., Reichard, P., and Widen, L. (1990). Cerebral blood flow and substrate utilization in insulin-treated diabetic subjects. *Am. J. Physiol.* **258,** E813–E820.

Hamilton, M. G., Tranmer, B. I., and Auer, R. N. (1995). Insulin reduction of cerebral infarction due to transient focal ischemia. *J. Neurosurg.* **82**(2), 262–268.

Harris, M. I., and Eastman, R. C. (2000). Early detection of undiagnosed diabetes mellitus: A US perspective. *Diabetes Metab Res. Rev.* **16**(4), 230–232.

Harris, R. J., Wieloch, T., Symon, L., and Siesjö, B. K. (1984). Cerebral extracellular calcium activity in severe hypoglycemia: relation to extracellular potassium and energy state. *J. Cereb. Blood Flow Metab.* **4,** 187–193.

Hauser, E., Strohmayer, C., Seidl, R., Birnbacher, R., Lischka, A., and Schober, E. (1995). Quantitative EEG in young diabetics. *J. Child. Neurol.* **10**(4), 330–334.

Havel, P. J., Hahn, T. M., Sindelar, D. K., Baskin, D. G., Dallman, M. F., Weigle, D. S., and Schwartz, M. W. (2000). Effects of streptozotocin-induced diabetes and insulin treatment on the hypothalamic melanocortin system and muscle uncoupling protein 3 expression in rats. *Diabetes* **49**(2), 244–252.

Helgason, C. M. (1988). Blood glucose and stroke. *Stroke* **19,** 1049–1053.

Heller, S. R., and Cryer, P. E. (1991). Reduced neuroendocrine and symptomatic responses to subsequent hypoglycemia after 1 episode of hypoglycemia in nondiabetic humans. *Diabetes* **40,** 223–226.

Hingorani, V., and Brecher, P. (1987). Glucose and fatty acid metabolism in normal and diabetic rabbit cerebral microvessels. *Am. J. Physiol.* **252,** E648–E653.

Hirsch, B. R., and Shamoon, H. (1987). Defective epinephrine and growth hormone responses in type 1 diabetes are stimulus specific. *Diabetes* **36,** 20–26.

Holmes, C. S., Hayford, J. T., Gonzalez, J. L., and Weydert, J. A. (1983). A survey of cognitive functioning at difference glucose levels in diabetic persons. *Diabetes Care* **6,** 180–185.

Hoyer, S. (1994). Oxidative metabolism deficiencies in brains of patients with Alzheimer's disease. [Review] [90 refs]. *Acta Neurol. Scand.* (Suppl.), 16–24.

Inui, K., Sannan, H., Ota, H., Uji, Y., Nomura, S., Kaige, H., Kitayama, I., and Nomura, J. (1998). EEG findings in diabetic patients with and without retinopathy. *Acta Neurol. Scand.* **97**(2), 107–109.

Jakobsen, J., Nedergaard, M., Aarslew-Jensen, M., and Diemer, N. H. (1990). Regional brain glucose metabolism and blood flow in streptozocin-induced diabetic rats. *Diabetes* **39,** 437–440.

Johnson, P. C., Brendel, K., and Meezan, E. (1982). Thickened cerebral cortical capillary basement membranes in diabetics. *Arch. Pathol. Lab. Med.* **106,** 214–217.

Jones, T. W., McCarthy, G., Tamborlane, W. V., Caprio, S., Roessler, E., Kraemer, D., Starick Zych, K., Allison, T., Boulware, S. D., and Sherwin, R. S. (1990). Mild hypoglycemia and impairment of brain stem and cortical evoked potentials in healthy subjects. *Diabetes* **39,** 1550–1555.

Kalimo, H., Auer, R. N., and Siesjö, B. K. (1985). The temporal evolution of hypoglycemic brain damage. III. Light and electron microscopic findings in the rat caudoputamen. *Acta Neuropathol.* (*Berl.*) **67,** 37–50.

Kalmijn, S., Feskens, E. J., Launer, L. J., Stijnen, T., and Kromhout, D. (1995). Glucose intolerance, hyperinsulinaemia and cognitive function in a general population of elderly men. *Pharmacol. Ther.* **38**(9), 1096–1102.

Kastrup, J., Rrsgaard, S., Parving, H. H., and Lassen, N. A. (1986). Impaired autoregulation of cerebral blood flow in long-term type I (insulin-dependent) diabetic patients with nephropathy and retinopathy. *Clin. Physiol.* **6,** 549–559.

Kety, S. S., Polis, B. D., Nadler, C. S., and Schmidt, C. F. (1948). Blood flow and oxygen consumption of the human brain in diabetic acidosis and coma. *J. Clin. Invest.* **27,** 500–510.

Kinnala, A., Suhonen-Polvi, H., Aarimaa, T., Kero, P., Korvenranta, H., Ruotsalainen, U., Bergman, J., Haaparanta, M., Solin, O., Nuutila, P., and Wegelius, U. (1996). Cerebral metabolic rate for glucose during the first six months of life: an FDG positron emission tomography study. *Arch. Dis. Child. Fetal Neonatal Ed.* **74**(3), F153–F157.

Klepper, J., Garcia-Alvarez, M., O'Driscoll, K. R., Parides, M. K., Wang, D., Ho, Y. Y., and De Vivo, D. C. (1999a). Erythrocyte 3-*O*-methyl-D-glucose uptake assay for diagnosis of glucose-transporter-protein syndrome. *J. Clin. Lab. Anal.* **13**(3), 116–121.

Klepper, J., Wang, D., Fischbarg, J., Vera, J. C., Jarjour, I. T., O'Driscoll, K. R., and De Vivo, D. C. (1999b). Defective glucose transport across brain tissue barriers: A newly recognized neurological syndrome. *Neurochem. Res.* **24**(4), 587–594.

Koranyi, L., Bourey, R. E., James, D., Mueckler, M., Fiedorek, G. T., Jr., and Permutt, M. A. (1991). Glucose transporter gene expression in rat brain: Pretranslational changes associated with chronic insulin-induced hypoglycemia, fasting, and diabetes. *Mol. Cell. Neurosci.* **2,** 244–252.

Korol, D. L., and Gold, P. E. (1998). Glucose, memory, and aging. *Am. J. Clin. Nutr.* **67**(4), 764S–771S.

Korytkowski, M. T., Mokan, M., Veneman, T. F., Mitrakou, A., Cryer, P. E., and Gerich, J. E. (1998). Reduced beta-adrenergic sensitivity in patients with type 1 diabetes and hypoglycemia unawareness. *Diabetes Care* **21**(11), 1939–1943.

Kovatchev, B., Cox, D., and Summeers, K. (2001). Do people with T1DM recognize Hyper glycemia? *Diabetes* **50**(Suppl. 2), Abstr. #1643.

Kovatchev, B. P., Cox, D. J., Farhy, L. S., Straume, M., Gonder-Frederick, L., and Clarke, W. L. (2000). Episodes of severe hypoglycemia in type 1 diabetes are preceded and followed within 48 hours by measurable disturbances in blood glucose. *J. Clin. Endocrinol. Metab.* **85**(11), 4287–4292.

Kuller, L. H. (1995). Stroke and diabetes. *In* "Diabetes in America" National Institutes of Health, pp. 449–456.

Kumagai, A. K. (1999). Glucose transport in brain and retina: Implications in the management and complications of diabetes. *Diabetes/Metab. Res. Rev.* **15**(4), 261–273.

Kumagai, A. K., Kang, Y. S., Boado, R. J., and Pardridge, W. M. (1995). Upregulation of blood-brain barrier GLUT1 glucose transporter protein and mRNA in experimental chronic hypoglycemia. *Diabetes* **44**(12), 1399–1404.

LaManna, J. C., and Harik, S. I. (1986). Regional studies of blood-brain barrier transport of glucose and leucine in awake and anesthetized rats. *J. Cereb. Blood Flow Metab.* **6,** 717–723.

Lass, P., Knudsen, G. M., Pedersen, E. V., and Barry, D. I. (1989). Impaired beta-adrenergic mediated cerebral blood flow response in streptozotocin diabetic rats. *Pharmacol. Toxicol.* **65,** 318–320.

Lawrence, M. S, Sun, G. H., Kunis, D. M., Saydam, T. C., Dash, R., Ho, D. Y., Sapolsky, R. M., and Steinberg, G. K. (1996). Overexpression of the glucose transporter gene with a herpes simplex viral vector protects striatal neurons against stroke. *J. Cereb. Blood Flow Metab.* **16**(2), 181–185.

Lewis, L. D., Ljunggren, B., Norberg, K., and Siesjö, B. K. (1974). Changes in carbohydrate substrates, amino acids and ammonia in the brain during insulin-induced hypoglycemia. *J. Neurochem.* **23,** 659–671.

Lins, P.-E., and Adamson, U. (1994). Neurological manifestations of hypoglycaemia. *In* "Hypoglycaemia and Diabetes: Clinical and Physiological Aspects" (B. Frier and M. Fisher, eds.), pp. 347–354. Edward Arnold, London.

Lozovsky, D., Saller, C. F., and Kopin, I. J. (1981). Dopamine receptor binding is increased in diabetic rats. *Science* **214,** 1031–1033.

Malmberg, K. (1997). Prospective randomised study of intensive insulin treatment on long term survival after acute myocardial infarction in patients with diabetes mellitus. DIGAMI (Diabetes Mellitus, Insulin Glucose Infusion in Acute Myocardial Infarction) Study Group. *BMJ* **314**(7093), 1512–1515.

Manjarrez-Gutierrez, G., Rocio Herrera-Marquez, J. R., Bueno-Santoyo, S., Gonzalez-Ramirez, M., and Hernandez, J. (2000). Changes in brain serotonin biosynthesis in rats with diabetes mellitus induced by streptozocin: effect of insulin treatment. [Spanish] *Revista de Investigacion Clinica* **52,** 509–516.

Manjarrez-Gutierrez, G., Rocio, J. R., Herrera-Marquez, S., Bueno-Santoyo, M., Gonzalez-Ramirez, and Hernandez, J. (2000). Changes in brain serotonin biosynthesis in rats with diabetes mellitus induced by streptozocin: effect of insulin treatment. [Spanish]. *Rev. Investi. Clin.* **52**(5), 509–516.

Manson, J. E., Colditz, G. A., Stampfer, M. J., Willett, W. C., Krolewski, A. S., Rosner, B., Arky, R. A., Speizer, F. E., and Hennekens, C. H. (1991). A prospective study of maturity-onset diabetes mellitus and risk of coronary heart disease and stroke in women. *Arch. Intern. Med.* **151**(6), 1141–1147.

McCall, A. L. (1998). New findings and treatment strategies for hypoglycemia and hypoglycemia unawareness. *Curr. Opin. Endocrinol. Diabetes* **5,** 136–143.

McCall, A. L., Van Bueren, A. M., Huang, L., Stenbit, A., Celnik, E., and Charron, M. J. (1997). Forebrain endothelium expresses GLUT4, the insulin-responsive glucose transporter. *Brain Res.* **744**(2), 318–326.

McCall, A. L., and Figlewicz, D. P. (1997). How does diabetes mellitus produce brain dysfunction?. *Diabetes Spect.* **10**(1), 25–32.

McCall, A. L. (1997). IDDM, counterregulation, and the brain. *Diabetes Care* **20**(8), 1228–1230.

McCall, A. L. (1993). Effects of glucose deprivation on glucose metabolism in the central nervous system *In* "Hypoglycaemia and Diabetes Mellitus" (B. M. Frier and M. Fisher, eds.), pp. 56–71. Edward Arnold, Oxford.

McCall, A. L. (1992a). The impact of diabetes on the CNS. *Diabetes* **41**(5), 557–570.

McCall, A. L. (1992b). Cerebral microvascular transport and metabolism: Implications for diabetes. *In* "Hyperglycemia, Diabetes, and Vascular Disease" (N. Ruderman, J. Williamson, and M. Brownlee, eds.), pp. 59–103. Oxford University Press, New York.

McCall, A. L., Sussman, I., Tornheim, K., Cordero, R., and Ruderman, N. B. (1988). Effects of hypoglycemia and diabetes on fuel metabolism by rat brain microvessels. *Am. J. Physiol.* **254** (End. Metab. 17), E272–E278.

McCall, A. L., Fixman, L. B., Fleming, N., Tornheim, K., Chick, W., and Ruderman, N. B. (1986). Chronic hypoglycemia increases brain glucose transport. *Am. J. Physiol.* **251,** E442–E447.

McCall, A. L., Gould, J. B., and Ruderman, N. B. (1984). Diabetes-induced alterations of glucose metabolism in rat cerebral microvessels. *Am. J. Physiol.* 247 (End. Metab. 10), E462–E467.

Meissner, T., Beinbrech, B., and Mayatepek, E. (1999). Congenital hyperinsulinism: molecular basis of a heterogeneous disease. [Review] [80 refs]. *Hum. Mutat.* **13**(5), 351–361.

Mitrakou, A., Ryan, C., Veneman, T., Mokan, M., Jenssen, T., Kiss, I., Durrant, J., Cryer, P., and Gerich, J. (1991). Hierarchy of glycemic thresholds for counterregulatory hormone secretion, symptoms, and cerebral dysfunction. *Am. J. Physiol.* **260**(1:Pt 1), E67–74.

Mooradian, A. D. (1997a). Central nervous system complications of diabetes mellitus—a perspective from the blood-brain barrier. *Brain Res. Brain Res. Rev.* **23**(3), 210–218.

Mooradian, A. D. (1997b). Pathophysiology of central nervous system complications in diabetes mellitus. *Clin. Neurosci.* **4**(6), 322–326.

Mooradian, A. D., Perryman, K., Fitten, J., Kavonian, G. D., and Morley, J. E. (1988a). Cortical function in elderly non-insulin dependent diabetic patients. Behavioral and electrophysiologic studies. *Arch. Intern. Med.* **148,** 2369–2372.

Mooradian, A. D. (1988b). Diabetes and the central nervous system. *Endocrinol. Rev.* **9**(3), 346–356.

Mukai, N., Hori, S., and Pomeroy, M. (1980). Cerebral lesions in rats with streptozotocin-induced diabetes. *Acta Neuropathol. (Berlin)* **51,** 79–84.

Nagamatsu, S., Sawa, H., Inoue, N., Nakamichi, Y., Takeshima, H., and Hoshino, T. (1994). Gene expression of GLUT3 glucose transporter regulated by glucose in vivo in mouse brain and in vitro in neuronal cell cultures from rat embryos. *Biochem. J.* **300,** 125–131.

Neaton, J. D., Wentworth, D. N., Cutler, J., Stamler, J., and Kuller, L. (1993). Risk factors for death from different types of stroke. Multiple Risk Factor Intervention Trial Research Group. *Ann. Epidemiol.* **3**(5), 493–499.

Neil, H. A., Gale, E. A., Hamilton, S. J., Lopez Espinoza, I., Kaura, R., and McCarthy, S. T. (1987). Cerebral blood flow increases during insulin-induced hypoglycaemia in type 1 (insulin-dependent) diabetic patients and control subjects. *Diabetologia* **30,** 305–309.

Norberg, K., and Siesjö, B. K. (1976). Oxidative metabolism of the cerebral cortex of the rat in severe insulin-induced hypoglycemia. *J. Neurochem.* **26,** 345–352.

Nordli, D. R., Jr., and De Vivo, D. C. (1997). The ketogenic diet revisited: Back to the future [editorial; comment] [see comments]. [Review] [41 refs]. *Epilepsia* **38**(7), 743–749.

Ohtani, N., Ohta, M., and Sugano, T. (1997). Microdialysis study of modification of hypothalamic neurotransmitters in streptozotocin-diabetic rats. *J. Neurochem.* **69**(4), 1622–1628.

Pardridge, W. M. (1994). Glucose transport and phosphorylation: which is rate limiting for brain glucose utilization? [editorial; comment] [see comments]. *Ann. Neurol.* **35**(5), 511–512.

Pardridge, W. M., Triguero, D., and Farrel, C. R. (1990). Downregulation of blood-brain barrier glucose transporter in experimental diabetes. *Diabetes* **39,** 1040–1044.

Pardridge, W. M. (1983). Brain metabolism: a perspective from the blood-brain barrier. *Physiol. Rev.* **63,** 1481–1535.

Pelligrino, D. A., LaManna, J. C., Duckrow, R. B., Bryan, R. M., Jr., and Harik, S. I. (1992). Hyperglycemia and blood-brain barrier glucose transport. *J. Cereb. Blood Flow Metab.* **12,** 887–899.

Pelligrino, D. A., and Albrecht, R. F. (1991). Chronic hyperglycemic diabetes in the rat is associated with a selective impairment of cerebral vasodilatory responses. *J. Cereb. Blood Flow Metab.* **11,** 667–677.

Pelligrino, D. A., Segil, L. J., and Albrecht, R. F. (1990). Brain glucose utilization and transport and cortical function in chronic vs. acute hypoglycemia. *American J. Physiol.* **259,** E729–E735.

Pelligrino, D., Almquist, L. O., and Siesjö, B. K. (1981). Effects of insulin-induced hypoglycemia on intracellular pH and impedance in the cerebral cortex of the rat. *Brain Res.* **221,** 129–147.

Pelligrino, D., and Siesjö, B. K. (1981). Regulation of extra- and intracellular pH in the brain in severe hypoglycemia. *J. Cereb. Blood Flow Metab.* **1,** 85–96.

Pennebaker, J. W., Cox, D. J., Gonder-Frederick, L., Wunsch, M. G., Evans, W. S., and Pohl, S. (1981). Physical symptoms related to blood glucose in insulin-dependent diabetics. *Psychosom. Med.* **43**(6), 489–500.

Perlmuter, L. C., Nathan, D. M., Goldfinger, S. H., Russo, P. A., Yates, J., and Larkin, M. (1988). Triglyceride levels affect cognitive function in noninsulin-dependent diabetics. *J. Diab. Complic.* **2,** 210–213.

Perlmuter, L., Hakami, M., Harrington, C., Ginsberg, J., Katz, J., Singer, D., and Nathan, D. (1984). Decreased cognitive function in aging non-insulin-dependent diabetic patients. *Am. J. Med.* **77,** 1043–1048.

Perros, P., Deary, I. J., Sellar, R. J., Best, J. J., and Frier, B. M. (1997). Brain abnormalities demonstrated by magnetic resonance imaging in adult IDDM patients with and without a history of recurrent severe hypoglycemia. *Diab. Care* **20,** 1013–1018.

Perros, P., and Frier, B. M. (1997). The long-term sequelae of severe hypoglycemia on the brain in insulin-dependent diabetes mellitus. *Horm. Metab. Res.* **29,** 197–202.

Perros, P., Sellar, R. J., and Frier, B. M. (1994). Chronic pontine dysfunction following insulin-induced hypoglycemia in an IDDM patient. *Diabetes Care* **17**(7), 725–727.

Pozzessere, G., Valle, E., de Crignis, S., Cordischi, V. M., Fattapposta, F., Rizzo, P. A., Pietravalle, P., Cristina, G., Morano, S., and di Mario, U. (1991). Abnormalities of cognitive functions in IDDM revealed by P300 event-related potential analysis. Comparison with short-latency evoked potentials and psychometric tests. *Diabetes* **40,** 952–958.

Pramming, S., Thorsteinsson, B., Theilgaard, A., Pinner, E. M., and Binder, C. (1986). Cognitive function during hypoglycaemia in type I diabetes mellitus. *Br. Med. J.* (*Clin. Res. Ed.*) **292**(6521), 647–650.

Pulsinelli, W. A., Levy, D. E., Sigsbee, B., Scherer, P., and Plum, F. (1983). Increased damage after ischemic stroke in patients with hyperglycemia with or without established diabetes mellitus. *Am. J. Med.* **74,** 540–544.

Pyorala, M., Miettinen, H., Halonen, P., Laakso, M., and Pyorala, K. (2000). Insulin resistance syndrome predicts the risk of coronary heart disease and stroke in healthy middle-aged men: the 22-year follow- up results of the Helsinki Policemen Study. *Arterioscler. Thromb. Vasc. Biol.* **20**(2), 538–544.

Rapoport, S. I. (1997). Discriminant analysis of brain imaging data identifies subjects with early Alzheimer's disease. *Int. Psychogeriatr.* **9,** Suppl 1, 229–35.

Rattarasarn, C., Dagogo-Jack, S., Zachwieja, J. J., and Cryer, P. E. (1994). Hypoglycemia-induced autonomic failure in IDDM is specific for stimulus of hypoglycemia and is not attributable to prior autonomic activation. *Diabetes* **43**(6), 809–818.

Reagan, L. P., Gorovits, N., Hoskin, E. K., Alves, S. E., Katz, E. B., Grillo, C. A., Piroli, G. G., McEwen, B. S., and Charron, M. J. (2001). Localization and regulation of GLUTx1 glucose transporter in the hippocampus of streptozotocin diabetic rats. *Proc. Natl. Acad. Sci. USA* **98**(5), 2820–2825.

Reaven, G. M., Thompson, L. W., Nahum, D., and Haskins, E. (1990). Relationship between hyperglycemia and cognitive function in older NIDDM patients. *Diabetes Care* **13,** 16–21.

Reske-Nielsen, E., Lundback, K., and Rafaelsen, O. J. (1965). Pathological changes in the central and peripheral nervous system of young long-term diabetics. *Diabetologia* **1,** 233–237.

Riddle, M. C., and Hart, J. (1982). Hyperglycemia, recognized and unrecognized, as a risk factor for stroke and transient ischemic attacks. *Stroke* **13,** 356–359.

Ruderman, N. B., Ross, P. S., Berger, M., and Goodman, M. N. (1974). Regulation of glucose and ketone-body metabolism in brain of anaesthetized rats. *Biochem. J.* **138,** 1–10.

Ryan, C., Longstreet, C., and Morrow, L. (1985). The effects of diabetes mellitus on the school attendance and school achievement of adolescents. *Child Care Health Dev.* **11,** 229–240.

Sawka, A. M., Burgart, V., and Zimmerman, D. (2000). Loss of hypoglycemia awareness in an adolescent with type 1 diabetes mellitus during treatment with fluoxetine hydrochloride. *J. Pediatr.* **136**(3), 394–396.

Segel, S. A., Fanelli, C. G., Dence, C. S., Markham, J., Videen, T. O., Paramore, D. S., Powers, W. J., and Cryer, P. E. (2001). Blood-to-brain glucose transport, cerebral glucose metabolism, and cerebral blood flow are not increased after hypoglycemia. *Diabetes* **50**(8), 1911–1917.

Seidner, G., Alvarez, M. G., Yeh, J. I., O'Driscoll, K. R., Klepper, J., Stump, T. S., Wang, D., Spinner, N. B., Birnbaum, M. J., and De Vivo, D. C. (1998). GLUT-1 deficiency syndrome caused by haploinsufficiency of the blood-brain barrier hexose carrier. *Nat. Genet.* **18**(2), 188–191.

Shah, G. N., Giddings, S. J., and Mooradian, A. D. (1997). Shortening of poly (A) tail of glucose transporter—one mRNA in experimental diabetes mellitus. *Brain Res.* **754**(1–2), 213–220.

Sharma, N., Crane, A., Gonzalez, G., Bryan, J., and Aguilar-Bryan, L. (2000). Familial hyperinsulinism and pancreatic beta-cell ATP-sensitive potassium channels. [Review] [30 refs]. *Kidney Int.* **57**(3), 803–808.

Shepherd, P. R., Gibbs, E. M., Wesslau, C., Gould, G. W., and Kahn, B. B. (1992). Human small intestine facilitative fructose/glucose transporter (GLUT5) is also present in insulin-responsive tissues and brain. Investigation of biochemical characteristics and translocation. *Diabetes* **41**(10), 1360–1365.

Siesjö, B. K. (1988). Hypoglycemia, brain metabolism, and brain damage. *Diabetes/Metab. Rev.* **4,** 113–144.

Simpson, I. A., Appel, N. M., Hokari, M., Oki, J., Holman, G. D., Maher, F., Koehler-Stec, E. M., Vannucci, S. J., and Smith, Q. R. (1999). Blood-brain barrier glucose transporter: effects of hypo- and hyperglycemia revisited. *J. Neurochem.* **72**(1), 238–247.

Simpson, I. A., Chundu, K. R., Davies-Hill, T., Honer, W. G., and Davies, P. (1994). Decreased concentrations of GLUT1 and GLUT3 glucose transporters in the brains of patients with Alzheimer's disease. *Ann. Neurol.* **35**(5), 546–561.

Simpson, I. A., and Davies, P. (1994). Reduced glucose transporter concentrations in brains of patients with Alzheimer's disease [letter; comment]. *Ann. Neurol.* **36**(5), 800–801.

Sipols, A. J., Baskin, D. G., and Schwartz, M. W. (1995). Effect of intracerebroventricular insulin infusion on diabetic hyperphagia and hypothalamic neuropeptide gene expression. *Diabetes* **44**(2), 147–151.

Soltesz, G., and Acsadi, G. (1989). Association between diabetes, severe hypoglycaemia, and electroencephalographic abnormalities. *Arch. Dis. Child.* **64**(7), 992–996.

Sue, C. M., Hirano, M., DiMauro, S., and De Vivo, D. C. (1999). Neonatal presentations of mitochondrial metabolic disorders. *Semin. Perinatol.* **23**(2), 113–124.

Thompson, S. C., Woods, S. C., Hendricks, S., Bell, S. M., and Figlewicz, D. P. (2000). Intraventricular insulin suppresses the acoustic startle response in rats. *Physiol. Behav.* **69,** 433–437.

Tkac, I., Troscak, M., Javorsky, M., Petrik, R., and Tomcova, M. (2001). Increased intracranial arterial resistance in patients with type 2 diabetes mellitus. *Wien. Klin. Wochenschr.* **113**(22), 870–873.

Trulson, M. E., Jacoby, J. H., and MacKenzie, R. G. (1986). Streptozotocin-induced diabetes reduces brain serotonin synthesis in rats. *J Neurochem.* **46**(4), 1068–1072.

Tun, P. A., Nathan, D. M., and Perlmuter, L. C. (1990). Cognitive and affective disorders in elderly diabetics. *Clin. Geriatr. Med.* **6,** 731–746.

Tun, P. A., Perlmuter, L. C., Russo, P., and Nathan, D. M. (1987). Memory self-assessment and performance in aged diabetics and non-diabetics. *Exp. Aging Res.,* **13,** 151–157.

Tuomilehto, J., Rastenyte, D., Jousilahti, P., Sarti, C., and Vartiainen, E. (1996). Diabetes mellitus as a risk factor for death from stroke: Prospective study of the middle-aged Finnish population. *Stroke* **27**(2), 210–215.

Uehara, Y., Nipper, V., and McCall, A. L. (1997). Chronic insulin hypoglycemia induces GLUT-3 protein in rat brain neurons. *Am. J. Physiol. (Endocrinol. Metab.)* **272**(4), E716–E719.

Vannucci, R. C., and Vannucci, S. J. (2001). Hypoglycemic brain injury. *Semin. Neonatol.* **6**(2), 147–155.

Vannucci, S. J., Willing, L. B., Goto, S., Alkayed, N. J., Brucklacher, R. M., Wood, T. L., Towfighi, J., Hurn, P. D., and Simpson, I. A. (2001). Experimental stroke in the female diabetic, db/db, mouse. *J. Cereb.Blood Flow Metab.* **21**(1), 52–60.

Vannucci, S. J., Clark, R. R., Koehler-Stec, E., Li, K., Smith, C. B., Davies, P., Maher, F., and Simpson, I. A. (1998). Glucose transporter expression in brain: relationship to cerebral glucose utilization. *Dev. Neurosci.* **20**(4–5), 369–379.

Vannucci, S. J., Maher, F., and Simpson, I. A. (1997). Glucose transporter proteins in brain: delivery of glucose to neurons and glia. [Review] [140 refs]. *Glia* **21**(1), 2–21.

Voll, C. L., and Auer, R. N. (1991). Insulin attenuates ischemic brain damage independent of its hypoglycemic effect. *J. Cereb. Blood Flow Metab.* **11,** 1006–1014.

Wakisaka, M., Nagamachi, S., Inoue, K., Morotomi, Y., Nunoi, K., and Fujishima, M. (1990). Reduced regional cerebral blood flow in aged noninsulin-dependent diabetic patients with no history of cerebrovascular disease: evaluation by N-isopropyl-123I-p-iodoamphetamine with single-photon emission computed tomography. *J. Diabetic Complications* **4,** 170–174.

White, N. H., Skor, D. A., Cryer, P. E., Levandoski, L. A., Bier, D. M., and Santiago, J. V. (1983). Identification of type I diabetic patients at increased risk for hypoglycemia during intensive therapy. *N. Engl. J. Med.* **308,** 485–491.

Wieloch, T. (1985). Hypoglycemia-induced neuronal damage prevented by an *N*-Methyl-D-Aspartate antagonist. *Science* **230,** 681–683.

Wieloch, T., Harris, R. J., Symon, L., and Siesjö, B. K. (1984). Influence of severe hypoglycemia on brain extracellular calcium and potassium activities, energy, and phospholipid metabolism. *J. Neurochemi.* **43,** 160–168.

Woo, E., Ma, J. T., Robinson, J. D., and Yu, Y. L. (1988). Hyperglycemia is a stress response in acute stroke. *Stroke* **19,** 1359–1364.

Wortis, J., Bowman, K. M., Orenstein, L. I., and Rosenbaum, I. J. (1939). Further experiences at Bellevue Hospital with the hypoglycemic insulin treatment of schizophrenia. *Am. J. Psychiatry* **94,** 153–158.

DIABETES, THE BRAIN, AND BEHAVIOR: IS THERE A BIOLOGICAL MECHANISM UNDERLYING THE ASSOCIATION BETWEEN DIABETES AND DEPRESSION?

Alan M. Jacobson, Jacqueline A. Samson, Katie Weinger, and Christopher M. Ryan[1]

Harvard Medical School,
Joslin Diabetes Center,
Boston, Massachusetts 02215

I. Introduction

Over the past 20 years, a large body of research has examined the relationship of diabetes to changes in behavior, social and psychological functioning, psychiatric status, cognition, and brain structure and function. While far from definitive, there is now increasing evidence of important relationships between diabetes and these aspects of human function. Several other reviews have examined many facets of this research area, including psychosocial impacts of diabetes and the types of interventions that can be useful for integration into diabetes treatment (Gavard *et al.*, 1993; Lezak, 1995;

[1]Current affiliation: Department of Psychiatry, University of Pittsburgh, School of Medicine, 15213.

0074-7742/02 $35.00

Jacobson, 1996; Jacobson and Weinger, 1998; de Groot *et al.*, 2000; Gonder-Frederick *et al.*, 2000; Talbot and Nouwen, 2000; Anderson *et al.*, 2001, 2002), Such studies have demonstrated that, in general, diabetes in children does not lead to profound psychological disturbances, although changes in development as well as influences on family life have been found (Jacobson and Weinger, 1998; Hauser *et al.*, 1990, 1993; Jacobson *et al.*, 1994, 1997a,b; Kovacs *et al.*, 1997). Studies of adults with type 1 diabetes suggest that there is an increased risk of psychiatric disorders, such as depression and eating disorders (Wells *et al.*, 1989a; Gavard *et al.*, 1993; Kovacs *et al.*, 1997). Among adults with type 2 diabetes, it is also reasonable to assume, from smaller literature, that, like type 1 diabetes, it also influences the quality of individual lives and changes the thematic concerns of patients and their families (Jacobson *et al.*, 1994; Jacobson, 1996; Ritholz and Jacobson, 1998; Fisher *et al.*, 2000; Ryan and Greckle, 2000). Intervention studies have demonstrated the benefit of a wide variety of educational programs and counseling techniques to improve self-care behaviors, even though the translation of these interventions to actual clinical practice has *not* been demonstrated to lead to marked improvements in patient motivation or adherence. Psychiatric treatments for depression and anxiety disorder appear to benefit patients with diabetes as would be extrapolated from nondiabetic samples.

There is also a small body of evidence suggesting that diabetes and its metabolic consequences may lead to changes in brain structure and function and cognition (Dejgaard *et al.*, 1991; Bjorgaas *et al.*, 1996; Drevets *et al.*, 1997; Leibson *et al.*, 1997; Ryan, 1997; Rovet and Ehrlich, 1999; Jacobson *et al.*, 2001). However, there has been minimal attention paid to the interrelationships between these structural and functional changes in the brain, cognitive effects of diabetes, and the aforementioned influences of diabetes on the development of psychiatric conditions (Dejgaard *et al.*, 1991; Jacobson *et al.*, 2001). Of particular relevance to this trilateral set of relationships is the growing research base addressing the biological basis for the pathogenesis of depression in nondiabetic individuals and the evidence that diabetes, in particular type 1 diabetes, leads to an increase in the incidence and prevalence of unipolar major depression (Wells *et al.*, 1989a; Gavard *et al.*, 1993; Kovacs *et al.*, 1997). The purpose of this chapter is to examine this more narrow set of relationships based on research findings from our group and a focused review of relevant studies of other investigators.

Fig. 1 presents a model of potential ways in which diabetes may influence development of depression. Of note, both psychosocial factors and biological mechanisms may play a role in the increased incidence and prevalence of depression. In this chapter, we will present information regarding what is known about the influences of diabetes and its metabolic alterations on brain structure and function and explore links between changes in brain

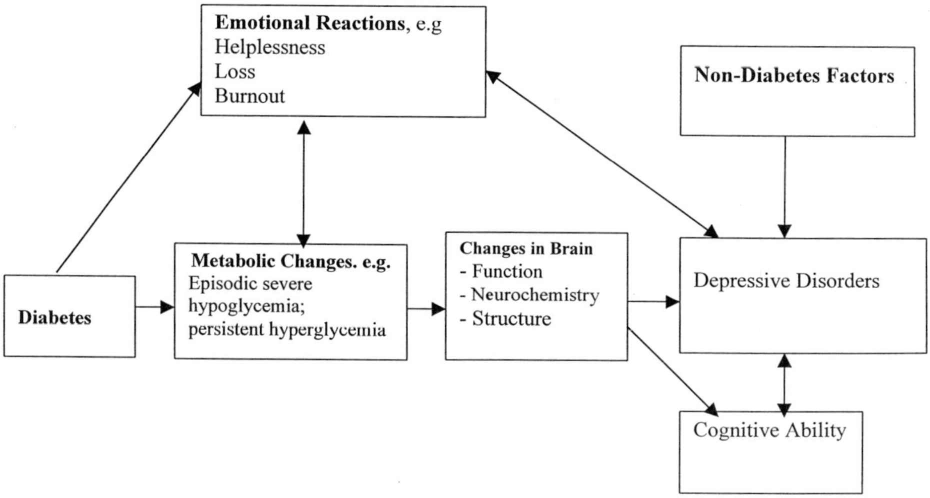

FIG. 1. Hypothesized model of relationship of diabetes and depression.

structure and function among diabetic patients to the increased risk for depression.

II. Background

Type 1 diabetes is now recognized as a disease set in motion primarily by an autoimmune process, which leads to the destruction of the pancreatic β-cell and, therefore, loss of insulin secretory capability. The genetic preconditions for this process are, as yet, not well understood. It is assumed that it is a polygenic disease in which genetic changes condition the impact of environment on the individual. Therefore, the genetics remain complex and many individuals with type 1 diabetes appear to develop it without any family history.

Type 2 diabetes is thought to be an illness of relative insulin deficiency brought on by multiple risk factors (e.g., obesity and lack of physical activity), which lead to insulin resistance, added stress to the β-cell's insulin producing capability and gradual decrease in production of insulin, resulting in diabetes. The precise mechanism causing this decrease in secretory capacity is as yet unknown. It is clear that there are important differences across ethnic groups so that South Asian, African-American, American Indian, Middle Eastern, and Latino populations appear to be at higher risk for type 2 diabetes than Northern European Caucasian populations. This relative risk of illness is, in fact, reversed in comparison with type 1 diabetes where the highest risk is among Northern European, Scandinavian populations. The growing problem of obesity and physical inactivity appears to uncover a genetic susceptibility for type 2 diabetes. Type 2 diabetes, itself, appears to occur as part of a more encompassing syndrome which includes insulin resistance associated with hypercoagulability, hyperlipidemia, and hypertension. In general, type 1 diabetes is more commonly found in younger individuals with peak incidence in early childhood, around puberty, and with later peaks during adult years. Type 2 diabetes increases in incidence with age.

Because of the absolute deficiency in insulin production, patients with type 1 diabetes require insulin from diagnosis and are more susceptible to both extreme elevations in blood glucose (hyperglycemia) and sudden and dangerous drops in blood glucose (hypoglycemia). Moreover, patients with type 1 diabetes, when their blood glucose levels are elevated, are more susceptible to states of diabetic ketoacidosis (DKA). These are far less likely to occur in those with type 2 diabetes because of some insulin production capability persisting over long periods of time. Serious hypoglycemia is accompanied by symptomatic changes associated with autonomic nervous system reactivity (e.g., nervousness, irritability, sweating, and headaches) and also acute changes in cognitive function. Severe hypoglycemic episodes can lead to loss of consciousness, seizures, and in extreme cases, death. Because of the changes in cognition during severe hypoglycemic episodes, patients may have difficulty with decision making and with carrying out tasks such as driving. Over time, patients with type 1 diabetes become more susceptible to "hypoglycemia unawareness," a state in which patients lose their autonomic symptom responses to a hypoglycemic episode and, therefore, experience the episode as a change in cognition or loss of consciousness. The risk factors for developing hypoglycemia unawareness are not fully understood, but it is known that tight control of diabetes, i.e., the achievement of glycemic control levels that approximate those of nondiabetic individuals, and associated recurrent hypoglycemic episodes, increase the likelihood of hypoglycemia unawareness. As will be discussed, both recurrent severe hypoglycemia and persistent blood glucose elevations (persistent hyperglycemia) may have *chronic effects* on brain structure and function and cognition and, therefore, may play a role in the added risk for depression in diabetic patients.

III. Diabetes and Psychiatric Illness

A small but increasing number of studies suggest rates of psychiatric illnesses may be higher in diabetic populations than in community samples (Cassileth *et al.*, 1984; Lustman *et al.*, 1986; Popkin *et al.*, 1988; Wells *et al.*,

1988, 1989a; Weyerer *et al.,* 1989; Mayou *et al.,* 1990, 1991; Bryden *et al.,* 1999), particularly rates of depressive and anxiety-related disorders (Wells *et al.,* 1988, 1989a; Robinson *et al.,* 1988; Jacobson, 1993; Lustman *et al.,* 1998; Jacobson and Weinger, 1998; Anderson *et al.,* 2001). Questions remain as to whether these elevations develop in response to biological mechanisms specific to diabetes or are manifestations of being chronically ill. Moreover, it is not clear whether these elevations are characteristic of *both* type 1 and 2 patients, or if rates are equally elevated in males and females (Anderson *et al.,* 2001).

Contributing to this lack of clarity is the variability in methods used to examine comorbidity across studies (Anderson *et al.,* 2001). Some studies have examined assessment using self-reported psychiatric symptom inventories in samples selected from medical treatment centers. In general, medically ill patients show self-reported psychiatric symptom severity scores that are higher than those found in medically healthy patients and lower than those found in clinically depressed patients. Diabetic patients appear to show psychiatric symptom severity scores that are similar in magnitude to those found in other groups of medically ill patients. Other studies (Anderson *et al.,* 2001) have examined rates of specific DSM-III psychiatric disorders among diabetic patients using structured diagnostic interviews. For example, using data from the Epidemiologic Catchment Area (ECA) study, Wells *et al.* (1988) found that respondents with diabetes (type 1 or 2 unspecified) reported significantly higher lifetime rates of affective illnesses and anxiety disorders compared with healthy subjects. However, prevalence rates for affective illnesses and anxiety disorders were comparably elevated across four other groups of nondiabetic medically ill subjects. Other studies using standardized interview methods have also suggested that diabetic individuals have higher prevalence rates of psychiatric disorders (especially affective disorder and that these appear to occur relatively early in the course of diabetes, i.e., prior to the onset of disabling complications (Lustman *et al.,* 1986, 1997; Kovacs *et al.* 1995, 1997; Jacobson and Weinger, 1998; Anderson *et al.,* 2001).

A. Joslin Study of Psychiatric Illness and Diabetes

Our own research exemplifies these findings. In one study, adult diabetic subjects were outpatients receiving treatment in the medical department of the clinic at the Joslin Clinic. Demographic and clinical characteristics for both the screened and interviewed sample are given in Table I.

Prevalence rates were estimated on the basis of a two-stage case identification process that consisted of an initial screening followed by a Structured Clinical Interview for DSM-IV (SCID). Patients scoring at or above a SCL-90 Global Severity Index (GSI) T-Score of 63 were considered as likely cases

TABLE I
SOCIAL AND DEMOGRAPHIC CHARACTERISTICS OF PATIENTS IN PHASE I AND PHASE II OF DATA COLLECTION

Variable	Screening phase I (n = 240)		Interview phase II (n = 143)	
	n	%	n	%
Number male	118	49.2	78	54.6
Number female	122	50.8	65	45.5
Education level				
High school or less	79	32.9	40	27.9
More than high school	161	67.1	103	72.0
Diabetes type				
Type 1–Insulin	111	100	62	100
Type 2–Insulin	68	53	40	50
Type 2–Oral agent	49	38	34	42.5
Type 2–Diet only	12	9.3	7	8.6
	Mean	SD	Mean	SD
Age in years	52.9	16.0	52.1	15.7
Years of diabetes	15.4	10.3	15.8	10.8

following the guidelines of Derogatis *et al.* (1983) and, therefore, interviewed. A total of 85 patients or approximately one-third of the sample met the criterion for caseness. All patients with a GSI T-score of 63 or greater were contacted by phone to arrange for a follow-up SCID diagnostic interview. To screen for false-negative results, 40% (n = 68) of the remaining patients (with GSI T-Scores less than 63) were selected at random for follow-up SCID diagnostic interviews (Spitzer *et al.*, 1992).

A total of 57% of the type 1 patients and 46% of the type 2 patients showed a positive history of at least one DSM-III-R diagnosis (Table II). The majority of disorders reported by both groups were in the category of affective disorder (38% type 1 and 32% type 2 patients), with major depressive disorder (29% type 1 and 15% type 2 patients), and depression NOS (11% type 1 and 13% type 2 patients) as the most frequently found diagnoses. Anxiety disorders were found in 33% of the type 1 and 12% of the type 2 patients, with no one specific anxiety diagnosis predominating. Psychoactive substance abuse was found in 21% of the type 1 and 19% of the type 2 patients. No psychotic disorders were found in the type 1 patients and only one type 2 patient met criteria for schizophreniform disorder. Regression analyses were performed to examine diabetes type on the presence or absence of

TABLE II
LIFETIME PREVALENCE ESTIMATES[a] FOR DSM-III-R DIAGNOSES FOR PATIENTS WITH TYPE 1 AND TYPE 2 DIABETES MELLITUS

	Type 1 (n = 111)			Type 2 (n = 129)		
Disorder	Actual cases	Estimated prevalence(%)	S.E. (%)	Actual cases	Estimated prevalence(%)	S.E. (%)
Any psychiatric disorder	40	57	7.0	43	46	5.6
Any affective disorder	29	38	6.4	32	32	4.9
Bipolar	2	2	1.5	2	2	1.2
Major depression	22	29	5.9	15	15	3.7
Depression NOS	8	11	4.3	12	13	3.7
Dysthymia	3	3	1.8		3	1.7
Adjust. disorder depressed Mood	3	6	3.7	3	3	1.5
Any anxiety disorder	23	33	6.5	12	12	3.5
Any psychoactive substance	13	21	5.8	17	19	4.3
Any psychotic disorders[b]	0	0	0	1	1	.9

[a] Prevalence estimates are based on diagnoses assigned with no exclusion for comorbidity.
[b] Psychotic disorders includes one lifetime case of schizophreniform disorder.

major depression, any affective disorder, and any anxiety disorder. Results of these analyses showed that affective and anxiety diagnoses did not differ by diabetes type, once age was taken into account.

B. GENDER DIFFERENCES

While rates of affective illnesses are comparable across gender (approximately 35%), rates for anxiety disorders are higher for females (31% females, 14% males, $p = 0.02$), whereas rates of substance abuse and dependence tend to be higher in males (14% females, 25% males; $p = 0.10$) (Table III). The comparability of male and female rates of depressive illnesses in our diabetic patients is at odds with the data reported in community surveys showing depression in females to exceed that in males by a ratio of approximately 2:1. Rates for anxiety disorders, as expected, show females outnumber males by about 2:1. The higher rate of substance abuse

TABLE III
LIFETIME PREVALENCE ESTIMATES[a] FOR DSM-III-R DIAGNOSES FOR ALL MALE AND FEMALE PATIENTS

	Males (n = 118)			Females (n = 122)		
Disorder	Actual cases	Estimated prevalence (%)	S.E. (%)	Actual cases	Estimated prevalence (%)	S.E. (%)
Any psychiatric disorder	42	48	5.6	41	53	6.8
Any affective disorder	32	36	5.3	29	33	5.7
Bipolar I	0	0	0	0	0	0
Bipolar NOS	2	2	1.4	2	2	1.3
Major depression	20	23	4.8	17	17	3.9
Depression NOS	11	13	3.7	9	12	4.3
Dysthymia	3	3	1.7	4	4	1.7
Adjust. disorder depressed mood	2	2	1.4	4	6	3.4
Any anxiety disorder	13	14	3.5	22	31	6.2
Any psychoactive substance	21	25	4.9	9	14	4.7
Any psychotic disorders[b]	1	1	1.0	0	0	0

[a] Prevalence estimates are based on diagnoses assigned with no exclusion for comorbidity.
[b] Psychotic disorders includes one case of schizophreniform disorder.

in males is also comparable to what would be expected based on community reports.

Male and female samples did not differ in age for either the type 1 (male mean age = 41.90 years, SD = 14.80; female mean age = 45.98 years, SD = 16.46) or type 2 groups (male mean age = 59.27 years, SD = 11.30; female mean age = 61.84 years, SD = 12.18).

C. DEPRESSION AND MEDICAL COMPLICATIONS

To address the possibility that the high rates of major depression could be explained by the presence of severe medical complications, chi square analyses were performed that compared patients with zero to three complications and presence of major depression. Medical records were reviewed to establish the presence of major diabetes complications (neuropathy, nephropathy, and retinopathy). No significant associations were found between the presence of major depression and number of complications in either type 1 (chi square = 0.71, DF = 3, p = ns) or type 2 patients (chi square = 1.95, DF = 3, p = ns).

D. Age of Onset

One explanation for the absence of a gender difference in rates of depression might be that depressive illness in diabetic patients occurs as a result of the experience of the illness, and thus, these are secondary rather than primary depressions. To examine for this possibility, we examined the data reflecting the age of onset of diabetes against the age of onset of major depressive disorder.

Examined by diabetes type, type 1 diabetic patients showed an earlier onset of diabetes than depression (paired t = 3.41, $p = 0.003$, n = 22) whereas type 2 diabetic patients did not (paired t = 0.69, p = ns, n = 15). Moreover, other investigators have suggested that depression itself could be a risk factor for type 2 diabetes, or both conditions could reflect underlying risk factors both have in common. These data showing that the age of onset of major depression is later than the age of onset of diabetes in type 1 patients is not surprising given that the age of risk for onset of major depression is typically between the early twenties and mid-thirties, whereas (by definition) the age of onset for type 1 diabetes is typically younger.

To further understand the age of onset results found in type 1 patients, we rank ordered our SCID interview data by age of onset of diabetes and examined the relationship between the presence of lifetime major depression and age of diabetes onset. These listings were limited to those patients who had undergone a SCID interview. Examined by decades, 77% of type I patients we interviewed with onset of diabetes before the age of 10 years showed a positive lifetime history of major depressive disorder, compared to 20% of patients with diabetes onset between the ages of 10 and 19 (chi square = 9.07, $p < 0.001$), 22% with diabetes onset between 20 and 29 (chi square = 9.12, $p < 0.001$), 43% with diabetes onset between 30 and 39 (chi square = 2.32, $p = 0.13$), and 22% with diabetes onset after age 40 (chi square = 6.42, $p = 0.01$). Thus, patients with diabetes onset before the age of ten years showed elevated rates of lifetime major depression compared with patients who showed a later age of diabetes onset.

This vulnerability of the patients with earliest onset of diabetes (before age 10) may reflect the high level of familial distress engendered by very early onset diabetes. Alternatively, the onset of diabetes in young children may pose danger to the developing brain. As we will discuss, severe hypoglycemia or persistent hyperglycemia may be the reason that children with very early onset of diabetes experience these changes.

Many studies have not routinely reported prevalence estimates separately for males and females. Given the fact that community and psychiatric surveys consistently demonstrate gender differences in rates of psychiatric disorder (women show higher rates of depression and anxiety disorders and lower rates of alcohol or substance abuse/dependence than men) (Gater *et al.*, 1998), it is likely that the processes that govern the onset of specific psychiatric illnesses differ for men and women.

E. Conclusions

In sum, there is growing evidence that patients with diabetes, especially adults with type 1 diabetes, have an increased risk for psychiatric morbidity. This increased risk has been noted in particular for depression (Popkin *et al.*, 1988; Wells *et al.* 1989a,b; Mayou *et al.*, 1991; Gavard *et al.*, 1993; Kovacs *et al.*, 1997; Jacobson and Weinger, 1998). Depending on the study, diabetic adults have rates of current depression that range between 8.5 and 28% with the average across studies of 19% and rates of symptomotology as high as 39%. Moreover, lifetime histories from these samples suggest rates of lifetime history in the 30–40% range. It is reasonable to assume that an illness with these complex lifestyle demands as well as the clear risk for several life-threatening complications would lead patients to feel greater distress and, therefore, increased risk for depression. This would be expected to be found among patients with long-term serious complications of diabetes. If this psychosocial mechanism leads to added risk for depression, then the risk should increase with time and specifically rise after onset of complications. There are no data to controvert this model, but studies of adolescents and young adults suggest that depressive disorders appear to develop at higher than expected rates in patients with type 1 diabetes before the onset of serious complications of diabetes (Lustman *et al.*, 1986, 1997; Kovacs *et al.*, 1995, 1997; Jacobson and Weinger, 1998).

Prior studies have suggested that patients with poorly controlled diabetes have alterations in the hypothalamic-pituitary axis similar to those found in patients with depression (Cameron *et al.*, 1984; Hudson *et al.*, 1984; Geringer, 1990). This suggests the possibility that alterations in a hypothalamic-pituitary axis could play a role in the development of depression in diabetic patients. However, no studies have explicitly examined this issue. Studies of streptozotocin-induced diabetes in rats suggest that neurotransmitters (serotonin and norepinephrine) are reduced (Crandall *et al.*, 1981; Trulson and Himmel, 1985). Moreover, rats with streptozotocin-induced diabetes subjected to uncontrollable stress are more likely than nondiabetic rats to develop a syndrome of "learned helplessness" and neurochemical changes similar to those found in depression (Trulson and Himmel, 1985; Bellush *et al.*, 1991). These studies suggest that metabolic disturbances of diabetes could lead to changes in the brain that alter susceptibility to social stressors, thereby leading to depression-like syndromes. Thus, these few animal studies are consistent with both biological and psychosocial mechanisms. Extrapolating from the animal data, the metabolic alteration of diabetes may lead to changes in neuronal function, and these changes, whether permanent or temporary, could lead to increased susceptibility to stress. Our assumption is that diabetes-induced changes in neuronal function and structure impact on areas of the brain that regulate affect and, therefore, increase risk for depression.

We now review research examining the effects of diabetes on cognition, brain structure, and brain function to more fully develop our working model linking diabetes to the unexpectedly high incidence and prevalence of comorbid depression.

IV. Diabetes and Cognition

Cognitive ability provides a measure of brain performance and as such a marker, albeit an indirect measure of the changes in brain functioning wrought by diabetes. This literature, therefore, is useful in this discussion to the extent it provides evidence of changes in CNS integrity secondary to diabetes. We can use such findings heuristically as a "place holder" for more direct findings of changes in brain structure and function. It is important to note that such changes may well understate such influences of diabetes on brain function for two reasons: (1) reserve capacity may hide actual changes in the brain, which are not of sufficient magnitude to overcome cognitive capacity; and (2) brain damage could occur in regions of the brain that have little influence on cognition.

A. Hypoglycemia and Cognitive Dysfunction

From a public health perspective, especially as intensive diabetes therapy is implemented more widely, a more critical safety question is whether repeated episodes of less profound hypoglycemia contribute to the development of cognitive dysfunction in diabetic adults. Cross-sectional neuropsychological studies have suggested that diabetic adults with multiple lifetime episodes of severe hypoglycemia—defined as requiring external assistance or triggering a seizure or coma—tend to perform more poorly than those without prior hypoglycemia (Gold *et al.*, 1993; Perros and Deary, 1999). A wide variety of cognitive domains are reported to be affected, although there is little consistency of findings from one study to another. Visuospatial deficits and psychomotor slowing appear to occur in diabetic patients who have had five or more lifetime episodes of severe hypoglycemia (Wredling *et al.*, 1990), as are performance decrements on tasks requiring rapid decision making and adaptive problem solving skills ("fluid intelligence") (Langan *et al.*, 1991; Deary *et al.*, 1992; Lincoln *et al.*, 1996). Given the sensitivity of the hippocampus to profound hypoglycemia, one would have predicted that memory skills would also be affected, yet only one study has found evidence of mnestic deficits (Sachon *et al.*, 1992). Other studies, using learning and memory tests, have failed to find a consistent and robust association between recurrent hypoglycemia and learning or memory deficits (Langan *et al.*, 1991; The Diabetes Control and Complications Trial Research Group, 1996; Hershey *et al.*, 1997). Clinically significant deterioration in cognitive functioning and in personality has been reported in a group of five older patients with longstanding diabetes and recurrent severe hypoglycemia, but because these individuals had multiple diabetes-associated complications—including peripheral neuropathy—it is impossible to identify the etiology of these deficits (Gold *et al.*, 1994).

All studies reporting hypoglycemia-associated cognitive dysfunction have been cross-sectional, evaluated small numbers of subjects, and relied on retrospective recall to estimate prior hypoglycemic events. The two large-scale longitudinal clinical trials that incorporated measures of cognitive function did *not* find a relationship between prospectively ascertained episodes of severe hypoglycemia and neuropsychological test performance (see DCCT reference for a complete summary of the neuropsychological test results) (The Diabetes Control and Complications Trial Research Group, 1996; Austin and Deary, 1999). A similar conclusion was reached by the Stockholm Diabetes Intervention Study group, which conducted a much smaller longitudinal study (n = 91) that employed a very different, computer-based test battery (Reichard *et al.*, 1991, 1996).

Nevertheless, it is premature to conclude from those carefully conducted clinical trials that severe hypoglycemia is entirely benign. Examining all of the extant neuropsychological data leads us to agree with the conclusion, articulated several years ago by Deary and Frier (1996), that the link between severe hypoglycemia and cognitive impairment in diabetes is, of yet, "not proven." In many ways, the failure to prove (or disprove) the existence of such a link reflects the methodological shortcomings of the extant research: samples are small; severe hypoglycemic episodes are poorly ascertained; other diabetes-associated biomedical complications are rarely considered in the analyses; and cognitive assessments tend to be quite limited.

B. Persistent Hyperglycemia and Cognition

A few studies suggest that persistent hyperglycemia may be associated with decreased cognitive functioning (Meuter *et al.*, 1980; Franceschi *et al.*, 1984; Skenazy and Bigler, 1984). For example, The work of Ryan *et al.* (1992, 1993) in Pittsburgh demonstrated a relatively circumscribed pattern of impairments, with deficits most apparent on measures of attention and psychomotor efficiency, and on spatial information processing. Poor metabolic control, as indexed by a clinical diagnosis of peripheral neuropathy, was the best predictor of those deficits. In contrast, these diabetic subjects performed as well as nondiabetic subjects on measures of learning and memory, deductive reasoning, and simple motor speed. We now believe that the psychomotor slowing is *the* defining neuropsychological characteristic of chronic hyperglycemia, insofar as it is also evident in children and adolescents who developed type 1 diabetes after 5 or 6 years of age (Ryan *et al.*, 1984; Hershey *et al.*, 1999), and also in middle-aged (Ryan and Greckle, 2000) or elderly adults with type 2 diabetes (Strachan *et al.*, 1997; Vanhanen *et al.*, 1998).

V. Diabetes, Brain Structure, and Brain Function

Some research suggests that changes in EEG and in evoked potential can be found in patients with type 1 diabetes and that these are associated with duration of diabetes and presence of peripheral neuropathic and microvascular complications (Soltesz and Acsadi, 1989; Nakamura *et al.*, 1991; Lunetta *et al.*, 1994; Seidl *et al.*, 1996). For example, Dejgaard *et al.* (1991), compared 20 long-duration type 1 patients with recognized microvascular and peripheral neuropathic complications to 19 gender matched short-duration type 1 patients without clinical evidence of neuropathy or microangiopathy. They found abnormal brain stem auditory-evoked responses in 40% of the long-duration, but only 5.3% of the short-duration patients. Other studies of patients with type 1 diabetes similarly demonstrate these effects (Nakamura *et al.*, 1991; Lunetta *et al.*, 1994; Seidl *et al.*, 1996). It has been suggested that the changes found in evoked potential studies may reflect alterations due to microangiopathy as well as metabolic disturbances (Nakamura *et al.*, 1991) and that they reflect a central neuropathy. Studies evaluating patients experiencing acute episodes of severe hypoglycemia suggest that such episodes may play an important causative role in the development of changes in the brain (Chalmers *et al.*, 1991; Lunetta *et al.*, 1994; Seidl *et al.*, 1996; Jacobson *et al.*, 2000).

Some case reports demonstrate that a single episode of profound hypoglycemia *may* produce permanent structural brain damage, with persisting neurological and neuropsychological sequelae. A single episode of profound hypoglycemia can trigger significant brain damage in adults, with lesions distributed bilaterally in the basal ganglia, hippocampus, and cerebral cortex—particularly mesial temporal regions (Kalimo and Olsson, 1980; Auer *et al.,* 1989; Patrick and Campbell, 1990; Chalmers *et al.,* 1991; Boeve *et al.,* 1995; Fujioka *et al.,* 1997)—although damage has also been noted in brain stem regions (Perros *et al.,* 1994). Serial MR and CT imaging has suggested that these lesions represent tissue degeneration that is characterized by selective neuronal loss and a proliferation of astrocytic glial cells (Fujioka *et al.,* 1997). Cases of profound hypoglycemia are remarkably infrequent and quite atypical of most adults with diabetes insofar as the blood glucose values are extremely low (<1.0 mmol) for an indeterminate, but quite prolonged time.

Significant neuronal necrosis, particularly in the hippocampus, has been found in rodents and primates following profound hypoglycemia accompanied by an isoelectric EEG (Brierley *et al.,* 1971a,b; Auer *et al.,* 1984). When neuronal necrosis occurs, it tends to be most pronounced in hippocampal structures and is correlated with the duration of the hypoglycemic event (Kahn and Myers, 1971; Auer *et al.,* 1984, 1985; Siesjo and Bengtsson, 1989). On the other hand, there is no evidence from animal research that recurrent bouts of severe hypoglycemia, *without* coma or isoelectric EEG, necessarily produce structural brain changes (Auer and Siesjo, 1993). Indeed, a recent study of diabetic patients who experienced severe hypoglycemia (requiring assistance but without subsequent development of neurological sequelae) failed to find increases in two proteins, neuron-specific enolase and protein S-100, that are known to be elevated following neuronal injury in humans. Elevations were only apparent in two of the three patients who experienced profound hypoglycemia, i.e., were comatose at admission and subsequently died (Strachan *et al.,* 1999).

Increased rate of brain atrophy has been found in patients with type 1 diabetes (Woods *et al.,* 1995). These findings of brain atrophy have been interpreted as signs of the increased rate of "aging" in the brains of patients with diabetes, i.e., increased frequency of cerebral vascular changes. Similarly, white matter hyperintensities have been found in both patients with type 1 and type 2 diabetes (Dejgaard *et al.,* 1991). For example, in Dejgaard's (1991) study of type 1 patients, fully 69% of long-duration patients were found to have abnormalities on MRI with white matter hyperintensities being the most frequent type of lesion. Unfortunately, in this study, only a subset of the diabetic patients were studied using MRI. However, age-matched controls were studied. In these control subjects, only 12% were found to have abnormalities on the MRI and these were only seen in subjects greater than 40 years of age. Dejgaard's findings of a rate of 69% of long-duration patients with clear diabetic complications having white matter hyperintensities is quite consistent with our preliminary study findings of 44% of a long-term diabetic population without clear complications having MRI abnormalities (Dejgaard *et al.*, 1991; Jacobson *et al.*, 2000, 2001). One would expect that our sample, in which patients with recognized neuropathy and microangiopathy were excluded and who were somewhat younger, would have lower rates of MRI abnormality. The sample of Dejgaard *et al.* was heavily weighted toward patients with complications. This may account for Dejgaard *et al.*, not finding correlations of MRI abnormalities with metabolic control. This contrasts with our findings of a significant relationship between prior history of HbA1c level and MRI changes. Moreover, our findings indicate that changes in brain structure consistent with a central neuropathy can occur in the absence of clinically evident peripheral and autonomic neuropathy (Jacobson *et al.*, 2000). One study of diabetic patients using MRSI found increased myoinositol and choline metabolites in both gray and white matter compared to healthy controls (Kreis and Ross, 1992), suggesting that diabetes could lead to changes in brain detectable by neurochemical markers.

While these studies provide an important base of scientific information guiding current understanding, they also demonstrate the significant gaps in the current literature. For example, past studies have utilized long-term history of severe hypoglycemia as a predictor of brain structure and function, but measures of long-term glycemic control have not typically been utilized as a predictor of brain structure and function. This may be the reason that prior studies fail to show relationships (Dejgaard *et al.*, 1991). This failure to demonstrate a relationship is also a reflection of the methodologic limitations of the studies in which a small number of patients as recruited with very specific hypotheses in mind. For example, the Perros' study (1997) compared patients with and without histories of severe hypoglycemia and included, therefore, two separate patient groups with very similar levels of glycemic control. The study by Dejgaard *et al.* (1991) compared patients with very short-duration diabetes with those with long-duration diabetes who already had significant microangiopathy and neuropathy. Therefore, it's not surprising that Dejgaard *et al.* failed to find a relationship between hyperglycemia and change in brain.

However, there is minimal research, which has examined the relationship of these abnormalities to the development of depressive disorders, which are now thought to occur at higher than expected rates among type 1 diabetic patients (The Diabetes Control and Complications Trial Research Group, 1993). As Talbot and Nouwen (2000) suggest, the etiology of depression in type 1 diabetic patients is not well understood. Given the findings from patient studies of nondiabetic depressed subjects, which show higher than expected levels of white matter hyperintensities, these findings suggest a mechanism by which diabetes may lead to depressive disorder at higher rates than found in non-ill populations.

In summary, findings from structural MRI, spectroscopic imaging studies, EEG, and evoked potential in patients with diabetes and other chronic medical illnesses provide support for one set of key assumptions in our proposed model accounting for the excess risk of depression among younger adult diabetic individuals, i.e., neuronal disruption in parts of the brain that regulate affect leading to depression. If the metabolic abnormalities of type 1 diabetes accelerate changes in the brain, then these metabolic abnormalities could play an important role in the development of depression among diabetic patients.

VI. Depression, Brain Structure, and Function

Moreover, there is now growing literature on the structural and functional correlates of psychiatric disorders (Buchsbaum *et al.*, 1986; Coffey *et al.*, 1988–1990, 1993; Dupont *et al.*, 1990; Figiel *et al.*, 1991; Krishnan *et al.*, 1992; Brown *et al.*, 1992; Breteler *et al.*, 1994a,b; Hickie *et al.*, 1995; Persaud *et al.*, 1997; McDonald *et al.*, 1999; Lyoo *et al.*, 2001). Of particular relevance for this discussion, a series of studies in adults, and more recently children, with affective disorder suggests that patients with both bipolar disorder and unipolar depression are more likely to have MRI abnormalities, i.e., white matter signal hyperintensities as compared with patients without psychiatric conditions (Coffey *et al.*, 1988–1990, 1993; Dupont *et al.*, 1990; Figiel *et al.*, 1991; Krishnan *et al.*, 1992; Brown *et al.*, 1992; Hickie *et al.*, 1995; Persaud *et al.*, 1997; McDonald *et al.*, 1999; Lyoo *et al.*, 2001). In addition, some studies of schizophrenia (Breteler *et al.*, 1994a) and obsessive compulsive disorder also suggest greater than expected frequencies of white matter changes. Patients with psychiatric disorders are also more likely to have white matter changes if they are older (Araki *et al.*, 1994).

MR volumetric studies also suggest that patients with depression history, compared to patients with other psychiatric disorders, have decreased volume of frontal lobes (Dupont *et al.*, 1990; Drevets *et al.*, 1997, 1999; Schurhoff *et al.*, 2000). These findings are also consistent with the hypothesis that depression may develop in some individuals secondary to tissue disruption. Both white matter hyperintensities (WMH) and volume changes reflect tissue disruption caused by unknown factors. Diabetes could trigger or accelerate this process. Among individuals genetically at risk or those not already at risk, such changes could, if they occurred in brain regions that regulate affect, lead to depression. Moreover, MRSI studies, which are more sensitive to changes in neuronal function and cell membrane integrity, suggest that the brains of both adolescent and adult subjects with current depression show increased choline content of the frontal lobe and basal ganglia (Pillay *et al.*, 1998; Drevets, 1999; Lai *et al.*, 2000). Other MRSI studies of *N*-acetyl aspartate (NAA) suggest that this neurochemical marker is sensitive for detecting early evidence of neuronal white matter damage even before WMH can be identified by MRI. For example, NAA changes have been found in illnesses, such as lupus erythematosus (Axford *et al.*, 2001) and multiple sclerosis (Simone *et al.*, 2001) and are also associated with cognitive declines in healthy elderly individuals (Valenzuela *et al.*, 2000) and outcomes following traumatic brain injury (Garnett *et al.*, 2000). To date, decreased NAA has not been found in primary depression but is common in neurological disorders associated with secondary depression. Since frontal cortex-basal ganglia circuitry is involved in affect regulation, these studies support the linkage of neuronal damage as measured by three methods (MRSI, WMH, and volume) playing a role in development of depression.

A recent large study by Lyoo *et al.* (2001) underlines the potential importance of white matter abnormalities in the etiology of psychiatric conditions among children and adolescents. In this study, child patients with bipolar disorder, unipolar depression, and conduct disorders/attention deficit disorders were all more likely to have white matter hyperintensities than a nonpsychiatric comparison group. Of note, child patients with schizophrenia, unlike findings from adult studies, had relatively low rates of white matter hyperintensities. This suggested that in younger patients with severe psychiatric problems, there may be a stronger association of changes in white matter with the development of affective and anxiety disorders. This is consistent with findings in type 1 diabetes in which affective and anxiety disorder rates are elevated, while there is no evidence of increased rates of schizophrenia. Moreover, in this large study, the investigators were able to examine location of white matter hyperintensities and found that frontal abnormalities were more likely to be found in patients with affective disorders.

VII. Testing the Model: Preliminary Evidence

In order to examine our hypothesized model, linking diabetes changes in brain structure and function and depression, we studied 16 type 1 diabetic patients who were ages 20–40 and had a duration of diabetes of 10–34 years. All patients were diagnosed before the age of 20; 7 were male and 9 were female. As a group, the patients were all educated (75% college graduates). Patients were identified from the roster of patients followed at the Joslin Clinic. Histories were obtained from patient self-report and from chart review. Specifically, information was gathered regarding the history of severe, hyperglycemic episodes defined as those causing unconscious reactions, coma, or seizures. This criterion was the same as that used for the narrow definition of severe hypoglycemia in the Diabetes Control and Complications Trial. Moreover, information was gathered regarding results from prior glycoserated hemoglobin tests in order to characterize the patient's history of glycemic control. All patients were examined using SCID, MRI, and single photon emission computed tomography (SPECT).

We found 9 patients with a history of 0–2 severe hypoglycemic events; while the remaining 7 patients had 5 or more such events. Ten of the patients studied (63%) had a lifetime history of psychiatric illness and 6 (38%) had a lifetime history of depressive disorder. Seven patients had abnormal MRIs (44%), typically characterized by having white matter hyperintensities. The SPECT exam showed hypofrontality in 14 of 16 patients (87.5%). Patients with 5 or more severe hypoglycemic episodes had lower full-scale IQs than those with 0–2 such episodes (1 of 4 ± 11 versus 122 ± 13). In addition, patients with abnormal MRIs had higher HbA1c levels on average in comparison with those with normal MRIs. Patients with 5 or more unconscious reactions versus those with 0–2 were more likely to have abnormal MRIs (71% versus 22%) as well as more positive psychiatric histories (86% versus 44%). When we examined this subset of 12 patients who had either a history of depression or no psychiatric history, we found that those with 5 or more unconscious reactions versus those with 0–2 such reactions were more likely to have a history of depression (80% versus 28%). We concluded from this study that there was preliminary evidence linking functional and structural changes found using SPECT and MRI to type 1 diabetes and that recurrent severe hypoglycemia and/or persistent hyperglycemia could play a role in the pathogenesis brain structure changes and, therefore, lead to the development of depressive disorder in this population.

VIII. Summary

In summary, our review of the literature suggests that diabetes, especially type 1 diabetes, may place patients at risk for a depressive disorder through a biological mechanism linking the metabolic changes of diabetes to changes in brain structure and function. Further studies are warranted examining these relationships in order to better understand the impact of diabetes on brain functioning and structure as well as one potential manifestation of such changes—affective disorder. Moreover, such studies could play a useful role in better understanding mechanisms that commonly underlie the development of depression in individuals without diabetes but with other medical problems or conditions.

References

Anderson, B. A., Goebel-Fabbri, A. E., and Jacobson, A. M. (2002). Behavioral research and psychological issues in diabetes: Progress and prospects. "Joslin Diabetes Mellitus" 14th Edition. (C. R. Kahn and G. C. Weir, eds.). Lea & Febinger, Malvern, PA, (in press).

Anderson, R., Freedland, K. E., Clouse, R. E., and Lustman, P. J. (2001). The prevalence of comorbid depression in adults with diabetes. *Diabetes Care* **24,** 1069–1078.

Araki, Y. *et al.* (1994). MRI of the brain in diabetes mellitus. *Neuroradiology* **36,** 101–103.

Auer, R. N., Hugh, J., Cosgrove, E., and Curry, B. (1989). Neuropathologic findings in three cases of profound hypoglycemia. *Clin. Neuropathol.* **8,** 63–68.

Auer, R. N., Kalimo, H., Olsson, Y., and Siesjo, B. K. (1985). The temporal evolution of hypoglycemic brain damage. I. L. *Acta Neuropathol.* (*Berl.*) **67,** 13–24.

Auer, R. N., Olsson, Y., and Siesjo, B. K. (1984a). Hypoglycemic brain injury in the rat. Correlation of density of brain damage with the EEG isoelectric time: A quantitative study. *Diabetes* **33,** 1090–1098.

Auer, R. N., and Siesjo, B. K. (1993). Hypoglycaemia: Brain neurochemistry and neuropathology. *Baillieres Clin. Endocrinol. Metab.* **7,** 611–625.

Auer, R. N., Wieloch, T., Olsson, Y., and Siesjo, B. K. (1984b). The distribution of hypoglycemic brain damage. *Acta Neuropathol.* (*Berl.*) **64,** 177–191.

Austin, E. J., and Deary, I. J. (1999). Effects of repeated hypoglycemia on cognitive function: a psychometrically validated reanalysis of the Diabetes Control and Complications Trial data. *Diabetes Care* **22,** 1273–1277.

Axford, J. S., Howe, F. A., Heron, C., and Griffiths, J. R. (2001). Sensitivity of quantitative (1)H magnetic resonance spectroscopy of the brain in detecting early neuronal damage in systemic lupus erythematosus. *Ann. Rheum. Dis.* **60,** 106–111.

Bellush, L. L., Reid, S. G., and North, D. (1991). The functional significance of biochemical alterations in streptozotocin-induced diabetes. *Physiol. Behav.* **50,** 973–981.

Bjorgaas, M., Sand, T., and Gimse, R. (1996). Quantitative EEG in type 1 diabetic children with and without episodes of severe hypoglycemia: A controlled, blind study. *Acta. Neurol. Scand.* **93,** 398–402.

Boeve, B. F., Bell, D. G., and Noseworthy, J. H. (1995). Bilateral temporal lobe MRI changes in uncomplicated hypoglycemic coma. *Can. J. Neurol. Sci.* **22,** 56–58.

Breteler, M. M. *et al.* (1994a). Cognitive correlates of ventricular enlargement and cerebral white matter lesions on magnetic resonance imaging. The Rotterdam Study. *Stroke* **25,** 1109–1115.

Breteler, M. M. *et al.* (1994b). Cerebral white matter lesions, vascular risk factors, and cognitive function in a population-based study: The Rotterdam Study. *Neurology* **44,** 1246–1252.

Brierley, J. B., Brown, A. W., and Meldrum, B. S. (1971a). The nature and time course of the neuronal alterations resulting from oligaemia and hypoglycaemia in the brain of Macaca mulatta. *Brain Res.* **25,** 483–499.

Brierley, J. B., Brown, A. W., and Meldrum, B. S. (1971b). "Brain Hypoxia" (J. B. Brierley and B. S. Meldrum, eds.), pp. 225–230. Spastics International Medical Publications, London.

Brown, F. W., and Lewine, R. J. *et al.* (1992). White matter hyperintensity signals in psychiatric and nonpsychiatric subjects. *Am. J. Psychiatry* **149,** 620–625.

Bryden, K. S. *et al.* (1999). Eating habits, body weight, and insulin misuse. A longitudinal study of teenagers and young adults with type 1 diabetes. *Diabetes Care* **22,** 1956–1960.

Buchsbaum, M. S., Wu, J. *et al.* (1986). Frontal cortex and basal ganglia metabolic rates assessed by positron emission tomography with [18F]2-deoxyglucose in affective illness. *J. Affect. Disord.* **10,** 137–152.

Cameron, O., Kronfol, Z., Greden, J., and Carroll, B. (1984). Hypothalamic-pituitary-adrenocortical activity in patients with diabetes mellitus. *Arch. Gen. Psychiatry* **41,** 1090–1095.

Cassileth, B. R. *et al.* (1984). Psychosocial status in chronic illness. A comparative analysis of six diagnostic groups. *N. Engl. J. Med.* **311,** 506–511.

Chalmers, J. *et al.* (1991). Severe amnesia after hypoglycemia. *Diabetes Care* **14,** 922–925.

Coffey, C. E., Figiel, G. S. *et al.* (1988). Leukoencephalopathy in elderly depressed patients referred for ECT. *Biol. Psychiatry* **24,** 143–161.

Coffey, C. E., Figiel, G. S. *et al.* (1989). Subcortical white matter hyperintensity on magnetic resonance imaging: Clinical and neuroanatomic correlates in the depressed. *J. Neuropsychiatry* **1,** 135–144.

Coffey, C. E., Figiel, G. S. *et al.* (1990). Subcortical hyperintensity on magnetic resonance imaging: A comparison of normal and depressed elderly subjects. *Am. J. Psychiatry* **147,** 187–189.

Coffey, C. E., Wilkinson, W. E. *et al.* (1993). Quantitative cerebral anatomy in depression: A controlled magnetic resonance imaging study. *Arch. Gen. Psychiatry* **50,** 7–16.

Crandall, E. A., Gillis, M. A., and Fernstrom, J. D. (1981). Reduction in brain serotonin synthesis rate in streptozotocin diabetic rats. *Endocrinology* **109,** 310–312.

Deary, I. J., and Frier, B. M. (1996). Severe hypoglycaemia and cognitive impairment in diabetes: Link not proven. *Br. Med. J.* **313,** 767–768.

Deary, I. J., Langan, S. J., Graham, K. S., Hepburn, D., and Frier, B. M. (1992). Recurrent severe hypoglycemia, intelligence, and speed of information processing. *Intelligence* **16,** 337–359.

de Groot, M., Anderson, R. J., Freedland, K. E., Clouse, R. E., and Lustman, P. J. (2000). Association of diabetes complications and depression in type 1 and type 2 diabetes: A meta-analysis. *Diabetes* **49** (Suppl. 1), A63–A64.

Dejgaard, A. *et al.* (1991). Evidence for diabetic encephalopathy. *Diab. Med.* **8,** 162–167.

Derogatis, L. R. (1983). SCL-90-R administration, scoring and procedures manual—II for the revised version. *Towson, MD, Clin. Psychometr. Res.*

Drevets, W. C. (1999). Prefrontal cortical-amygdalar metabolism in major depression. *Ann. N.Y. Acad. Sci.* **877,** 614–637.

Drevets, W. C. *et al.* (1997). Subgenual prefrontal cortex abnormalities in mood disorders [see comments]. *Nature* **386,** 824–827.

Dupont, R. M. *et al.* (1990). Subcortical abnormalities detected in bipolar affective disorder using magnetic resonance imaging. Clinical and neuropsychological significance. *Arch. Gen. Psychiatry* **47,** 55–59.

Figiel, G. S., Krishnan, K. R. *et al.* (1991). Subcortical hyperintensities on brain magnetic resonance imaging: A comparison of normal and bipolar subjects. *J. Neuropsychiatry Clin. Neurosci.* **3,** 18–22.

Fisher, L. *et al.* (2000). The family and disease management in Hispanic and European-American patients with type 2 diabetes. *Diabetes Care* **23,** 267–272.

Franceschi, M. *et al.* (1984). Cognitive processes in insulin-dependent diabetes. *Diabetes Care* **7,** 228–231.

Fujioka, M. *et al.* (1997). Specific changes in human brain after hypoglycemic injury. *Stroke* **28,** 584–587.

Garnett, M. R. *et al.* (2000). Early proton magnetic resonance spectroscopy in normal-appearing brain correlates with outcome in patients following traumatic brain injury. *Brain* **123,** 2046–2054.

Gater, R. *et al.* (1998). Sex differences in the prevalence and detection of depressive and anxiety disorders in general health care settings: report from the World Health Organization Collaborative Study on Psychological Problems in General Health Care. *Arch. Gen. Psychiatry* **55,** 405–413.

Gavard, J. A., Lustman, P. J., and Clouse, R. E. (1993). Prevalence of depression in adults with diabetes: An epidemiological evaluation. *Diabetes Care* **16,** 1167–1178.

Geringer, E. D. (1990). Affective disorders and diabetes mellitus. *Neuropsychol. Behav. Aspects Diabetes* 239–272.

Gold, A. E., Deary, I. J., and Frier, B. M. (1993). Recurrent severe hypoglycaemia and cognitive function in type 1 diabetes. *Diab. Med.* **10,** 503–508.

Gold, A. E. *et al.* (1994). Severe deterioration in cognitive function and personality in five patients with long-standing diabetes: A complication of diabetes or a consequence of treatment? *Diab. Med.* **11,** 499–505.

Gonder-Frederick, L., Cox, D., Clarke, W., and Julian, D. (2000). "Psychology in Diabetes Care" (F. J. Snoek and T. C. Skinner, eds.), pp. 169–205. Wiley, New York.

Hauser, S. T., Borman, E. H., Powers, S. I., Jacobson, A. M., and Noam, G. G. (1990). Paths of adolescent ego development: Links with family life and individual adjustment. *Psychiatr. Clin. North Am.* **13,** 489–510.

Hauser, S. T., DiPlacido, J., Jacobson, A. M., Willett, J., and Cole, C. (1993). Family coping with an adolescent's chronic illness: An approach and three studies. *J. Adolesc.* **16,** 305–329.

Hershey, T., Bhargava, N., Sadler, M., White, N. H., and Craft, S. (1999). Conventional vs. intensive diabetes therapy in children with type 1 diabetes: Effects on memory and motor speed. *Diabetes Care* **22,** 1318–1324.

Hershey, T., Craft, S., Bhargava, N., and White, N. H. (1997). Memory and insulin dependent diabetes mellitus (IDDM): Effects of childhood onset and severe hypoglycemia. *J. Int. Neuropsychol. Soc.* **3,** 520.

Hickie, I. *et al.* (1995). Subcortical hyperintensities on magnetic resonance imaging: clinical correlates and prognostic significance in patients with severe depression. *Biol. Psychiatry* **37,** 151–160.

Hudson, J. *et al.* (1984). Abnormal results of dexamethasone suppression test in non-depressed patients with diabetes mellitus. *Arch. Gen. Psychiatry.* **41,** 1086–1089.

Jacobson, A. M. (1993). Depression and diabetes *Diabetes Care* **16,** 1621–1623.

Jacobson, A. M. (1996). Psychological care of patients with insulin-dependent diabetes mellitus. *N. Engl. J. Med.* **334,** 1249–1253.

Jacobson, A. M. (1997). Quality of life in patients with diabetes mellitus. *Semin. Clin. Neuropsychiatry* **2,** 82–93.

Jacobson, A. M., de Groot, M., and Samson, J. A. (1994). The evaluation of two measures of quality of life in patients with type I and type II diabetes. *Diabetes Care* **17,** 267–274.

Jacobson, A. M. *et al.* (1994). Family environment and glycemic control: A four-year prospective study of children and adolescents with insulin-dependent diabetes mellitus. *Psychosomat. Med.* **56,** 401–409.

Jacobson, A. M. *et al.* (1997a). Psychological adjustment to IDDM: 10-year follow-up of an onset cohort of child and adolescent patients. *Diabetes Care* **20,** 811–818.

Jacobson, A. M. *et al.* (1997b). Social relationships among young adults with insulin-dependent diabetes mellitus: ten-year follow-up of an onset cohort. *Diab. Med.* **14,** 73–79.

Jacobson, A. *et al.* (2000). Brain functioning, cognition and psychiatric disorders in patients with type 1 diabetes. *Diabetes* **49**(Suppl. 1), A132.

Jacobson, A. M. *et al.* (2001). Factors related to the development of MRI abnormalities in type 1 diabetic patients. Unpublished observations.

Jacobson, A. M., and Weinger, K. (1998). Treating depression in diabetic patients: Is there an alternative to medications? *Ann. Intern. Med.* **129,** 656–657.

Kahn, K. J., and Myers, R. E. (1971). "Brain Hypoxia" (J. B. Brierley and B. S. Meldrum, eds.), pp. 185–194. Spastics International Medical Publications, London.

Kalimo, H., and Olsson, Y. (1980). Effects of severe hypoglycemia on the human brain. Neuropathological case reports. *Acta Neurol. Scand.* **62,** 345–356.

Kovacs, M. *et al.* (1997). Major depressive disorder in youths with IDDM: A controlled prospective study of course and outcome. *Diabetes Care* **20,** 45–51.

Kovacs, M., Ho, V., and Pollock, M. H. (1995). Criterion and predictive validity of the diagnosis of adjustment disorder: A prospective study of youths with new-onset insulin-dependent diabetes mellitus. *Am. J. Psychiatry* **152,** 523–528.

Kreis, R., and Ross, B. D. (1992). Cerebral metabolic disturbances in patients with subacute and chronic diabetes mellitus: Detection with proton MR spectroscopy. *Radiology* **184,** 123–130.

Krishnan, K. R., McDonald, W. M. *et al.* (1992). Magnetic resonance imaging of the caudate nuclei in depression: Preliminary observations. *Arch. Gen. Psychiatry* **49,** 553–557.

Lai, T., Payne, M. E., Byrum, C. E., Steffens, D. C., and Krishnan, K. R. (2000). Reduction of orbital frontal cortex volume in geriatric depression. *Biol. Psychiatry* **48,** 971–975.

Langan, S. J., Deary, I. J., Hepburn, D. A., and Frier, B. M. (1991). Cumulative cognitive impairment following recurrent severe hypoglycaemia in adult patients with insulin-treated diabetes mellitus. *Diabetologia* **34,** 337–344.

Leibson, C. L. *et al.* (1997). Risk of dementia among persons with diabetes mellitus: A population-based cohort study. *Am. J. Epidemiol.* **145,** 301–308.

Lezak, M. D. (1995). "Neuropsychological Assessment" Oxford University Press, New York.

Lincoln, N. B., Faleiro, R. M., Kelly, C., Kirk, B. A., and Jeffcoate, W. J. (1996). Effect of long-term glycemic control on cognitive function. *Diabetes Care* **19,** 656–658.

Lunetta, M. *et al.* (1994). Evidence by magnetic resonance imaging of cerebral alterations of atrophy type in young insulin-dependent diabetic patients. *J. Endocrinol. Invest.* **17,** 241–245.

Lustman, P. J., Griffith, L. S., Clouse, R. E., and Cryer, P. E. (1986). Psychiatric illness in diabetes mellitus. Relationship to symptoms and glucose control. *J. Nerv. Ment. Dis.* **174,** 736–742.

Lustman, P. J., Griffith, L. S., Freedland, K. E., and Clouse, R. E. (1997). The course of major depression in diabetes. *Gen. Hosp. Psychiatry* **19,** 138–143.

Lustman, P. J., Griffith, L. S., Freedland, K. E., Kissel, S. S., and Clouse, R. E. (1998). Cognitive behavior therapy for depression in type 2 diabetes mellitus: a randomized, controlled trial. *Ann. Intern. Med.* **129,** 613–621.

Lyoo, I. K., Lee, H. K., Jung, J. H., Noam, G. G., and Renshaw, P. F. (2002). White matter hyperintensities on brain MRI in children with psychiatric disorders. *Compr. Psychiatry* (in press).

Mayou, R., Bryant, B., and Turner, R. (1990). Quality of life in non-insulin-dependent diabetes and a comparison with insulin-dependent diabetes. *J. Psychosom. Res.* **34,** 1–11.

Mayou, R., Peveler, R., Davies, B., Mann, J., and Fairburn, C. (1991). Psychiatric morbidity in young adults with insulin-dependent diabetes mellitus. *Psychol. Med.* **21,** 639–645.

McDonald, W. M., Tupler, L. A. *et al.* (1999). Hyperintense lesions on magnetic resonance images in bipolar disorder. *Biol. Psychiatry* **45,** 965–971.

Meuter, F., Thomas, W., Gruneklee, D., Gries, F. A., and Lohmann, R. (1980). Psychometric evaluation of performance in diabetes mellitus. *Horm. Metab. Res. Suppl.* **9,** 9–17.

Nakamura, Y. *et al.* (1991). Abnormal brainstem evoked potentials in diabetes mellitus: Evoked potential testings and magnetic resonance imaging. *Electromyogr. Clin. Neurophysiol.* **31,** 243–249.

Patrick, A. W., and Campbell, I. W. (1990). Fatal hypoglycaemia in insulin-treated diabetes mellitus: Clinical features and neuropathological changes, *Diab. Med.* **7,** 349–354.

Perros, P., and Deary, I. J. (1999). "Hypoglycaemia in Clinical Diabetes" (B. M. Frier and B. M. Fisher, eds.), pp. 187–210. John Wiley & Sons Ltd., Chinchester, UK.

Perros, P., Deary, I. J, Sellar, R. J., Best, J. J. K., and Frier, B. M. (1997). Brain abnormalities demonstrated by magnetic resonance imaging in adult IDDM patients with and without a history of recurrent severe hypoglycemia. *Diabetes Care* **20,** 1013–1018.

Perros, P., Sellar, R. J., and Frier, B. M. (1994). Chronic pontine dysfunction following insulin-induced hypoglycemia in an IDDM patient. *Diabetes Care* **17,** 725–727.

Persaud, R., Russow, H. *et al.* (1997). Focal signal hyperintensities in schizophrenia. *Schizophrenia Res.* **27,** 55–64.

Pillay, S. S. *et al.* (1998). A quantitative magnetic resonance imaging study of caudate and lenticular nucleus gray matter volume in primary unipolar major depression: relationship to treatment response and clinical severity. *Psychiatry Res. Neuroimaging* **84,** 61–74.

Popkin, M. K., Callies, A. L., Lentz, R. D., Colon, E. A., and Sutherland, D. E. (1988). Prevalence of major depression, simple phobia, and other psychiatric disorders in patients with long-standing type 1 diabetes mellitus. *Arch. Gen. Psychiatry* **45,** 64–68.

Reichard, P., Berglund, A., Britz, A., Levander, S., and Rosenqvist, U. (1991). Hypoglycaemic episodes during intensified insulin treatment: increased frequency but no effect on cognitive function. *J. Intern. Med.* **229,** 9–16.

Reichard, P., Britz, A., and Rosenqvist, U. (1991). Intensified conventional insulin treatment and neuropsychological impairment. *Br. Med. J.* **303,** 1439–1442.

Reichard, P., Pihl, M., Rosenqvist, U., and Sule, J. (1996). Complications in IDDM are caused by elevated blood glucose level: The Stockholm Diabetes Intervention Study (SDIS) at 10-year follow up. *Diabetologia* **39,** 1483–1488.

Ritholz, M. D., and Jacobson, A. M. (1998). Living with hypoglycemia. *J. Gen. Intern. Med.* **13,** 799–804.

Robinson, N., Fuller, J. H., and Edmeades, S. P. (1988). Depression and diabetes. *Diab. Med.* **5,** 268–274.

Rovet, J. F., and Ehrlich, R. M. (1999). The effect of hypoglycemic seizures on cognitive function in children with diabetes: A 7-year prospective study. *J. Pediatr.* **134,** 503–506.

Ryan, C. M. (1997). Effects of diabetes mellitus on neuropsychological functioning: A lifespan perspective. *Semin. Clin. Neuropsychiatry* **2,** 4–14.

Ryan, C. M., and Greckle, M. (2000). Why is learning and memory dysfunction in type 2 diabetes limited to older adults. *Diabetes Metab. Res. Rev.* **16,** 308–315.

Ryan, C., Vega, A., Longstreet, C., and Drash, A. (1984). Neuropsychological changes in adolescents with insulin-dependent diabetes. *J. Consult. Clin. Psychol.* **52,** 335–342.

Ryan, C. M., Williams, T. M., Finegold, D. N., and Orchard, T. J. (1993). Cognitive dysfunction in adults with type 1 (insulin-dependent) diabetes mellitus of long duration: effects of recurrent hypoglycaemia and other chronic complications. *Diabetologia* **36,** 329–334.

Ryan, C. M., Williams, T. M., Orchard, T. J., and Finegold, D. N. (1992). Psychomotor slowing is associated with distal symmetrical polyneuropathy in adults with diabetes mellitus. *Diabetes* **41,** 113.

Sachon, C. *et al.* (1992). Cognitive function, insulin-dependent diabetes and hypoglycaemia. *J. Int. Med.* **231,** 471–475.

Schurhoff, F., Bellivier, F. *et al.* (2000). Early and late onset bipolar disorders: Two different forms of manic-depressive illness. *J. Affect. Disord.* **58,** 215–221.

Seidl, R. *et al.* (1996). Brainstem auditory evoked potentials and visually evoked potentials in young patients with IDDM. *Diabetes Care* **19,** 1220–1224.

Siesjo, B. K., and Bengtsson, F. (1989). Calcium fluxes, calcium antagonists, and calcium-related pathology in brain ischemia, hypoglycemia, and spreading depression: A unifying hypothesis. *J. Cereb. Blood Flow Metab.* **9,** 127–140.

Simone, I. L. *et al.* (2001). Axonal damage in multiple sclerosis plaques: a combined magnetic resonance imaging and 1H-magnetic resonance spectroscopy study. *J. Neurol. Sci.* **182,** 143–150.

Skenazy, J. A., and Bigler, E. D. (1984). Neuropsychological findings in diabetes mellitus. *J. Clin. Psychol.* **40,** 246–258.

Soltész, G., and Acsádi, G. (1989). Association between diabetes, severe hypoglycemia, and electroencephalographic abnormalities. *Arch. Dis. Child.* **64,** 992–996.

Spitzer, R. L., Williams, J. B., Gibbon, M., and First, M. B. (1992). The structured clinical interview for DSM-III-R (SCID). I: History, rationale and description. *Arch. Gen. Psychiatry* **49,** 624–629.

Strachan, M. W. J., Deary, I. J., Ewing, F. M. E., and Frier, B. M. (1977). Is type 2 (non-insulin dependent) diabetes mellitus associated with an increased risk of cognitive dysfunction? *Diabetes Care* **20,** 438–445.

Strachan, M. W. *et al.* (1999). Evaluation of serum markers of neuronal damage following severe hypoglycaemia in adults with insulin-treated diabetes mellitus. *Diabetes Metab. Res. Rev.* **15,** 5–12.

Talbot, F., and Nouwen, A. (2000). A review of the relationship between depression and diabetes in adults: Is there a link? *Diabetes Care* **23,** 1556–1562.

The Diabetes Control and Complications Trial Research Group. (1993). The effect of intensive treatment of diabetes on the development and progression of long-term complications in insulin-dependent diabetes mellitus. *N. Engl. J. Med.* **329,** 977–986.

The Diabetes Control and Complications Trial Research Group. (1996). Effects of intensive diabetes therapy on neuropsychological function in adults in the Diabetes Control and Complications Trial. *Ann. Intern. Med.* **124,** 379–388.

Trulson, M. E., and Himmel, C. D. (1985). Effects of insulin and streptozotocin-induced diabetes on brain norepinephrine metabolism in rats. *Neurochemistry* **44,** 1873–1876.

Valenzuela, M. J. *et al.* (2000). Dual voxel proton magnetic resonance spectroscopy in the healthy elderly: Subcortical-frontal axonal *N*-acetylaspartate levels are correlated with fluid cognitive abilities independent of structural brain changes. *Neuroimage* **12,** 747–756.

Vanhanen, M. *et al.* (1998). Cognitive function in an elderly population with persistent impaired glucose tolerance. *Diabetes Care* **21,** 398–402.

Wells, K. B., Golding, J. M., and Burnam, M. A. (1988). Psychiatric disorder in a sample of the general population with and without chronic medical conditions. *Am. J. Psychiatry* **145,** 976–981.

Wells, K. B., Bolding, J. M., and Burnam, M. A. (1989a). Affective, substance use, and anxiety disorders in persons with arthritis, diabetes, heart disease, high blood pressure, or chronic lung conditions. *Gen. Hosp. Psychiatry* **11,** 320–327.

Wells, K. B., Golding, J. M., and Burnam, M. A. (1989b). Chronic medical conditions in a sample of general population with anxiety, affective, and substance use disorders. *Am. J. Psychiatry* **146,** 1440–1446.

Weyerer, S., Hewer, W., Pfeifer-Kurda, M., and Dilling, H. (1989). Psychiatric disorders and diabetes–results from a community study. *J. Psychosomat. Res.* **33,** 633–640.

Woods, B. T., Yurgelun-Todd, D. *et al.* (1995). Age-related MRI abnormalities in bipolar illness: A clinical study. *Biol. Psychiatry* **38,** 846–847.

Wredling, R., Levander, S., Adamson, U., and Lins, P. E. (1990). Permanent neuropsychological impairment after recurrent episodes of severe hypoglycaemia in man. *Diabetologia* **33,** 152–157.

SCHIZOPHRENIA AND DIABETES

David C. Henderson and Elissa R. Ettinger

Schizophrenia Program, Massachusetts General Hospital, Harvard Medical School, Boston, Massachusetts 02114

I. Introduction

Recently, a great deal of attention has been given to the role of antipsychotic agents in the development of diabetes mellitus (DM) in patients with neuropsychiatric disorders, including schizophrenia. Evidence long before antipsychotic agents were developed suggests a higher rate of DM in schizophrenia patients compared to the general population. This chapter will review the literature on schizophrenia, DM, and antipsychotic agents and discuss several implications.

II. Schizophrenia and DM

Studies conducted 20–30 years ago suggested that the first antipsychotic medications—perhaps even untreated schizophrenia itself—were associated with an increased risk of diabetes. Several studies have found increased rates of diabetes or hyperglycemia in schizophrenia with and without association with conventional antipsychotic agents. (Braceland and Meduna *et al.*, 1945; Waitzkin, 1966a,b; Schwartz and Munoz, 1968; Franzen, 1970; Marinow, 1971; Schimmelbusch *et al.*, 1971; Brambilla *et al.*, 1976; Yaryura-Tobias

0074-7742/02 $35.00

et al., 1978; McKee *et al.,* 1986; Mukherjee *et al.,* 1989, 1996). It is not clear if the increased rates are secondary to increases in insulin-dependent or noninsulin-dependent DM in these studies. Earlier studies also suffer problems with diagnostic criteria for schizophrenia. Mukherjee and colleagues (1989) found diabetes to be more common in neuroleptic-free schizophrenia patients compared to conventional antipsychotic-agent-treated patients. The rates were considered higher than the general population.

Studies that are more recent suggest that patients with schizophrenia may have an increased risk of developing diabetes. Mukherjee and colleagues (1989) found a 15.8% overall prevalence of diabetes in 95 schizophrenia patients. There were no cases of diabetes in patients younger than 50 years old; 12.9% in patients age 50–59 years; 18.9% in patients age 60–69 years; and 16.7% in patients age 70–74 years. Diabetes was significantly more common in neuroleptic-free patients than patients treated with neuroleptics, after controlling for age, sex, and cumulative duration of neuroleptic treatment. These results exceed expected rates of type 2 diabetes in the United States general population which are 5.1% in people age 20 years or less with 2.7% undiagnosed diabetes; and 12.3% age 40–74. Similar rates for diabetes were seen in men and women, but non-Hispanic blacks and Mexican-Americans were 1.6 and 1.9 times the rate of non-Hispanic whites (Harris *et al.,* 1998). All patients had extensive neuroleptic exposure (mean 27.04 ± 8.3 years for diabetic patients and 27.1 ± 7.1 years for nondiabetic patients) and were thought to have a sedentary lifestyle. Among the entire sample, 15.8% of the patients had diabetes (all type 2 diabetes), a prevalence that is considerably higher than among the general population. Bruno *et al.* (1997) reported the overall estimates for DM in the general population, in Italy, was 2.1% while 3% from age 50–59 and 5.5% from age 60–69.

III. Genetics of Diabetes and Schizophrenia

Given that individuals with schizophrenia appear to have higher rates of diabetes or hyperglycemia than the general population, the question arises whether genetic similarities exist in the two disorders. Unfortunately, the etiology of both disorders is complex and not fully understood. Nevertheless, progress is being made in understanding the genetic basis of these disorders (Hyman, 1999; National Institute of Mental Health's Genetics Workgroup 1999; Hy, 2000; Meltzer, 2000).

Twin and adoption studies have established a hereditary component in schizophrenia and the schizophrenia spectrum disorders (Gottesman,

1991; National Institute of Mental Health's Genetics Workgroup, 1999). Gottesman examined European family and twin studies performed between 1920 and 1987, and found the lifetime recurrence risk for schizophrenia among the offspring of one affected parent was approximately 8–10%. This risk is approximately 10 times higher than the risk in the general population.

Gottesman and several others also showed that no single gene causes schizophrenia (Gottesman, 1991; Tienari *et al.,* 1994; Tsuang and Faraone, 1995; National Institute of Mental Health's Genetics Workgroup, 1999; Tsuang, 2000). Instead, schizophrenia appears to be caused by the interaction of multiple genes and environmental factors, such as pregnancy and delivery complications (e.g., fetal hypoxia, infections, and malnutrition), residence in an urban area, and dysfunctional family communication (Tsuang, 2000). Indeed, if the schizophrenic phenotype were determined entirely by genetic factors, monozygotic (MZ) twins, who are genetically identical, would be expected to show a concordance rate that approaches 100%. Dizygotic (DZ) twins, siblings and parents, who share approximately 50% of their genes, would be expected to have concordance rates approaching 50%. In schizophrenia, however, this is not the case. Specifically, the studies aforementioned found concordance rates of approximately 50% for MZ twins and 15% for DZ twins (Tsuang, 2000).

Currently, many researchers are conducting linkage and association studies to identify which genes, and specifically which loci, have dominant effects on the development and progression of schizophrenia. To date, genetic linkage analysis has revealed susceptibility regions on chromosomes 13 (Blouin *et al.,* 1998; Pulver, 2000), 8 (Blouin *et al.,* 1998; Pulver, 2000), 22 (Blouin *et al.,* 1998; Pulver, 2000), and 6 (Kaufmann *et al.,* 1998; Pulver, 2000). In addition, preliminary findings suggest that chromosomes 10 (Faraone *et al.,* 1998; Pulver, 2000), 5 (Straub, 1997; Pulver, 2000), and others may be involved (Pulver, 2000). Of note, of these linkage findings, only the locus 1q21–q22 meets the significant linkage criteria of Lander and Kruylak (Blouin *et al.,* 1998; Pulver, 2000).

Like schizophrenia, DM is a disease with mutifactorial inheritance. Researchers are confident that no single gene causes type 1 or 2 DM, and that development of this heterogeneous group of disorders is likely the result of multiple genes and environmental factors (Atkinson and Mclaren, 1994). Siblings of individuals with type 1 DM face a substantial elevation in risk: approximately 6% as opposed to a risk of about 0.3–0.5% in the general population (Atkinson and Mclaren, 1994). Genetic susceptibility to type 1 DM may be located on chromosomes 2, 6, 11, and 15. Of these, the primary genetic site in humans is believed to be located in the histocompatibility locus on the short arm of chromosome 6 (Atkinson and Mclaren, 1994), which, as noted above, may be a susceptibility region for schizophrenia.

Type 2 diabetes is more common than type 1, accounting for 80–90% of all diabetes cases in the United States, and it more frequently exhibits familial aggregation with MZ twin concordance rates substantially higher than in type 1 diabetes, often exceeding 90% (Atkinson and Mclaren, 1994). In addition, the empirical recurrence risks for first-degree relatives in type 2 diabetes cases are higher than those for type 1, generally ranging from 10–15% (Atkinson and Mclaren, 1994). Even though there appears to be a high degree of genetic involvement in type 2 diabetes, specific genes have not yet been identified and linkage analysis studies continue.

IV. Tardive Dyskinesia

Schizophrenia patients with antipsychotic-agent-associated tardive dyskinesia (TD) appear to have higher rates of DM. The overall mean prevalence of TD among chronically neuroleptic-treated patients is approximately 24%. The annual incidence in younger adults is 4–5%. Aging is a major risk factor for TD (Jeste *et al.*, 1995). Likely risk factors include female gender, mood disorders, "organic" brain dysfunction or damage, DM, and early extrapyramidal side effects. Metoclopramide, a D2-receptor blocker commonly used in nonpsychiatric medical patients, can also produce persistent TD.

Woerner *et al.* (1993) examined 160 elderly individuals who were beginning neuroleptic treatment; 24 had diabetes and 136 did not. After 43 weeks of neuroleptic exposure, the cumulative incidence rates of TD were 54.1% (SE = 5.6%) for the diabetics and 25.6% (SE = 16.1%) for the nondiabetics. In a study examining the relationships among CSF concentrations of substrates of mitochondrial energy metabolism, neuroleptic medication, and neurological side effects, Goff *et al.* (1995) found CSF concentrations of alanine were significantly elevated in the medicated patients when TD status was controlled for and were significantly correlated with total scores on the Abnormal Involuntary Movement Scale (AIMS). The results were consistent with a model linking neuroleptic-induced neurological side effects with impairment of mitochondrial energy metabolism, possibly mediated by inhibition of complex 1 of the electron transport chain.

Ganzini *et al.* (1992) reported two studies that examined the prevalence of TD in neuroleptic-treated diabetic patients. Study 1 compared 38 diabetic patients with 38 nondiabetic patients treated for psychotic disorders with low to moderate doses of neuroleptics (mean chlorpromazine equivalents = 300 mg/day) for an average of 18 years. Study 2 compared 24 diabetic and 27 nondiabetic patients treated for an average of 2.6 years with a mean 31 mg/day of metoclopramide for gastrointestinal disease.

There were no differences between the diabetic and nondiabetic groups in age, sex, type of psychiatric illness, and dose and duration of neuroleptic treatment or severity of parkinsonism. In both studies, the diabetic patients had significantly greater prevalence and severity of TD. No measures of diabetes severity were associated with TD in either study. Mukherjee and colleagues (1989) investigated the relationship of persistent TD to glucose tolerance and family history of type 2 diabetes by performing fasting glucose and glucose tolerance tests (GTT) on 22 schizophrenic patients treated with haloperidol (Haldol), comparing those with and without TD. Fifteen patients also underwent a GTT when drug free. The investigators found that of the haloperidol-treated patients, those with TD exhibited significantly elevated fasting blood glucose and decreased glucose tolerance compared to non-TD patients. The authors concluded that a positive family history of type 2 diabetes was significantly associated with the presence of TD and with higher drug-free fasting blood glucose.

Schultz *et al.* (1999) found patients with impaired glucose tolerance had higher mean abnormal movement scores than those without glucose intolerance, but this difference was not statistically significant. There was, however, an association between the magnitude of the fasting insulin level and abnormal movements after the authors controlled for fasting glucose level. Additionally, the fasting glucose level predicted abnormal movements after the authors controlled for age. The authors concluded that hyperinsulinemia and hyperglycemia associated with insulin resistance may potentially contribute to the pathogenesis of TD.

Sumiyoshi *et al.* (1997) examined the effect of streptozotocin (STZ)-induced diabetes and a combination of chronic treatment with haloperidol (HPD) on dopamine(DA)D2, serotonin (5-HT) 5-HT1A and 5-HT2A receptors was investigated in rat brain. Striatal D2 receptor density of the STZ-HPD and the STZ-vehicle were not significantly different from the vehicle-vehicle group. A significant increase in cortical 5-HT2A receptor density was observed only in the group of STZ-vehicle. The results suggest that a diabetic state might affect brain serotonergic activity via an increase in the density of 5-HT2A receptors. The lack of an effect of the combined chronic treatment with STZ and HPD on the D2 receptor density may correspond to the increased risk to develop TD in patients with diabetes.

V. Conventional Antipsychotic Agents and Diabetes

The introduction of antipsychotic agents led to an increase in reports of DM in psychotic patients. Thonnard-Neumann in 1968 reported 92 of

528 (17.2%) of the psychotic women population at their institute were diabetic (Thonnard-Neumann, 1968). The incidence of diabetes among 277 patients who had not been treated with phenothiazines was 24 (8.6%) and 68 (27%) among 252 phenothiazine-treated patients. Of these patients, 40% identified a family member with DM. It is often difficult to relate these older studies to contemporary clinical practice because of inconsistencies in the ways in which schizophrenia or diabetes were diagnosed and because relatively little attention was given to other factors that influence the risk of diabetes, such as a high body mass index (BMI) or sustained obesity, race, age, physical activity level, and diet, however. It was also not always clear whether these increases were due to insulin-dependent (type 1) or noninsulin-dependent (type 2) diabetes. However, there were no reports of diabetic ketoacidosis (DKA) or deaths related to hyperglycemia. Tollefson and Lesar (1983) reported a case of nonketotic hyperglycemia associated with the introduction of loxapine that resolved with discontinuation of the drug. Later, amoxapine was reinitiated and resulted in hyperglycemia.

Keskiner and colleagues (1973) reported a greater incidence of blood glucose levels >120 mg/dl in a 2-h postprandial screening for diabetes in patients treated with thioridazine (11.59%) and chlorpromazine (10.63%) than haloperidol (2.90%) and fluphenazine (4.35%) treated patients. The increased incidence of blood glucose values >120 mg/dl was more significant for impaired glucose tolerance than diabetes. However, this study and many of the early studies examining antipsychotic agents and diabetes in schizophrenia, had major design flaws, particularly in the diagnostic criteria for schizophrenia and diabetes, and attention to other factors that impact on glucose metabolism.

VI. Atypical Antipsychotic Agents and DM

Recent case reports have linked clozapine to DKA (Koval *et al.,* 1994; Kostakoglu *et al.,* 1996; Peterson and Byrd, 1996; Ai *et al.,* 1998; Colli *et al.,* 1999; Mohan *et al.,* 1999; Smith *et al.,* 1999) and hyperglycemia (Kamran *et al.,* 1994). Popli and colleagues (1997) described four cases in which clozapine was associated with exacerbation of DM or new onset DM not attributable to weight gain. Wirshing and colleagues (1988) described four cases of clozapine-associated diabetes and two cases of olanzapine-associated diabetes. Hagg and colleagues (1988) recently reported, using an oral glucose tolerance test (OGTT), that 12% of patients treated with clozapine developed type 2 DM and 10% developed impaired glucose tolerance (IGT) compared to 6 and 3%, respectively, for patients treated with conventional depot antipsychotic agents. Although the increased rates of diabetes and IGT with clozapine did not achieve statistical significance, patients in the conventional antipsychotic agent group were older, which may have influenced the results. Melkersson *et al.* (1999) compared 28 schizophrenia patients on therapy with conventional agents to 13 patients treated with clozapine and found elevated fasting insulin levels and reduced growth-hormone-dependent insulin-like growth factor-I, suggesting insulin resistance with secondary increased insulin secretion.

We reported a 5-year naturalistic study to examine, in 82 clozapine patients with schizophrenia, the incidence of treatment-emergent DM in relation to other factors including weight gain, lipid abnormalities, age, clozapine dose, and treatment with valproate (Henderson *et al.*, 2000). The mean age at the time of clozapine initiation of the 82 patients studied was 36.4 years (SD 7.8) and 22 (27%) were women. Seventy-five (91%) were Caucasian, 3 (4%) African-American, 3 (4%) Hispanic, and 1 (1%) Asian. The baseline weight was 175.5 lb (SD 34.0) and BMI was 26.9 kg/m^2 (SD 5.0). During the 60-month study, 43 (52.44%) of the 82 patients experienced at least one elevated fasting blood glucose (FBG >140 mg/dl). Thirty of 82 patients (36.6%) were actually diagnosed with diabetes during the 5-year follow-up. Weight gain, valproate, and clozapine total daily dose were not significant risk factors for developing DM. Consistent with reports in the literature, some patients gained weight and developed diabetes, while others did not gain weight but also developed diabetes. Also, several patients developed a cluster of medical disorders similar to a metabolic syndrome X (diabetes requiring insulin, elevated serum lipids, abdominal obesity, and hypertension).

Finally, a group from the Food and Drug Administration (FDA) Research program identified 384 reports, with 54 cases of exacerbation of existing diabetes and 242 cases of new-onset diabetes in clozapine-treated patients from the FDA's Medwatch surveillance program (Koller *et al.*, 2001). There were 80 cases of probable DKA and 25 deaths. The majority of DKA episodes occurred in the first six months of treatment with clozapine. One patient actually developed diabetes immediately following an accidental 500 mg dose of clozapine.

Reports of olanzapine-associated DM have also emerged in the literature. Goldstein *et al.* (1999) reported several cases of new-onset DM and DKA associated with olanzapine treatment. In the majority of cases, diabetes developed within the first three months of treatment with olanzapine. However, several cases were complicated by the additional use of valproate which is thought to be associated with polycystic ovaries, centripetal obesity with associated hyperinsulinemia, and unfavorable serum lipid profiles (Isojarvi *et al.*, 1993, 1998). In another report from Germany, the authors report one case of olanzapine-associated elevated glucose concentration that returned to normal following olanzapine withdrawal while another case resulted in lethal DKA. (Von Hayek *et al.*, 1999). Wirshing *et al.* (1998) reported four cases of clozapine-associated diabetes and two cases of olanzapine-associated diabetes where five of the six had risks factors for diabetes and four experienced substantial weight gain after starting the particular agent.

Additionally, Lindenmayer and Patel (1999) reported a case of *de novo* DM eight months after initiating olanzapine, which resolved completely after discontinuation of olanzapine. Similarly, Fertig and colleagues (1998) reported the onset of diabetes in an obese African-American male within six weeks of olanzapine initiation, which resolved after olanzapine discontinuation, and returned within 8 days of rechallenge with olanzapine. Gatta and colleagues (1999) reported a case of olanzapine-associated DKA that required the discontinuation of olanzapine and the use of insulin. The diabetes resolved completely within 15 days of olanzapine discontinuation. Smith (Smith and Lindenmayer *et al.*, 1999) recently reported a naturalistic study of 52 patients before and after olanzapine initiation. She found 33%, (compared to 1% at baseline, before olanzapine initiation), demonstrated a serum glucose above 140 ng/ml. Twenty-five percent of the patients had diabetes before olanzapine-initiation and 77% of these exhibited abnormal glucose levels, while 18% of the nondiabetic patients (before olanzapine initiation) exhibited increased serum glucose values (>140 ng/ml) after olanzapine-initiation. Finally, Melkersson *et al.* (2000), who examined 14 olanzapine-treated patients, found weight gain and elevated levels of insulin, leptin, and blood lipids, with 3 patients developing diabetes compared to baseline.

There have been fewer reports of risperidone- or quetiapine-associated DM. A recent case report (Sobel *et al.*, 1999) described a 42-year-old male with a history of bipolar disorder admitted to the hospital for new-onset diabetes, requiring insulin, one month after beginning treatment with quetiapine. Quetiapine was discontinued and the insulin requirement steadily declined until it was discontinued five months after hospitalization. Additionally, Wilson *et al.* (2000) described 14 patients that developed diabetes and DKA (N = 5) while receiving treatment with either clozapine, olanzapine, or quetiapine. Bettinger *et al.* (2000) reported a case of severe exacerbation of type 2 DM in a 54-year-old African-American woman treated with olanzapine and quetiapine. In this case, glucose regulation improved following discontinuation of both olanzapine and quetiapine. Procyshyn *et al.* (2000) described a case of new-onset DM, which resulted in insulin and oral hypoglycemic agent therapy, in an obese, Ethiopian male following treatment with quetiapine.

Another case report (Croarkin *et al.*, 2000) described a 42-year-old, Caucasian, HIV-positive male with a history of chronic depression with psychotic features admitted to the hospital for DKA while taking risperidone, fluoxetine, and trazadone. Patients receiving the HIV-1 protease inhibitors often develop impaired glucose tolerance or diabetes, most likely due to an induction of insulin resistance (Schutt *et al.*, 2000). However, this patient was not receiving treatment with protease inhibitors (Croarkin and Bain, 2001). Risperidone was discontinued, the DKA resolved, and the patient was discharged with quetiapine and subcutaneous insulin. Recently, Newcomer and colleagues (1999) reported at the American College of Neuropsychopharmacology that clozapine and olanzapine but not risperidone used on normal control subjects experienced higher baseline and 75-min post 50 g dextrose challenge glucose levels. Comparison of insulin values was not statistically significant ($p = 0.07$). Subjects were well matched for age and BMI. Finally, Newcomer *et al.* using a modified oral GTT, reported significant elevations in fasting glucose in clozapine- and olanzapine-treated subjects compared to normal controls and subjects treated with typical antipsychotic agents (Newcomer and Fucetola *et al.*, 2000, 2001). Risperidone-treated subjects also showed elevation in fasting glucose when compared to untreated, healthy controls, but not typical agents. A homeostatic model assessment suggested effects on insulin resistance.

VII. Mechanisms of Glucose Metabolism Impairment by Antipsychotic Agents

Few medications have been linked to DKA. The reports of DKA with clozapine and olanzapine and possible increased incidence of type 2 DM may represent two separate populations with varying risks. The majority of clozapine- and olanzapine-DKA cases reported have been in the first six months of treatment. Several cases of DKA resolved after clozapine or olanzapine were discontinued, only to have hyperglycemia return upon reinstitution of the drug. By virtue of DKA cases taken off the drug and achieving remission, strongly suggests that insulin secretion is effected. Individuals who developed DKA associated with the use of the antipsychotic agents may be a different, poorly understood, subpopulation of schizophrenia with different risk factors. Insulin resistance is a characteristic feature of patients with impaired glucose tolerance and particularly with type 2 DM, as these patients have an even greater degree of insulin resistance. Insulin resistance can be due to abnormalities at any step in the entire insulin action sequence, broadly categorized as receptor defects or postreceptor defects in insulin action. It is possible that atypical antipsychotic agents cause a decrease of insulin-sensitive glucose transporters (GLUT) or an inability to stimulate recruitment of GLUT4 transporter from microsomes to the plasma membrane. GLUT4 is a transporter that mediates the bulk of insulin-stimulated transport activity (Taniguchi *et al.*, 1994). With a greater degree of insulin resistance, there is a greater need for effective glucose transport in clearing glucose from the serum. Chronic exposure to high concentrations of glucose and insulin reduces the subsequent ability of insulin to maximally stimulate glucose transport by inhibiting transporter translocation (Kopelman and Albon, 1997). In fact, a reduction in glucose utilization or effectiveness (S_G) is a common finding in type 2 DM patients regardless of insulin sensitivity. Alternatively, it may be that antagonism at serotonin 5-HT1A receptors by atypical antipsychotic agents decrease pancreatic β-cell responsiveness to blood glucose levels resulting in impairment of glucose metabolism (Wirshing *et al.*, 1998). However, in light of the findings of hyperinsulinemia in clozapine- and olanzapine-treated subjects, β-cell responsiveness, at least early in treatment, does not appear to be the problem (Melkersson *et al.*, 1999, 2000).

We (Henderson *et al.*, 2000) presented preliminary results of a cross-sectional study comparing clozapine-, olanzapine- and risperidone-treated, non-obese (BMI <30 mg/k^2) schizophrenia subjects with a frequent sampled intravenous GTT using Bergman's Minimal Model Analysis (Bergman *et al.*, 2000). There were no differences between the three groups for age, sex, race, family history of DM, BMI, percent body fat, fasting glucose, fasting insulin, cortisol, and letpin. Mean age was 40.5 years (range 22–60 years) and the mean BMI was 26.88 (range 19–30 kg/m^2). There was no difference between the three groups for age, race, BMI, fasting glucose, and fasting insulin. There was significant difference between treatment groups for the insulin sensitivity index (SI) (F value = 7.64, DF = 2, $p = 0.0057$). There was a significant difference in SI, comparing clozapine (mean 1.944) to risperidone (mean 12.843) ($p = 0.0017$) and olanzapine (mean 4.257) to risperidone (mean 12.843) ($p = 0.0085$). Normal values for SI with the minimal model analysis in the adult male population are $7.56 +/- 1.13 \times 10^{-4^*}$ min$^{-1^*}$ ml^{-1} and for SG are 0.026 +/− 0.008 min^{-1}. The normal values for healthy women are SI $5.61 \times 10^{-4^*}$ min$^{-1^*}$ ml^{-1}, and SG 0.023 min^{-1}, suggesting differences in SI between men and women (Bergman, 1989; Bergman *et al.*, 2000). There was no significant difference between the three groups for glucose effectiveness (SG), though there was a significant trend comparing clozapine (mean 0.014) to risperidone (mean 0.024) ($p = 0.062$). SG represents the ability of glucose itself, independent of insulin, to increase glucose uptake and suppress hepatic glucose output. This is a component of glucose disposal that is dependent on glucose transporters. All seven subjects treated with clozapine exhibited abnormal SI (all below $6.0 \times 10^{-4*}$ min^{-1*} ml^{-1}) and four of the five SGs that could be calculated were considered abnormal (less than 0.020 min^{-1}). For olanzapine, five of seven exhibited abnormal SI (below 6.0) and four of seven exhibited abnormal SG (below 0.020). Preliminary results suggest that there is a significant difference between the three groups for insulin sensitivity with clozapine and olanzapine exhibiting significant impairment. Larger sample sizes are needed to elucidate any effect on SG.

Dwyer *et al.* (1999a) reported that, *in vitro,* pimozide, fluphenazine, chlorpromazine, and clozapine inhibited glucose tranport in rat pheochromocytoma cells, while haloperidol and sulpiride did not. Dwyer *et al.* (1999b) also reported fluphenazine, chlorpromazine, and clozapine inhibited glucose transporter 3 (GLUT3) while haloperidol effect was significantly less. Glucose is essential for energy metabolism in neurons, and drugs which cause hyperglycemia in patients appear to inhibit neuronal glucose transport which may partly explain their effects (Dwyer *et al.,* 2001).

VIII. Cognition and Schizophrenia

Postmortem studies suggest that chronic, poorly controlled DM might be associated with degenerative changes in the central nervous system (Prescott *et al.,* 1990). However, the evidence that these changes are reflected in any impairment of cognitive performance is equivocal. Prescott *et al.* (1990) evaluated the effects of duration of illness and glycemic control upon memory function in 40 cases of insulin-dependent diabetes mellitus (IDDM). Duration of illness was inversely related to performance, but degree of glycemic control had no effect at all. This suggests that any impairment of cognitive function associated with IDDM is more likely to be attributable to the impact of suffering from a life-threatening disease rather than to organic neurological damage. The effect of chronicity was somewhat more pronounced in the case of concrete material than in the case of abstract material, and it is suggested that chronic type 1 DM might be associated with a specific impairment in elaborative encoding.

Up until recently, treatment of schizophrenia concentrated on elimination of positive and negative symptoms. Beginning in the 1990s, however, psychiatrists realized the importance of treating cognitive dysfunction, a widely recognized main feature of schizophrenia. The impairment in the ability to recognize and process complex tasks adequately is seen early in the illness, is usually mild initially, and increases in severity in some chronic patients (Goldberg, 1990; Hoff, 1992; Sharma, 1999). The cognitive deficits persist even when positive and negative symptoms improve (Sharma, 1999). To assess cognitive function, a wide variety of neuropsychological tests are used. In the process of testing, distinguishing syndrome-related impairment from cognitive impairment caused by side effects of antipsychotic and anticholinergic medications has proven difficult (Sharma, 1999). The presence of psychotic symptoms, depression, anxiety, lack of volition, long-term institutionalization, and inappropriate demographic control measures may also confound the results of neuropsychological tests. Despite methodological problems, studies with first-episode and antipsychotic-naïve patients have shown that even before the onset of other symptoms, cognitive deficit is a feature of schizophrenia (Sharma, 1999). Additionally, some studies have found that the cognitive deficit is less severe in first-episode patients than in chronic patients, thus suggesting deterioration in cognitive functioning (Nelson *et al.*, 1990; Bilder *et al.*, 1992).

Schizophrenia patients show cognitive impairments in attention, memory, and executive function. Attention is often tested using the Continuous Performance Test (CPT), and schizophrenic patients persistently perform poorly on the CPT because of inability to remain vigilant (Neuchterlein *et al.*, 1994; Sharma, 1999). Neuropsychologists have also used tests of immediate attention span, selective attention, executive control of attention, sustained attention, reaction time, and visual search and tracking to demonstrate attentional dysfunction in individuals with schizophrenia (Kremen *et al.*, 1992).

Tests of learning of lists, memory for stories, and paired associate learning and memory designs have been used to assess memory function in schizophrenic patients (Sharma, 1999). From these tests, neuropsychologists have come to believe memory impairment exists throughout encoding, storage, and retrieval (Gold *et al.*, 1992), and it is more significant than the overall intellectual deficit (Sharma, 1999).

Without doubt, cognitive impairment contributes substantially to the disability seen in schizophrenic patients. In particular, problems with attention, memory, and executive function are present to varying degrees in all patients with schizophrenia (Sharma, 1999), and this cognitive impairment has been shown to greatly affect social and vocational abilities (Green, 1996). These same studies recognize that cognitive impairments are more strongly related to functional outcome than positive symptom severity (Green, 1996). As a result, there has been an increasing emphasis on the treatment of cognitive dysfunction in addition to positive and negative symptoms in schizophrenic patients.

The conventional antipsychotic agents improve psychotic symptoms in schizophrenic patients, but they have had little effect on patients' cognitive deficits and chronic problems in self-care, social relations, and ability to work. In particular, while it has been shown that D2 antagonists decrease positive symptoms in schizophrenic patients, most psychiatrists acknowledge that these antipsychotic agents are not very effective against negative symptoms (King, 1994; Sharma, 1999). Indeed, given that these drugs have sedative effects and induce secondary negative symptoms, they may worsen cognitive function in patients.

The introduction of the atypical antipsychotics in the 1990s, however, brought promise to the treatment of cognitive dysfunction in schizophrenic patients. Indeed, a number of reports have indicated that clozapine, risperidone, ziprasidone, aripiprazole, olanzapine, and quetiapine improve cognitive functioning as compared to treatment with traditional antipsychotics (Sharma, 1999). However, these beneficial effects were assessed in studies with much variation in methodology, and it is therefore necessary to discriminate between high- and low-quality studies when considering these data and the implications for clinical relevance.

Several early studies have indicated that some measurements of cognitive function were reduced in elderly patients with type 2 diabetes compared with control subjects of similar age (Reaven *et al.*, 1990). These studies were replicated in 1990 by Reaven *et al.* (1990) who revealed that type 2 diabetes patients specifically performed more poorly than control subjects on cognitive measures of learning, abstract reasoning, and complex psychomotor functioning. However, performance on simple motor tasks and verbal tasks was similar for both groups. In addition, studies conducted on young adults with type 1 DM with poor long-term glycemic control, as evidenced by abnormally high HbA1c levels, showed decline in cognitive function (Reaven *et al.*, 1990).

IX. The Role of Neurotransmitter Systems

Since the 1970s, neurophysiologists and psychiatrists have proposed that either an excess or paucity of dopaminergic activity in the brain is responsible for the symptoms of schizophrenia. To explain their theories, these researchers have pointed to evidence that dopamine agonists, such as amphetamines, produce psychosis in individuals without schizophrenia (Bell, 1973; Goff *et al.*, 2001). In addition, dopamine D2 receptor antagonists have proven to be very effective in improving psychosis (Creese *et al.*, 1976; Snyder, 1981). The most current theory touting dopaminergic dysregulation as the primary cause of schizophrenia holds that there is decreased dopamine activity in the frontal cortex, thus leading to the negative symptoms of schizophrenia, and increased dopamine activity in the mesolimbic area, producing the positive symptoms (Davis *et al.*, 1991; Goff *et al.*, 2001).

Until recently, these theories positing dopaminergic dysregulation as the primary causal factor in the pathophysiology of schizophrenia dominated thinking about schizophrenia in psychiatric circles. Problems existed with these theories, however, because there was no direct evidence for an abnormality in the regulation of dopamine in the schizophrenic brain (Goff *et al.*, 2001). In particular, abnormalities in dopamine receptor density have not been found consistently, and when comparing the blood and CSF of schizophrenic patients and controls, no differences in concentrations for dopamine and its metabolites have been found (Goff *et al.*, 2001).

Currently, a new model for the pathophysiology of schizophrenia involving glutamatergic dysregulation is receiving attention (Goff and Coyle, 2001) and antagonism of the glutamatergic *N*-methyl-D-aspartic acid (NMDA) receptor complex, produces behavioral and cognitive deficits in normal subjects resembling those in schizophrenic patients. It has been demonstrated that agents that enhance NMDA receptor activity selectively improve negative symptoms in schizophrenic patients (Goff and Coyle, 2001). Furthermore, postmortem studies have identified changes in density and in subunit composition of NMDA and alpha-amino-3 hydroxy-5-methyl-4-isoxazolepropionic acid (AMPA) subunits of glutamatergic receptors in the prefrontal cortex and hippocampal regions (Kerwin *et al.*, 1990; Harrison *et al.*, 1991; Akbarian *et al.*, 1995, 1996; Eastwood *et al.*, 1995; Goff *et al.*, 2001). Similarly, other researchers have found decreased glutamate concentrations in the frontal cortex and hippocampus of schizophrenic brains (Tsai *et al.*, 1995), as well as abnormalities of glutamate receptor density in the prefrontal cortex, thalamus, and temporal lobe (Goff and Coyle, 2001). Functional imaging of these latter CNS regions demonstrated impaired activation when schizophrenic patients performed cognitive tasks (Heckers *et al.*, 1999, 2000). Finally, studies showing that the dissociative anesthetics phencyclidine and ketamine, drugs that noncompetitively block NMDA-receptor-gated cation channels, produce positive, negative, and cognitive symptoms very similar to those seen in the schizophrenic patient, further support the emerging glutamate model (Goff *et al.*, 2001).

While the glutamate model is gaining popularity, the dopamine model and the glutamate model are not mutually exclusive. Rather, the two models complement each other because the two neurotransmitter systems reciprocally regulate each other (Goff *et al.*, 2001). Specifically, studies have shown that firing of ventral tegmental dopamine neurons is disrupted during the administration of NMDA antagonists (Goff *et al.*, 2001). At the same time, the NMDA antagonists decrease dopamine concentrations in the frontal cortex and increase mesocortical dopamine activity (Goff *et al.*, 2001).

Lapidot and Haber (2001) examined the effect of hyperglycemia and hyperketonemia, on glucose metabolism, the glucose flux via the pyruvate dehydrogenase (PDH) and pyruvate carboxylase (PC) pathways and subsequent metabolism in the tricarboxylic acid cycles in neurons and glia in the diabetic brain in young adult rabbits. Significantly elevated brain glucose levels were found. In the hyperketonemic rabbits, elevated cerebral levels of β-hydroxybutyrate (β-HBA) were found. Alterations in the labeling patterns of glutamine in the hyperketonemic group led to the conclusion that the elevated β-HBA levels inhibit glucose metabolism, mostly in glia. This results in accumulation of glucose in the diabetic brain. In addition, altered levels of glutamine, glutamate, and GABA were also attributed to the effect of β-HBA on brain metabolism.

Di Luca *et al.* (1999) studied the effect of experimental diabetes on the expression of postsynaptic glutamate NMDA ionotropic receptors and of other key proteins regulating synaptic transmission at the postsynaptic compartment in rats. NR1 and NR2A as well as GluR1, GluR2/3, PSD–95, and alphaCaMKII protein concentrations in postsynaptic densities were the same in both control and diabetic rats, whereas the immunoreactivity for NR2B was reduced by about 40%. In addition, the activity of alphaCaMKII on exogenous substrates, such as syntide-2, and the phosphorylation of NR2A/B subunits of NMDA receptor were reduced in hippocampal postsynaptic densities of STZ-diabetic rats as compared with control rats. Furthermore, an insulin intervention three months after diabetic duration partially restored both alphaCaMKII activity and NR2B levels. The authors suggested that NMDA receptor expression and phosphorylation are possibly involved in behavioral and electrophysiological abnormalities observed in STZ-diabetic rats.

X. Conclusions

Schizophrenic patients appear to be at increased risk for DM and hyperglycemia. Antipsychotic agents, particularly atypical antipsychotic agents, may further increase this risk. There is little genetic overlap with diabetes and schizophrenia. Understanding the mechanisms of glucose metabolism impairment by antipsychotic agents is critical to develop prevention of life-threatening illnesses such as DKA.

The advantages of atypical agents, including improvement in cognitive symptoms in schizophrenia patients, must be balanced with data that suggest patients with diabetes or hyperglycemia demonstrate a cognitive decline. Thus, there exists a paradox. Why do the patients with schizophrenia and antipsychotic-agent-associated DM improve cognitively while diabetes and hyperglycemia has the opposite effect in nonschizophrenic patients with diabetes? Further research in the area of schizophrenia and diabetes is warranted.

References

Ai, D., Roper, T. A., and Riley, J. A. (1998). Diabetic ketoacidosis and clozapine. *Postgrad. Med. J.* **74,** 493–494.

Akbarian, S., Kim, J. J., Potkin, S. G., Hetrick, W. P., Bunney, W. E. Jr., and Jones, E. G. (1996). Maldistribution of interstitial neurons in prefrontal white matter of the brains of schizophrenic patients. *Arch. Gen. Psychiatry* **53,** 425–436.

Akbarian, S., Kim, J. J., Smith, M. A., and Jones, E. G. (1995). Editing for an AMPA receptor subunit RNA in prefrontal cortex and striatum in Alzheimer's disease, Huntington's disease and schizophrenia. *Brain Res.* **699,** 297–304.

Atkinson, M. A., and Maclaren, N. K. (1994). The pathogenesis of insulin-dependent diabetes mellitus. *N. Engl. J. Med.* **331,** 1428–1436.

Bell, D. S. (1973). The experimental reproduction of amphetamine psychosis. *Arch. Gen. Psychiatry* **29,** 35–40.

Bergman, A. J., Silverman, J. M., Harvey, P. D., Smith, C. J., and Siever, L. J. (2000). Schizotypal symptoms in the relatives of schizophrenia patients: an empirical analysis of the factor structure. *Schizophr. Bull.* **26,** 577–586.

Bergman, R. (1989). Lilly Lecture 1989. Toward Physiological understanding of glucose tolerance. *Diabetes* **38,** 1512–1527.

Bettinger, T. L., Mendelson, S. C., Dorson, P. G., and Crismon, M. L. (2000). Olanzapine-induced glucose dysregulation. *Ann. Pharmacother.* **34,** 865–867.

Bilder, R. M., Lipschutz-Broch, L., Reiter, G., Geisler, S. H., Mayerhoff, D. I., and Lieberman, J. A. (1992). Intellectual deficits in first-episode schizophrenia: Evidence for progressive deterioration. *Schizophr. Bull.* **18,** 437–448.

Blouin, J. L., Dombroski, B. A., Nath, S. K., Lasseter, V. K., Wolyniec, P. S., Nestadt, G., Thornquist, M., Ullrich, G., McGrath, J., Kasch, L., Lamacz, M., Thomas, M. G., Gehrig, C., Radhakrishna, U., Snyder, S. E., Balk, K. G., Neufeld, K., Swartz, K. L., DeMarchi, N., Papadimitriou, G. N., Dikeos, D. G., Stefanis, C. N., Chakravarti, A., Childs, B., and Pulver, A. E. (1998). Schizophrenia susceptibility loci on chromosomes 13q32 and 8p21. *Nat. Genet.* **20,** 70–73.

Braceland, F. J., and Meduna, L. J. (1945). Delayed action of insulin in schizophrenia. *Am. J. Psychiatry* **102,** 108–110.

Brambilla, F., Guastalla, A., Guerrini, A., Riggi, F., Rovere, C., Zanoboni, A., and Zanoboni-Muciaccia, W. (1976). Glucose-insulin metabolism in chronic schizophrenia. *Dis. Nerv. Syst.* **37,** 98–103.

Bruno, G., F. Merletti, F., De Salvia, A., Lezo, A., Arcari, R., and Pagano, G. (1997). Comparison of incidence of insulin-dependent diabetes mellitus in children and young adults in the Province of Turin, Italy, 1984–91. Piedmont Study Group for Diabetes Epidemiology. *Diabetes Med.* **14,** 964–969.

Colli, A. M., Cocciolo, M., Francobandiera, F., Rogantin, F., and Cattalini, N. (1999). Diabetic ketoacidosis associated with clozapine treatment. *Diabetes Care* **22,** 176–177.

Creese, I., Burt, D. R., and Snyder, S. H. (1976). Dopamine receptor binding predicts clinical and pharmacological potencies of antischizophrenic drugs. *Science* **192,** 481–483.

Croarkin, P. E., and Bain, B. K. (2001). Reply to Drs. Haupt and Newcomer. *Psychosomatics* **42,** 280.

Croarkin, P. E., Jacobs, K. M., and Bain, B. K. (2000). Diabetic ketoacidosis associated with risperidone treatment? *Psychosomatics* **41,** 369–70.

Davis, K. L, Kahn, R. S., Ko, G., and Davidson, M. (1991). Dopamine in schizophrenia: A review and reconceptualization. *Am. J. Psychiatry* **148,** 1474–1486.

Di Luca, M., Ruts, L., Gardoni, F., Cattabeni, F., Biessels, G. J., and Gispen, W. H. (1999). NMDA receptor subunits are modified transcriptionally and post-translationally in the brain of streptozotocin-diabetic rats. *Diabetologia* **42,** 693–701.

Dwyer, D. S., Bradley, R. J., Kablinger, A. S., and Freeman, A. M., III (2001). Glucose metabolism in relation to schizophrenia and antipsychotic drug treatment. *Ann. Clin. Psychiatry* **13,** 103–113.

Dwyer, D. S., Liu, Y., and Bradley, R. J. (1999a). Dopamine recepetor antagonists modulate glucose uptake in rat pheochromocytoma (PC12) cells. *Neurosci. Lett.* **274,** 151–154.

Dwyer, D. S., Pinkofsky, H. B., Liu, Y., and Bradley, R. J. (1999b). Antipsychotic drugs affect glucose uptake and the expression of glucose tranporters in PC12 cells. *Prog. Neuropsychopharmacol. Biol. Psychiatry* **23,** 69–80.

Eastwood, S. L., McDonald, B., Burnet, P. W., Beckwith, J. P., Kerwin, R. W., and Harrison, P. J. (1995). Decreased expression of mRNAs encoding non-NMDA glutamate receptors GluR1 and GluR2 in medial temporal lobe neurons in schizophrenia. *Brain Res. Mol. Brain Res.* **29,** 211–223.

Faraone, S. V., Matise, T., Svrakic, D., Pepple, J., Malaspina, D., Suarez, B., Hampe, C., Zambuto, C. T., Schmitt, K., Meyer, J., Markel, P., Lee, H., Harkavy, Friedman, J., Kaufmann, C., Cloninger, C. R., and Tsuang, M. T. (1998). Genome Scan of European-American Schizophrenia Pedigrees: Results of the NIMH Genetics Initiative and Millenium Consortium. *Am. J. Med. Genet.* **81,** 290–295.

Fertig, M. K., Brooks, V. G., Shelton, P. S., and English, C. W. (1998). Hyperglycemia associated with olanzapine. *J. Clin. Psychiatry* **59,** 687–689.

Franzen, G. (1970). Plasma free fatty acids before and after an intravenous insulin injection in acute schizophrenic men. *Br. J. Psychiatry* **116,** 173–177.

Ganzini, L., Casey, D. E., Hoffman, W. F., and Heintz, R. T. (1992). Tardive dyskinesia and diabetes mellitus. *Psychopharmacol. Bull.* **28,** 281–286.

Gatta, B., Rigalleau, V., and Gin, H. (1999). Diabetic ketoacidosis with olanzapine treatment. *Diabetes Care* **22,** 1002–1003.

Goff, D. C., and Coyle, J. T. (2001). The emerging role of glutamate in the pathophysiology and treatment of schizophrenia. *Am. J. Psychiatry* **158,** 1367–1377.

Goff, D. C., Heckers, S., and Freudenreich, O. (2001). Schizophrenia. *Med. Clin. North Am.* **85,** 663–689.

Goff, D. C., Tsai, G., Beal, M. F., and Coyle, J. T. (1995). Tardive dyskinesia and substrates of energy metabolism in CSF. *Am. J. Psychiatry* **152,** 1730–1736.

Gold, J. M., Randolph, C., Carpenter, C. J., Goldberg, T. E., and Weinberger, D. R. (1992). Forms of memory failure in schizophrenia. *J. Abnormal Psychol.* **101,** 487–494.

Goldberg, T. E., Ragland, J. D., Torrey, E. F., Gold, J. M., Bigelow, L. B., and Weinberger, D. R. (1990). Neuropsychological assessment of monozygotic twins discordant for schizophrenia. *Arch. Gen. Psychiatry* **47,** 1066–1072.

Goldstein, L. E., Sporn, J., Brown, S., Kim, H., Finkelstein, J., Gaffey, G. K., Sachs, G., and Stern, T. A. (1999). New-onset diabetes mellitus and diabetic ketoacidosis associated with olanzapine treatment. *Psychosomatics* **40,** 438–443.

Gottesman, I. I. (1991). "Schizophrenia Genesis: The Origins of Madness." Freeman New York,

Green, M. F. (1996). What are the functional consequences of neurocognitive deficits in schizophrenia? *Am. J. Psychiatry* **153,** 321–330.

Hagg, S., Joelsson, L., Mjorndal, T., Spigset, O., Oja, G., and Dahlqvist, R. (1998). Prevalence of diabetes and impaired glucose tolerance in patients treated with clozapine compared with patients treated with conventional depot neuroleptic medications. *J. Clin. Psychiatry* **59,** 294–299.

Harris, M. I., Flegal, K. M., Cowie, C. C., Eberhardt, M. S., Goldstein, D. E., Little, R. R., Wiedmeyer, H. M., and Byrd-Holt, D. D. (1998). Prevalence of diabetes, impaired fasting glucose, and impaired glucose tolerance in U.S. adults. The Third National Health and Nutrition Examiniation Survey, 1988–1994. *Diabetes Care* **21,** 475–476.

Harrison, P. J., McLaughlin, D., and Kerwin, R. W. (1991). Decreased hippocampal expression of a glutamate receptor gene in schizophrenia. *Lancet* **337,** 450–452.

Heckers, S., Curran, T., Goff, D., Rauch, S. L., Fischman, A. J., Alpert, N. M., and Schacter, D. L. (2000). Abnormalities in the thalamus and prefrontal cortex during episodic object recognition in schizophrenia. *Biol. Psychiatry* **48,** 651–657.

Heckers, S., Goff, D., Schacter, D. L., Savage, C. R., Fischman, A. J., Alpert, N. M., and Rauch, S. L. (1999). Functional imaging of memory retrieval in deficit vs. nondeficit schizophrenia. *Arch. Gen. Psychiatry* **56,** 1117–1123.

Henderson, D., Cagliero, E., Borba, C. P., Nathan, D., and Goff, D. C. (2000). Atypical antipsychotic agents and glucoe metabolism: Bergman's Minimal Model analysis. NCDEU Annual Conference, Boca Raton, FL.

Henderson, D. C., Cagliero, E., Gray, C., Nasrallah, R. A., Hayden, D. L., Schoenfeld, D. A., and Goff, D. C. (2000). Clozapine, diabetes mellitus, weight gain, and lipid abnormalities: A five-year naturalistic study. *Am. J. Psychiatry* **157,** 975–981.

Hoff, A. L., Riordan, H., O'Donnell, D. W., Morris, L., and DeLisi, L. E. (1992). Neuropsychological functioning of first-episode schizophreniform patients. *Am. J. Psychiatry* **149,** 898–903.

Hy, M. (2000). Genetics and etiology of schizophrenia and bipolar disorder. *Biol. Psychiatry* **47,** 171–173.

Hyman, S. E. (1999). Introduction to the complex genetics of mental disorders. *Biol. Psychiatry* **45,** 518–521.

Isojarvi, J. I., Rattya, J., Myllyla, V. V., Knip, M., Koivunen, R., Pakarinen, A. J., Tekay, A., and Tapanainen, J. S. (1998). Valproate, Lamotrigine, and insulin-mediated risks in women with epilepsy. *Ann. Neurol.* **43,** 446–451.

Isojarvi, J. I., Laatikainen, T. J., Pakarinen, A. J., Juntunen, K. T., and Myllyla, V. V. (1993). Polycystic ovaries and hyperandrogenism in women taking valproate for epilepsy. *N. Engl. J. Med.* **329,** 1383–1388.

Jeste, D. V., Caligiuri, M. P., Paulsen, J. S., Heaton, R. K., Lacro, J. P., Harris, M. J., Bailey, A., Fell, R. L., and McAdams, L. A. (1995). Risk of tardive dyskinesia in older patients. *Arch. Gen. Psychiatry* **52,** 756–765.

Kamran, A., Doraiswamy, P. M., Jane, J. L., Hammett, E. B., and Dunn, L. (1994). Severe hyperglycemia associated with high doses of clozapine. *Am. J. Psychiatry* **151,** 1395.

Kaufmann, C. A., Suarez, B., Malaspina, D., Pepple, J., Svrakic, D., Markel, P. D., Meyer, J., Zambuto, C. T., Schmitt, K., Matise, T. C., Harkavy,, Friedman, J. M., Hampe, C., Lee, H., Shore, D., Wynne, D., Faraone, S. V., Tsuang, M. T., and Cloninger, C. R. (1998). NIMH

Genetics Initiative Millenium Schizophrenia Consortium: Linkage Analysis Of African-American Pedigrees. *Am. J. Med. Genet.* **81,** 282–289.

Kerwin, R., Patel, S., and Meldrum, B. (1990). Quantitative autoradiographic analysis of glutamate binding sites in the hippocampal formation in normal and schizophrenic brain post mortem. *Neuroscience* **39,** 25–32.

Keskiner, A., El-Toumi, A., and Bousquet, T. (1973). Psychotropic drugs, diabetes and chronic mental patients. *Psychosomatics* **14,** 176–181.

King, D. J. (1994). Psychomotor impairment and cognitive disturbances induced by neuroleptics. *Acta Psychiatr. Scand.* (Suppl. 380), 53–58.

Koller, E., Schneider, B., Bennett, K., and Dubitsky, G. (2001). Clozapine associated diabetes. *Am. J. Med.* **111,** 716–723.

Kopelman, P., and Albon, L. (1997). Obesity, non-insulin-dependent diabetes mellitus and the metabolic syndrome. **53,** 322–340.

Kostakoglu, A. E., Yazici, K. M., Erbas, T., and Guvener, N. (1996). Ketoacidosis as a side-effect of clozapine: a case report. *Acta Psychiatr. Scand.* **93,** 217–218.

Koval, M. S., Rames, L. J., and Christie, S. (1994). Diabetic ketoacidosis associated with clozapine treatment. *Am. J. Psychiatry* **151,** 1520–1521.

Kremen, W. S., Tsuang, M. T., Faraone, S. V., and Lyons, M. J. (1992). Using vulnerability indicators to compare conceptual models of genetic heterogeneity in schizophrenia. *J. Nerv. Mental Dis.* **180,** 141–152.

Lapidot, A., and Haber, S. (2001). Effect of endogenous beta-hydroxybutyrate on glucose metabolism in the diabetic rabbit brain: A (13)C-magnetic resonance spectroscopy study of [U-(13)C]glucose metabolites. *J. Neurosci. Res.* **64,** 207–216.

Lindenmayer, J. P., and Patel, R. (1999). Olanzapine-induced ketoacidosis with diabetes mellitus. *Am. J. Psychiatry* **156,** 1471.

Marinow, A. (1971). Diabetes in chronic schizophrenia. *Dis. Nerv. Syst.* **32,** 777–778.

McKee, H. A., D'Arcy, P. F., and Wilson, P. J. (1986). Diabetes and schizophrenia—a preliminary study. *J. Clin. Hosp. Pharmacol.* **11,** 297–299.

Melkersson, K. I., Hulting, A. L., and Brismar, K. E. (1999). Different influences of classical antipsychotics and clozapine on glucose-insulin homeostasis in patients with schizophrenia or related psychoses. *J. Clin. Psychiatry* **60,** 783–791.

Melkersson, K. I., Hulting, A. L., and Brismar, K. E. (2000). Elevated levels of insulin, leptin, and blood lipids in olanzapine-treated patients with schizophrenia or related psychoses. *J. Clin. Psychiatry* **61,** 742–749.

Meltzer, H. Y. (2000). Genetics and etiology of schizophrenia and bipolar disorder. *Biol. Psychiatry* **47,** 171–173.

Mohan, D., Gordon, H., Hindley, N., and Barker, A. (1999). Schizophrenia and diabetes mellitus. *Br. J. Psychiatry* **174,** 180–181.

Mukherjee, S., Decina, P., Bocola, V., Saraceni, F., and Scapicchio, P. L. (1996). Diabetes mellitus in schizophrenic patients. *Compr. Psychiatry* **37,** 68–73.

Mukherjee, S., Roth, S. D., Sandyk, R., and Schnur, D. B. (1989). Persistent tardive dyskinesia and neuroleptic effects on glucose tolerance. *Psychiatry Res.* **29,** 17–27.

National Institute of Mental Health's Genetics Workgroup (1999). Report of the National Institue of Mental Health's Workgroup. *Biol. Psychiatry* **45,** 559–602.

Nelson, H. E., Pantelis, C., Carruthers, K., Speller, J., Baxendale, S., and Barnes, T. R. (1990). Cognitive Functioning and Symptomatology in Chronic Schizophrenia. *Psychol. Med.* **20,** 357–365.

Neuchterlein, K. H., Dawson, M. E., and Green, M. F. (1994). Information-processing abnormalities as neuropsychological vulnerability indicators for schizophrenia. *Acta Psychiatr. Scand.* 71–79.

Newcomer, J. W., and Fucetola, R. (2000). Atypical antipsychtoic-induced differences in glucose regulation in schizophrenia independent of differences in adiposity. *Soc. Neurosci. Abstr.* **26,** 275.

Newcomer, J. W., and Fucetola, R. (1999). Abnormalities in glucose regulation during antipsychotic treatment. ACNP Nashville, TN.

Newcomer, J. W., and Fucetola, R. (2001). Glucose metabolism during antipsychotic treatment in schizophrenia. *Schizophrenia Res.* **49,** Suppl. 288.

Peterson, G., and Byrd, S. (1996). Diabetic ketoacidosis from clozapine and lithium cotreatment. *Am. J. Psychiatry* **153,** 737–738.

Popli, A. P., Konicki, P. E., Jurjus, G. J., Fuller, M. A., and Jaskiw, G. E. (1997). Clozapine and associated diabetes mellitus. *J. Clin. Psychiatry* **58,** 108–111.

Prescott, J. H., Richardson, J. T., and Gillespie, C. R. (1990). Cognitive function in diabetes mellitus: the effects of duration of illness and glycaemic control. *Br. J. Clin. Psychol.* **29,** 167–175.

Procyshyn, R. M., Pande, S., and Tse, G. (2000). New-onset diabetes mellitus associated with quetiapine. *Can. J. Psychiatry* **45,** 668–669.

Pulver, A. E. (2000). Search for schizophrenia susceptibility genes. *Biol. Psychiatry* **47,** 221–230.

Reaven, G. M., Thompson, L. W., Nahum, D., and Haskins, E. (1990). Relationship between hyperglycemia and cognitive function in older NIDDM patients. *Diabetes Care* **13,** 16–21.

Schimmelbusch, W. H., Mueller, P. S., and Sheps, J. (1971). The positive correlation between insulin resistance and duration of hospitalization in untreated schizophrenia. *Br. J. Psychiatry* **118,** 429–36.

Schultz, S. K., Arndt, S., Ho, B. C., Oliver, S. E., and Andreasen, N. C. (1999). Impaired glucose tolerance and abnormal movements in patients with schizophrenia. *Am. J. Psychiatry* **156,** 640–642.

Schutt, M., Meier, M., Meyer, M., Klein, J., Aries, S. P., and Klein, H. H. (2000). The HIV-1 protease inhibitor findinavir impairs insulin signalling in HepG2 hepatoma cells. *Diabetologia* **43,** 1145–1148.

Schwartz, L., and Munoz, R. (1968). Blood sugar levels in patients treated with chlorpromazine. *Am. J. Psychiatry* **125,** 253–255.

Sharma, T. (1999). Cognitive effects of conventional and atypical antipsychotics in schizophrenia. *Br. J. Psychiatry* Suppl. 38, 44–51.

Smith, H., Kenney-Herbert, J., and Knowles, L. (1999). Clozapine-induced diabetic ketoacidosis. *Aust. N. Z. J. Psychiatry* **33,** 120–121.

Smith, R. C., and Lindenmayer, J. P. (1999). Hyperglycemia in schizophrenic patients treated with olanzapine and clozapine. ACNP, Nashville, TN.

Snyder, S. H. (1981). Dopamine receptors, neuroleptics and schizophrenia. *Am. J. Psychiatry* **138,** 460–464.

Sobel, M., Jaggers, E. D., and Franz, M. A. (1999). New-onset diabetes mellitus associated with the initiation of quetiapine treatment. *J. Clin. Psychiatry* **60,** 556–557.

Straub, R. E., MacLean, C. J., O'Neill, F. A., Walsh, D., and Kendler, K. S. (1997). Support for a possible schizophrenia vulnerability locus in a region of 5q22–31 in Irish families. *Mol. Psychiatry* **2,** 148–155.

Sumiyoshi, T., Ichikawa, J., and Meltzer, H. Y. (1997). The effect of streptozotocin-induced diabetes on dopamine2, serotonin1A and serotonin2A receptors in the rat brain. *Neuropsychopharmacology* **16,** 183–190.

Taniguchi, A., Nakai, Y., Fukushima, M., Imura, H., Kawamura, H., Nagata, I., Florant, G. L., and Tokuyama, K. (1994). Insulin sensitivity, insulin secretion, and glucose effectiveness in subjects with impaired glucose tolerance: A minimal model analysis. *Metabolism* **43,** 714–718.

Thonnard-Neumann, E. (1968). Phenothiazines and diabetes in hospitalized women. *Am. J. Psychiatry* **124,** 978–982.

Tienari, P., Wynne, L. C., Moring, J., Lahti, I., Naarala, M., Sorri, A., Wahlberg, K. E., Saarento, O., Seitamaa, M., and Kaleva, M. (1994). The Finnish adoptive family study of schizophrenia. Implications for family research. *Br. J. Psychiatry* 163, Suppl. 23, 20–26.

Tollefson, G., and Lesar, T. (1983). Nonketotic hyperglycemia associated with loxapine and amoxapine: case report. *J. Clin. Psychiatry* **44,** 347–348.

Tsai, G., Passani, L. A., Slusher, B. S., Carter, R., Baer, L., Kleinman, J. E., and Coyle, J. T. (1995). Abnormal excitatory neurotransmitter metabolism in schizophrenic brains. *Arch. Gen. Psychiatry* **52,** 829–836.

Tsuang, M. T. (2000). Schizophrenia: Genes and Environment. *Biol. Psychiatry* **47,** 210–220.

Tsuang, M. T., and Faraone, S. V. (1995). The case for heterogeneity in the etiology of schizophrenia. *Schizophrenia Res.* **17,** 161–175.

Von Hayek, D., Huttl, V., Reiss, J., Schweiger, H. D., and Fuessl, H. S. (1999). Hyperglycemia and ketoacidosis associated with olanzapine. *Nervenarzt* **70,** 836–837.

Waitzkin, L. (1966a). A survey of unknown diabetes in a mental hospital. I. *Diabetes* **15,** 97–104.

Waitzkin, L. (1966b). A survey of unknown diabetics in a mental hospital. II. Men from age fifty. *Diabetes* **15,** 164–172.

Wilson, R. W., and D'Souza, L. (2000). New onset diabetes and ketoacidosis with atypical antipsychotic agents. NCDU Annual Conference, Boca Raton, FL.

Wirshing, D. A., Spellberg, B. J., Erhart, S. M., Marder, S. R., and Wirshing, W. C. (1998). Novel antipsychotics and new onset diabetes. *Biol. Psychiatry* **44,** 778–783.

Woerner, M. G., Saltz, B. L., Kane, J. M., Lieberman, J. A., and Alvir, J. M. (1993). Diabetes and development of tardive dyskinesia. *Am. J. Psychiatry* **150,** 966–968.

Yaryura-Tobias, J. A., Chang, A., and Neziroglu, F. (1978). A study of relationships of serum glucose, insulin, free fatty acids, and free and total tryptophan to mental illness. *Biol. Psychiatry* **13,** 243–254.

PSYCHOACTIVE DRUGS AFFECT GLUCOSE TRANSPORT AND THE REGULATION OF GLUCOSE METABOLISM

Donard S. Dwyer, Timothy D. Ardizzone, and Ronald J. Bradley

Departments of Psychiatry and Pharmacology, LSU Health Sciences Center
Shreveport, Louisiana 71130

I. Introduction

In many tissues, the transport of glucose into cells is the rate-limiting step in glucose metabolism. In order to increase glucose uptake in muscle and fat, additional glucose transporter (GLUT) proteins are recruited to the cell surface through the actions of insulin (Simpson and Cushman, 1986). The extent of the contribution of the glucose transport step to regulation of glucose metabolism in the brain is not known. However, a 50% reduction in GLUT1 at the blood-brain barrier has profound consequences for brain development and function (Seidner *et al.*, 1998; Klepper *et al.*, 1999b). In glucose transporter protein syndrome (GTPS), there is a reduced number of GLUTs as a result of mutations in one allele of the GLUT1 gene (Seidner *et al.*, 1998; Klepper *et al.*, 1999b). This syndrome is characterized by microcephaly, seizures and retarded growth and development. Thus, any

0074-7742/02 $35.00

condition or substance that reduces glucose transport in the brain may potentially alter neuronal function. This review will summarize the findings concerning various psychoactive drugs that affect glucose uptake and/or utilization. Although these drugs are derived from distinct classes of medications, many share a common feature—they can produce tranquilization or sedation in patients. In addition, some of the drugs have been observed to induce hyperglycemia and even diabetes as adverse effects. It is not known precisely how the drugs interfere with glucose transport or produce hyperglycemia. Several possible mechanisms will be discussed here, together with the clinical implications of these findings.

II. Regulation of Glucose Transport

Before describing the actions of the various drugs on glucose uptake, it is worthwhile to discuss general aspects of the regulation of sugar transport. The glucose transport capacity of a cell can be modulated in one of two ways. First, the rate of glucose uptake will be determined by the balance between recruitment of GLUTs to the plasma membrane and internalization of GLUTs to intracellular compartments. Synthesis of new GLUTs in response to a signal would generally be expected to drive more transporters to the cell surface. Secondly, glucose transport can be regulated through changes in the intrinisic activity of GLUTs that reside in the plasma membrane. Each of these mechanisms will be discussed in greater detail below with special emphasis on regulation of glucose uptake in the brain.

A. Translocation of GLUTs

In peripheral tissues, such as muscle and fat, basal levels of GLUTs are expressed at the cell surface, while a significant reserve of transporters is localized in intracellular vesicles. Insulin (and other signals) induce translocation of the intracellular GLUTs to the cell surface in order to increase glucose uptake by the cells (Simpson and Cushman, 1986; Nystrom and Quon, 1999; Watson and Pessin, 2001). The signaling pathway involved in this response is complex and beyond the scope of this paper; for recent reviews on this subject see Avruch (1998) and Farese (2001). GLUT4 is the major insulin-responsive GLUT isoform (Watson and Pessin, 2001). There is a dileucine motif near the carboxy terminus of the protein that constitutes a targeting signal involved in trafficking (Verhey and Birnbaum, 1994; Corvera *et al.*, 1994). GLUT4 is expressed in the brain in neurons, endothelial cells, and ependymal cells of the ventricles (Vannucci *et al.,* 1997; Dwyer *et al.,* 2002). It is not currently known if the cell surface expression of GLUT4 in brain is modulated by insulin or other signals. A dileucine motif is also present in GLUT8 (also known as GLUTx1) (Doege *et al.,* 2000b; Ibberson *et al.,* 2000) and GLUT6 (formerly GLUT9; Doege *et al.,* 2000a), which are prominently expressed in brain. Although GLUT8 is located at the surface of neuronal cell bodies and apical dendrites, significant levels of the protein are also found in the intracellular compartment (Reagan *et al.,* 2001). Therefore, translocation may be an important mechanism for the regulation of glucose transport in the brain.

B. Intrinsic Activity

Another potential mechanism for regulating glucose uptake in brain cells involves modulation of the intrinsic activity of GLUT proteins. The intrinsic activity of a GLUT is defined as the catalytic turnover number or Vmax/the concentration of GLUT proteins at the cell surface (Kristiansen *et al.,* 1996). A number of studies have provided evidence that the intrinsic activity of GLUTs is variable and subject to regulation. The strongest support for this notion is derived from findings of a discordance between the number of GLUTs at the cell surface and the rate of glucose transport, i.e., an increase in glucose uptake with no change in GLUT expression. A variety of treatments or drugs have been reported to alter the intrinsic catalytic activity of GLUT1 and GLUT4 (Czech *et al.,* 1992). For example, cadmium has been shown to enhance glucose transport in 3T3–L1 cells as a result of an increase in the affinity of GLUT1 for substrate (a decrease in the *K*m) (Harrison *et al.,* 1991; Lachaal *et al.,* 1996), whereas adenosine reduces transport by lowering the affinity of GLUT1 for substrate (Lachaal *et al.,* 2001).

III. Modulation of Glucose Transport by Psychoactive Drugs

A. Barbiturates and Anesthetics

Pentobarbital was one of the first psychoactive drugs that was found to affect glucose metabolism in the brain (Crane *et al.,* 1978; Gjedde and Rasmussen, 1980). In general, the barbiturates decrease cerebral blood flow and the uptake of glucose in brain tissue. In addition, these drugs inhibit glucose transport in various cell lines and in human erythrocytes (Honkanen *et al.*, 1995; El-Barbary *et al.*, 1996; Naftalin and Arain, 1999). The IC_{50}s for inhibition of transport are generally in the range of 1–5 mM and the rank order of potency is secobarbital > pentobarbital > phenobarbital ~ barbital (Honkanen *et al.*, 1995; Haspel *et al.*, 1999). Studies of the intrinsic fluorescence of the GLUT1 protein reveal that the barbiturates bind directly to the transporter protein to affect its function (Honkanen *et al.*, 1995). The drugs inhibit glucose transport via a noncompetitive interaction that decreases the intrinsic activity of GLUT1 (Hokanen *et al.*, 1995; Naftalin and Arain, 1999; Klepper *et al.*, 1999a). It is widely agreed that the barbiturates decrease the Vmax of transport and some studies have shown an appreciable change in the *Km* (Otsuka *et al.*, 1991; Naftalin and Arain, 1999), whereas other studies did not (El-Barbary *et al.*, 1996; Klepper *et al.*, 1999a). Haspel *et al.* (1999) have reported that barbiturates inhibit glucose transport mediated by GLUT1, GLUT2, and GLUT3, but have little effect on transport by GLUT4. The overall picture that emerges from these studies is that the barbiturates bind to a noncatalytic site on GLUTs and prevent conformational changes that are necessary to reorient the protein for transport of substrate.

Additional anesthetic compounds have been reported to affect glucose transport in the brain. Nitrous oxide inhibits glucose metabolism in the CNS with a profile that is similar to that of the barbiturates (Alexander *et al.*, 1975). On the other hand, halothane (and other halogenated inhalation anesthetics) typically have very modest effects on glucose transport (Gjedde and Rasmussen, 1980; Honkanen *et al.*, 1995). Ethanol inhibits glucose uptake only at very high concentrations (>50 mM) (Singh *et al.*, 1990), however, other alcohols, e.g., benzylalcohol, block transport at concentrations similar to the low-potency barbiturates (Lacko *et al.*, 1978). Thus, a number of different anesthetic drugs decrease the uptake and utilization of glucose in brain cells. It is possible that this interference with energy metabolism contributes to the overall depression of neuronal activity during anesthesia.

These findings may have significant ramifications in the clinical setting. In GTPS, the existing deficiency in GLUT expression may make the use of barbiturates problematic because the combination may severely compromise glucose transport in affected individuals (Klepper *et al.*, 1999a). Furthermore, energy metabolism in neurons may be especially sensitive to the effects of barbiturates (and other anesthetics) because glucose flux in these cells is nearly unidirectional and is, therefore, more susceptible to inhibition by the drugs than the equilibrium exchange of glucose that occurs in most cells (El-Barbary *et al.*, 1996). This might explain, in part, why energy metabolism in the brain is reduced to a greater degree than in peripheral tissues during anesthesia. Finally, ethanol and anesthetics may produce greater injury to the brain at certain developmental stages that are highly dependent upon glucose. There is a strict requirement for glucose early in fetal development (Shepard *et al.*, 1970) and consumption of alcohol by the mother at this stage may increase the risk for fetal alcohol syndrome (Streissguth *et al.*, 1980; West and Hamre, 1985) due to interference with glucose metabolism in embryonic brain tissue.

B. Antidepressants

The antidepressant drugs increase the levels of norepinephrine and serotonin at synapses by blocking reuptake via monoamine transporters (Svensson, 2000). The older tricyclic medications inhibit reaccumulation of the major monoamines in brain, whereas the newer drugs, such as fluoxetine and sertraline, are serotonin selective reuptake inhibitors (SSRIs) (Garattini *et al.*, 1998; Masand and Gupta, 1999). The common theme is that the antidepressants block specific transporter proteins in the brain. Recently, it has been shown that the tricyclic drugs also inhibit glucose transport in human erythrocytes (Pinkofsky *et al.*, 2000). These results confirmed earlier findings that imipramine and nortriptyline blocked glucose uptake in human platelets (Gaut, 1974). Thus, it appears that antidepressants can block the transport of several different substrates.

The initial studies of the effects of antidepressant drugs on glucose transport in neuronal cells focused on the tricyclic compounds. The K_i for inhibition of glucose uptake ranged from 0.5–1.5 mM; imipramine was the most potent of the drugs tested (Pinkofsky *et al.*, 2000). In addition, the drugs blocked the binding of cytochalasin B to GLUT1 in red blood cell membranes (Pinkofsky *et al.*, 2000). Cytochalasin B is a specific inhibitor of glucose transport and binds to GLUTs with high affinity. Taken together, the data suggest that the tricyclic antidepressant drugs inhibit glucose transport by binding directly to GLUT proteins. More recently, the ability of antidepressants to inhibit glucose uptake in neuronal cells has been evaluated. The results are shown in Fig. 1. Imipramine and fluoxetine both inhibited glucose uptake by about 50% at concentrations of 20 μM. From these data, we can conclude that imipramine inhibited glucose transport in PC12 cells much more effectively than in human erythrocytes. One difference might be that GLUT3 is the major transporter isoform of PC12 cells, whereas GLUT1 is the isoform found in human red blood cells. Some GLUT isoforms may be more sensitive than others to the inhibitory effects of the tricyclic antidepressants. These drugs inhibit the reuptake of monoamines at somewhat

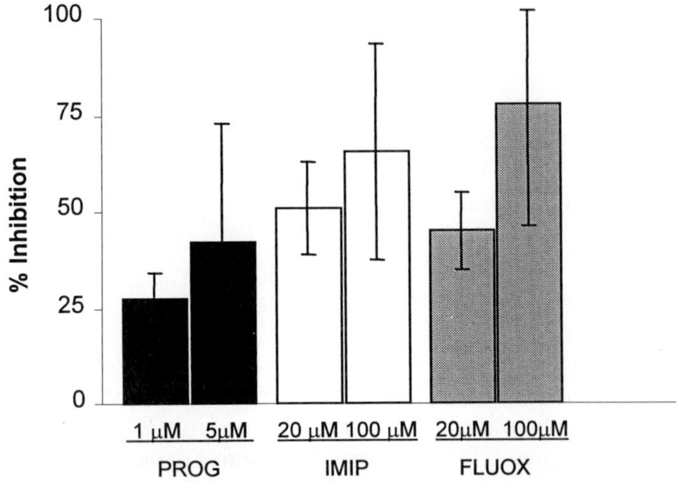

FIG. 1. Inhibition of glucose transport by progesterone and antidepressant drugs. PC12 cells were incubated with vehicle (DMSO) or drug—progesterone (PROG), imipramine (IMIP), or floxetine (FLUOX)—for 30 min. The uptake of ^{3}H-2-deoxyglucose (0.5 μCi) was then measured over 5 min at room temperature. Within experiments, duplicate samples were evaluated and each experiment has been repeated at least three times. These data have been expressed as the percent inhibition of transport by the drug in comparison to the control (vehicle) condition. Mean values and standard deviations are presented here.

lower concentrations, nevertheless, they appear to have a significant affinity for GLUTs as well.

It is noteworthy that antidepressant drugs have also been reported to affect the regulation of glucose metabolism *in vivo.* In the case of overdose with tricyclic antidepressants, marked hyperglycemia of a transient nature was noted (Royds and Knight, 1970). Under normal treatment conditions, hyperglycemia has been observed in some patients who have concurrent depression and diabetes (Katz *et al.,* 1991; Goodnick *et al.,* 1995), however, a very early study by Kaplan *et al.* (1960) found no such effect of the antidepressants. Acute administration of the tricyclic drugs in mice and rabbits was accompanied by significant hyperglycemia within about 1 h of the injection of drug (Wannarka *et al.,* 1979; Gupta *et al.,* 1992). Similar results were obtained with fluoxetine (Jacoby and Bryce, 1979; Yamada *et al.,* 1999). However, numerous studies in patients have shown that fluoxetine improved glucose disposition in diabetic patients with depression (Breum *et al.,* 1995; Connolly *et al.,* 1995; Maheux *et al.,* 1997). Therefore, the relationship between antidepressant therapy and altered regulation of glucose utilization is complex and may depend upon genetic factors that determine insulin responsiveness.

C. Steroids

Although steroidal drugs are not primarily psychoactive agents, they have significant biological effects on the brain. The steroid hormones shape behavior during development and glucocorticoids can induce behavioral disturbances and even psychosis with prolonged administration (McEwen *et al.*, 1986). In addition, adrenal steroids are intimately involved in learning and cognition, especially the attentional and emotional aspects of these processes (Lupien and McEwen, 1997). In peripheral tissues, glucocorticoids stimulate gluconeogenesis, lipolysis, and amino acid metabolism (Munck and Naray-Fejes-Toth, 1995). Under normal circumstances, these physiological responses prepare the organism to face sudden environmental changes or stresses that are potentially harmful. However, glucocorticoids also play a major role in the deleterious stress response of the brain. Thus, excessive and chronic stress can lead to memory impairment and even neuronal loss as a consequence of elevated levels of circulating glucocorticoids (Sapolsky *et al.*, 1990; Sapolsky, 1999). It is, therefore, important that the organism maintain optimum concentrations of glucocorticoids to avoid their deleterious effects.

At the molecular level, steroids have been reported to affect glucose transport in neuronal cells. Horner and colleagues (1990) showed that dexamethasone and corticosterone inhibited glucose transport in cultured neurons and glial cells. The effect required more than 4 h to manifest and appeared to involve internalization of GLUTs from the plasma membrane. Sakoda *et al.* (2000) studied the effects of dexamethasone on glucose transport in adipocytes and found that this drug prevented the increase in glucose transport normally produced in response to insulin. Dexamethasone affected the translocation of GLUTs rather than insulin signaling pathways. In a similar vein, progesterone has been reported to inhibit glucose transport in oocytes stimulated by insulin (Hainaut *et al.*, 1991). We have evaluated progesterone for more direct effects on glucose uptake and, as seen in Fig. 1, this steroid inhibits glucose transport in the PC12 cell line. Additional studies will be necessary to characterize the nature of the inhibition of transport produced by progesterone and to evaluate additional steroids for their effects. Estradiol has been reported to inhibit the exit of glucose from human erythrocytes and a putative binding region has been identified on an intracellular face of the GLUT1 protein (Afzal *et al.*, 2001). In contrast to these observations, estradiol and tamoxifen (an estrogen receptor modulator) increased the expression of GLUT3 and GLUT4 in monkey brain tissue (Cheng *et al.*, 2001). The animals were treated with estradiol for three days in order to elicit these effects. Similarly, estradiol has been reported to enhance GLUT1 expression in an animal model of stroke injury, especially in penumbral regions of the infarct (Shi *et al.*, 1997). The positive effects of estradiol on GLUT expression could conceivably represent an adaptive response to initial downmodulation of glucose transport in brain cells. Alternatively, steroid hormone receptors may participate in the regulation of GLUT transcription. Interestingly, the production of GLUT8, which is expressed in the brain and testes, is downregulated by estrogen (Doege *et al.*, 2000b). Thus, steroids may modify glucose transport in the brain through several different mechanisms, including direct actions on GLUT proteins, modulation of the surface expression of GLUTs, and transcriptional regulation of GLUT synthesis. Further investigation of these phenomena will clarify the physiological role of glucocorticoids in the regulation of glucose metabolism and neuronal survival in the brain.

D. Antipsychotic Drugs

Antipsychotic drugs are used to treat schizophrenia and other psychotic disorders. The older drugs such as chlorpromazine, fluphenazine, and haloperidol are now sometimes referred to as conventional or typical antipsychotics, whereas the newer medications such as clozapine, olanzapine, and risperidone have been termed atypical antipsychotics. The nomenclature refers to both pharmacological (less selectivity for D2 dopamine receptors) and clinical properties of the drugs (Remington and Kapur, 2000). More specifically, novel atypical drugs are viewed as having a high ratio of antagonism of 5–HT2/D2 receptors and actions at multiple neurotransmitter receptors in comparison to the conventional antipsychotic drugs. In addition, the atypical drugs appear to be more effective in refractory schizophrenia and provide more relief from the negative symptoms of this disease (Remington and Kapur, 2000). Although the major targets of these medications are dopamine (and to a lesser degree serotonin) receptors, some of the drugs show appreciable affinity for other proteins, including monoamine transporters (Tatsumi *et al.*, 1999).

An early report described inhibition by the phenothiazine antipsychotic medications of glucose transport in human erythrocytes (Baker and Rogers, 1973). More recently, a number of antipsychotic drugs (including the newer atypical agents) have been shown to inhibit glucose transport in neuronal and muscle cells (Dwyer *et al.*, 1999a,b; Ardizzone *et al.*, 2001). The initial studies demonstrated inhibition of glucose uptake in PC12 cells by the older drugs, chlorpromazine and fluphenazine, and by the atypical, clozapine (Dwyer *et al.*, 1999b). Subsequently, a role for dopamine receptors in the response was ruled out because haloperidol and sulpiride, both potent D2 antagonists, had little effect on glucose uptake in this system (Dwyer *et al.*,

1999a). The drugs inhibited glucose transport after only a short incubation with the cells (1–2 min) indicating a rapid onset of the effects. The time course profile suggested that events proximal to the transport step were affected rather than gene transcription and protein synthesis. Analysis of a more extensive panel of antipsychotic drugs has revealed several important points (Ardizzone *et al.*, 2001):

1. Atypical antipsychotic drugs such as clozapine, risperidone, and loxapine inhibited glucose transport in PC12 cells (Ardizzone *et al.*, 2001), whereas amoxapine, a structurally related antidepressant, was a very weak inhibitor of transport. Loxapine has mixed typical and atypical properties (Meltzer and Jayathilake, 1999; Herrick-Davis *et al.*, 2000).
2. One major metabolite of clozapine, desmethylclozapine, was far more potent than the parent compound, whereas another metabolite, clozapine N-oxide, was essentially inactive. These data provide evidence for a structure-activity relationship among the dibenzazepine compounds that is especially sensitive to changes in the piperazinyl substituent (Ardizzone *et al.*, 2001).
3. The inhibition of glucose transport by the antipsychotic drugs is sensitive to the level of substrate in the medium and there is a noncompetitive relationship between the drugs and glucose.
4. The drugs block glucose uptake in additional cell types, namely the L6 muscle cell line. Thus, a variety of antipsychotic medications can inhibit glucose transport in diverse cell types.

The antipsychotic drugs produce additional changes in cells related to glucose metabolism. The expression of GLUT proteins, both GLUT1 and GLUT3, was upregulated (about two- to three-fold) by incubation of PC12 cells with the drugs for 24 h (Dwyer *et al.*, 1999b). The increase in GLUT protein was detected by both Western blotting and by analysis of cell surface expression with monoclonal antibodies directed against extracellular epitopes of GLUT3 (Ardizzone and Dwyer, 2002). It is possible that this cellular response was elicited by drug-induced glucose deprivation. In other words, the cells may have upregulated the expression of GLUTs as part of a normal response to the glucose-deficient state that resulted from inhibition of sugar transport. The antipsychotic drugs were also cytotoxic for PC12 cells (and other cell lines) with longer exposure times (48–72 h) (Dwyer *et al.*, 1999b; Dwyer *et al.*, 2001), which might be consistent with the development of a glucose-deprived state in the cells. However, additional factors probably contribute to the cytotoxicity of the drugs *in vitro.* It is not known how these *in vitro* observations relate to the adverse effects of chronic administration of the antipsychotic drugs. For example, many of the antipsychotic medications, particularly the newer atypical drugs, have been implicated in the development of type 2 diabetes mellitus in some psychotic patients (Dwyer *et al.,* 2001). This issue will be discussed more thoroughly in Section IV.

Clues have recently been obtained about how the antipsychotic drugs might inhibit glucose transport. The possibility that blockade of dopamine receptors accounted for the inhibition of transport was ruled out by studies of haloperidol and sulpiride (Dwyer *et al.,* 1999a). Alternatively, the drugs might act on calcium channels, especially in view of the fact that some antipsychotics are potent inhibitors of L-type channel proteins (Gould *et al.,* 1983; Galizzi *et al.,* 1986). This issue has been discussed at length in the next section. From a theoretical perspective, the antipsychotics could ultimately inhibit glucose transport by one of several mechanisms.

1. The drugs could reduce the number of GLUTs at the cell surface by increasing the internalization of transporter proteins.
2. Antipsychotic drugs could bind directly to GLUTs and either block the pore or alter the intrinsic activity of the GLUT protein (i.e., its kinetic properties).
3. The drugs might inhibit downstream events in glucose metabolism, such as hexokinase activity.

This last possibility seems unlikely because the antipsychotic drugs inhibit the uptake of 3-*O*-methylglucose, which is not phosphorylated by hexokinase (Dwyer *et al.,* 1999b). Recent studies cast doubt on the role of the first hypothetical mechanism. It has been shown that the antipsychotic drugs do not reduce cell surface expression of GLUT3 in PC12 cells under conditions where glucose transport is inhibited by 50% or more (Ardizzone and Dwyer, 2002), which is contrary to expectations based on the first mechanism. On the other hand, evidence is accumulating in support of the second possibility. Thus, the antipsychotic drugs inhibit the transport of substrate in a glucose-sensitive manner (Ardizzone *et al.,* 2001). This is similar to the situation with cytochalasin B, which blocks glucose transport by binding directly to GLUTs in a glucose-displaceable manner (Taverna and Langdon, 1973; Lin and Spudich, 1974). Both cytochalasin B and the antipsychotic drugs are noncompetitive inhibitors of glucose transport (Taverna and Langdon, 1973; Krupka and Deves, 1986; Ardizzone *et al.,* 2001). Preliminary studies have shown that the antipsychotic drugs appear to affect the binding of cytochalasin B to GLUTs in membrane preparations. Moreover, the effects of the drugs on the kinetic properties of GLUTs are being investigated. Overall, the data suggest that GLUTs should be considered as an additional target of the antipsychotic drugs. It is not known whether inhibition of glucose transport plays any role in the therapeutic effects of these medications.

E. Calcium Channel Antagonists

Calcium channel antagonists are primarily used to treat angina and hypertension (Klein and Weiner, 1987; Scriabine, 1987), however, some of these drugs are also used in migraine headaches and epilepsy (Scriabine, 1987; Jain, 2000), and they have been tried in patients with psychiatric illness (Reiter *et al.,* 1989; Schwartz *et al.,* 1997). The drugs are also being evaluated as potential therapeutic agents for chronic pain (Atanassoff *et al.,* 2000) and traumatic brain injury (Verweij *et al.,* 2000). There are various calcium channel antagonists that specifically block particular classes of channel protein. For instance, verapamil inhibits L-type calcium channels, flunarizine is specific for T-type channels, and ω-conotoxin GVIA is highly selective for N-type calcium channel proteins (Jones, 1998).

In view of the fact that some antipsychotic medications (e.g., fluphenazine, chlorpromazine, and pimozide) are potent inhibitors of calcium channels (Gould *et al.,* 1983; Galizzi *et al.,* 1986), the effects of these drugs on glucose transport might possibly be explained by inhibition of calcium flux into cells. Others have previously shown that calcium plays an important role in the overall regulation of glucose metabolism (Bihler, 1988). Therefore, a number of calcium channel antagonists have been evaluated for their effects on glucose transport. These findings have recently been reviewed elsewhere (Ardizzone and Dwyer, 2002) and are summarized here.

Representative drugs specific for each of these classes were tested for inhibition of glucose transport in PC12 cells and the L6 muscle cell line. L-type channel blockers, including verapamil, nifedipine, nimodipine and diltiazem, inhibited glucose transport in a dose-dependent fashion (Ardizzone *et al.,* 2002) in both neuronal (PC12) and muscle (L6) cells. By contrast, flunarizine and ω-conotoxin were inactive in the glucose uptake assay. Time-course studies revealed that a 2-min incubation with the drugs was sufficient to produce maximum inhibition of transport. Several findings suggested that the effects of the drugs on glucose transport were not mediated by interference with calcium channel function (Ardizzone *et al.,* 2002).

1. Inhibition of glucose transport was sensitive to the level of glucose in the assay buffer, whereas blockade of calcium channels by these drugs was not.
2. S(–)–Bay K8644, a calcium channel activator, also inhibited glucose uptake.
3. The calcium channel antagonists blocked glucose transport even in the absence of calcium in the uptake buffer.

Thus, it appears that the calcium channel blockers interfere with glucose transport via a novel, calcium-independent mechanism. Perhaps, the calcium channel drugs block glucose uptake in the same way as the antipsychotics, i.e., through direct interaction with GLUT proteins. Preliminary data suggest that nimodipine, like imipramine, competes with cytochalasin B for binding to GLUTs in PC12 cell membranes. This drug decreases the Vmax for transport but not the *K*m. All the available evidence points to direct inhibition by the calcium channel antagonists of glucose transport. Some of these drugs have also been implicated in the development of hyperglycemia in overdose situations as discussed in the next section.

IV. Inhibition of Glucose Transport and Hyperglycemia

Most of the psychoactive drugs that inhibit glucose transport *in vitro* have also been reported to produce hyperglycemia *in vivo*. In the most striking case, treatment with antipsychotic drugs has been linked to the development of type 2 diabetes in some schizophrenic patients (Dwyer *et al.*, 2001). The initial reports that antipsychotic medications produced hyperglycemia and even diabetes in patients treated for psychosis appeared shortly after the introduction of the first antipsychotic drug, chlorpromazine, to clinical practice (Hiles, 1956; Arneson, 1964). The hyperglycemia resolved with discontinuation of the drug which argued convincingly against a natural onset of the diabetes. A number of studies confirmed that the phenothiazine drugs induced hyperglycemia in some schizophrenic patients (Thonnard-Neumann, 1968; Erle *et al.*, 1977; Price and Giannini, 1983; Hagg *et al.*, 1998). However, there was greater concern among clinicians about other adverse effects of these drugs, namely the induction of extrapyramidal syndrome (EPS)—a spectrum of serious movement disorders. The newer atypical medications produce much less EPS; therefore, attention has now turned to the weight gain and metabolic disturbances that are associated with these drugs. Nevertheless, it is worth emphasizing that these same types of metabolic disturbances were produced by the older antipsychotic drugs.

At about the same time that Hiles (1956) described drug-induced hyperglycemia in patients, Norman and Hiestand (1955) reported that chlorpromazine produced hyperglycemia in mice, hamsters, and rats. Moreover, chlorpromazine protected mice against insulin-induced coma and death suggesting that its main effect was to counteract insulin rather than to prevent insulin secretion. A number of other investigators confirmed that acute administration of chlorpromazine elicited hyperglycemia in animals (Ryall, 1956; Chagovets and Shtutman, 1963; Jori *et al.*, 1964). The mechanism by which chlorpromazine elevates blood glucose levels has not been established. In addition, the newer atypical antipsychotic drugs have not been evaluated in the same way, nor have other drugs that block glucose transport *in vitro* (e.g., the calcium channel antagonists) been tested.

Recently, our group has begun evaluating the atypical antipsychotic drugs for their ability to induce hyperglycemia in mice. For comparison, some animals were injected with chlorpromazine (a positive control), haloperidol (a conventional antipsychotic), or nimodipine, which potently inhibits glucose transport *in vitro*. The results of these studies are shown in Fig. 2. As anticipated, chlorpromazine produced a significant elevation of blood glucose when compared to the control (vehicle) mice. Clozapine and nimodipine also induced a striking increase in blood glucose values; many of the mice had glucose levels that would normally indicate the presence of diabetes. On the other hand, haloperidol did not produce hyperglycemia in the mice, which is consistent with its lack of effect on glucose metabolism in schizophrenic patients. An additional group of mice that were injected with cytochalasin B—a specific inhibitor of glucose transport—also developed marked hyperglycemia (Fig. 2). This last observation is important because it demonstrates that a compound, whose only relevant activity is to block glucose uptake, can significantly increase blood glucose levels. Thus, direct

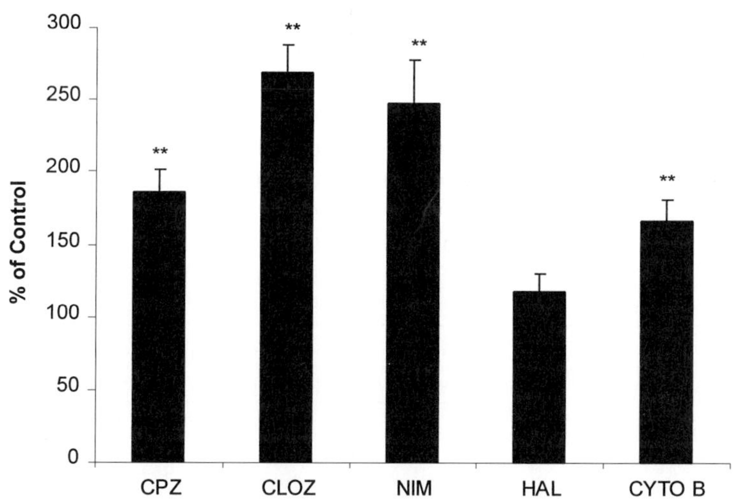

FIG. 2. Induction of hyperglycemia by psychotropic drugs. C57B1/6 mice (8 mice per group) were given a single intraperitoneal injection of drug: chlorpromazine (CPZ, 5 mg/kg), clozapine (CLOZ, 20 mg/kg), nimodipine (NIM, 20 mg/kg), haloperidol (HAL, 2 mg/kg), or cytochalasin B (CYTO B, 20 mg/kg). Control mice (N = 8) received vehicle alone. The mice were bled 30 min later and serum glucose levels were determined by the glucose Trinder method. These data have been expressed as the percentage of the control value in the Trinder assay. Significant elevations of serum glucose levels are indicated by the asterisks ($^{**}p < 0.01$).

interference by a drug with glucose transport is at least one possible mechanism for developing acute hyperglycemia. Perhaps, chronic treatment with a drug that acutely elevates blood glucose levels will eventually lead to insulin resistance, particularly in susceptible individuals. Of course, additional factors are likely to contribute to the metabolic effects of the antipsychotic drugs.

As mentioned earlier, the tricyclic antidepressants produce hyperglycemia in animals and in some patients who have overdosed on the drugs. Likewise, the calcium channel antagonists induce hyperglycemia in cases of accidental poisoning (Herrington *et al.*, 1986; Thomas *et al.*, 1995; Kline *et al.*, 1997). However, there have also been reports that these drugs can produce a significant increase in blood glucose values in some patients even under normal treatment conditions (Giugliano *et al.*, 1980; Charles *et al.*, 1981; Collins *et al.*, 1987; Ferguson *et al.*, 1989). Finally, hyperglycemia is a common outcome of treatment with the glucocorticoids (Munck and Naray-Fejes-Toth, 1995).

There has generally been a strong correlation between the antipsychotic drugs that inhibit glucose transport *in vitro* and those that produce hyperglycemia and diabetes *in vivo*. A summary of the relevant data is presented in Table I. The drugs have been ranked in order of citations of case reports and larger studies of drug-induced hyperglycemia and diabetes. Clozapine and chlorpromazine have the most literature citations and have collectively affected the greatest number of patients. Interestingly, these drugs are also used at the highest daily doses of the medications listed here. A recent study found that 36% of patients on clozapine for five years developed clinical diabetes that was not related to weight gain (Henderson *et al.*, 2000).

TABLE I

CORRELATION BETWEEN *in vivo* VERSUS *in vitro* EFFECTS OF THE ANTIPSYCHOTIC DRUGS

Drug	# Reports	# Patients	Dose range (mg/day)	Serum glucose[a] (% of CON)	EC_{50}[b] (μM)
1. Clozapine	16	64	100–500	268	30
2. Chlorpromazine	8	70	300–800	163	35
3. Olanzapine	8	18	10–15	NT	ND
4. Fluphenazine	4	19	10–25	NT	15
5. Risperidone	2	2	1–6	NT	~25
6. Haloperidol	1	1	5–20	118	>200

[a] These data were taken from the 30 min time point of the acute studies in mice described in Fig. 2 and are expressed as the percentage of the control group.

[b] Estimated from the glucose transport assay.

NT—not tested. ND—not determined.

Clearly, these findings suggest a significant problem that has not been widely appreciated.

Many questions remain concerning the possible causes of the hyperglycemia. For the most part, the number of patients who develop diabetes while treated with antipsychotic drugs is low. It is not known why a particular patient develops drug-induced hyperglycemia. One possibility might be that the drugs induce diabetes only in patients who are genetically predisposed to insulin resistance in the first place. According to this scenario, a genetic tendency toward defective insulin secretion and/or low insulin responsiveness would be exacerbated by the drugs. Patients with normal regulation of glucose metabolism may experience transient hyperglycemia in response to drug therapy, however, they would not develop frank diabetes. On the other hand, drug metabolism may be a significant factor in the emergence of diabetes during treatment, especially in light of data concerning the divergent effects of clozapine analogs on glucose transport (Ardizzone *et al.*, 2001). There are two major metabolites of clozapine, desmethylclozapine and clozapine N-oxide (Shen, 1999). Desmethylclozapine is far more potent than the parent drug at inhibiting glucose transport, whereas clozapine N-oxide is inactive. Thus, an individual who produces more of the desmethyl metabolite relative to the N-oxide species may be at greater risk for developing hyperglycemia than a patient who produces mainly the inactive N-oxide metabolite. Both metabolites are found in significant amounts in the plasma of patients treated with clozapine (Shen, 1999; Spina *et al.*, 2000). Therefore, the severity of an adverse effect may depend upon the balance of these species in circulation or accumulated in tissues.

Alternatively, the brain may play a major role in determining which patients develop diabetes. Some, but not all, patients treated with the older antipsychotic drugs develop movement disorders, such as akathisia and even tardive dyskinesia, as a consequence of drug treatment. The movement disorders result from drug-induced changes in neurotransmission in the nigrostriatal pathway. The fact that only a subpopulation of patients develop severe EPS suggests that there is significant heterogeneity in the brain's response to these drugs. It is possible that other brain regions are also affected in a highly idiosyncratic manner. The hypothalamus is an important brain area involved in the regulation of peripheral glucose levels (Oomura, 1983). Certain neurons in the hypothalamus are sensitive to glucose and increase or decrease their activity as a function of the ambient level of sugar (Oomura, 1983; Silver and Erecinska, 1998; Levin *et al.*, 1999). Perhaps, the antipsychotic drugs affect such neurons in the brains of certain patients more than others leading to dysregulation of glucose metabolism in peripheral tissues. The drugs accumulate in the brain to levels that are 20- to 30-fold higher than in plasma (Baldessarini *et al.*, 1993; Kornhuber *et al.*, 1999). At these levels, there may be significant interference with glucose uptake and glucose sensing in critical brain regions.

The antipsychotic drugs, the antidepressants, and the calcium channel antagonists have all been shown to inhibit glucose transport in cell lines. As outlined here, there have been numerous case reports of these drugs causing hyperglycemia (and even diabetes), especially in overdose situations. Perhaps, there is a common mechanism involved in the development of hyperglycemia. As shown here, inhibition of glucose transport (by cytochalasin B) is sufficient to produce a transient elevation of glucose levels in mice. Other possible mechanisms might include altered regulation of glucose metabolism mediated by inhibition of serotonergic or adrenergic receptors. Direct effects of the drugs on β-cells of the islets (and thus on insulin secretion) may also play a role. Certainly those patients who develop overt diabetes while on these drugs must have impaired responsiveness to insulin.

V. Drug-Binding Site on GLUTs

The results of various studies suggest that a number of drugs may bind directly to GLUT proteins to affect glucose transport (Pinkofsky *et al.*, 2000; Ardizzone *et al.*, 2001; Ardizzone *et al.*, 2002). It has already been established that the antipsychotic drugs and calcium channel antagonists inhibit transport in a glucose sensitive manner (Ardizzone and Dwyer, 2002). Cytochalasin B also binds to GLUTs in a glucose displaceable fashion. Another GLUT ligand, forskolin, inhibits glucose transport and competes with cytochalasin B for binding to the transporter protein (Sergeant and Kim, 1985). These compounds are thought to bind to an intracellular site on the protein and inhibit transport in a noncompetitive manner. Photoaffinity versions of cytochalasin B and forskolin label the C-terminal half of the protein in the vicinity of transmembrane (TM) segments TM8–TM11 (Cairns *et al.*, 1984; Holman and Rees, 1987; Shanahan *et al.*, 1987; Wadzinski *et al.*, 1990). A putative binding site for forskolin and cytochalasin B has recently been identified in a molecular model of GLUT3 (Dwyer, 2001). The binding pocket is formed by the intracellular loops that connect TM8 and TM9 and TM10 and TM11; additional residues also contribute to the site. The model is consistent with mutagenesis studies (Garcia *et al.*, 1992; Schuermann *et al.*, 1993) and photoaffinity labeling experiments aimed at mapping the binding site for cytochalasin B and forskolin. We propose that a wide array of lipophilic molecules bind to this site, including cytochalasin B, forskolin, antipsychotic drugs, barbiturates, antidepressants, calcium channel antagonists, and steroids such as progesterone. These various compounds may interact with distinct subsites in the pore region, however, there must be some overlap in their binding mode because all of the compounds inhibit glucose transport. Glucose can also displace most or all of these drugs from the proposed binding site.

The list of drugs that bind to and inhibit glucose transport is very similar to the list of compounds that block the transport of xenobiotics by P-glycoprotein (Gottesman and Pastan, 1993). P-glycoprotein is a product of a multidrug resistance gene that confers drug resistance to tumors by enhancing the efflux of antitumor medications from tumor cells (Chen *et al.*, 1986; Gros *et al.,* 1986). This protein consists of 12 TM segments with a similar overall topology to the GLUTs (Gottesman and Pastan, 1993). The major target of the antidepressant drugs (monoamine transporters [MATs]), is also comprised of 12 TM helices (Erickson *et al.,* 1992). Once again, there is significant overlap in the types of compounds that bind to the MAT proteins. For example, certain antipsychotic drugs, calcium channel antagonists, and steroids block the uptake of monoamines via MATs (Yelin and Schuldiner, 1995; Wu *et al.,* 1998; Tatsumi *et al.,* 1999). Finally, there appear to be some similarities between these various transporter proteins and calcium channels, especially in the putative drug-binding region of the L-type channel (Mitterdorfer *et al.,* 1998). These similarities include: structural features (conducting pores composed of multiple TM helices), pharmacology (all of the proteins are affected by calcium channel antagonists, antipsychotics, and antidepressants), and function (the transport of small molecules across membrane barriers). In view of the similar pharmacological profiles of these various proteins, we analyzed the amino acid sequences to determine if there might be homologous regions that define the proposed promiscuous binding site. The results of this analysis are summarized in Fig. 3. For comparison, the segments from the L-type calcium channel that are believed to interact with antagonist drugs are shown. There are a number of common features, such as the secondary structure (α-helix), amphipathic helical segments, and conserved residues or motifs (e.g., IAF and (Y/V)AXC in IIIS6 and IVS6 of the calcium channel and their counterparts in the other proteins). We suggest that the overall topology of the proteins is also similar, including a common vestibule or mouth (formed by the bundle of TM helices) that constitutes the promiscuous drug-binding site. Moreover, aromatic amino acid residues may line the vestibule and provide stabilizing π interactions to a variety of molecules, such as cytochalasin B and the phenothiazines, whose common features include multiple rings and highly delocalized electrons. Others (Martell *et al.*, 1997) have suggested that GLUTs and the multidrug resistance-associated protein (MRP) share a common structural motif involved in cytochalasin B binding. Thus, the presence of a

PgP	TM7	P Y L L V G V L C A V I N G C - - I Q P V F
GLUT3	TM7	P L L I S V V L Q L S Q Q F S - G I N A V F
MAT	TM7	I L I A A G S I C F A N M G I A M L E P A L
LTCC	IIIS5	I V I V T T L L Q F M - F A C I G V Q - L F
PgP	TM8	S L F F L V M G - L I S F V - - T Y F F Q G F
GLUT3	TM8	T I G A G V V N - T I F T V V - S L F L V E R
MAT	TM8	Q L G V A F L P A S I S Y L I G T N I F G I L
LTCC	IIIS6	S I F F I I Y I I I I A F F M M - N I F V G F
PgP	TM11	F T Q A M M Y F S Y A A C F R F G A - Y L V A
GLUT3	TM11	G P R P A A M A V A G C S N W T S N F L V G
MAT	TM11	Y G S V Y A I A D V A F C M G Y - A I G P S A
LTCC	IVS6	F A V F Y F I S F Y M L C A F L I I N L F V A

FIG. 3. Amino acid sequence homology between GLUT3 (from Dwyer, 2000, with modifications), P-glycoprotein (PgP; from Gros *et al.*, 1986, with modifications), the vesicular MAT (from Erickson *et al.*, 1992, with modifications), and L-type calcium channels (LTCC; from Mitterdorfer *et al.*, 1998, with modifications). The amino acids have been represented according to the single letter code. Identical residues have been enclosed by solid lines, whereas conserved residues have been indicated by dashed lines. The location of these transmembrane (TM) segments in the respective proteins has also been indicated.

similar drug-binding site on a diverse set of cellular transporters may reflect a common evolutionary origin of these proteins.

VI. Endogenous Molecules That Regulate Glucose Transport

The preliminary evidence for a promiscuous drug-binding site on GLUTs suggests the possibility that endogenous molecules may also bind to this site and regulate glucose transport under normal conditions. ATP is an established modulator of glucose transport (Levine *et al.*, 1998). Fatty acids and steroids would be additional promising candidates as regulatory molecules. Studies have already shown that GLUTs undergo covalent modification by fatty acids and lipid peroxidation products (Pouliot and Beliveau, 1995; Reagan *et al.*, 2000). Thus, GLUT1 in brain capillaries is palmitoylated and it has been suggested that this step provides a rapid mechanism for downregulating glucose transport (Pouliot and Belivau, 1995). Likewise, conjugation of GLUT3 in the hippocampus is associated with a decrease in glucose utilization in the brain (Reagan *et al.*, 2000). As we have shown here, progesterone also inhibits glucose transport, however, it is unlikely that this occurs by covalent modification of GLUT proteins. We propose that the fatty acids and steroids interact with the promiscuous binding site on GLUTs and serve as endogenous modulators of glucose transport. This type of regulation may be important in a number of situations. For instance, the competition among neurons for limiting amounts of glucose may be regulated by reversible inhibition of glucose transport. Essential neurons may continually maintain their GLUTs in a transport-competent state. By contrast, the rate of glucose uptake in less critical neurons may fluctuate significantly in relation to the amount of available energy substrate and the number of active transporters. Fatty acids, ATP, and other endogenous mediators, such as NAD or steroids, might provide feedback control of GLUT activity in these neurons. Finally, transient inhibition of glucose efflux may be a useful adaptation to hypoglycemic conditions in order to prevent further loss of glucose from neuronal and glial cells in the brain.

Additional studies are needed to better characterize the putative binding site for endogenous regulatory ligands on GLUT proteins. Mutagenesis studies could provide the first clues about amino acid residues that are involved in this binding site. A more detailed picture of the drug-binding domain will await the solution of the three-dimensional structure of a GLUT protein by X-ray crystallography or nuclear magnetic resonance.

VII. Future Directions

A. Drug-Induced Hyperglycemia

As discussed here, there has generally been a good correspondence between the ability of a drug to block glucose transport *in vitro* and its hyperglycemic effects in patients. However, little is currently known about the possible mechanisms involved in this response. A worthwhile goal would be to determine why a particular patient develops drug-induced hyperglycemia, while most do not. A number of different genes contribute to the development of noninsulin-dependent diabetes mellitus in the general population (Velho and Froguel, 1997; Ehm *et al.*, 2000). Perhaps, those patients who develop diabetes on the drugs have an elevated risk for hyperglycemia due to inheritance of one or more of these susceptibility genes. Genomics approaches (e.g., gene arrays) could be used to evaluate the effects of the drugs that cause hyperglycemia on the expression of genes linked to diabetes susceptibility or to the regulation of glucose utilization. Alternatively, individual differences in drug metabolism may contribute to the emergence of serious side effects, including hyperglycemia. Patients who mainly metabolize the drugs to inactive species (i.e., with respect to interference with glucose transport) may have a lower risk of hyperglycemia than patients who generate metabolites, such as desmethylclozapine, that are even more active than the parent compound. Studies of drug metabolism in patients in relation to the development of hyperglycemia and diabetes could provide valuable information concerning this possibility. It would also be useful to establish animal models of drug-induced hyperglycemia. As shown here, there is significant elevation of blood glucose levels in mice following acute administration of antipsychotic drugs. This is a starting point, however, long-term studies should also be conducted with the goal of producing sustained hyperglycemia in the animals. It may then be possible to pinpoint the mechanisms involved in the hyperglycemia response to the drugs. By studying the response of mice to the drugs under euglycemic clamp conditions, it should be possible to determine whether hyperglycemia results from impaired insulin secretion, failure to respond adequately to insulin, or from some combination of these processes. Similar approaches could also be used to evaluate insulin responsiveness in patients who have developed diabetes during drug therapy.

B. Drug Design

The weight gain and diabetes induced by the antipsychotic drugs are becoming serious limitations to effective therapy in some patients. There is special concern with clozapine because about one-third of patients may develop diabetes while on this drug (Henderson *et al.*, 2000), yet clozapine often provides benefit to patients who have failed to improve on other medications. Ideally, new antipsychotic drugs would be developed that treat psychosis without causing adverse metabolic effects. This will first require detailed characterization of the pharmacology of the inhibition of glucose transport *in vitro* and the induction of hyperglycemia *in vivo.* At this stage, the data reveal an emerging structure-activity relationship among the antipsychotic drugs that inhibit glucose transport (Ardizzone *et al.*, 2001). Clozapine and its desmethyl derivative block glucose uptake, whereas clozapine-N-oxide is inactive. The structurally related drug, loxapine, also inhibits glucose transport, however, amoxapine (the desmethyl version of loxapine) is much less active. Thus, changes to substituent groups on the piperazinyl ring of these drugs are associated with gain or loss of activity in the glucose transport assay. It is not clear if these same structural changes determine whether a drug produces hyperglycemia *in vivo.* This important line of research should be pursued in animal studies with the goal of establishing an animal model for evaluating the adverse effects of antipsychotic drugs on glucose metabolism.

Such a model could then be used to screen new drugs for their ability to produce hyperglycemia *in vivo.* Promising candidates for clinical development should have minimal effects on glucose metabolism in the animal model.

Inhibition of glucose transport and induction of hyperglycemia are not inevitable properties of antipsychotic drugs. Haloperidol and sulpiride do not inhibit glucose uptake *in vitro.* Furthermore, there has only been one case report of diabetes in a patient treated with haloperidol (Balzan and Cacciottolo, 1992) despite its long and widespread use; this patient had also received tricyclic antidepressants and trifluoperazine, which complicates the interpretation of the findings. However, a major drawback to haloperidol is the induction of EPS, including tardive dyskinesia, in a significant number of patients. Clearly, more research is needed in order to design a new generation of antipsychotic drugs that have fewer serious side effects. It will be important to evaluate lead compounds for their ability to block glucose transport as part of the selection process for identifying drug candidates for clinical development.

References

Afzal, I., Cunningham, P., and Naftalin, R. J. (2001). Interactions of estradiol, genistein, and the anti-estrogens, ICI 182,780 (Faslodex^{TM}) and tamoxifen with the human erythrocyte glucose transporter, GLUT1. Pharmconference, Interlaken, Switzerland.

Alexander, S. C., Helmer, P. H., Ramirez, O., and Chang, J. (1975). Effects of general anesthesia on canine blood-brain barrier glucose transport. *In* "Blood Flow and Metabolism in the Brain" (M. Harper, B. Jennett, O. Miller, and J. Rowan, eds.), pp. 9.37–9.41. Churchill Livingstone, Edinburgh.

Ardizzone, T. D., Bradley, R. J., Freeman, A. M., III, and Dwyer, D. S. (2001). Inhibition of glucose transport in PC12 cells by the atypical antipsychotic drugs risperidone and clozapine, and structural analogs of clozapine. *Brain Res.* **923,** 82–90.

Ardizzone, T. D., and Dwyer, D. S. (2002). Inhibition of neuronal glucose transport by calcium channel antagonists. *Rec. Res. Dev. Neurochem.* (in press).

Ardizzone, T. D., Lu, X.-H., and Dwyer, D. S. (2002). Calcium-independent inhibition of glucose transport in PC12 and L6 cells by calcium channel antagonists. *Am. J. Physiol. Cell Physiol.* **283,** C579–C586.

Arneson, G. A. (1964). Phenothiazine derivatives and glucose metabolism. *J. Neuropsychiatry* **5,** 181–185.

Atanassoff, P. G., Hartmannsgruber, M. W., Thrasher, J., Wermeling, D., Longton, W., Gaeta, R., Singh, T., Mayo, M., McGuire, D., and Luther, R. R. (2000). Ziconotide, a new N-type calcium channel blocker, administered intrathecally for acute postoperative pain. *Reg. Anesth. Pain Med.* **25,** 274–278.

Avruch, J. (1998). Insulin signal transduction through protein kinase cascades. *Mol. Cell. Biochem.* **182,** 31–48.

Baker, G. F., and Rogers, H. J. (1973). Further effects of chlorpromazine on the hexose permeability of the human erythrocyte. *J. Physiol.* **232,** 597–608.

Baldessarini, R. J., Centorrino, F., Flood, J. G., Volpicelli, S. A., Huston-Lyons, D., and Cohen, B. M. (1993). Tissue concentrations of clozapine and its metabolites in the rat. *Neuropsychopharmacology* **9,** 117–124.

Balzan, M., and Cacciottolo, J. M. (1992). Neuroleptic malignant syndrome presenting as hyperosmolar non-ketotic diabetic coma. *Br. J. Psychiatry* **161,** 257–258.

Bihler, I. (1988). The 1987 Upjohn award lecture. The role of membrane transport in the control of glucose metabolism and its coupling to cellular function. *Can. J. Physiol. Pharmacol.* **66,** 549–560.

Breum, L., Bjerre, U., Bak, J. F., Jacobson, S., and Astrup, A. (1995). Long-term effects of fluoxetine on glycemic control in obese patients with non-insulin-dependent diabtes mellitus or glucose intolerance: influence on muscle glycogen synthase and insulin receptor kinase activity. *Metab. Clin. Exp.* **44,** 1570–1576.

Cairns, M. T., Elliott, D. A., Scudder, P. R., and Baldwin, S. A. (1984). Proteolytic and chemical dissection of the human erythrocyte glucose transporter. *Biochem. J.* **221,** 179–188..

Chagovets, R. V., and Shtutman, T. M. (1963). Changes in glucose synthesis and utilization in rats treated with chlorpromazine. *Fed. Proc.* **22,** T935–T937.

Charles, S., Ketelslegers, J.-M., Buysschaert, M., and Lambert, A. E. (1981). Hyperglycemic effect of nifedipine. *Br. Med. J.* **283,** 19–20.

Chen, C., Chin, J. E., Ueda, K., Clark, D. P., Pastan, I., Gottesman, M. M., and Roninson, I. B. (1986). Internal duplication and homology with bacterial transport proteins in the *mdr*1 (P-glycoprotein) gene from multidrug-resistant human cells. *Cell* **47,** 381–389.

Cheng, C. M., Cohen, M., Wang, J., and Bondy, C. A. (2001). Estrogen augments glucose transporter and IGF1 expression in primate cerebral cortex. *FASEB J.* **15,** 907–915.

Collins, W. C., Cullen, M. J., and Feely, J. (1987). Calcium channel blocker drugs and diabetic control. *Clin. Pharmacol. Ther.* **42,** 420–423.

Connolly, V. M., Gallagher, A., and Kesson, C. M. (1995). A study of fluoxetine in obese elderly patients with type 2 diabetes. *Diabetic Med.* **12,** 416–418.

Corvera, S., Chawla, A., Chakrabarti, R., Joly, M., Buxton, J., and Czech, M. P. (1994). A double leucine within the GLUT4 glucose transporter COOH-terminal domain functions as an endocytosis signal. *J. Cell Biol.* **126,** 979–989..

Crane, P. D., Braun, L. D., Cornford, E. M., Cremer, J. E., Glass, J. M., and Oldendorf, W. H. (1978). Dose dependent reduction of glucose utilization by pentobarbital in rat brain. *Stroke* **9,** 12–18.

Czech, M. P., Clancy, B. M., Pessino, A., Woon, C.-W., and Harrison, S. A. (1992). Complex regulation of simple sugar transport in insulin-responsive cells. *Trends Biochem. Sci.* **17,** 197–201.

Doege, H., Bocianski, A., Joost, H.-G., and Schuermann, A. (2000a). Activity and genomic organization of human glucose transporter 9 (GLUT9), a novel member of the family of sugar-transport facilitators predominantly expressed in brain and leucocytes. *Biochem. J.* **350,** 771–776.

Doege, H., Schuermann, A., Bahrenberg, G., Brauers, A., and Joost, H.-G. (2000b). GLUT8, a novel member of the sugar transport facilitator family with glucose transport activity. *J. Biol. Chem.* **275,** 16275–16280.

Dwyer, D. S. (2001). Model of the 3-D structure of the GLUT3 glucose transporter and molecular dynamics simulation of glucose transport. *Prot. Struct. Funct. Genet.* **42,** 531–541.

Dwyer, D. S., Bradley, R. J., Kablinger, A. S., and Freeman, A. M., III (2001). Glucose metabolism in relation to schizophrenia and antipsychotic drug treatment. *Ann. Clin. Psychiatry* **13,** 103–113.

Dwyer, D. S., Liu, Y., and Bradley, R. J. (1999a). Dopamine receptor antagonists modulate glucose uptake in rat pheochromocytoma (PC12) cells. *Neurosci. Lett.* **274,** 151–154.

Dwyer, D. S., Pinkofsky, H. B., Liu, Y., and Bradley, R. J. (1999b). Antipsychotic drugs affect glucose uptake and the expression of glucose transporters in PC12 cells. *Prog. Neuropsychopharmacol. Biol. Psychiatry* **23,** 69–80.

Dwyer, D. S., Simpson, I. A., and Vannucci, S. J. (2002). Expression, regulation and functional role of glucose transporters (GLUTs) in brain. *Int. Rev. Neurobiol,* this volume.

Ehm, M. G., Karnoub, M. C., Sakul, H., Gottschalk, K., Holt, D. C., Weber, J. L., Vaske, D., Briley, D., Briley, L., Kopf, J., McMillen, P., Nguyen, Q., Reisman, M., Lai, E. H., Joslyn, G., Shepherd, N. S., Bell, C., Wagner, M. J., and Burns, D. K. American Diabetes Association GENNID Study Group. (2000). Genomewide search for type 2 diabetes susceptibility genes in four American populations. *Am. J. Hum. Genet.* **66,** 1871–1881.

El-Barbary, A., Fenstermacher, J. D., and Haspel, H. C. (1996). Barbiturate inhibition of GLUT-1 mediated hexose transport in human erythrocytes exhibits substrate dependence for equilibrium exchange but not unidirectional sugar flux. *Biochemistry* **35,** 15222–15227.

Erickson, J. D., Eiden, L. E., and Hoffman, B. J. (1992). Expression cloning of a reserpine-sensitive vesicular monoamine transporter. *Proc. Natl. Acad. Sci. USA* **89,** 10993–10997.

Erle, G., Basso, M., Federspil, G., Sicolo, N., and Scandellari, C. (1977). Effect of chlorpromazine on blood glucose and plasma insulin in man. *Eur. J. Clin. Pharmacol.* **11,** 15–18.

Farese, R. V. (2001). Insulin-sensitive phospholipid signaling systems and glucose transport. Update II. *Exp. Biol. Med.* **226,** 283–295.

Ferguson, J. E., II, Dyson, D. C., Holbrook, H., Jr., Schutz, T., and Stevenson, D. K. (1989). Cardiovascular and metabolic effects associated with nifedipine and ritodrine tocolysis. *Am. J. Obstet. Gynecol.* **161,** 788–795.

Galizzi, J.-P., Fosset, M., Romey, G., Laduron, P., and Lazdunski, M. (1986). Neuroleptics of the diphenylbutylpiperidine series are potent calcium channel inhibitors. *Proc. Natl. Acad. Sci. USA* **83,** 7513–7517.

Garattini, S., Barbui, C., and Saraceno, B. (1998). Antidepressant agents: From tricyclics to serotonin uptake inhibitors. *Psychol. Med.* **28,** 1169–1178.

Garcia, J. C., Strube, M., Leingang, K., Heller, K., and Mueckler, M. M. (1992). Amino acid substitutions at tryptophan 388 and tryptophan 412 of the HepG2 (Glut 1) glucose transporter inhibit transport activity and targeting to the plasma membrane in *Xenopus* oocytes. *J. Biol. Chem.* **267,** 7770–7776.

Gaut, Z. N. (1974). Influence of various substances which induce and inhibit aggregation on the uptake of deoxyglucose by human blood platelets. *J. Pharmacol. Exp. Ther.* **190,** 180–186.

Giugliano, D., Torella, R., Cacciapuoti, F., Gentile, S., Verza, M., and Varricchio, M. (1980). Impairment of insulin secretion in man by nifedipine. *Eur. J. Clin. Pharmacol.* **18,** 395–398.

Gjedde, A., and Rasmussen, M. (1980). Pentobarbital anesthesia reduces blood-brain glucose transfer in the rat. *J. Neurochem.* **35,** 1382–1387.

Goodnick, P. J., Henry, J. H., and Buki, V. M. (1995). Treatment of depression in patients with diabetes mellitus. *J. Clin. Psychiatry* **56,** 128–136.

Gottesman, M. M., and Pastan, I. (1993). Biochemistry of multidrug resistance mediated by the multidrug transporter. *Annu. Rev. Biochem.* **62,** 385–427.

Gould, R. J., Murphy, K. M., Reynolds, I. J., and Snyder, S. H. (1983). Antischizophrenic drugs of the diphenylbutylpiperidine type act as calcium channel antagonists. *Proc. Natl. Acad. Sci. USA* **80,** 5122–5125.

Gros, P., Croop, J., and Housman, D. (1986). Mammalian multidrug resistance gene: Complete cDNA sequence indicates strong homology to bacterial transport proteins. *Cell* **47,** 371–380.

Gupta, B., Awasthi, A., and Jaju, B. P. (1992). Effect of acute and chronic imipramine treatment on glucose homeostasis. *Ind. J. Med. Res.* **96,** 65–71.

Hagg, S., Joelsson, L., Mjorndal, T., Spigset, O., Oja, G., and Dahlqvist, R. (1998). Prevalence of diabetes and impaired glucose tolerance in patients treated with clozapine compared with patients treated with conventional depot neuroleptic medications. *J. Clin. Psychiatry* **59,** 294–299.

Hainaut, P., Kowalski, A., Le Marchand-Brustel, Y., Giorgetti, S., Gautier, N., and Van Obberghen, E. (1991). Effects of insulin, insulin-like growth factor-I (IGF-I) and progesterone on glucose and amino acid uptake in Xenopus oocytes. *Mol. Cell. Endocrinol.* **75,** 133–139.

Harrison, S. A., Buxton, J. M., Clancy, B. M., and Czech, M. P. (1991). Evidence that erythroid-type glucose transporter intrinsic activity is modulated by cadmium treatment of mouse 3T3-L1 cells. *J. Biol. Chem.* **266,** 19438–19449.

Haspel, H. C., Stephenson, K. N., Davies-Hill, T., El-Barbary, A., Lobo, J. F., Croxen, R. L., Mougrabi, W., Koehler-Stec, E. M., Fenstermacher, J. D., and Simpson, I. A. (1999). Effects of barbiturates on facilitative glucose transporters are pharmacologically specific and isoform selective. *J. Membr. Biol.* **169,** 45–53.

Henderson, D. C., Cagliero, E., Gray, C., Nasrallah, R. A., Hayden, D. L., Schoenfeld, D. A., and Goff, D. C. (2000). Clozapine, diabetes mellitus, weight gain, and lipid abnormalities: A five-year naturalistic study. *Am. J. Psychiatry* **157,** 975–981.

Herrick-Davis, K., Grinde, E., and Teitler, M. (2000). Inverse agonist activity of atypical antipsychotic drugs at human 5-hydroxytryptamine2C receptors. *J. Pharmacol. Exp. Ther.* **295,** 226–232.

Herrington, D. M., Insley, B. M., and Weinmann, G. G. (1986). Nifedipine overdose. *Am. J. Med.* **81,** 344–346.

Hiles, B. W. (1956). Hyperglycemia and glycosuria following chlorpromazine therapy. *JAMA* **162,** 1651.

Holman, G. D., and Rees, W. D. (1987). Photolabelling of the hexose transporter at external and internal sites: Fragmentation patterns and evidence for a conformational change. *Biochim. Biophys. Acta* **897,** 395–405.

Honkanen, R. A., McBath, H., Kushmerick, C., Callender, G. E., Scarlata, S. F., Fenstermacher, J. D., and Haspel, H. C. (1995). Barbiturates inhibit hexose transport in cultured mammalian cells and human erythrocytes and interact directly with purified GLUT-1. *Biochemistry* **34,** 535–544.

Horner, H. C., Packan, D. R., and Sapolsky, R. M. (1990). Glucocorticoids inhibit glucose transport in cultured hippocampal neurons and glia. *Neuroendocrinology* **52,** 57–64.

Ibberson, M., Uldry, M., and Thorens, B. (2000). GLUTX1, a novel mammalian glucose transporter expressed in the central nervous system and insulin-sensitive tissues. *J. Biol. Chem.* **275,** 4607–4612.

Jacoby, J. H., and Bryce, G. F. (1979). The acute effects of 5HTP, fluoxetine and quipazine on insulin and glucagon release in the intact rat. *Horm. Metab. Res.* **11,** 90–94.

Jain, K. K. (2000). An assessment of zonisamide as an anti-epileptic drug. *Expert Opin. Pharmacother.* **1,** 1245–1260.

Jones, S. W. (1998). Overview of voltage-dependent calcium channels. *J. Bioenerg. Biomembr.* **30,** 299–312.

Jori, A., Bernardi, D., and Garattini, S. (1964). Chlorpromazine and glucose metabolism. *Int. J. Neuropharmacol.* **3,** 553–558.

Kaplan, S. M., Mass, J. W., Pixley, J. M., and Ross, W. D. (1960). Use of imipramine in diabetics. Effects on glycosuria and blood sugar levels. *JAMA* **174,** 119–125.

Katz, L. M., Fochtmann, L. J., and Pato, M. T. (1991). Clomipramine, fluoxetine and glucose control. *Ann. Clin. Psychiatry* **3,** 271–274.

Klein, M. D., and Weiner, D. A. (1987). Treatment of angina pectoris and hypertension with sustained-release calcium channel-blocking drugs. *Circulation* **75** (Suppl. 5), V-110–V-113.

Kline, J. A., Raymond, R. M., Schroeder, J. D., and Watts, J. A. (1997). The diabetogenic effects of acute verapamil poisoning. *Toxicol. Appl. Pharmacol.* **145,** 357–362.

Klepper, J., Fischbarg, J., Carlos Vera, J., Wang, D., and De. Vivo, D. C. (1999a). GLUT-1 deficiency: barbiturates potentiate haploinsufficiency in vitro. *Pediatr. Res.* **46,** 677–683.

Klepper, J., Wand, D., Fischbarg, J., Carlos Vera, J., Jarjour, I. T., O'Driscoll, K. R., and De Vivo, D. C. (1999b). Defective glucose transport across brain tissue barriers: A newly recognized neurological syndrome. *Neurochem. Res.* **24,** 587–594..

Kornhuber, J., Schultz, A., Wiltfang, J., Meineke, I., Gleiter, C. H., Zoechling, R., Boissl, K., Leblhuber, F., and Riederer, P. (1999). Persistence of haloperidol in human brain tissue. *Am. J. Psychiatry* **156,** 885–890.

Kristiansen, S., Youn, J., and Richter, E. A. (1996). Effect of vanadate on glucose transporter (GLUT4) intrinsic activity in skeletal muscle plasma membrane giant vesicles. *Biochim. Biophys. Acta* **1282,** 71–75.

Krupka, R. M., and Deves, R. (1986). Looking for probes of gated channels: studies of the inhibition of glucose and choline transport in erythrocytes. *Biochem. Cell Biol.* **64,** 1099–1107.

Lachaal, M., Liu, H., Kim, S., Spangler, R. A., and Jung, C. Y. (1996). Cadmium increases GLUT1 substrate binding affinity in vitro while reducing its cytochalasin B binding affinity. *Biochemistry* **35,** 14958–14962.

Lachaal, M., Spangler, R. A., and Jung, C. Y. (2001). Adenosine and adenosine triphosphate modulate the substrate binding affinity of glucose transporter GLUT1 in vitro. *Biochim. Biophys. Acta* **1511,** 123–133.

Lacko, L., Wittke, B., and Lacko, I. (1978). Inhibition of glucose transport in human erythrocytes by benzylalcohol. *J. Cell. Physiol.* **96,** 199–202.

Levin, B. E., Dunn-Meynell, A. A., and Routh, V. H. (1999). Brain glucose sensing and body energy homeostasis: Role in obesity and diabetes. *Am. J. Physiol.* **276,** R1223–R1231.

Levine, K. B., Cloherty, E. K., Fidyk, N. J., and Carruthers, A. (1998). Structural and physiologic determinants of human erythrocyte sugar transport regulation by adenosine triphosphate. *Biochemistry* **37,** 12221–12232.

Lin, S., and Spudich, J. A. (1974). Biochemical studies on the mode of action of cytochalasin B. Cytochalasin B binding to red cell membrane in relation to glucose transport. *J. Biol. Chem.* **249,** 5778–5783.

Lupien, S. J., and McEwen, B. S. (1997). The acute effects of corticosteroids on cognition: integration of animal and human model studies. *Brain Res. Rev.* **24,** 1–27.

Maheux, P., Ducros, F., Bourque, J., Garon, J., and Chiasson, J. L. (1997). Fluoxetine improves insulin sensitivity in obese patients with non-insulin-dependent diabetes mellitus independently of weight loss. *Int. J. Obesity Rel. Metab. Dis.* **21,** 97–102.

Martell, R. L., Slapak, C. A., and Levy, S. B. (1997). Effect of glucose transport inhibitors on vincristine efflux in multidrug-resistant murine erythroleukemia cells overexpressing the multidrug resistance-associated protein (MRP) and two glucose transport proteins, GLUT1 and GLUT3. *Br. J. Cancer* **75,** 161–168.

Masand, P. S., and Gupta, S. (1999). Selective serotonin-reuptake inhibitors: an update. *Harv. Rev. Psychiatry* **7,** 69–84.

McEwen, B. S., De Kloet, E. R., and Rostene, W. (1986). Adrenal steroid receptors and actions in the nervous system. *Physiol. Rev.* **66,** 1121–1180.

Meltzer, H. Y., and Jayathilake, K. (1999). Low-dose loxapine in the treatment of schizophrenia: is it more effective and more "atypical" than standard-dose loxapine. *J. Clin. Psychiatry* **60 Suppl.,** 47–51.

Mitterdorfer, J., Grabner, M., Kraus, R., Hering, S., Prinz, H., Glossmann, H., and Striessnig, J. (1998). Molecular basis of drug interaction with L-type Ca^{2+} channels. *J. Bioenerg. Biomembr.* **30,** 319–333.

Munck, A., and Naray-Fejes-Toth, A. (1995). Glucocorticoid action. *In* "Endocrinology" (L. J. DeGroot, M. Besser, H. G. Burger, J. L. Jameson, D. L. Loriaux, J. C. Marshall, W. D. Odell, J. T. Potts, Jr., and A. H. Rubenstein, eds.), pp. 1642–1656. W. B. Saunders Company, Philadelphia.

Naftalin, R. J., and Arain, M. (1999). Interactions of sodium pentobarbital with D-glucose and L-sorbose transport in human red cells. *Biochim. Biophys. Acta* **1419,** 78–88.

Norman, D., and Hiestand, W. A. (1955). Glycemic effects of chlorpromazine in the mouse, hamster and rat. *Proc. Soc. Exp. Biol. Med.* **90,** 89–91.

Nystrom, F. H., and Quon, M. J. (1999). Insulin signalling: metabolic pathways and mechanisms for specificity. *Cell. Signal.* **11,** 563–574.

Oomura, Y. (1983). Glucose as a regulator of neuronal activity. *Adv. Metab. Dis.* **10,** 31–65.

Otsuka, T., Wei, L., Bereczki, D., Acuff, V., Patlak, C., and Fenstermacher, J. (1991). Pentobarbital produces dissimilar changes in glucose influx and utilization in brain. *Am. J. Physiol.* **261,** R265–R275.

Pinkofsky, H. B., Dwyer, D. S., and Bradley, R. J. (2000). The inhibition of GLUT1 glucose transport and cytochalasin B binding activity by tricyclic antidepressants. *Life Sci.* **66,** 271–278.

Pouliot, J.-F., and Beliveau, R. (1995). Palmitoylation of the glucose transporter in blood-brain barrier capillaries. *Biochim. Biophys. Acta* **1234,** 191–196.

Price, W. A., and Giannini, A. J. (1983). Thioridazine and diabetes. *J. Clin. Psychiatry* **44,** 469.

Reagan, L. P., Magarinos, A. M., Yee, D. K., Swzeda, L. I., Van Bueren, A., McCall, A. L., and McEwen, B. S. (2000). Oxidative stress and HNE conjugation of GLUT3 are increased in the hippocampus of diabetic rats subjected to stress. *Brain Res.* **862,** 292–300.

Reagan, L. P., Gorovits, N., Hoskin, E. K., Alves, S. E., Katz, E. B., Grillo, C. A., Piroli, G. G., McEwen, B. S., and Charron, M. J. (2001). Localization and regulation of GLUT × 1 glucose transporter in the hippocampus of streptozotocin diabteic rats. *Proc. Natl. Acad.Sci. USA* **98,** 2820–2825.

Reiter, S., Adler, L., Angrist, B., Peselow, E., and Rotrosen, J. (1989). Effects of verapamil on tardive dyskinesis and psychosis in schizophrenic patients. *J. Clin. Psychiatry* **50,** 26–27.

Remington, G., and Kapur, S. (2000). Atypical antipsychotics: Are some more atypical than others? *Psychopharmacology* **148,** 3–15.

Royds, R. B., and Knight, A. H. (1970). Tricyclic antidepressant poisoning. *Practitioner* **204,** 282–286.

Ryall, R. W. (1956). Some actions of chlorpromazine. *Br. J. Pharmacol.* **11,** 339–345.

Sakoda, H., Ogihara, T., Anai, M., Funaki, M., Inukai, K., Katagiri, H., Fukushima, Y., Onishi, Y., Ono, H., Fujishiro, M., Kikuchi, M., Oka, Y., and Asano, T. (2000). Dexamethasone-induced insulin resistance in 3T3-L1 adipocytes is due to inhibition of glucose transport rather than insulin signal transduction. *Diabetes* **49,** 1700–1708.

Sapolsky, R. M. (1999). Glucocorticoids, stress, and their adverse neurological effects: Relevance to aging. *Exp. Gerontol.* **34,** 721–732.

Sapolsky, R. M., Uno, H., Rebert, C. S., and Finch, C. E. (1990). Hippocampal damage associated with prolonged glucocorticoid exposure in primates. *J. Neurosci.* **10,** 2897–2902.

Schuermann, A., Keller, K., Monden, I., Brown, F. M., Wandel, S., Shanahan, M. F., and Joost, H. G. (1993). Glucose transport activity and photolabelling with 3-[^{125}I]iodo-4-azidophenethylamido-7-*O*-succinyldeacetyl (IAPS)-forskolin of two mutants at tryptophan-388 and -412 of the glucose transporter GLUT1: Dissociation of the binding domains of forskolin and glucose. *Biochem. J.* **290,** 497–501.

Schwartz, B. L., Fay-McCarthy, M., Kendrick, K., Rosse, R. B., and Deutsch, S. I. (1997). Effects of nifedipine, a calcium channel antagonist, on cognitive function in schizophrenic patients with tardive dyskinesia. *Clin. Neuropharmacol.* **20,** 364–370.

Scriabine, A. (1987). Current and potential indications for Ca^{2+} antagonists. *Ration. Drug Ther.* **21,** 1–7.

Seidner, G., Garcia-Alvarez, M., Yeh, J. I., O'Driscoll, K., Klepper, J. R., Stump, T. S., Wand, D., Spinner, N. B., Birnbaum, M. J., and De Vivo, D. C. (1998). Glut-1 deficiency syndrome caused by haploinsufficiency of the blood-brain barrier hexose carrier. *Nat. Genet.* **18,** 1–4.

Sergeant, S., and Kim, H. D. (1985). Inhibition of 3-*O*-methylglucose transport in human erythrocytes by forskolin. *J. Biol. Chem.* **260,** 14677–14682.

Shanahan, M. F., Morris, D. P., and Edwards, B. M. (1987). [^{3}H]Forskolin. Direct photoaffinity labeling of the erythrocyte D-glucose transporter. *J. Biol. Chem.* **262,** 5978–5984.

Shen, W. W. (1999). The metabolism of atypical antipsychotic drugs: an update. *Ann. Clin. Psychiatry* **11,** 145–158.

Shepard, T. H., Tanimura, T., and Robkin, M. A. (1970). Energy metabolism in early mammalian embryos. *Dev. Biol. Suppl.* **4,** 42–58.

Shi, J., Zhang, Y. Q., and Simpkins, J. W. (1997). Effects of 17β-estradiol on glucose transporter 1 expression and endothelial cell survival following focal ischemia in the rats. *Exp. Brain Res.* **117,** 200–206.

Silver, I. A., and Erecinska, M. (1998). Glucose-induced intracellular ion changes in sugar-sensitive hypothalamic neurons. *J. Neurophysiol.* **79,** 1733–1745.

Simpson, I. A., and Cushman, S. W. (1986). Hormonal regulation of mammalian glucose transport. *Annu. Rev. Biochem.* **55,** 1059–1089.

Singh, S. P., Snyder, A. K., and Eman, S. (1990). Effects of ethanol on hexose transport by cultured rat brain cells. *Alcohol Clin. Exp. Res.* **14,** 741–745.

Spina, E., Avenosi, A., Facciola, G., Scorda, M. G., Ancione, M., Madia, A. G., Ventimiglia, A., and Perucca, E. (2000). Relationship between plasma concentrations of clozapine and norclozapine and therapeutic response in patients with schizophrenia resistant to conventional neuroleptics. *Psychopharmacology* **148,** 83–89.

Streissguth, A. P., Landesman-Dwyer, S., Martin, J. C., and Smith, D. W. (1980). Teratogenic effects of alcohol in humans and laboratory animals. *Science* **209,** 353–361.

Svensson, T. H. (2000). Brain noradrenaline and the mechanisms of action of antidepressant drugs. *Acta Psychiat. Scand. Suppl.* **402,** 18–27.

Tatsumi, M., Jansen, K., Blakely, R. D., and Richelson, E. (1999). Pharmacological profile of neuroleptics at human monoamine transporters. *Eur. J. Pharmacol.* **368,** 277–283.

Taverna, R. D., and Langdon, R. G. (1973). Reversible association of cytochalasin B with the human erythrocyte membrane. Inhibition of glucose transport and the stoichiometry of cytochalasin binding. *Biochim. Biophys. Acta* **323,** 207–219.

Thomas, S. H., Stone, C. K., and May, W. A. (1995). Exacerbation of verapamil-induced hyperglycemia with glucagon. *Am. J. Emer. Med.* **13,** 27–29.

Thonnard-Neumann, E. (1968). Phenothiazines and diabetes in hospitalized women. *Am. J. Psychiatry* **124,** 978–982.

Vannucci, S. J., Maher, F., and Simpson, I. A. (1997). Glucose transporter proteins in brain: delivery of glucose to neurons and glia. *Glia* **21,** 2–21.

Velho, G., and Froguel, P. (1997). Genetic detrminants of non-insulin-dependent diabetes mellitus: strategies and recent results. *Diabetes Metab.* **23,** 7–17.

Verhey, K. J., and Birnbaum, M. J. (1994). A Leu-Leu sequence is essential for COOH-terminal targeting signal of GLUT4 glucose transporter in fibroblasts. *J. Biol. Chem.* **269,** 2353–2356.

Verweij, B. H., Muizelaar, J. P., Vinas, F. C., Peterson, P. L., Xiong, Y., and Lee, C. P. (2000). Improvement in mitochondrial dysfunction as a new surrogate efficiency measure of preclinical trial: dose-response and time-window profiles for administration of the calcium channel blocker Ziconotide in experimental brain injury. *J. Neurosurg.* **93,** 829–834.

Wadzinski, B. E., Shanahan, M. F., Seamon, K. B., and Ruoho, A. E. (1990). Localization of the forskolin photolabelling site within the monosaccharide transporter of human erythrocytes. *Biochem. J.* **272,** 151–158.

Wannarka, G. L., Fletcher, H. P., and Schnell, R. C. (1979). Variations of chlorpromazine-related isosteres on the glycemic response in mice. *Gen. Pharmacol.* **10,** 169–172.

Watson, R. T., and Pessin, J. E. (2001). Intracellular organization of insulin signaling and GLUT4 translocation. *Rec. Prog. Horm. Res.* **56,** 175–193.

West, J. R., and Hamre, K. M. (1985). Effects of ethanol during different periods of development: changes in hippocampal mossy fibers. *Dev. Brain Res.* **17,** 280–284.

Wu, X., Kekuda, R., Huang, W., Fei, Y. J., Leibach, F. H., Chen, J., Conway, S. J., and Ganapathy, V. (1998). Identity of the organic cation transporter OCT3 as the extraneuronal monoamine transporter (uptake 2) and evidence for the expression of the transporter in brain. *J. Biol. Chem.* **273,** 32776–32786.

Yamada, J., Sugimoto, Y., and Inoue, K. (1999). Selective serotonin reuptake inhibitors fluoxetine and fluvoxamine induce hyperglycemia by different mechanisms. *Eur. J. Pharmacol.* **382,** 211–215.

Yelin, R., and Schuldiner, S. (1995). The pharmacological profile of the vesicular monoamine transporter resembles that of multidrug transporters. *FEBS Lett.* **377,** 201–207.

INDEX

D

E

F

G

I

M

N

Q

R

CONTENTS OF RECENT VOLUMES

Volume 39

Volume 40

Volume 41

Volume 42

Volume 43

Volume 44

Volume 45

Volume 46

Volume 47

Volume 48

Volume 49

Volume 50